9th AUN/SEED-Net Regional Conference on Natural Disaster

IOP Conference Series: Earth and Environmental Science
Volume 1091

Online
15-16 December 2021

Editors:

Mohd Hakim Ab Halim
Aznah Nor Anuar
Muhammad Ali Muhamad Yuzir

ISBN: 978-1-7138-6437-0
ISSN: 1755-1307

Printed from e-media with permission by:

Curran Associates, Inc.
57 Morehouse Lane
Red Hook, NY 12571

Institute of Physics
Dirac House, Temple Back
Bristol BS1 6BE UK

Phone: 44 1 17 929 7481
Fax: 44 1 17 920 0979

techtracking@iop.org

Additional copies of this publication are available from:

Curran Associates, Inc.
57 Morehouse Lane
Red Hook, NY 12571 USA
Phone: 845-758-0400
Fax: 845-758-2633
Email: curran@proceedings.com
Web: www.proceedings.com

TABLE OF CONTENTS

DISASTER SCIENCE AND MANAGEMENT

Ground Surface and Building Settlement Monitoring using Internet of Things (IoT) Approach 1
N A Mohd Razip, N N Nik Daud

Vibration Effects on Soil Slope Detected by using Arduino System 16
M A A Azharudin, N N Nik Daud

Predicting and avoiding hazardous occurrences of stuck pipe for the petroleum wells at offshore Vietnam using machine learning techniques 29
Quang Khanh Do, Trong Quang Hoang, Thang Nguyen, Vu Khanh Phat Ong

Strategic framework of using drone in cities disaster response 39
H. Zahir, M. S. Fathi, A. F. Tharima

Development of Fragility Curves for the Underground PVC Water Distribution Pipes in Quezon City and Damage Estimation under a Magnitude 7.2 Scenario Earthquake 49
C M Soriano, P E Quinay

Development of A Two-storage Reservoir Operation Rules for Flood Control Case Study of Wonogiri Multipurpose Reservoir, Central Java, Indonesia 59
R Jayadi, Istiarto, E P A Pratiwi, K Windianita

Communicating landslide risks during the COVID-19 pandemic: A baseline study for PHIVOLCS-Dynaslope stakeholders 69
Eunice R. Decena, Heinrich Joyce S. Domingo, Ardeth Joy E. Maligon, Arturo S. Daag, Renato U. Solidum

Seismic screening and structural investigation of heritage buildings for adaptive reuse: a survey study at Iloilo City, Philippines 79
Richard De Jesus, Andres Winston Oreta, Mary Earl Grio, Rodolfo Mendoza, Lessandro Garciano, Rhodella Ibabao, Giacomo Sevieri, Carmine Galasso

Effect of Disaster Preparedness Training on Knowledge Regarding Flood and Landslides Disaster Management among Family with Elderly 89
Fatmah

Climate Change Induced Landslide Susceptibility Assessment - for Aiding Climate Resilient Planning for Road Infrastructure: A Case Study in Rangamati District, Chittagong Hill Tracts, Bangladesh 100
I P K Wijaya, A Joshi, M N Alam, S Jayasinghe, N Laila

Heritage Community Resilience: The Experience of Stakeholders in Calle Real, Iloilo City, Philippines 110
Rhodella A. Ibabao, Vicente Balinas, Jerilee Camena, Rene Trance, Gay Defiesta, Mary Earl Grio, Andres Winston Oreta, Suzette Penaredondo

The Effect of Clay-Soil on Landslide: Case Study from Central Java, Indonesia 119
W Budianta, H Ohta, J Takemura

Numerical simulation of the 1994 tsunami in Calapan, Mindoro, Philippines using Delft3D 125
Marjorie David, Christian Paul Allan, Eric Cruz

Physical Vulnerability of Surabaya due to Earthquake Based On Rapid Visual Screening.............................. 134
A. B. Habieb, D. Iranata, S. Nurlaela, R. Anggoro, R. A. W. D. Syafitri

Fear of the fall of space objects in people of rural communities of the district of Azangaro, Peru 144
A P Calatayud-Mendoza, H N Chui-Betancur, G Belizario-Quispe, E Huaquisto-Ramos, R Alfaro-Alejo

Using the TELEMAC model for analyzing the hydrodynamic regime in the Vam Nao River, Vietnam 154
Nga N Q Tra, Kim T Tran, Nghia H Ho, Hoai C Huynh, Bay T Nguyen

Flash Flood Susceptibility Mapping in Sungai Pinang catchment using Weight of Evidence.......................... 164
A Saleh, N Sabtu, M R Bunmi

Alternative Design of Post-Earthquake Permanent Housing in Indonesia using Cold Formed Steel............... 174
R. Alghiffary, Data Iranata, D. Irawan, B. Suswanto, H. Kristijanto

Relationship of Building Form with Damage Pattern and Damage Level... 182
Rita Laksmitasari Rahayu, Triyadi Sugeng S, Lily Tambunan

Generating City-Scale 3D Finite Element Models from GIS Features and Heuristic-based Framing Models using Feed-Forward Artificial Neural Networks 191
Jamie Alea B Ramos, Pher Errol B Quinay, Karlo Daniel Q Colegio

Development of a k-Nearest Neighbor (kNN) Machine Learning Model to Estimate the SPT N-Values of Valenzuela City, Philippines 200
J Galupino, J Dungca

Attitude and flood preparedness of households in Pasir Mas, Kelantan, Malaysia 210
F A Noor Diyana

Finding an Optimal Use of Tides for Leaching of Pond Waters in Channel Networks with Various Size and Simultaneous Opening Number of Gates, Case Study of Sei Teras Fishpond Irrigation Area, Central Kalimantan... 216
R Ciptadi, A P Rahardjo, B Kamulyan

Application of Delft3D-WAVE in the investigation of the typhoon-induced high wave in Central Vietnam 227
L T Anh, D D T Trang, N D Thao, N Kiet

The condition of upper soil layer that causes severe damage to land due to liquefaction 237
H Muhanifah, A D Adi, F Faris

Slope stability analysis during an earthquake in flow-slide affected area of Jono Oge.................................. 247
A N Andiny, F Faris, A D Adi

Mitigation plan against liquefaction potential on Gumbasa Irrigation Canal around Jono Oge area after 2018 Central Sulawesi Earthquake.. 257
A Pratama, T F Fathani, I Satyarno

Isolation and characterization of filamentous biological soil crust forming algae from Malaysia 267
P Chheang, F T Y Min, J Md Khudzari, K Iwamoto

Liquefaction Potential Analysis and Mitigation at The Sediment Control Structures on Bangga River, Central Sulawesi 274
Fauzan, A Rifa'I, S Ismanti

Hydrological Drought Responding to Meteorological Drought in Prek Thnot River Basin............................ 284
Chhordaneath Hen, Ty Sok, Sophal Try, Ratboren Chan, Ilan Ich, Kimleang Chum, Chantha Oeurng

Assessing Flood Hazard Index using Analytical Hierarchy Process (AHP) and Geographical Information System (GIS) in Stung Sen River Basin 294
Chanseyma Khoeun, Ty Sok, Ratboren Chan, Sotheanea Khe, Ilan Ich, Kimsan Chan, Chantha Oeurng

River Water Quality Implication after the Earthquake 304
Leangkim Kong, Aznah Nor, Mohd Hakim, Nf. Said, J. Asanuma

BSC (biological soil crust) Method for soil erosion control and vegetation recovery: Case studies in Japan.. 316
T C T Nguyen, J Md Khudzari, F Imamura, S Sethi, K Iwamoto

Assessment of Reservoir and Canals Revitalization to Improve The Reclamation Process at Unit Tamban Tidal Irrigation Area, Central Kalimantan 328
S S Rahajeng, R Jayadi, A Maas

Behavior of Flow of Channel Network of Dadahup Swampy Irrigation Area 338
M Ihsan Arif, Djoko Legono, Djoko Luknanto, Budi Wignyosukarto, Maruddin F Marpaung

Climate change adaptation measures in the agricultural sector in Southeast Asia: A mini-review.................. 348
S Sha'Arani, Nsa Sabri, N.A Hamdi, F.A Riyadi

Utilizing Google Street View for Rapid Seismic Vulnerability Assessment: Case Study in the City of Manila, Philippines 355
M K Evangelista, D H Aquino, J A Ramos

The Effect of Various Peak Ground Acceleration Level in Petobo Irrigation Canal Area Post 2018's Liquefaction Disaster... 363
S Sauri, A Rifa'I, H C Hardiyatmo

Analysis of the Effective Length of Handil Channel for Tidal Irrigation at Palingkau Irrigation Area in Central Kalimantan Province for National Food Estate Program... 370
E. E. Ramadhani, J. Sujono, Taryono

Review of flood prediction hybrid machine learning models using datasets.. 380
Ainaa Hanis Zuhairi, Fitri Yakub, Sheikh Ahmad Zaki, Mohamed Sukri Mat Ali

Flood River Water Level Forecasting using Ensemble Machine Learning for Early Warning Systems.. 390
Amrul Faruq, Shamsul Faisal Mohd Hussein, Aminaton Marto, Shahrum Shah Abdullah

Community based index of coastal erosion using ahp analysis 398
Wan Norshuhada Binti Wan Khairuddin, Shuib Bin Rambat, Aizul Nahar Bin Harun

Lightweight deep CNN models for identifying drought stressed plant 409
Mh. Kamarudin, Zool H. Ismail

ENVIRONMENTAL SCIENCE

Determination of Critical Environmental, Social, Infrastructure, and Economic Indicators as Evaluation Checklist for Pre-Project Abandonment Plan Assessment in-line with Sustainable Development Goal Plan 418
Sn Abdullah, A Aznah, Mohd Hakim, Mf Md Din, M Ponraj, S Mat-Taib

High cadmium tolerance in Stichoccocus-like microalgae (Tetratostichoccocus sp. P1) from Malaysia 428
E Sahabudin, N Othman, I Suzuki

The effect of LEDs on Galdieria sulphuraria cultivation at indoor and outdoor condition. 435
N Jusoh, Y S Lee, Y C Teo, F S Othman, S E Mohamad

Potential Application and Future Prospects of Algal-Bacterial Aerobic Granular Sludge for Wastewater Treatment 441
L D A Purba, K Shimizu, J Hermana

Analysis of Rock Mass Surface Colour using CIELAB and Munsell Chart 451
M. Razali, M. A. M Ismail, I. N. Yusoff, H. Tobe, T. Miyoshi, K. Date, Y. Yokota

Rapid method of aerobic granular sludge bioreactor start-up for domestic wastewater treatment 464
Karn Tanavarotai, Aznah Nor Anuar, Alijah Mohd Aris, Zhongfang Lei, Mohd Hakim Ab Halim

Development of Microalgae-Bacteria Aerobic Granular Sludge for Landfill Leachate Treatment 471
C Khen, D Ilmasari, A Yuzir

The applications of aerobic granular sludge for leachate treatment: A review 480
D Ilmasari, A Yuzir

Development and characteristics of microalgae-bacteria aerobic granules treating low strength municipal wastewater 490
S A Zahra, L D A Purba, N Abdullah

One-way flow system for improvement of the acid sulfate soil reclamation process in the Belanti II tidal swamp irrigation network, Central Kalimantan, Indonesia. 499
A Setiawan, B S Wignyosukarto, A P Rahardjo

Management of Waste Cooking Oil and its Potential for Value-added Materials: A Mini Review 507
N. A. Hamdi, S. Sha'Arani, N. F. Azman, S. B. Mohd. Rafi, E. Norsin, N. Othman

Backswamp Development's Effects on Water Management System of Tidal Swamp Irrigation Palingkau SP1 SP2 SP3, Kalimantan, Indonesia 515
Athanasius Donny Firstyadi, Budi Santosa Wignyosukarto, Istiarto Istiarto, Yakubson Yakubson

Energy efficiency and nutrient removal performance: comparison between several types of activated sludge process 524
Wan Mohd Bukhary W Muzaffar, A. Aznah, Hakim Halim

Sustainable indicators for designing the wastewater treatment plant 534
Azila Sembuil, Aznah Nor, Md. Hakim A-Halim

Physical Characteristics and Removal Performance of Aerobic Granular Sludge in Biological Treatment of Domestic Wastewater 544
Mohd Hakim Ab Halim, Aznah Anuar, Nur Syahida Abdul Jamal, Siti Izaidah Azmi, Wan Nur Atiqah Wan Hassan, Zaini Ujang, Mustafa M Bob

To Optimize Sequencing Batch Reactor (SBR) and Monitor Aerobic Granular Sludge (AGS) Development for Domestic Wastewater Treatment 551
Norazah Mohd Kassim, Aznah Nor Anuar, Hakim Halim, Noor Atiqah Sharani, Alijah Md Aris, Khor Bee Chin

Current state and the application of energy conversion of sewage sludge treatment in Malaysia 561
L M Quan, I Warmadewanthi, Y Helmut

Anammox bacterium 'Candidatus Kuenenia stuttgartiensis': a review 572
Thilagavathi Arumugham, Shaza Eva Mohamad

Isolation and characterization of dye-degrading bacteria from batik wastewater 585
Rakhmania, Z Ibrahim, M Abba

Production of Ligninolytic Enzymes from Thermophilic Bacterial Strains Isolated from Palm Oil Wastes 596
F A Riyadi, N F Azman, N Yusof, F N M Akhir

Preparation Method of Titanium Dioxide Nanoparticles and Its Application: An Update 603
N Seman, Z I Tarmizi, R R Ali, S H M Taib, M S N Salleh, J C Zhe, S N A Mohamad Sukri

Modified Surface Polymer Prepared by using Radiation-Induced Grafting Co-polymerization Method and Its Application: A Mini Review 613
N Seman, Z I Tarmizi, R R Ali, M S N Salleh

Author Index

IOP Conf. Series: Earth and Environmental Science 1091 (2022) 011001 doi:10.1088/1755-1315/1091/1/011001

Abstract

The 9th AUN/SEED-Net Regional Conference on Natural Disaster (RCND) 2021 is a continuing series of regional conferences on natural disaster. RCND2021 was held on December 15th-16th, 2021 on virtual format. It aims to bring policymakers, scientists, engineers, researchers and students to the areas of Geo-Spatial Knowledge and Intelligence.

Through RCND2021 theme "Disaster Resilience for Sustainable Development" this international conference is part of the continuing series of regional conference to provide a forum to collect and disseminate the most updated technology and research of regional issues and public interests in the field of Natural Disaster in order to contribute to the community and to draw support from the industry and government sectors.

RCND2021 features broad issues on natural disaster, including but not limited to the following topics - Disaster Risk Reduction, Disaster Prevention and Mitigation, Disaster Resilience Community, Disaster Rehabilitation and Recovery, Information Management for Disaster Risk Reduction (DRR), Internet of Things (IoT) in Disaster Risk Management, Electronic Sensor for Disaster, Machine Learning and Artificial Intelligence (AI) on Disaster Management, Fluvial and Coastal Disaster, Water-Related Disaster, Watershed Management, Flood Management, Water Supply and Sanitation, Tsunami Hazard Earthquake Disaster, Geological Hazard, Volcanic Hazard, Early Warning System and Environmental Science. The conference provides a seminar and special session for sharing experiences and original research contributions on those topics.

We received many paper submissions from various parts of the world. The International Program Committee worked very hard to have all papers peer-peer reviewed before the review deadline. There are four (4) keynote speeches from Japan (2) and Malaysia (2). All the keynote speakers are internationally recognized leading experts in disaster research fields, who have demonstrated outstanding proficiency and have achieved distinction in their profession. Japan and Malaysia are also well-known as efficient countries in disaster risk management towards resilient society.

We would like to express our sincere gratitude to all the members of the International Program Committee and organizers for their enthusiasm, time, and expertise. Our deepest gratitude also goes to all volunteers and staff for the long hours and hard work they have generously given to RCND2021. We are very grateful to Prof. Dr. Kaoru Takara, Dr. Senro Kuraoka, Prof. Ir. Dr. Azlan Adnan, Assoc. Prof. Dr. Aznah Nor Anuar, and Prof. Dr. Muhammad Ali Muhamad Yuzir for their support in making RCND2021 possible. Finally, we would like to thank all the authors, speakers, and participants of this conference for their contributions to RCND2021.

Editors

The 9th AUN/SEED-Net Regional Conference On Natural Disaster 2021 (RCND 2021)

Main Editor: Dr. Mohd Hakim Ab Halim
Co-Editors: Assoc. Prof. Dr. Aznah Nor Anuar and Prof. Dr. Muhammad Ali Muhamad Yuzir

Funding Acknowledgement

The 9th AUN/SEED-Net Regional Conference on Natural Disaster 2021 would like to thank AUN/SEED-Net and Malaysia-Japan International Institute of Technology (MJIIT), Universiti Teknologi Malaysia (UTM) for funding the conference under grant number A.K430000.5600.06369.

IOP Conf. Series: Earth and Environmental Science 1091 (2022) 011001 doi:10.1088/1755-1315/1091/1/011001

Peer Review Statement

All papers published in this volume of IOP Conference Series: Earth and Environmental Science have been peer-reviewed through processes administered by the proceedings Editors. Reviews were conducted by expert referees to the professional and scientific standards expected of a proceedings journal published by IOP Publishing.

IOP Conf. Series: Earth and Environmental Science 1091 (2022) 011001 doi:10.1088/1755-1315/1091/1/011001

THE 9TH AUN/SEED-NET REGIONAL CONFERENCE ON NATURAL DISASTER 2021 (RCND 2021)

ORGANIZING COMMITTEE

PATRON
Professor Dr. Ali Selamat, Dean of Malaysia-Japan International Institute of Technology, Universiti Teknologi Malaysia

ADVISOR
Associate Professor Dr. Siti Hamidah Mohd Setapar, Deputy Dean of Research and Innovation Centre of Malaysia-Japan International Institute of Technology, Universiti Teknologi Malaysia

CHAIRPERSON
Professor Dr. Muhamad Ali Muhammad Yuzir, Department of Chemical and Environmental Engineering, Malaysia-Japan International Institute of Technology, Universiti Teknologi Malaysia

DEPUTY CHAIRPERSON
Dr. Sumiaty Ambran, Electronic Systems Engineering, Malaysia-Japan International Institute of Technology, Universiti Teknologi Malaysia

SECRETARIES
Dr. Mohd. Hakim Ab Halim, Disaster Preparedness and Prevention Centre, Department of Chemical and Environmental Engineering, Malaysia-Japan International Institute of Technology, Universiti Teknologi Malaysia

Anis Fariha Mahmad, Malaysia-Japan International Institute of Technology, Universiti Teknologi Malaysia

Syamimi Mohd Zul, Malaysia-Japan International Institute of Technology, Universiti Teknologi Malaysia

TREASURERS
Akmalaini Ahmad, Malaysia-Japan International Institute of Technology, Universiti Teknologi Malaysia

Mazliza Aini Abdul Majid, Malaysia-Japan International Institute of Technology, Universiti Teknologi Malaysia

SECRETARIATS
Masafumi Goto, Disaster Preparedness and Prevention Centre, Malaysia-Japan International Institute of Technology, Universiti Teknologi Malaysia

Azlan Adnan, Earthquake Engineering, Universiti Teknologi Malaysia

Norhayati Abdullah, Chemical Process Engineering, Malaysia-Japan International Institute of Technology, Universiti Teknologi Malaysia

Koji Iwamoto, Chemical Process Engineering, Malaysia-Japan International Institute of Technology, Universiti Teknologi Malaysia

Aznah Nor Anuar, Chemical Process Engineering, Malaysia-Japan International Institute of Technology, Universiti Teknologi Malaysia

Azura Hamzah, Electronic Systems Engineering, Malaysia-Japan International Institute of Technology, Universiti Teknologi Malaysia

Faizah Che Ros, Chemical Process Engineering, Malaysia-Japan International Institute of Technology, Universiti Teknologi Malaysia

Fazrena Nadia Md. Akhir, Chemical Process Engineering, Malaysia-Japan International Institute of Technology, Universiti Teknologi Malaysia

Shuib Rambat, Chemical Process Engineering, Malaysia-Japan International Institute of Technology, Universiti Teknologi Malaysia

Aizul Nahar Harun, Malaysia-Japan International Institute of Technology, Universiti Teknologi Malaysia

Zatil Izzah Ahmad Tarmizi, Malaysia-Japan International Institute of Technology, Universiti Teknologi Malaysia

Mohd. Sharudin Mat Deli @ Abdul Manan, Malaysia-Japan International Institute of Technology, Universiti Teknologi Malaysia

Nurasmawisham Alel, Malaysia-Japan International Institute of Technology, Universiti Teknologi Malaysia

Zamri Ramli, Malaysia-Japan International Institute of Technology, Universiti Teknologi Malaysia

Muhammad Syafiq Yazid, Malaysia-Japan International Institute of Technology, Universiti Teknologi Malaysia

Mohd. 'Afiq Murad, Malaysia-Japan International Institute of Technology, Universiti Teknologi Malaysia

Muhammad Islah Sulaiman, Malaysia-Japan International Institute of Technology, Universiti Teknologi Malaysia

Huzaimah Yahya, Malaysia-Japan International Institute of Technology, Universiti Teknologi Malaysia

Ahmad Fitri Ahmad Faozi, Malaysia-Japan International Institute of Technology, Universiti Teknologi Malaysia

Nur Fatehah Othman, Malaysia-Japan International Institute of Technology, Universiti Teknologi Malaysia

Siti Nabihah Johari, Malaysia-Japan International Institute of Technology, Universiti Teknologi Malaysia

Nurain Abd Jalil, Malaysia-Japan International Institute of Technology, Universiti Teknologi Malaysia

Ruzanna Abd Samad, Malaysia-Japan International Institute of Technology, Universiti Teknologi Malaysia

Siti Nur Murni Mazli, Malaysia-Japan International Institute of Technology, Universiti Teknologi Malaysia

Nur Faezatul Hartini Radzi, Malaysia-Japan International Institute of Technology, Universiti Teknologi Malaysia

Megat Farizul Auzan Zuhairi, Malaysia-Japan International Institute of Technology, Universiti Teknologi Malaysia

Laila Dina Amalia Purba, Malaysia-Japan International Institute of Technology (MJIIT), Universiti Teknologi Malaysia (UTM KL)

Thilagavathi Arumugham, Malaysia-Japan International Institute of Technology (MJIIT), Universiti Teknologi Malaysia (UTM KL)

Eri Sahabudin, Malaysia-Japan International Institute of Technology (MJIIT), Universiti Teknologi Malaysia (UTM KL)

Le Minh Quan, Malaysia-Japan International Institute of Technology (MJIIT), Universiti Teknologi Malaysia (UTM KL)

Sasmitha Aulia Zahra, Malaysia-Japan International Institute of Technology (MJIIT), Universiti Teknologi Malaysia (UTM KL)

Dhaneswara Ilmasari, Malaysia-Japan International Institute of Technology (MJIIT), Universiti Teknologi Malaysia (UTM KL)

Nguyen Thi Cam Tu, Malaysia-Japan International Institute of Technology (MJIIT), Universiti Teknologi Malaysia (UTM KL)

Khen Chanlyda, Malaysia-Japan International Institute of Technology (MJIIT), Universiti Teknologi Malaysia (UTM KL)

Karn Tanavarotai, Malaysia-Japan International Institute of Technology (MJIIT), Universiti Teknologi Malaysia (UTM KL)

Chheang Pothiphimean, Malaysia-Japan International Institute of Technology (MJIIT), Universiti Teknologi Malaysia (UTM KL)

Kong Leangkim, Malaysia-Japan International Institute of Technology (MJIIT), Universiti Teknologi Malaysia (UTM KL)

IOP Conf. Series: Earth and Environmental Science 1091 (2022) 011001 doi:10.1088/1755-1315/1091/1/011001

THE 9TH AUN/SEED-NET REGIONAL CONFERENCE ON NATURAL DISASTER 2021 (RCND 2021)

EDITORS

Dr. Mohd Hakim Ab Halim, Main Editor, Disaster Preparedness and Prevention Centre, Department of Chemical and Environmental Engineering, Malaysia-Japan International Institute of Technology, Universiti Teknologi Malaysia

Assoc. Prof. Dr. Aznah Nor Anuar, Co-Editor, Department of Chemical and Environmental Engineering, Malaysia-Japan International Institute of Technology, Universiti Teknologi Malaysia

Prof. Dr. Muhammad Ali Muhamad Yuzir, Co-Editor, Department of Chemical and Environmental Engineering, Malaysia-Japan International Institute of Technology, Universiti Teknologi Malaysia

SCIENTIFIC COMMITTEE

Dr. Sumiaty Ambran, Malaysia-Japan International Institute of Technology (MJIIT), Universiti Teknologi Malaysia (UTM KL)

Ts. Dr. Azlinda Saadon, Department of Civil Engineering, Faculty of Engineering and Technology Infrastructure, IUKL Geo Disaster Research Center (I-GEO) University Kuala Lumpur (IUKL)

Dr. Suriani Mohd Sam, Razak Faculty of Technology and Informatics (RFTI), Universiti Teknologi Malaysia (UTM KL)

Dr. Sri Suhartini, Universitas Brawijaya, Indonesia

Assoc. Prof. Dr. Chart Chiemchaisri, Department of Environmental Engineering, Faculty of Engineering, Kasetsart University, Thailand

Dr. Mohd Hakim Ab Halim, Disaster Preparedness and Prevention Centre, Department of Chemical and Environmental Engineering, Malaysia-Japan International Institute of Technology, Universiti Teknologi Malaysia (UTM KL)

Dr. Chan Ratboren, Water Resource and Infrastructure Engineering, Faculty of Hydrology and Water Resources Engineering, Institute of Technology of Cambodia (ITC)

Prof. Joni Hermana, Teknik Lingkungan, Fakultas Teknik Sipil, Perencanaan dan Kebumian, Institut Teknologi Sepuluh Nopember (ITS)

Mdm. Ansita Gupitakingkin Pradipta, Department of Agriculture Engineering, Universitas Gadjah Mada (UGM)

Prof. Hitoshi Morikawa, Department of Civil and Environmental Engineering, School of Environment and Society, Tokyo Institute of Technology (Tokyo Tech)

Dr. Anurak Sriariyawat, Department of Water Resources Engineering, Faculty of Engineering, Chulalongkorn University

Dr. Mohammad Zakri Tarmidi, Faculty of Built Environment and Surveying, Universiti Teknologi Malaysia (UTM JB)

Dr. Carolyn Melissa Payus, Faculty of Science & Natural Resources, Universiti Malaysia Sabah (UMS)

Dr. Try Sophal, Disaster Prevention Research Institute (DPRI), Kyoto University (KyotoU)

Assoc. Prof. Dr. Koji Iwamoto, Malaysia-Japan International Institute of Technology (MJIIT), Universiti Teknologi Malaysia (UTM KL)

Prof. Ryosuke Uzuoka, Department Civil and Earth Resources Engineering, Division of Engineering, University of Tsukuba

Dr. Intan Sazrina Saimy, Razak Faculty of Technology and Informatics (RFTI), Universiti Teknologi Malaysia (UTM KL)

Dr. Davin Sang, Water Resource and Infrastructure Engineering, Faculty of Hydrology and Water Resources EngineeringInstitute of Technology of Cambodia (ITC)

Dr. Wan Mohd Muhiyyudin Wan Ibrahim, School of Humanities, Universiti Sains Malaysia

Assoc. Prof. Dr. Hirofumi Hara, Department of Biotechnology, Graduate School of Agricultural and Life Sciences / Faculty of Argiculture, The University of Tokyo

Dr. Kyohei Ueda, Department civil and Earth Resources Engineering, Division of Engineering, University of Tsukuba

Dr. Fazreena Nadia Md Akhir, Malaysia-Japan International Institute of Technology (MJIIT), Universiti Teknologi Malaysia (UTM KL)

Dr. Seri Adelila Sari, State University of Medan (UNIMED), Indonesia

Prof. Tetsuya Sumi, Disaster Prevention Research Institute (DPRI), Kyoto University (KyotoU)

Assoc. Prof. Dr. Uchida Taro, School of Life and Environmental Sciences, University of Tsukuba

Dr. Nurashikin Ihsan, Faculty of Science, Universiti Teknologi Malaysia

Prof. Dr. Montri Choowong, Department of Geography, Faculty of Arts, Chulalongkorn University

Dr. Nuruol Syuhadaa Mohd, Department of Civil Engineering, Faculty of Engineering, Universiti of Malaya (UM)

Dr. Siti Uzairiah Mohd Tobi, Razak Faculty of Technology and Informatics (RFTI), Universiti Teknologi Malaysia (UTM KL)

Mdm. Marlina Ali, School of Education, Faculty of Social Science and Humanities, Universiti Teknologi Malaysia (UTM JB)

Dr. Chanita Duangyiwa, Department of Geography, Faculty of Arts, Chulalongkorn University

Assoc. Prof. Dr. Anil Wijeyewickrema, Department of Civil and Environmental Engineering, School of Environment and Society, Tokyo Institute of Technology (Tokyo Tech)

Assoc. Prof. Ir. Dr. Fadzli Mohamed Nazri, School of Civil Engineering, Universiti Sains Malaysia

Dr. Mariyana Aida Ab. Kadir, School of Civil Engineering, Faculty of Engineering, Universiti Teknologi Malaysia (UTM JB)

Dr. Rabieahtul Abu Bakar, Razak Faculty of Technology and Informatics (RFTI), Universiti Teknologi Malaysia (UTM KL)

Assoc. Prof. Behzad Elahifar, Department of Geoscience and Petroleum, Faculty of Engineering, Norwegian University of Science and Technology (NTNU)

Prof. Ir. Ts. Dr. Azman Kassim, School of Civil Engineering, Faculty of Engineering, Universiti Teknologi Malaysia (UTM JB)

Prof. Mayank Mishra, School of Infrastructure, Indian Institute of Technology Bhubaneswar

Sasmitha Aulia Zahra, Malaysia-Japan International Institute of Technology (MJIIT), Universiti Teknologi Malaysia (UTM KL)

Dhaneswara Ilmasari, Malaysia-Japan International Institute of Technology (MJIIT), Universiti Teknologi Malaysia (UTM KL)

Nguyen Thi Cam Tu, Malaysia-Japan International Institute of Technology (MJIIT), Universiti Teknologi Malaysia (UTM KL)

Khen Chanlyda, Malaysia-Japan International Institute of Technology (MJIIT), Universiti Teknologi Malaysia (UTM KL)

Karn Tanavarotai, Malaysia-Japan International Institute of Technology (MJIIT), Universiti Teknologi Malaysia (UTM KL)

Chheang Pothiphimean, Malaysia-Japan International Institute of Technology (MJIIT), Universiti Teknologi Malaysia (UTM KL)

Kong Leangkim, Malaysia-Japan International Institute of Technology (MJIIT), Universiti Teknologi Malaysia (UTM KL)

IOP Conf. Series: Earth and Environmental Science 1091 (2022) 011001 doi:10.1088/1755-1315/1091/1/011001

CONFERENCE PHOTOS OF THE 9TH AUN/SEED-NET REGIONAL CONFERENCE ON NATURAL DISASTER 2021 (RCND 2021)

RCND 2021 MJIIT Website Photos

https://mjiit.utm.my/rcnd2021/

IOP Conf. Series: Earth and Environmental Science 1091 (2022) 011001 doi:10.1088/1755-1315/1091/1/011001

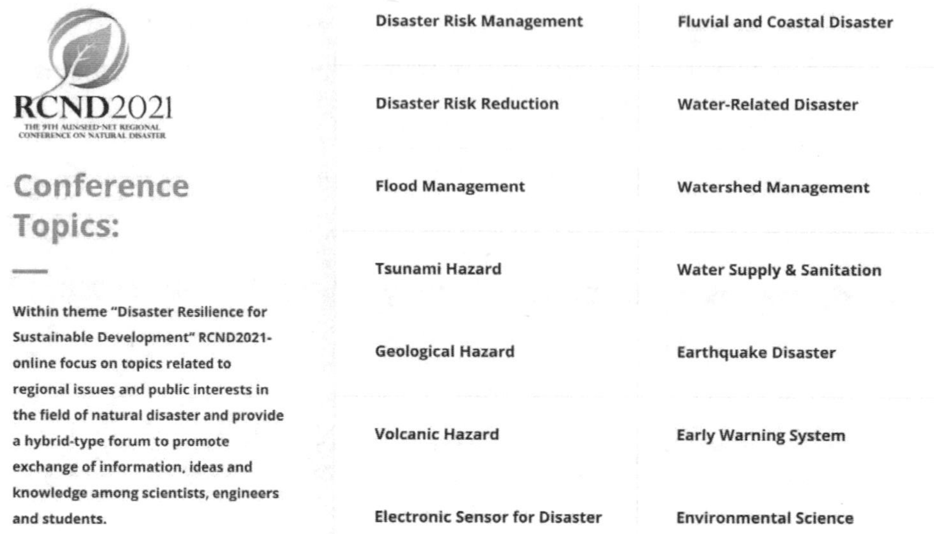

RCND 2021 Morressier Platform Photos

https://www.morressier.com/o/event/6155b046bcf459001bc87cc6

RCND 2021 Presentation Photos at Morressier Platform

IOP Conf. Series: Earth and Environmental Science 1091 (2022) 011001 doi:10.1088/1755-1315/1091/1/011001

The opening ceremony

The keynote speech 1

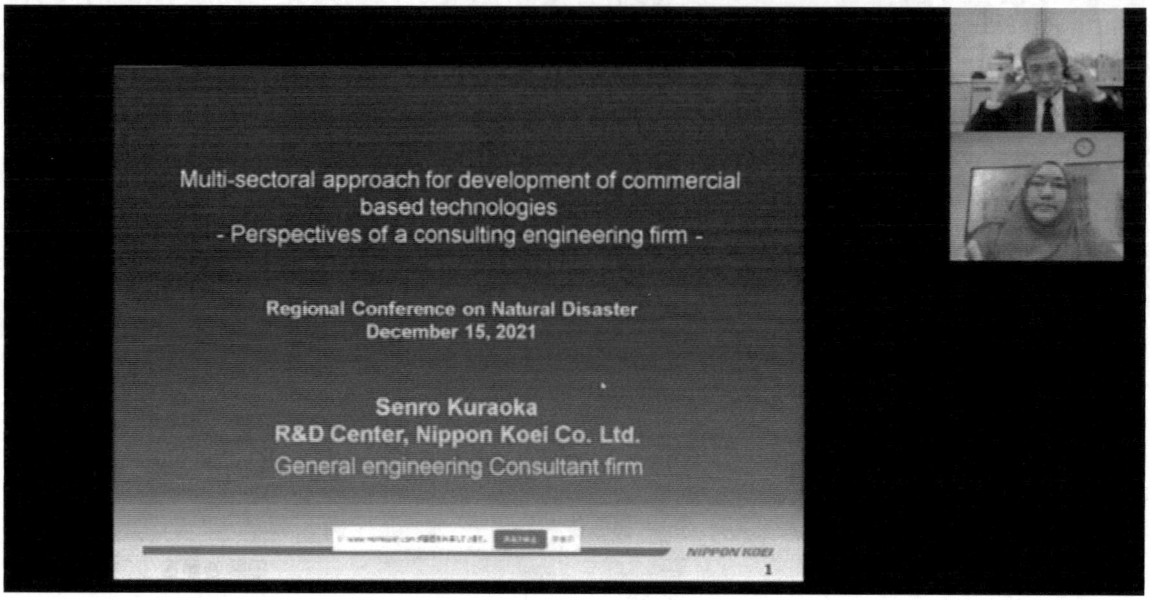

The keynote speech 2

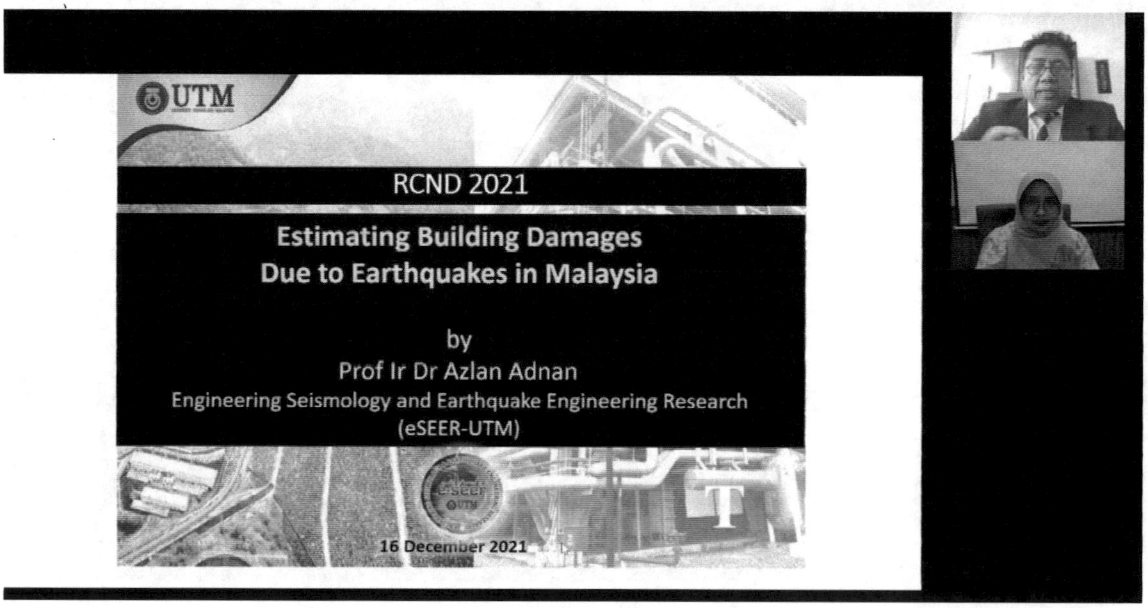

The keynote speech 3

IOP Conf. Series: Earth and Environmental Science 1091 (2022) 011001 doi:10.1088/1755-1315/1091/1/011001

The keynote speech 4

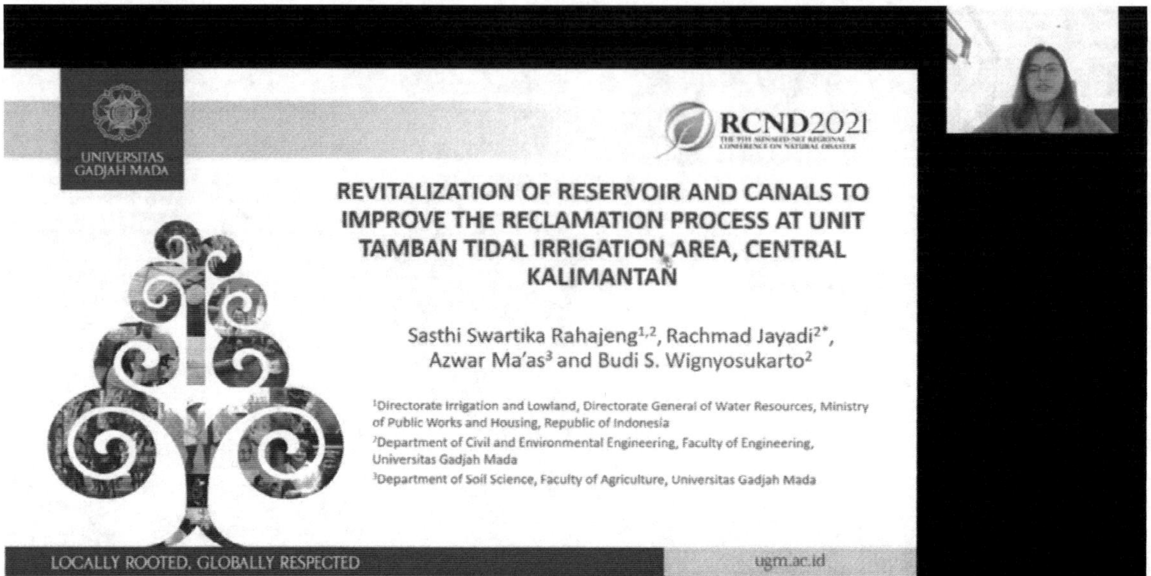

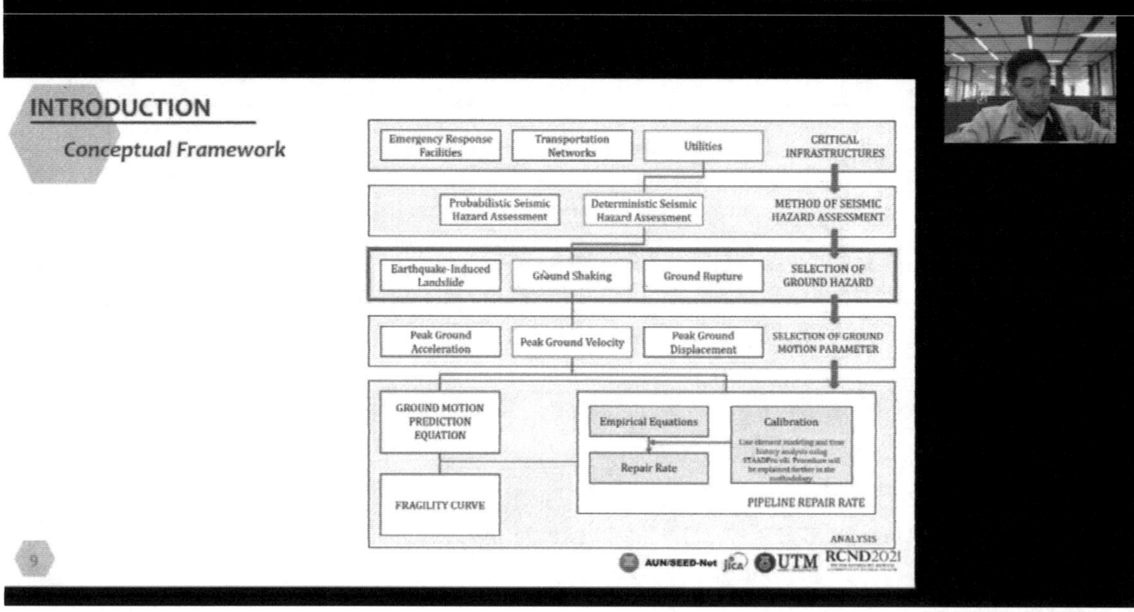

The parallel sessions

IOP Conf. Series: Earth and Environmental Science 1091 (2022) 011001 doi:10.1088/1755-1315/1091/1/011001

The closing and the award ceremony

IOP Conf. Series: Earth and Environmental Science 1091 (2022) 011002 doi:10.1088/1755-1315/1091/1/011002

Peer Review Statement

All papers published in this volume have been reviewed through processes administered by the Editors. Reviews were conducted by expert referees to the professional and scientific standards expected of a proceedings journal published by IOP Publishing.

- Type of peer review: Single anonymous
- Conference submission management system: Morressier
- Number of submissions received: 84
- Number of submissions sent for review: 84
- Number of submissions accepted: 65
- Acceptance Rate (Submissions Accepted / Submissions Received x 100): 77%
- Average number of reviews per paper: 2.00
- Total number of reviewers involved: 51
- Contact person for queries:
  Name: Dr. Mohd Hakim Ab Halim
  Affiliation: Universiti Teknologi Malaysia
  Email: mohdhakim@utm.my

  *means value has been edited

Ground Surface and Building Settlement Monitoring using Internet of Things (IoT) Approach

N A Mohd Razip[1] and N N Nik Daud[1*]

[1] Department of Civil Engineering, Faculty of Engineering, Universiti Putra Malaysia, Serdang, Selangor, Malaysia

*Corresponding author e-mail: niknor@upm.edu.my

Abstract. The ground settlement caused big concern to the engineers and the residents as the settlement can cause dangers and lead to building failure if no remedial actions are done. This paper aims to monitor building settlement related to the ground surface behavior by using the Internet of Things (IoT) approach. Two phases of laboratory works were outlined; Phase 1 – Geotechnical properties of sample collected at the site and Phase 2 – Develop a monitoring system using Arduino incorporating ultrasonic and tilt sensors for checking ground settlement and building tilt movement. All data collected using sensors will analyze based on ground settlement and building tilt behavior of the fabricated physical model. Blynk application was used to create a warning system for the user when there is settlement detected. Geotechnical properties showed that soil consists of coarse and fine particles which are 64% and 36%, respectively, and was classified as low clay (CL). The Arduino system showing the reading on tilt and ultrasonic sensors set on a fabricated model on the ground surface near to a slope under vibration condition had recorded a movement and showed the highest percentage of settlement which is 20%, respectively. In conclusion, both ultrasonic and tilt sensors have proved system ability in monitoring the building and ground surface settlement.

1. Introduction

Building and ground surface settlement is quite common to occur during or post-construction. It has always been a concern to engineers as it can negatively impact the safety, environment, and the economy. Settlement is where the downward movement and displacement of soil happen. Settlement and the downward movement of the ground can occur when there is an increase in the vertical strain of the soil [1]. It can happen when the effective stress acting on the soil increase. In real-life situations, almost all buildings will undergo settlement after years of initial construction. Settlement can be happened due to several factors such as loss of strength of the soil, maturing trees, changes in the value of moisture content which will then cause consolidation, and due to vegetation or poor compaction of the soil. Settlement of the ground surface will eventually and indirectly lead to the building settlement [2]. It is predicted that the building will settle due to its weight and the sudden decrease of soil stiffness beneath the foundation.

There are several ground surface settlement cases reported in Malaysia which has led to severe structural failures. The cause of the settlement varies but the focus of this study is the settlement due to the existence of groundwater with the assistance of load applied and due to the vibration effect. Back in 2017, in Penang, a few houses have been reported to shows significant cracks due to soil movement. The affected houses were located along an old riverbank in Hong Seng Estate as shown in **Figure 1 (a)** and **Figure 1 (b)**. The soil movement happened due to the heavy downpour in the area for a few days straight. The heavy downpour causes the soil to be loosened and decreases its strength of the soil. The loosened of the soil caused the shifting of the house structures and losing their structural integrity. The houses were badly damaged showing significant cracks on the wall, ground surface, and obvious tilting of structure [3]. According to The Star (2017), the site was inspected by the professionals, and the residents were ordered to evacuate their homes as the houses were severely affected and might cause danger to them [4].

(a) (b)

Figure 1. a) Apron slab detached from the ground beam, b) major cracks on the wall of houses [3]

Another ground settlement case has been recorded in Ranau, Sabah due to the earthquake. It happened on 5 June 2015 at 5.9 magnitudes. The strong earthquake magnitude showed seismic intensity in that area with loose soil properties. Even though the earthquake magnitude was not that high, there were still a few buildings that have been affected and severely damaged as shown in **Figure 2 (a)** and **Figure 2 (b)**. The seismic design was not considered during the designing of all the affected buildings. The earthquake has generated higher ground motions and peak ground acceleration which causes the soil particles to move aggressively during the earthquake. The movement of soils especially soft soil tends to amplify the shaking of the ground which will then leads to ground movement. A few severe damages have been detected and some repairs and rehabilitation works have been conducted by [5].

(a) (b)

Figure 2. a) Severe buckling of link reinforcement, b) severe damages detected on the beam-column joints [5]

The severe damage of the building structure happened as the settlement occurred exceeded the maximum permissible settlement. The allowable settlement of a structure depends on several factors such as the foundation type, soil type, and the usage of the structure.

The usage of Internet of Things (IoT) has been used widely in a lot of sectors. The IoT monitoring system has shown its ability and validity to provide continuous and real-time measurements with the same accuracy [6]. The real-time monitoring by IoT system helps to detect any abnormal changes of structure faster. Thus, appropriate warning signals and remedial actions can be conducted in time. In addition, monitoring of ground and building settlements using IoT is important to be conducted, especially in the risk areas so that even a small magnitude of settlement can be detected immediately. It is also important to monitor if the ground and building are behaving according to the expected design criteria.

2. Methodology

Figure 3 shows the flow chart for this study. It consists of two phases. The first phase was divided into two which is soil laboratory testing and the development of ground and building settlement monitoring system. Next, is the second phase which is analyzing the settlement behavior of the soil sample on different conditions, based on the data obtained from the Arduino software.

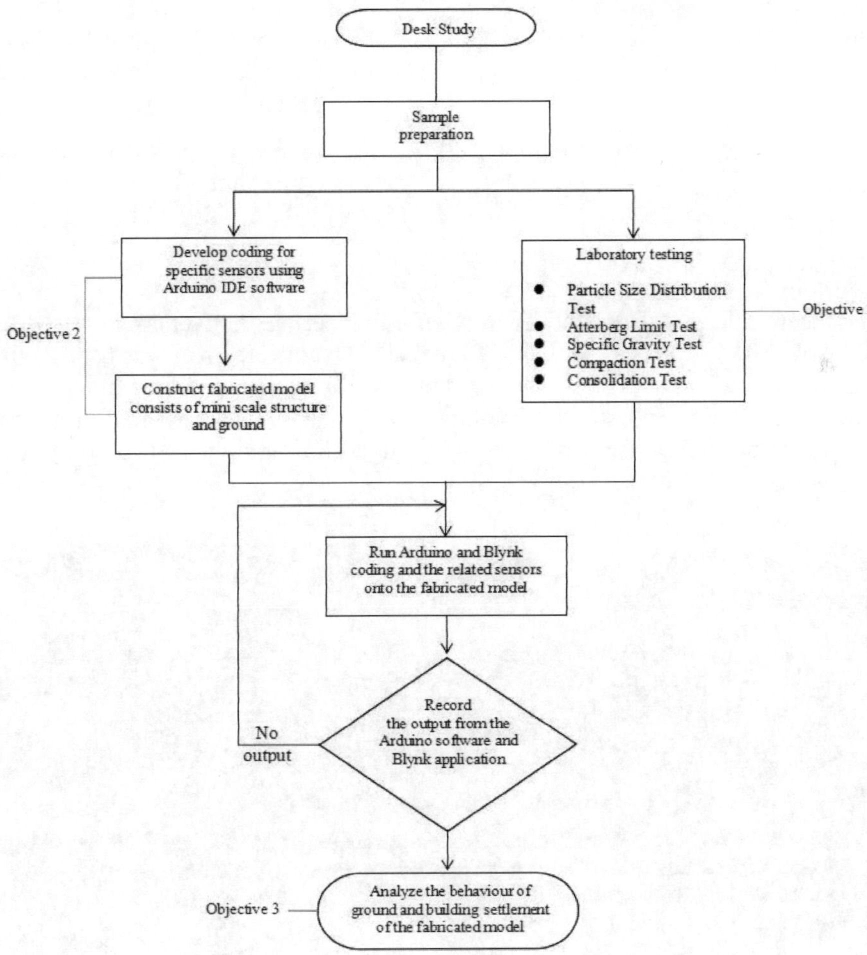

Figure 3. Flowchart of the methodology

IOP Conf. Series: Earth and Environmental Science 1091 (2022) 012001 doi:10.1088/1755-1315/1091/1/012001

2.1 Apparatus and Standard

The apparatus used in this study were tabulated in **Table 1** according to each of the tests that have been conducted. The standard that has been used in conducting the tests was the American Society for Testing and Materials (ASTM).

Table 1. List of apparatus and standards used in the study

Apparatus	Laboratory Testing	Standard
Sieve shaker, hydrometer, sedimentation cylinder	Grain Analysis Test	ASTM D 422 - Standard Test Method for Particle-Size Analysis of Soils [7]
Desiccator, vacuum pump, funnel	Specific Gravity Test	ASTM D 854 - Standard Test Method for Specific Gravity of Soils [8]
Casagrande device, flat grooving tool	Atterberg Limit Test	ASTM D 4318 - Standard Test Method for Liquid Limit, Plastic Limit, and Plasticity Index of Soils [9]
Compaction mould, manual rammer, graduated cylinder	Compaction Test	ASTM D 698 - Standard Test Method for Laboratory Compaction Characteristics of Soil Using Standard Effort [10]
Consolidation load frame, consolidation cell	Consolidation Test	ASTM D 2435 - Standard Test Method for One-Dimensional Consolidation Properties of Soils [11]

2.2 Studied Location

Soil sample has been collected from the Faculty of Engineering; Universiti Putra Malaysia (GPS coordinate: 3.005957, 101.722087) as shown in **Figure 4**. The collected soil was used as the ground of the fabricated model and to analyze the settlement behavior by conducting laboratory testing. The site has been chosen because there are building development around it. This is related to the main purpose of the study which is to monitor the settlement of ground and building and to study the settlement behavior.

Figure 4. The site location (Source: GoogleMap, 2022)

2.3 Work Program

There are several parts of the work program for this study, i) laboratory test, ii) setup of physical fabricated model and iii) development of warning system using Blynk application. **Part 1:** A laboratory testing was conducted for this paper to determine the physical and mechanical properties of soil samples and to predict the behavior of the soil.

Part 1.1: Physical soil testing such as grain size analysis, hydrometer analysis, Atterberg limit, specific gravity, and compaction test has been conducted to investigate the physical properties of the soil sample (**Figure 5**).

(a) (b) (c)

Figure 5. a) Compaction test, b) soil samples prepared from compaction test for moisture content, c) soil samples prepared from Atterberg limit test

Part 1.2: A consolidation test was conducted to study the settlement behavior of the soil samples (**Figure 6**). The test was conducted by referring to the standard reference of ASTM D 2435 - Standard Test Method for One-Dimensional Consolidation Properties of Soils.

Figure 6. Consolidation test

Part 2: Physical fabricated model setup

A fabricated model includes mini building and ground set up was constructed in laboratory scale as shown in **Figure 7**. The purpose of using fabricated model is to test the functionality of the ultrasonic and tilt sensor in monitoring settlement of ground and building.

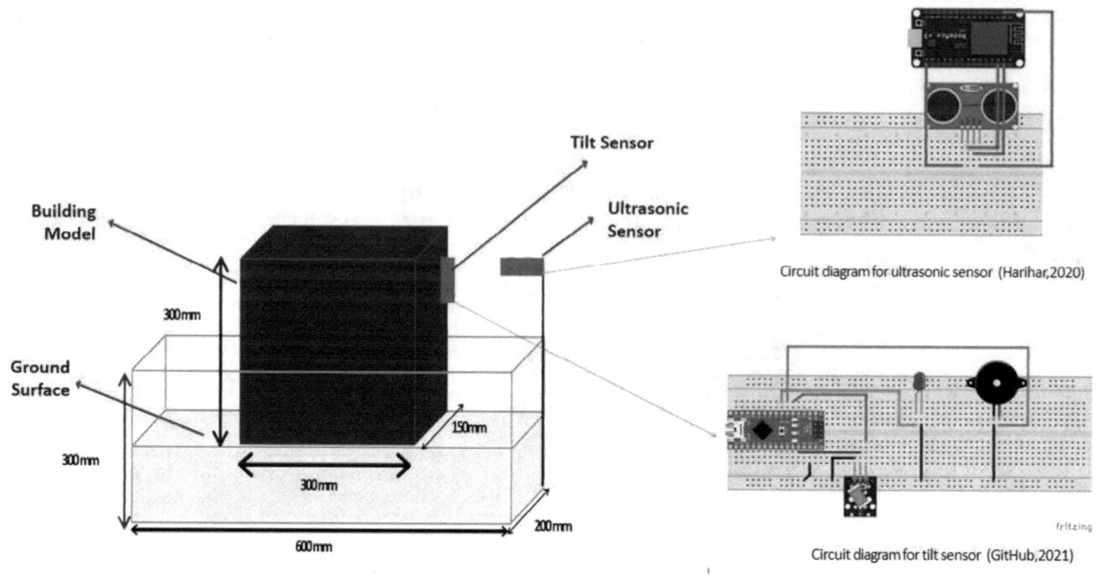

Figure 7. Setup for physical fabricated model

The ground and building settlement monitoring were done by using the IoT approach which is the Arduino programming. For this project, the settlement was monitored using a few sensors such as an ultrasonic sensor and a tilt sensor by the Arduino program. The circuit diagram for both sensors was referred to GitHub, 2021 [12]. The ultrasonic sensor was used to determine the magnitude of settlement of the ground surface and tilt sensor functions to detect if there is any downward movement occurring in the building. All these sensors will be combined using Arduino programming to form a system.

The ultrasonic sensor was set up at a certain height from the surface of the fabricated model vertically. The sensor was adjusted accordingly until it can measure the vertical distance of the ultrasonic sensor to the ground surface of the model. If the distance of the ultrasonic from the ground surface increases more than the initial distance, thus it indicates that there is a downward movement of the ground surface. The tilt sensor was attached to the structure at one of the corners. This is done to detect if there is any tilt of the structure which may be the consequence of the ground settlement. The Arduino's sensors have been developed to function for a lab-scale reading. Hence, there is no limitation on the period of the device can function with monitoring. Some improvements can be conducted to able the devices to function in real settlement monitoring situations.

Part 3: Development of a warning system using Blynk application

Blynk is an open-source IoT platform that allows quick interfaces for monitoring hardware projects from a mobile phone. As for this project, the Blynk application is used to display the output from Arduino and create a warning system to the user when there is a settlement detected. The user can monitor the ground and building settlement through mobile phone and will receive a notification instantly when settlement occurs. Hence, a regular site visit for only settlement monitoring purposes is no longer necessary and more time and cost can be saved.

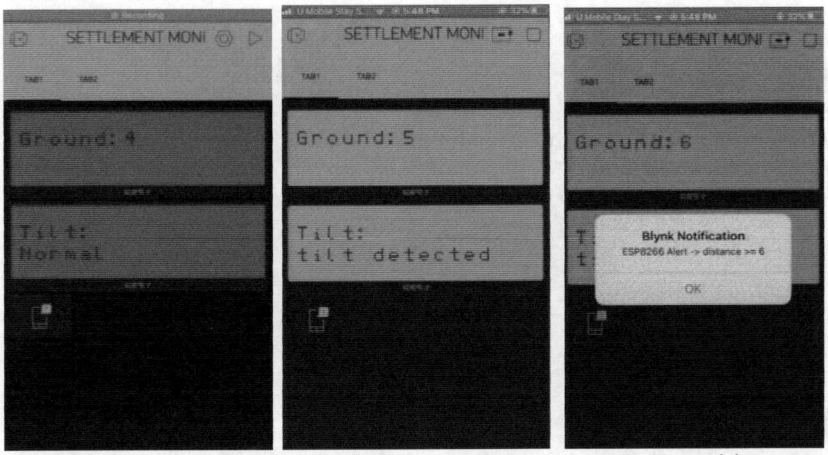

(a) (b) (c)

Figure 8. a) the Blynk application displays the output from the Arduino when there is no settlement (when the initial height of ultrasonic sensor = 4cm), b) the Blynk application displays the output from the Arduino when there is settlement occur, c) the warning alert from Blynk application when there is settlement detected

Figure 8 (a) shows the output displays on the Blynk application when there is no settlement occur. It is the situation when there is no movement of ground and tilt of building. The tilt shown as 'Normal' indicates that there is no tilt of the building, Besides, if the distance of the ground to the ultrasonic sensor remains the same, it shows that there is no ground settlement. **Figure 8 (b)** shows the output displays on the Blynk application when there is settlement occurs. It is the situation when there is a movement of ground and tilt of building. The tilt shown is 'tilt detected' indicates that there is a tilt of the building. An increase of distance between the ground and ultrasonic sensor indicates that the ground is moving away from the sensor and there is ground settlement has occurred. While **Figure 8 (c)** shows the warning system created through the Blynk application when there is a settlement occurs. The user will receive notifications instantly as shown by the figure.

2.4 Testing Program

To create the settlement in the fabricated model, two different situations which can trigger and lead to the settlement were applied to it. The first situation is by increasing the groundwater level with the assistance of increasing loads onto the structure. Increasing groundwater level causes soil to be loosened and decreases the strength of the soil. Extra loads can lead to the increase of vertical strain in the soil which will then causes vertical movement in the soil. Hence, a settlement can be generated. To analyze the settlement behavior and the effect of the loads and groundwater level on the structure, various amounts of loads corresponding to the level of groundwater were applied. The loads applied to the model were various in the range of 0.1kg to 1.8kg. Next, to generate the settlement, the vibration was applied to the model. The vibration will trigger more movements between soil particles which can lead to the settlement of the ground surface and the building. In real-life situations, the vibration in the soil can be generated by an earthquake, construction works, or seismic response. But, for this project since the model is built on a small scale, a shaking table was used to create the vibration situation to the fabricated model. All related figures are shown in **Figure 9 (a) – (c)**.

(a)

(b)

(c)

Figure 9. a) Fabricated model under increasing groundwater level and load condition, b) Fabricated model on ground surface on shaking table, c) Fabricated model on the ground surface near to a slope on shaking table

3. Results and Discussion

All results that have been obtained from the soil laboratory testing (physical and mechanical properties of soil sample) and software output from Arduino software using ultrasonic sensor and tilt sensor will discuss in this part. The first result is the geotechnical properties of the soil sample. The second results are the settlement values and tilt detection that can be obtained from the ultrasonic sensor and tilt sensor using Arduino software on the fabricated model. The settlement behavior of the fabricated model for each of the different situations will be analyzed in this chapter.

3.1 Physical and Mechanical Properties of Soil Sample

The physical properties of the soil sample were tabulated in **Table 2**. The results have been obtained from the soil laboratory testing and been analyzed to study the properties and behavior of the soil sample.

Table 2. Physical Properties of Soil

Properties		Unit	Values
Grain Size Distribution			
	Gravel	%	20.8
	Sand	%	43.5
	Fines	%	35.7
Specific Gravity, G_s		-	2.60

Liquid Limit, *LL*	%	36
Plastic Limit, *PL*	%	22
Plasticity Index, *PI*	%	14
Optimum Moisture Content, *OMC*	%	11.11
Maximum Dry Density, *MDD*	kg/m^3	1638

The physical properties value of soil samples has been obtained by conducting physical laboratory tests. From grain size analysis as shown in **Figure 10**, the soil sample can be classified as fine soils as the percentage of fine soils is slightly more than 35% which is 35.7%. The specific gravity of the soil is 2.60 which indicates the soil contains a large amount of organic matter. While for the values of the liquid limit, plastic limit and plasticity index are 36%, 22%, and 14% respectively. Using the plasticity index chart, the soil can be classified as clay of low plasticity (**Figure 11**). The compaction test showed the optimum moisture content of soil and maximum density obtained are 11.11% and 1638 kg/m^3 respectively.

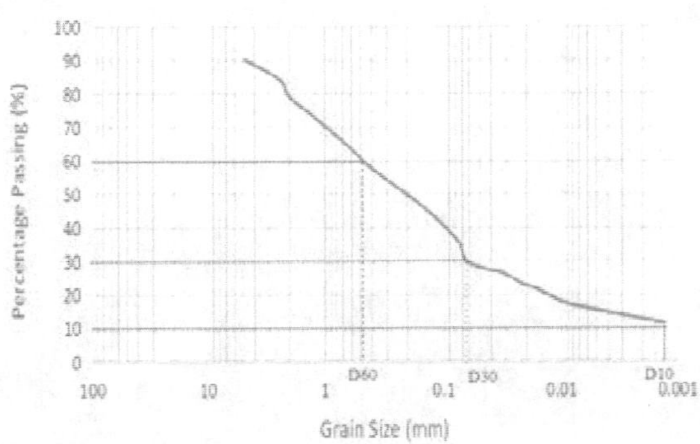

Figure 10. Particle size distribution of the soil sample

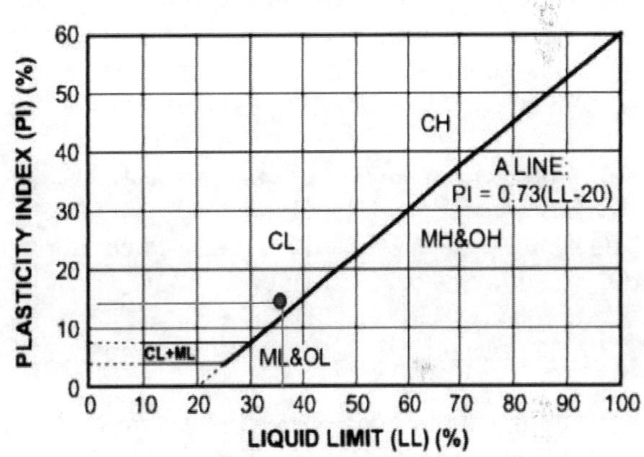

Figure 11. Plasticity Index chart

A consolidation test has also been conducted to study the settlement behavior of the soil. The pressure applied to the sample has been incremented after every 24 hours which is from 2.55 kPa to 5.10 kPa and 10.20 kPa. The permeability of soils at 2.55 kPa, 5.10 kPa and 10.20 kPa pressures are 1.2026 x 10^{-9} ms^{-1}, 3.0516 x 10^{-10} ms^{-1} and 7.7321 x 10^{-11} ms^{-1}, respectively. The values of coefficient permeability of soil obtained have indicated that the permeability of the soil sample is low since the range of permeability is lower than 10^{-8} ms^{-1} (**Table 3**).

Table 3. Mechanical Properties of Soil

Parameters	Pressure Applied (kPa)		
	2.55	5.10	10.2
Coeff. of consolidation, c_v(m^2/year)	0.381	0.173	0.067
Coeff. of vol. compressibility, m_v (m^2/kN)	0.0996	0.0557	0.0365
Coeff. of permeability, *k* (m/s)	1.203 x 10^{-9}	3.052 x 10^{-10}	7.732 x 10^{-11}
Compression index, C_c		0.138	

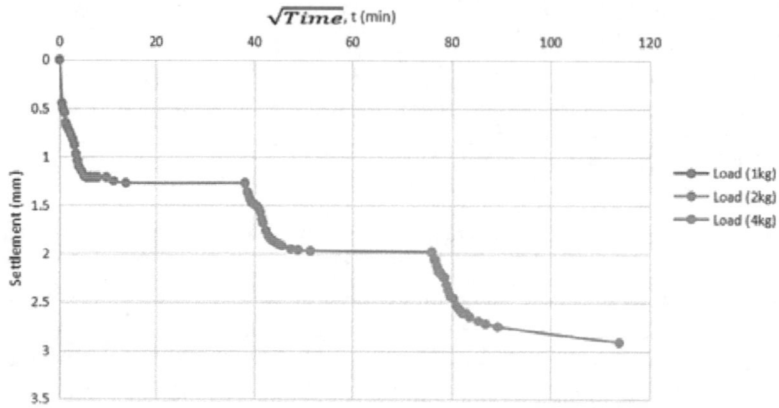

Figure 12. Settlement behavior of the soil sample with variation of pressure increments

Figure 12 shows the settlement behavior of the soil sample. It can be observed that the samples undergo immediate settlement as the load is applied and added. At pressure 2.55kPa, the primary settlement occurs up to the 20th minute where the settlement increases rapidly. Passing the 20th minute, secondary settlement can be observed as the rate of settlement slowly decreases over time.

3.2 Settlement Behaviour of Mini-Scaled Fabricated Model using Arduino Software

The settlement behavior of the mini-scaled fabricated model by using Arduino software will be discussed in this section.

3.2.1 Settlement Behavior under variation Loads and Groundwater Condition

The settlement behavior under increasing loads and groundwater existence condition of the fabricated model was analyzed based on **Table 4** and **Figure 13.**

Table 4. Settlement values and tilt detection using ultrasonic sensor and tilt sensor

Load (g)	Groundwater Table Level	Initial Distance Ultrasonic Sensor to Ground (cm)	Distance (cm)	Settlement (cm)	Tilt Sensor
0	0.1H	14	14	0	Normal
1000	0.3H	14	15	1	Tilt Detected
1500	0.6H	14	16	2	Tilt Detected
1800	1.0H	14	16	2	Tilt Detected

* H = total height of soil ground

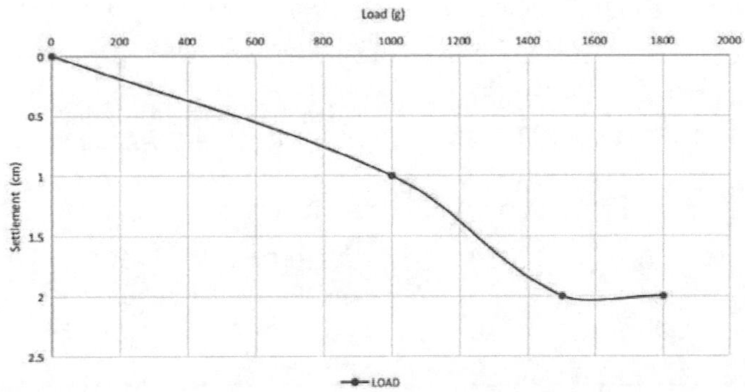

Figure 13. Settlement behavior of soil sample under variation of loads using ultrasonic sensor

An ultrasonic sensor has been used to measure the difference in distance between the sensor itself to the ground surface. The changes of distance between the ultrasonic sensor to the ground surface indicate that there are soil movements. While for the tilt sensor, it was used to detect the tilt movement of the building model.

A different number of loads has been applied to the building model and the groundwater table has been increased to different levels. Based on the results output from the ultrasonic sensor and the tilt sensor, there is no settlement and tilt movement detected when there are no extra loads applied to the building and no increase of initial groundwater table high. The settlement of the ground surface at 1cm and tilt movement of the building started to occur when the load applied to the building model has been added at 1000g and the groundwater table increased to 0.3H. The settlement value increased to 2cm when the load applied increased from 1000g to 1500g and groundwater table high increased from 0.3H to 0.6H. This shows that the increase of load applied, and groundwater level may lead to settlement.

Changes of groundwater level may lead to differential settlement of structures and foundations as it can affect the mechanical properties and bearing capacity of the soil [13]. The increase of extra loads onto the building model causes the vertical strain of soil to increase and leads to the downward movement of the ground surface [1]. There is no further settlement recorded once the load applied onto the building model is more than 1500g and the groundwater table rise until it reached 1.0H. It indicates that the soil sample has been fully saturated with water.

It can be observed that there is a similarity in the settlement behavior of the soil sample during consolidation test shown in **Figure 12** and the settlement behavior of soil samples under increasing load and groundwater table condition shown in **Figure 13**. In both situations the soil sample undergoes immediate settlement once the first load is applied resulting the decreased in settlement rate as the load applied is incremented.

3.2.2 Settlement Behaviour under Vibration Condition

The settlement behavior under vibration conditions of the fabricated model placed on flat ground surface and near to slope was analyzed as show in **Table 5**, **Table 6**, **Figure 14**, and **Figure 15**.

Table 5. Settlement values and tilt detection using ultrasonic sensor and tilt sensor of fabricated building model placed on flat ground

Shaking Table Excitation (V)	Time (s)	Initial Distance (cm)	Distance (cm)	Settlement (cm)	Tilt
	0	20	20	0	Normal
	30	20	20	0	Normal
1	60	20	20	0	Normal
	90	20	20	0	Normal
	120	20	20	0	Normal
	0	20	20	0	Normal
	30	20	20	0	Normal
1.5	60	20	20	0	Normal
	90	20	20	0	Normal
	120	20	21	1	Tilt detected
	0	20	20	0	Normal
	30	20	20	0	Normal
2	60	20	21	1	Tilt detected
	90	20	21	1	Tilt detected
	120	20	22	2	Tilt detected

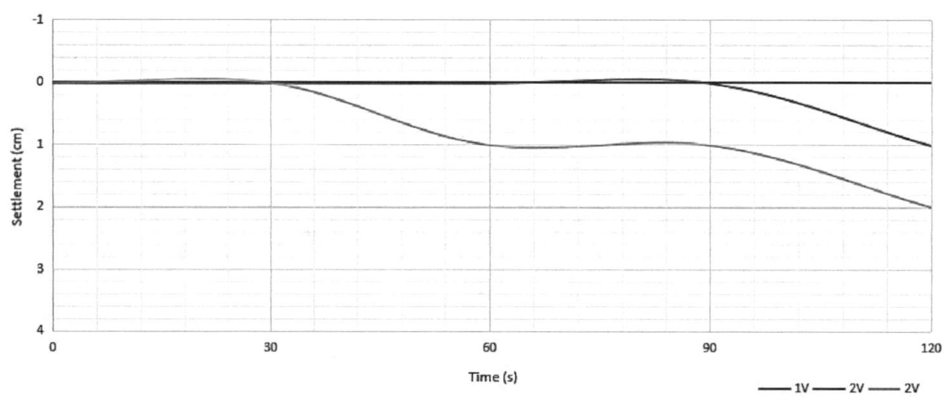

Figure 14. Settlement behavior of soil sample under various shaking table voltage on flat ground

Table 6. Settlement values and tilt detection using ultrasonic sensor and tilt sensor of fabricated building model placed on flat ground near to slope

Shaking Table Excitation (V)	Time (s)	Initial Distance (cm)	Distance (cm)	Settlement (cm)	Tilt
	0	20	20	0	Normal
	30	20	20	0	Normal
1	60	20	20	0	Normal
	90	20	20	0	Normal
	120	20	21	1	Tilt detected

1.5	0	20	20	0	Normal
	30	20	20	0	Normal
	60	20	21	1	Tilt detected
	90	20	22	2	Tilt detected
	120	20	22	2	Tilt detected
2	0	20	20	0	Normal
	30	20	21	1	Tilt detected
	60	20	21	2	Tilt detected
	90	20	22	2	Tilt detected
	120	20	23	3	Tilt detected

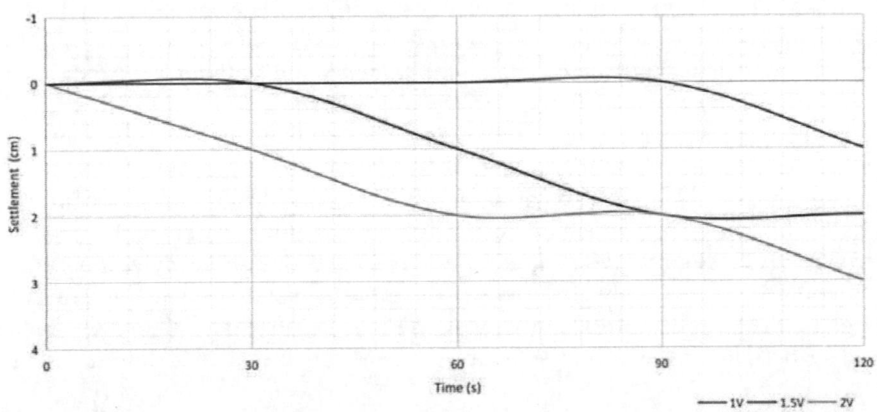

Figure 15. Settlement behavior of soil sample under various shaking table voltage located on flat ground near to slope

The settlement behavior due to seismic response can be analyzed through **Figure 14** and **Figure 15**. The settlement behavior is different depending on the respective shaking table voltage. The difference between these two graphs is that **Figure 14** shows the settlement behavior when the fabricated building model is put on flat ground. While for **Figure 15**, the fabricated building model is placed on flat ground but near to a slope.

Each graph shows that the settlement increases as the shaking table voltage increase. Increasing of shaking table voltage indicates that the vibration of the shaking table also increases. Referring to **Figure 14**, building model is placed on flat ground, it can be observed that there is no settlement and tilt movement of the building model recorded when the voltage of the shaking table is at 1V. It indicates that voltage produces lower vibration which does not affect much the stability of the building model. The settlement started to occur when the shaking table voltage is 1.5V. The settlement at 1cm and the tilt movement of the building model were recorded at 120th seconds. Moreover, the settlement rate is faster at the higher voltage, 1.5V as the first settlement detected was at the 30th second after running the shaking table. The settlement increases from 1cm at 30th seconds to 2cm at 120th seconds. This is due to a higher voltage of the shaking table, vibration, and speed produced by the shaking table is higher causing the stability of the building model to be affected. In certain situations, the combination of ground vibration and cohesive soils may be the cause of soil settlements [14]. It can lead to damage to the structures due to the interaction between soil structure and waves propagated from the vibration movement.

While for the other situation, the building model was put onto a flat ground but near to a slope. The settlement behavior of the building model can be observed and concluded much faster compared to the flat ground without a slope. This can be proved as there is a 1cm settlement and tilt movement of the building model recorded at 1V voltage of shaking table. Compared to the other situation when there is no slope, no settlement was recorded at the same voltage, 1V. As the voltage of the shaking table increases, the rate, and magnitude of settlement increase. When the voltage of the shaking table is 1.5V, settlement and tilt movement of the building model started to occur at 60 seconds (1cm) and the maximum settlement recorded was 2cm at 120 seconds. As the voltage increases to 2.0V, the first settlement and tilt movement recorded are earlier which is at 30 seconds and the maximum settlement recorded was 3cm at 120 seconds.

Based on the results and discussions above, it can be concluded that the vibration produced by the shaking table can lead to building model settlement. This is because the vibration can trigger movements between the soil particles. The higher voltage of the shaking table means higher vibration produced. Hence the movement between soil particles becomes more aggressive. Comparing the two situations where the building model was put onto a flat ground with and without slope, there is a different observation regarding the settlement behavior. The rate of settlement and tilt movement with the presence of slope is higher and faster compared to the one without slope. The presence of slope causes the stability of soil to be affected by increasing the shear stress and reducing the shear strength [15]. More shear strength such as frictional resistance, gravity forces, and cohesion between soil particles are needed to resist the downward movement of soil particles and building model with the presence of slope near to the structure. The slope movement can be slower and decreased by adding embedded piles. According to Varco d'Izzo case study [16], the movement of landslide in clay soil and a railway tunnel can be decreased by the presence of sheet pile walls. This is because the embedded piles influence the earth pressure distribution transferred to the tunnel and can increase the landslide-tunnel interaction.

4. Conclusion

This study focuses on investigating the geotechnical properties of soil samples and analyzing the settlement and tilt behavior of the soil sample in a mini-scaled fabricated model. Based on the analysis of laboratory testing results, the soil is categorized as clay of low plasticity. Results obtained from both sensors show that the settlement behavior of the soil sample varies depending on the situation. The highest percentage of settlement was recorded from the fabricated model on the ground surface near a slope under vibration conditions, which is 20%. The other two conditions (fabricated model under increasing groundwater level and load condition, and fabricated model on ground surface under vibration condition) recorded the same percentage of settlement which is 13.3%. Finally, the settlement behavior of soil samples has been successfully obtained and analyzed proving that the ultrasonic and tilt sensor using the Arduino program can monitor the ground surface and building settlement.

Acknowledgments

Authors would like to acknowledge and give the warmest thanks for the technical support offered by the laboratory staff of Geotechnical and Geological Engineering Laboratory of Civil Engineering Department, Engineering Faculty, Universiti Putra Malaysia.

References

[1] Day C D 2018 Differential settlement in buildings. Retrieved from https://vertexeng.com/insights/differential-settlement-in-buildings/

[2] Seok J, Kim O, Chung C and Kim M 2001 Evaluation of ground and building settlement near braced excavation sites by model testing *Canadian Geotechnical Journal* **38(5)** 1127–1133

[3] Mohamed B 2017 Penang Homes Suffer Structural Damage as Soil Sinks Following Downpour. New Straits Times

[4] The Star 2017 Cracks in Homes Force Seven Penang Families to Flee Retrieved from https://www.thestar.com.my/news/nation/2017/10/01/cracks-in-homes-force-seven-penang-families-to-flee/

[5] Khoiry M A, Hamzah, N, Osman S, Mutalib A and Hamid R 2018 Physical damages effect on residential houses caused by the earthquake at Ranau, Sabah Malaysia *International Journal of Engineering and Technology* **10** 414-418 10.7763/IJET.2018.V10.1094.

[6] Lam R C Y, Junus A, Cheng W M Y, Li X and Lam L C H 2017 IoT Application in Construction and Civil Engineering Works *International Conference on Computational Science and Computational Intelligence (CSCI)* 1320-1325

[7] ASTM D422-63 2007 Standard Test Method for Particle-Size Analysis of Soils (Withdrawn 2016) *ASTM International* West Conshohocken

[8] ASTM D854-00 2000 Standard Test Methods for Specific Gravity of Soil Solids by Water Pycnometer *ASTM International* West Conshohocken

[9] ASTM D4318-17e1 2007 Standard Test Methods for Liquid Limit, Plastic Limit, and Plasticity Index of Soils *ASTM International* West Conshohocken

[10] ASTM D698-12 2021 Standard Test Methods for Laboratory Compaction Characteristics of Soil Using Standard Effort (12,400 ft-lbf/ft^3 (600 kN-m/m^3)) *ASTM International* West Conshohocken

[11] ASTM D 2435 2006 Standard Test Method for One-Dimensional Consolidation Properties of Soils *ASTM International* West Conshohocken

[12] GitHub 2021 The Tilt Sensor *Experimenting-with-Arduino* Retrieved from https://github.com/

[13] Lizhu M, Zhonghua Z and Yue T 2019 The Influence of Groundwater Level Rise on Bearing Capacity of Sand Foundation *IOP Conference Series: Earth and Environmental Science* 304 022089. doi:10.1088/1755-1315/304/2/022089

[14] Svinkin, M R 2015 Tolerable Limits of Construction Vibrations *Practice Periodical on Structural Design and Construction* **20(2)** 04014028. doi:10.1061/(asce)sc.1943-5576.0000223

[15] Jiang C, Wu W, He J and Chen L 2020 Computation method for the settlement of a vertically loaded pile in sloping ground *Advances in Civil Engineering* 1–10 doi:10.1155/2020/2109535

[16] Vassallo R, Mishra M, Santarsiero G and Masi A 2019 Modeling of Landslide–Tunnel Interaction: the Varco d'Izzo Case Study *Geotechnical and Geological Engineering*, (), – . doi:10.1007/s10706-019-01020-x

IOP Conf. Series: Earth and Environmental Science 1091 (2022) 012002 doi:10.1088/1755-1315/1091/1/012002

Vibration Effects on Soil Slope Detected by using Arduino System

M A A Azharudin[1], N N Nik Daud[1]*

[1]Department of Civil Engineering, Faculty of Engineering, Universiti Putra Malaysia, Serdang, Selangor, Malaysia.

*Corresponding author e-mail: niknor@upm.edu.my

Abstract. Slope failure is a phenomenon occurs when a slope rupture due to several factors and often happen in many countries including our country, Malaysia. These factors lead to slope failure which are underground movement due to vibration, erosion, soil characteristics, heavy rainfall, and others. Thus, monitoring process is important to conduct at risky slope. The purpose of this study is to develop a system that can monitor the slope before a failure can happen using an Internet of Things (IoT), Arduino programming. The process involved by developing a vibration effect using the mini-shaking table and sensors (accelerometer, tilt) will use by incorporating it with Arduino UNO kit. The slope of physical model was prepared in lab scale by using the soil sample collected at specific slope at varsity area. Basically, the whole works were covered a lab-based experiment and slope simulation when different vibration of the force was applied on the slope, the Arduino sensor will detect the value due to induced vibration. The created force causes movement of slope which will detected by accelerometer applied and the acceleration values will be recorded using Arduino kit. Based on the graph of acceleration against time, the values of acceleration will be analysed. Particle size distribution of soil sample results shown that the soil consists more fine particles which is 54% and low silt (ML) also agreeable by plasticity index chart. The optimum water content and maximum dry density for the soil sample are 13.35% and 16.95 g/cm^3. The strength parameters indicated a cohesion value, c is 18.9 kPa and internal friction angle, ϕ is 32.3°. From physical model observations, it is shown that when the vibration force is applied with certain voltage, the slope tends to fail and there is significant reading was recorded. It can be summarized that the response of the soil is dependent on the type of soil, as well as the amplitude of the vibration.

1. Introduction

For the past decade, Malaysia has experienced a prompt infrastructure evolution due to the population growth. Nowadays, people favoured to build their buildings and houses at mountainous and hilly area which can be categorised as an unsafe area. As a result, slope stability problems have become the most concern part in the construction sector. Slope failure occurs if the shear strength of the ground is surpassed, resulting in slope collapse [1]. Hence, to avoid the unexpected collateral damage and to overcome rapid population growth, engineers must come out with a solution using a simple structure model at the hilly area to observe the condition of the slope at the area.

Slope failure is a main risk in geotechnical engineering where tragedies and fatalities relating the damage of the properties and death always occur. Even it does not cause accidental injury, there is a high possibility due to production interruption, lost invention, and structure damages. Numerous landslides that have had a significant effect on Malaysian society have been recorded. One of the incidents that is still give a big impact is the Highland Tower disaster back on December 11, 1993, that located at Ulu Klang, Selangor, Malaysia. The tragedy had causing deaths of 48 peoples when the Tower Block 1 collapsed (**Figure 1**). Based on the investigation, the collapse is caused by the slope failure due to heavy rain that burst diversion pipes [2].

Figure 1. Highland Tower Collapse due to slope failure [3]

Early warning is the important thing to alert authorities and citizen about the incoming danger. Sometimes, evacuation to shelters are the most significant things when having a disaster. In general, people send warning signals based on rainfall features, which assume that the danger of a slope failure is assessed by the rainfall intensity. The method is basically giving a proper assumption and the early warning system was being implemented nowadays. The action must be taken to stabilize an occurrence and minimize the possible damage. Actions taken immediately in the first minutes following an emergency will significantly affect what happens afterward.

There are several factors that can take into the consideration that contributed to the slope failure, such as the underground movement, erosion, soil characteristics, rainfall characteristics etc. For example, the presence of surface run-off water when at the rainy condition will penetrate the ground and occur the underground soil movement. Throughout history, stability of slope issues has been faced when the construction of building has interrupted the delicate equilibrium of natural slopes [4]. Others, the slope failure will occur from the vibration that come from the earthquake. Hence, it will be decreasing the overall strength of slope failure. According to Omar et al. (2007), another hazardous site with a greater possibility of slope failure is on highways or roads that have undergone extensive excavation [5]. Engineers must take careful the precautions before starting any development and construction to ensure that the slope design is stable. Slope stability monitoring is often performed at the beginning, and sometimes during the life cycle of a project and construction design, maintenance, and rehabilitation.

It is important to analyse and understand the process involved on the slope failure where the vibration effect that comes from the load, whether it's a dynamic; vibration from the transportation, to ensure the safer urban development on the hilly area. Thus, the studies of the slope monitoring are very important to learn from the case studies or histories by develop some device or sensor to monitor the slope failure. This has been agreed by Bunawan et al., (2016) which has developed a slope monitoring device using accelerometer. The application and operation of this system does not interrupt by physical obstacles, different climate change conditions and the construction works at site [6].

Vibration monitoring is comprehensive system that can observe vibration signals. Internet of Things (IoT) instruments is used to detect and record the movement. This project is to develop the slope monitoring based on effect of the vibration by using IoT where some several sensors are being used to measure some parameter of the slope such as acceleration and tilting angle of the slope. For the formula of acceleration, the measurement of the change in velocity or speed divided by time was being used. In this paper, the setup of some several sensors such as accelerometer and tilt sensor by using Arduino was developing to monitor the slope failure. When an of force is applied to the slope model, change in the values of *x*, *y* and *z-axis* sensor was occurring. This change of the value is due to vibrations that comes from the force that has been applied to the slope model. When the vibrations are detected, its values can be. It can be analysed through a user-friendly device, Arduino IDE software.

2. Materials and Methodology

This section will explain and described in detail about the sampling process and the design of laboratory works carried out for this study.

2.1. Studied Location

The location of the soil sample taken is near Bukit Ekspo, Putra Agriculture Centre, UPM with the coordinate of GPS: 2.987981, 101.712280 as shown in (**Figure 2**).

Figure 2. Studied site located near to Bukit Ekspo, Universiti Putra Malaysia (Google Map, 2022)

Figure 3. Conditions of failure at studied slope

The purpose of collecting soil sample at this specific site location due to slope failure observed at site as shown in **Figure 3**. The slope was failed due to vibration that comes from the agriculture truck machines. It is an ideal place to test for the new develop system using Arduino kits. However, before the site test is carry out, it is important to study the type of soil presents and the condition which contribute to the failure. In this study, there is no in situ testing work have been done due to the time constraint and works limitation.

The samples were taken at failure plane to determine its shear strength parameters as well as to determine the type of soil which contribute to the failure. Samples were kept in a sealed plastic sampling bag to prevent excessive moisture loss. The collected soil sample will be analyse for their properties which are physical and mechanical behaviour focused on strength behaviour. The soil physical tests and mechanical test were performed within three days after the sample referring to ASTM International Standard [7-11]. The soil sample also will be used for constructing the soil slope model in laboratory.

2.2. Works Program

The study focused on the application of the several sensors (i.e., accelerometer and tilt) to monitor the slope movement/ failure when the vibration applied to the slope model. The works program for this study is shown in flow chart (**Figure 4**).

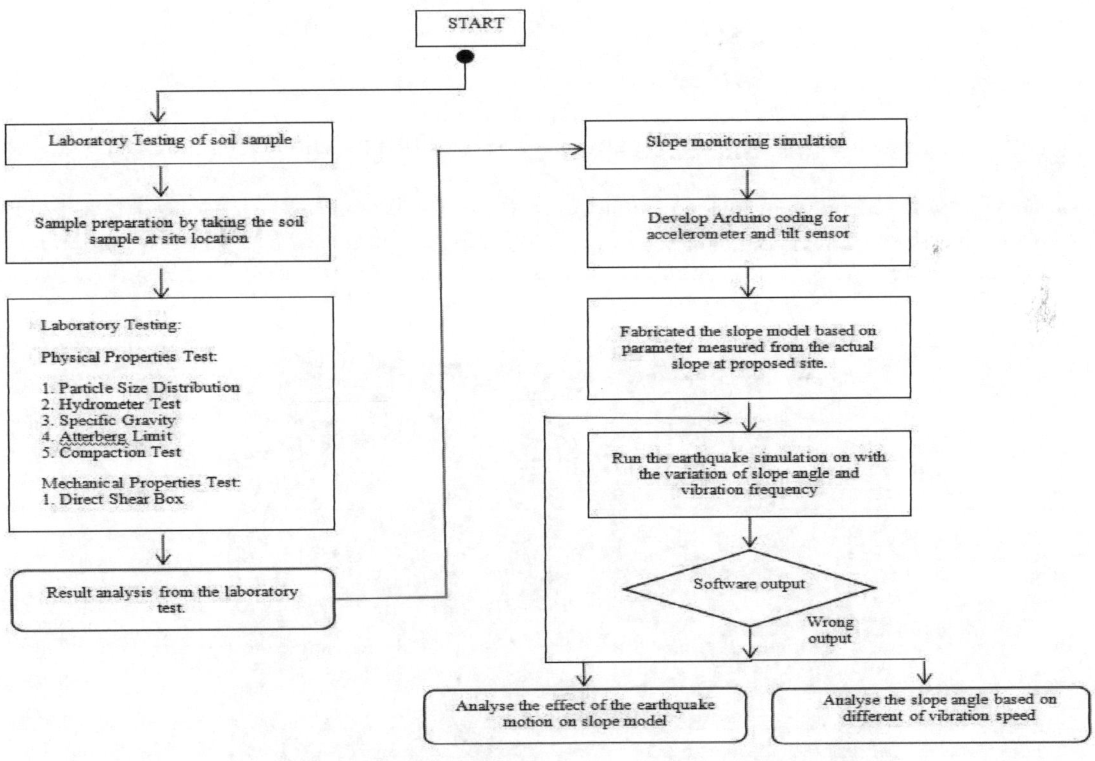

Figure 4. Works program flow chart

Soil laboratory testing was conducted after the soil sample was being taken at the study area. The laboratory works are divided into two stages. **Stage 1**: laboratory testing conducted to investigate the physical and mechanical properties of soil. The soil physical properties including particle size distribution test, hydrometer test, specific gravity test, Atterberg limit test and compaction. While for soil mechanical property; specifically on strength behaviour of soil, direct shear box test was conducted. For **Stage 2**: results from the slope simulation were being analysed which is the relationship between acceleration of the shaking table with those parameters; angle of slope, height of slope, depth of slope and ground surface.

2.3. Soil Slope Physical Model Setup

The drawing of the slope model completes with dimensions using an AutoCAD is prepared before the physical model was fabricated in laboratory scale (**Figure 5**). The boundary conditions do not consider in this study since there is no computed analysis for factor of safety will carry out.

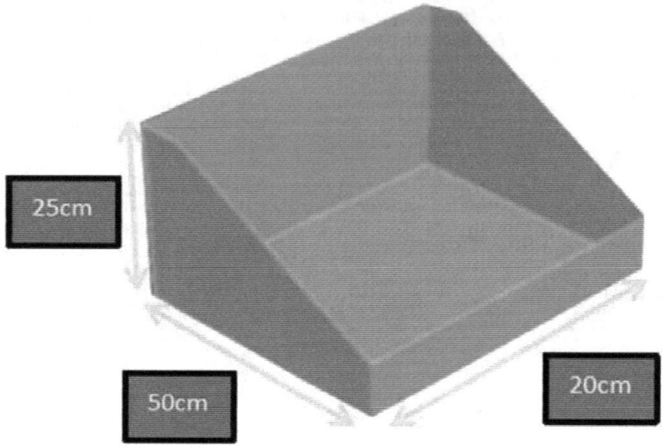

Figure 5. The drawing of soil slope physical model with dimension using AutoCAD

The vibration sensors were placed on slope surface of soil slope physical model fabricated in laboratory to detect vibration force (**Figure 6**). The soil is place within the Perspex box model; in the slope shape (i.e., slope angle, θ, and height, H, details with ratio to the real slope dimension) reflecting the slope at site.

Figure 6. Placement of Arduino sensors (tilt and accelerometer) on the slope model

The sensor is positioned within the soil and the position of *x, y, z-axis* as shown in (**Figure 7**). Soil samples were taken at 3 location points in the slope area where the boundary area was measured and modelled according to the actual dimensions and being constructed using AutoCAD. Tests are carried out with respect to each orientation and the force is applied to the model to produce vibrations. Basically, the vibration force is created by using the shaking table equipment available in laboratory.

The data from the sensor was sent to Arduino IDE device and processed for further analysis. The frequencies are plotted on a graph of acceleration against time.

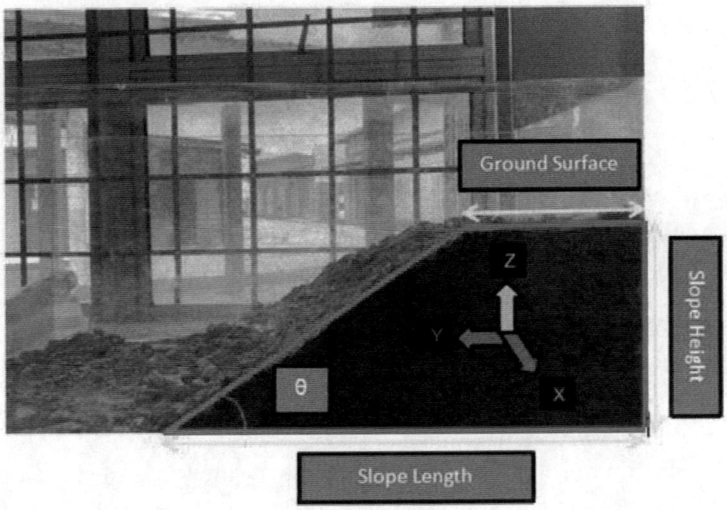

Figure 7. Soil slope physical model place in Perspex box for simulating the vibration effect

There are two separate setups was run in this study: i) vibration detection and ii) for measuring the angle of slope failure based on tilt sensor. Tilt sensor was used to detect the tilt of the slope once the vibration force was applied to the slope. The accelerometer ADXL 335 is used for detecting vibrations. This sensor is placed inside the soil slope surface physical model. The values are calculated at static state. Once the force is applied using shaking table to represent the vibration from earthquake as an example; the values are being detected and shown in the Arduino IDE Software. For this study, the earthquake acceleration was measured by assuming the forces caused by the shaking table as a percentage of gravity, or percent of g.

3. Results and Discussion

This section will describe in detail the results and analysis for each testing carried out in laboratory.

3.1 Soil Physical Properties

The results of soil physical properties are shown in **Table 1**. There are five tests were analysed which are particle size distribution, specific gravity, hydrometer, Atterberg limit, and compaction tests by referring to the ASTM International Standards.

Based on soil properties results, the percentage of silt/clay is about 54% that indicates the soils from the failed slopes are generally fined grain soil with more than 50% are classified as silt or clay. The angle of a slope, which defines its steepness threshold before failure, is influenced by the grain size of soil. A slope that has a greater proportion of coarser grain material is more stable than one that contains a higher percentage of fine grain material. A slope with a higher fine grain material may experience landslides [12].

Table 1. Physical properties of soil

Physical Properties	Unit	Values
Particle Size Distribution		
Gravel	%	5
Sand	%	41
Silt/ Clay	%	54
Specific Gravity, G_s	-	2.67
Liquid Limit, *LL*	%	14.3
Plastic Limit, *PL*	%	13.3
Plasticity Index, *PI*	%	1.1
Optimum Moisture Content, *OMC*	%	13.4
Maximum Dry Density, *MDD*	g/cm^3	16.95

The particle size distribution analysis was split into two categories of size particle of soil. One is the coarse-grain, which includes gravel and sand, and others is fine-grain, which includes silt and clay (**Figure 8**).

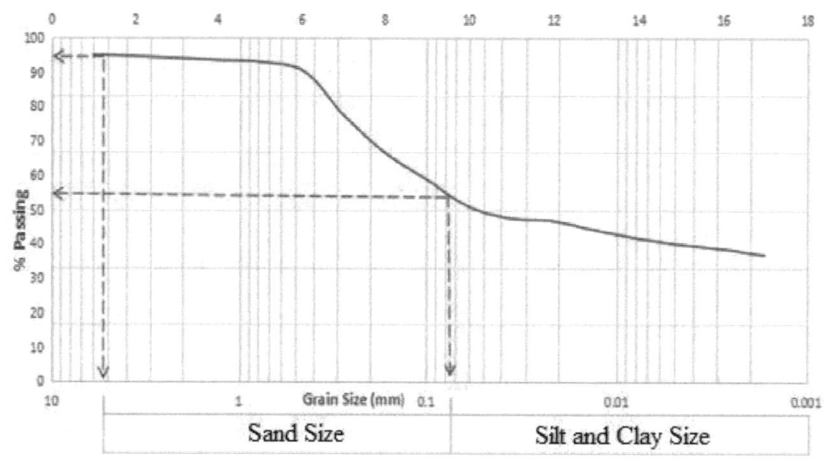

Figure 8. Particle size distribution curve of soil sample collected at site

The specific gravity can be referred as mass of solid in the soil compared to mass of the water at the same volume. The average of specific gravity test is about 2.67. Basically, the value of specific gravity test for the sample collected which is 2.67 was in the range of 2.65 to 2.70. Based on **Table 2**, the specific gravity value for the soil sample is categorised as silt.

Table 2. Typical value and classification for Specific Gravity [13]

Soil Type	Specific Gravity
Sand	2.63 – 2.67
Silt	2.65 – 2.70
Clay and Silty Clay	2.67 – 2.90
Organic Soil	Less than 2

IOP Conf. Series: Earth and Environmental Science 1091 (2022) 012002 doi:10.1088/1755-1315/1091/1/012002

From Plasticity Index chart (**Figure 9**), the soil sample with 1.1 value falls in low plasticity of silt category. Hence, the failure might be occurred for the studied slope due to low plasticity silt condition.

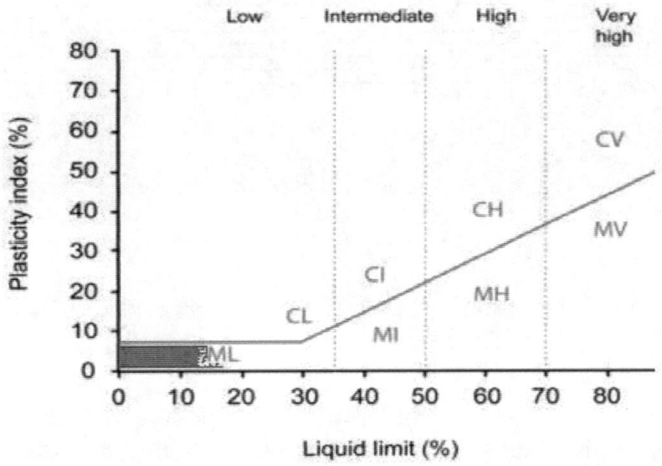

Figure 9. Soil sample on plasticity index chart

Furthermore, the slopes in the study area may exposed to the vibration that comes from the transportation such as heavy machinery and this condition may affect the stability of the slopes. Based on the **Table 3**, the value of the liquidity and consistency index is 1.620 and 1.537. Soils having a liquidity index of one or more are closer to becoming liquid. In other hands, if the soil consistency index is low, the shear strength of the soil will reduce.

Table 3. Liquidity Index and Consistency Index

Index	Value
Liquidity Index	1.620
Consistency Index	1.537

Figure 10 shown the plotted points of dry density and moisture content of soil sample collected as site. Dry density was increased as well as moisture content value, allowing the soil to be consolidated into a denser condition. Followed by decreased of dry density values as the water content increases after approximately the third point was plotted and reaches a point where it displaces soil particles in each volume. This means that the optimum moisture content has been exceed. The value of maximum dry density is about 16.95 g/cm^3 and optimum moisture content is 13.35%.

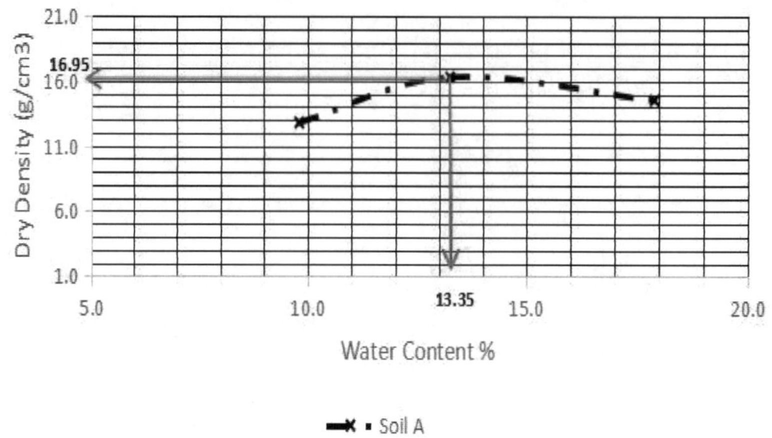

Figure 10. Compaction curve of soil sample collected at studied site

3.2 Strength Behaviour of Soil

In contrast to other materials such as concrete or metals, soil failure typically occurs on a particular surface (shear plane). When the shear stress operating parallel to the surface exceeds the shear strength, failure will occur. Shear strain refers to the deformation that occurs when shear stress is applied to undisturbed soil. Shear strain is quantified in direct shear testing as the distance between two sections of the soil specimen.

Based on the direct shear testing, shear stress increases with horizontal displacement and reaches a maximum before start decreasing (**Figure 11**). Shear stress will reduce as shearing and horizontal displacement in fined grained soil sample increase, until it achieves a nearly constant value. When a soil's shear stress exceeds its peak shear strength, the maximum stress that soil sample can sustain is equal to the residual stress.

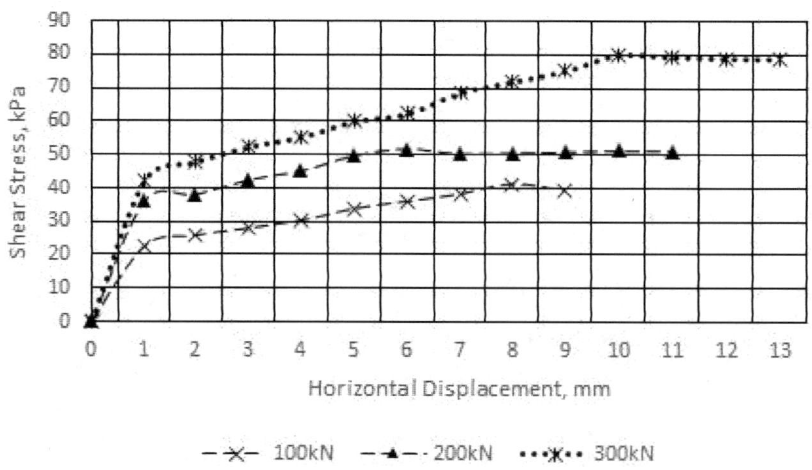

Figure 11. Graph of shear stress against horizontal displacement

Based on Normal Stress – Shear Stress curves shown in (**Figure 12**), the peak and normal stress values from each curve are shown. Normal stress and peak shear stress graph have linear correlations. The lines that cross the y-axis is indicated the soil cohesion, *c*. The friction angle, ϕ, is determined by the angle of each line to the x-axis. Based on the result, the value of cohesion is 18.9 kPa, while the angle of friction is 32.3°. The value of soil friction angle is a typical value indicated for silty sand. The most common analytic method to monitor the slope instability is by using the factor of safety, *FS* of the slope with respect to the limit equilibrium condition. **Equation 1** is used to calculate the *FS*.

$$\text{Factor of safety} = \frac{\text{Average shear strength of soil}}{\text{Average shear stress}} \quad \textbf{Equation 1}$$

From the result the average of the shear strength of soil is 49.59 kPa, while the average of shear stress, τ, is 56.67 kPa. Hence the factor of safety is lesser than 1, the slope is considered in a state of failure slope.

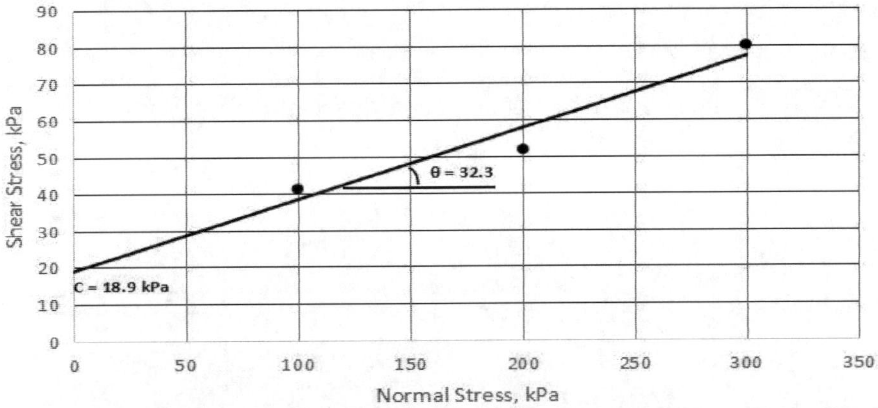

Figure 12. Shear stress against normal stress curve

Table 4. Strength properties of soil

Properties	Values	
	Normal Stress, σ	Max. Shear Stress, τ_{max}
Shear properties	100 kN	41.3 kPa
	200 kN	51.8 kPa
	300 kN	80.1 kPa
Cohesion, *c*	18.9 kPa	
Friction Angle, ϕ	32.3°	
Shear Strength	49.59 kPa	
Factor of Safety, *FS*	0.875	

3.3 Analysis of Acceleration Related to Slope Angle

The results indicate that the vibrations are cumulative, which implies that the readings from the accelerometer sensor play a significant part in the present operation, as shown by the graphs in (**Figures 13-15**). The grey color indicates the y-axis, and the direction is parallel to the movement of shaking table. The vibration from the shaking table was being measured after applied force is 1 Voltage. This is because there is no significant change in the value of the vibration when applied load is less than 1 Volt. Based on the curves show, the significant impact is on the y axis which is the acceleration at y axis is higher. Urvashi (2018) stated that the impact will become higher towards the axis that parallel to the direction of the vibration movement [14]. For example, when the vibration speed increasing, from 1 to 2 voltage, the values of acceleration in y-axis are increase significantly.

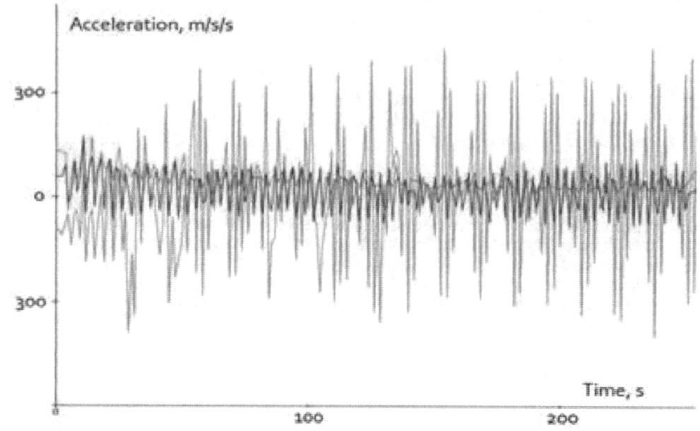

Figure 13. Acceleration against time curves for 1 Volt

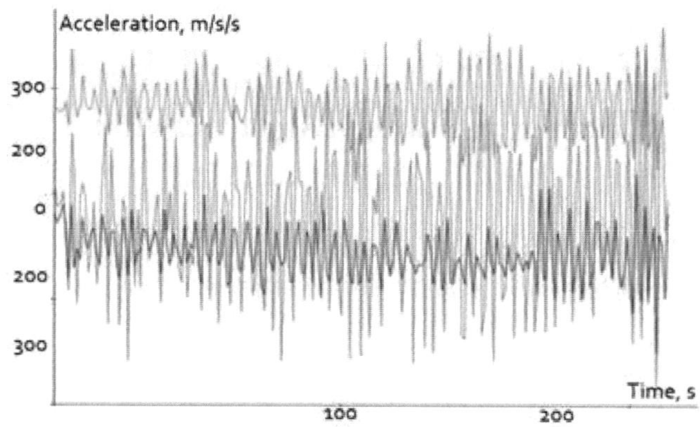

Figure 14. Acceleration against time curves for 1.5 Volt

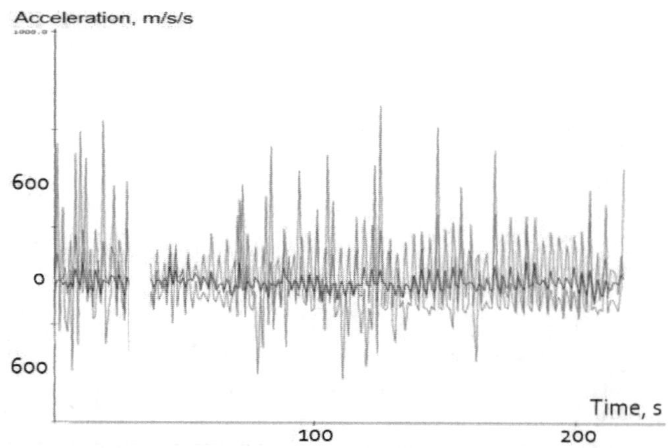

Figure 15. Graph of acceleration against time for 2 Volt

Figure 16 shown the slope parameters based on the variation of vibration speed. For the initial slope, which is in normal condition, the parameters of the slope including the angle of slope, height of slope, slope depth and ground surface was being measured from the actual site, Bukit Ekspo, Universiti Putra Malaysia. The effect of frequency and amplitude is very apparent in these results. The significant changes in the angle of slope occurred as the amplitude of vibration was increased. In normal condition, the angle of slope is 46.0°, the slope depth is 33.8 cm, slope height is 18.1 cm and ground surface are 15.3 cm. When increasing the vibration speed of 1 Volt, the result shows that the parameter of the slope tend to decrease except the slope depth will be increase.

The parameter of the slope after increasing to 1 Volt which is the angle of slope is decrease to 43.2˚, slope height is 14.8 cm and ground surface is 15.2 cm, but the slope depth increases to 35.4 cm. This is because when the sliding of the slope occurs, the slope depth will become increase. When increase the maximum vibration speed to 2 Volt, the parameter of slope will totally fail when the slope of the angle decreases to 24.4°, the height of slope is 14.1 cm and the slope depth increase significantly to 49.7 cm.

It can be summarized that the response of the soil is dependent on the type of soil, as well as the amplitude of the vibration. However, it is reasonable to assume that soil particles have a resonance condition that is a function of the gradation, effective size, and type of soil. Therefore, a study was made of the effect of vibration on a soil that would yield results dependent only on the properties of the soil. Other factors to be consider include material deterioration, which is accompanied by an increase in fine grained material between large particles, resulting in a decrease in shear strength, bearing capacity, and shear modulus that leading to slope failure.

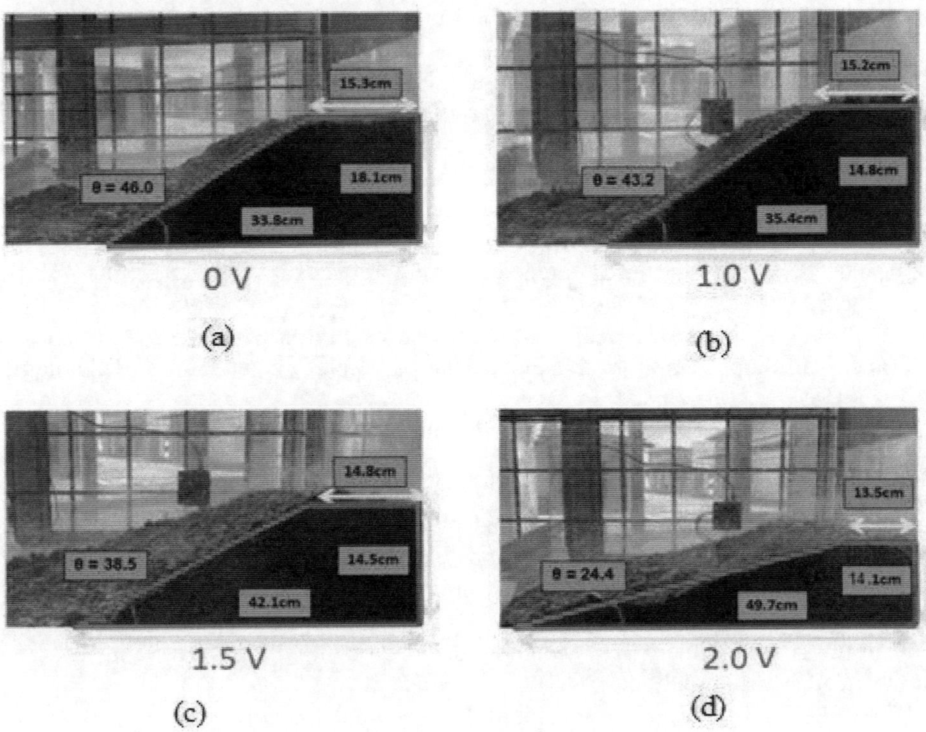

Figure 16. Slope parameters based on variation of vibration speed a) 0 volt, b) 1.0 volt, c) 1.5 volt and d) 2.0 volt

4. Conclusion

In conclusion, the soil sample categorized as silt with low plasticity and the shear strength of the soil is lower than shear stress of soil. The value from the Arduino output was being analysed such as the acceleration of the accelerometer and the angle of slope. From the simulation, the value of acceleration from the y-axis which is directly parallel to the direction movement of the shaking table will give the high value than x-axis and z-axis. The angle of slope will be reduced when the speed of shaking table increases. The changes in acceleration values resulting from the variations of applied vibration on the slope during observation. The variation of vibration can be assumed as the earthquake motion when comes during the real case.

Acknowledgments

Authors would like to acknowledge and give the warmest thanks for the technical support offered by the laboratory staff of Geotechnical and Geological Engineering Laboratory of Civil Engineering Department, Engineering Faculty, Universiti Putra Malaysia.

References

[1] Dewedree, S. and Jusoh, S. N. (2019). Slope Stability Analysis Under Different Soil Nailing Parameters Using the Slope/W Software. Journal of Physics: Conference Series, Volume 1174, 2nd International Conference on Women in Science, Engineering and Technology 17– 19 July 2018, Kuala Lumpur Convention Centre, Malaysia.
[2] MPAJ (1994) Report of the inquiry committee into the collapse of block 1 and the stability of blocks 2 and 3 Highland Towers Condominium, Hulu Klang, Selangor Darul Ehsan.
[3] New Straits Times (2019) https://www.nst.com.my/opinion/letters/2019/12/551073/have-we-really-learnt-lessons-highland-towers-collapse.
[4] Das, D., Roy, D. and Singh, R., 2007. A Correlation for Permanent Earthquake - Induced Deformation of Earth Embankments. Engineering Geology. Vol. 90. Pp: 174-185.
[5] Omar, H., Ibrahim, A. L. and Hashim, M. (2007) Slope Stability Analysis using Remote Sensing and GIS. 28th Asian Conference on Remote Sensing (ACRS), Putra World Center, Kuala Lumpur.
[6] Bunawan, A. R., Mahin, M. A. H., Hezmi, M. A. Kassim A. and Rashid, A. S. A. (2016) Development of Slope Monitoring Device Using Accelerometer. Jurnal Teknologi, vol. 78, no. 8, Pp:59-65.
[7] ASTM D422-63(2007) e2. 2007. Standard Test Method for Particle-Size Analysis of Soils (Withdrawn 2016), ASTM International, West Conshohocken.
[8] ASTM D4318-17e1. 2007. Standard Test Methods for Liquid Limit, Plastic Limit, and Plasticity Index of Soils, ASTM International, West Conshohocken.
[9] ASTM D698-12. 2021. Standard Test Methods for Laboratory Compaction Characteristics of Soil Using Standard Effort (12,400 ft-lbf/ft^3 (600 kN-m/m^3)), ASTM International, West Conshohocken.
[10] ASTM D854-00. 2000. Standard Test Methods for Specific Gravity of Soil Solids by Water Pycnometer, ASTM International, West Conshohocken.
[11] ASTM D3080 / D3080M-11. 2011. Standard Test Method for Direct Shear Test of Soils Under Consolidated Drained Conditions (Withdrawn 2020). ASTM International, West Conshohocken.
[12] Temesgen, B., Mohammed, M. U. and Korme, T. (2001) Natural Hazard Assessment using GIS and Remote Sensing Methods with Particular Reference to the Landslides in the Wondogenet Area, Ethiopia. Phys. Chem. Earth (C). 26:665-615.
[13] Bowles, J. (2012) Foundation Analysis and Design. 5th Edition. McGraw Hill. India.
[14] Urvashi J. V. G. (2018) Vibration Detection Instrument Based on Internet of Things. International Journal of Advance Research, Ideas and Innovation In technology, 4(6), 91-99.

IOP Conf. Series: Earth and Environmental Science 1091 (2022) 012003 doi:10.1088/1755-1315/1091/1/012003

Predicting and avoiding hazardous occurrences of stuck pipe for the petroleum wells at offshore Vietnam using machine learning techniques

Quang Khanh Do[*,1,2], Trong Quang Hoang[1,2], Thang Nguyen[1,2] and Vu Khanh Phat Ong[1,2]

[1] Faculty of Geology and Petroleum Engineering, Ho Chi Minh City University of Technology (HCMUT),
268 Ly Thuong Kiet Street, District 10, Ho Chi Minh City, Vietnam
[2] Vietnam National University - Ho Chi Minh City (VNU-HCM), Linh Trung Ward, Thu Duc District, Ho Chi Minh City, Vietnam

*Corresponding author: dqkhanh@hcmut.edu.vn; ORCID: 0000-0003-2872-5743

Abstract. Over many years, the petroleum industry has been facing the hazardous occurrences associated with drilling stuck pipe. It is actually one of the most challenging and costly problems at offshore drilling rigs, especially in the high-risk, high-cost environments as the ultra-deep or deep water. Nowadays, with the development of Artificial Intelligence these problems can be predicted and avoided by applying machine learning techniques. The purpose of study is to apply the two most powerful machine learning techniques including Artificial Neural Network (ANN) and Support Vector Machine (SVM) to predict and avoid stuck pipe occurrences for the petroleum wells at offshore Vietnam. Modeling ANN and SVM using Matlab programming language based on many actual drilling parameters to estimate the risk of stuck pipe. The output of the models of both ANN and SVM machine learning techniques is the probability of non-stuck or stuck pipe at different measured depth points. Predictive results from both ANN and SVM techniques are compared to real measurements with the accuracy of all models above 85%. These machine learning techniques can help predicting stuck pipe occurrences as well as avoiding them by considering the changes of some important drilling operational parameters.

1. Introduction

In the petroleum industry, stuck pipe is one of the hazardous occurerences affecting seriously the drilling process and costly problem. The severity of stuck pipe issues can vary from a minor inconvenience to major complications such as loss of the drill string or complete loss of the well. These occurrences usually happen everywhere in the drilling world and estimate to cost to the petroleum industry hundreds of millions of dollars per year. In some areas, they can be responsible for over 40% the total well cost. Moreover, stuck pipe is actually one of the most challenging and costly problems at offshore drilling rigs, especially in the high-risk, high-cost environments as the ultra-deep or deep water [1]; [2]. Therefore, predicting and avoiding hazardous occurrences of stuck pipe for the petroleum wells is very necessary in the well planning and efficient drilling performance approach.

Traditionally, stuck pipe situations have been resolved through trial and error by using standard methods and techniques available that have been developed in the field. Stuck pipe problems are divided into two categories: mechanical sticking and differential sticking [3]. If the case is a mechanical sticking, the first action most commonly taken is the jarring operation. In the event the jarring process is unsuccessful, finding some chemical materials is another solution to dissolve cuttings and shale caving in the annulus around the pipe, especially if the drill string is stuck in formations such as limestone, chalk and dolomite. In case of differential sticking, use spotting a lighter drilling fluid to reduce the difference between hydrostatic pressure and formation pressure to treat the problem.

Parameters affecting the possibility of stuck pipe include drilling operation parameters, properties of drilling fluid and formation characteristics. The input and output data are highly complex and nonlinear, so conventional mathematical models cannot be used to predict a stuck pipe. Machine learning methods such as Artificial Neural Networks (ANNs) and Support Vector Machines (SVMs) are essential tools to predict stuck pipe parameters. Features of machine learning methods based on the ability to learn and recognize complex data with reasonable accuracy over 85%; models are built with generalization, approximation and interpolation in a specific problem [2].

In this paper, we conducted a study on applying two most popular machine learning techniques, artificial neural networks and support vector machines to build models that predicts stuck pipe through the previous drilling parameters. The study compares machine learning models and chooses the effective model for drilling engineers as well as considering the changes of some important drilling operational parameters to predict and avoid hazardous occurrences of stuck pipe.

2. Methodology

The pipe is considered to be "stuck" while it is dangling state and cannot be removed out of the hole. There are many mechanisms of stuck pipe incident, two major hazardous stuck pipe categories: Mechanical sticking and differential sticking used to clearly understand about stuck pipe in this study [4].

2.1. Mechanical sticking

Mechanical sticking occurs when parts of the drillstring or bottom hole assembly are mechanically trapped downhole due to the causes of poor hole cleaning, cementing accumulation and wellbore instability. Poor hole cleaning that causes cutting accumulation is one of the main causes of mechanical sticking. The factor to evaluate the cleanliness of the well is the amount of mud in annulus. Failure can occur when the circulation is interrupted, causing the cuttings return to form plugs that cause stuck pipe [4].

2.2. Differential sticking

Differential sticking is a problem that occurs when the hydraulic pressure is greater than the formation pressure. The drillstring is pressed and stuck to the well wall so part of its circumference can only see the reservoir pressure, the rest will continue to be pushed by the pressure in the well. As a result, the drillstring gets stuck in the well formation and can take millions pounds of force to release, but this is not possible. In many emergency cases, the density of drilling fluid is reduced thereby reducing the pressure differential and releasing the drillstring [5].

2.3. Machine learning

Machine learning techniques are based on mathematical models. These models can learn trends and behavior of experimental or real data to create a suitable model. These methods are called machine learning due to their ability to learn from certain data sets of previous experience [6]. Based on capability learning of machines, the models of this method can generate generalizations, approximations, or interpolations in a particular problem, including a certain number of datasets with output which give results of the occurrence problem. Nowaday, the most powerful machine learning methods are Artificial Neural Network (ANN) and Support Vector Machine (SVM). Both of these tools are designed primarily for pattern recognition and classification.

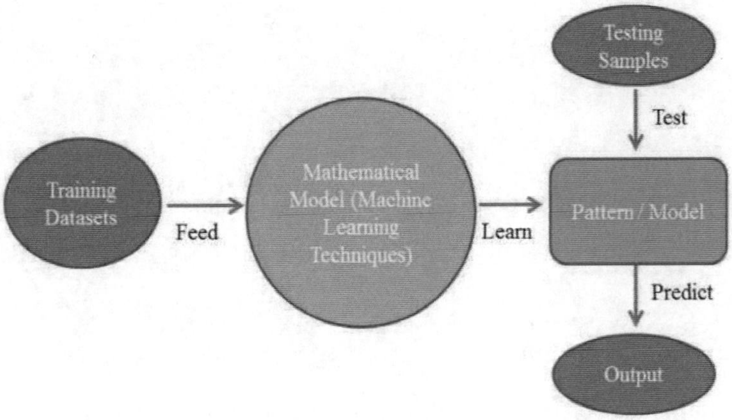

Figure 1. A schematic of how machine learning techniques work [2].

The learning and generalization ability of the machine learning model is evaluated on the basis of two statistical parameters including Mean Square Error (MSE) and Correlation Coefficient Square (R^2):

$$R^2 = \frac{\sum_{i=1}^{N}(y_i-\bar{y_i})(y_i^*-\overline{y_i^*})}{\sqrt{\sum_{i=1}^{N}(y_i-\bar{y_i})^2 \sum_{i=1}^{N}(y_i^*-\overline{y_i^*})^2}} \quad (1)$$

$$MSE = \frac{\sum_{i=1}^{N}(y_i-y_i^*)^2}{N} \quad (2)$$

Where y_i and y_i^* are the real and predicted values.

2.3.1. Artificial Neural Network (ANN)

Artificial Neural Network (ANN) is inspired by human brain function. An artificial neural network consists of many neurons connected to each other through weights similar to synapses connected to biological neurons in the system in the human nervous system. These weights represent how important the input data is to information processing (the process of converting data from one layer to another). Currently, the most popular type of artificial neural network is the multilayer perceptron [2].

These models consist of three layers: input, hidden and output layer. Each of these layers consists of neurons (nodes). The input layer is the first layer whose neurons represent the input parameters. The second layer is the hidden layer where the computation is started. Depending on the complexity and optimization of each problem, an ANN model can have one or more hidden layers. Each hidden layer consists of certain neurons connected by weights. The output layer represents the conclusions or outputs of the problem and includes certain neurons based on the nature of the problem.

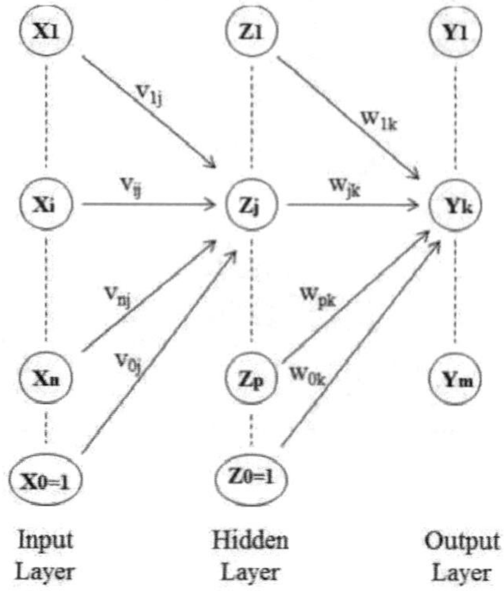

Figure 2. Typical structure of the Perceptron network model [2].

2.3.2. Support Vector Machine (SVM)

Support Vector Machine (SVM) is considered as one of the main used machine learning techniques for classification and nonlinear regression [7]. Basically, the SVM mappings and nonlinear inputs are transformed into a very high performance space where the inputs are linearly separated by a hyperplane. This transformation can be achieved by applying some nonlinear mappings such as polynomial or Radial Bases Functions. The major concept behind SVMs is to classify the data into two groups with the maximum margin between the boundaries that separate the data. The points where the decision lines intervene are called support vectors. SVM techniques have been applied in many critical fields, outperforming ANNs in some instances [2].

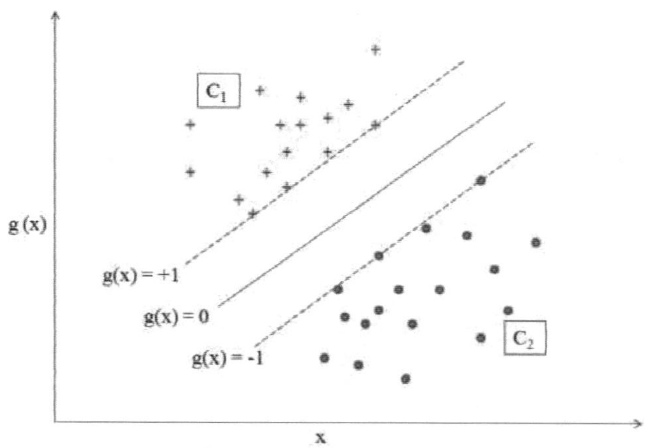

Figure 3. Linear classification of SVM [2].

3. Building a machine learning model for the drilling stuck pipe problem

3.1. Data Selection and Preprocessing

Input data to build machine learning models, including artificial neural network method and SVM method from well A of field X at Nam Con Son basin, offshore Vietnam. Input parameters include: depth (MD), rate of penetration (ROP), weight on bit (WOB), revolution per minute (RPM), torque (TORQUE), standpipe pressure (SPP), flow in (Q), mud weight (MW), the type of drilling fluid used is water-based mud (WBM) and the drill bit type is PDC. All parameters were recorded during the 8-1/2" hole section.

Input parameters need to be stable and normalized, and related to output parameters. Since the output parameter of the model is the probability of stuck pipe or non-stuck between [0,1], so convert all the input parameters to a dimensionless form with a range between [0,1]. This makes the network model optimal and has less error [8].

$$x_j = \frac{x_i - x_{min}}{x_{max} - x_{min}} \quad (3)$$

Where x_j is the dimensionless value, x_i is the initial value of the data point.

The network construction data consists of a set of 83 data points including input parameters and stuck pipe data to occur with stuck (= 1) and non-stuck (= 0). The input data for network construction is divided into 3 data groups at the ratio of 70:15:15, including:

- Training data: accounts for 70% of the dataset corresponding to 59 data points.
- Validation data: accounts for 15% of the dataset corresponding to 12 data points.
- Testing data: accounts for 15% of the dataset corresponding to 12 data points.

3.2. Building ANN models using Matlab

For the network training dataset of well A, the network that gives the lower error results and the higher correlation coefficient square is the 8-11-1 network with the corresponding structure of 8 neurons in input layer, 11 neurons in hidden layer, and 1 neuron in output layer.

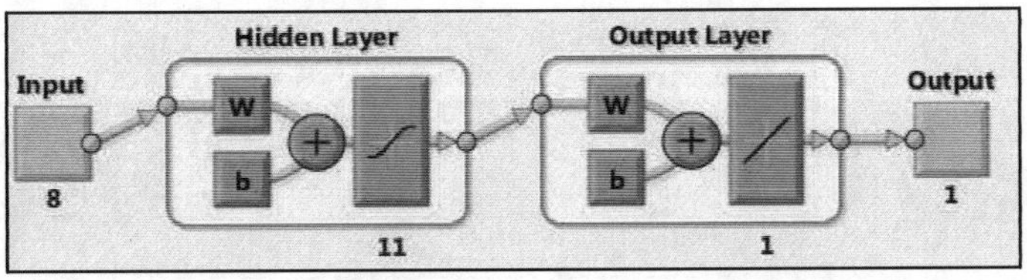

Figure 4. Structure of well network training model A.

After training the network with a total of 83 samples, the network gives a small mean square error MSE = 0.020883 and a high correlation coefficient square R^2 = 0.999495. The difference between the validation error and the training error is not large and lower, besides the error in the last epoch is stable, from which the network does not occur overfitting.

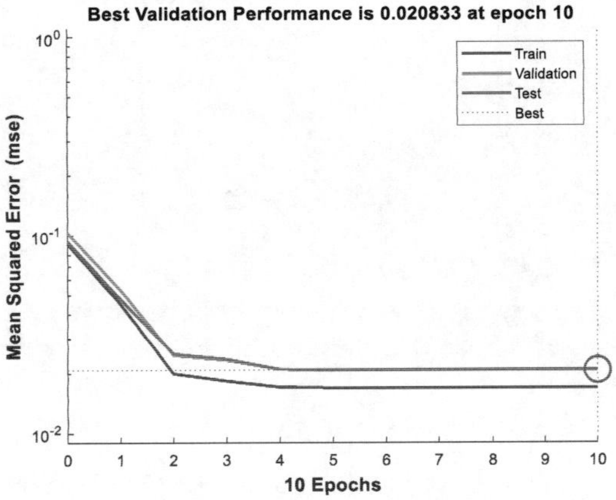

Figure 5. The graph shows MSE of train error, validation error and test error of the 8-11-1 network.

3.3. Building SVM models using Matlab

Similarly, the input data also includes 8 parameters during the drilling process that are converted to [0,1] (as in equation (3)) and parameters of the stuck or non-stuck (1 or 0). The output data of the model is the result of 1 or 0 corresponding to the incident stuck or non-stuck. The following figure (**Figure 6**) shows the relationship between the input and output data of the Linear SVM model.

Figure 6. Input data to build the Linear SVM model.

After training the Linear SVM model, the accuracy of the model is calculated by the following formula:

$$Accuracy\% = \frac{\text{Correctly classified testing samples}}{\text{Total number of testing samples}} \times 100 \tag{4}$$

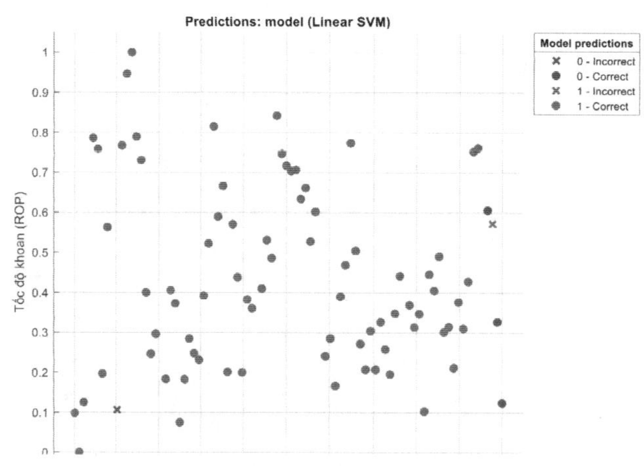

Figure 7. Linear SVM construction results for well A.

After building the Linear SVM model by using the Classification Learner tool for well A, it is shown in **Figure** 7. From **Figure 7,** we see that the predicted data points deviating from reality are represented by slashes and the model Linear SVM has 02 wrong prediction data points out of 83 data points including one non-stuck data point and one stuck point. From equation (4), the accuracy of the model is 97.59% $(=\frac{81}{83} \times 100)$.

4. Result and discussion of the stuck pipe prediction study

4.1. Stuck pipe prediction data

The input data of the predicted well is a dataset of 43 data points according to the measured depth (MD) along with the drilling parameters all recorded in the same 8-1/2" hole section of well B of field X at Nam Con Son basin, offshore Vietnam. The predicted input data of well B also includes: depth (MD), rate of penetration (ROP), weight on bit (WOB), revolution per minute (RPM), torque (TORQUE), standpipe pressure (SPP), flow in (Q), mud weight (MW). The parameters non-processing and reprocessing are shown in **Table 1** and **Table 2**:

Table 1: Input parameters of predicted well when non-processing.

No.	Depth	ROP	WOB	RPM	TQ	SPP	Flow in	MW
	m	m/hr	tons	rpm	m.kg	atm	l/s	kg/l
1	4365	13.62	5.55	170	2303.189	190.5967	19.4318	2.1
2	4366	7.49	5.5	170	2276.506	192.6381	19.4318	2.1
3	4367	6.24	5.85	170	2253.141	193.3866	19.4318	2.1
...								
42	4406	9.8	12.75	160	2382.409	191.5494	18.9270	2.088
43	4407	10.53	15.05	160	2816.806	205.2266	18.9270	2.088

Table 2: Input parameters of predicted well when pre-processing.

No.	Depth	ROP	WOB	RPM	TQ	SPP	Flow in	MW
	m	m/hr	tons	rpm	m.kg	atm	l/s	kg/l
1	0	1	0.2830	0.7027	0.3387	0.6571	0.9767	1
2	0.0238	0.3050	0.2792	0.7027	0.3044	0.7050	0.9767	1
3	0.0476	0.1632	0.3057	0.7027	0.2743	0.7225	0.9767	1
...								
42	0.9762	0.5669	0.8264	0.4324	0.4407	0.6794	0.7907	0
43	1	0.6497	1	0.4324	1	1	0.7907	0

4.2. Results and discussion of stuck pipe prediction of two machine learning models

Applying the optimal network trained according to ANN and SVM with the same input data set of predicted well B (as shown in **Table 2** with pre-processing data between [0,1]), the output of the stuck pipe occurence parameters is shown in **Table 3**. The predictive calculation results from the machine learning models are also compared with the actual results of the stuck pipe cases during the drilling of well B later.

Table 3. Results of stuck pipe prediction for well B.

Depth (m)	ANN prediction		SVM prediction	Actual result	Depth (m)	ANN prediction		SVM prediction	Actual result
	Probability ANN	Encode ANN				Probability ANN	Encode ANN		
4365	0.1311234	0	0	0	4387	0.222467	0	0	0
4366	0.1625848	0	0	0	4388	0.112781	0	0	0
4367	0.1239887	0	0	0	4389	0.047824	0	0	0
4368	0.3505118	0	0	0	4390	0.097248	0	0	0
4369	0.1127809	0	0	0	4391	0.050938	0	0	0
4370	0.1008683	0	0	0	4392	0.056588	0	0	0
4371	0.4950237	0	0	0	4393	0.0001	0	0	0

4372	0.4904522	0	0	0	4394	0.067372	0	0	0
4373	0.4145204	0	0	0	4395	0.191832	0	0	0
4374	0.4116193	0	0	0	4396	0.065355	0	0	0
4375	0.4101664	0	0	0	4397	0.088651	0	0	0
4376	0.3919696	0	0	0	4398	0.065734	0	0	0
4377	0.2861628	0	0	0	4399	0.073926	0	0	0
4378	0.1999853	0	0	0	4400	0.016748	0	0	0
4379	0.2327771	0	0	0	4401	0.097889	0	0	0
4380	0.6072238	1	1	0	4402	0.149231	0	0	0
4381	0.0796233	0	0	0	4403	0.187809	0	0	0
4382	0.1595436	0	0	0	4404	0.108215	0	0	0
4383	0.64935	1	0	0	4405	0.072671	0	0	0
4384	0.6764131	1	0	0	4406	0.944653	1	1	1
4385	0.6393361	1	0	0	4407	0.950028	1	1	1
4386	0.244723	0	0	0					

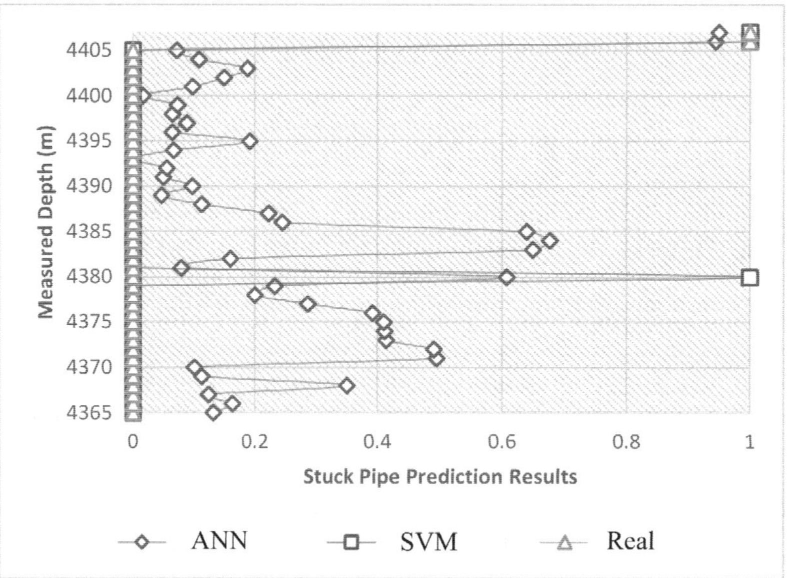

Figure 8. Graph of stuck pipe prediction for well B.

From **Figure 8,** the predictive graph of stuck pipe occurrence of both models can be seen that the predicted line quite accurately reflects the actual trend of stuck pipe cases. Although there are still biases, both models have very high reliability because the mean square error is very low (from equation (2), the ANN model has a mean square error MSE = 0.0441 and SVM model is MSE = 0.0465).

The output results of both models are values with stuck or non-stuck (1 or 0). In addition, the ANN model also shows the probability of stuck pipe. However, the SVM model shows the ability to predict the stuck pipe more accurately and better than the ANN model. Stuck pipe prediction results from **Table 3** show that the number of incorrect prediction points of the ANN model (at 04 depth points of 4380 m, 4383 m, 4384 m, 4385 m) more than the SVM model (only at 01 point at a depth of 4380 m). From equation (4), the prediction accuracy by SVM model is 97.67% ($=\frac{42}{43}\times 100$) higher than when using ANN model is 90.70% ($=\frac{39}{43}\times 100$).

By increasing or decreasing drilling operation parameters then the trend of stuck pipe incident can be clearly seen. As shown in **Figure 9**, when increasing torque index (percentage compared to the initial value), the tendency to stuck pipe increases because more stuck pipe points appear at depth values about 4371-4374 m. This also proves that when drilling at stuck points, the torque parameter will increase dramatically. A recommendation to reduce the torque index to the reasonable one while drilling may help to avoid the stuck pipe. Therefore, during drilling operations for petroleum wells when the torque index increases suddenly, it is necessary to take timely handling measures to avoid hazardous stuck pipe occurrence.

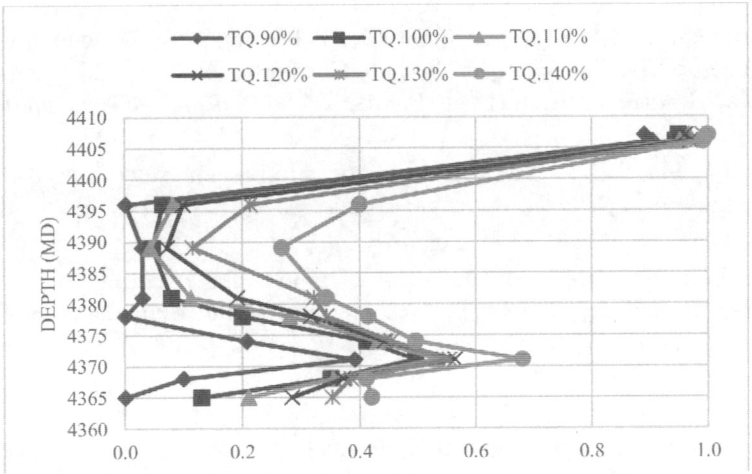

Figure 9. Effect of input parameters as the torque index to stuck pipe.

5. Conclusion

- Stuck pipe is one of the most costly and challenging occurrences in the drilling process of the oil and gas industry, especially in high-risk, high-cost environments as ultra-deep or deep water. Machine learning methods which include artificial neural network and support vector machine have been studied and applied successfully in the problem of stuck pipe, showing the most powerful and efficient prediction to help recognize and prevent the hazards of stuck pipe.
- The results of stuck pipe prediction in actual wells by both ANN and SVM methods have high accuracy above 85%. However, the SVM model has higher accuracy (97.67%) than the ANN model (90.70%), so the SVM model is more useful. Therefore, the machine learning method is a potential tool in predicting the hazardous occurrences of stuck pipe cases more efficiently.
- Reducing the torque index while drilling to the reasonable value help to avoid the hazardous occurrences of stuck pipe for the petroleum wells.

Acknowledgement: We would like to thank Simulation Lab (SIMLAB), Faculty of Geology and Petroleum Engineering (GEOPET), Ho Chi Minh City University of Technology (HCMUT) – Vietnam National University – Ho Chi Minh City (VNU-HCM) for the support of time and facilities for this study. Sincere thanks for the support and help of PetroVietnam Exploration Production Corporation (PVEP) and Bien Dong Petroleum Operating Company (BIENDONG POC).

References

[1] C. Siruvuri, S. Nagarakanti and R. Samuel. Stuck Pipe Prediction and Avoidance: A Convolutional Neural Network Approach. *IADC/SPE Drilling Conference; 2006. DOI:*https://doi.org/10.2118/98378-MS .

[2] I. Al-Baiyat and L. Heinze. Implementing Artificial Neural Networks and Support Vector Machines in Stuck Pipe Prediction. SPE Kuwait International Petroleum Conference and Exhibition; 2012. DOI:https://doi.org/10.2118/163370-MS .

[3] R. Jahanbakhshi and R. Keshavarzi. Intelligent Prediction of Differential Pipe Sticking by Support Vector Machine Compared With Conventional Artificial Neural Networks: An Example of Iranian Offshore Oil Fields. *SPE Drilling & Completion; 2012. DOI:* 10.2118/163062-PA

[4] M. J. Aljubran and A. S. Al-Yami. Best Practices to Prevent Stuck Pipe Incidents in Offshore Shaly Formations. *SPE Middle East Oil & Gas Show and Conference; 2017.* DOI:https://doi.org/10.2118/183717-MS.

[5] Abrar A. Alshaikh and Md. Amanullah. A Comprehensive Review of Differential Sticking, Spotting Fluids, and the Current Testing and Evaluation Methods. *SPE Kingdom of Saudi Arabia Annual Technical Symposium and Exhibition; 2018. DOI:*https://doi.org/10.2118/192169-MS.

[6] P. Skalle and A. Aamodt. Case-Based Reasoning, a method for gaining experience and giving advise on how to avoid and how to free stuck drill strings. *IADC Middle East Drilling Conference; 1998.*

[7] A. K. Abbas and R. Flori. Intelligent Prediction of Stuck Pipe Remediation Using Machine Learning Algorithms. *SPE Annual Technical Conference and Exhibition; 2019. DOI:* https://doi.org/10.2118/196229-MS.

[8] S. R. Shadizadeh and F. Karimi. Drilling Stuck Pipe Prediction in Iranian Oil Fields: An Artificial Neural Network Approach. *Iranian Journal of Chemical Engineering; 2010.*

[9] R.Miri and J. Sampaio. Development of Artificial Neural Networks to Predict Differential Pipe Sticking in Iranian Offshore Oil Fields. *International Oil Conference and Exhibition in Mexico; 2007. DOI:* https://doi.org/10.2118/108500-MS .

IOP Conf. Series: Earth and Environmental Science 1091 (2022) 012004 doi:10.1088/1755-1315/1091/1/012004

Strategic framework of using drone in cities disaster response

Zahir H.[1*], Fathi M.S.[2], and Tharima A.F.[3]

[1]Malaysia-Japan International Institute of Technology, Universiti Teknologi Malaysia, Jalan Sultan Yahya Petra, 54100 Kuala Lumpur, Malaysia
[2]Razak Faculty of Technology and Informatics, Universiti Teknologi Malaysia, Level 7, Razak Tower, Jalan Sultan Yahya Petra, 54100 Kuala Lumpur, Malaysia
[3]Fire and Rescue Department of Malaysia, Planning and Research Division Fire Research Centre, Lengkok Teknologi Kawasan Perindustrian, Enstek, 71760 Nilai, Negeri Sembilan, Malaysia

*Corresponding author e-mail: zahirhassan@gmail.com

Abstract. Drone technology has been around the world more than a decade and the application has becomes more vital, especially, during the disaster event such as COVID-19 pandemic. However, some of the countries were not ready in implementating the technology due to various reasons. Thus, this study intended to identify the technological and legislative aspects of drone application for disaster response in cities environment and to propose strategic framework on optimization of drone capabilities in cities disaster response. The literature search has been conducted in order to explore the applicability of drone technology in disaster management and disaster response. Later, further investigations were carried out using the Participatory Action Research (PAR) approach. In the PAR method, the data were collected in two phases; 1) observation and discussions and 2) interview sessions. The observation and discussions were conducted to obtain the implementation process of using the drone during the disaster event by the rescue team via recorded video. To triangulate the information gathered from the discussion, a series of interview had been made. Where, six selected stakeholders from end-user, regulatory agency and supplier were interviewed. It is found that the implementation of drone technology framework in disaster consists of six perspectives, namely; actions, internal organization, external environment, needs specification and feasibility analysis. To make the framework works, stakeholders are required to function in a cohesive and collaborative manner as illustrated in the Harmonization Triangle. The strength of cohesion and collaboration has also opened up future study for a full scope disaster management cycle phases, known as response, recovery, mitigation and preparedness.

1. Introduction

Disaster response requires accurate,quick and fast decisions and actions. The complexity and strenuous process working in a very uncertain and dynamic environment complicates the situation even further. Although disasters may have similarities but each disaster has its own uniqueness due the circumstances, geographical area and the extent of the devastation. Whatever the circumstances and situation, it may be, immediate and real time assessment of the disaster site conditions has to be undertaken. However, under some situations the assessment is hampered due to limited or no physical accessibility and/or safety risk considerations.

To overcome the constraints and limitations, one of the fast developing sectors is to deploy drones in disaster management including to support situational awareness [1]. In this review when drone is mentioned, it is to include Unmanned Aerial Vehicles (UAVs), the Unmanned Aircraft(UA), Unmanned Aerial System(UAS), Remotely Piloted Aircraft(RPA) or Remotely Piloted Aircraft System(RPAS)[2]. Cybersecurity, privacy, public safety and forensic are some of the challenges in urban setting[3].

The steep development in manufacturing and use of drones has become a major challenge in setting up the regulations of its usage. This is indeed challenging for regulators, policy makers and legislators[4]. Thus, positioning correctly human values in the age of technology and filling or closing in the gaps between admiration of technology and deep disorder[5]. The constraints in using this technology in disaster response is dependent on external and internal factors[6].

The objectives of the study was to identify the technological and legislative aspects of drone application and to propose strategies on optimization of drone capabilities in cities disaster response. Various questions were addressed including what are the advantages and constraints in using drone technology in disaster response; what are the laws/regulations/policies in place in Malaysia that hinders and/or facilitate the use of drones and what are the drone technologies available for disaster management, especially for disaster response purposes?.

Hence the scope of this paper is threefold; to analyze the current state, to identify and analyze the cause and effect for each and every limitation and constraint and to propose strategies optimizing drone technological potential for disaster response in cities.

2. Literature review

Five broad areas of Literature review were carried out namely disaster, drone technology, disaster management, disaster response, and Drone Application for flood disaster.

2.1. Disaster

The United Nations (UN) International Strategy for Disaster Reduction Secretariat (UNISDR) define disaster is a dangerous, unbearable and serious disruption of community functionality during normal circumstances and unable to deal the situation with their own readily available resources may it be human, economic, material or environmental losses and impacts.

Drones have been used to evaluate the impact of the natural disaster for decades. After the 2008 Wenchuan Earthquake, drones equipped with sensors were used by disaster responders to assess the extent of the damage caused by the earthquake. Drones were deployed to inspect the damaged at the nuclear reactor plant during the 2011 Fukushimi Dalichi crisis in Japan; Typhoon Haiyan in the Philippines in 2013 [7], [8]; the aftermath of the 2015 Cyclone Pam in Vanuatu and the 2016 Nepal Earthquake[9]. The use of drone drones in disaster response in Malaysia also greatly reduce the risks to responders and expense of the missions. The use of drones for disaster, collecting spatial data compared to other remote sensing platforms is much lower , with a greater ability to fly under cloud cover and able to collect an extremely high resolution both video and imagery data[10],[11]. In a study conducted[12] has concluded that majority of the public seems to have a good understanding of drones and their capabilities, however, the acceptance levels of drones did significantly differ depending on the context of use.

This can be achieved by having first responder drones deployed prior to the arrival of the human first responder to the crisis site or location. Thus, the first responders are equip with critical data and information upon arrival to the accident or disaster area. This would enable precision decision and action taken by the search and rescue team[13]. Drones can also provide communication gateways between first responders during disaster response at various locations of the disaster affected areas[14].

2.2. Drone technology

Significant development in drone technology and capability in recent years. The features of drones for emergency responses should have the following capabilities depending on the disaster specific, condition and geographical locations. Technologies have the potential to facilitate disaster responders with more precise damage assessment and reconnaissance. Technologies lessen the time to process the data and the analytical work of the data collected. Various types of sensory or non-sensory payloads can be attached to the drones. The maximum altitude, flight duration, flight range, and maximum payload are generally determined and influence by the weight, the model, and the energy source of a drone[15].

Data and information security and also stability and reliability of the internet connectivity as are some of the issues that requires countermeasures, a high and safe proof encrypted and protected server and platform is required. The failure of network infrastructure or the internet connection may lead communication gap, which in turn may lead to catastrophic results and affects[16]. There are three major developments in drone technology, namely miniaturization, autonomy and swarms[15]. The most significant development is miniaturization that are smaller, lighter and cheaper. Furtherance to the development of a more autonomous drones, the next level of development is the capability of drones operating in swarm or operating as a squadron of drones[15].

2.3. Disaster management

Moving forward from Hyogo Framework for Action (HFA) 2005-2015, Integrated Disaster Management(IDM) has moved forward to increase its capacity and reduce vulnerability through the The Sendai Framework for Disaster Risk Reduction(SFDRR)2015-2030 by introducing 3 positives changes compared to HFA**.** 3 positive changes are from saving lives to saving lives and livelihoods, investment for disaster risk reductioand in case of unfortunate disaster, Build Back Better

SFDRR has set out and identify 7 targets for disaster risk reduction (1) reduce number of death, (2) reduce the number of affected people, (3) reduce economic losses, (4) reduce damage to critical infrastructure, (5) more countries formulate national and local strategies, (6) develop and increase international cooperation, and (7) enhance access to disaster information. These positives changes and 7 intents of SFDRR, will be empowered further and achieve societal goal shared by all relevant stakeholders with the integration of research into implementation via Transdisciplinary Approach(TDA). The challenge is to get all relevant stakeholders from all disciplines and sectors working together. To achieve this, it is prime important to get the buying in from all stakeholders the need for a holistic and transformative solution possible.

On a broad framework, disaster includes earthquake, tsunamis, volcanic eruptions, floods, cyclones, hurricanes, landslides, bushfires, droughts, pandemic, epidemic, accidents and civil unrests. Each of the disaster has its own extent of impact pending on the respective circumstances and community preparedness, vulnerability and resilience.

Having a multi stakeholder involvements in every phase of disaster management(Mitigation, Preparedness, Response and Recovery) will ensure effectiveness in reducing the impact of the disaster by reducing our vulnerability and increasing our capacity. The Disaster Management Phases model helps to address issues related to disaster preparedness, and also business and economic recovery after a disaster. Each phase is unique but yet it is connected and has particular needs, requires definite and specific tools, strategies, and resources under various challenges.

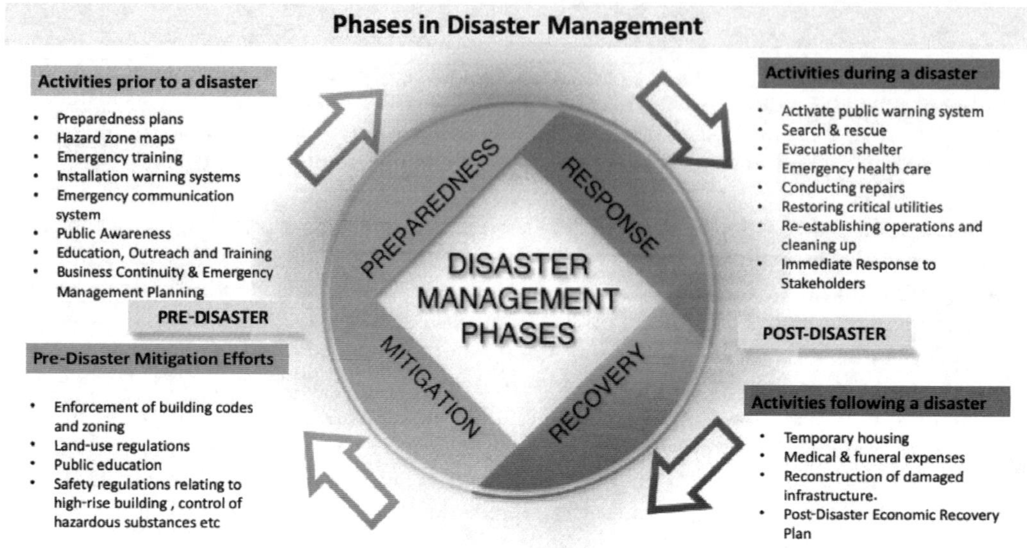

Figure 1. Disaster management phases

The Disaster Management Phases model as in Figure 1 facilitate to understand the relationship amongst the multi stakeholders, multi-dimensional and multi facet scenario. In our effort to increase resilience and reduce vulnerability of the community in facing disaster, mitigation measures requires to be indentified and implemented. This involves making changes and improvements in the non-structural and the structural measures. The next Disaster Management phase is the preparedness phase. The preparedness phase focuses on understanding how a disaster might impact the community and how education, outreach and training can build capacity to respond to and recover from a disaster. This may include engagement with the business community, joint pre-disaster strategic planning, and other logistical readiness activities. As for the response phase, as being the core and focus subject matter, it will be elaborated further in the later part of the paper. Recovery phase of disaster management cycle inclides restoration works on damaged and impacted livelihood on a community are to be undertaken as soon as disaster subsided and to build back better.

2.4. Disaster response

This phase immediate after disaster occurs, typically characterized by the need of a large set of tasks to be undertaken and coordinated with multiple resource requirements and time pressure[17].The scientific and technological input in decision making in line with the TDA is crucial for an accurate and effective decision making during the critical moments during the disaster.

Humanitarian logistics or logistic activities in disaster response comprise of planning, executing, control, monitoring and coordinating in the most efficient and cost-effective arrangement to receive and distribute relevant information, goods and materials from the point of origin to the rightly beneficiaries at the specified time and location. Thus, drone is preferred and applicable as the last-mile means of transport to overcome and made distribution possible to areas that have been cut-off and inaccessible. Although limitation of its lifting capacity, the delivery of light-weight multiple relief packages, including vaccine, water purification tablets are made possible via drones to remote or inaccessible locations within a disaster effected area[18].

2.5. *Drone application for flood disaster*

The application of drones combined with photogrammetry software has made it possible to collect high-performance spatial data and build-up spatial database, thus creating a high-resolution Digital Surface Model(DSM) of 22.6 cm per pixel[19]. Flood assessment can be carried out by using remote sensing technology, combining data and information obtained through satellite and Full Motion Video(FMV) via drone. Integrating FMV with appropriate software such as would be able to analyze the Geographic Information System(GIS) collected to produce critical and relevant outputs for disaster responders to consider and decide accurate, effective and efficient flood mitigation plan[20]. The 2014 worst flood disaster ever since 20 years in Malaysia has opened up the push and realisation to the need to improve the search and rescue operations by incorporating drone technology in the operation. The use of drone reduces cost and a shorter respond time and cost. Due to its relatively smaller size and ease of its deployment into disaster site very quickly as drones are expendable in relation to human life[21].

3. Research methodology

The research will be undertaken based on the methodology framework as shown in Figure 2 below. The study began with introducing the subject matter by identifying and establishing the problem statements, objectives and scopes of the research. The next phase will be the desktop literature review. In this research the Participatory Action Research (PAR) approach adopted. Next, critical evaluation, analysis of findings and identifying gaps for future study will be discussed before making recommendations and conclusion.

3.1. *Research framework*

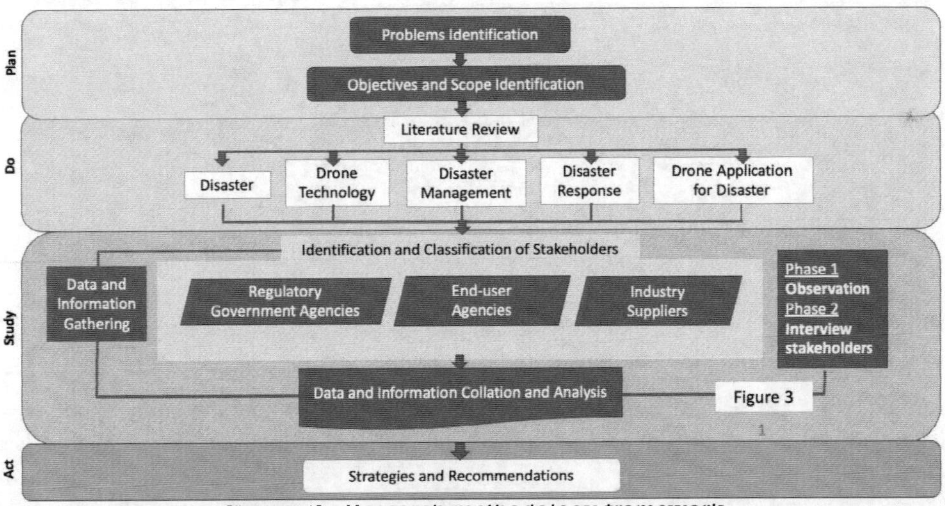

Figure 2. Research methodology framework

It is thought that the qualitative method is most suited as the method and techniques that integrate the oberservation, collecting, collating, documenting, analyzing and interpreting features, forms, elements and values of human experiences[22].

3.2. *Participatory action research (PAR) approach*

Participatory Action Research (PAR) approach has been chosen since the expected outcome seems to be more sustainable by getting the participation of all relevant stakeholders especially the community in every phase of the study towards the formulation of practical policy and best practice. It is hope that equitable participation would lead to a social transformation and change for the better[23]. In this context, the approach will create close bonding relationship that will close the gap between the interviewer and interviewee, the researcher and researched[24]. PAR approach attempt to bridge between science and practice, combining the best of the two. PAR facilitates people in a community to co-discover, co-design and implement solutions to the problems and challenges of their community.

3.3. Respondent criteria

The assessment was based on interviews and review data and information made available by regulatory body, government departments and industry suppliers. The findings were then systematically and effectively evaluated.

3.4. Stakeholder engagement

The stakeholders participated in this study were from end-user agencies, regulatory body and industry suppliers. The end-users are disaster response agencies that provide services to various types of disaster. The regulator is the government agency that regulate all civil aviation matters under the Ministry of Transport Malaysia. While the industry suppliers are supplier of drones and also all related technology that comes with it.

3.5. Two(2) phases research approach

Figure 3.3 below shows, the two phases of the research approach. The approach has been adopted as an alternative or second best option to the required oberservation and discussion on actual operation, training and simulation drill. In the first phase, observations and discussion were made on the selected videos. While the second phase were a series of interviews conducted via Zoom video conference.

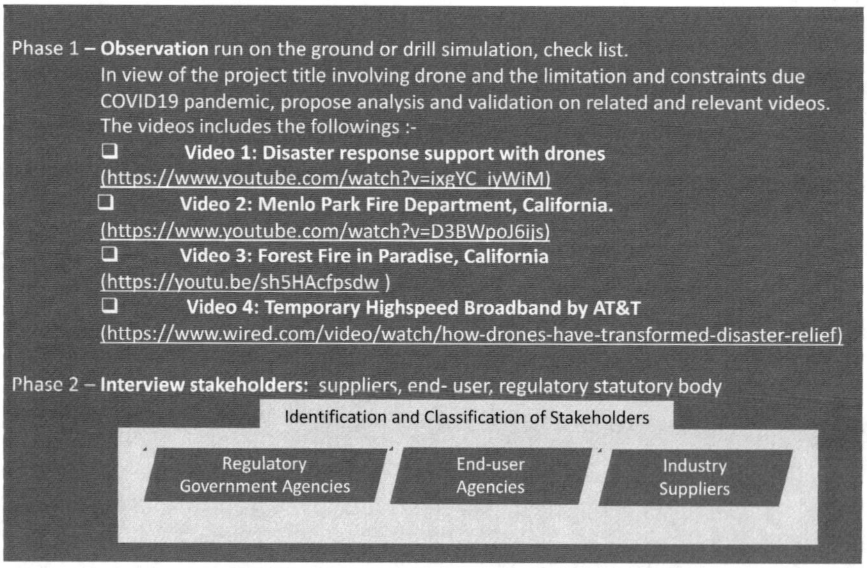

Figure 3. Phases in research approach

Each interviewee were asked three main and central questions namely issues of the current state of drone used in disaster management, the causes and effects for each and every limitation and constraint in operationalizing drone and how to increase in awareness on the need and optimizing drone technological potential for cities disaster management.

4. Result and finding

4.1. Observation and Validation on Operational Matters

In view of the COVID19 pandemic and the Movement Control Order(MCO) which is still inforce at the time the research being written, an alternative approach is to make observation on relevant videos available on youtube. These videos were also shared with interviewees prior to their interview session for their comments and validations, wherever possible.

Similar experiences were reported on a colloborative SAR operations consist of multiple drones by multiple operators including hobbyist, media together with rescue agencies. Observation and validation on operational matters were shared by the four(4) end-users responders.

4.2. End-user perspective
This section discusses the current state of key end-user agencies implementing drones within the response phase of disaster management. From the data collected from the interviews conducted, it shows that the drone unit of these agancies was set-up quite recently, the ealiest was in 2016. The drone fleet of the agencies are quite basic having minimal and limited in terms of numbers and capabilities to just having camera for taking photos, videos and thermal detection payloads. While very limited budget allocation for the procurement of the drone fleet. Training of drone operation is limited that comes together with the procurement of the drones and conducted by the suppliers.

4.3. Drone technology
As from the industry players perspective, it is impotatnt to realized that drone based solution is 30% flying or operating drone while the other 70% is the analysis of the collected data by the drone. The combination and integrating these two(2) broad skill sets have been proven and agreed upon through the literature review, observations and interviews conducted.

4.4. Drone regulation : present and future
The custodian and regulatory authority to ensure the safety of Malaysia air space is the Civil Aviation Agency Malaysia(CAAM). CAAM has plan for future Regulations and Directives targeted to be enforced in 2022. The 2022 regulations and directives will be a migration from classification based on weight, classification base on the risk factor namely open(Low risk), Specific(Medium risk) and Certified(High Risk) categories. It will also include regulation for operator registration and direct remote identification.

5. Discussion
Based on the findings under the research framework, in every identified and relevant disaster response operation, firstly to identify critical actions required to incentive technological adoption of drone capabilities in disaster response and secondly to set strategic operation framework aims to guide disaster response agency throughout the adoption of drones.

5.1. The six facet perspectives
The dynamics to the discussion is summarised by the six facet perspectives namely actions, internal organization, external environment, needs specification, feasibility analysis and implementation, as shown in Figure 4. The core of all the the six perspectives lies with the three (3) main players namely the the end-user, technology provider and regulator. In turn, these three (3) players are required to function in a cohesive and collaborative manner in order to ensure their respective role is played effectively as illustrated by the Harmonization Triangle in Figure 5.

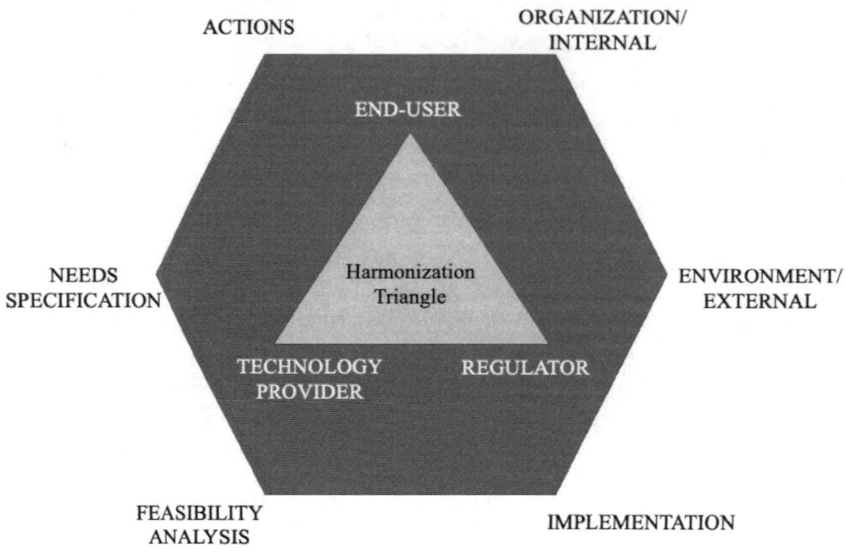

Figure 4. Six facet perspectives

Leadership quality, technology awareness and staff training to justify the procurement of drones are part of **internal organization** issues while statutory regulatory agency and industry players are some of the **external environmental** issues. In the **implementation** stage, is to ensure the most optimum drone capabilities and related technologies are identified, determine and procure correctly and accurately that leads to determination of the **needs specification.** Finally to carry out the **feasibility analysis**. All this requires **actions**.

5.2. Harmonization triangle

The Harmonization Triangle illustrate the interaction between all the three sectors on the various common interest and strategic issues. The inter and intra harmonization of the regulator, end-user and industry supplier would guarantee a brighter future in optimization of drone capabilities in disaster response.

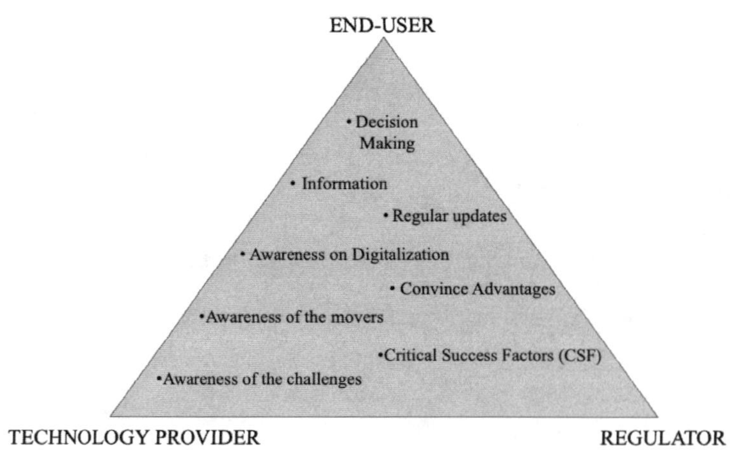

Figure 5. Harmonization triangle

5.3. Strategy management process

The outcome from the research, can then be used as basis to continue with the strategy management process as in Figure 6. Identified strategic issues and the strategic framework can then be used as basis to continue with the full and complete strategy management process.

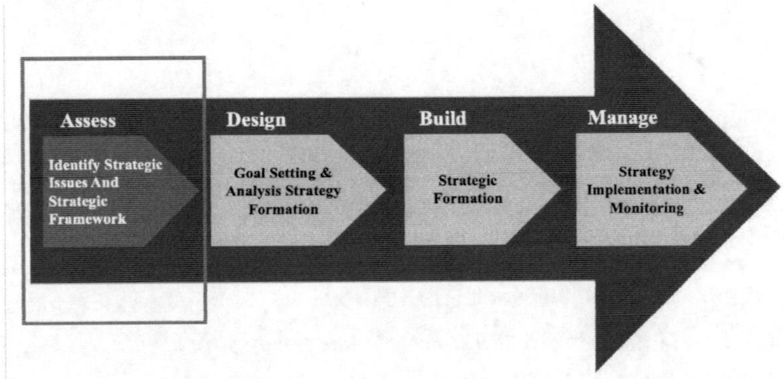

Figure 6. Strategy management process

6. Recommendation and conclusion

The proposed strategic framework consists both non-structural and structural framework. The strategic framework on the optimization of drone capabilities in cities disaster response lays the foundation to continue with the design, build and manage of the strategy management process. The inter and intra harmonization of the regulator, end-user and industry supplier would guarantee a brighter future in optimization of drone capabilities in disaster response.

This study focuses on the response phase of disaster management cycle. To evaluate the implementation of drones in the full scope of the 4-phase disaster management cycle namely response, recovery, mitigation and preparedness is an option for future research.

References

[1] Zwegliński T. *The use of drones in disaster aerial needs reconnaissance and damage assessment-Three-dimensional modeling and orthophoto map study. Sustain* 2020;12:1–20. https://doi.org/10.3390/su12156080.

[2] Sarrión Esteve J, Benlloch Domènech C. *Rights and Science in the Drone Era Actual Challenges in the Civil Use of Drone Technology.* Rights Sci R&S 2017:117–33.

[3] Vattapparamban et al. *Drones for Smart Cities: Issues in Cybersecurity, Privacy, and Public Safety* Edwin. IWCMC 2016 12th Int Wirel Commun Mob Comput Conf Sept 5-9, 2016, Paphos, Cyprus 2016:216–21.

[4] Nugraha RA, Jeyakodi D, Mahem T. *Urgency for Legal Framework on Drones: Lessons for Indonesia, India, and Thailand.* Indones Law Rev 2016;6:137. https://doi.org/10.15742/ilrev.v6n2.229.

[5] Katyal N. *Disruptive technologies and the law*. Georgetown Law J 2014;102:1685–9.

[6] Hristozov S, Zlateva P. *Concept model for drone selection in specific disaster conditions.* Int. Arch. Photogramm. Remote Sens. Spat. Inf. Sci. - ISPRS Arch., 2018. https://doi.org/10.5194/isprs-archives-XLII-3-W4-245-2018.

[7] techteam. New {Use} for {Drones} in the {Philippines} {Is} {Saving} {Lives}. Impact Hub 2017:1–9.

[8] Sjaak S. *Using Drone Imagery for real-time information after Typhoon Haiyan in The Philippines* 2016:9–12.

[9] Anhorn J, Herfort B, De Albuquerque JP. *Crowdsourced validation and updating of dynamic features in Open Street Map: An analysis of shelter mapping after the 2015 Nepal earthquake.* Proc Int ISCRAM Conf 2016.

[10] Consulting GW, Events GW, Media GW, Us A. *Did you know Drones can be used for Humanitarian Work ?* Drones for Humanitarian Work : The United Nations Approach Mapping 2021:1–4.

[11] Soesilo Denise. *How Drones Can Help in Humanitarian Emergencies 2016.* http://www.ictworks.org/2016/12/07/how-drones-can-help-in-humanitarian-emergencies/.

[12] Lin Tan LK, Lim BC, Park G, Low KH, Seng Yeo VC. *Public acceptance of drone applications in a highly urbanized environment.* Technol Soc 2021;64:101462. https://doi.org/10.1016/j.techsoc.2020.101462.

[13] Lemayian JP, Hamamreh JM. *First Responder Drones for Critical Situation Management.* Proc - 2019 Innov Intell Syst Appl Conf ASYU 2019 2019. https://doi.org/10.1109/ASYU48272.2019.8946353.

[14] Micheletto M, Petrucci V, Santos R, Orozco J, Mosse D, Ochoa SF, et al. F*lying real-time network to coordinate disaster relief activities in urban areas.* Sensors (Switzerland) 2018;18:1–20. https://doi.org/10.3390/s18051662.

[15] Vergouw B, Nagel H, Bondt G, Custers B. *Drone Technology: Types, Payloads, Applications, Frequency Spectrum Issues and Future Developments* 2016:21–45. https://doi.org/10.1007/978-94-6265-132-6_2.

[16] Sorsaniemi A. *Internet of Things and H2020* 2014:87–92.

[17] Schryen G, Rauchecker G, Comes T. *Resource Planning in Disaster Response: Decision Support Models and Methodologies.* Bus Inf Syst Eng 2015;57:243–59. https://doi.org/10.1007/s12599-015-0381-5.

[18] Rabta B, Wankmüller C, Reiner G. A *drone fleet model for last-mile distribution in disaster relief operations*. Int J Disaster Risk Reduct 2018;28:107–12. https://doi.org/10.1016/j.ijdrr.2018.02.020.

[19] TUROĞLU H, DOLO A. the *Advantages of Drone Technology in Urban Flood Studies.* Int. Symp. Geomorphol. 2017, 2017.

[20] Sharom MAAM, Fauzi MFA, Azmi MZM, Samsudin S, Rahman MHA, Fadzil MA, et al. *Assessment And Mitigation Of Monsoon Floods Via Satellite Imagery Data Extraction And Drone Full Motion Video (Fmv).* Def S T Tech Bull 2021;14:82–90.

[21] Sobri Hashim A, Syafiq Mohamad Tamizi M. *Development of Drone for Search and Rescue Operation in Malaysia Flood Disaster*. Int J Eng Technol 2018;7:9. https://doi.org/10.14419/ijet.v7i3.7.16195.

[22] Macdonald CD. *Understanding Participatory Action Research: A Qualitative Research Methodology Option.* Can J Action Res 2012;13:34–50.

[23] Benjamin-Thomas TE, Corrado AM, McGrath C, Rudman DL, Hand C. *Working Towards the Promise of Participatory Action Research: Learning From Ageing Research Exemplars.* Int J Qual Methods 2018;17:1–13. https://doi.org/10.1177/1609406918817953.

[24] Baum F, MacDougall C, Smith D. *Participatory action research. J Epidemiol Community Health* 2006;60:854–7. https://doi.org/10.1136/jech.2004.028662.

Development of Fragility Curves for the Underground PVC Water Distribution Pipes in Quezon City and Damage Estimation under a Magnitude 7.2 Scenario Earthquake

C M Soriano[*1,2], **P E Quinay**[1]

[1] Institute of Civil Engineering, University of the Philippines Diliman, T.H. Pardo de Tavera, Quezon City, Philippines
[2] Metropolitan Waterworks and Sewerage System – Corporate Office, Katipunan Road, Quezon City, Philippines

*Corresponding author e-mail: cfsoriano@up.edu.ph

Abstract. Quezon City is traversed by the West Valley Fault System (WVFS) that has the capability of generating a magnitude 7.2 earthquake known as "The Big One". It has an extensive water distribution network that is very susceptible to damages that will be caused by the ground shaking component of the M 7.2 earthquake. This study focuses on determining the behavior and estimating the damage of underground polyvinyl chloride (PVC) water pipes due to the M 7.2 earthquake by using appropriate empirical repair rates (RR) and developing fragility curves. The appropriate empirical RR equation was determined by comparing the results of selected PVC RR equations and the simulation using line-element modeling. The PGV ranges from 23.10 cm/s to 64.49 cm/s as determined using the Boore and Atkinson (2008) ground motion prediction equation. Using the results from the empirical and simulations methods, the equation by the American Lifelines Alliance (2001) was determined to be the appropriate empirical RR equation for the study area. The expected average repair rate of PVC is 0.05 repairs/km length of pipe or an estimated 84 total PVC pipe repairs in the city. Three fragility curves were generated showing the relationship of PGV and RR which is an important tool in estimating the underground pipe damages with respect to the M 7.2 earthquake and other future earthquakes of similar properties.

1. Introduction

1.1. Background

The Philippines experiences around 20 earthquakes per day, most of which have magnitudes that can't be felt according to the Department of Science and Technology - Philippine Institute of Volcanology and Seismology (DOST-DOST-PHIVOLCS). In the National Capital Region (NCR), 18 scenario earthquakes were determined to impact the region significantly. Of the scenario earthquakes, the Model 08 (West Valley Fault Model) will have the most significant impact on NCR [1] which can generate a magnitude 7.2 earthquake.

Quezon City is a highly urbanized city in the NCR. It is also the largest city in NCR and is served with water services by the Metropolitan Waterworks and Sewerage System (MWSS) through its Concessionaires (Maynilad Water Services, Inc. and Manila Water Company, Inc.). Water is distributed to the customers through a water distribution network composed of the water source, water treatment plants, reservoirs, and underground pipes. Water distribution pipes in the city can be classified into three groups based on their material – plastic (e.g. polyvinyl chloride (PVC), high-density polyethylene

(HDPE)), metallic (e.g. steel (SP), cast iron (CIP)), and cement (e.g. asbestos concrete (ACP), concrete (CON)) pipes. In Quezon City, most of the pipes used are PVC, HDPE, and SP, which comprise 66.08%, 18.93%, and 6.83%, respectively, while the remaining 8.16% is comprised of different materials.

The WVFS accounts for the highest hazard due to an earthquake in Quezon City based on a probabilistic seismic hazard assessment (PSHA) [2]. Also, the GMMA-RAP predicted that the epicenter of the M 7.2 earthquake will be near the border of Quezon City. Ground shaking is the most prevalent earthquake-related hazard in Quezon City based on the hazard maps produced by DOST-PHIVOLCS.

1.2. Objectives

The objectives of the research are:

1. To determine the appropriate empirical repair rate equation(s) that can be used to describe the possible damage of PVC pipes under a M 7.2 scenario earthquake by modeling the structural response of the selected study zones.
2. To determine the behavior and estimate the damage of underground PVC water pipes in the whole city due to the predicted magnitude 7.2 earthquake.
3. To develop fragility curves based on the pipe repair rates and peak ground velocity (PGV) as a function of distance from the predicted epicenter.

1.3. Significance

Critical infrastructures are of utmost importance during natural disasters such as earthquakes, hurricanes, and typhoons. Water distribution systems extend to large and complicated distribution systems that traverse different geologic formations, are exposed to different hazards, and experience widespread damage and shock due to an earthquake [3]. When subjected to an earthquake, buried pipes are subjected to the transient ground deformation is mainly caused by ground shaking, and the permanent ground deformation is caused by surface faulting or secondary effects leading to localized ground failures [4].

In preparation for the M 7.2 earthquake, several studies have been conducted to assess the extent of damage to structures. However, these studies mainly focused on buildings or above-ground structures. With the vast extent of the water distribution network and being a critical infrastructure, it is of utmost importance that a study is conducted to estimate its behavior and extent of the damage.

1.4. Limitations

The limitations of the research are:

1. Only underground PVC water pipes with diameters of 25 to 400 mm with standard thickness are focused on in this study. Appurtenances, valves, reservoirs, water treatment plants, and water sources are not included.
2. Damages to underground water pipes are based only on the ground shaking component of the M 7.2 scenario earthquake. Other components such as earthquake-induced landslide, liquefaction, ground rupture, and tsunami were not considered due to insignificant effects on the study area.
3. The method of the study is limited to deterministic seismic hazard assessment using peak ground velocity (PGV) as the independent variable for the ground motion parameter.
4. Corrosion effects, age of pipeline, manufacturing and construction conditions and defects, and history of repairs or previous pipe failures which may be of considerable effect in actual situations are not considered in this study.
5. The analysis is limited to the functionality of ArcGIS for geo-data processing and STAADPro for structural analysis.

2. Methodology

The study area is divided into a 250-m by 250-m grid called zones. These zones are used as representative areas in determining the actual behavior of the whole study area.

2.1. Division of Study Area

The Geographic Information System (GIS) datasets containing the maps of water lines and geographical boundaries of the study area are combined using ArcMap. The combined datasets are divided into smaller areas called zones that are used to describe the behavior of a group of PVC pipes. Clustering of pipe diameters is made in accordance with the groups mentioned in [5] to reduce the required analysis as a function of diameters. It is assumed that pipes of the same diameter behave similarly in the same group or cluster when corrosion, manufacturing, and construction effects are neglected. The distances of the zones from the predicted epicenter are computed using the haversine formula. The haversine formula measures the great-circle distance of two points using their respective longitudes and latitudes.

2.2. Computation of PGV Values

The Greater Metro Manila Area – Risk Analysis Project (GMMA-RAP) assesses the potential impact of flood, cyclone, and earthquake in the Greater Metro Manila Area by developing fundamental datasets and information on hazards, exposure, and vulnerability. In the said study, eight (8) ground-motion prediction equations (GMPEs) were used. The number of GMPEs was later narrowed to [6], [7], [8], and [9] based on the method of residuals. In this study, the GMPE by Boore & Atkinson (BA08) [6] will be used.

$$\ln Y = F_M(\boldsymbol{M}) + F_D(R_{JB}, \boldsymbol{M}) + F_S(V_{S30}, R_{JB}, \boldsymbol{M}) + \varepsilon\sigma_T \quad (1)$$

where Y is the ground motion parameter (e.g. PGA, PGV), $F_M(\mathbf{M})$ is the magnitude scaling component, $F_D(R_{JB}, \mathbf{M})$ is the distance function component, $F_D(V_{S30}, R_{JB}, \mathbf{M})$ is the site amplification factor, R_{JB} is the Joyner-Boore distance, V_{S30} is the average seismic shear wave velocity from the surface to a depth of 30 meters, and T is the aleatory uncertainty component equal to 0.560 (fault-type specified) corresponding to the combined intra-event and inter-event uncertainties.

The BA08 ground motion prediction equation (GMPE) (1) will be used to determine the PGV per corner. The corresponding PGV_{zone} is the geometric mean described in eqn. (2).

$$PGV_{zone} = \left((PGV_a)(PGV_b)(PGV_c)(PGV_d)\right)^{1/4} \quad (2)$$

where PGV_a, PGV_b, PGV_c, and PGV_d are the PGV per corner of the corresponding zone.

Using the PGV values obtained from eqn. (2), the range of the PGV values is obtained. The range is divided into groups (PGV_{group}) of specified equal intervals. Consequently, the zones with their respective PGV values are grouped into their respective PGV group.

2.3. Selection of Study Zones

The selection of study zones for simulation is done using stratified random sampling wherein the total number of zones per PGV group corresponds to its percentage from the total number of zones.

The respective number of zones per PGV group is counted and its corresponding percentage from the total number of zones is calculated. After calculating the corresponding percentage, the number of sample zones is calculated by multiplying the percentage by the total number of study zones. Then, the total number of pipe diameter groups per zone is counted and the zones are arranged in a descending order based on the number of pipe diameter groups represented.

The individual standard deviation per zone is calculated within the group of zones with the same number of pipe diameter groups represented. The zones with the highest priority are selected correspondingly based on the number of study zones per PGV group. After carefully selecting the study zones, each study zone is analyzed using both the empirical RR equations and the line-element modeling simulation.

2.4. Computation of Pipeline Repair Rate

In the Philippines, there is still no developed pipeline repair rate (RR) equation or a calibrated empirical equation. To validate its effectiveness in the Philippine setting, the simulation will be conducted for the study zones. For simplification, Table 1 lists the following abbreviations used for each empirical RR equation.

IOP Conf. Series: Earth and Environmental Science 1091 (2022) 012005 doi:10.1088/1755-1315/1091/1/012005

Table 1. Abbreviation of Empirical RR Equations

Empirical RR Equation	Abbreviation	Reference
O'Rourke & Ayala (1993), FEMA (1999)	ORA93	[10], [11]
Eidinger (1998)	EID98	[12]
Isoyama, Ishida, Yune & Shirozu (2000)	ISO00	[5]
American Lifelines Alliance (2001)	ALA01	[13]
Shih & Chang (2006)	SHI06	[14]

The common dimensions and specifications of PVC pipes are provided in different ASTM standards such as [15], [16], [17], [18], and [19]. To simplify the analysis, constant pipe diameters are used for every pipe diameter group for both methods. The concept of a weighted average is used in determining the common pipe diameter and are rounded up and down to the next commercially available pipe diameter and are used to determine the effect of changing the diameters on the RR of PVC.

2.4.1. Empirical Equations. Each pipe is analyzed based on their corresponding probability of failures as a function of the empirical RR equations given in eqn. (3). If the length of pipe repair is less than the length of a specific pipe, the said pipe failed ($L_{repair} < L_{pipe} \rightarrow$ Pipe Failure!).

$$L_{repair} = -\frac{1}{RR}\ln(1-u) \tag{3}$$

where L_{repair} is the length of pipe repair (unit depends on the unit length of repair rate), RR is the repair rate (unit in number of repairs/unit length of pipe), and u is the random variable which is uniformly distributed between 0 and 1.

The percentage of pipe that fails is computed at the end of every run (1 run = eqn. (3) is used in all pipes). The process is repeated 20 times per pipe per empirical RR equation. The average of 20 runs per empirical RR equation is compared to the results of the simulation method.

2.4.2. Simulation Method. Simulations are used to analyze and predict the behavior of a certain structure when subjected to various loading conditions. The effect of the M 7.2 earthquake on the study zones is simulated using a line-element modeling and time history analysis in STAADPro v8i software and is referred to as the "simulation method" in this study. The following are the loads applied:

a) Self-weight
b) Earthquake Loads. Three (3) earthquake time histories are applied to a whole study zone with amplitudes depending on its location. Actual time history is scaled accordingly using ratio and proportion based on its PGA values using the BA08 GMPE. Time history files are obtained from the Pacific Earthquake Engineering Research (PEER) Center database that has the same characteristics as the predicted M 7.2 earthquake in our country.
c) Weight of Fluid Conveyed
d) Overburden Load. Refers to the soil column placed on top of the buried pipe within the recommended trench width from the ground to the crown of the pipe. The soil unit weight is obtained using the allowable soil bearing capacities from [20] and the Terzaghi's Formula for strip footing using an angle of internal friction (ϕ) = 34° and cohesion (c) = 0 kPa. (USCS clayey gravel soils).

$$L_{overburden} = \gamma W^2 \left(\frac{1 - e^{-2kf(H/w)}}{2kf}\right) \tag{4}$$

where $L_{overburden}$ is the overburden load (unit in kN/m linear length of pipe), H is the depth of soil from the ground to crown of pipe (unit in m, a uniform soil cover of 0.75 m is applied throughout), w is the recommended trench width (unit in m, as specified in ASTM D-2321), and γ is the soil unit weight (unit in kN/m^3).

e) Live Load. Often due to vehicular loads transmitted through the ground, a uniform live load is applied along the length of the pipe based on the values provided by AASHTO H-20 (for diameters >100 mm) and AASHTO H-15 (for diameters ≤ 100 mm).
f) Lateral Soil Pressure

The pipes are supported by soil springs at specified intervals, depending on their diameters. Soil spring formulas for pipe-soil interactions were inferred from pile shaft interactions [21]. The total pipe span is divided into equal parts such that the individual length of pipe segments will not exceed the allowable span length. Thrust blocks, modeled as pinned supports, are added at bends and intersections.

Due to the lack of actual data on soil properties, other soil parameters were derived from previous researches in the study area. Allowable soil bearing capacity at 1-meter depth from [20] was used to derived soil undrained shear strength using the rules of thumb in geotechnical engineering practice [22].

2.5. *Selection of Appropriate Empirical Equation*
The selection of appropriate empirical equation(s) is made through the comparison of the percentage failure values from the simulation method and the empirical equations method. The percentage failure values are compared by plotting the percentage failure values of the simulation method on the horizontal axis and the percentage failure values of the empirical method on the vertical. The slope of the trendlines is used to determine what empirical RR equation is/are appropriate for the whole study area. Two (2) cases can happen in this comparison:

a) Case 1 (Uniform Case) – A single empirical equation best fits the simulated values for the whole range of PGV values; and
b) Case 2 (Variable Case) – Different empirical equations best fit the simulated values for a specific range of PGV values.

2.6. *Formulation of the Pipeline Fragility Curve*
For pipelines, two damage states are considered: leaks and breaks. For pipeline damage caused by seismic wave propagation or ground shaking, 80% are leaks and 20% are breaks [23].

After identifying the appropriate empirical equation, the corresponding repair rates of each zone and a scatter plot is made with the vertical axis as the repair rate (unit in the number of repairs/unit km length of pipe) and the horizontal axis as the PGV (unit in cm/s). Consequently, two (2) similar fragility curves are also generated representing its damage states.

3. Results and Discussion

3.1. *Study Area and Pipe Statistics*
The study area is divided into 2,475 zones of 250 meters by 250 meters. Out of the 2,475 zones, only 1,922 (77.65%) have PVC pipes laid. The determination of the study zones is based on this selection. The most common PVC diameter used is 100 mm with a cumulative pipe length of 669.03 kilometers and followed by 150 mm with a cumulative pipe length of 426.78 kilometers. The said diameters constitute 68.28% of the total PVC pipe length. The summary for the diameter groups and the corresponding common diameters is presented in Table 2.

Table 2. Summary of PVC Diameter Groups

Diameter Group	Total Length (km)	Percentage (%)	Common Diameters (mm) [Lower, Higher]
< 100 mm	242.05	15.08%	71.38 → [50, 75]
100 – 150 mm	1,095.82	68.28%	119.47 → [100,150]
200 – 450 mm	267.13	16.64%	229.48 → [200,250]
Total	1,605.00	100.00%	

IOP Conf. Series: Earth and Environmental Science 1091 (2022) 012005 doi:10.1088/1755-1315/1091/1/012005

3.2. Peak Ground Velocity

The peak ground velocities of the respective zones are calculated using the BA08 GMPE. The lowest and highest PGV calculated for the study area were 23.10 cm/s and 64.49 cm/s respectively with the highest PGV nearest the predicted epicenter and the lowest PGV being those farthest.

3.3. Determination of Pipeline Repair Rates

A total of 55 study zones are analyzed using the empirical and simulation methods. The selection of zone to be simulated are done using stratified random sampling and standard deviation. The number of study zones for PGV Groups 25 to <=30, 55 to <=60, and 60 to <= 65 have been adjusted to two (2) to have additional data points for the comparison of the results of the empirical and simulation methods.

3.3.1. Empirical Method. The comparison of results for the empirical method is shown in Figure 1. For the whole PGV range, the ALA01 exhibits the lowest probability of failure with the lowest variation from 4.11% to 10.62%. EID98 and ORA 93 exhibit almost the same probability of failure for the whole PGV range from 7.07% to 35.98%. ISO00 exhibits the largest variation in the probability of failure from 10.97% to 49.12% while the SHI06 exhibits a logarithmic variation from 14.85% to 34.76%. For PGV up to 35 cm/s, the SHI06 exhibits the largest probability of failure while for PGV of more than 35 cm/s, the ISO00 exhibits the largest probability of failure.

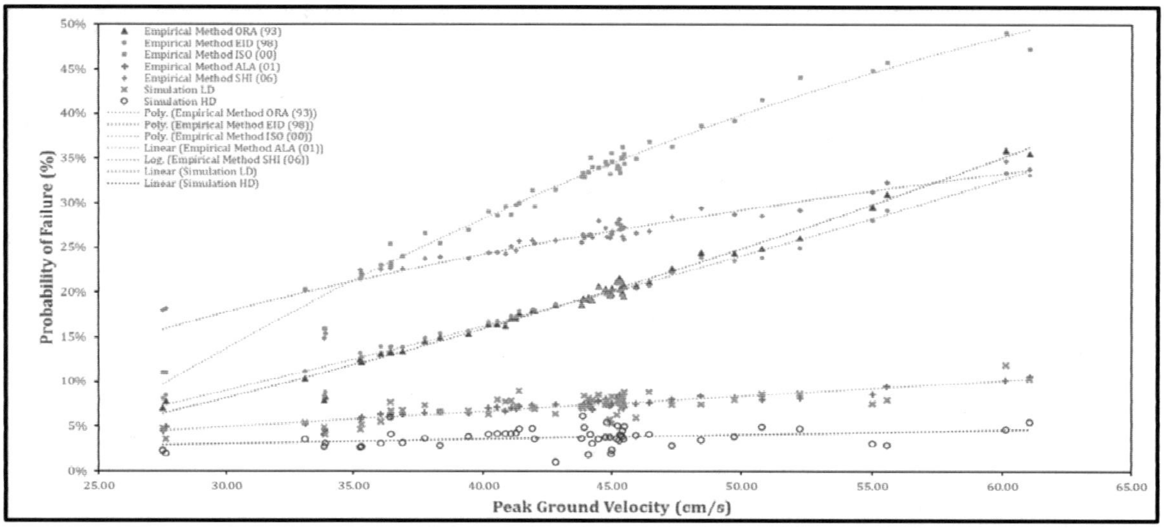

Figure 1. Results of the Empirical and Simulation Methods

3.3.2. Simulation Method. Figure 2 shows a sample study zone (AR-27) overlaid on a Google Earth Map. The study zone shows portions of Commonwealth Avenue and Don Mariano Marcos Avenue (right side) and Brgy. Holy Spirit (left side). The thick red lines indicate the zone boundaries and the thin green lines indicate the location of PVC lines.

Figure 3 shows a sample line-element model of study zone AR-27. The right figure shows the zoom extent of the line-element model for the whole study zone. The bottom-left figure shows an isometric view of the pin supports and soil springs. The top-left figure shows the 3D-rendered view of a portion of the study zone.

Figure 4 shows sample pipe stresses results of study zone AR-27. The left figure shows the pipe stress results of the same portion of AR-27 in Figure 3. The top-right figure shows the 3D-rendered view of a single pipe showing the pipe stress distribution. The axes indicated in the figure are in local coordinates. The bottom-right figure shows the 2D-view (front view) of a single pipe showing the pipe stress distribution. The maximum resulting pipe stress is 32.00 MPa (red, compression) and 34.14 MPa (blue, tension).

Figure 2. Sample Study Zone

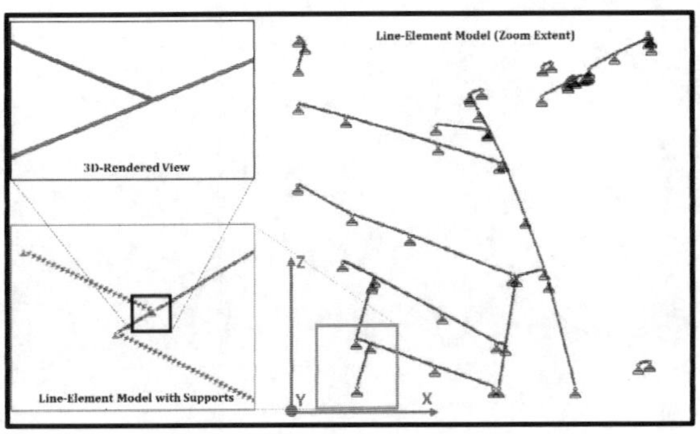

Figure 3. Sample Line-Element Modeling

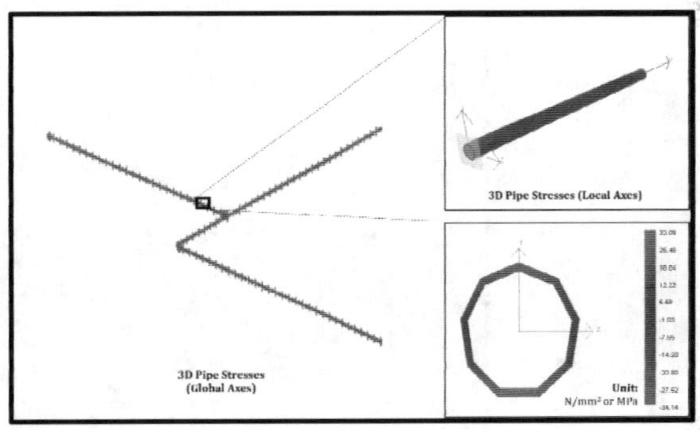

Figure 4. Pipe Stresses

The comparison of results for the simulation method is shown in Figure 1. The resulting probability of failure for using the lower diameter is higher than the probability of failure for using the higher diameter. The decrease of probabilities of failure is due to the increase in the geometrical properties of the pipe (e.g. diameter, wall thickness, etc.) and the values of the soil spring supports. The resulting pipe stresses in line-element modeling are compared with the allowable tensile stress of 51.36 MPa for PVC pipes since it is the lowest allowable stress when compared with compression (66.18 MPa) and flexure (99.62 MPa). The majority of the pipe failures occurred within the first diameter group (<100 mm) because it has the lowest geometrical properties resulting in lower moments of inertia and rigidity.

3.3.3. Comparison of Empirical and Simulation Method. Figure 1 shows the comparison of the results from the empirical and simulation methods. The results of the simulation using the lower diameter (denoted by "x") exhibit almost the same trend as the results using the ALA01 empirical RR equation (denoted by "+"). On the other hand, the results of the simulation using the higher diameter (denoted by "o") falls below all the results of the empirical RR equation and being closest to the ALA01 empirical RR equation. The trendlines show that ALA01 (denoted by "solid line") has a slope of 0.98 which is almost similar to the slope of the ideal line (slope = 1.00). The trendline for ISO00 (denoted by "dotted lines") exhibits the largest deviation from the ideal trendline with a slope of 4.36. Using this comparison, ALA01 is the appropriate empirical RR equation when using lower diameters.

On the other hand, the results of the simulation using the higher diameter yields the same conclusion with that of the lower diameter and are not presented in this summary to make it concise.

Given that the comparison for both the lower and higher diameters yield the same results, ALA01 is the appropriate empirical RR equation that can be used to describe the possible damage of PVC pipes under a M 7.2 earthquake for the whole study area.

3.3.4. Overall Pipeline Repair Rate. The ALA01 is the appropriate empirical RR equation for determining the overall PVC pipeline repair rate in the whole study area. The maximum RR is 0.07 repairs/km length of pipe and the minimum RR (excluding RR = 0) is 0.03 repairs/km length of pipe. The average RR per zone is 0.05 repairs/km length of pipe. Given the average RR, the minimum length to have at least one (1) repair is 20.00 km. For the whole study area, the estimated total number of repairs is 84.

*3.3.5. Fragility Curves.*Figure 5 shows the generated fragility curves for the study area. The solid line denotes the fragility curve for total damage and the broken lines represent its corresponding damage states. The fragility curves are only within the bounds of the peak ground velocities (23.10 cm/s to 64.49 cm/s) obtained in this study.

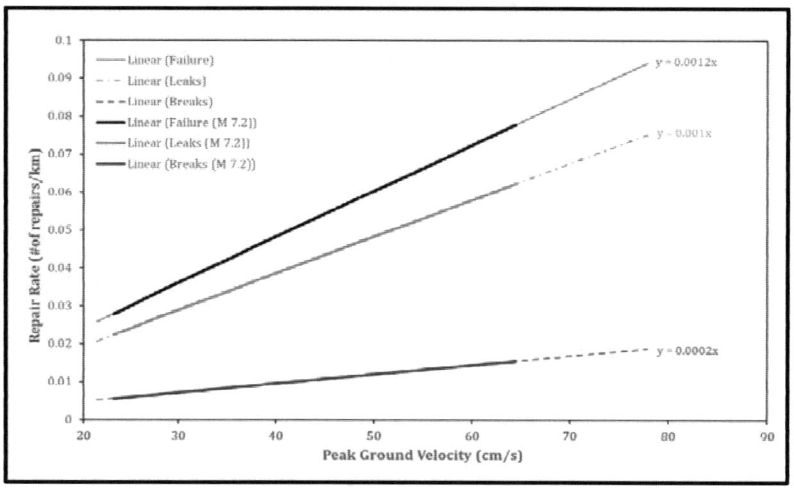

Figure 5. Fragility Curves (M 7.1 to M 7.5)

The generated fragility curves are tested on their applicability for other earthquake magnitudes. Earthquakes of magnitudes 7.1 and 7.5 are tested. Using the BA08 GMPE, the range of PGVs calculated are 21.36 cm/s to 60.26 cm/s and 28.77 cm/s to 77.84 cm/s for M 7.1 and M 7.5 respectively. The change in values of PGVs did not significantly affect the results of the simulation method since only the method of application of the time histories are PGV-dependent. Table 3 shows that there is no significant variation in the amplitude scaling factors for AR-27 for the three different magnitudes. The other loads such as dead and live loads are PGV-independent.

Table 3. Comparison of Amplitude Scaling Factor (AR-27)

Parameters	M 7.1	M 7.2	M 7.5
PGV for AR-27, cm/s	30.780	33.050	40.280
Maximum PGV, cm/s	60.260	64.490	77.840
Amplitude Scaling Factor	0.510	0.512	0.517
Amplitude Scaling Factor with g[a]	5.010	5.027	5.076

[a]Amplitude scaling factor multiplied by 9.81 m/s^2.

Having no significant difference for the other magnitudes, the generated fragility curves can be used for the magnitude range of 7.1 to 7.5 with ground shaking as the major component of the earthquake.

The generated fragility curves have two (2) major uses: inputs for computer software that directly calculates the probability of damage of a certain pipeline when subjected to an earthquake and cost estimation.

Developing computer software that directly calculates the probability of damage and identifies possible pipeline failure would be of great help in designing distribution networks that use complex pipeline structures. Restoration and hazard mitigation plans can also be developed using this software.

Cost estimation is important for budget preparation for potential repairs after any disasters such as earthquakes. Knowing that the average RR = 0.05 repairs/km length of PVC pipe and the total length of PVC pipe is 1,597.80 km, the estimated average total number of repairs is 83.05 ≅ 84.00. The calculated total number of repairs represents the average probable number of repairs and actual data may vary (increase or decrease). The ballpark estimate for the cost of repairs is PhP 363,084.62. The estimate covers the cost for PVC pipe replacement only and does not include associated joints and appurtenances, other pipe materials, and other components of the water distribution system.

4. Conclusions and Recommendations

4.1. Summary and Conclusions

From the study, it can be concluded that:

1. The American Lifelines Alliance 2001 (ALA01) is the appropriate empirical repair rate equation for determining pipeline damage caused by the ground shaking component of an earthquake.
2. The probability of failure for PVC pipes increases as the value of the PGV increases from 23.10 cm/s to 64.49 cm/s. In addition, the probability of failure is lowest when PVC pipe diameter is more than 200 mm and highest when PVC pipe diameter is less than 100 mm.
3. The generated fragility curves illustrate that the pipeline repair rate increases linearly from 0.02 to 0.07 number of repairs/km length of pipe as PGV increases as a consequence of the M 7.2 earthquake with the predicted epicenter at 14.65°N, 121.087°E and the major earthquake hazard is ground shaking.

The results of the empirical and simulation methods illustrate the increase in the probability of failure for PVC pipes as the value of the PGV increases. Furthermore, the results of the simulation method show that the majority of the pipe failures occur in the first diameter group (Ø < 100 mm) because it has the lowest geometrical properties (e.g. diameter, wall thickness, etc.). The probability of failure decreases when pipe diameter increases as illustrated in the results of the simulation using the next lower and the next higher commercially available PVC pipe diameters. The percentage decrease varies depending on the pipe diameters compared with the smallest reduction for the first diameter group (<100 mm) and the largest reduction for the third diameter group (200 mm to 450 mm).

This study developed fragility curves based on the pipeline repair rates and peak ground velocity (PGV) as a function of distance from the predicted epicenter. The generated fragility curves are tested on their applicability for other earthquake magnitudes (M 7.1 and M 7.5). From the further analysis and data post-processing conducted, it can be observed that negligible differences are observed (except for PGV values) when M 7.1 and M 7.5 are used as scenario earthquakes. The adjusted fragility curves can be used to estimate the PVC pipe repair rate for earthquakes with magnitude within the range of 7.1 to 7.5 subject that ground shaking is the major contributory component of the said earthquake and the maximum PVC pipe diameter is 450 mm. The fragility curves illustrate that the estimated PVC pipe repair rate linearly increases with the value of the PGV.

4.2. Recommendations

The methodology of this study can be replicated in the other areas by using their corresponding GIS data sets of the water lines and geographical boundaries considering the other earthquake-related hazards that may have significant contributions within the study area.

In addition, studies can be made for different types of pipe materials, especially for brittle pipes (e.g. asbestos concrete, reinforced concrete, cast iron, etc.) which have significantly higher repair rates than

PVC as observed in different literature. If different pipe materials are to be analyzed, additional empirical RR equations derived in past researches.

Other components of the water distribution system such as joints and appurtenances can be studied while the fragility curves for buildings can be applied to treatment plants as well. The generation of fragility curves for pipelines can be extended to other cities, municipalities, and regions while considering other earthquake scenarios as well.

Methods of cost estimation can be further improved by providing a more detailed cost-breakdown analysis and considering other factors to create a unit price. The repairs can be further broken down into its components – leaks and breaks and their corresponding associated repair costs can be integrated.

5. References

[1] JICA, MMDA and DOST-PHIVOLCS 2004 *Earthquake impact reduction study for Metropolitan Manila, Republic of the Philippines*

[2] Wong I, Dawson T, Dober M 2008 *Fourteenth World Conference on Earthquake Engineering* 8 pp

[3] Davis C A 2014 *Earthquake Spectra* **30** pp 1487-1509

[4] Tromans I 2004 *Behavior of buried water supply pipelines in earthquake zones* (London: Imperial College of Science, Technology, and Medicine)

[5] Isoyama R, Ishida E, Yune K and Shirozu T 2000 *Twelfth World Conference on Earthquake Engineering* **1762** 8 pp

[6] Boore D M and Atkinson G M 2008 *Notes on the equation to use for pga4nl*

[7] Chiou B and Youngs R R 2008 *Earthquake Spectra* **24** pp 173-216

[8] Sadigh K, Chang C Y, Egan J A and Makdisi F 1997 *Seismological Research Letters* **68** pp 180-9

[9] Fukushima Y and Tanaka T 1990 *Bulletin of the Seismological Society of America* **80** 757-83

[10] O'Rourke M and Ayala G 1993 *Journal of Geotechnical Engineering* **119** pp 1490-8

[11] FEMA 1999 *Earthquake loss estimation methodology, HAZUS 99*

[12] Eidinger J 1998 Water Distribution System – *The Loma Prieta, California, earthquake of October 17, 1989 – Lifelines* ed Schiff A J (Washington: United States Government Printing Office) pp 63-78

[13] American Lifelines Alliance 2001 *Seismic fragility formulations for water systems*

[14] Shih B J and Chang C H 2006 *Natural Hazards* **37** pp 71-85

[15] ASTM International 2014 *ASTM D638. Standard test method for tensile properties of plastics* 08.01 17 pp

[16] ASTM International 2015 *ASTM D695. Standard test method for compressive properties of rigid plastics* 08.01 8 pp

[17] ASTM International 2017 *ASTM D790. Standard test methods for flexural properties of unreinforced and reinforced plastics and electrical insulating materials* 08.01 12 pp

[18] ASTM International 2020 *ASTM D792. Standard test methods for density and specific gravity (relative density) of plastics by displacement* 08.01 6 pp

[19] ASTM International 2012 *ASTM D1785. Standard for poly(vinyl chloride) (PVC) plastic pipe, schedules 40, 80, and 120* 08.04 11 pp

[20] Dungca J R 2020 *International Journal of GEOMATE* **19** pp 42-7

[21] American Lifelines Alliance 2005 *Seismic Guidelines for Water Pipelines*

[22] Atkinson J 2008 *18th NZGS Geotechnical Symposium on Soil-Structure Interaction* **34** 13 pp

[23] FEMA 2003 *Multi-hazard lost estimation methodology earthquake model – HAZUS MR4 Technical manual* 712 pp

Development of A Two-storage Reservoir Operation Rules for Flood Control
Case Study of Wonogiri Multipurpose Reservoir, Central Java, Indonesia

R Jayadi1[1]*, Istiarto[1], E P A Pratiwi[1] and K Windianita[2]

[1] Department of Civil and Environmental Engineering, Universitas Gadjah Mada, Jalan Grafika No. 2, Kampus UGM, Yogyakarta 55281, Indonesia
[2] Jasa Tirta I Corporation, Jalan Surabaya 2A, Malang 65145, PO BOX 39, Indonesia

*Rachmad Jayadi: rjayadi@ugm.ac.id

Abstract. Wonogiri Reservoir, one of the large reservoirs in Indonesia, regulates water for flood control, irrigation, domestic use, and hydropower. It is located at the upper part of Bengawan Solo River Basin, Central Java. A closure dike has recently been constructed as a countermeasure against severe sedimentation that separates the reservoir into two parts, i.e., Sediment Storage Reservoir (SSR) and Main Reservoir (MR). A two-gate spillway was constructed in the SSR to facilitate sediment flushing and sluicing. During flood events, it controls reservoir water level jointly with the MR spillway. This brings the need for an update on the reservoir operation rules, especially on the gates opening of the new spillways to maintain the flood control performance. An assessment of reservoir routing simulation results was carried out by considering the overflow from SSR to MR, or vice versa with two options of Control Water Level (CWL), i.e. +135.8 m and +136.3 m. The flood control performance of the new reservoir operation rules was evaluated by considering the reservoir water level recovery, the maximum total outflow discharge, and the dam safety due to the risk of overtopping. These assessments give a foundation for the selection of CWL and the spillway operation to ascertain an optimal flood control of the reservoir.

Keywords: Multipurpose reservoir, sedimentation, closure dike, reservoir routing, dam safety

1. Introduction

Wonogiri Reservoir is located at the upper part of Bengawan Solo River basin with a 1343 km^2 watershed, a total storage of 735 million cubic meters, and a design life of 100 years. Its construction and impoundment had started in 1977 and 1981 before the reservoir has been operated since 1982. Wonogiri Reservoir was initially built as flood control. Due to a catastrophic flood in 1966, the Bengawan Solo River's management mainly focused on flood control until the Bengawan Solo River Development Master Plan in 1974 [1]. The construction of the Wonogiri Dam was one of the master plan's recommendations. The occurrence of Bengawan Solo floods, particularly for the Solo City, decreased significantly with the completion of the Wonogiri Dam. Eventually, Wonogiri Reservoir has also functioned as a multipurpose reservoir that serves irrigation water supply, domestic water supply, and hydropower.

Previous studies [2,3] show that the Wonogiri Reservoir faces serious sedimentation problems because of the rapid population growth and land-use change in the watershed. The sedimentation inflow rate, mainly from Keduang River, is high, so the reservoir's storage capacity decreases rapidly. It was found that the volume of sediment depositing in the dam was estimated to be 98 million m^3, with a sedimentation rate of 2.72 million m^3/year. As a result, the dam's service life is 44 years, much shorter compared to the dam's intended service life of 100 years [4]. The sediment deposit blockage in the front of intake structures also often affects reservoir operation for water supply. Short-term countermeasure, such as excavation, is insufficient for maintaining storage capacity. If no long-term countermeasure be taken for the sedimentation problem, Wonogiri Reservoir will lose its function shortly, and its economic life will not last as planned. The countermeasures to overcome the sedimentation problem must comprise structural and non-structural. Non-structural countermeasure is taken with watershed conservation activities. The recommended structural measures were closure dikes, overflow dikes, and a new spillway [3]. The construction of those structural countermeasures finished in 2019. The dikes split the reservoir into two compartments: Main Reservoir (MR) and Sediment Storage Reservoir (SSR). In the SSR, a two-gate spillway was constructed. The new spillway enables sediment flushing and sluicing. It works together with the MR spillway to maintain reservoir water levels during flood situations. The presence of structural countermeasures in the reservoir, consequently, changes the characteristics of the reservoir. Maintaining flood control effectiveness necessitates a modification to the reservoir operation regulations, particularly regarding the gates opening of the new spillways.

Jasa Tirta I Corporation operates the Wonogiri Reservoir during flood period in compliance with the operation policy of Bengawan Solo River Basin Development Master Project 1994 with the elevation of Control Water Level (CWL) at +135.3 m. The elevation of the new CWL should be evaluated because the condition and characteristics of the reservoir have changed a lot in 27 years due to sedimentation, the construction of structural sedimentation countermeasures, the change of hydrological state, and the change of water demand. To recover the storage capacity loss by the sedimentation, CWL can be raised so the water storage at the end of the rainy season increases. Raising CWL must be carefully conducted because it reduces the reservoir's capability to control the flood. This paper aims to assess the new reservoir operation rules with two options of CWL (+135.8 m and +136.3 m) on the flood control performance. Assessments were performed by evaluating the reservoir water level recovery, the maximum outflow, and the maximum reservoir water level. Apart from CWL, there were some adjustments to operational limits on the reservoir water level, i.e., Normal High-Water Level (NHWL), Surcharge Water Level (SWL), and Extra Flood Water Level (EFWL).

2. Methods

2.1 Data collection

Data used in this study comprises hydrology data, technical data of Wonogiri Reservoir, including the new and old spillways, reservoir characteristics, and the previous study regarding the operation of Wonogiri Reservoir. The Wongiri Reservoir inflow originates from the runoff in 1343 km^2 reservoir's watershed, consisting of 10 rivers, namely Pondok, Keduang, Kepuh, Wiroko, Temon, Solo Hulu, Alang Ngunggahan, Kedungguling, Wuryantoro, and Durensewu. Hourly and daily rainfall data are sourced from various rain gauges managed by Bengawan Solo River Basin Organization, Jasa Tirta I Corporation, Wonogiri Regency Public Works Office, and Bengawan Solo Watershed Management Office. After evaluating the comprehensiveness and quality of rainfall data, this study categorizes rainfall data to calculate the daily area rainfall of each watershed into two periods: from 1990-2005 using 14 rain gauges and from 2006-2018 using 18 rain gauges. Figure 1 shows a map of the reservoir's watershed and the location of the rain gauges. The data of hourly area rainfall and the corresponding hydrograph at the various occurrence of heavy runoff in Solo Hulu, Temon, Tirtomoyo, and Keduang watersheds, are used to determine unit hydrographs using the Collins method [5].

Primary technical data for Wonogiri Reservoir is shown in Table 1. Figure 2 shows a modified sketch of Wonogiri Reservoir storage zonation from a previous study [5] that shows the existing operational limits of reservoir water level prior to constructing a closure dike. Technical data for the closure dike, overflow dike, and new spillway is shown in Table 2. A picture of Wonogiri Reservoir's storage conditions after constructing the closure dike dividing storage into MR and SR is shown in Figure 3.

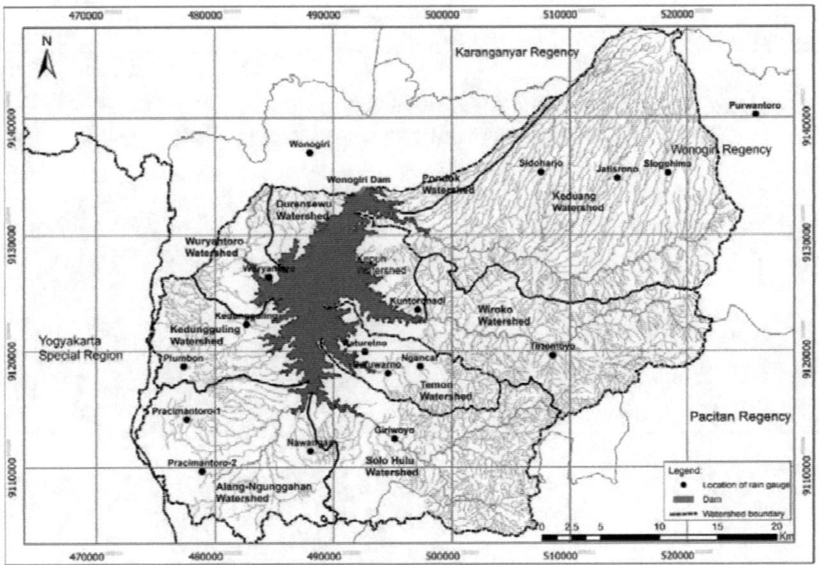

Figure 1. Map of Wonogiri Reservoir watershed and location of rain gauges.

Table 1. Primary technical data of the Wonogiri Reservoir.

Items	Size/description
Dam type	Rockfill with clay core
Elevation of the dam crest	+142.0 m
Length of the dam crest	830 m
Dam height	40 m
Existing flood control storage volume	220 million m^3
Existing effective storage volume	440 million m^3
Design sediment storage volume	120 million m^3
Existing total storage volume	735 million m^3
Existing spillway type	Ogee with 4 radial gates
Elevation of the existing spillway crest	+131.0 m
Existing spillway crest width	30 m
Existing spillway flow capacity	1360 m^3/s

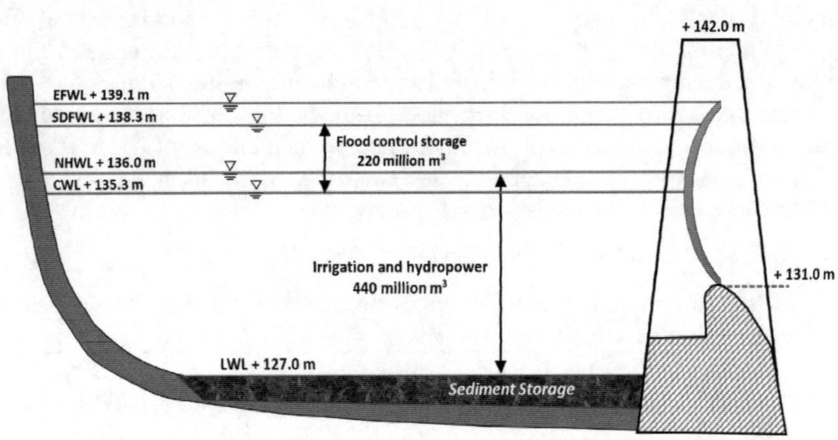

Figure 2. The existing Wonogiri Reservoir's operational storage zoning.

Table 2. Technical data for closure dike, overflow dike, and new spillway.

Component of structure	Items	Size/description
Closure dike	Type	Earth fill with counterweight
	Elevation of the dike crest	+139.40 m
	Foundation reinforcement	Deep mixing soil improvement
	Length of closure dike A	700 m
	Length of closure dike B	700 m
	Length of closure dike C	402 m
Overflow dike	Elevation of the dike crest	+137.0 m
	Length of the dike crest	250 m
New spillway	Type	Closed type control structure
	Width of crest	15 m with two roller gates
	Elevation of the crest	+127.0 m
	Design discharge	400 m^3/s

Figure 3. A photo of MR, closure dike, overflow dike, and SSR, taken in September 2020.

2.2 Design flood and operation rules for flood control

2.2.1 Design flood update results. The existing design flood as a reference for operating Wonogiri Reservoir complies with Minister of Public Works Decree No.229/KPTS/1986 regarding Guidelines for Exploitation and Maintenance of Wonogiri Multipurpose Dam. The peak discharges of existing design floods for the 60-year and 500-year return periods are 4000 m^3/s and 5100 m^3/s, respectively. The maximum discharge of PMF that is equivalent to a 5500-year return period is 9600 m^3/s. In this study, flood hydrograph calculation from 10 watersheds in the reservoir's watershed uses the observed unit hydrograph method for the three design floods. For the reservoir routing calculations, the design flood hydrographs are separated as inflow to the SSR, the sum of the Pondok and Keduang watersheds, and the summation of the flood hydrographs of the other eight watersheds as inflow to the MR [6]. Table 3 shows the maximum discharge of the updated design floods for MR and SSR. The design flood hydrograph of PMF for MR and SSR is shown in Figure 4.

Table 3. The updated design floods of MR and SSR.

Return period (Year)	Peak discharge into MR (m^3/s)	Peak discharge into SSR (m^3/s)	Total peak discharge (m^3/s)
Q_{60}	2326.6	1112.4	3186.0
Q_{500}	3451.3	1709.1	4790.1
Q_{5500} (Q_{PMF})	8091.3	2568.4	1015.8

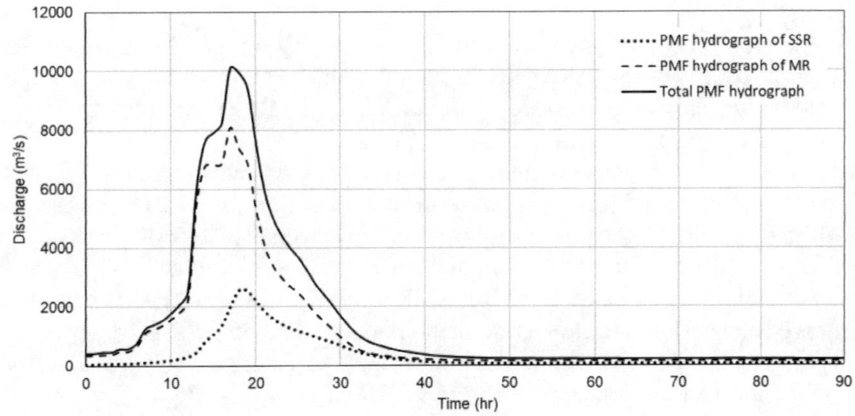

Figure 4. Design flood hydrograph of PMF for MR and SSR.

2.2.2 Reference operation rules of flood control. The reservoir operation rules for the flood period used for reservoir routing simulation comply with the recommendations of the Study Report on Operation and Maintenance of Wonogiri Multipurpose Dam and Related Structures 2016 [7]. The rules consider the new sedimentation countermeasures structures. The recommended reservoir operating rules also propose changes to the operational limits of the reservoir water level, as shown in Table 4.

Table 4. Changes in the operational reservoir water level.

Items	Existing rules 1994	Proposed rules 2016
Control Water Level (CWL)	+135.3 m	+136.3 m
Normal High-Water Level (NHWL)	+136.0 m	+137.0 m
Standard Highest Flood (SHF)	+137.7 m	+138.6 m
Surcharge Water Level (SWL)	+138.2 m	+139.1 m
Spillway Design Flood Water Level (SDFWL)	+138.3 m	+139.4 m
Extra Ordinary Flood Water Level (EFWL)	+139.1 m	+139.9 m

Listed below are key recommendation points for flood control operations at Wonogiri Reservoir at SHF conditions:

1 At the condition where the reservoir water level is below CWL +136.3 m, all spillway gates are closed, and release is only done through intake for hydropower at 72.6 m^3/s.
2 If the reservoir water level is +136.3 m $\leq$ CWL < 137.0 m, release is done from existing spillway gates no. 2 and 3, new spillway gates, and intake with a total discharge of 400 m^3/s.
3 If the reservoir water level is +137.0 m $\leq$ CWL < 138.2 m, existing spillway gates are closed and released from new spillway gates and intake with a total discharge of 400 m^3/s.
4 If the reservoir water level is +138.2 m $\leq$ CWL < 138.6 m, existing spillway gates are opened by keeping the top 0.50 m above the water level and release from new spillway gates and intake with a total discharge of 400 m^3/s.

The recommended rules for flood control operations at Wonogiri Reservoir during emergency conditions equivalent to the occurrence of PMF inflow are as follows:

1. If the reservoir water level is $\geq$ +138.6 m and abnormal flood is forecasted, release from existing spillway gates by keeping the top 0.50 m above the water level and from new spillway gates and intake with a total discharge of 1360 m^3/s.
2. At conditions like point (1), if new spillway gates are already fully opened but the total discharge cannot reach 1360 m^3/s, existing spillway gates are opened for the total discharge to reach 1360 m^3/s.

2.3 Reservoir routing modeling with a two-storage reservoir operation

Even with the sedimentation countermeasures structures where Wonogiri Reservoir storage is divided into MR and SR, reservoir operation has to be conducted as one reservoir with two storages. A previous study was conducted to develop an optimized model of Wonogiri Reservoir operation rules to minimize the sediment trap [8]. The optimization model used is Colorado State University Dynamic Programming (CSUDP), where SSR and MR are assumed as a serial two-reservoir system. With this approach, reservoir routing calculation to achieve optimal reservoir operation rules has not been simultaneously done because it does not consider the possibility of overflow from MR to SSR.

In this study, reservoir routing simulation calculation is done simultaneously for MR and SR storages while considering the possibility of overflow from MR to SSR and vice versa, depending on the water level in each storage. The bathymetric measurements data of MR and SSR storages received from the Jasa Tirta I Corporation has been processed to obtain the characteristics of the relationship between reservoir water level and reservoir storage volume at 10 cm intervals. Figure 5 shows the schematic procedure for calculating reservoir routing with the recommended reservoir operation rules. Reservoir routing simulation calculations were done for three inflow conditions, using 60-year and 500-year return periods of a design flood hydrograph and PMF. For the necessity of assessing flood control performance, the recommended operation rules are simulated with CWL of +135.8 m and +136.3 m, which are 0.5 m and 1.0 m, respectively, above the current CWL. The increase in CWL is intended to add reservoir storage for the water supply.

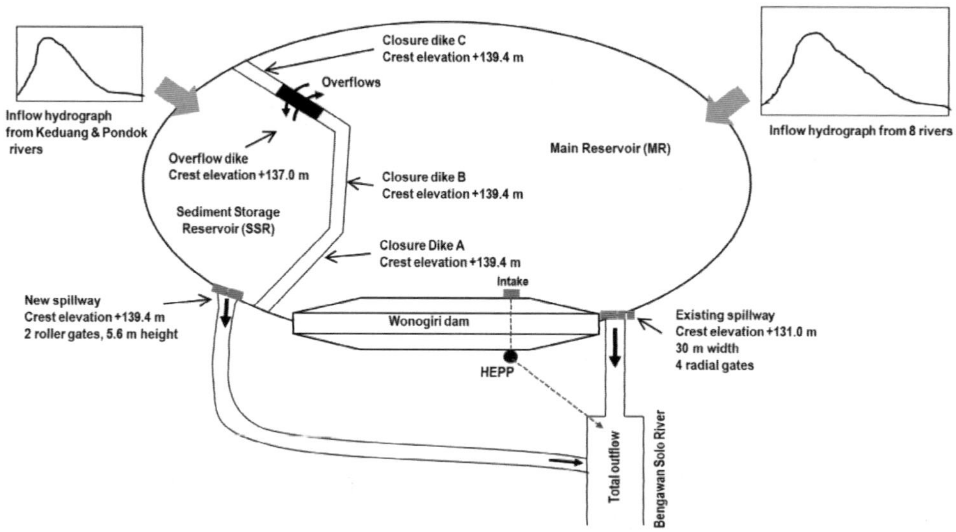

Figure 5. Schematic of the reservoir routing calculation with a two-storage reservoir.

Reservoir routing calculation uses a principal conservation approach, applying mass level pool routing method on MR and SSR using a base formula shown in equation (1) [9]. Various essential hydraulic formulae to calculate outflow discharge from existing spillway gates, new spillway gates, and overflow discharge from MR to SSR (or vice versa) is shown in equation (2) - (4) [5].

$$S_{j+1} - S_j = \frac{I_j + I_{j+1}}{2}\Delta t - \frac{O_j + O_{j+1}}{2}\Delta t \tag{1}$$

where S_j and S_{j+1} are the storage volumes at the time j and j+1, respectively, I_j and I_{j+1} are the inflow discharges at the time j and j+1, respectively, O_j and O_{j+1} are the outflow discharges at the time j and j+1, respectively, and Δt is the calculation time interval.

$$Q = \frac{2}{3}\sqrt{2g}CL\left({H_1}^{3/2} - {H_2}^{3/2}\right) \tag{2}$$

where Q is the outflow discharge from the spillway gate, C is the discharge coefficient which depends on the ratio between the height of the spillway gate opening and the upstream head, g is the gravity acceleration, L is the spillway crest length, H_1 is the reservoir water level above the spillway crest, and H_2 is the reservoir water level above the bottom of the spillway gate.

If free overflow occurs from SSR to MR, or vice versa, then the following equation can be used to calculate discharge.

$$Q = Cd\ (BH^{1,5} + tan\theta H^{2,5}) \tag{3}$$

where Q is the overflow discharge from SSR to MR or vice versa, Cd is the discharge coefficient which depends on the cross-section shape of the overflow dike, B is the length of the overflow dike crest, H is the reservoir water level above the overflow dike crest, and θ is the slope angle of the overflow dike slope above the crest.

If the overflow from SSR to MR or vice versa is submerge flow, then the following equation is used to calculate discharge.

$$Q = Cd\ [BH^{1,5} + tan\theta H^{2,5}]\left(1 - \left(\frac{h}{H}\right)^{1,5}\right)^{0,385} \tag{4}$$

where h and H are the reservoir water levels in two reservoirs above the overflow dike crest, and other symbols are the same as the description in equation (3).

By applying these hydraulic equations and the recommended reservoir operation rules for the flood period, reservoir routing calculation can be done to determine maximum reservoir water level and maximum total outflow discharge of MR and SSR for each design flood and CWL option.

3. Results and discussion

3.1 Results of reservoir routing simulation with CWL +135.8 m

Three performance indicators of flood control are maximum reservoir water elevation, maximum outflow, and the recovery time of reservoir water level to reach the CWL for SSR. For example, CWL +135.8 m with a design flood of 60-year return period inflow (Q_{60}) results in maximum water level for SSR and MR each at +138.28 m occurring at hour 14.25 and +137.3 m at hour 63.75. SSR water level can return to CWL at hour 35.50. MR water level can return to CWL at the end of hour 96, or after day 4, at an elevation of +137.29 m. Maximum outflow discharge from SSR and MR are 319.06 m^3/s and 90.52 m^3/s, with a maximum total outflow discharge of 400 m^3/s. These results are concerning relative to dam safety. If heavy rain occurs consecutively with a respite of fewer than four days, flood control storage volume cannot return to the expected state and controlling a maximum total outflow discharge of 400 m^3/s becomes difficult. If total outflow discharge is kept at 400 m^3/s, then reservoir water level elevation will be very high, and freeboard will not comply with dam safety requirements of a higher than 1.25 m. Figures 6 and 7 explain the results of the reservoir routing simulation, which shows the fluctuation of reservoir water level, inflow, and outflow discharges at SSR and MR.

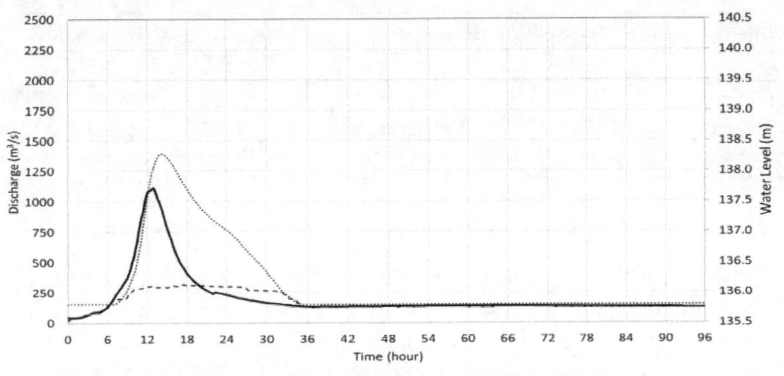

Figure 6. Results of reservoir routing at SSR for CWL +135.8 m with design flood of Q_{60}.

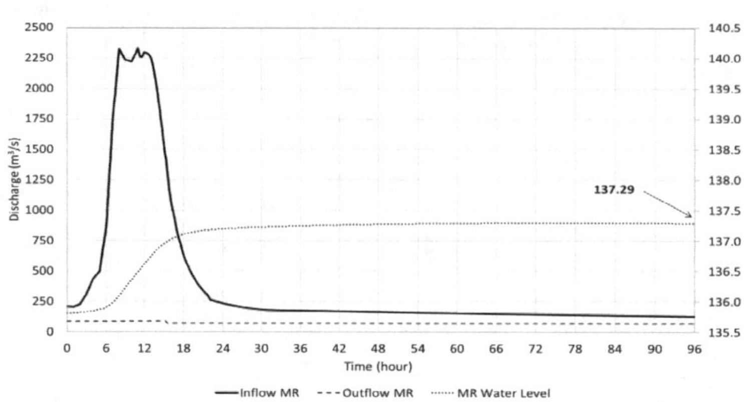

Figure 7. Results of reservoir routing at MR for CWL +135.8 m with design flood of Q_{60}.

The summary of results for all reservoir routing simulations for CWL +135.8 m can be seen in Table 5. From the following table, key findings can be observed: MR water level for inflow of Q_{60} and Q_{500} cannot return to CWL until the end of day 4. Aside from that, the water level of SSR and MR at Q_{PMF} exceeds EFWL (+139.9 m), and the freeboard does not comply with dam safety requirements.

Table 5. Summary of reservoir routing simulation results with CWL +135.8 m.

Inflow	Maximum outflow discharge (m³/s)			Maximum water level (m)		Freeboard
	SSR	MR	Total	SSR	MR	(m)
Q_{60}	319.0	90.5	400.0	+138.28	+137.30	3.72
Q_{500}	327.0	90.5	400.0	+139.06	+137.89	2.94
Q_{PMF}	835.9	1.719.0	2.490.8	+140.88	+140.88	1.12

3.2 Results of reservoir routing simulation with CWL +136.3 m

Flood control performance assessment applying CWL +136.3 m for three design flood parameters can be conducted in the same manner as evaluating reservoir routing simulation results for CWL +135.8. A summary of reservoir routing simulation results for CWL +136.3 m is provided in Table 6, which shows that for inflow Q_{60}, the water level of MR cannot return to CWL until the end of simulation calculation time. For inflow Q_{500}, water level recovery at SSR and MR cannot reach CWL. SSR and MR maximum water level for inflow Q_{PMF} exceeds EFWL, and freeboard does not meet a minimum requirement of 1.25 m for spillway with gates, thus endangering dam stability and risks overtopping.

Table 6. Summary of reservoir routing results with CWL +163.3 m.

Inflow	Maximum outflow discharge (m³/s)			Maximum water level (m)		Freeboard
	SSR	MR	Total	SSR	MR	(m)
Q_{60}	319.0	90.5	400.0	+138.28	+137.30	3.72
Q_{500}	325.9	112.5	400.0	+139.15	+138.31	2.85
Q_{PMF}	1166.6	1982.7	2831.5	+141.20	+141.20	0.80

3.3 Proposed changes to the rules for opening spillway gates

By considering MR water level recovery to very slowly reaching CWL, closure of existing spillway gates needs to be accelerated from an elevation of +137.0 m to +137.5 m. It does not require waiting until reaching an elevation of +138.2 m. This change is conducted to decrease MR water level to CWL more quickly. All existing spillway gates are opened to maintain a total maximum outflow discharge for inflow of Q_{60} and Q_{500} at 400 m³/s and 1100 m³/s.

The proposed update of reservoir operation rules for inflow Q_{PMF} or emergency conditions is to fully open the new spillway gates when the SSR water level reaches +137.5 m. It is unnecessary to wait until the water level reaches +138.6 m and the total outflow discharge is raised. The change in the rules for opening the spillway gates is to lower the SSR water level during peak inflow discharges. It is impossible to restrict the total outflow discharge to a maximum of 1360 m^3/s. Instead, the total outflow discharge is adjusted so that the maximum reservoir water level at SSR and MR provides a safe freeboard to avoid dam overflow. In other words, the consideration of Wonogiri Dam safety is prioritized to avoid the risk of a dam break. The consequence of this adjustment to the reservoir operation rules has indeed resulted in a high risk of flooding in the flood-prone area downstream of the Wonogiri Reservoir.

Table 7 summarizes the results of the reservoir routing simulation calculations for CWL +135.8 m using the new recommended reservoir operation rules. The important result obtained is that for inflow of Q_{60} and Q_{500}, SSR and MR water levels can return to CWL before the end of the fourth day, except for MR with Q_{60} inflow. SSR and MR maximum water level for inflow of Q_{PMF} exceeds EFWL (+139.90 m), which reaches +140.65 m. Freeboard for inflow of Q_{500} and Q_{PMF} meets dam safety requirements, which are 2.94 m and 1.35 m, respectively.

Table 7. Performance of the proposed new operation rules for flood control with CWL +135.8 m.

Inflow	Maximum outflow discharge (m^3/s)			Maximum water level (m)		Freeboard
	SSR	MR	Total	SSR	MR	(m)
Q_{60}	319.0	90.5	400.0	+138.28	+137.30	3.72
Q_{500}	561.6	1028.6	1100.0	+139.06	+138.31	2.94
Q_{PMF}	826.2	1723.0	2463.7	+140.65	+140.65	1.35

Figure 8 shows an example of reservoir routing simulation at MR with Q_{PMF} inflow using the proposed new operation rules for CWL +136.3 m. A summary of results applying the proposed new operating rules for reservoir routing simulation with CWL +136.3 m can be seen in Table 8. The significant result obtained is that SSR and MR water level recovery can return to CWL, except for inflow Q_{60}, where MR water level at the end of day four is still at +137.06 m. The freeboard is still safe for inflow of Q_{60} and Q_{500}, but for Q_{PMF} inflow, the SSR and MR water level exceeds EFWL with a freeboard less than the required minimum, i.e., only reaching 1.03 m. The maximum total outflow discharge of Q_{PMF} inflow reaches 2761.3 m^3/s, exceeding the existing design flood of 1360 m^3/s.

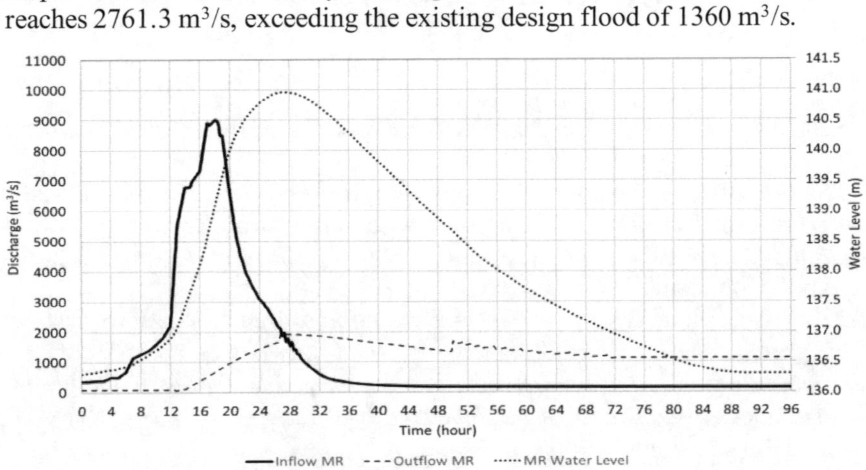

Figure 8. Performance of the proposed new operation rules for CWL +136.3 m with design flood of Q_{PMF}

Table 8. Performance of the proposed new operation rules with CWL +163.3 m.

Inflow	Maximum outflow discharge (m^3/s)			Maximum water level (m)		Freeboard
	SSR	MR	Total	SSR	MR	(m)
Q_{60}	307.4	110.5	400.0	+138.41	+137.63	3.59
Q_{500}	570.8	996.0	1100.0	+138.89	+137.97	3.11
Q_{PMF}	1150.2	1921.7	2761.3	+140.97	+140.97	1.03

4. Conclusions

The assessment results regarding flood control performance for CWL +135.8 m and +136.3 m show that updating reservoir operation rules is required. Failure for the reservoir water level recovery to reach CWL, freeboard that does not meet dam safety requirements, and maximum total outflow discharge control are all basis for considering the proposed new reservoir operation rules. The proposed new operation rules can improve flood control performance, especially in the aspect of reservoir water level recovery reaching CWL, as well as maximum water level of MR and SSR for inflow of Q_{60} and Q_{500} not exceeding EFWL. For inflow of Q_{PMF}, the maximum water level of MR and SSR still exceeds EFWL, but by using CWL +135.8 m, the freeboard still complies with dam safety requirements. It is very important to note that applying CWL of +135.8 m and +136.3 m, the maximum total outflow discharge for the Q_{PMF} inflow is impossible to be controlled so as not to exceed the design discharge of 1360 m^3/s. This condition needs to be considered by operators and all parties involved in mitigating flood disasters in the Bengawan Solo River basin because the increased risk of flood inundation and its impact in flood-prone areas will be even greater if there is a reservoir inflow equal to Q_{PMF}. This assessment provides the basis for determining the operation of the CWL and spillway gates to ensure optimal reservoir flood control. A comprehensive assessment of the results of this study with the results of studies regarding the reservoir operation performance to fulfil water requirements is needed for developing rule curves of future reservoir operation.

Acknowledgment
The author would like to express gratitude to the Faculty of Engineering Universitas Gadjah Mada and Jasa Tirta I Corporation for their financial support of this research.

ORCID iDs
R Jayadi https://orcid.org/0000-0003-1664-8103
Istiarto https://orcid.org/0000-0001-7753-7824
EPA Pratiwi https://orcid.org/0000-0001-6806-3700
K Windianita https://orcid.org/0000-0002-5434-3258

References

[1] Sarwono S 2013 Usaha Mereduksi Banjir di Bengawan Solo Hilir. *Jurnal Sumber Daya Air* **9** p 79–92
[2] Budi T J I 2016 Analysis of Sedimentation in Wonogiri Reservoir. *JCEF* **2** p 27-28
[3] Nippon Koei Co. Ltd. 2007 *The Study on Countermeasures for Sedimentation in Wonogiri Multipurpose Dam Reservoir in The Republic of Indonesia. Final Report* (Jakarta: Ministry of Public Works The Republic of Indonesia) **2** p 1-2
[4] Pamuji A, Sobriyah S and Ikhsan C 2017 *The Prediction of Wonogiri Dam's Service Life Using Point Integrated Sampling* IPTEK Journal of Proceedings Series **3** p 75–78
[5] Pradipta A G 2014 *Studi Pengaruh Pembangunan Closure Dike Waduk Wonogiri Terhadap Pola Operasi Waduk Perioda Banjir. A Master Program Thesis Report* (Yogyakarta: Universitas Gadjah Mada) p 24-31 and 39-40
[6] Faculty of Engineering Universitas Gadjah Mada 2020 *Kajian Perubahan Control Water Level (CWL) Bendungan Wonogiri. Final Report* (Malang: Jasa Tirta I Corporation) p 36-38
[7] Nippon Koei Co. Ltd 2016 *Study Report on Operation and Maintenance of Wonogiri Multipurpose Dam and Related Structures, The Republic of Indonesia. Final Report* (Jakarta: Ministry of Public Works and Housing The Republic of Indonesia, Directorate General of Water Resources) p 2-14
[8] Wulandari D A, Legono D and Suseno D 2014 Reservoir Operation to Minimize Sedimentation. *Int. J. Eng. Sci.* **6** p 16–23
[9] Chow V T, Maidment D R and Mays L W 1988 *Applied Hydrology* (New York: Mc.Graw-Hill Book Company) p 245-246

IOP Conf. Series: Earth and Environmental Science 1091 (2022) 012007 doi:10.1088/1755-1315/1091/1/012007

Communicating landslide risks during the COVID-19 pandemic: A baseline study for PHIVOLCS-Dynaslope stakeholders

Eunice R. Decena[1], **Heinrich Joyce S. Domingo**[1], **Ardeth Joy E. Maligon**[1], **Arturo S. Daag**[1], **Renato U. Solidum, Jr.**[1,2]

[1]Department of Science and Technology-Philippine Institute of Volcanology and Seismology, Manila, Philippines

[2]Department of Science and Technology (DOST)

Abstract. Dynaslope Project, a government-funded program implemented by the Philippine Institute of Volcanology and Seismology (PHIVOLCS), develops and deploys an early warning system (EWS) for deep-seated landslides. It uses landslide sensor technology and community involvement in its EWS implementation in 50 sites all over the country. As the pandemic altered people's modes of communication, it is important to understand how these changes relate to communicating landslide early warning information (LEWI). This research answered: How can the Dynaslope Project effectively communicate landslide risk during the COVID-19 pandemic? Specifically, it a) identified their preferred communication channels and tools during the pandemic, b) enumerated the stakeholders' perceived participation and communication barriers, and c) measured the degree of trust of the stakeholders in the Project, the LEWI it released, and other key messages. This research was guided by the Actor-Network Theory and the concept of co-orientation which posited that, instead of a single entity defining a network or a system, it was the interactions between people, objects, and institutions that created a collective network and continuously negotiated a coherent understanding. The study employed a descriptive quantitative methodology. It used stratified random sampling to select the participants. Among the major findings include the stakeholders' preference for offline communication like SMS and calls, the presence of communication and participation barriers including weak or no phone reception and limited access to up-to-date communication devices and the internet, and a high level of trust in the Dynaslope Project because it was implemented by a science-based national agency PHIVOLCS.

1. Introduction

The Dynaslope Project is a government-funded program implemented by the Department of Science and Technology (DOST) under its service institute, the Philippine Institute of Volcanology and Seismology (PHIVOLCS). The Project develops and deploys an early warning system (EWS) for deep-seated landslides in 50 sites all over the country. The EWS uses landslide sensor technology which detects subsurface movements. Also, it analyzes data on rainfall and earthquakes which are hazards that can cause landslides. Lastly, it involves the participation of the local stakeholders. Communities participate by providing on-site ground measurements and observations of landslide features. They do so by recording the displacement of the surficial markers and noting manifestations of earth or slope movement which they subsequently submit to Dynaslope.

Dynaslope analyzes and interprets these data to release landslide early warning information (LEWI) to help the communities make informed decisions and timely actions to reduce disaster risk.

The EWS uses a four-level alert system that describes each alert level and provides the recommended response for the landslide early warning committee (LEWC), local government unit (LGU), and the community. Dynaslope sends these alert levels to the community through text messaging and phone calls.

Each of the communities has an LEWC which is a volunteer organization composed of households at risk of landslides as well as other members of the community. They serve as the point-person in the communication process between Dynaslope and the community. Other EWS stakeholders are the local government units at the barangay (BLGU), municipal/city (MLGU/CLGU), and provincial (PLGU) levels. The municipal and provincial LGUs are represented by their respective disaster risk reduction and management offices (DRRMOs).

Aside from community members and LGUs, there are also external stakeholders of the EWS who participate in the operation and are invited to assessments, conferences, and other similar activities. These external stakeholders include PHIVOLCS field stations, DOST regional offices, Mines and Geosciences Bureau (MGB) regional offices, Office of the Civil Defense (OCD) regional offices, social media audience, academe, and civil society organizations (CSOs).

Before COVID-19, the community stakeholders and Dynaslope used text messaging and phone calls to report ground measurements and observations as well as updates on the status of the landslide sensor-data logger system. Meanwhile, Dynaslope staff regularly visited the sites for several purposes including research, landslide risk assessment, sensor maintenance, capacity building, and post-landslide alert assessment.

The series of lockdowns and community quarantines, which started on March 15, 2020, limited face-to-face site visits and community engagement of the Project staff. This meant that the Project had to shift to mostly online communication in dealing with its community stakeholders. For one, it revamped its Facebook page in July 2020 to offer another communication channel. The Project posted project activities, news, and updates on its Facebook page for the stakeholders to read and watch. Meanwhile, the communities sent out surficial marker measurements and other updates (e.g., site maintenance activities, LEWC meetings) to Dynaslope through Facebook Messenger.

1.1. Research Problem

To date, there were limited studies on risk communication during COVID-19 that were not health or pandemic-related. Natural hazards were essential to examine as they continued to pose risks even in the middle of a health pandemic. In the same sense, there was a need to study how to communicate natural hazard risks such as landslides during a global health crisis.

In the case of the Dynaslope Project, the pandemic posed challenges in communicating landslide risks to its stakeholders mainly because of restrictions on travel and mass gatherings. The convenient, albeit low-tech ways of communicating in the past (e.g., physical meetings, group consultations, interviews, etc.) had become challenging. Given the foregoing, this research answered the question: How can the Dynaslope Project effectively communicate landslide risk during the COVID-19 pandemic? Its general objective was to identify strategies and techniques how to communicate landslide risk during the COVID-19 pandemic effectively. Also, it specifically aimed to:

- identify channels and tools that the stakeholders prefer for communicating with the Dynaslope Project during the COVID-19 pandemic;
- enumerate the stakeholders' perceived participation and communication barriers; and
- measure the degree of trust of the stakeholders in the PHIVOLCS-Dynaslope Project, the landslide early warning information (LEWI) it released, and other key messages.

This research was valuable in understanding communication during the time of COVID-19. It provided insights as to how government agencies and DRR programs may be implemented amidst restrictions on travel and mass gatherings.

Also, this research was beneficial to the stakeholders of the Dynaslope Project. While the Project had been using online communication tools and channels to communicate with the stakeholders, it was essential to identify whether doing so was favorable to the stakeholders' current situations.

Determining the stakeholders' preferred means of communication as well as addressing the different communication barriers could decrease their risk of being severely affected by landslides.

1.2. Communication and the COVID-19 Pandemic
The physical restrictions set by COVID-19 had forced people across the globe to modify their mode of communication. Specifically, the pandemic had increased people's reliance on digital communication technology. A study conducted by communication and media researchers from the University of Zurich found that during the pandemic, there was a significant increase in the respondents' use of digital technology to communicate. Its data showed that 35% of the respondents used social media more often and there was a 30% increase in video calls and 25% more in email usage [1]. Most of the respondents had to resort to digital communication to stay connected with their family, friends, and work colleagues.

COVID-19 had also altered the way that people communicate with social institutions around them. A study about the social media use of three LGUs in the Philippines namely Quezon City, Iloilo City, and Davao City during the COVID-19 pandemic, noted that "majority (57%) of the respondents said that they usually receive/seek risk/crisis information from their LGUs through their social media accounts and pages, e.g. Facebook" [2]. Public offices had to offer these new modes of communication as more of their constituents are using them now.

Both studies note the increased usage of social media to communicate. They provided the basis for this study's inquiry on the tools and channels preferred by Dynaslope stakeholders in communicating during the pandemic. In this study, the researchers instead focused on disaster risk communication, specifically on landslide risks. It was noteworthy that many of the COVID-19 studies were about how COVID-19 was communicated to the public. This study added to the dearth of research about landslide risk communication during the COVID-19 pandemic.

1.3. Communicating Landslide Risk
The Philippines is one of the most at-risk countries for natural hazards because it is located along the Ring of Fire and the typhoon belt. This meant common occurrences of earthquakes and storms. With floods and storms came landslides which often resulted in significant loss of life and social, economic, and environmental assets [3]. Given the impact of landslides on the lives of Filipinos, it was important to examine how this hazard is communicated to the people.

Despite the critical importance of landslide communication and reporting, one international study noted how this hazard had often been misreported in the past. In researching landslide vulnerability, David Alexander (2005) collected situation reports and news briefings on natural hazards published by news sites, NGOs, UN agencies, and disaster workers between August 1993-May 2002. He found out that damage and casualties were underreported, causes were not identified, and landslide events were not differentiated between a disaster or a mere accident [4]. This often resulted in the failure of the affected communities to understand landslide risk. While changes and developments had been made since then, this study provided a glimpse of why there should be a conscious effort to improve landslide risk communication.

Meanwhile, studies had noted how proper and adequate communication could empower communities in preparing for landslide events. A research article by Klimeš, et al. (2019) discussed the landslide risk reduction efforts in a small community in the Rampac Grande of the Peruvian Andes, which experienced a landslide event in 2009 [5]. The researchers found that proper communication and adequate sharing of knowledge between the experts and the locals resulted in active community engagement and acceptance of the DRR project. Their study suggested that enhanced communication and the use of participative methods in a DRR project could increase community engagement.

Both studies emphasized the need to examine the accuracy of landslide communication. They suggested that the agencies or organizations reporting or communicating landslides be studied as well.

Notably, most studies on communicating landslide risks were made by and about non-Filipinos. Despite the high risk of many Filipino communities to landslides, few studies had been made about landslides and how they were communicated. This research was an attempt to fill in this research gap.

1.4. Dynaslope Project Communities

In the past years, several researchers published papers about the Dynaslope Project. In 2019, Project researchers assessed the capabilities and vulnerabilities of five of the project sites to support the establishment of a community-based early warning system for deep-seated landslides (CBEWS-L). One of the major findings of the study was that "information and communication technology (ICT) capacities are lacking and are outnumbered by vulnerabilities" [6]. In particular, they noted that most of the Project sites lack ICT materials like phones and computers, have poor ICT infrastructures such as cell sites and internet lines, and lack ICT skills, including computer, internet, and smartphone use.

A similar study was done in 2021 when researchers developed an action plan for the creation of a multi-stakeholder early warning system for landslides in a project site in Alimodian, Iloilo. Researchers examined the infrastructure, organization, training, and communication needs of the community. Much like the earlier study, this research noted the lack of communication infrastructures, equipment, and knowledge of the community [7]. It identified how these issues and limitations on communication protocol and equipment contribute to the challenges of disaster response.

One common limitation of the studies mentioned above was that their analysis only involved a small sample of the Project sites. The first research examined five sites while the second one only examined one. To date, there are 50 Project sites, which this research studied. Also, all these studies were done before COVID-19, which meant that the social and economic contexts of these sites had become different in the past two years.

While both the 2019 and 2021 studies have emphasized the limitations of communication protocols and equipment in their examination of the selected sites, they were unable to identify what specific tools and channels were preferred by the stakeholders in the community. In contrast, this research focused on the preferred communication tool and channels of the community stakeholders in all the Project sites.

2. Materials and Methods

2.1. Research Framework

This research was guided by the Actor-Network Theory (ANT) developed by science and technology scholars John Law, Bruno Latour, and Michael Callon. The group was interested in examining what human and nonhuman objects contributed to the process of the scientific enterprise [8]. They wanted to better understand the process of creation of knowledge in the field of science and technology studies (STS). The basic principle of ANT was to examine how actors (human or nonhuman) were gathered together to form stable, diverse networks with similar interests [9]. It posited that, instead of a single entity defining a network or a system, it was the interactions of people, objects, and institutions that created a collective network. This collective network was dynamic in nature as a result of interaction among actors.

Aside from the ANT, this research employed James Taylor's concept of *co-orientation*. It was from the Montreal School Approach to Communicative Constitution of Organizations which was interested in looking at how organizations were created and continuously evolved. The process of co-orientation happened when "two people orient to a common object (a topic, issue, concern)" [10]. In some other cases, an entire network could co-orient with another network as well.

In the case of an organization, co-orientation happened as members negotiated to come up with agreeable goals, objectives, practices, informal rules, etc. Different perspectives might arise given the context of each individual, and so the process of negotiating and co-orienting each other happened as they reached a mutual understanding.

This research combined the ANT and the concept of co-orientation. ANT was helpful in identifying the (human and nonhuman) actors in an organization while co-orientation helped in finding out how these actors worked together to arrive at a common understanding. They both tackled the creation and recreation of shared meanings in an organization by different actors.

Figure 1 showed a network of human and nonhuman actors interacting to achieve co-orientation. These actors were independent of one another but were connected by their shared interests and goals. As seen here, an actor could be connected with all other actors in the network while in other cases, an actor was only connected to one or two other actors. Also, it was possible that an actor in this network could be connected to a totally different network. In discussing dynamic entities like organizations, it was likely to see actors and networks intermingling with one another and forming convoluted connections.

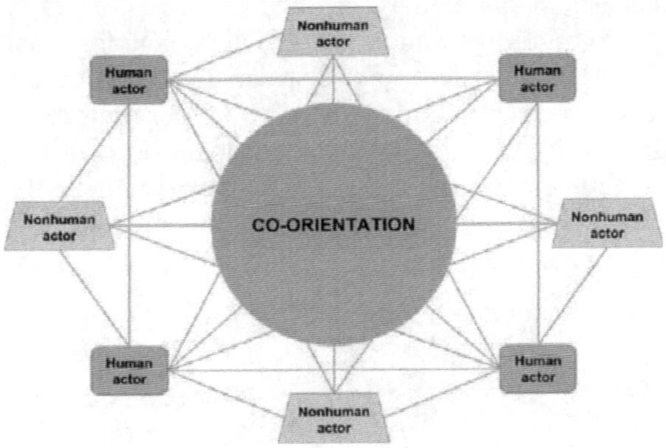

Figure 1. This showed how the human and nonhuman actors co-orient as they form a network.

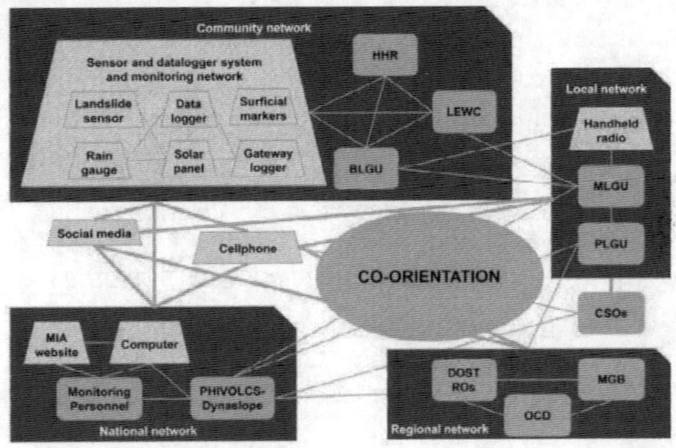

Figure 2. This showed how the various networks in the Dynaslope Project co-orient towards the same goals.

Applying these concepts in this study, the researchers identified four main networks in the Dynaslope Project, mainly the Community, Local, Regional, and National networks. In the community network was a subset of nonhuman actors forming the sensor and datalogger system and monitoring network. They interacted with the human actors namely the households-at-risk, LEWC and, BLGU. These actors were connected by a thin line which means they seldom interact with one another. In contrast, the thicker lines that connect them to social media and cellphone indicated more frequent interactions. Notably, social media and cellphone were outside the four networks as they were related to all the human actors. Meanwhile, CSOs were outside any other network as they only interacted with the communities they serve, although not necessarily a Dynaslope Project site.

The Community network was connected to the Local network composed of municipal and provincial local governments as well as the nonhuman actor, handheld radio. Unlike the Community network, the Local network was connected to all the other networks. The third network in Figure 2 was the Regional network composed of government institutions, namely regional offices of DOST, OCD, and MGB. Lastly, there was the National network composed of the Dynaslope staff and the tools they use in their monitoring, research, community engagement, and other works.

In the middle of all these interactions was the process of co-orientation. Here, human and non-human actors negotiated to find equilibrium towards the same goal of building an effective and efficient EWS for deep-seated landslides. While these actors bore their own interests and objectives, they co-oriented for the achievement of the Projects' goal that is to contribute to empowerment and resilience of Filipino communities by using science and technology in the development of community-based early warning system for landslides.

2.2. Research Design and Methods

The researchers conducted a cross-sectional survey with the human actors in the community, local, and regional network of the Dynaslope Project (Figure 2). The survey was divided into four sections namely: 1) demographics, 2) preferred channels and tools for communication, 3) perceived participation and communication barriers, and 4) trust in sources of information. These sections followed the order of the research questions and objectives. The items contained in the questionnaire were combinations of close-ended, rating, matrix, and Likert scale questions. However, due to the fixed and strict format of a questionnaire, intensive and in-depth analysis was impossible.

Stratified random sampling was used to select the participants. Employing the ANT and the concept of co-orientation, this research followed the three identified networks in grouping its sample. Stratum 1 was the Community Network, 2 was the Local Network, and 3 was the Regional Network.

There were two criteria in selecting the research participants namely: a.) being a landslide early warning information (LEWI) recipient, or b.) being a primary/secondary contact person for their stakeholder group. After grouping the participants into strata and identifying participants based on the main criteria, the population size was computed. The sample size was calculated at a 95% confidence level with a 5% margin of error through an online sample size calculator. A number identification was then attached to each name in the population. The number ID was used to identify which of the population was selected to proceed in the survey. An online randomizer was used for this method. The population size and sample size were summarized in the table below.

Table 1. Table showing the research population and sample size.

Strata	Population size	Sample size
Stratum 1	96	76
Stratum 2	71	59
Stratum 3	9	8
Total	176	143

Data was collected in the second half of 2021. A total of 143 participants were approached. Among them, 114 agreed to participate in the survey. To respond to the survey, the researchers provided two options to the participants namely a.) through a computer-assisted telephonic interview (CATI) or b.) answer the survey on their own.

Among the 114 surveyed, only six preferred to answer via Google Forms, all of which belonged to Stratum 2 (MLGU and PLGU). The rest was done via CATI. The response rates were as follows. For Stratum 1, a total of 58 participants took part in the survey, translating to a 76.32% response rate. Stratum 2 had the highest turnout with an 84.75% response rate and 50 participants. Stratum 3, which targeted 8 participants, was able to reach the acceptable response rate of 75% or 6 participants. On average, the survey garnered a 78.69% response rate.

Some participants were unable to participate due to weak to null cellphone signals on the site and differences in priorities at the time of data collection (e.g., typhoon). During the data collection, a typhoon hit a number of Dynaslope Project sites which restrained the researchers from proceeding with the activity. Most of the affected sites experienced power outages with no specified schedule of power restoration.

3. Results and Discussion

3.1. Demographics

A total of 114 individuals participated in the survey with 68% (77) males and 33% (37) females. As seen in Figure 3, the participant turnout for females and males was equal for Stratum 3, slightly different in Stratum 1, and a huge gap in Stratum 2. It was noteworthy that Stratum 2 was composed of the M/CLGUs and PLGUs where most of those who held leadership positions in DRRM Offices or for Operations and Warning units were male.

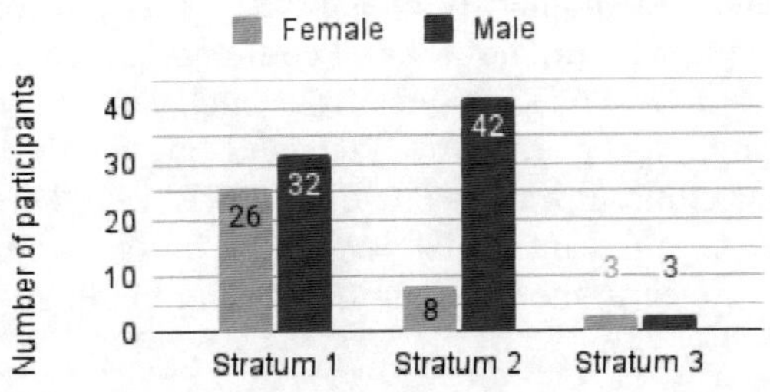

Figure 3. This showed the sex of surveyed participants per stratum.

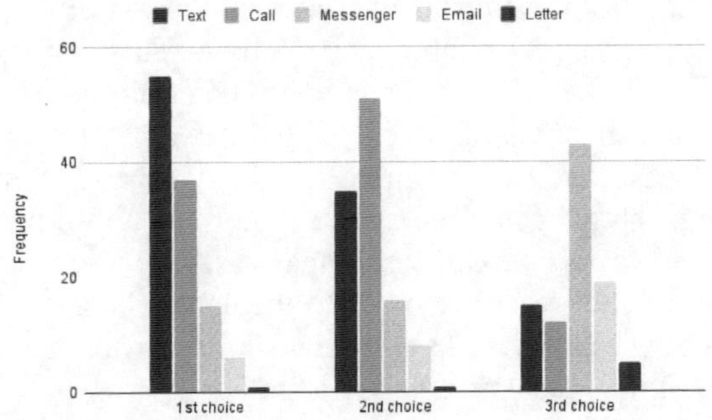

Figure 4. This showed the participants' preferred communication channels with Dynaslope.

Apropos of age, almost all of the participants (97.37%) belonged to the working-age population of 15-64 years old [11] while the remaining (2.63%) were 65 years old and above. Classified according to generation [12], most of the participants were Millennials (42.98%) followed by Generation X (28.95%). Close to the participant turnout of Generation X were the Boomers (25.44%). And the least were the Zoomers (2.63%).

3.2. Channels and Tools of Communication

Based on the gathered responses, the main channels of communication used by the participants were call and text through cellphone (96%), followed by social media (75%), and then television (64%). When it comes to the most common communication tools used, cellphones (97%), radio (64%), TV (60%), and Internet (58%) topped the list. Traditional media such as TV and radio, through the years, were the most accessible. However, social media was now the up-and-coming channel and played a prominent role in today's communication, even among the Project stakeholders. In addition, only a third of the participants (31%) answered word-of-mouth (face-to-face). During the CATI, participants expressed that before the pandemic, they used to do face-to-face interaction more often but due to COVID-19, they opted to use other means of communication.

When asked how they primarily share information, text, call, and messenger placed first, second, and third respectively. Some of them said that they preferred sending/receiving texts because it was easier to get back to once they got the time. Most of them expressed that calls were preferred for urgent matters. Messenger, on the other hand, was preferred over text and calls, especially when the participant had no load credit. They shared that Messenger had a "free data" feature that allowed users to send and receive messages without incurring data charges. Notably, handheld radio was also preferred by some, particularly in areas with poor or no signal at all.

The participants were also asked which social media platforms they had been using. Facebook (86%) and Messenger (82%) came out to be the two most popular. Most of them said that due to not being able to meet face-to-face, creating group chats to facilitate discussions and meetings became a solution to the problem. This affirmed the Project's decision to revamp its Facebook page in July 2020.

Meanwhile, when it comes to communicating with the Project, the participants were asked to rank their top three preferred methods. As seen in Figure 4, text messaging was the number one choice among the participants. Call and Messenger came in second and third, respectively. This was an indication that in the Project's future communication endeavors, Messenger could be considered a channel to disseminate landslide information and communicate with its stakeholders.

3.3. Participation and Communication Barriers

Of all the participants surveyed, 19 unique languages (Figure 5) were used. They chose Filipino/Tagalog as both their primary (28%) and secondary (30%) language. This was despite most of the Project sites being located in rural areas in the province. Other languages commonly spoken were Ilocano, Bisaya/Cebuano, and Waray. This was partly because, in many communities, different enclaves spoke different languages and could not understand one another. Hence, there was a need for a lingua franca, a language used by different groups of people who did not share a common language. In the case of the site in Sitio Labey, Brgy. Ambuklao in Bokod, Benguet, two enclaves spoke different languages. One spoke Ibaloi and the other Kalanguya. In order to understand one another, they had to use Ilokano or Tagalog (A. Cupido, personal communication, October 12, 2021). This suggested that when it comes to the use of language in communicating landslide risk, the information did not necessarily need to be translated to each language spoken in the communities. Using the lingua franca in the area could be sufficient.

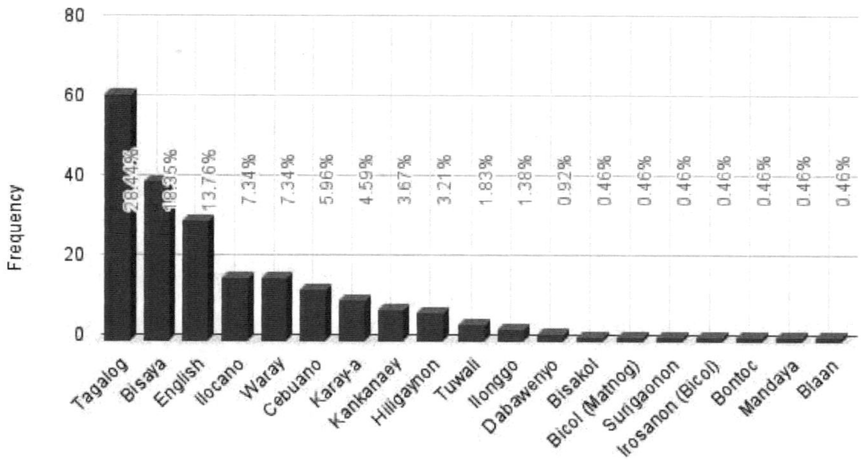

Figure 5. This showed the primary languages of the research participants.

When it comes to the mastery of using cellphones and navigating the internet, most of the participants rated themselves four (4 or highly knowledgeable) on a scale of one to five (1-5 or *least knowledgeable, slightly knowledgeable, moderately knowledgeable, highly knowledgeable, or extremely knowledgeable*). Notably, during the CATI, participants qualified their answers by indicating that they would give themselves a lower score for their mastery in navigating the internet than in using a cellphone. Also, some mentioned that they were using feature phones, the ones with press-button inputs and small non-touch screens, instead of smartphones. Feature phones were far more affordable than standard smartphones, which could cost around thrice as much. Most of these feature phones could not connect to the internet.

This indicated that familiarity with using cell phones did not necessarily equate to easy access to online information. For the Dynaslope Project, it meant that while some of the stakeholders had access to the internet, others were still unfamiliar with navigating this platform. Thus, in communicating landslide risk, non-Internet-based materials should also be made.

Moreover, 40% of the participants answered that they had weak mobile phone signals. Five percent of them said that their mobile phone reception was dependent on weather conditions. This showed that communication should not rely entirely on cell phones (i.e., SMS, calls, and the internet). For an EWS project like the Dynaslope, which was dependent on fast and reliable information dissemination, other communication channels must be navigated.

Some participants identified no or weak phone signal as their major obstacle in their participation in online trainings or seminars (26%), in interviews via phone call and surveys via text message (36%), and in their access to general information (54%). This indicated that to capacitate the human actors in a network, they must have access to reliable phone and internet connections, up-to-date communication devices, and other nonhuman actors.

3.4. Trust

The top three trusted sources of general information were the government (76 mentions), punong barangay or barangay captain (56 mentions), and family (51 mentions). When it comes to landslide information, PHIVOLCS-Dynaslope was the most trusted source with a mean score of 4.80. More than 80% of the participants gave the Project a perfect (5) rating.

The participants were also asked to rate the PLGU/DRRMO, MLGU/DRRMO, BLGU, family/relatives, and neighbors from 1 as not trusted and 5 as the most trusted source of landslide information. For the participants, the second most trusted source was the MLGU with an average score of 4.35. With a minuscule difference of 0.03 points came the BLGU in third, with an average of 4.32. Not far behind was the PLGU with a mean rating of 4.19. Family and relatives had a 3.90 mean score. And lastly, neighbors had an average rating of 3.01.

According to the participants, PHIVOLCS-Dynaslope Project was a trusted source of information because it was a national agency mandated by the Philippine Government to develop an early warning system for landslides. Moreover, its use of science and evidence-based methods such as the collection of rainfall data via rain gauge, subsurface movement through landslide sensors, and other instruments gave reliability and accuracy to the information they released. Another common reason pertained to the credibility of the Project researchers or staff as they were seen as qualified professionals with pertinent background and experience.

Affinity to the stakeholders was also a factor in the trust given to the PHIVOLCS-Dynaslope Project. The participants mentioned that they personally knew the staff because of site visits, courtesy calls, and trainings. The regular sending of early warning information also gave reassurance to the stakeholders that their site was being monitored.

4. Conclusions and Recommendations

The Dynaslope Project can effectively communicate landslide risk during the COVID-19 pandemic by maximizing the use of the preferred tools and channels by the stakeholders. The top preference of the stakeholders in receiving and sharing landslide early warning information was through cellphones via text messaging. However, it should be noted that there was no common preference for all the stakeholders. Because of this, the Project needs to diversify its communication strategies by using the available channels identified by the stakeholders.

At the start of the research, the researchers hypothesized that the use of different mother tongues was one of the communication and participation barriers and that the Project needs to translate materials to each of the stakeholder's vernacular. However, this study revealed that *lingua franca* existed among networks. Aside from using Filipino/Tagalog, which was the country's national language, the Project could use Ilocano, Bisaya, and Waray among others. These were the *lingua franca* in North Luzon, most of Visayas and Mindanao, and Eastern Visayas, respectively.

Also, given that weak or no phone signal was the stakeholders' main barrier to their participation in Project activities, Dynaslope has to be strategic in scheduling its online seminars and trainings, phone interviews, and text surveys. For one, during bad weather conditions that greatly affect phone signals, according to the respondents, scheduled activities may be postponed to a later date.

While the survey indicated a high level of trust (mean score of 4.8 out of 5 points) among stakeholders in the PHIVOLCS-Dynaslope Project, there is also a need to build trust between the community and the LGU. LGUs are the first responders in disaster risk reduction efforts. Hence, the existing capacity-building initiatives for these agencies must be furthered.

For future studies, the researchers recommend examining other networks in the Dynaslope Project. This research only identified four major networks. Examining other networks can result in the inclusion of more human and nonhuman actors in the process of co-orientation. By doing so, the Project can create ways to strengthen these previously unidentified actors.

In terms of the research design, it is recommended to include semi-structured research instruments such as open-ended surveys, interviews, and focus group discussions as they allow for more nuanced answers. As this research employed a descriptive quantitative approach, further studies must delve into correlation and causation research designs. This is to deepen the understanding of the relationships of the actors and networks in the Dynaslope Project.

Acknowledgments
We would like to express our most sincere appreciation and thanks to the Dynaslope Community for participating in this research and their continuous involvement in the early warning system. We also extend our recognition to our colleagues and our mother institute, PHIVOLCS, for their insights and critique in this research. And most importantly, our heartfelt gratitude to our family and loved ones for their support and understanding.

ORCID iDs
E Decena https://orcid.org/0000-0003-3277-6551
H Domingo https://orcid.org/0000-0001-8465-1203
A Maligon https://orcid.org/0000-0002-4850-9506

5. References

[1] Nguyen M, Gruber J, Fuchs J, Marler W, Hunsaker A, Hargittai E 2020 Changes in digital communication during the COVID-19 global pandemic: implications for digital inequality and future research *Sage Journals* **6**(3) 1-6
[2] Flores R, Asuncion X 2020 Toward an improved risk/crisis communication in this time of COVID-19 pandemic: a baseline study for Philippine local government units *J. Science Communication* **19**(7) 1-16
[3] Acosta L, Eugenio E, Macandog P, Macandog D, Lin E, Abucay E, Cura A, Primavera M 2016 Loss and damage from typhoon-induced floods and landslides in the Philippines: community perceptions on climate impacts and adaptation options *Int. J. Global Warming* **9**(1) 33-65
[4] Alexander D 2005 Vulnerability to landslides *Landslide hazard and risk* ed T Glade, M Anderson, MJ Crozier (England: John Wiley & Sons Ltd) Chapter 5 pp 175-98
[5] Klimeš J, Rosario A, Vargas R, Raška P, Vicuña L, Jurt C 2019 Community participation in landslide risk reduction: a case history from Central Andes, Peru *Landslides* **16** 1763-77
[6] Gumiran B, Moncada F, Gasmen H, Boyles-Panting N, Solidum R 2019 Participatory capacities and vulnerabilities assessment: towards the realisation of community-based early warning system for deep-seated landslides *Jàmbá* **11**(1) 1-11
[7] Gumiran B, Daag A 2021 Negotiated participatory action research for multi-stakeholder implementation of early warning systems for landslides *Int. J Disaster Risk Reduction* **58** 1-12
[8] Littlejohn S, Foss K, Oetzel J 2017 *Theories of human communication* 11th ed (Illinois: Waveland Press, Inc) p 498
[9] Law J 1992 Notes on the theory of the actor-network: ordering, strategy, and heterogeneity. *Systems Practice* **5**(4) 379-93
[10] Littlejohn S, Foss K, Oetzel J 2017 *Theories of human communication* 11th ed (Illinois: Waveland Press, Inc) p 498
[11] Philippine Statistics Authority 2003 *Working age population* (Quezon City PH: Philippine Statistics Authority) Available from: https://psa.gov.ph/content/working-age-population-1
[12] WJSchroer *Generations X, Y, Z and the others* (Michigan USA: WJSchroer) Available from: http://socialmarketing.org/archives/generations-xy-z-and-the-others/

IOP Conf. Series: Earth and Environmental Science 1091 (2022) 012008 doi:10.1088/1755-1315/1091/1/012008

Seismic screening and structural investigation of heritage buildings for adaptive reuse: a survey study at Iloilo City, Philippines

Richard De Jesus[1,5], Andres Winston Oreta[1], Mary Earl Grio[2], Rodolfo Mendoza, Jr. [1], Lessandro Garciano[1], Rhodella Ibabao[3], Giacomo Sevieri[4] and Carmine Galasso[4]

[1]De La Salle University – Manila, 2401 Taft Avenue, 1004 Manila, Philippines
[2]Central Philippine University – Lopez Jaena Street, Jaro, Iloilo City, 5000 Iloilo, Philippines
[3]University of the Philippines Visayas – New Administration Building, Miagao, 5023 Iloilo, Philippines
[4]University College London – Gower Street, London WC1E 6BT, United Kingdom
[5] Corresponding author e-mail: richard.dejesus@dlsu.edu.ph

Abstract. Building adaptive reuse has been a popular strategy for cultural heritage conservation of structures for it serves two basic objectives of preserving heritage structures and promoting cultural heritage and tourism. Adaptive reuse is the adaptation of an old structure to allow the introduction of a modern function into the old structure. Old buildings are not demolished but recycled into modern uses. This strategy, however, needs a careful and thorough assessment of these existing heritage buildings to assure their resilience and sustainability to future hazards. Heritage buildings for adaptive reuse due to age and structural deterioration are highly vulnerable to hazards like earthquakes. Moreover, these buildings when used for commercial purposes may be subjected to additional loads due to changes in function. Considering the age of the structure, conformity to new design codes, additional loads imposed by unit modification, and the new function brought about by adaptive reuse of these structures, there is a need to assess these buildings to assure their safety and continuous use. Promoting adaptive reuse of heritage buildings and tourism is most appropriate in heritage zones like Calle Real in Iloilo City, the site for the case study. This paper presents a rapid seismic screening of buildings to prioritize a population of heritage buildings in a heritage zone for further detailed inspection. A site survey of the buildings was conducted at the heritage zone of Calle Real, Iloilo City in terms of their current use, and current condition to determine potential structural, maintenance, and functional issues related to resilience and to recommend future improvements in the implementation of adaptive reuse of heritage structures in cities and towns to assure their sustainability.

1. Introduction

Heritage buildings pose as important landmarks not only to add character with their beauty and traditional designs but also to display the history of a society long past and now inherited by the current generation. Heritage buildings contribute greatly to the surrounding society in different aspects by simply existing in the modern era. Heritage buildings, because of their importance with respect to a country's culture and history must be preserved. Conservation of heritage structures allows them to retain their significance as a historical landmark while proving their worth economically and environmentally through reuse. However, heritage buildings suffer deterioration, alteration, and other forms of degradation and changes throughout its life like any other existing structure.

There are several ways to conserve and preserve heritage structures. Preserving a population of heritage structures is best applied in heritage zones where adaptive reuse is promoted. Adaptive reuse is utilizing a heritage structure for a different purpose from what it was originally built as to not become unproductive. This method allows old buildings to be recycled instead of having to build new structures

(Commonwealth of Australia, 2004). Aside from using the heritage buildings as museums, these buildings can be used for various purposes such as commercial, educational, or office. However, before adaptive reuse can be used on structures such as heritage buildings, rules and regulations set by a country's heritage laws about these buildings must be observed and followed.

In the Philippines, the Republic Act 10066 (RA 10066), or the National Cultural Heritage Act of 2009, defines adaptive reuse as using the buildings, built-in structures, or sites for a different purpose than originally intended. This would allow buildings or sites with intrinsic value to be conserved and maintain their engineering integrity and design. Any conservation attempts must go through the appropriate channels, and the cultural agency is responsible for the supervision and approval of the procedures and materials. Adaptive reuse is a common conservation option for buildings marked as cultural property, either being over fifty (50) years old or stated by the Commission. Aside from buildings, adaptive reuse is also implemented on an entire site marked as Heritage Zones, sites holding cultural significance. These Heritage Zones are maintained by the local government. The culture agency is responsible for issuing a Cease-and-Desist order should any significant modification on the cultural property pose a threat to the integrity of the property (Philippines, 2009). Iloilo City has a City Ordinance No. 00-054, declaring certain building structures or sites in Iloilo City (e.g. Calle Real) as a local heritage zone. Any form of modification (repair, rehabilitation, restoration, or construction) is prohibited without a recommendation from the Iloilo City Cultural Heritage Conservation Council (ICCHC).

Conservation of heritage structures allows them to retain their significance as a historical landmark while proving their worth economically and environmentally through reuse. However, heritage buildings suffer through deterioration, alteration, and other forms of degradation and changes throughout its life like any other existing structure. Assessment of existing structures has become a regular occurrence to determine their value economically, environmentally, and socio-politically. However, these assessments are in accordance with advanced methods that have yet to be realized in the traditional codes that the heritage buildings complied with. Due to this update in several provisions, heritage buildings must be assessed and verified for their structural reliability using current codes, with the old codes only present as guidance. Any change brought on the structure must abide by the current codes. Additionally, structural behavior should be investigated for ultimate and serviceability limit states to foresee the degree of deterioration on the building (Holicky & Sykora, 2012). According to their study, several issues that are evaluated are the following: new structural members added to existing load-carrying system due to rehabilitation; load resistance upon any change in use of the facility, operational changes, or extension of its design working life; the damage suffered from time-dependent deterioration, earthquake, and other effects; and other concerns that affect actual reliability of the heritage structure.

Preserving heritage structures in heritage zones is a challenge to the building owners, cultural advocates, and local government officials. This study was conceptualized to support the local government's policies in preserving heritage structures. Issues that need to be addressed include assessing the asset value of the heritage buildings, investigating the structural vulnerability of the buildings to hazards like earthquakes, wind, fire, and strengthening and retrofitting of these buildings to assure their resilience and sustainability. To accomplish these tasks, there is a need to conduct prioritization and screening of the existing buildings in the heritage zone using rapid screening and risk assessment of the site and the buildings. For this paper, seismic risk assessment and structural investigation were conducted at Calle Real, Iloilo City to demonstrate the challenges and issues on heritage preservation of buildings for adaptive reuse considering seismic hazards.

2. Multi-hazard risk assessment

Heritage buildings because of age and deterioration are at great risk of being damaged due to their vulnerable state and the hazards surrounding them. With limited budget and resources, it is a problem to preserve a great number of buildings simultaneously. Hence, a rapid risk assessment of these buildings due to natural hazards (earthquake, wind, flood) and man-made hazards (fire, terrorism) will guide decision-makers like LGUs, cultural heritage councils, architects, and engineers to rank and prioritize the buildings that must be given immediate attention for detailed structural assessment and retrofitting.

The multi-hazard risk assessment methodology developed for this study is based on the relationship among the parameters: Asset (A), Hazard (H), Vulnerability (V), and Risk (R) as shown in figure 1.

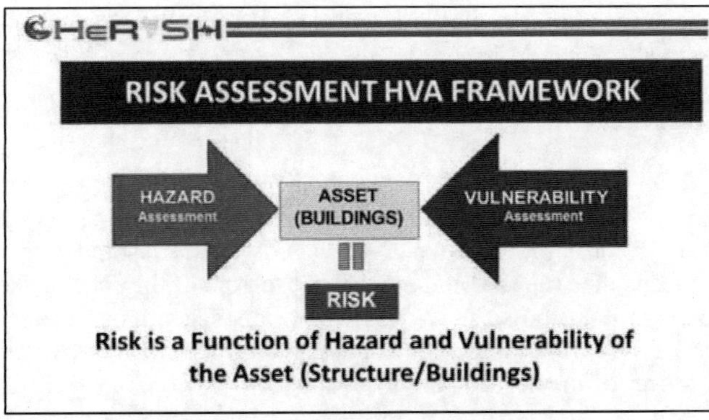

Figure 1. Risk Assessment Framework.

At the center of the risk assessment framework is the "Asset" which is the heritage structure or building in this study. An asset has a value that can be determined qualitatively or quantitatively by considering the relative importance of the heritage building in terms of architectural, historical, commercial, tourism, and adaptive reuse values.

"Hazard" is a "potentially damaging physical event, phenomenon or human activity that may cause losses in life or injury, property damage, social and economic disruption or environmental degradation" (UNISDR 2015). Natural hazards include earthquakes, typhoons, flood, tsunami, landslide, storm surge, drought, volcanic eruption, while human-induced hazards include terrorism, fire, war, and industrial catastrophe. Hazard assessment aims to determine the impact of a hazard to an asset by considering quantitative parameters that contribute to the degree of severance of the hazard like the size of the hazard (e.g. earthquake magnitude, peak ground acceleration, wind speed, precipitation, flood depth) and amplifying factors of the hazard (e.g. nearness to the fault, rivers or slopes, soil type).

"Vulnerability" refers to the "conditions determined by physical, social, economic and environmental factors or processes, which increase the susceptibility of a community to the impact of hazards" (UNISDR 2015). With respect to structures, vulnerability is the degree of susceptibility of the structural and non-structural elements of the building to specific hazards. A qualitative and semi-quantitative vulnerability assessment aims to assess if a building has vulnerable elements to a specific hazard. A building that possesses many vulnerable elements has a greater "risk" to perform poorly to the specific hazard. All in all, the risk (R) is determined using the general relationship below,

$$R = (H \times V) \times A \tag{1}$$

where R, H, V, and A are indices corresponding to the risk, hazard, vulnerability, and asset value, respectively; the latter three indices are referred to as *sub-indices*.

3. Qualitative/semi-quantitative risk assessment

The sub-indices are each defined by a cluster of key parameters, rated through a rapid visual screening (RVS) procedure. RVS is a first-level screening method that involves a quick visual inspection of key critical parameters of the structure. The key parameters for each index are rated high, medium, or low risk, corresponding to a score of 3, 2, or 1, respectively. The *weighted sum* of the relevant parameters will return the value of the sub-index in the study. Equation 1 is then used to compute the main risk index.

The methodology which is a modified version from Yu and Oreta (2014) uses the conventional technique of risk assessment (Vicente, Mendes Da Silva, & Varum, 2005). This technique returns a vulnerability index without estimating the corresponding damage. The corresponding weights for each key parameter under hazard, vulnerability, and asset can be determined through the conduct of an analytic hierarchy process (AHP) survey which is a pairwise comparison of all possible parameter combinations. Experts in the field are asked to choose the relative weight of each pair of parameters over a nine-point scale then they are aggregated by geometric mean before the calculation of weights.

This method is known as aggregation by individual judgments (Angiz, Mustafa, Noraida, & Kamil, 2012).

4. Seismic risk assessment

4.1. Seismic hazard parameters

The strength of the seismic event is measured through three key parameters: (1) distance of the study location to the nearest fault, (2) peak ground acceleration (PGA), and (3) soil conditions. Altogether, the three parameters aim to encapsulate the size and amplifying factors of the earthquake ground shaking at the site. The earthquake magnitude and seismicity are important seismic parameters, but the PGA best represents the local effect of the earthquake ground shaking at a site. The hazard rating is the sum of the weighted product of the scores of the hazard parameters.

4.2. Seismic vulnerability and modifier parameters

The vulnerability parameters, including its modifiers, aim to measure the relative capacity of a structure through the identification of "weak points". The number of weak points will determine the structure's capacity. The key vulnerability parameters are code year which is related to the age of structure (V1), plan irregularity (V2), vertical irregularity (V3), building proximity which is a provision for seismic gap (V4), number of floors (V5), structural system material (V6), number of bays in the transverse axis (V7), and column-to-column spacing (V8).

The vulnerability rating is the sum of the weighted product of the scores of the vulnerability parameters. The vulnerability rating is then further modified through time deterioration and structural modifiers (M). The modifiers, based on a report from the Architectural Institute of Japan (n.d.), adjust the vulnerability index through consideration of cracks, deflections, building age, continuity of the force load path, and material deterioration among others. Hence, the sub-index vulnerability, V, is multiplied by the modifying factor (M).

4.3. Seismic hazard risk index

The product between the hazard rating (H) and structural vulnerability rating (V), given as HxV, represents a hazard-based risk index. The seismic risk classification of low, medium, and high is used to group the buildings based on seismic risk.

The risk index of the heritage building considering the asset value can then be multiplied by the hazard risk index to obtain the final risk seismic risk index as R = A x H x V.

Figure 2. Villanueva Building (01-008), one of the CH structures in Calle Real

5. Pilot study in Calle Real, Iloilo City

The City of Iloilo is situated in the central part of the Philippines. It is one of the oldest cities and has witnessed one of the first establishments of Spanish colonial rule. Calle Real, the downtown central business district (CBD) of the city at present hosts a total of thirty-one (31) identified heritage buildings. These structures were houses, of art deco architectural type (see figure 2), of the Illustrados (educated Filipino class), and trading bazaars of the entrepreneurial Chinese during the Spanish era (Table 1). Calle Real bounded by J. M. Basa, Aldeguer, Guanco and Iznart, and Mapa streets was declared by the National Historical Commission of the Philippines as a district and a heritage zone on August 8, 2014. It is by Ordinance No. 00-054, also known as the Local Cultural Heritage Conservation Ordinance which established the Iloilo City Cultural Heritage Conservation Council (ICCHC). The government and the Iloilo Cultural Heritage Foundation, Inc. (ICHFI) partnered for the restoration of the heritage structures. Recent efforts from the city government have demarcated the CBD as a cultural hub and regulated the preservation of such structures. The city has also instituted a cultural heritage conservation council to oversee preservation regulations (Yu, Oreta, Ibabao, & Hechanova, 2013). Through the council, several renovations of the heritage buildings have been undertaken to preserve the buildings. In line with these efforts, the identified heritage buildings were subjected to the developed methodology for heritage buildings. Only twenty-three buildings of the thirty-one (31) buildings listed in Table 1 were assessed for seismic risk since two of the buildings were burned down while the six (6) others were not listed yet at the time of this research. The number of identified heritage buildings on the CBD Heritage Zone will continue to increase as many of the buildings in that area will turn 50 years old in years to come.

Table 1. Calle Real Cultural Heritage Structures' List.

#	Building Code	Building Name	Vulnerability rating							
			V1	V2	V3	V4	V5	V6	V7	V8
1	01-001	Celso Ledesma	0.16	0.11	0.12	0.32	0.27	0.44	0.13	0.36
2	01-003	Pilar	0.16	0.11	0.12	0.32	0.27	0.44	0.25	0.18
3	01-004	Javellana	0.16	0.11	0.12	0.32	0.27	0.44	0.13	0.18
4	01-005	Iloilo Lucky Supply	0.16	0.11	0.12	0.32	0.27	0.44	0.38	0.36
5	01-006	S. Villanueva	0.16	0.11	0.12	0.32	0.27	0.44	0.25	0.18
6	01-007	Villanueva	0.16	0.33	0.12	0.32	0.27	0.44	0.13	0.18
7	01-008	Villanueva	0.16	0.33	0.12	0.32	0.27	0.44	0.25	0.18
8	01-009	Divinagracia	0.05	0.11	0.12	0.32	0.27	0.44	0.25	0.18
9	01-010	S. Javellana	0.16	0.33	0.35	0.32	0.27	0.44	0.13	0.18
10	01-011	Iloilo Central Trading	0.16	0.11	0.23	0.32	0.27	0.44	0.25	0.18
11	01-012	Regent Arcade	0.16	0.11	0.35	0.32	0.27	0.44	0.13	0.18
12	01-013	Elizalde	0.16	0.22	0.23	0.21	0.27	0.22	0.25	0.18
13	01-014	Magdalena	0.16	0.11	0.12	0.32	0.27	0.44	0.25	0.18
14	01-015	S. Villanueva	0.16	0.11	0.35	0.11	0.27	0.44	0.25	0.18
15	01-016	S. Villanueva	0.16	0.33	0.12	0.21	0.27	0.44	0.38	0.18
16	01-018	Villanueva	0.16	0.33	0.23	0.32	0.27	0.44	0.13	0.18
17	02-001	Villanueva	0.16	0.33	0.12	0.32	0.27	0.65	0.38	0.18
18	02-002	Villanueva	0.16	0.33	0.35	0.32	0.27	0.44	0.38	0.18
19	02-003	S. Villanueva	0.16	0.11	0.12	0.32	0.27	0.44	0.13	0.18
20	02-004	Iloilo Central Market	0.16	0.22	0.23	0.32	0.27	0.44	0.38	0.18
21	02-005	J. Melliza	0.16	0.33	0.23	0.32	0.27	0.44	0.38	0.18
22	02-007	LJ Hormillosa	0.16	0.11	0.23	0.32	0.27	0.44	0.38	0.18
23	02-008	L.J. Hormillosa	0.16	0.22	0.23	0.32	0.27	0.44	0.13	0.36
Average			0.16	0.20	0.19	0.30	0.27	0.44	0.25	0.20

5.1. Seismicity of Iloilo City

The Philippine Institute of Volcanology and Seismology (PHIVOLCS) lists the city under seismic zone 4 in the seismic map of the Philippines (National Structural Code of the Philippines, 2010). A fault map from the PHIVOLCS shows that there are around three active faults in the proximity of the city.

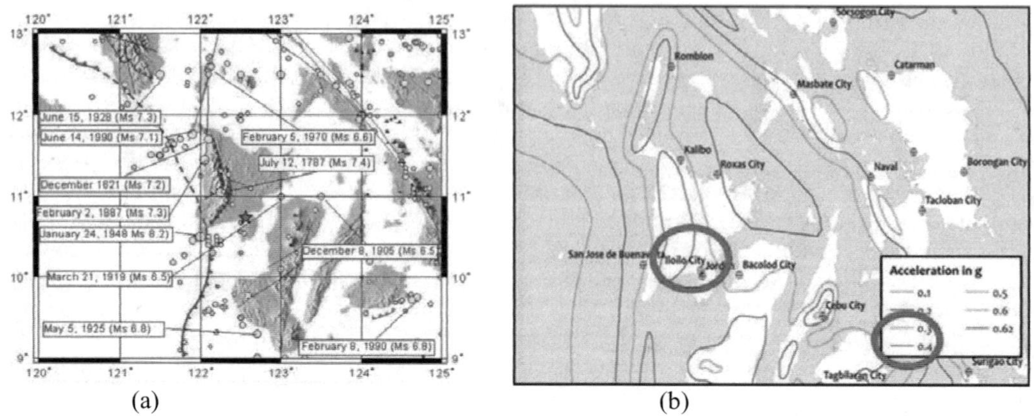

Figure 3. (a) Documented seismic activities in the Western Visayas Region. (b) PGA map from the Philippine Earthquake Model (PHIVOLCS).

Historical earthquakes of the city with magnitudes six and above are plotted on a map; see Figure 3a (Bautista M. et al, 2012). The red star is the location of the city. Figure 3b shows a PGA map for a 500-year return period on a rock site determined by PHIVOLCS. The PGA for Iloilo ranges from 0.3g to 0.4g.

5.2. Seismic Rapid Assessment Result at Calle Real, Iloilo City

The seismic hazard ratings in this research are categorized into fault distance (H1), earthquake peak ground acceleration (H2), and soil foundation (H3). Earthquake peak ground acceleration (H2), and soil foundation (H3) are the highest factors that contribute to the hazard and fault distance (H1) being the least factor. The hazard ratings for these parameters are assumed to be the same for all the 23 cultural heritage (CH) structures in Calle Real since these buildings are relatively close to each other in the heritage zone. In the seismic risk assessment tool, the level of risk values for H1, H2, and H3 are 1, 2, and 2 respectively for each building. With equal weight given to these 3 parameters, a total hazard rating (H) of 1.67 is computed and thus used for all assessed structures.

Table 1 shows the structural vulnerability results under seismic vulnerability assessment. Specifically, the table presents the vulnerability rating and its average value for each parameter for all buildings. The structural system material (V6), pounding effect (V4), and building height (V5) respectively, were revealed to be the main parameters of the buildings that are most vulnerable to earthquake.

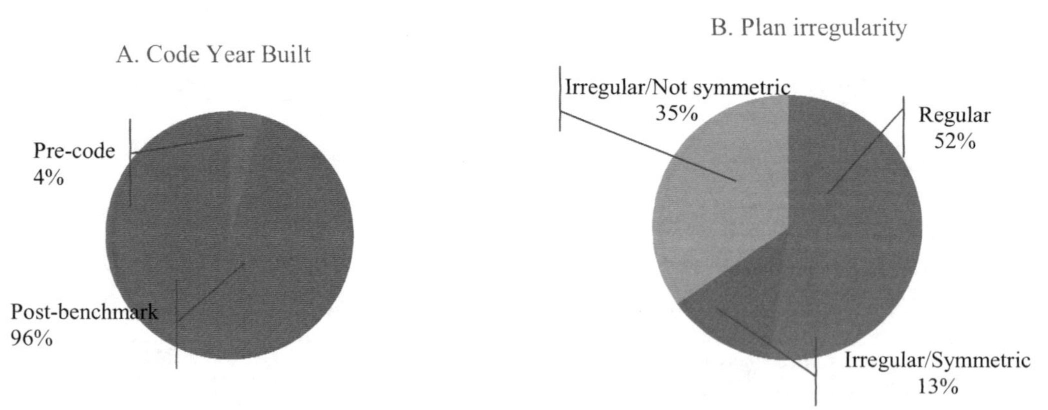

IOP Conf. Series: Earth and Environmental Science 1091 (2022) 012008 doi:10.1088/1755-1315/1091/1/012008

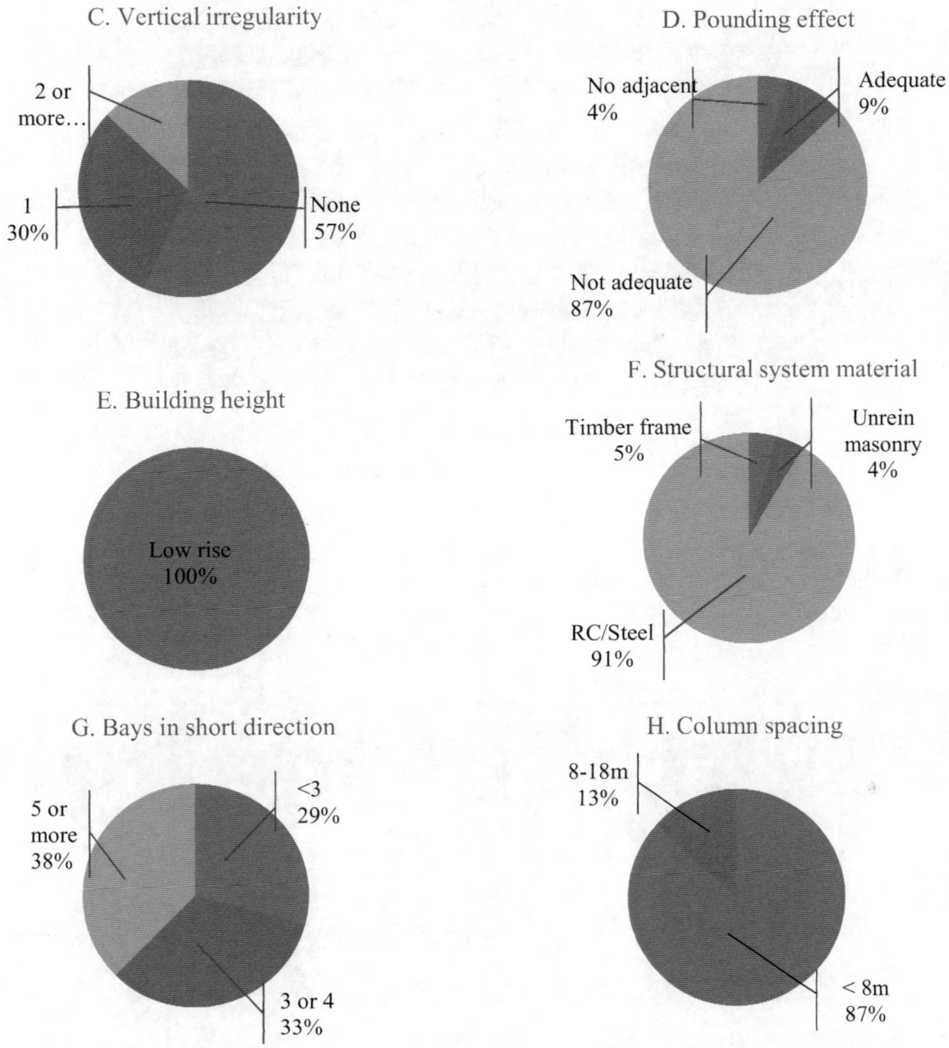

Figure 4. (A-H) Compiled pie charts for seismic vulnerability factors (Sevieri et al, 2019).

The distribution of seismic vulnerability factors for CH structures in Calle Real is graphically represented in figure 4 (compiled seismic vulnerability factor charts). All structures can be considered low-rise with 91% being constructed from reinforced concrete material. Moreover, 95.7% of structures were constructed before the adoption of building code with 47.8% of structures having plan irregularity, and 44.4% having vertical irregularity. Most of the buildings were subjected to a pounding effect (87%).

Figure 5 shows the statistics for seismic vulnerability modifiers. All buildings did not exhibit significant inclination as judged through a street survey. Also, all twenty-three (23) buildings had not been exposed to fire. However, all buildings somehow had used the whole or a portion of flooring for storage. Many of the structures had visible deformations, had shown leaks on walls and columns, and had exhibited cracks on structural members. Severe deterioration can also be observed from the finishing of both the exterior and interior parts of the structures. Finally, many of these structures had installed an additional floor/mezzanine for additional storage.

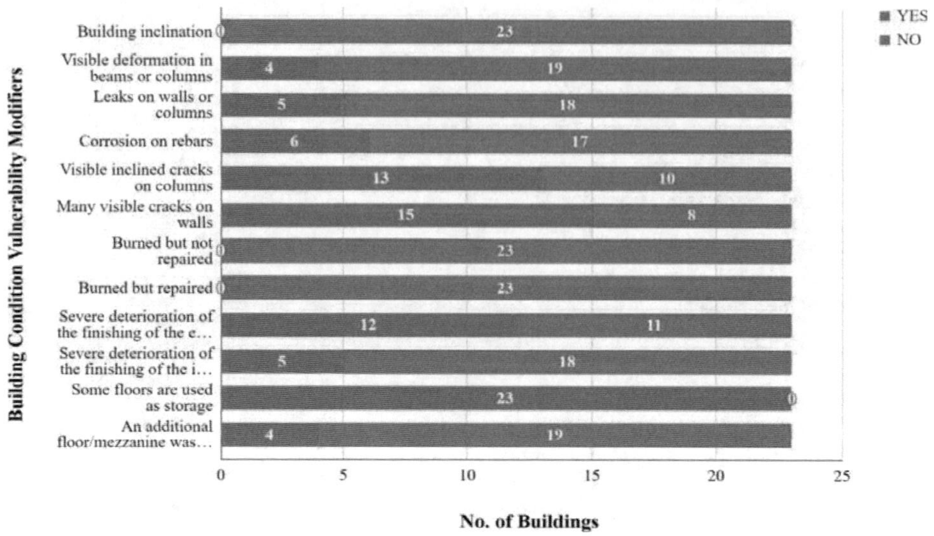

Figure 5. Building Condition Modifiers.

In the seismic risk index of figure 6 and seismic risk map in figure 7, it is seen that the 02-002 Villanueva is the most at-risk building with a seismic index of 6.78 classified under high risk. On the other hand, the building with the lowest risk as its index is only 3.38 is the 01-003 Pilar building classified as medium risk.

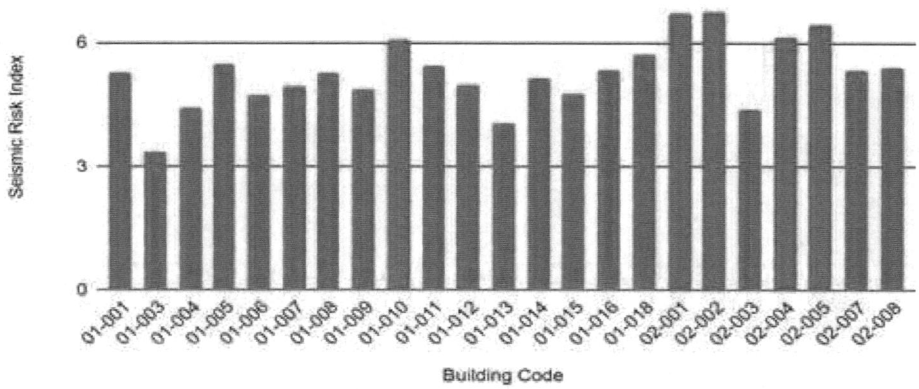

Figure 6. Seismic risk index.

Among the 23 assessed buildings, five (5) heritage structures are at high risk while the rest (18) are at medium risk. The seismic risk index ranges from 3.38 to 6.78. Five buildings that are found to be at high risk shared the same highest rating criteria which is the structural system material (V6). Twenty-one (21) out of twenty-three (23) buildings scored 0.44 or medium risk since the beams, columns, and girders of the buildings are mostly made up of reinforced concrete. This is estimated to contribute to 21.89% of the average vulnerability rating.

6. Site survey of Calle Real

During the site survey, several functional and structural issues were found in cultural heritage structures in Calle Real. Structure owners, to maximize earnings from their property with the least expense, put their property up for commercial lease based on the as-is condition of the structure. Since Calle Real is one of the busiest commercial areas in Iloilo City, rental fees are significantly high and as a result, tenants have imposed modifications on the structure to maximize the usable commercial space of the structure. Some of these modifications are conversion of spaces for storage, installation of the mezzanine for additional storage area (figure 8a), removal of a building component (e.g., column) to increase commercial space, installation of billboards on top of the structures (figure 8b), conversion of entire 2nd floor to storage, and cutting off a portion of roof for other purposes.

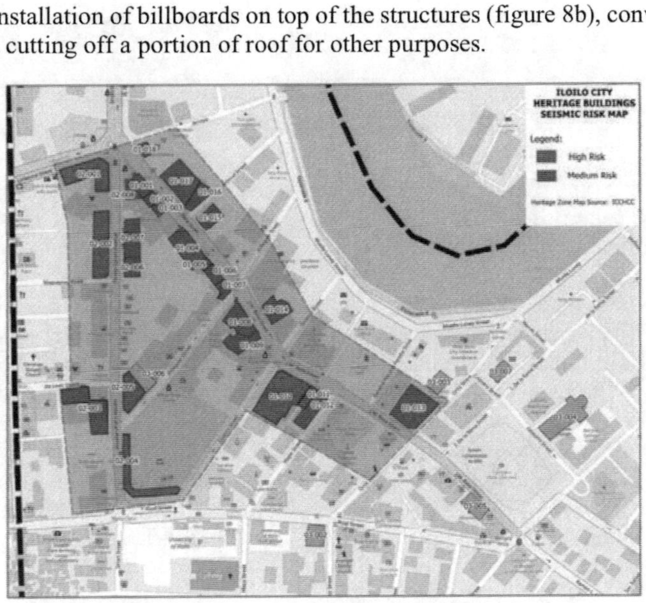

Figure 7. Seismic Risk Map of Calle Real

Figure 8. (a) Installation of mezzanine; (b) installation of billboards on top of the structure

Issues about structural integrity were also discovered during the site survey. Based on the preliminary survey, although a truss system was installed to provide support for the billboard structure placed on the roof deck of the building, no evidence is seen on the improvement of the structural elements of this building to carry the additional load. Roof alterations were also observed in some structures to accommodate the installation of a water supply tank. Other structural integrity issues found in some buildings are the conversion of entire floors as storage spaces and also the installation of mezzanines for additional storage. These building use alterations have caused excessive sagging of the floor due to excessive weight brought about by stored supplies.

Some structural issues were brought about, not by modification of the structure, but by its original design. These are plan irregularity, vertical irregularity, mass irregularity, and pounding (Sevieri et al., 2019). Out of 25 structures, 36% require an in-depth structural investigation, as its condition is deemed poor based on the preliminary survey.

Many of the structures also have exhibited cracks. Fourteen (14) cultural heritage structures included in this research have shown visible cracks that range from mild to severe. Some cracks are so severe that it has already led to partial spalling and exposure of some portion of the reinforcing steel which resulted in corrosion of steel bars. A building that seemed to have cut off some of the structural elements (i.e. columns) to provide more commercial space, also has its external columns exhibiting severe wide and lengthy cracks.

7. Conclusion

Using the seismic risk assessment tool, five (5) out of twenty-three (23) CH structures in Calle Real turned out to have a high-risk seismicity rating while the rest of the structures have a medium risk rating. The year built, pounding, and building height, are the vulnerability factors that are commonly found in most of the structures (>20 structures). Cracks on walls, columns, and severe signs of deterioration of the finishing are the most common building condition modifiers, with at least half of the structures manifesting these conditions. Along with hazard rating, these vulnerability factors and building code modifiers were used to assess the seismic risk indices of CH structures in Calle Real.

The seismic risk assessment was conducted with a preliminary site survey. This preliminary site survey has identified functional and structural issues in the CH structures in Calle Real. Some of the issues were a direct result of adaptive reuse since commercial tenants of the structures had made some alterations in the structure to either add commercial space or add storage spaces. An indirect result of adaptive reuse had to do with the structure being old and had survived several calamities already like typhoons and earthquakes. These indirect results were manifested through cracks on walls, columns, and beams of the structures.

The result of this study fairly indicates that the use of cultural heritage structures through adaptive reuse must need an in-depth structural investigation to ensure the reliability and safety of the structures.

Acknowledgments

The study would like to acknowledge the senior Civil Engineering students of Central Philippine University for conducting the seismic risk assessment for this study, the Iloilo City Government for field research support, and CHED-Newton Institutional Link Program for funding support.

References

[1] Commonwealth of Australia 2004 Adaptive Reuse: Preserving our past, building our future (Australia: Department of the Environment and Heritage) https://www.awe.gov.au/sites/default/files/documents/adaptive-reuse.pdf
[2] Philippines 2009 *Republic Act No. 10066: National Cultural Heritage Act of 2009*
[3] ICCHC 2009 *Implementing Rules and Regulations for the Downtown Central Business District Heritage Zone* (Philippines: Iloilo City Cultural Heritage Conservation Council)
[4] Holicky M and Sykora M 2012 Structural assessment of heritage buildings *WIT Transactions on the Built Environment* (Southhampton, UK: WIT Press) Vol 123 pp 69-80
[5] Yu K K and Oreta A W C 2014 Seismic risk assessment of heritage buildings in Iloilo city, Philippines *Proc. 5th Asia Conf. on Earthquake Engineering* October 2014 Taipei
[6] Vicente R, Mendes Da Silva J and Varum H 2005 Seismic vulnerability assessment of buildings in the old city centre of Coimbra *250th Anniversary of the 1755 Lisbon Earthquake Int. Conf.* 1-4 November 2005 Lisbon
[7] Angiz L, Mustafa A, Noraida A and Kamil A 2012 group decision via usage of analytic hierarchy process and preference aggregation method *Sains Malaysiana* **41** 361-66
[8] Architectural Institute of Japan. n.d. Time deterioration modifiers *In Seismic Screening Procedure*
[9] Yu K K, Oreta A W C, Ibabao R and Hechanova N 2013 Supporting local initiatives in preserving heritage buildings in Iloilo (Philippines) through risk assessment *Int. Conf. on Cultural Heritage and Disaster Risk Reduction* 18-20 November 2013 Bangkok
[10] Association of Structural Engineers of the Philippines 2015 *National Structural Code of the Philippines* Vol 1, ed 6th (Manila: Association of Structural Engineers of the Philippines)
[11] Bautista ML, Bautista B, Narag I, Lanuza A, Deocamppo J, Papiona K, Atando L, Solidum R, Allen T, Jakab M, Ryu H, Edwards M, Nadimpalli K, Leonard M and Dunford M 2012 Strengthening natural hazard risk assessment in the Philippines: an earthquake impact pilot study for Iloilo City, Western Visayas Geoscience Australia: Canberra
[12] Sevieri G, Di Stasio S, O'Sullivan K and Galasso C 2019 The CHeRiSH project: towards a multilevel, multi-hazard risk assessment framework for cultural heritage assets in the Philippines *Conf. Paper in 18th Convegno Anidis* September 2019 Italy

Effect of Disaster Preparedness Training on Knowledge Regarding Flood and Landslides Disaster Management among Family with Elderly

Fatmah*[1]

[1] Disaster Management Study Program at School of Environmental Science Universitas Indonesia, Indonesia

*Corresponding author e-mail: ffatmah@yahoo.com

Abstract. Elderly population is among vulnerable populations during disasters. This study aimed to assess the effect of disaster preparedness training on knowledge on flood and landslide disaster preparedness and management in families with elderly. This was a quasi-experimental pre-post study on 30 respondents participating in natural disaster preparedness training. A significant change in general knowledge on disaster, flood, and landslide (12.9, 20, and 12.4 points, respectively). Disaster preparedness practice was good as reflected in actions performed before, during, and after disaster. Before flood occurs, families prepare a disaster preparedness bag for important documents as well as logistics (food) for emergencies and equipment for evacuation while also store valuable goods in a safe place. During flood, families prioritize elderly evacuation while seeking information on the flood through neighbors, handy talky, handphone, television, and radio and ensure the availability of sufficient logistics. After disaster, families clean their properties and house; provide clean water and toilet facilities for the family; check the health of family members that may be impacted by the flood; and make sure that all electrical panels safe. Disaster training affects the knowledge of flood and landslide disaster management in families with elderly.

1. Introduction

Indonesia has a fairly high potentials for natural disasters and other disaster threats. Since the occurrence of a disaster cannot be predicted with certainty, people must always be prepared to deal with disasters. Since early 2021 to the beginning of September 2021, 1,829 natural disasters have occurred in Indonesia with flood (750 cases) and landslide (346 cases) as the dominant types of disaster[1]. The flood that struck the Greater Jakarta area on January 1, 2020 is among the most extreme and highest floods experienced since 154 years ago. The flood was triggered by heavy rains that submerged some parts of the nation's capital and its surrounding areas.

Law No. 24 of 2007 on Disaster Management has declared the elderly as one of the groups that are vulnerable to disasters [2]. During disaster, most of the victims of disaster consist of children and other vulnerable groups, such as women and the elderly. Elderly can be categorized into facing moderate to high levels of hazards during a disaster due to their limited physical ability, reduced sensory ability, and presence of degenerative diseases [3]. Therefore, it is important to improve knowledge and behavior of the elderly who live in flood-prone areas to have better preparedness in dealing with the disaster.

Disaster preparedness is defined as activities performed prior to disasters in order to develop the operational capacity and facilitate effective responses when a disaster strikes [4]. Through increased knowledge in preparedness, the elderly will understand the actions to take when facing a disaster and act accordingly. Knowledge is the key factor that can influence the attitude and awareness of the community to be ready and prepared in dealing with disasters. Elderly with good attitude are the elderly who are ready to deal with disasters based on good knowledge of the signs of disaster, making them always ready to evacuate at any time. Disaster preparedness has the aim of ensuring appropriate and rapid efforts in dealing with disaster events. One of the stages of disaster preparedness is education and training.

Disaster management comprises efforts or activities carried out in the context of disaster prevention, mitigation, preparedness, emergency response, and recovery that are performed before, during, and after a disaster [5]. Meanwhile, disaster mitigation is a series of efforts to reduce disaster risk through physical development, increased awareness, and ability to deal with disaster threats. One form of disaster mitigation is education, counseling, and training to enable people to achieve the objectives of disaster mitigation. Mitigation and preparedness should be established before a disaster occurs (pre-disaster) and training on disaster preparedness as a form of disaster mitigation and disaster preparedness effort is needed to increase public knowledge and awareness of the disaster risks in the area where they live [6]. It is expected that people who are well trained in disaster preparedness and mitigation will have good public awareness to reduce disaster risks and to collaborate well with agencies that are responsible for disaster management and other external parties before, during, and after a disaster.

Indonesia is a country that is prone of disasters. As an archipelagic country known as a country with the ring of fire, this country has a risk for various types of disaster including, among others, volcanic eruption, earthquakes, flood, landslide, and storms. West Java Province, as the most populous island in the country, is prone to earthquakes, flood, landslides, and storms, and Depok City, as one of the cities in the province, has even declared a disaster alert status for flood, landslide, and strong winds for a period of 14 days from January 1 to January 14, 2020 as a response to increased disaster events. From 2007 to 2015, Depok City had experienced 12 floods and 6 landslides with no fatalities [7]. In the last three years, there is an increased incidence of floods and landslides in Pancoran Mas and Mampang Sub-districts in Depok City due to poor drainage systems and high rainfall. These areas have a high potential for flooding caused by a high intensity of rain which leads to the overflow of river water triggered by inadequate slope inclination and river flow capacity. This is worsen by the fact that the two areas do not have good drainage system, leading to a potential flood every year. Some areas in Pancoran Mas Sub-district are also prone to landslides, especially those located on the river side of Ciliwung and Pesanggrahan rivers, as well as the area of Situ Pedongkelan (Figure 1).

Based on the data in the flood emergency contingency TWG of Depok City Fire and Disaster Management Agency, several flood and landslide incidences have been recorded in Pancoran Mas Sub-district: a 1.5 meter flood in Marine Housing Complex, RW 6, Rangkapan Jaya Baru Urban Village with 250 people affected; a landslide in Maharaja Housing Complex in Rangkapan Jaya Urban Village, flood caused by the overflow of Krukut stream in Arsip Nasional Housing Complex of Mampang urban village, and land slide in Rangkapan Jaya Urban Village [8][9][10]. To improve the preparedness in dealing with disasters, especially among those with higher vulnerability, a training session on flood and landslide disaster for families with elderly was conducted on August 30, 2021 in 30 *posbindu* (integrated service post for elderly) by involving sub-district and village cadres and elderly working groups. The purpose of the training is to increase the knowledge of families with elderly in flood and landslide preparedness at before, during, and after facing natural disaster.

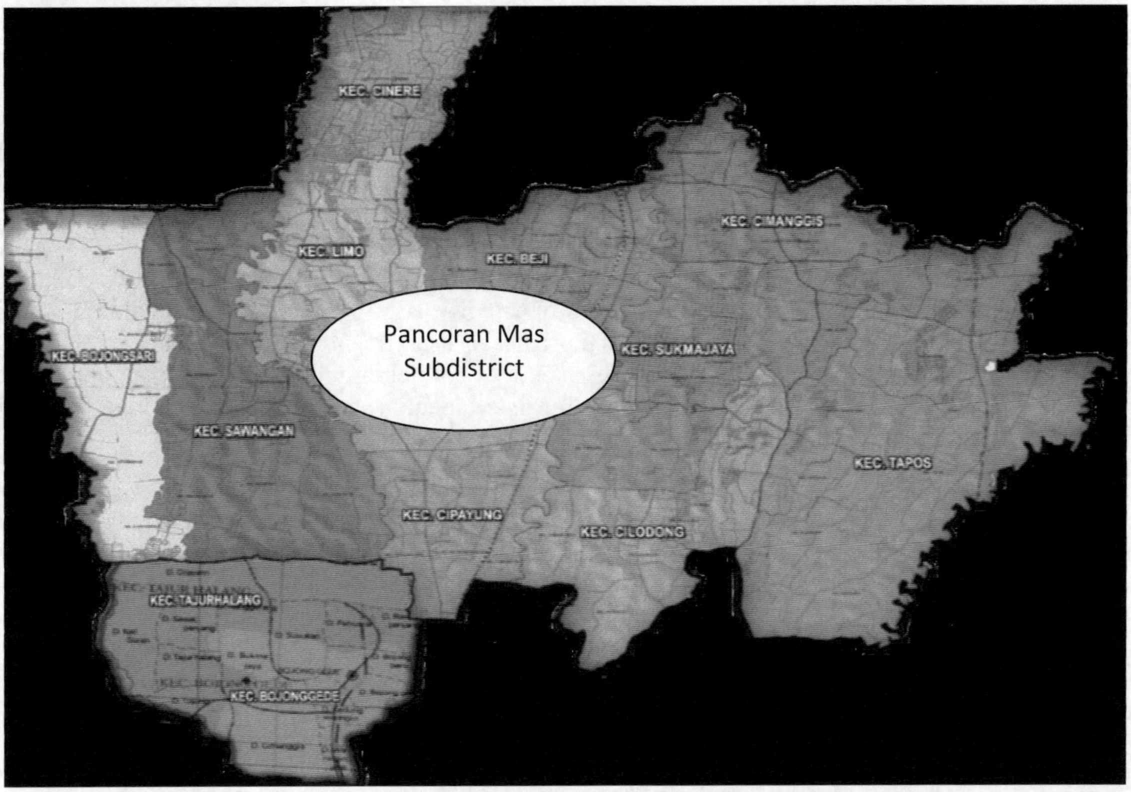

Figure 1. Map of Depok City flood emergency contingency

2. Material and methods

2.1 Study Design

This study used a quasi-experimental design without a control group and a purposive sampling technique [11]. The ethical clearance was obtained from the Health Research Ethics Commission of the National Institute of Health Research and Development, Ministry of Health of the Republic of Indonesia. The informed consent form was signed by the respondents in the third week of August 2021.

2.2 Population and Sample

The intervention of this study was a one-day training on flood and landslide preparedness for families with elderly held in August 30, 2021. The training was attended by 30 *posbindu* cadres, sub-district elderly working group members, and urban village elderly working group members who have elderly family members. They come from six villages at Pancoran Mas Subdistrict, Depok City, West Java Province, Indonesia (Depok Jaya, Depok, Pancoran Mas, Mampang, Rangkap Jaya, and Rangkap Jaya Baru). The training participants consisted of 3 men and 27 women. The resource persons for the training were the research team from the School of Environmental Sciences, University of Indonesia, and relevant partners from the Pancoran Mas Sub-district of Depok City, West Java Province, Indonesia.

2.3 Data collection

The independent variables of the study were socio-demographic characteristics (marital status, age, education background, primary job, length of time as a cadre, and whether or not they attended training), general knowledge of natural disasters (definition, types, examples of disasters, impacts, natural disaster-prone groups, preparedness for families and the elderly), specific knowledge about floods (definition; causes; impacts; actions taken before, during, and after floods; and flood prevention and mitigation efforts), specific knowledge about landslides (definitions; causes; impacts; actions taken before, during, and after landslides; as well as efforts to prevent and control landslides), and practices regarding preparedness for the elderly before, during, and after floods. The training was held at the Pancoran Mas Sub-district Office. The mean score for knowledge was measured twice; first, during the pre-test and, second, during the post-test. The purpose of the pre- and post-tests was to assess the changes in knowledge of families with elderly in flood and landslide disaster preparedness. The questionnaire used consisted of a total of 50 items with 4 items on socio-demographic characteristics, 9 items on general knowledge about natural disasters, 7 items on flood knowledge, 9 items on land landslide, and 20 items on practices of elderly before, during, and after flood. Preparedness education was carried out using printed media (leaflets and flipcharts) and visual media (video) developed by the study team.

2.4 Data analysis
Data were analyzed univariately for frequency distribution, minimum and maximum values, and study variable means. Scores of preparedness knowledge and practice before and after training were analyzed using the independent t-test and $p<0.05$ was considered significant. Data analysis was performed using the SPSS version 21 for Windows.

3. Results
Table 1 presents the sociodemographic characteristics of respondents including marital status, gender, age, highest education, current employment status, and participation in nutrition and health training. Most of the respondents were *posbindu* cadres (83.3%) and the remaining were members of the sub-district and urban village elderly working groups. By gender, most of the training participants were women, added by 3 male respondents. The majority of respondents were pre-elderly with the oldest being 73 years and the youngest being 30 years (mean, 51.9 years old). More than three-quarters of the respondents were married and most had completed senior high school or equivalent. Almost all respondents no longer worked, but a small proportion still worked as traders and entrepreneurs. More than half of the total respondents had attended nutrition and health-related training including training on elderly; COVID-19; stunting; nutrition aware family; food and nutrition; Clean and Healthy Lifestyle; nutrition for under-five children; hypertension; Alzheimer; Diabetes; *posbindu* cadre training; solid food introduction to infants; elderly exercise; cervical cancer; and health care.

3.1 Changes in knowledge of disaster, flood, and landslide
The increase in the mean score of respondents' knowledge on disasters in general, as well as knowledge on flood and landslide drawn in Table 2. The mean change in knowledge on flood ranked first, followed by knowledge of general disasters and landslide. The three types of knowledge presented a significant relationship before and after intervention. The mean changes in knowledge of disasters, flood, and landslide by socio-demographic characteristics are shown in Table 3. There was no significant difference between the increase in respondents' knowledge on disasters in general, flood, and landslide based on marital status, gender, age, education background, employment status, and participation in previous health and non-health related training ($p < 0.01$). However, respondents who have attended previous health and non-health related training tend to have a slightly higher score compared to those who had never participated in such training.

The comparison of the proportion of correct answers on general knowledge of disasters, flood, and landslide before and after the training is reflected in Table 4. The largest increase in the proportion of general knowledge of disasters was found in the item on the types of disasters (40%), followed by items on impact of disasters and examples of disasters. The items that were mostly answered correctly in the pre-test were items on elderly, including that elderly is one of the vulnerable groups during

Table 1. Socio-demografi characteristic

Variable	Person (n)	Proportion (%)	Mean ± SD (Min – Max)
Marital status			
Married	24	80	
Widow/widower	6	20	
Gender			
Female	27	90	
Male	3	10	
Age (years old)			51.9 ± 9.8 (30-73)
< 60	23	76.7	
>= 60	7	23,3	
Final education level			
Low (<=9 years)	5	16.7	
Medium (> 9 year)	25	83.3	
Working status	3	10.0	
Entrepreneur	2	6.7	
Trader	1	3.3	
Did not work	27	90.0	
Ever attended training before: Yes	20	66.7	
No	10	33.3	

Table 2. Changes in knowledge about disasters in general, floods, and landslides

Variable	Score Pre -test Mean ± SD	Post-test Mean ± SD	Difference Mean ± SD	*p*
Knowledge on general disaster	69.6 ± 19.9	82.5 ± 11.0	12.9 ± 18.2	0.001
Knowledge on flood	52.4 ± 18.9	72.4 ± 15.8	20.0 ± 14.1	0.001
Knowledge on landslide	47.6 ± 16.5	60.0 ± 18.6	12.4 ± 15.6	0.001

disaster and that the elderly needs to prepare themselves for disasters. For knowledge on flood, the item that had the highest proportion of correct answers was actions that should be taken by family when flood occurred. The item with the lowest proportion of correct answers was the item on four flood prevention and mitigation efforts. In general, the percentage of correct answers in the flood knowledge section was still less than 40%. For landslide, the item with the highest proportion of correct answers, compared to the other 8 items, was the item on factors causing landslide. Knowledge on the signs of an imminent landslide is the item that most respondents answered correctly during the pre-test. The item on actions need to be taken by families and elderly before, during, and after landslide was only answered correctly by less than 20% of respondents during the post-test. More than half of the respondents already knew before the training that the elderly is one of the vulnerable groups during a landslide.

3.2 Practices of families with elderly in dealing with flood

The first thing the respondents did before and when facing a sudden flood was evacuating electronic goods and household appliances to a higher place, evacuating themselves and family members to a higher floor of the house, evacuating to a relative's house or someone else's house who is located outside the flood area. When the flood suddenly entered the house, the first action taken by the respondent's family was to evacuate the elderly to a safer place. When asked whether the respondent felt anxious or worried during the previous flood, more than half of the total respondents felt anxious because there were elderly and toddlers in the house, worried that the house would be completely submerged, and panicked.

Table 3. Mean change in knowledge of general disasters, floods, and landslides based on the socio-demographic characteristic

Variable		Knowledge on general disaster				Knowledge on flood				Knowledge on landslides			
	n	Mean	±	SD	*p*	Mean	±	SD	*p*	Mean	±	SD	*p*
Marital status:													
Married	24	13.4	±	19.9	0.791	21.1	±	13.3	0.412	14.0	±	15.2	0.270
Widow/widower	6	11.1	±	9.4		15.7	±	17.7		6.1	±	16.9	
Gender:													
Male	3	2.8	±	4.8	0.316	11.8	±	15.6	0.295	15.2	±	34.4	0.755
Female	27	14.0	±	18.8		20.9	±	14.0		12.1	±	13.3	
Age (years old):													
< 60	23	15.0	±	19.8	0.253	22.5	±	13.5	0.078	9.5	±	13.3	0.059
>= 60	7	6.0	±	9.3		11.8	±	14.0		22.1	±	19.5	
Education level:													
Low (< 9 years)	5	22.5	±	14.9	0.201	23.5	±	23.5	0.550	16.4	±	7.6	0.545
Medium (>=9 years)	25	11.0	±	18.4		19.3	±	12.1		11.6	±	16.7	
Working status:													
Work	7	11.9	±	7.8	0.870	23.5	±	16.3	0.460	11.7	±	10.1	0.889
Does not work	23	13.2	±	20.4		18.9	±	13.6		12.6	±	17.1	
Had ever participated in the previous training													
Yes	20	14.2	±	20.1	0.603	18.2	±	14.8	0.342	13.6	±	17.1	0.556
No	10	10.4	±	14.1		23.5	±	12.7		10.0	±	12.5	

Actions taken by respondents before the flood were preparing disaster preparedness bag containing important/valuable documents, preparing logistics (food items) for emergencies, preparing equipment for evacuation, and planning/placing valuables in a relatively safe place. The respondents also sought information related to flood through neighbors, handy talkies, cellphones, television, and radio. They also ensured the availability of sufficient food/logistics supplies; ensured that all family members were safe; and evacuated to a safer place from when a flood struck.

Almost all respondents stated that they prioritized the elderly during flood due to the elderly's physical limitations that prevent the elderly to do quick actions to save themselves from the flood. When the flood had subsided, most respondents cleaned their properties and houses from mud, dirt, and waste carried by flood; provided clean water and toilet facilities for family members; checked the health of the family members that may be affected by flood; and ensured that all electrical panels in the house were safe to use after the flood. Usually, families evacuated to the second floor of their own house or went to the neighbor's house and/or relative's house that was located outside the flooding area or went to a public building that was safe from the flood.

Table 4. Comparison of the correct answers proportion regarding knowledge of general natural disasters, floods, and landslides before and after training

Knowledge on general natural disaster, flood, and landslides	Proportion of correct answer (%)		
	Pre	Post	Difference
Knowledge on general natural disaster			
Able to mention three definitions of disaster	30.0	70.0	40.0
Able to mention three types of disaster	36.7	50.0	13.3
Able to mention at least three examples of natural disaster	56.7	76.7	20.0
Able to mention three types of disaster impact	66.7	96.7	30.0
Elderly are vulnerable to disaster	93.3	100.0	6.7
Elderly, children, pregnant mothers, and lactating mothers are vulnerable group to natural disaster	76.7	86.7	10.0
Elderly need to prepare for natural disaster	93.3	100.0	6.7
Preparedness of elderly and their families for natural disasters	63.3	73.3	10.0
Flood:			
Able to mention four definitions of disaster	20.0	33.3	13.3
Able to mention three causal factors of flood	60.0	80.0	20.0
Able to mention three impacts of flood	66.7	90.0	23.3
Able to mention four actions before the flood	13.3	46.7	33.4
Family actions taken during flood	40.0	93.3	53.3
Family actions taken at post-flood	60.0	93.3	33.3
Able to mention four flood prevention and control	26.7	30.0	3.3
Landslides:			
Definition	60.0	60.0	0.0
Causal factors	23.3	60.0	36.7
Impacts	46.7	53.3	6.6
Elderly are vulnerable group to landslides	60.0	63.3	3.3
Know the signs of an impending landslide	86.7	96.7	10.0
Actions taken before landslides occurs (pre)	20.0	36.7	16.7
Actions taken during landslides occurs (at)	26.7	43.3	16.6
Actions taken after landslides occurs (post)	73.3	93.3	20.0
Three efforts of landslides prevention and control	10.0	26.7	16.7

The emergency response plan prepared by the respondent in the event of a flood was to prepare medicines for first aid at home; prepare equipment for self-rescue/emergencies; and prepare sufficient food stocks to survive during the flood. The majority of respondents received flood early warning information from the neighborhood level that facilitated the people to receive the latest flood information. The media used for sharing information were WhatsApp (WA) group; announcement from mosque/Islamic prayer room speakers; Instagram, Facebook, and other internet media. Only a small proportion of respondents had attended training on natural disaster preparedness at the sub-district level in 2019 and 2020. The majority of respondents expected that all people in their respective neighborhood cleaned up the waste from rivers/streams. The Depok City Government regularly normalizes/dredges shallow rivers, and repairs roadside waterways, including providing flood education to the community and preparing better infrastructure for anticipating future disasters.

4. Discussion

The elderly is one of the vulnerable groups during disaster, in addition to adolescent girls, pregnant women, breastfeeding women, children, and persons with disabilities. The elderly are classified as vulnerable to natural disasters due to their decreased physical abilities and health status; reduced ability to move/mobility; decreased sensory awareness; and increased risk of morbidity in pre-, during, and post-disaster situations [12][13][14]. Specific protection and measures need to be implemented to the vulnerable groups during disasters to enable them to survive the disaster and post-disaster situations. One of the specific efforts to reduce disaster risk is increasing the preparedness and mitigation in dealing with disasters. Preparedness comprises of organized real and effective steps to anticipate disasters and is included in the pre-disaster cycle [15].

To have a good preparedness level in the community, it is necessary to increase public knowledge and awareness on disasters. This good knowledge and awareness will influence how the people deal with disaster situations before, during, and after disaster for themselves, their families, the elderly, and other vulnerable groups. The ability of the family to assist elderly in disaster situation holds an utmost importance. Several studies have shown a significant relationship between family support and the elderly's preparedness in dealing with natural disasters [16][17]. Family preparedness in assisting the elderly in dealing with disasters is a form of family care [18]. The flood preparedness practice for elderly was assessed at the beginning of the study. The aim was to obtain an overview of the practice of the elderly before, during, and after the flood. The assessment observed in this study is flood disaster because floods occupy the highest order of natural disasters in Depok City. There are quite a lot of flood-prone areas in Depok City during 2018-2020. Depok City's flood vulnerability index includes low (46.4%) and moderate (36.5%) [7].

In general, the practice of preparedness of families with elderly in dealing with flood is adequately good from the perspective of the disaster cycle (before, during, and after disaster). Families with elderly have taken appropriate action to evacuate the elderly to a safer place. Information support is also provided by the family to the elderly by explaining actions that need to be taken before the flood. The family also prepares equipment that may be needed if a flood strikes, such as disaster preparedness bag and food logistics. The disaster preparedness bag prepared by families with elderly also contains elderly medications, first aid kits, clothing, blankets, emergency lamps/torches, batteries, et cetera [16]. The findings of this study are in line with a study on the preparedness of families with elderly in facing potential volcanic eruption in Klaten [18] [19]. Information needs to be given repeatedly to the elderly considering potential age-related changes in the cognitive function of the elderly [20].

Efforts to improve preparedness of the elderly to face flood is carried out by involving them or their family members in regular training and seminars, as well as updating knowledge and information that can be directly accessed by the elderly and their family members. The increased knowledge of families with elderly will become an important part in the efforts to increase disaster preparedness [21]. Disaster warning systems used by families with elderly takes the form of WA groups at the neighborhood level; announcements from mosque/Islamic prayer room speakers; and information from internet sources such as Instagram, Facebook, and other media that can be used to warn people regarding the signs of the imminent emergencies or flood. The purpose of the early warning is to notify and quickly respond to potential disasters and threats, which will lead to faster evacuation of affected residents and prevention and reduced impact of disasters [22]. This is also emphasized in findings of a study on flood preparedness in Bekasi City [23].

General knowledge of disasters and knowledge of flood and landslide has a significant relationship before and after the intervention. No significant relationship was found between the changes in the three types of knowledge and all socio-demographic characteristics variables. This may be due to the homogeneity of respondents in the education background, gender, and age group variables with most respondents graduated from secondary education level, are female, and are in pre-elderly group. This finding is supported by several studies that show a relationship between age, education, gender, and knowledge [24][25][26]. The more mature a person is, the higher the level of maturity and ability in dealing with issues. A higher education also facilitates better information absorption which, in turn, will impact the level of knowledge. Women consider disaster events or threats as a more serious and risky situation than men. Women are more involved in disaster mitigation and preparedness activities than men, especially in activities centered at home [27]. Knowledge is the result gained after a person uses his/her senses to identify a specific object through sight, hearing, smell, taste, and touch [28]. Changes in knowledge of disaster, flood, and landslide as a result of the training are in the cognitive domain of knowing and understanding. It has not yet reached the domain of application, analysis, synthesis, and evaluation because it has not been applied in the family and communities [29].

In general, the respondent's knowledge about floods and landslides is fairly good. This can be seen from the fairly high proportion of correct answers (score higher than 60) to several items, namely: elderly is one of the vulnerable groups in disaster; elderly needs to be prepared for disaster; three causes of flood; three effects of flood; family actions taken after flood; knowledge of landslide; elderly is a one of the vulnerable groups in landslide; signs of a landslide; and actions taken after a landslide. This fairly good knowledge may be due to the experience of dealing with floods that frequently strike the areas where the respondents live, leading to a better understanding of flood. However, this is not very obvious for landslide. In addition, all respondents are involved in the urban village and sub-district elderly working groups, as well as in the *posbindu* that they have daily exposure to issues faced by the elderly. Good knowledge of flood and landslide also links to the practice of disaster preparedness in families with elderly in disaster management phases (before, during, and after flood). The findings of this study support those of two previous studies that prove knowledge has a positive direction with community preparedness practices for flood and landslides [30][31]. Knowledge of flood and landslide preparedness can influence the attitude and attention of the community so that they are always alert in anticipating potential disasters, especially for people who live in areas prone to natural disasters [32].

5. Conclusion

This study has limitations that it does not include a control group for comparison with the intervention group. In addition, the small sample size with a homogeneous educational background, age, and gender may also influence the results. Knowledge and practice of cadres and member of elderly working group who have elderly as family member regarding flood and landslide disaster management is quite good, especially in the efforts to evacuate the elderly first during natural disaster. The knowledge on flood, disasters in general, and landslide increases significantly after the training. Meanwhile, no significant relationship is seen between respondents' socio-demographic characteristics and the increase in knowledge of flood and landslide and whether respondents have attended any previous training. Nevertheless, there is a tendency that respondents who have attended previous training have a higher score compared to those who do not attend such training.

It is necessary to hold flood preparedness training for families with elderly to increase flood disaster literacy, especially among the elderly and other groups that are vulnerable to disasters in the community. The objective is to increase alertness in anticipating flood that may occur at any time. This training also needs to be replicated in other areas prone to disasters by involving the community and disaster management authorities at the subdistrict, district/city, and national levels.

Acknowledgement

Authors would like to thank the Ministry of Research and Technology of the Republic of Indonesia for the PDUPT Grant in 2021 under the contract of NKB-221/UN2.RST/HKP.05.00/2021. We also would like to thank all parties who have supported the implementation of this study, including Pancoran Mas Sub-district Office of Depok City, urban village elderly working groups, and *posbindu* of 6 urban villages in Pancoran Mas Sub-district, Depok City, Indonesia, as well as all *posbindu* cadres who participated in this study.

ORCHID iDs

Fatmah https://orchid.org./0000-0002-2763-0219

References

[1] Since the beginning of the year, there have been 1829 natural disasters in Indonesia. https://nasional.kontan.co.id/news/sejak-awal-tahun-telah-terjadi-1829-bencana-alam-di-indonesia

[2] The Indonesian National Board for Disaster Management. The Indonesian Constitution No. 2 Year 2007. https://bnpb.go.id/ppid/file/UU_24_2007.pdf

[3] Cornell VJ, Cusack L, Arbon P 2012 Older people and disaster preparedness:a literature review. *The Australian Journal of Emergency Management*. **27** 49-53

[4] The Indonesian National Board for Disaster Management 2017 *Disaster Preparedness Handbook Building Awareness, Alertness, and Preparedness in Facing Disasters* (Jakarta: BNPB)

[5] The Indonesian Minister for Public Works and Human Settlements 2017 *Modul of Disaster Management Flood Disaster Management Training (*Bandung: Water Resources and Construction Education and Training Center)

[6] The Indonesian National Board for Disaster Management 2016 *The Indonesian Disaster Risk (*Jakarta: BNPB) p.-

[7] The Indonesian Ministry of Agraria and Spatial Planning 2006 *Depok City Towards a Disaster Resilient and Climate Change-Resilient City* (Jakarta: Directorate General of Spatial Planning)

[8] Dadan R 2021 *Presentation of Proposals for Handling Urgent Needs for FY 2021 in the Water Resources Sector* (Depok: Depok City Public Works and Spatial Planning Service)

[9] Depok City Fire and Disaster Management Service 2020 *Flood Emergency Contingency TWG Map FY 2020* (Depok: Depok City Fire and Disaster Management Service)

[10] Depok District of Public Works and Human Settlements 2020 *Landslide and Inundation Handling Efforts in the Field of Natural Resources FY 2020* (Depok: Depok District of Public Works and Human Settlements)

[11] Sastroasmoro S, Ismael S 2014 Fundamentals of Clinical Research Methodology (Jakarta: CV Sagung Seto) p 149-150.

[12] Hildayanto A 2020 Knowledge and preparedness of the community against flood disasters *Higeia Journal of Public Health Research and Development* 577-586

[13] Gill J C, Ekbal H, Bruce D M 2020 *Workshop Report: Multi-Hazard Risk Scenarios for Tomorrow's Cities* Tomorrow's Cities Urban Risk in Transition (London: UK Research and Innovation GRCF)

[14] Kar N 2016 Care of older persons during and after disasters: meeting the challenge. *Journal of Geriatric Care and Research* **3** 7-12

[15] Lesley G 2017 Social determinants of health, disaster vulnerability, severe and morbid obesity in adults: triple jeopardy? *International Journal of Environmental & Public Health* **14** 1-10.

[16] Erita, Donny M, Adventus M R L B 2019 *Emergency and Disaster Management Learning Materials Book (*Jakarta: UKI) p. 6.

[17] Dodon 2013 Indicators and behavior of community preparedness in traditional settlements in anticipation of various phases of flood disaster http://www.sappk.itb.ac.id/jpwk/wp-content/uploads/2014/02/Jurnal-9-

[18] Nurhidayati I, Bahar K 2018 Family support increases the preparedness of the elderly in the face of volcanic disasters . *Jurnal Keperawatan Respati Yogyakarta* **5** 302-308

[19] Lenawida 2011 *The Effect of Knowledge, Attitudes, and Support of Family Members on Household Preparedness in Facing Earthquake Disasters in Deyah Raya Village, Syiah Kuala District, Banda Aceh City* Tesis (Medan:Universitas Sumatra Utara Medan)

[20] Fatmah 2010 *Geriatric Nutrition* (Jakarta: Penerbit Erlangga) p 15

[21] Jannah I, Daniah, Nur A 2021 Analysis of elderly preparedness to face flood disaster in Kebalen Village, Jambi *STIKES Mitra Ria Husada* X 1-11

[22] The Indonesian National Board for Disaster Management 2012 *Community Based Early Warning System Guidelines* (Jakarta: BNPB)

[23] Akhirianto N A 2018 Community knowledge and preparedness for flood disaster in Bekasi City (case study: Pondok Gede Permai Housing) *Jurnal Alami* **2** 63-72.
[24] Islam FMA, Chakrabarti R, Dirani M 2014 Knowledge, attitudes and practice of diabetes in rural Bangladesh: The Bangladesh Population Based Diabetes and Eye Study (BPDES). *Plos One* **9** 1-11
[25] Oktarina, Fachrudi H, Made A B 2009 The relationship between respondent characteristics, regional conditions and knowledge, attitudes towards HIV/AIDS in Indonesian society *Buletin Penelitian Sistem Kesehatan* **12** 362–369
[26] Nastiti R P, Rafiah M P, Acim H I 2021 Factors related to community preparedness in dealing with flood disasters in Kebon Pala Village, East Jakarta *Poltekita: Jurnal Ilmu Kesehatan* 48-56
[27] World Health Organization 2002 *Gender and Health in Disasters* (Geneve: WHO)
[28] Notoatmojo S 2007 *Health Behavioral Science* (Jakarta: Rineka Cipta)
[29] Anderson, LW and Krathwohl DR 2017 *Foundation Framework for Learning, Teaching, and Assessment* (Yogyakarta: Pustaka Pelajar)
[30] Firmansyah I, Rasni H, Rondhianto 2014 The relationship between knowledge and preparedness behavior in dealing with floods and landslides in adolescents 15-18 years old at Al-Hasan Kemiri High School, Panti District, Jember Regency https://repository.unej.ac.id/bitstream/handle/123456789/60652/Iman%20Firmansyah.pdf?sequence=1
[31] Puspitasari P, Lukita D, Rismayanti 2021 Cadre development in an effort to increase cadre readiness to face flood disasters in one of the villages in Baleendah District Bina Sehat Masyarakat **1** 14-18.
[32] Teja M, Research Center of the Indonesian House of Representatives Expertise Board 2021 Community preparedness for vulnerable groups in dealing with natural disasters in Lombok https://berkas.dpr.go.id/sipinter/files/sipinter-773-016-20200708131842.pdf

Climate Change Induced Landslide Susceptibility Assessment - for Aiding Climate Resilient Planning for Road Infrastructure: A Case Study in Rangamati District, Chittagong Hill Tracts, Bangladesh

I P K Wijaya*[1], A Joshi[1], M N Alam[1], S Jayasinghe[1], and N Laila[1]

[1] Asian Disaster Preparedness Center, Bangkok, Thailand

*Corresponding author e-mail: putu.wijaya@adpc.net

Abstract. Climate change induced extreme hydro-meteorological conditions have increased the prevalence of landslides in the hilly and geologically fragile region of Chittagong Hill Tract districts (including Rangamati) in Bangladesh. These landslides have attributed to significant damages to transportation infrastructures such as roads and bridge. In this study, we investigated the susceptibility of landslides due to extreme rainfall events under different climate change scenarios in Rangamati district. We developed high-resolution 1km x 1km downscaled extreme rainfall projections under RCP 4.5 and RCP 8.5 scenarios for baseline period 1976-2005 and for future time horizons 2030s, 2050s, and 2080s. Based on these extreme rainfall scenarios, the combination of the Frequency Ratio (FR) and Analytical Hierarchy Process (AHP) techniques were applied to map and analyse the landslide susceptibility maps. Nine multi-variate factors contributing to the landslides were considered including terrain slope, aspect, elevation, lithology, soil, distance from the lineaments, distance from the stream, land use and mean annual rainfall in four different time periods for scenario RCPs. Further, an Area Under the Curve (AUC) approach was used to evaluate the quality of the model A total of seven landslide susceptibility maps were developed and classified into five susceptible classes. The models were validated using the Receiver Operating Characteristic curve (ROC) approach, which showed a satisfactory result of 80-86 percent accuracy.

1. Introduction

In 2017, a landslide in Rangamati district caused severe damage and degraded the region's geomorphology, increasing the danger of future slope failure. The risk of slope failure has been further exacerbated due to the extreme rainfall events caused by climate change. We developed high-resolution 1km x 1km downscaled extreme rainfall projections under selected extreme GCMs from the Coupled Model Inter-comparison Project Phase 5 (CMIP5), Global Climate Models (GCMs), with representative concentration pathways (RCP 4.5 and RCP 8.5) scenarios for baseline period 1976-2005 and future time horizons 2030s, 2050s and 2080s. Based on these extreme rainfall scenarios, we modelled and mapped landslide susceptibility by merging the Frequency Ratio (FR) and Analytical Hierarchy Process (AHP) techniques. The developed approach applies Geographic Information System (GIS) as a tool along with open data to develop the landslide susceptibility models. The study aims to obtain a better understanding of the controlling and triggering factors of landslides and the mutual relationships among controlling, triggering factors and the spatial distribution of landslide susceptibility level for current climate condition and future time horizons, i.e., the 2030s, 2050s and 2080s with RCP 4.5 and RCP 8.5 scenarios.

1.1. Description of the study area

Rangamati district is situated at the South-eastern side of Bangladesh. Rangamati is located in between latitudes of 22°27′ and 23°44′ N, and the longitudes of 91°56′ and 92° 33′ E. Rangamati district encompasses 6116.11 km^2 which is considered to be the biggest district in the country, in relation to land size [1]. Geologically, the study area is part of folded belt area and regionally is dominated by two major geological group formations: Surma Group and Tipam Group [2]. Rangamati district (zila) consists of 10 sub-district (upazila, Figure 1). The rural roads in Bangladesh are divided into three types i.e., upazila, union, and village roads, and the spatial distribution of rural roads are shown in Figure 1. There are two types of pavements for rural roads in Rangamati District, 4310.27km of earthen roads and 629.92km of paved roads with grand total of 4940.19km [3].

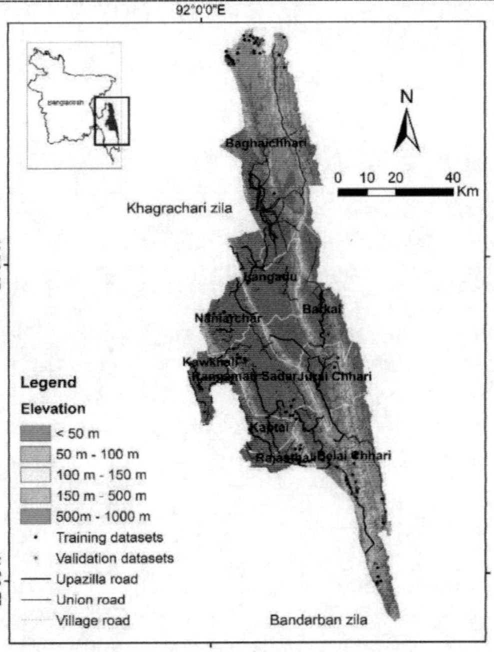

Figure 1. The digital map shows the location, topography and distribution of rural roads of study area.

1.2. General description of the landslides in Rangamati District

The majority of landslides in Bangladesh occurred during the monsoon season as a result of heavy rainfall (> 40 mm/day) in a short period of time (2–7 days, [4]). Based on [1], ± 40% of landslide type in Rangamati was the debris flow. [2] mentioned that there were two dominant types of landslides in this area, which were slide and flow. Based on [1], 60 percent of the landslide's size in Rangamati district was 100 m^2. In accordance with Hungr's categorization [5], most of the landslide types in Rangamati district can be categorized into slides and flows.

2. Datasets and Method

2.1. Landslide inventory map

The preparation of inventory map is an important part for landslide susceptibility mapping. The landslide inventory map aids in determining the link among previous landslide occurrences and various controlling and triggering parameters. In this study, the observations and digitation of landslide areas were collected through visual interpretations of existing landslides on high resolution satellite imagery, using Google Earth images and references from available studies including [1] and [6].

The change of morphological surface and the presence of debris were considered to identify landslides. 237 landslides were mapped in study area. The distribution of landslides in landslide inventory was randomly divided into two subsets, the training set area (80 percent) and the test/validation set area (20 percent) as suggested by [7], Figure 3. The landslide size in study area varied from 163 m^2 to 0.04 km^2. The existing landslides mapped through visual interpretation covered an area of about 0.62 km^2.

2.2. Extreme rainfalls projections under climate change scenarios

The process of proposed climate impact modeling for identification of extreme events at the study area comprise of six methodological stage as shown in the Figure 2.

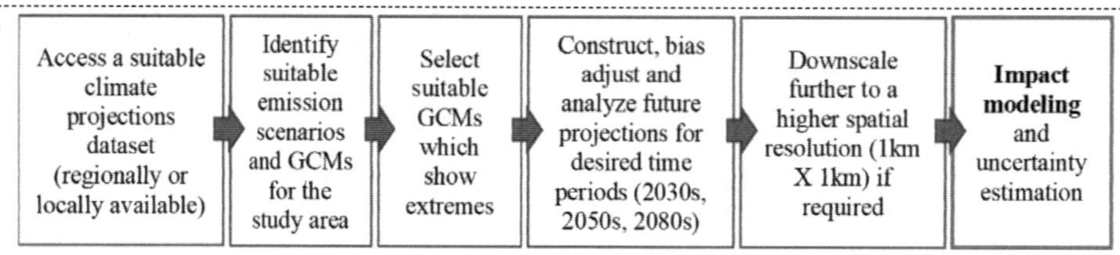

Figure 2. Schematic of climate impact modeling for assessing risk from extreme landslides at the watershed scale using downscaled GCMs.

Future climate scenarios for 2030 (averaging 2016-2045), 2050 (averaging 2036-2065), and 2080 (averaging 2066-2095) are developed in the research area based on present climate (rainfall and mean temperature from 1976-2005) across the same study region during the wet season. Because landslides are more common during the rainy (monsoon) season, it is found to be an excellent time to pick suitable GCMs. The pattern of rainfall projections under different future time horizon and climate change scenarios are shown in Figure 3.

2.3. Landslide controlling and triggering factor maps

With the combination of the aforementioned work of literatures and the nature of the study area, nine different thematic layers together with terrain slope (°), distance from lineaments (m), distance from streams (m), elevation (m), lithology, soil, land use, aspect, and mean annual rainfall in different climate change scenarios were taken into consideration. The landslide controlling and triggering factor's maps were rasterized with 12.5 m x 12.5 m pixels and each raster was reclassified into appropriate thematic classes.

2.4. Frequency ratio method

Frequency ratio (FR) values represent the relationship between landslide event and the classes of every single controlling and triggering parameters. The landslide susceptibility can be evaluated from the spatial relationship between the controlling, triggering factors and landslides occurrences. The greater the FR ratio, the stronger the link among the landslide causing parameter and the incidence of the landslide [8]. The frequency ratio value is computed like described in the following formula:

IOP Conf. Series: Earth and Environmental Science 1091 (2022) 012010 doi:10.1088/1755-1315/1091/1/012010

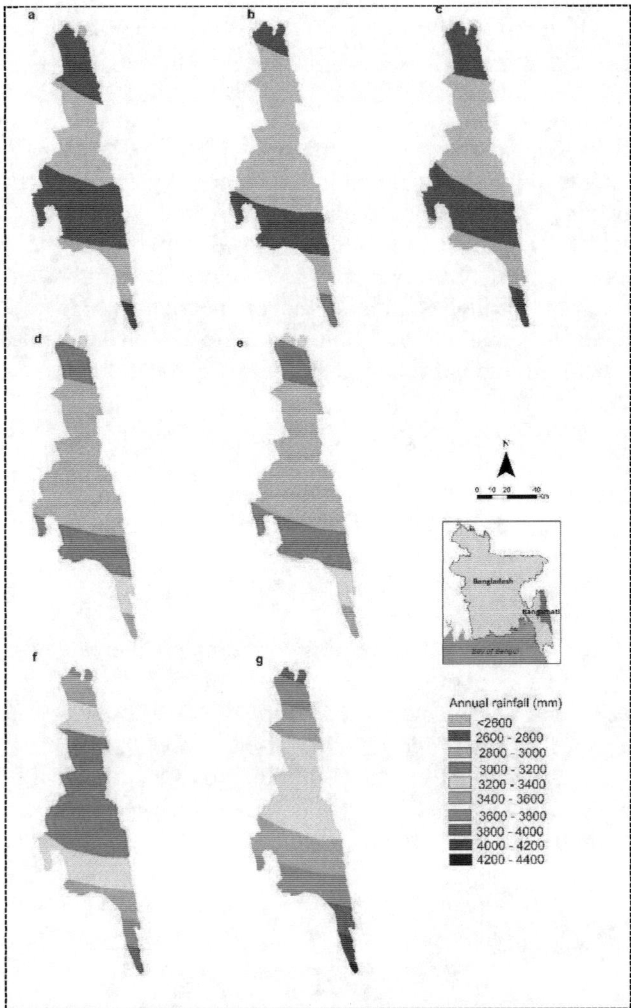

Figure 3. Mean annual rainfall for baseline (1976-2005, a), future time horizon 2030s with RCP 4.5 (b) RCP 8.5 (c). Time horizon 2050s, with RCP 4.5 (d) and 8.5 (e). Time horizon 2080s, with RCP 4.5 (f) and 8.5 (g).

$$FRi = \frac{Ncell\ (Si)/Ncell\ (Ni)}{\sum Ncell\ (Si)/\sum Ncell\ (Ni)}$$

where Ncell (Si) is the amount of grid cells in class I that have been recognized as landslides and Ncell (Ni) is the total amount of grid cells in class I in the area. The overall amount of grid cells identified as landslides in the entire region is ∑Ncell (Si), whereas the overall amount of grid cells in the entire region is ∑Ncell (Ni).

2.5. Analytical and hierarchy process method

The analytical hierarchy process avoids the difficulties associated with random weights and ratings systems by allowing judgment to influence relative importance features or weights rather than randomly assigning those features [9]. In this study, the AHP method was used for creating a map of landslide susceptibility zonation.

The AHP technique used the consistency ratio (CR), that is a ratio between the matrix's consistency index and the random index. CR stands for the likelihood that the matrix judgments were created at random, as follows:

$$CR = \frac{\mathrm{CI}}{\mathrm{RI}}$$

based on the arrangement of Malczewski's matrix [10], RI is the average of the resultant consistency index, and CI is the consistency index.

2.6. Integrated weighted index

There are three important processess in the integrated weighted index method. The first step is to determine the relative importance of landslide controlling and triggering parameters by using AHP technique (Figure 4). The second step is by describing the mutual relationship between the location of landslide and the controlling & triggering parameters using the FR method. The third step, by using the weighted overlay tool in GIS environment, the landslide susceptibility maps were created. By combining the FR and AHP techniques, the integrated weighted model examined the connection among the conditioning parameters as well as the effect of each landslide conditioning parameter on landslide frequency (Figure 4). The following formula can be used to obtain the integrated weighted index:

$$I = \sum_{i}^{m} (\mathrm{W}i \times \mathrm{FR}i)$$

where m is the number of controlling and triggering factors, Wi denotes the weight assigned to each conditioning factor using the AHP method, and FRi denotes the FR value assigned to the conditioning factor using the FR method.

2.7. Validation method

Validation of the model predictions is critical for landslide susceptibility mapping. The receiver operating characteristics curve (ROC) approach [11] was adopted to evaluate the performance of the integrated weighted index model. The ROC curve is a plot of the model prediction's sensitivity (percent of true positives) against the complement of its specificity (proportion of false positives). The true positive rate (TPR) was plotted in contradiction of the false positive rate (FPR) on the ROC curve, with TPR on the y-axis and FPR on the x-axis (Figure 6). The area under ROC curve AUC ranged between 0.5 and 1.0, where value of 1.0 suggested that the model performed perfectly, but a value near 0.5 indicated that the model performed poorly.

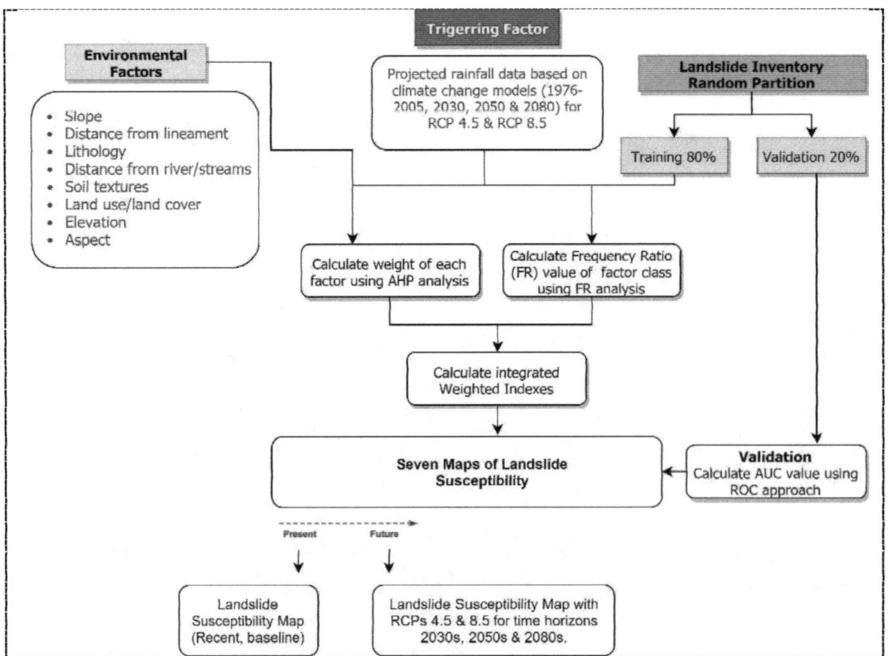

Figure 4. The flow chart depicted the study approaches.

3. Results

3.1. Landslide susceptibility maps

Based on the pairwise comparison matrix of AHP, the weight of slope was the highest (0.205 suggesting that slope has the greatest effect on the frequency of landslides, the weight of distance from lineaments and rainfall the second and third highest (0.198 and 0.174, respectively), followed by lithology (0.134), soil (0.067), distance from streams (0.061) and finally the weights of aspect, elevation, and land use were the lowest (0.054), This suggested that these three parameters had the lowest impact on the prevalence of the landslide. The consistency ratio (CR) of the matrix of paired comparisons between the 9 controlling and triggering factors in susceptibility map is 0.012. A consistency ratio of 0.10 or less is a reasonable degree of consistency [12].

Based on FR analysis, the highest frequency of landslide occurred on the slope class 30° to 60° (2.216), distance from lineaments class <500 m (1.075), mean annual rainfall class 3800-3850 (7.734), Lithological type is the yellowish-brown to brown, fine to medium grained pebbly and cross-bedded sandstone (2.503), soil type is the deep brown soil (3.307), distance from stream class <500 m (1.154) the east terrain aspect (2.410), elevation range 100-150 m (2.317), and the land use type is trees (1.137). With the help of weighted overlay tools in GIS environment, a total of seven landslide susceptibility maps were generated from two RCP 4.5 and RCP 8.5 for time horizons 2030s, 2050s, and 2080s and baseline (1976-2005). The landslide susceptibility maps were classified into five susceptible classes using the natural breaks classification in GIS environment *viz.* very low , low, moderate, high, and very high (Figure 5). The total area of landslide susceptibility was quantified by calculating the area geometry in GIS environment. Based on the acquired landslide susceptibility maps, the area for each level of susceptibility in scenarios shows different values (Table 1). Based on RCP scenario 4.5 for time horizon 2030, 2050, and 2080, the very high and high susceptibility level zones, occupied approximately 42%, 49%, and 53 % of study area, respectively. The highest area value (3477 km^2) of the high to very high susceptibility level was achieved by the susceptibility map of RCP scenario 8.5 for time horizon 2080. Concerning the susceptibility map of RCP 4.5 and 8.5 for time horizon 2030, 2050, and 2080, most parts of high to very high-level susceptibility zones are mainly distributed in the North part of Baghaichhari Upazila, East part of Naniarchar, Kawkhali and Rangamati sadar Upazila, Northeast part of Kaptai Upazila, center, North, Southeast part of Belai Chhari Upazila, West part of Rajasthali and Jura Chhari upazila, East and North part of Barkal Upazila and the almost entire area of Rajasthali and Bella Charri upazila (only for time horizon 2080 with RCP 4.5 and 8.5 cases, Figure 5).

3.2. Validation of landslide susceptibility maps

A sample subset of 44 landslides were used as a validation subset from the total (237 landslides sample set) to validate the reliability of the landslide susceptibility model. To examine the integrated weighted index model's predictive accuracy, a validation dataset (i.e., 44 landslides) was used to to calculate predictions percentages. The AUC values for landslide susceptibility map (current climate and future projected scenarios) varied from 80 – 86%, which indicated a reliable predicting capability of the integrated weighted index model adopted for this study. Figure 6 showed two examples of ROC of landslide susceptibility map for time horizon 2080s with RCP 4.5 and RCP 8.5.

Table 1. The total area of landslide susceptibility in Rangamati district for current climate (baseline), future time horizons with different RCPs scenarios.

Susceptibility area	Baseline (km^2)	RCP 4.5 (km^2)			RCP 8.5 (km^2)		
	1976-2005	2030s	2050s	2080s	2030s	2050s	2080s
Very low	261	261	213	210	261	213	173
Low	533	529	405	384	526	403	330
Moderate	2511	2496	2241	2023	2483	2236	1668
High	1791	1808	2068	2116	1824	2061	2294
Very high	540	554	721	914	553	733	1183

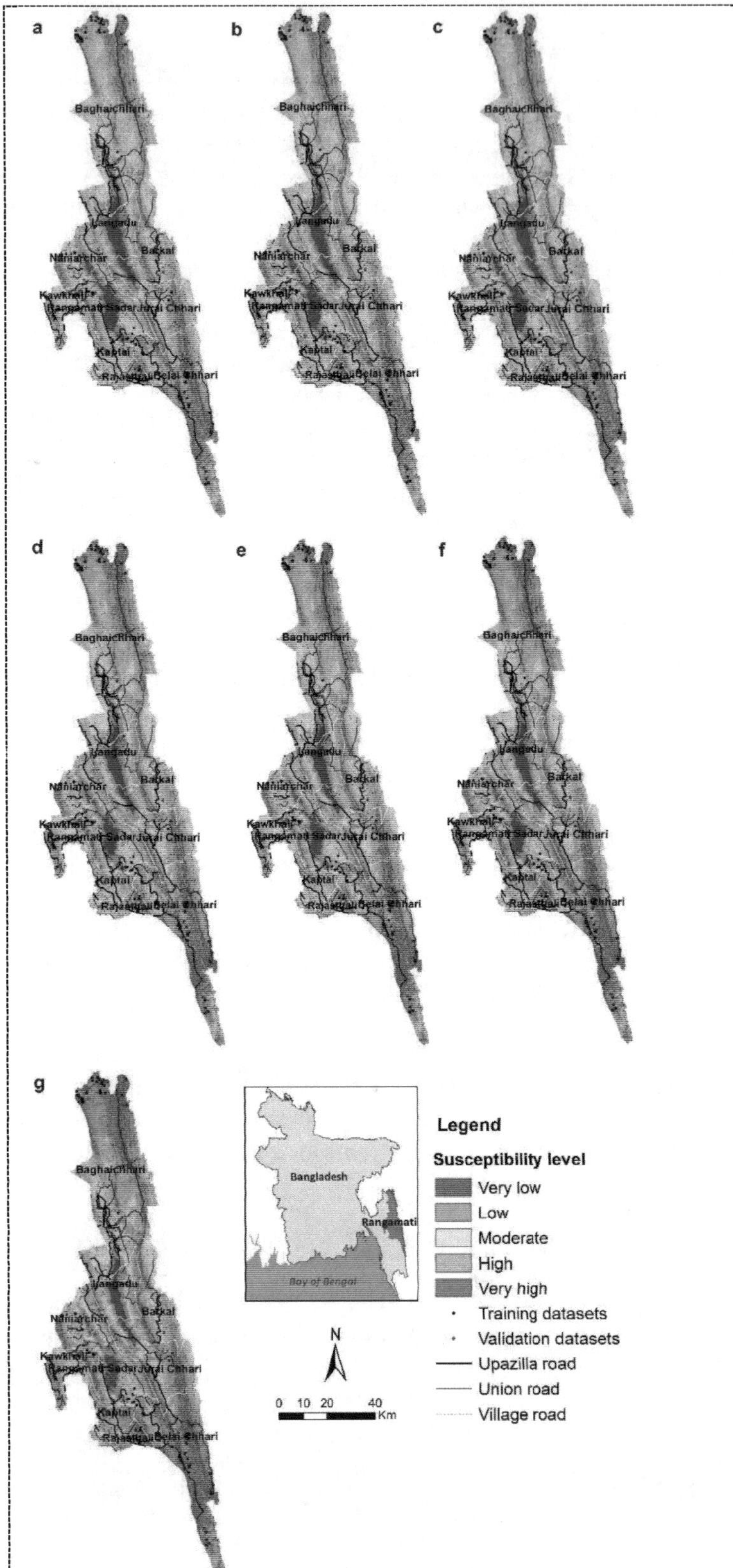

Figure 5. a) Recent landslide susceptibility map, projected landslide susceptibility map for time horizon 2030s, with RCP 4.5 (b) and 8.5 (c). Time horizon 2050s, with RCP 4.5 (d) and 8.5 (e). Time horizon 2080s, with RCP 4.5 (f) and 8.5 (g).

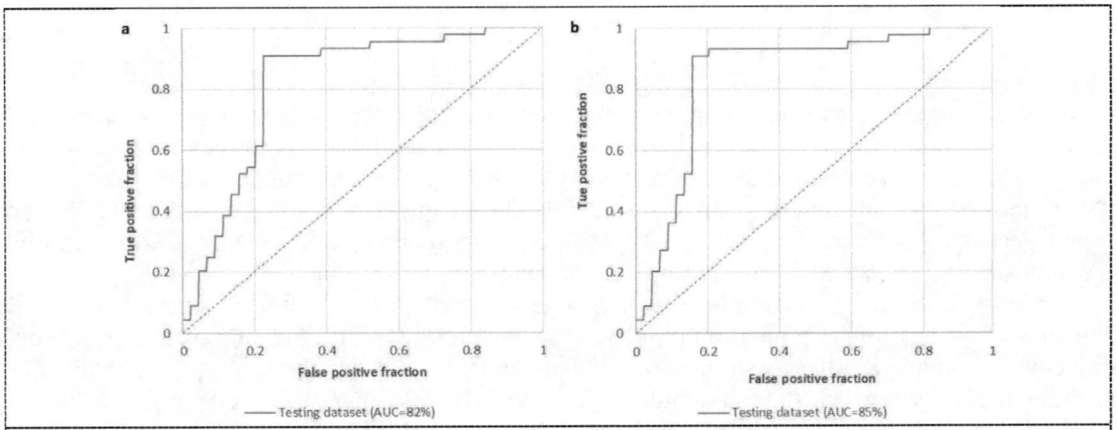

Figure 6. The integrated weighted index model's forecast effectiveness. Two examples of ROC of landslide susceptibility map for time horizon 2080s with RCP 4.5 (a) and RCP 8.5 (b).

4. Discussion

Compared to the landslide susceptible areas modelled for the current baseline climate conditions (1976-2005), the landslide susceptible areas are significantly increased in extreme rainfall conditions for the future climate change scenarios for both RCP 4.5 and RCP 8.5. Results revealed that the regions with "Very High" landslide susceptibility increased from 374km^2 to 643 km^2 in the 2080s. Thus, it is predicted that landslide susceptibility will increase in future time horizons (in 2030s, 2050s, and 2080s). RCP 8.5 showed larger susceptible areas than RCP 4.5 for the future time horizons, as RCP 8.5 models predicted heavier rainfall than RCP 4.5. In future periods, the southern part of Rangamati is likely to have higher landslide susceptibility area compared with the same under current condition. In particular, it is predicted that landslide susceptibility in the northern area would increase in the 2050s (2036-2065) and 2080s (2066-2095) in both RCP scenarios. In both scenarios, the landslide susceptible areas for 2080s were predicted to expand significantly to the area adjacent to Bella Charri upazila. In the case of future time horizon 2080 with RCP 8.5, significant landslide susceptibility was expected to occur at Baghaichhari Upazila.

The susceptibility map for time horizon 2080 with RCP 8.5 represents the worst-case scenario for the landslide. Although the majority of the study area (43-44%) was predicted to be located on "Moderate" susceptible to future landslides, (32%) was indeed on "High" susceptible category. The unsafe areas mentioned in the previous section, require immediate mitigation action. Reactivation of existing landslide sites and new landslide may occur particularly along lineaments. The findings revealed the dependability (indicated by the AUC value that showed a satisfactory result of 80 – 86 percent accuracy) and practicality of the integrated weighted index model in regional landslide susceptibility mapping.

Based on those results, it showed that under climate change scenarios, exposures to landslides would substantially increase for all rural roads in Rangamati district. Thus, the next step, further studies will be conducted in 2022. The overall maximum kilometers (km) of rural road networks exposed to landslide susceptibility levels under climate change scenarios will be analysed. A detailed landslide hazard assessment was planned for future studies, including landslide magnitudes analysis along representative road segments to better understand the landslide types and their physical processes (volume, flow direction, size, run out distance, & impact area of the landslide) to support designing the landslide mitigation measures.

Finally, the landslide susceptibility maps produced in the study is useful for planners and engineers for planning and designing roads and other infrastructures to obtain a better understanding of future climate-induced landslide and their potential impacts on road networks for planning new roads and rehabilitation/retrofitting of the existing roads in Rangamati district.

5. Conclusion

The landslide susceptibility modelling and mapping spatially locates the regions susceptible to landslides due to multi-variate factors under the extreme rainfall conditions for the current and projected future time horizons. Seven landslide susceptibility maps were developed representing scenarios for RCP 4.5 and RCP 8.5 for future time horizons 2030s, 2050s, and 2080s and the current (1976-2005) baseline period. The significant increase between 374km^2 to 643 km^2 in "Very High" landslide susceptible area between the baseline period and the long-term horizon in 2080 (for RCP 8.5 scenario) indicates that the extreme rainfall due to climate change will attribute to substantial increase in landslides at Rangamati district. Higher proportion of the land falling under "High" landslide susceptibility under future climate change scenarios also indicate limited available land for planning and development of road infrastructure. These landslide susceptibility models and maps enables decision making in planning of road infrastructure through relatively safer areas indicated by "Very Low", "Low" and/or "Moderate" susceptible areas. Planning, designing and construction of road infrastructure in the "Moderate" susceptible areas may require considerations of road design parameters to adapt and/or mitigate the impacts of landslide hazards. The "Moderate" susceptible areas may also be managed through climate adaptive measures including green-grey infrastructures such as incorporation of Nature based Solutions (NbS) along with engineering solutions. These landslide susceptibility maps will further support vulnerability assessment of rural roads in the district and identification of critical road segments for planning and implementing climate adaptation measures and activities to enhance the climate resilience of the road infrastructure in the district.

Acknowledgment

This study was done under Climate Adaptation and Resilience (CARE) for South Asia Project's Component 2 being implemented by Asian Disaster Preparedness Center (ADPC) with the support of World Bank. We advise disclaimer that the political/administrative boundary representation on the maps is not necessarily authoritative.

References

[1] Abedin J, Rabby Y W, Hasan I 2020 An investigation of the characteristics, causes, and consequences of June 13, 2017, landslides in Rangamati District Bangladesh *Geoenviron Disasters* **7** 20
[2] Rabby Y W, Li Y 2019 An integrated approach to map landslides in Chittagong Hilly Areas, Bangladesh, using Google Earth and field mapping *Landslides* **16** 633–645
[3] LGED Road and Market Database: ICT Unit LGED 2020 Available from: https://oldweb.lged.gov.bd/ViewRoad2.aspx [Accessed 20th December 2021]
[4] Khan Y A, Lateh H, Baten M A 2012 Critical antecedent rainfall conditions for shallow landslides in Chittagong City of Bangladesh *Environ Earth Sci* **67** 97–106
[5] Hungr O, Leroueil S, Picarelli, L 2014 The Varnes classification of landslide types an update *Landslides*
[6] Sifa S F, Mahmud, T, Tarin, M A, & Haque, D M E 2020 Event-based landslide susceptibility mapping using weights of evidence (WoE) and modified frequency ratio (MFR) model: a case study of Rangamati district in Bangladesh *Geology Ecology and Landscapes **4:3**, 222-235*
[7] Getachew N, Meten M 2021 Weights of evidence modeling for landslide susceptibility mapping of Kabi-Gebro locality, Gundomeskel area, Central Ethiopia *Geoenviron Disasters* **8** 6
[8] Lee S, Pradhan B, 2007 Landslide hazard mapping at Selangor Malaysia using frequency neural network model at Cameron Highland Malaysia *Landslides* **7** 13–30
[9] Yalcin A 2008 GIS-based landslide susceptibility mapping using analytical hierarchy process and bivariate statistics in Ardesen (Turkey): Comparisons of results and confirmations *Catena* **72**
[10] Malczewski J 1999 *GIS and multicriteria decision analysis* (New York: Wiley)

[11] Brenning A 2005 Spatial prediction models for landslide hazards: review, comparison and evaluation *Nat. Hazards Earth Syst* Sci **5** 853–862
[12] Yi Y, Zhang Z, Zhang W, Xu Q, Deng C, and Li Q 2019 GIS-based earthquake-triggered-landslide susceptibility mapping with an integrated weighted index model in Jiuzhaigou region of Sichuan Province, China Nat. Hazards Earth Syst Sci **19**

IOP Conf. Series: Earth and Environmental Science 1091 (2022) 012011 doi:10.1088/1755-1315/1091/1/012011

Heritage Community Resilience: The Experience of Stakeholders in Calle Real, Iloilo City, Philippines

Rhodella A. Ibabao*[1], Vicente Balinas[1], Jerilee Camena[1], Rene Trance[1], Gay Defiesta[1], Mary Earl Grio[2], Andres Winston Oreta[3], & Suzette Penaredondo[1]

[1]University of the Philippines Visayas Iloilo – New Administration Building, Miagao, 5023 Iloilo, Philippines
2Central Philippine University – Lopez Jaena Street, Jaro, Iloilo City, 5000 Iloilo, Philippines
[3]De La Salle University – Manila, 2401 Taft Avenue, 1004 Manila, Philippines 2Central Philippine University – Lopez Jaena Street, Jaro, Iloilo City, 5000 Iloilo, Philippines

*Corresponding author e-mail: raibabao@up.edu.ph

Abstract. Cultural heritage is a significant part of the UN Agenda 2030 in making cities safe and resilient. Yet, the role of cultural heritage on resilience has not been sufficiently investigated. The study contributes to the discussion of the role of knowledge and practices in the prevention and recovery from risks within the frameworks of heritage preservation and community-based disaster management approach. The research aims to describe the knowledge about the heritage structures as well as the heritage-related practices of stakeholders employed in Calle Real, Philippines, an urban heritage zone in Iloilo City, Philippines. Through a social survey with 200 stakeholders, key results showed that respondents are aware that the unique character of Calle Real is worth preserving. Moreover, the multiplicity of uses of the buildings makes these prone to rapid wear and tear. Also, a majority of the respondents are aware of the buildings' susceptibility to fire, seismic hazards, and strong winds. There is a gap, however, between the respondents' awareness of the risks of the buildings and their heritage preservation practices. Few respondents reported being affiliated with heritage advocacy groups or to take part in actual conservation activities. There are possible ways to address the disparities in knowledge and practices with the goal of enhancing heritage resilience to natural & man-made hazards.

1. Introduction

The past decade has seen the occurrence of large-scale disasters in many countries that have caused not only loss of lives but also widespread damage to the cultural heritage of these towns and cities [1]. The Great East Japan Earthquake and Christchurch, New Zealand Earthquake in 2011 and the Bohol earthquake in 2013 are just some examples of the extreme vulnerability of cultural heritage and the lack of resources and planning in place to protect it [1, 2]. Cultural heritage is also vulnerable to the adverse impacts of rapid urbanization and the inherent risks faced by dense urban areas so that it is necessary to develop specialised approach to risk management of cultural heritage in urban areas [3]. Moreover, local communities are often excluded from heritage reconstruction processes, which are frequently led by external donors and agencies and national authorities. Yet, local knowledge and engagement is a unique resource for resilience and long-term heritage conservation and management [4]. In academic debates, the question of the relationship between heritage and resilience has been evolving [5]. One way to address the debate is through studies that analyse the results of the processes that occurred before and after the disasters and have highlighted the contribution of the local material culture in prevention and recovery from risks ([6]. In the prevention phase, for example, the role of knowledge of traditional construction techniques or traditional prevention strategies resulting from subsequent trial and error in the management of known and expected risks is underlined [6]. In light of these challenges, making use of intangible heritage knowledge and practices apart from technical and structural solutions to disaster recovery is of importance within the overall planning and management frameworks [4].

This paper contributes to the understanding of stakeholders' awareness of heritage structures and of their heritage-related practices for the overall improvement of the outcome of disaster management in a heritage zone. The paper shows the results of a social survey among stakeholders whose main employment is in Calle Real, Iloilo City which is an urban heritage zone in one of the highly urbanized cities in the Philippines.

2. Linking Community Resilience and Cultural Heritage

The scientific literature [5, 6] has reinvigorated the discussion on the significance of cultural heritage in disaster risk-reduction, post disaster rehabilitation and recovery. Discussions have focused on the need to safeguard cultural heritage and its important role for the well-being and quality of life of people. There are also studies that have highlighted the contribution of the local material culture in prevention and recovery from risks. In the prevention phase, for example, the role of knowledge of traditional construction techniques or traditional prevention strategies resulting in subsequent trial and error in the management of known and expected risks is emphasized [6]. There are also researches on the capacity of cultural heritage to make and innovate communities in a proactive process aimed at preventing, coping with and recovering from disturbances and/ or disasters [5,6]. An example is the partnership with community heritage groups on creating digital archives of cultural artefacts and of information system to help develop more resilient communities [6]. Still, there is little reference in the programmes of governments to the role of cultural heritage in building community resilience [5]. In the context of heritage places, tourism studies show very little is known about public awareness of the management, importance, or designation of historic sites [7].

Community residents have a strong stake in a local heritage site and the Community-based Disaster Risk Management (CBDRM) framework has been used to mobilize residents to participate in heritage conservation. CBDRM can be defined as inclusive and owned community driven processes aimed at the direct and continued involvement of at-risk communities in the decision-making process of disaster risk reduction. It assumes that local knowledge is comparable to scientific knowledge and the community works together with in unison with other outside organizations and agencies. The Philippines has recognized the importance of building community resilience to disasters to reduce loss of lives and other community assets. This is shown through the National Disaster Risk Reduction and Management Plan (NDRRMP) for 2011-2018 and the creation of a National Disaster Risk Reduction and Management Council [8]. At the local level, there are equivalent plans and councils that align with higher level plans and bodies.

3. The Study Area

Calle Real is located in Iloilo City which is one of the oldest cities in the country and is a Highly Urbanized City (HUC) with a population of 457,626 residents in the 2020 census [9]. Calle Real is the old Central Business District (CBD) of the city and was declared by the National Historical Commission of the Philippines as a heritage zone on August 8, 2014. The heritage buildings found in the heritage site reflect the architectural designs of those times: the Beaux Arts style, Art Deco, art modern, post-art modern, & neo-classical designs [10]. The Iloilo City Hall and commercial banks are located in the old CBD that contributes to the everyday characteristics of the place where people engage in government & commercial transactions. Moreover, the buildings in Calle Real are low-rise (5 storeys and below) and have mixed uses. The lower floors are for commercial activities and the upper floors are used as residential and storage facilities. In general, the development and design guidelines of the old CBD are embodied in the Implementing Rules & Regulations (IRR) thru EO 46-2009 [11]. The Iloilo City

Cultural Heritage and Conservation Council (ICCHCC) implements the IRR, a body that was created thru Ordinance No. 2000-054 [12]. There are currently 23 buildings of heritage value that can still be found in Calle Real. The heritage zone is just a few meters away from Iloilo River and is a 20-minute boat ride away from Guimaras Island province from the Parola wharf at For San Pedro.

Figure 1. Calle Real is the Old Central Business District of Iloilo City

4. Data Collection Methods

A pre-test survey of the stakeholders' perceptions was first conducted by the authors during the period of March 2021. The actual data collection was conducted at a time when Iloilo City was on Modified General Quarantine Status (MGCQ) in April 2021. The MGCQ has the most relaxed quarantine guidelines in terms of movements of people and business services allowed to operate. Data were collected with 200 stakeholders: 75 city hall, 50 employees of shops, 50 tenants or property managers, and 25 vendors. Purposive and convenient sampling methods were used in the selection of respondents.

The questionnaire is divided into six major sections. Section A is a description of the stakeholders who work and/or reside in Calle real, Iloilo City, particularly their socio-demographic characteristics. Section B is a description of the stakeholders' general activities in the heritage site. Section C is the stakeholders' perceptions about the buildings in Calle Real on strength, stability, and security; about how the buildings represent the Ilonggo culture and history; and the stakeholders' knowledge of cultural heritage and of the buildings' risks. Section D is the stakeholders' willingness to help in the preservation of heritage buildings in the Calle Real. These include educating the public about the value of these buildings, giving money, and cleaning the building among other things. Section E describes the current or earlier practices of stakeholders related to the preservation of heritage buildings in Calle Real and of similar structures. Section F consists of data about the tenants/property managers.

Table 1. Frequency and Percentage Distribution of Respondents by Sex and by Sectoral Membership

Sectoral Membership	Male		Female		Total	
	f	%	f	%	f	%
City LGU Employee	36	36.7	39	38.2	75	37.5
Employee of the Shops	23	23.5	27	26.5	50	25.0
Tenants/Property Managers	26	26.5	24	23.5	50	25.0
Vendors	13	13.3	12	11.8	25	12.5
Total	98	100.0	102	100.0	200	100.0

The survey was supplemented by desk review and informal face-to-face interviews to provide context to the answers. Materials read were the old CBD catalogue, newspaper articles, earlier works of some authors on Calle Real done in 2014 and 2019, and undergraduate thesis.

5. Key Results

5.1. Stakeholders' Socio-Economic Characteristics

The socio-economic profile showed that there are slightly more females than males and majority of the stakeholders have relatively high educational attainment (college education and college degree holders. The mean age is 42 years old, the computed monthly mean income is around Php 29,000.00, and the monthly median income of Php 13,000.00. Males have a higher mean monthly income (Php around Php 13, 700.00) than the females (Php 27, 200.00) but the latter has a higher median monthly of Php 15,000.00 than the former which is Php 12,000.00. The majority of the stakeholders reside outside Calle Real but still within Iloilo City. There are those who live in adjacent towns with a few residing in another island (Guimaras Island Province).

Table 2. Frequency and Percentage Distribution of Respondents by Residence

Current Residence	f	%
Iloilo City	180	90.0
Iloilo Province	16	8.0
Others		
Guimaras	4	2.0
Total	200	100.0

5.2. Calle Real and I

- Majority (61.0%) have full-time employment at the time of the interview. Thirty-six (35%) percent reported to being self-employed while a handful of the stakeholders (3.0%) have part-time employment.
- There are respondents who have worked in Calle Real in less than 5 years while there are those who have worked there for 9 years.
- The nature of economic activities of vendors, tenants, and employees of business establishments are mostly on food, retail, and photocopying/printing services.
- Males have higher reported educational attainment (college graduates) than females (college level). Of the 5 stakeholders who reported to be members of heritage groups, 4 of them are

females. Both male and female stakeholders go to Calle Real 2.6 times in a week during the pandemic period.

- Doing some shopping and eating/dining are relatively popular activities compared to other reported activities among the stakeholders. All (100%) vendors reported to dine/eat-in Calle Real and a high percentage of them do their marketing in Iloilo Central Market (Tienda Mayor). Very few reported to doing bank transactions in Calle Real. The stakeholders did not say where they conduct their bank transactions outside of Calle Real.
- Places visited frequently: the Iloilo Centra Market, drug stores, department stores, stores that sell general merchandise, restaurants, hardware, bank, certain streets, and the plaza/park.

Table 3. Activities of Respondents

Activities	City LGU Employee (n=75)		Employee of Shops (n=50)		Tenants / Property Managers (n=50)		Vendors (n=25)	
	f	%	f	%	f	%	f	%
Shopping	60	80.0	26	52.0	22	44.0	9	36.0
Do government transactions	25	33.3	2	4.0	4	8.0	1	4.0
Grocery shopping	42	56.0	25	50.0	24	48.0	19	76.0
Marketing at Tienda Mayor	29	38.7	15	30.0	18	36.0	21	84.0
Eating/ Dining	45	60.0	27	54.0	16	32.0	25	100.0
Others:								
Buy Medicine	0	0.0	0	0.0	3	6.0	1	4.0
Coffee	1	1.3	0	0.0	0	0.0	0	0.0
Bank transactions	2	2.7	2	4.0	0	0.0	0	0.0
No Response (Did not want to answer)	3	4.0	9	18.0	13	26.0	0	0.0

5.3. Perceptions of Calle Real

- Majority of the stakeholders believe that the buildings in Calle Real are strong, stable, secure, and safe.
- Majority of the stakeholders 'agree' or 'strongly agree' that Calle Real represents the "Ilonggo" culture through the products sold in the area, the architectural styles, and the general pride the place gives them. Moreover, most of the stakeholders 'agree' or 'strongly agree' that the designs of old buildings complement with the newer developments in the area, and that the buildings tell important stories of Iloilo's history. For most stakeholders, the buildings are sources of income and are essential for tourism activities.
- Majority of the stakeholders 'agree' and 'strongly agree' that the buildings can be lost due to fire, earthquake, strong winds during a tropical cyclone. There are a few stakeholders, however, who 'don't know' of the risks of buildings to natural hazards.
- Stakeholders were asked about certain information about Calle Real. The majority of the stakeholders 'don't know' about specific details of the buildings and of incentives to owners of heritage buildings in Calle Real. However, this statement does not hold true about the Iloilo Central Market (Tienda Mayor) where an overwhelming majority agree on information about the structure.

Table 4. Respondents and Their General Perceptions about Calle Real

Perceptions	City LGU Employee		Employee of Shops		Tenants / Property Managers		Vendors	
	f	%	f	%	f	%	f	%
Strength & Stability								
Yes	55	73.3	39	78.0	35	70.0	16	64.0
No	20	26.7	11	22.0	15	30.0	9	36.0
Total	75	100.0	50	100.0	50	100.0	25	100.0
Security								
Yes	57	76.0	37	74.0	40	80.0	16	64.0
No	18	24.0	13	26.0	10	20.0	9	36.0
Total	75	100.0	50	100.0	50	100.0	25	100.0

5.4. Attitudes Towards Cultural Heritage

- Majority of the participants reported to 'agree' and 'strongly agree' of their willingness to advocate for the preservation of heritage buildings and to study, document, or do research about the buildings. Yet, most of the stakeholders (specifically the City LGU employees) provided 'neutral' responses about giving money for the preservation of heritage buildings in Calle Real and their willingness to spend time or to volunteer for the maintenance work for the buildings.
- An overwhelming number of stakeholders 'agree' and 'strongly agree' on the importance of the following characteristics to preserve the buildings:
  - ✓ The appearance and design qualities
  - ✓ Streetscapes and views
  - ✓ Availability of goods and services
  - ✓ Rarity or unique characteristics such as landmarks
  - ✓ Tourism activities (e.g. tours of buildings and of food)

5.5. Practices Related to the Preservation of Heritage Buildings

- Majority of the stakeholders reported 'sometimes', or' rarely', or 'never' in terms of engagement to a variety of heritage preservation activities.

Table 5. Frequency and Percentage Distribution of Respondents by Attitudes to Cultural Heritage

Practices	City LGU Employee		Employee of the Shops		Tenants / Property Managers		Vendors	
	f	%	f	%	f	%	f	%
I encourage the local and international tourists to visit the heritage buildings in Calle Real.								
Never	7	9.3	21	42.0	7	14.0	12	48.0
Rarely	6	8.0	11	22.0	7	14.0	9	36.0
Sometimes	25	33.3	9	18.0	14	28.0	2	8.0
Often	19	25.3	8	16.0	16	32.0	2	8.0
Always	18	24.0	1	2.0	6	12.0	0	0.0
I attend public gatherings or online webinars about Calle Real and the Ilonggo built heritage.								

Practices	City LGU Employee		Employee of the Shops		Tenants / Property Managers		Vendors	
	f	%	f	%	f	%	f	%
Never	38	50.7	41	82.0	38	76.0	8	32.0
Rarely	18	24.0	3	6.0	5	10.0	9	36.0
Sometimes	15	20.0	4	8.0	5	10.0	0	0.0
Often	3	4.0	2	4.0	2	4.0	5	20.0
Always	1	1.3	0	0.0	0	0.0	3	12.0
I participate in conservation and preservation projects for the Ilonggo built heritage.								
Never	42	56.0	45	90.0	35	70.0	20	80.0
Rarely	17	22.7	3	6.0	7	14.0	5	20.0
Sometimes	10	13.3	2	4.0	4	8.0	0	0.0
Often	5	6.7	0	0.0	1	2.0	0	0.0
Always	1	1.3	0	0.0	3	6.0	0	0.0

5.6. Information about the Tenants/Property Managers

- The reported uses of the buildings are of mixed uses: commercial shops/storage/warehouse/ and residential and commercial purposes or residential/commercial/storage.
- The mean no. of employees is 4 and the median is 3 per shop.
- The reported monthly income ranges between Php 5,000.00-Php 12,000,000.00 with a mean income of Php 1,422.00 and a median of Php 570.00. It can be implied that the figures reflect the condition since the pandemic started. The figures could have been higher pre-pandemic period.
- Most of the business establishments have employees that go home after work while a few reported to having staff that resides in the same building.

5.7. Desk Review: Hazards of Calle Real

Based on secondary accounts, Calle Real experienced a great fire on February 7, 1966. It destroyed nearly 75% of the central business district of Iloilo City at that time. As a result, the Ilonggo Chinese Community established the Filipino Chinese Fire Prevention Association Incorporated in the aftermath of the event. Prior to 1966, a supreme court decision (GR No. L-12695, Mar 23, 1959) laid down the rules for hotels, tenements, lodging/boarding houses to avoid fire. Up to now, fire remains to be the most destructive man-made disaster in the city. In 2015, fire assessment was conducted on certain heritage structures in Calle Real. Results showed that buildings do not have a passive fire prevention system. If there are incidents of fire, the usual cause is faulty electrical wiring [12]. Apart from fire, Calle Real is also susceptible to seismic hazard and strong winds brought about by typhoons.

6. Preliminary Insights

There is a disparity between knowledge and practices among the stakeholders particularly on the following matters: heritage value of Calle Real and the buildings, the risks of the buildings, the stakeholders' current practices on heritage activities, and their willingness to engage in preservation efforts. The gap between knowledge and practices can be enhanced through extensive and intensive information dissemination program that the city government can undertake together with the academe and other sectors. By publicizing information using appropriate materials, this can help improve what people know about the built structures. In addition, materials to be produced are targeted to various groups and highlight the distinct characteristics of Calle Real and its buildings. One of the distinct characters are those areas that were reported to be frequented by the stakeholders, particularly for eating/dining and for doing some shopping. The production of such materials, however, implies that thorough and regular inventory and documentation of the engineering, architectural, and socio-cultural conditions of Calle Real should be undertaken. The academe and other sectors working on heritage preservation can be of great assistance in providing science-based and practical knowledge on this aspect.

Survey results showed that at present there appears to have limited engagement of stakeholders to practices that demonstrate the principles of community-based disaster risk management. The stakeholders reported to 'rarely' or to have 'never' participated in heritage preservation and disaster management activities. While there is a community that appreciates heritage preservation and know about disaster management in Calle Real, their responses demonstrate a lack of strategies and operational tools aimed at community-based disaster risk management and community resilience.

Some buildings have no passive fire prevention in place. This maybe because the tenant or building owner does not have sufficient money to install elements to slow or impede the spread effects of fire or smoke. Heritage structures have a different set of guidelines compared to other buildings. It is difficult for the owner/tenant to do repairs or retrofitting without complying to the procedures set for heritage structures. Mechanisms can be set in place to allow some flexibility in doing such types of works without compromising the safety of buildings. At the same time, such mechanisms will not give too much inconvenience to the owners or tenants of the buildings.

The onset of the pandemic has affected the work arrangement of the stakeholders. On a positive note, vehicle traffic has been lessened given the limited mobility of the people who go to Calle Real and the public vehicles that ply the area. However, the stakeholders have reported that the pandemic has affected their income and revenues due to the number of people allowed in the workplace at certain times of the day and the travel and health restrictions imposed by the national government. Thus, there is a need to review and implement work and travel arrangements in the new normal that address income and revenue concerns of the stakeholders.

7. Policy and Research Methods Recommendations

It is necessary to identify factors and strategies that will actually engage the stakeholders to heritage preservation and disaster management activities. Certain strategies used in the co-management of natural heritage resources may be applicable in the context of managing tangible immovable structures. It is also important to integrate heritage resiliency to the overall framework of Community-based Disaster Risk Management (CBRM) to increase the stakeholders' knowledge and enhance activities aimed at heritage resiliency through community-based disaster response. A systematic program or projects to reinvigorate activities in Calle Real with the goal of improving the income and revenues should be put in place. For tenant/heritage tenants that engage in reactive maintenance especially against fire, incentives and loans can be explored to motivate them to invest on preventive maintenance measures.

In terms of methods, adjustment have to be made resulting in the quarantine status. This means that the methods and the types of data that can be collected have to be adjusted by the researchers following the quarantine guidelines. Moreover, ethical considerations have to be discussed by the researchers and the stakeholders depending on the quarantine level and when there is no quarantine status anymore.

Acknowledgments

The Cultural Heritage Resilience & Sustainability to multiple Hazards (CHeRiSH) project was funded by the CHED-Newton Institutional Links Program. The Iloilo City government is further acknowledged for the creation of a Task Force for the field research.

References

[1] RitsDMUCH or the Research Center for Disaster Mitigation of Urban Cultural Heritage, Ritsumeikan University n.d. *Disaster Risk Management of Cultural Heritage in Urban Areas*: *A Training Guide* https://www.preventionweb.net/files/44208_trainingguide1.pdf

[2] Garciano E and Taclibon P 2013 *The October 15, 2013 Bohol Island, Philippines earthquake and its damaging effects: An investigative report* https://committees.jsce.or.jp/disaster/system/files/FSID28_Garciano_DisasterMitigation.pdf

[3] World Bank & GFDR 2017 *Promoting Disaster Resilient KNOWLEDGE NOTE Cultural Heritage* https://documents1.worldbank.org/curated/en/696061511882383371/pdf/121709-WP-P161985-PUBLIC-DisasterResilientCulturalHeritageKnowledgeNoteENWEB.pdf#:~:text=In%20post-disaster%20situations%2C%20cultural%20heritage%20can%20also%20play,from%20a%20variety%20of%20risks%2C%20including%20disaster%20risks.

[4] Heritage and Our Sustainable Future 2021 8 6th December 2021 https://reliefweb.int/sites/reliefweb.int/files/resources/Heritage-Disaster-Response-and-Resilience-Report.pdf

[5] *The National Disaster Risk Reduction and Management Plan NDRRMP 2011-2018* https://www.dilg.gov.ph/PDF_File/reports_resources/DILG-Resources-2012116-420ac59e31.pdf

[6] Beel DE, Wallace CD, Webster G, Nguyen H, Tait E, MacLeod M, Mellish C 2017 Cultural resilience: the production of rural community heritage, digital archives and the role of volunteers J Rural Stud 54:459–468 https://eprints.whiterose.ac.uk/92273/

[7] Nyaupane G & Dallen T 2016 *Public Awareness and Perceptions of Heritage Buildings in Arizona, USA*. Travel and Tourism Research Association: Advancing Tourism Research Globally 32 https://scholarworks.umass.edu/ttra/2009/Presented_Papers/32

[8] Fabbricatti K, Boissenin L, & Citoni M 2020 *Heritage Community Resilience: towards new approaches for urban resilience and sustainability* City, Territory and Architecture 7 17 2020 https://cityterritoryarchitecture.springeropen.com/articles/10.1186/s40410-020-00126-7

[9] Census of Population 2020 *Region VI (Western Visayas) Total Population by Province, City, Municipality and Barangay* Philippine Statistics Authority

[10] Executive Order No. 46-2009 *Implementing Rules & Regulations (IRR) of the old CBD*

[11] Ordinance No. 2000-054 *The Creation of the Iloilo City Cultural Heritage and Conservation Council (ICCHCC)*

[12] Tormis G. et. al. 2015 Fire Risk Assessment of Heritage Buildings in Iloilo City An undergraduate thesis for the BS in Civil Engineering program Dept. of Civil Engineering, Central Philippine University, Philippines

IOP Conf. Series: Earth and Environmental Science 1091 (2022) 012012 doi:10.1088/1755-1315/1091/1/012012

The Effect of Clay-Soil on Landslide: Case Study from Central Java, Indonesia

W Budianta[1,3], H Ohta[2], J Takemura[2]

[1] Department of Geological Engineering, Jl. Grafika 2 Kampus UGM, Universitas Gadjah Mada, Yogyakarta, Indonesia
[2] Tokyo Institute of Technology, 2-12-1 Ookayama, Meguro, Tokyo, Japan
[3]Corresponding author e-mail: wbudianta@ugm.ac.id

Abstract. Landslides are caused by triggering factors such as rainfall, geological factor, geotechnical properties of surface rock, and human activity. Weathering is an essential phenomenon that generates clay minerals and influences the landslide mechanism because the process will reduce rock mass resistance. In the study area, intensive rain contributed to the weathering and increased the water content in the clay soil that affected the stability of the slopes. The clay mineral in the soil in the study area was investigated. The type of clays was determined by plasticity index and liquid limit value and X-Ray diffraction analysis. Some geotechnical properties also were determined, including grain size, shear strength, and atterberg limit. As a result, the clay minerals within the study area's volcanic rock formations play an important role in causing landslides because of illite and smectite, which reduce shear strengths and have high swelling potentials.

1. Introduction

Landslide is a process of mass movement of soil and rock that usually occurs in areas with steep slopes and is very common in Indonesia [1]. Landslide is one of the geological disasters that cause considerable material losses and death tolls. Landslides usually occur in areas with steep slopes and are triggered by several factors such as rainfalls, earthquakes, and the effect of human activities [2]. The rainfall factor, where water enters the soil on steep slopes, will affect shear strength decreasing caused by a decrease soil cohesion on the slip plane, which is closely related to the intensity of rainfall at a landslide location. The potential for landslides depends on many factors, including the land slopes, geotechnical properties of the soil and rock constituents, cohesive forces, and the presence of discontinuities presence [3]. The water that enters the soil on the slopes is also influenced by several factors, namely vegetation cover, soil, and rock types. Anthropogenic activities are also an important factor in making it a landslide vulnerability factor, usually due to changes in land use. Mechanically, landslides often occur due to the slope balance's disturbance between stress and the strength of its materials. This relationship can be seen from many factors such as the degree of slope and soil density as well as the cohesion and angle of friction of the soil, which make the landslide vulnerability higher. There will be balance's disturbance when the maximum material resistance exceeds the stress of the downslope. In theory, the material's shear strength depends on the internal cohesion generated by the particles interlocking that makes up the soil and the material's resistance depending on the overburdened weight. Landslides triggered by high rainfall are often found in Indonesia, and one of them is located at the research location in Tawangmangu, Central Java, as shown in Figure 1 [4].

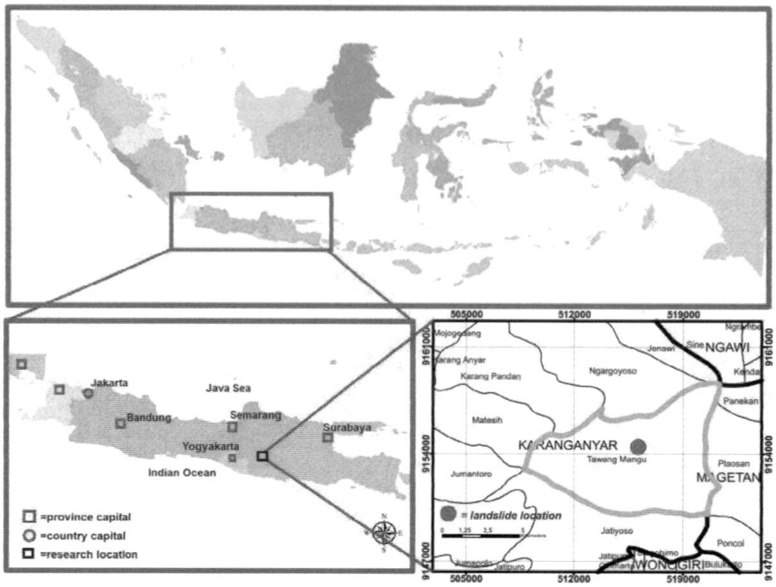

Figure 1. Location of study area [4]

In addition to the high rainfall factor, the geological materials that make soil and rock also determine the occurrence of landslides, especially in areas with very intensive rock weathering. Weathering and alteration are the processes of changing primary minerals into secondary ones caused by both internal and external factors. The secondary mineral is formed in clay. Soil with a high clay content will contribute to landslides that are accelerated by a large amount of rainwater infiltration into the clay [5]. This research aims to examine how the clay content in the soil at the research location contributes to the occurrence of landslides.

2. Research methodology

The clay soil samples used in this research were obtained from the landslide location in Figure 2, which were named S1, S2, S3, S4, and S5.

Figure 2. Overview of landslide location

All clay soil samples obtained were then brought to the laboratory for analysis, including grain size analysis, consistency test, specific gravity, and direct shear. The grain size analysis was carried out using sieve and hydrometer methods [6]. Mineralogical composition of the samples was determined by X-Ray diffraction method. The consistency test was carried out to obtain the Atterberg limit value where the soil taken from the research location was clay soil which is a cohesive soil. The Atterberg limit value is required because it provides information for soil strength, stability, behaviour, and soil organic or inorganic clay types. The values between the plastic and liquid limits were low and provided information about the rapid change of the soil clay sample. The semi-solid to liquid state occurs as a change which results in a cohesion decreasing of soil, internal angle of friction, and also bearing capacity after percolation of rainwater. Pore water pressure will theoretically contribute to slope failure and lead to landslides. Effective stress will decrease with increasing water pressure and consequently decrease shear resistance [7]. Thus, the direct shear stress test carried out on clay samples taken from the research site is important to get the effective shear strength value. Direct stress tests were carried out on undisturbed samples obtained at the study site and sealed with paraffin in situ. A consolidated shear strength test was carried out at 0.05 millimeters per minute in a shear box at a constant rate with controlled deformation conditions and recorded the stress.

3. Result and discussion

Geologically, the study area as part of quaternary rock consists of volcanic breccia, andesite lava, and lahar [8]. The lithology of the study area consists of Lawu lahar (Qlla), Sidoramping Lava (Qvsl), and other Lawu volcanic rock. Lawu lahar consists of an andesitic component, basaltic and tuff mixed with volcanic sand. Sidoramping lava consists of andesitic lava containing plagioclase, quartz, and feldspar. Other Lawu volcanic rocks consist of tuff and volcanic breccia with lava interlayer. Apart from tuff has been altered to clay and chlorite [8]. Soil conditions in the study area consist of mixed sand-clay, also found for boulder as part of altered breccia. Figure 2 shows that the sampling location as part of the lahar deposit consists of mixed boulder-gravel-sand volcanic material. The result of the grain size analysis of all clay soil was determined more than 40% larger than the sieve size of no. 200. This grain size analysis result revealed that all clay soil samples have the clay size in high perçentage. Table 1 shows that the limit consistency values in clay samples taken at the research site result from weathering of the volcanic breccia rocks of Mount Lawu, which are quaternary-aged.

Table 1. Consistency limits

Sample	Consistency limits	Sample				
		S1	S2	S3	S4	S5
Weathered volcanic rock	LL (liquid limit %)	52,36	54,44	64,56	59,67	63,56
	PI (plasticity index %)	16,74	19,73	23,45	18,56	24,78

The clay samples had medium to high liquid limit values ranging from 40% - 57% and had 49% on average. On the other hand, the plasticity index ranged from 17% - 26%, with an average of 19%, whereas the average plastic limit had an average of 28% in the range of 23% - 31%. The result also shows that more than 40% of the clay samples obtained at the site are placed in the high plasticity-inorganic clay or CH group based on the plasticity chart [9]. On the other hand, the others soil clay samples are included in the low plasticity-inorganic clay or CL group, as shown in Figure 3.

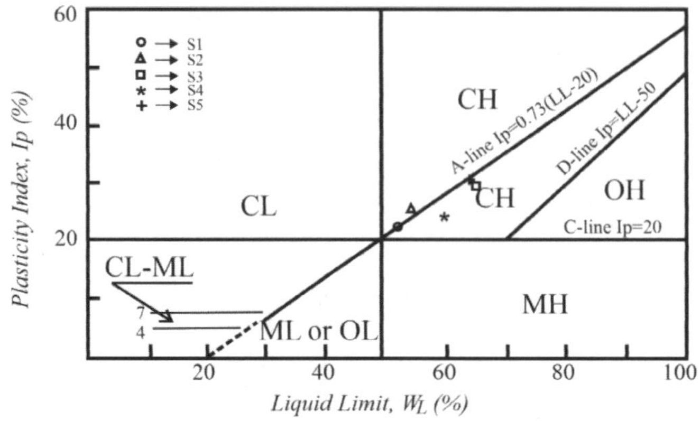

Figure 3. Plasticity chart [9]

As additional information, the Atterberg limits position value plotted is closely related to the type of clay mineral in the soil sample [10,11]. In this research, mineralogical interpretation is obtained by plotting the plasticity index and liquid limit value for the soil clay samples taken at the research site on a plasticity chart. The plotting results showed that the clay samples taken at the study site were in the illite, kaolinite area (Figure 4). The result of by X-Ray diffraction revealed that the samples consists of kaolinite and illite mineral in dominant and a few of smectite was found.

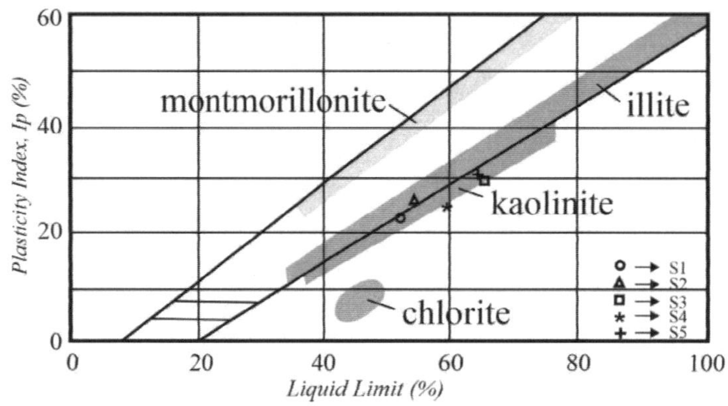

Figure 4. Plasticity chart of mineralogical interpretation [9]

The results of the measurement of the specific gravity value of clay samples taken from the research site showed a value of 26.48, with a value ranging between 2.4 - 3.6 depending on the condition of the constituent minerals. The property index of the clay samples taken from the research site can be seen in table 2.

Table 2. Soil clay index properties

Sample		γ_s	porosity (%)	void ratio
Weathered volcanic rock	S1	2.65	96.5	1.41
	S2	2.65	76.16	1.56
	S3	2.64	89	1.55
	S4	2.65	78	1.43
	S5	2.64	79	1.60

The measurement results showed that the saturated unit weight average value is 19.91, with a natural value of the saturated weight of 18.04. The porosity ranged from 37% - 48%, with an average of 44.7%, while the void ratio value ranged from 58% to 80%, with an average of 74%. Figure 5 shows the normal stress and maximum shear stress correlation were found from the graphics of the parameters of shear displacement and shear strength.

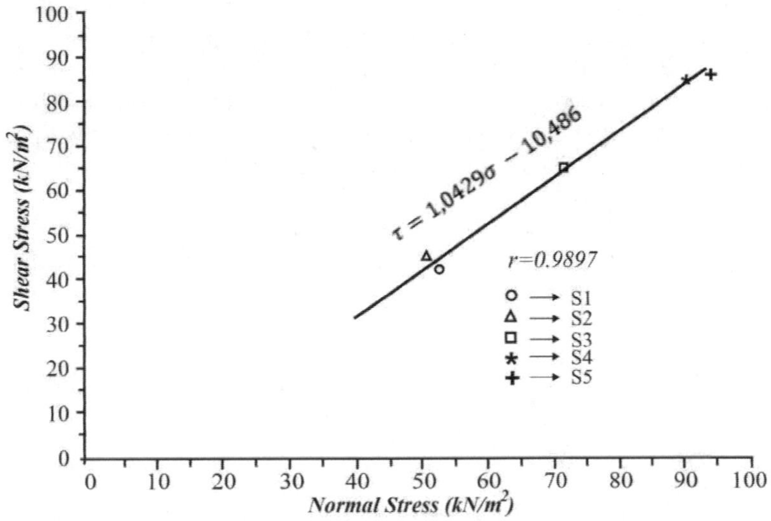

Figure 5. Shear strength parameter

Percolation will make the soil layer become saturated with water and, theoretically, will reduce the value of shear strength. The drainage process was produced by pore water pressure that develops in clay soils which immediately decreases the soil clay shear strength. Changes in the value of shear strength can also be seen from the relationship between moisture content, with the aim of experimentally studying the changes in question, which are mechanical changes that are influenced by water content in soil samples. This can be seen in Figure 5, which shows a correlation where a small amount of water can significantly increase the value of shear stress, which is closely related to rainfall in the research site. The lithological condition of the research site was very strong weathered rock, and there were many changes in land use. This is the influencing factor of landslides. The presence of high rainfall also contributed to the occurrence of landslides, which would reduce slope stability due to an increase in pore water pressure in the soil layer. Clay soil at the research site is the result of weathering of volcanic rocks that play a role in the occurrence of landslides at the research site. The mineralogical interpretation obtained from the plasticity chart revealed that the soil sample is consists of clay minerals of illite and kaolinite, which are commonly found in soils resulting from weathering of rocks. This clay mineral has low shear strength and has the potential for swelling, which contributes to landslides [12]. The results of property index calculations and laboratory analysis for the Atterberg limit of clay samples show that the plasticity index and liquid limit increase with the increase in the void ratio value [13]. Clay soil at the research site includes CL with high plasticity - inorganic and CH as low high plasticity - inorganic. The plasticity of the clay soil obtained from the study area was influenced by the void ratio value. The clay soil at the study site had a low shear strength value when it was saturated with water as shear strength parameters.

4. Conclusion

Based on the results of the research, it can be concluded that weathering and human activities at the landslide site in the research location contributed significantly to the occurrence of landslides. Water that enters the soil layer, in this case, high rainfall, was the dominant factor causing landslides during the rainy season. In this case, the construction of good drainage channels to drain the surface water and reduce percolation is very necessary. The research site is tropical with high rainfall, causing frequent landslides. High rainfall, steep slopes, and intensive weathering of rock are natural factors that contribute to the reduced slope stability at the research site. The geotechnical investigations that have been carried out showed that the movement of the soil layer that caused the landslide was triggered by sliding along the tension crack, which was caused by pore water pressure increases along the boundary of bedrock and weathered rock. The slip plane above the clay soil at the site is associated with saturated clay pore and deep subsurface flow, which results in pore water pressure of the slope-composing materials.

Acknowledgments

Author thanks to Tokyo Institute of Technology, Japan for laboratory facility. Author thanks to JICA Project for support to have research in Japan.

References

[1] Karnawati D, Fathani T F, Ignatius S, Andayani B, Legono, D, Burton PW 2011 *J. Mt. Sci.* **8(2)** 149-153
[2] Davies TR and Shroder JF (Eds.) 2015 *Landslide hazards, risks, and disasters* (Amsterdam, The Netherlands: Elsevier)
[3] Alimohammadlou Y, Najafi A, Yalcin A 2013 *Catena* **104** 219-232
[4] Badan Informasi Geospasial 2016 *Peta RBI* 1:25.000
[5] Yalcin A 2007 *Appl. Clay Sci.* **38** 77–85
[6] Craig R F 1987 *Soil Mechanics* (ELBS, England)
[7] Lambe T W and Whitman RV 1979 *Soil Mechanics* (Wiley, New York)
[8] Sampurno and Samodra 1997 *Geological Map Ponorogo Area* P3G Bandung
[9] Bain J A 1971 *Clay Minerals* **9(1)** 1-17
[10] Casagrande A 1948 *Trans. Am. Soc. Civ. Eng.* **113** 901–930
[11] Holtz R D, Kovacs W D 1981 *An Introduction to Geotechnical Engineering* (Prentice-Hall, Englewood Cliffs, New Jersey, USA)
[12] Ohlmacher G C 2000 *Curr. Res. Earth Sci.* **244 (3)** 1–16.
[13] Vacondios I, Konstantopoulou G, Karadassi St 2007 *Proc. of the 11th Int. Congress* Athens May 2007

IOP Conf. Series: Earth and Environmental Science 1091 (2022) 012013 doi:10.1088/1755-1315/1091/1/012013

Numerical simulation of the 1994 tsunami in Calapan, Mindoro, Philippines using Delft3D

Marjorie David*[1], Christian Paul Allan[2] and Eric Cruz[1]

[1] Institute of Civil Engineering, University of the Philippines Diliman, Quezon City, Philippines
[2] Public-Private Partnership Center of the Philippines, Brgy Pinyahan, Quezon City, Philippines

*Corresponding author e-mail: mcturiano@up.edu.ph

Abstract. A tsunami is a long-period sea wave caused by an earthquake, submarine landslide, or other disturbance such as meteor impact. A 7.1-magnitude earthquake on November 15, 1994 caused a tsunami that hit the coastal areas of Batangas, Quezon, and Mindoro in the Philippines, resulting to 78 deaths. In this study, the 1994 Mindoro tsunami was numerically simulated by modelling of the initial water surface generated by the ground rupture and propagating the tsunami wave towards the coastal areas. Based on the simulations, the generated tsunami wave has a height not exceeding 1.5 m above mean sea level and caused water runup heights in the coast of Calapan, Mindoro reaching 2.6 m. The results of the numerical simulation locally agree with the post-disaster measurements of Imamura et al. (1995) for the coast of Calapan, Mindoro, but the maximum runup levels recorded in the nearby Baco Islands were not obtained with acceptable accuracy. Further calibration of the numerical model is underway to improve the performance of the model.

1. Introduction

A tsunami is a long-period sea wave caused by an earthquake, submarine landslide, or other disturbance such as meteor impact. It is a hazard that poses high risk to coastal communities of the Philippines, due to the country's exposure to several active faults and trenches that cause earthquakes. Numerous accounts of tsunami incidences have been recorded in various parts of the Philippines from the 1600s to the present. Examples of tsunami events that recorded significantly high levels of water level runup on coasts are summarized in Table 1. The 1976 Moro Gulf tsunami recorded the highest runup of 9 meters above mean sea level (MSL) and greatest number of deaths of about 4,000 to 8,000 people [1] while the latest severe event was the 1994 Mindoro tsunami.

Table 1. Historical tsunami events in the Philippines with significant maximum runup (Source: https://www.ngdc.noaa.gov/hazard/tsu_db.shtml).

Tsunami Event (Year and Location)	Source Earthquake Magnitude	Maximum Runup (meters above MSL)
1897 Sulu Sea	8.7	7
1918 Celebes Sea	8.3	5.5
1976 Moro Gulf	8.0	9
1992 Southern Mindanao	7.3	6
1994 Mindoro	7.1	7.29

This paper reports the methodology, results, and findings of the numerical simulation of the 1994 Mindoro tsunami, focusing on the coastline of Calapan, Mindoro. It also presents the implementation of the generation of the initial conditions of the water level disturbance caused by the submarine earthquake as well as the propagation of the tsunami from the source to the affected coastal areas. The results of the numerical model are validated against measurements and post-event surveys.

2. The 1994 Mindoro Tsunami

The 1994 Mindoro tsunami hit the coastal areas of Batangas, Quezon and Mindoro, Philippines on November 15, 1994, and resulted to 78 deaths and about 7,800 damaged houses [2]. It had a magnitude of 7.1 in the Richter scale, with epicenter at 13.5° latitude and 121.1° longitude as shown in Figure 1 [3]. Figure 1 also shows the nearest fault systems in the vicinity of the epicenter of the earthquake: the Lubang Fault and the Aglubang River Fault. The earthquake was associated with the movement of the Aglubang River Fault, which is a right lateral strike slip fault traversing perpendicular to the northern coast of Calapan, Mindoro [2].

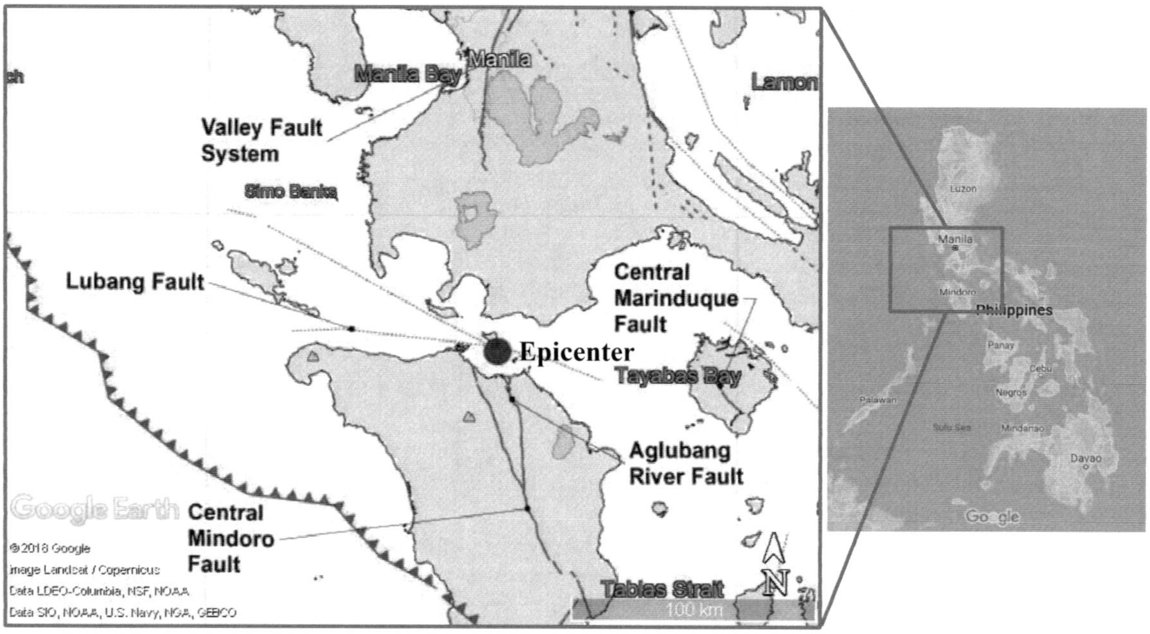

Figure 1. Study area showing epicenter of 1994 Mindoro earthquake. Map also shows the distribution of active faults and trenches in the vicinity of the study area. [4]

A post-event survey was conducted by the group of Imamura [3] a few days after the tsunami event, and they recorded a total of 54 tsunami runup locations, with the maximum runup of 7.29 m above MSL in Baco Island.

3. Methodology

The numerical modeling method to determine the tsunami runup is divided into two major parts: the generation model and the propagation model. The generation model describes the initial water surface conditions at the location of the source based on the seismic characteristics of the earthquake. The developed disturbed water surface at the source is then propagated onto the coastal areas.

The generation model used in this study is based on the work of Okada [5]. It is an analytical method of determining the ground displacement based on the seismic characteristics. It is assumed that the displacement of the water surface is identical to the displacement of the ground.

The propagation model uses the hydrodynamic model of the Delft3D modeling suite [6,7]. To calibrate the hydrodynamic model, astronomic tide conditions were first modeled (i.e., non-earthquake conditions). The astronomic tide forcing was applied at the boundaries, and the water levels were propagated to the Calapan Port tide station where the simulated water levels were compared with forecast levels which are based on a long-time history of actual tide measurements at the station. The calibration parameters were the Manning's roughness coefficient of the seabed, and some corrections in tide fluctuation amplitude and phase at the open boundaries of the model. Since the study area is not located in very deep and open ocean, it is anticipated that there may be errors in the public-domain astronomic tide database such as TPXO 7.2 [8]. The calibrated Manning's roughness coefficients and open boundary corrections were used in the succeeding tsunami runup simulations.

For the tsunami runup simulations, the generated initial ocean surface at the earthquake source were propagated to the coastlines. The simulated water levels were compared with the recorded runup based on the post-tsunami survey.

4. Numerical Model Setup

4.1. Generation Model

The generation model of the tsunami describes the initial sea surface conditions based on the seismic characteristics of the earthquake source. It is often assumed that the initial water surface is identical to the vertical deformation of the ocean bottom faulting, based on three assumptions: (1) the tsunamis are long waves, i.e., the wavelength is much larger than the water depth; (2) the vertical deformation of the ocean bottom is assumed to be instantaneous; and (3) the horizontal movement of the ocean bottom due to faulting is assumed to be negligible in tsunami generation [9].

Hammack [10] suggested a criterion using a dimensionless size ratio that must be significantly greater than unity, and a dimensionless time ratio that must be significantly less than unity to qualify if a tsunamigenic earthquake scenario is large and fast. If so, the assumptions for the claim that the vertical deformation of the ground surface equals the initial water surface hold.

For this study, the deformation of the ground surface was calculated analytically using the model of Okada [5]. The implementation of the analytical solutions of Okada was aided by the open-source program Delft Dashboard developed by Deltares [11].

The alignment of the Aglubang River Fault beyond the land mass was projected based on the report of the PHIVOLCS Quick Response Team [2]. The seismic characteristics of the 1994 Mindoro earthquake was obtained from the Global Centroid-Moment-Tensor database (https://www.globalcmt.org/) [12] and are summarized in Table 2.

Table 2. Earthquake parameters of 1994 Mindoro earthquake based on the Global CMT database (https://www.globalcmt.org/).

Parameter	Value
Depth (km)	15
Strike (degrees)	339
Dip (degrees)	70
Rake (degrees)	-172
Magnitude	7.1

4.2. Propagation Model

The initial water surface level generated by the ruptured fault at the source was propagated to the coastal areas using the hydrodynamic model of the Delft3D modeling suite. The hydrodynamic model, Delft3D-FLOW, solves the unsteady shallow water equations (depth-averaged) in two dimensions [6,7].

The propagation model consisted of a single computational domain covering the general study area of the Verde Island Passage and has a uniform grid resolution of about 150 m. The applied bathymetry was based on the GEBCO-2008 dataset [13] supplemented with depth information from 1:50,000 scale topographic map and 1:20,000 scale bathymetric map of Calapan published by the National Mapping and Resource Information Authority (NAMRIA). The computational domain and the modeled bathymetry are presented in Figure 2. The open boundary conditions applied in the FLOW model for astronomic tide simulations was based on the TPXO 7.2 global tide model [8]. Riemann-type boundary conditions were used to account for weakly reflective boundaries for tsunami simulations.

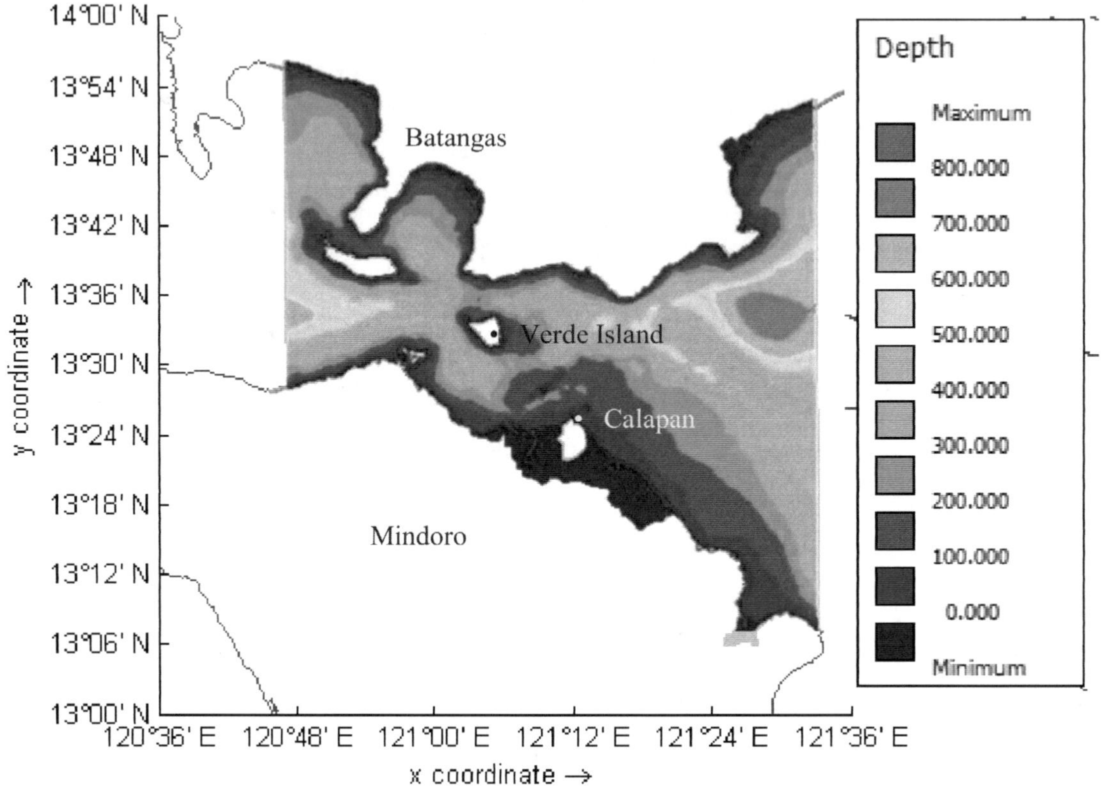

Figure 2. Computational domain and modelled bathymetry of the study area.

5. Results and Discussions

5.1. Initial Water Surface at Source

The model was checked against the Hammack [10] criterion to validate the assumption of the model that an equal deformation of seabed and water surface were made. The dimensionless width has value of 98.148, which is in the order of hundreds, while the dimensionless duration has a value of 0.044, which is significantly less than unity. This indicated that the earthquake event was large and fast, hence the deformation of the seabed can be translated to the free water surface.

The initial water level at the source of ground rupture was generated and produced a maximum height of about 1.5 m above the mean sea level, assuming that the slip length is 10 m. The distribution of the initial water surface is shown in Figure 3.

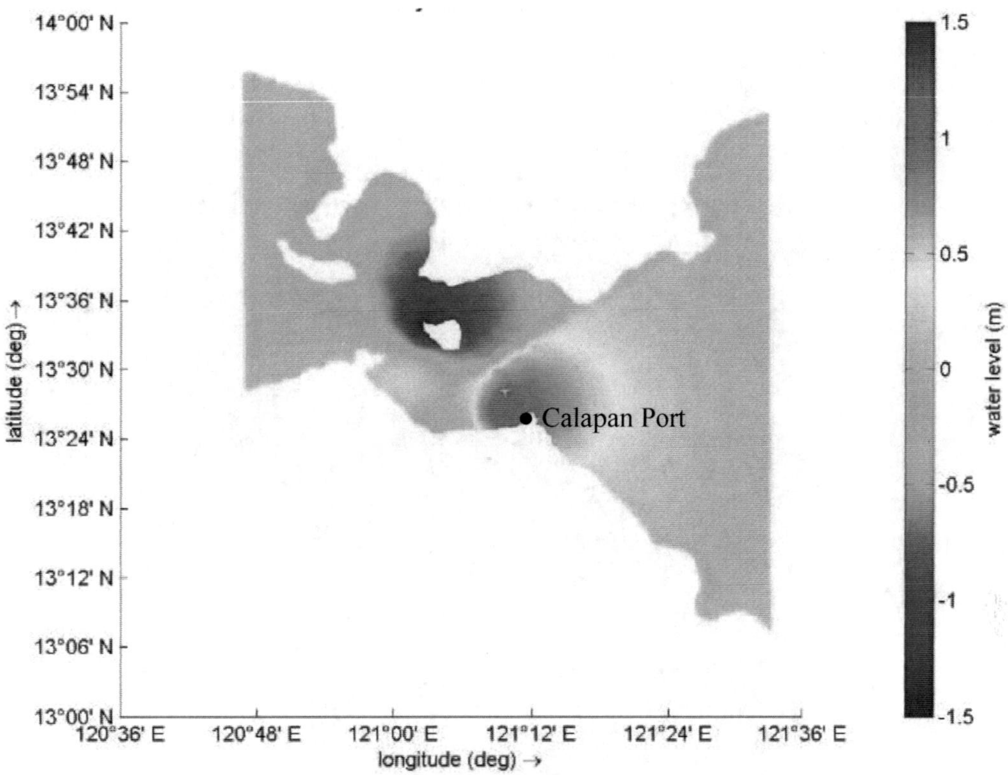

Figure 3. Initial water surface due to 1994 Mindoro Earthquake.

5.2. Astronomic tide calibration

The hydrodynamic model was calibrated by introducing amplitude correction factors in the applied open boundary conditions. Figure 4 shows the comparison of the simulated and forecast astronomic tides for a period of two weeks in Calapan Port after calibration. The root mean square error is 4.6 cm, and the coefficient of determination, R^2, is 0.9866, which makes the calibration highly acceptable. This implies that the hydrodynamic model has the capability to properly propagate water levels from the boundaries of the model up to the coastlines of interest.

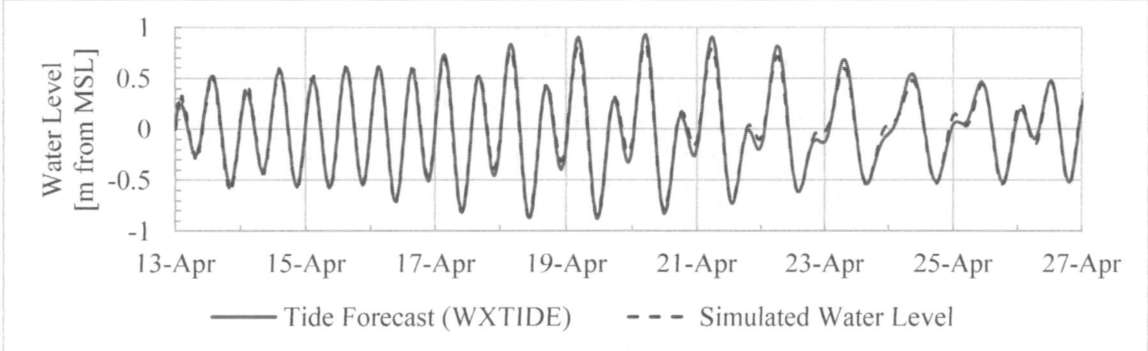

Figure 4. Forecast vs simulated astronomic tide levels in Calapan Port.

5.3. Tsunami runup simulation

The propagated tsunami wave caused runups on the nearby coastlines. A comparison of the recorded and simulated maximum tsunami runup along the coastline of Calapan is shown in Figure 5 and is summarized in Table 3. It is observed that the model gave acceptable simulation results, with percent relative errors ranging between 7.11% and 39.42%.

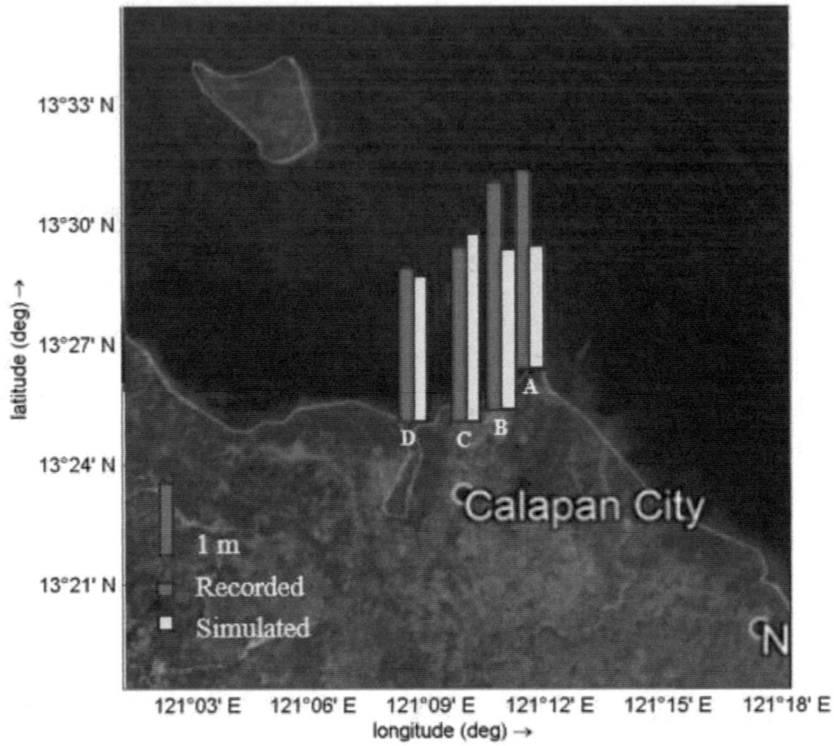

Figure 5. Simulated vs recorded tsunami runup in Calapan coast.

Table 3. Comparison of simulated vs recorded maximum runup in Calapan.

Point	Recorded Maximum Runup (m)	Simulated Maximum Runup (m)	% Relative Error
A	2.74	1.66	-39.42
B	3.14	2.21	-29.62
C	2.41	2.61	8.30
D	2.11	1.96	-7.11

Time histories of the simulated water levels for points A, B, C, and D along the coastline of Calapan are presented in Figure 6. The measured maximum tsunami runup levels per station are also indicated.

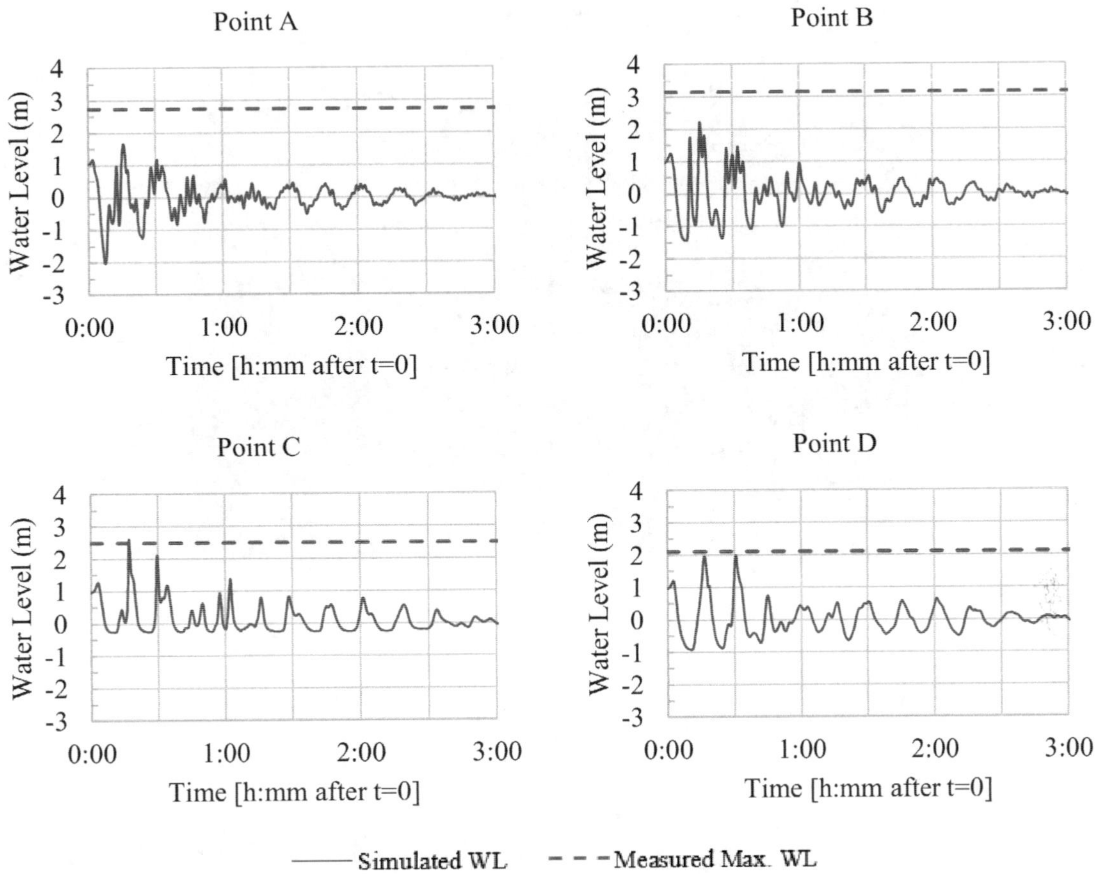

Figure 6. Water level time histories vs reported maximum tsunami runup level for points A, B, C, and D.

The simulated tsunami runup in Baco Islands located just north of Calapan, however, did not show acceptable agreement with the measurements of the post-disaster survey. Figure 7 shows the locations of four (4) survey points in Baco Islands, while Figure 8 show the time history of water levels compared to the maximum runup based on the survey. Table 4 summarizes the simulated and maximum tsunami runup heights in Baco Islands, with the computed percent deviation. Percent errors range between 6% and 82.99%, which are unacceptable. The large error may be attributed to the coarseness of the computational grid relative to the size of the small islands.

Table 4. Comparison of simulated vs recorded maximum runup in Baco Islands.

Point	Recorded Maximum Runup (m)	Simulated Maximum Runup (m)	% Relative Error
E	7.29	1.24	-82.99
F	3.54	1.10	-68.92
G	1.64	0.94	-42.68
H	0.12	0.84	600

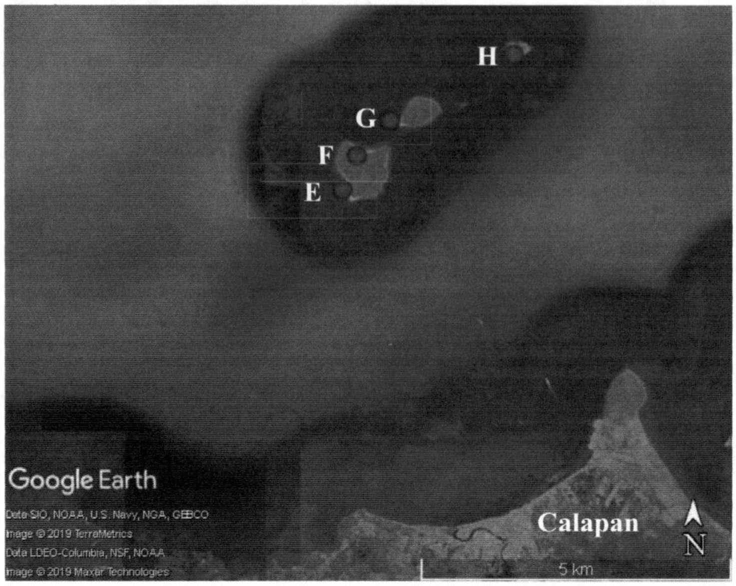

Figure 7. Survey Points in Baco Islands.

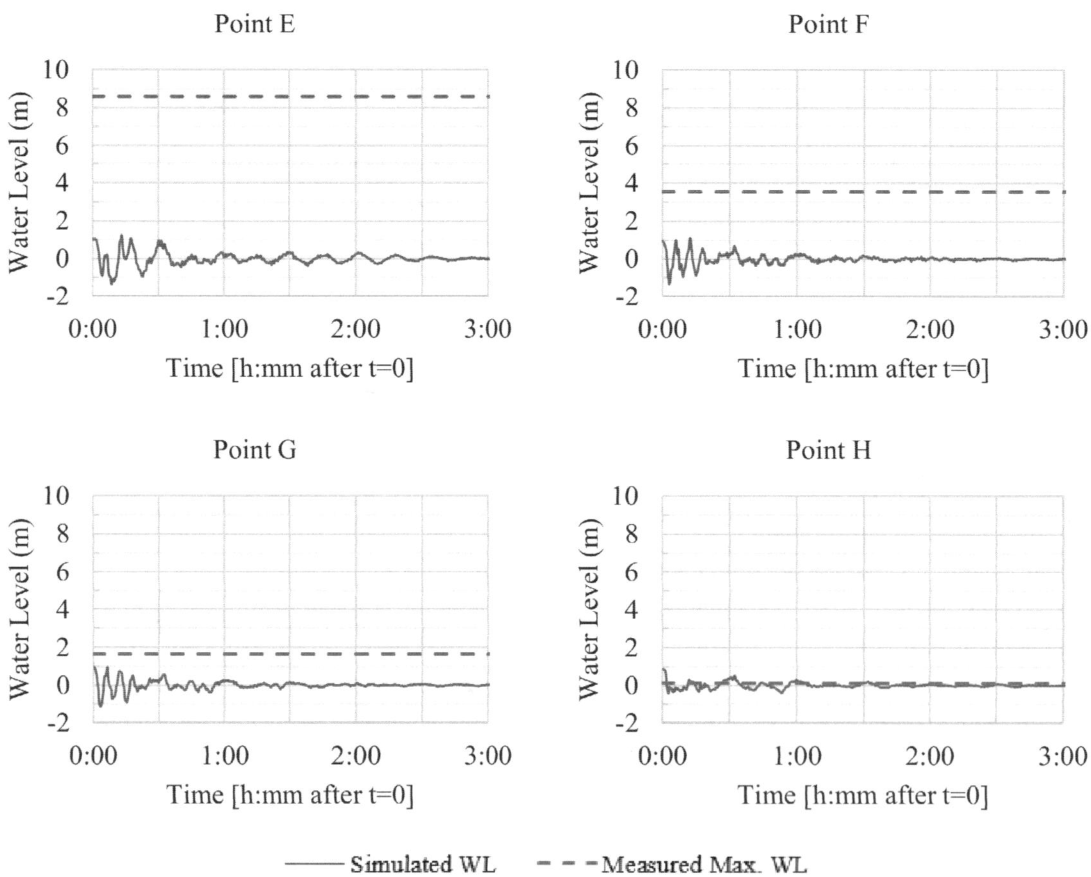

Figure 8. Water level time histories vs reported maximum tsunami runup level for points E, F, G, and H.

6. Conclusions

Based on the numerical simulations, the 1994 Mindoro earthquake created a tsunami wave that has a height of about 1.5 m above mean sea level, which then propagated to the nearby coastal areas. The propagation model was able to produce tsunami runup levels reaching 2.6 m in the coast of Calapan. In comparison with the measurements of the post-disaster survey, the model was able to produce acceptable results for the coast of Calapan, Mindoro only. Further calibration of the numerical model is underway to improve the performance of the model in simulating tsunami runup for the other survey points with available observations.

ORCID iDs

M David https://orcid.org/0000-0002-1351-0494

References

[1] National Centers for Environmental Information. NGDC/WDS Global Historical Tsunami Database [Internet]. National Oceanic and Atmospheric Administration. [cited 2018 May 01]. Available from https://www.ngdc.noaa.gov/hazard/tsu_db.shtml DOI:10.7289/V5PN93H7

[2] Punongbayan RS, Listanco EL, Tayag JC. 15 November 1994 MINDORO Earthquake: Preliminary Report of Investigation. Phivolcs Special Report No. 2. Department of Science and Technology - Philippine Institute of Volcanology and Seismology; 1994. 36 p.

[3] Imamura F, Synolakis CE, Gica E, Titov V, Listanco E, Lee H J. Field Survey of the 1994 Mindoro Island, Philippines Tsunami. In: Tsunamis: 1992-1994. Their generation, dynamics, and hazard. Pure and Applied Geophysics. 1995; 144(3-4):875–890.

[4] Philippine Institute of Volcanology and Seismology. Distribution of Active Faults and Trenches in the Philippines [Internet]. Department of Science and Technology. [cited 2018 May 01] Available from https://gisweb.phivolcs.dost.gov.ph/gisweb/earthquake-volcano-related-hazard-gis-information

[5] Okada, Y. Surface Deformation due to Shear and Tensile Faults in a Half-space. Bulletin of the Seismological Society of America. 1985;75(4): 1135–1154.

[6] Lesser, GR, Roelvink JA, van Kester JATM, Stelling GS. Development and Validation of a Three-dimensional Morphological Model. Coastal Engineering. 2004;51(8–9): 883–915. DOI:10.1016/j.coastaleng.2004.07.014.

[7] Deltares. Delft3D-FLOW: Simulation of Multidimensional Hydrodynamic Flow and Transport Phenomena, Including Sediment User Manual Ver 3.15,686. Delft: Deltares. 2014.

[8] Egbert, GD, Erofeva, SY. Efficient inverse modeling of barotropic ocean tides. Journal of Atmospheric and Oceanic Technology. 2002;19: 183–204.

[9] Tanioka, Y, Satake K. Tsunami generation by horizontal displacement of ocean bottom. Geophysical Research Letters. 1996;23(8): 861–864.

[10] Hammack, J. A note on tsunamis: their generation and propagation in an ocean of uniform depth. J. Fluid Mechanics. 1973;60(4): 769–799.

[11] Nederhoff, K, van Dongeren A, van Ormondt M. [Internet] Delft Dashboard: a MATLAB-based rapid tool for setting up coastal and estuarine models – Tool description and model verification. 2016. [cited 2018 May 01] Available from https://publicwiki.deltares.nl/display/DDB/Validation+Cases

[12] Dziewonski, AM, Chou TA, Woodhouse JH. Determination of earthquake source parameters from waveform data for studies of global and regional seismicity, J. Geophys. Res. 1981;86: 2825–52. DOI:10.1029/JB086iB04p02825

[13] Weatherall, P, Marks KM, Jakobsson M, Schmitt T, Tani S, Arndt JE, et al. A New Digital Bathymetric Model of the World's Oceans. Earth and Space Science. 2015;2: 331–345. DOI:10.1002/2015EA000107

Physical Vulnerability of Surabaya due to Earthquake Based On Rapid Visual Screening

A.B. Habieb[1], D. Iranata*[1], S. Nurlaela[2], R. Anggoro[3], and R.A.W.D. Syafitri[2]

[1] Department of Civil Engineering, Institut Teknologi Sepuluh Nopember (ITS), Surabaya, Indonesia
[2] Department of Urban Planning, Institut Teknologi Sepuluh Nopember (ITS), Surabaya, Indonesia
[3] Department of Informatics, Institut Teknologi Sepuluh Nopember (ITS), Surabaya, Indonesia

*Corresponding author e-mail: iranata80@gmail.com

Abstract. Finding of new faults in Surabaya city has attracted many researchers to evaluate the seismic resilience of the city. A major earthquake in Surabaya may cause a huge losses for the country due to its massive population and large business activities. In this research, a physical vulnerability map of buildings in Surabaya was developed, particularly due to the updated potential earthquake in Surabaya. The results are expected to be used as part of a mitigation and risk reduction strategy for earthquake disasters in the city of Surabaya. The sampled buildings were aggregated in the scale of regional blocks and urban villages in Surabaya, considering the structural type and typology of the buildings. The aggregation process to capture the distribution of existing buildings uses various inputs from landsat maps, google maps, google street view, CKTR Surabaya (Cmap), and openstreetmap Surabaya. Spatial mapping in the scale of regional and village blocks with more precise data will be an essential information in earthquake risk reduction strategies.

1. Introduction

Major earthquake in Surabaya may trigger huge losses for the country due to its massive population and large business activities. In this study, a physical vulnerability map of Surabaya due to earthquake was developed in order to provide an essential information in earthquake risk reduction strategies.
Physical vulnerability analysis due to potential earthquake in Surabaya has been performed in this study. The analyses is based on the Rapid Visual Screening (RVS) as presented in FEMA 154 [1] on around seven hundred buildings in Surabaya, including: housing and settlement; commercial buildings; public facility; and critical Facility.
The building score is calculated to determine the potential damage level of building due to a maximum considered earthquake. Some parameters that affects the final score of building are: structural type, number of storey, vertical and plan irregularity, use of seismic design code, and soil type. In table 1, the distribution of 724 sampling buildings were presented according to building types and ages.

2. RVS Survey Results

2.1 Result on Vulnerability assessment for housing and settlement

In general, family housing in Indonesia were built as either single or two storeys confined masonry or reinforced concrete structures. The housing can be classified as informal and formal housing. The first were generally built without proper design by engineer. Meanwhile, the later were generally designed and built by housing companies. Therefore, it can be stated that informal housing is more vulnerable than the formal ones. Figure 1 shows the typology of single and two storeys informal and formal housing in Surabaya city.

Figure 1. Typical image typology of (a,b) single and two-storey informal housing, (c,d) single and two-storey formal housing, (e) walk-up flat, and (f) appartment building

Source: Author (2019)

Based on the rapid survey (RVISITS) by ITS Team, the vulnerability level of sampled Housing and Settlement buildings in Surabaya is presented in the table 1. The classification of potential damage is estimated based on the literatures.

Table 1. The Results of RVISITS from sampled Housing and Settlement Buildings in Surabaya

Building types	Total number and percentage of buildings in high damage category	Total number and percentage of buildings in medium damage category	Total number and percentage of buildings in low damage category	Total number and percentage of buildings in no damage category
Real estate	0	0	2 (11.11%)	16 (88.89%)
Apartment	2 (15.38%)	0	3 (23.08%)	8 (61.54%)
Permanent housing	37 (17.96%)	2 (0.97%)	29 (14.08%)	138 (66.99%)
Semi-permanent housing	20 (60.61%)	0	2 (6.06%)	11 33.33%)

2.2 Result on Vulnerability assessment for commercial Building

Vulnerability of commercial buildings according to physical characteristics is presented in Figure 2. Noted that in general, commercial buildings pose strong structure and usually are resilient to earthquake and potentially may only experience light damage on the time of earthquake.

Figure 2. Typical image typology of commercial buildings in Surabaya

Source: Google Maps Street View, (2020)

Based on the rapid survey (RVISITS) by ITS Team, the vulnerability level of sampled commercial buildings in Surabaya is presented in the table 2.

Table 2. The Results of RVISITS from sampled Commercial Buildings in Surabaya

Building types	Total number and percentage of buildings in high damage category	Total number and percentage of buildings in medium damage category	Total number and percentage of buildings in low damage category	Total number and percentage of buildings in no damage category
Shop House	0	0	0	3 (100%)
Mall	0	1 (25%)	1 (25%)	2 (50%)
Market	1 (25%)	1 (25%)	0	2 (50%)
Hotel	5 (11.36%)	8 (18.18%)	11 (25%)	20 (45.45%)

2.3 Result on Vulnerability assessment for Public Facilities Building

An additional indicator of vulnerability in terms of the capacity of the educational facilities, consisted of:

- Total number students and staffs in Elementary school facilities are 402.276, or on average 474 people per facilities,
- Total number students and staffs in Junior school facilities are 173.720, or on average 425 people per facilities,
- Total number students and staffs in High school facilities are 134.412, or on average 491 people per facilities,
- Total number students and staffs in Universities facilities are 168.806, or on average 1.986 people per facilities.

Vulnerability of public facilities according to physical characteristics is presented as shown in Figure 3. Noted that in general, public facilities buildings pose strong structure and usually are resilient to earthquake and potentially may only experience light damage on the time of earthquake, except for some of the religious buildings with weak structure that may invite medium damage to earthquake.

Figure 3. Typical image typology of public facilities in Surabaya

Based on the rapid survey (RVISITS) by ITS Team, the vulnerability level of sampled Public Services buildings in Surabaya is presented in the following table.

Table 3. The Results of RVISITS from sampled Public Facility Buildings in Surabaya

Building types	Total number and percentage of buildings in high damage category	Total number and percentage of buildings in medium damage category	Total number and percentage of buildings in low damage category	Total number and percentage of buildings in no damage category
Civil Services	4 (9.76%)	0	4 (9.76%)	33 (80.49%)
Hospital	4 (36.36%)	1 (9.09%)	1 (9.09%)	5 (45.45%)
School	17 (5.11%)	52 (15.62%)	115 (34.53%)	149 (44.74%)
Religious Building	2 (9.09%)	0	3 (13.64%)	17 77.27%)

2.4 Result on Vulnerability assessment for Critical Facility Building

Vulnerability of critical facilities according to physical characteristics is presented as shown in Figure 4. Noted that in general, critical facilities are very vulnerable and are not resilient to earthquake and potentially may experience primary and secondary impacts of the earthquake. Damage on the time of earthquake, except for some of buildings types such as ports and train stations.

Figure 4. Typical image typology of critical facilities (port and train station) in Surabaya

Based on the rapid survey (RVISITS) by ITS Team, the vulnerability level of sampled Critical facility buildings in Surabaya is presented in the table 4.

Table 4. The Results of RVISITS from sampled Critical Facility Buildings in Surabaya

Building types	Total number and percentage of buildings in high damage category	Total number and percentage of buildings in medium damage category	Total number and percentage of buildings in low damage category	Total number and percentage of buildings in no damage category
Port	0	1 (20%)	3 (60%)	1 (20%)
Train station	0	1 (33%)	2 (66%)	0

3. Potential Physical Losses from Damaged Buildings

As described in the methodology section, the physical loss percentage due to physical damages on buildings after earthquake is estimated based on indonesian disaster management code Permen PUPR No.22/2018, as reported in table 5.

Table 5. Estimation of the economic loss

Damage Category	Estimated Economic Loss*
High	65% -100%
Moderate	45%-<65%
Low	30%-<45%
No Damage	-

* refer to *Permen PUPR No.22 / 2018* [3]

The percentage in the previous table refers to the total construction cost of the building. The estimation of construction cost (total cost/m^2) of each building typology is presented in the following table. The cost can be estimated either through local HSPK (unit cost in province/city) or national regulation [2]

In order to calculate the potential physical loss in a subdistrict (kelurahan), the total floor area of each building typology in a subdistrict is calculated based on the population data from Bapeko. In table 6, for instance, the total floor area of several building tipologies in Kelurahan Asemrowo is presented.

Table 6. The estimation of Potential physical Loss in a Village

	Total Floor Area (m²)						
Kelurahan Asemrowo	Apartment	Real Estate	Informal Housing	Public service	Hospital	School	Religious Building
	0	2153.6	1317311.127	1841.4	4369.435	11793.6	16384

Based on the rapid building survey (RVISITS) by ITS team, the distribution of damage level per building typology in Surabaya is obtained, as presented in previous section. Thus, the total floor area of each building typologi can be divided based on the damage level distribution, as seen in the table 7.

Table 7. The estimation of damage level per building typology according to the RVISITS Results

	Damage level	Total Floor Area (m²)						
		Apartment	Real Estate	Informal Housing	Public service	Hospital	School	Religious Building
Kelurahan Asemrowo	High	0.00	0.00	237116.00	179.72	1588.73	602.65	1489.31
	Moderate	0.00	0.00	12777.92	0.00	397.18	1842.16	0.00
	Low	0.00	236.90	184423.56	179.72	397.18	4068.79	2234.78
	No damage	0.00	1916.70	882598.46	1482.33	1985.91	5276.46	12659.92

The total physical loss due to building damages can be estimated by multiplying the corresponding total floor area, construction cost, and percentage of physical loss. For instance, for School Building in moderate damage in Kelurahan Asemrowo, the corresponding physical loss is calculated as following:

Physical loss due to school in moderate damage = (1842.16 m²) x (IDR 3,588,845/m²) x (65%)

= IDR 4,310,655,149

Finally, the estimation of physical loss per building typology per villages in Surabaya can be calcuated.

The indonesian code for economic loss estimation, PERKA [4], has assigned the interval cut-off values for physical loss from three building classification at a very small amount as compared to the typical amount of loss from potential damaged/collapsed buildings in big cities like Surabaya. For example, PERKA assigned the IDR 1 Billion as a high physical loss for damaged/collapsed public facilities. In Surabaya, the smallest amount of physical loss for public facilities is IDR 3.3 Billion which is in Kelurahan Tambak Oso Wilangun. Total losses from public facilities is around IDR 4.5 Trillion, and each village on average will experience potential loss at around IDR 29.4 Billion.

Similarly, the loss standard for damaged/collapsed houses is set at IDR 800 million as the high loss interval cut-off, whereas the smallest amount of physical loss for houses is IDR 113.7 Billion in Kelurahan Perak Barat. Total losses from damaged/collapsed houses (including apartment, informal houses, and is around IDR 103,7 Trillion, and each village on average will experience potential loss at around IDR 673.5 Billion.

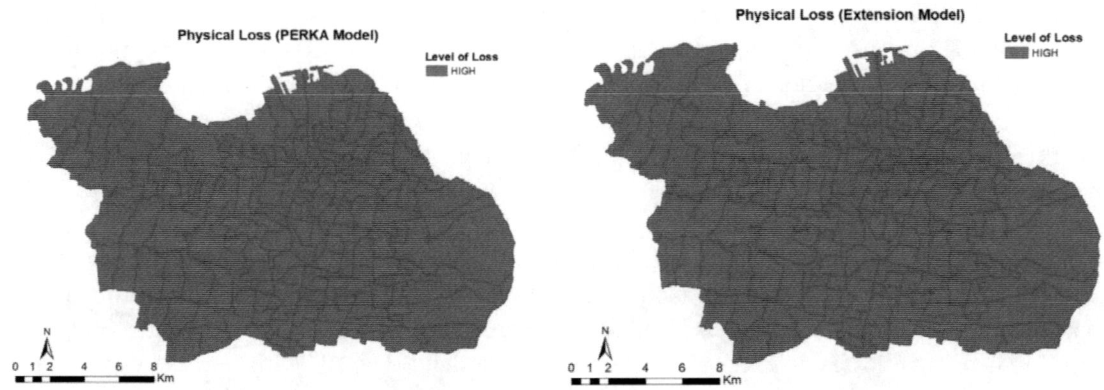

(a) (b)

Figure 5. Overall physical vulnerability level for the (a) PERKA model, and (b) extended model

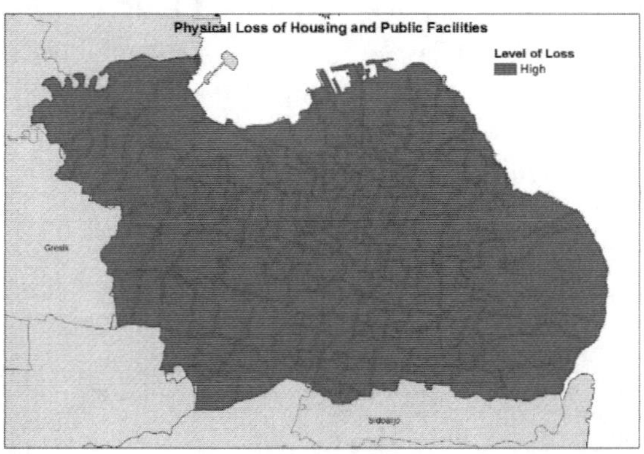

(a)

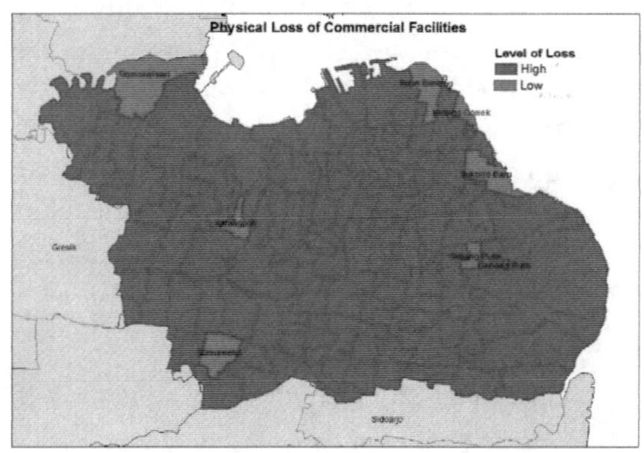

(b)

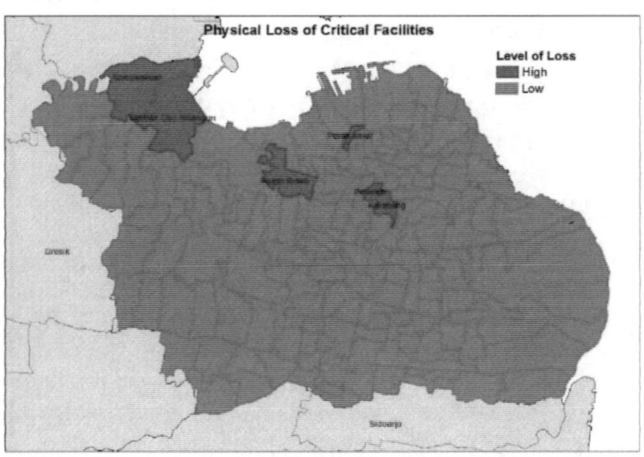

(c)

Figure 6. Potential physical loss of (a) housing/public facilities, (b) commercial facilites, and (c) critical facilities according to the extended model

On the other, PERKA has not assigned commercial building typologies as one determinant factor of physical vulnerability. Therefore, this study is to extend and propose to add the classification of commercial facilities in calculation of potential physical loss in order to fit into the Surabaya context as a metropolitan city. There are 7 (seven) villages in which the commercial facilities in terms of Hotel, Supermarket, and Traditional markets were not available in the vicinities, hence there will be no loss or the level of physical loss will be low in these 7 villages. These are: (1). Kelurahan Romokalisari, (2). Kedung Cowek, (3). Sukolilo Baru, (4). Bulak Banteng, (5). Sumurwelut, (6). Gebang Putih, and (7). Karangpoh. Meanwhile, all other villages will be similarly, experience high level of potential physical loss from damaged/collapsed commercial facilities. The lowest amount of potential loss is around IDR 4 Billion, which is in Kelurahan Tambak Wedi, while the highest is IDR 615,4 Billion in Kelurahan Embong Kaliasin. Total losses from commercial facilties is around IDR 10.3 Trillion, and each village on average will experience potential loss at around IDR 66.7 Billion. In relates to the critical facilities, the Extension model has calculated the physical loss for the critical facilities in terms of Port and Stations. Only 6 villages have these facilities, therefore the remaining villages will be zero in terms of physical loss. The average loss from the 6 villages is around IDR 6.4 Billion per villages, whereas the total loss at around IDR 9,9 Trillion. Figure 5(a) and 5(b) presented the distribution of physical loss from damaged/collapsed buildings according to total loss based on PERKA model and extended model, respectively. New weighted has applied in the calculation of loss index for the extended model according to the weight from AHP survey. Figure 6a,b, and c also presented the distribution of physical loss from damaged/collapsed building according to housing/public facilities, commercial facilities and critical facilities based on the extended model.

4. Conclusion

Physical vulnerability map and economic loss of Surabaya due to earthquake have been developed in this study. The vulnerability map was developed based on the result of Rapid Visual Screening (RVS) on 724 buildings in Surabaya. A local guideline for loss estimation after disaster, PERKA, and an extended model were used to estimate the economic loss due to earthquake. The lowest amount of potential loss is around IDR 4 Billion, which is in Kelurahan Tambak Wedi, while the highest is IDR 615,4 Billion in Kelurahan Embong Kaliasin. Total losses from commercial facilties is around IDR 10.3 Trillion, and each village on average will experience potential loss at around IDR 66.7 Billion. In relates to the critical facilities, the Extension model has calculated the physical loss for the critical facilities in terms of Port and Stations. Only 6 villages have these facilities, therefore the remaining villages will be zero in terms of physical loss. The average loss from the 6 villages is around IDR 6.4 Billion per villages, whereas the total loss at around IDR 9,9 Trillion. The vulnerability map and the estimated total loss can be an essential information in earthquake risk reduction strategies.

References

[1] FEMA 154. Rapid Visual Screening of Buildings for Potential Seismic Hazards: A Handbook. 2015.
[2] Wahyuni, E., Iranata, D., Suswanto, B., Nurcahyo, C. B., & Sutrisno, W. (2018). Assessment of vulnerable buildings due to earthquake loading using Rapid Visual Screening smartphone application. Int. J. Adv. Sci. Eng. Inf. Technol, 8(2), 567-572.
[3] Peraturan Menteri Pekerjaan Umum dan Perumahan Rakyat Nomor 22/PRT/M/2018 Tahun 2018 tentang Pembangunan Bangunan Gedung Negara.

[4] Peraturan Kepala Badan Nasional Penanggulangan Bencana Nomor 02 Tahun 2012 Tentang Pedoman Umum Pengkajian Risiko Bencana.

Acknowledgments
Authors wishing to acknowledge financial support from Institut Teknologi Sepuluh Nopember, Surabaya, Indonesia.

IOP Conf. Series: Earth and Environmental Science 1091 (2022) 012015 doi:10.1088/1755-1315/1091/1/012015

Fear of the fall of space objects in people of rural communities of the district of Azangaro, Peru

A P Calatayud-Mendoza[1], H N Chui-Betancur[1], G Belizario-Quispe[2], E Huaquisto-Ramos[2] and R Alfaro-Alejo*[2]

[1] Natural Sciences Research Institute, Universidad Nacional del Altiplano, Av. Floral 1153. C.P. 21001 Puno, Peru
[2] Center for Research in Water Resources, Universidad Nacional del Altiplano, Av. Floral 1153. C.P. 21001 Puno, Peru

Email: apcalatayud@unap.edu.pe

Abstract. The present study reports the fear scales, in adults over 18 years of age, caused by the fall of propellant tanks from outer space, considered as space debris, that occurred on January 27, 2018, in the communities of Larancahuani, Ninanuyo, Tiramaza, and Bellavista in the Azangaro district, Peru. 49 surveys with 6 items were applied to the residents surrounding the affected areas, using the fear inventory, in addition, the principal component analysis technique was applied to describe a set of information in terms of other variables that are not correlated in terms of least squares. The results show a reliability index of the items of 0.78, which implies the existence of internal consistency of the items. The first factor on affective symptoms: fear of staying overnight during the following nights is associated with a coefficient of 0.855; while the item Fear of the appearance of new diseases in people is positively related with a coefficient of 0.777; finally, the item fear of headaches and dizziness and fear of anxiety, depression, and panic are positively associated with 0.865 and 0.887 with factor 2, which are physical symptoms, which can even alter the development of daily activities of the inhabitants.

1. Introduction

The aerospace region is made up of airspace and outer space near the earth; in this space, there are several operational satellites in orbit with more than 60 years of total operational time [1,2], one of the concerns and debates are centered on the existence of satellites that have fulfilled their operational capacity and they are considered as space debris [3]. In the last ten years, the commercial communications satellite industry has grown tremendously, even with greater capacity and longer life in orbit, the fuel tanks of these satellites have different designs, which are working and others stopped. To operate in orbit [4], the shell of the propellant tanks are constructed of annealed titanium alloy 6AL-4V and commercially pure titanium material [5].

Some launches or space activities can generate collisions in space and turn create millions of pieces of space debris [6]. This space debris, attracted by Earth's gravity, can fall anywhere on Earth, causing damage from the impacts produced; worldwide a large amount of space debris is known to return to earth, it is estimated that one piece of debris per day on average falls on earth [7]. In the Arabian Peninsula using radar measurements and other techniques, it has been found that the re-entry rate of space debris is increasing, requiring a tracking system [8].

In this context, the falls of satellite tanks or space debris on the earth have caused fear in human beings, fear is a fundamental and universal emotion [9] with which human beings have been naturally programmed to act and react to danger based on the survival of the species [10], which is why it belongs to the defensive system of human nature that originates from the perception of danger and a series of physiological changes [9]. Fear can be both objective and subjective, real or imaginary, rational or irrational, and is based on the perception of the subject, and the system of beliefs and resources that can serve as a security system [11], feeling insecurity and anguish [12].

Fear affects emotional performance [13], the decrease in their labor productivity, a family dynamic with a greater probability of suffering a fatal outcome [14], also economic loss, and organizational culture [15]. Fear takes a heavy toll on organizations as it strips workers of their pride, hurts them, and deprives them of the opportunity to contribute to organizational goals [15]. The space objects that fell on January 27, 2018, in the rural communities of the province of Azangaro, Peru, would be components of a rocket, satellite, or spacecraft [16]. The fall was described as an "incandescent fireball". The impact did not cause human or material damage, with a hole just over 30 centimeters deep.

Falling dangerous objects can lead to worry, despair, tension, fear, fatigue, cognitive difficulties, and disorders in people's moods and affection [17]. In addition, they can leave stress and psychological trauma in the population and especially in children [18]. Fear is studied from different disciplinary approaches such as socio-cultural [13] psychological [19], neurological [13], and biological [20]. Each of these disciplinary approaches provides relevant information when interpreting fear in human beings. Therefore, the purpose of this study is to evaluate the fear of falling propellant tanks in residents of the community of Larancahuani, Ninanuyo, Tiramaza, and Bellavista of the district of Azangaro, Peru.

2. Materials and methods

2.1. Participants

The unit of analysis is made up of 49 inhabitants 33 men (67.35%) and 16 women (32.65%) belonging to 86 families from the communities of Laranccahuani (-14.837751°, -70.062537°), Ninanuyo (-14.821301°, -70.059359°), Tiramaza (-14.851687°, -70.034900°) and Bellavista (-14.845976°, -70.046327°) of the Azangaro district (Figure 1), whose main activity is livestock, complemented by agriculture, the trade of products of the area, crafts and regional and interregional consumer goods, as well as transport services.

2.2. Measurement

For the characterization of fears in the inhabitants of the communities: Laranccahuani, Ninanuyo, Tiramaza, and Bellavista, the fear inventory was used [21], for the application of principal components analysis, a technique used to describe a set of data in terms of other variables that are not correlated, it seeks a greater representation of the data in terms of least squares.

From the statistical point of view, the factorial score (X_i) is the linear combination of the factorial loads (F_{im}) weighted by the respective standardized multiple regression coefficients, that is, the eigenvectors, composed of the coefficients that correspond to each variable, are used to calculate principal component scores. The coefficients indicate the relative weighting of each variable in the component (Equation 1).

$$X_i = A_{i1}F_1 + A_{i2}F_2 + ... + A_{im}F_{im} + V_iU_i \tag{1}$$

Where: X_i: Standardized variable; A_{ij}: Standardized multiple regression coefficients; F_i: Common factor or factorial load; V_i: Standardized regression coefficient of variable i in a single factor i; U_i: Unique factor for variable i; m: Number of common factors.

The estimate of the *i-th* factor (F_i) is the linear combination of variables or items weighted by the respective factor coefficients (W_i), as detailed in equation 2.

$$F_i = W_{i1}X_1 + W_{i2}X_2 + \ldots + W_{ik}X_k \qquad (2)$$

Where: F_i: Estimate of the i-th factor; W_i: Factor coefficient; k: Number of variables.

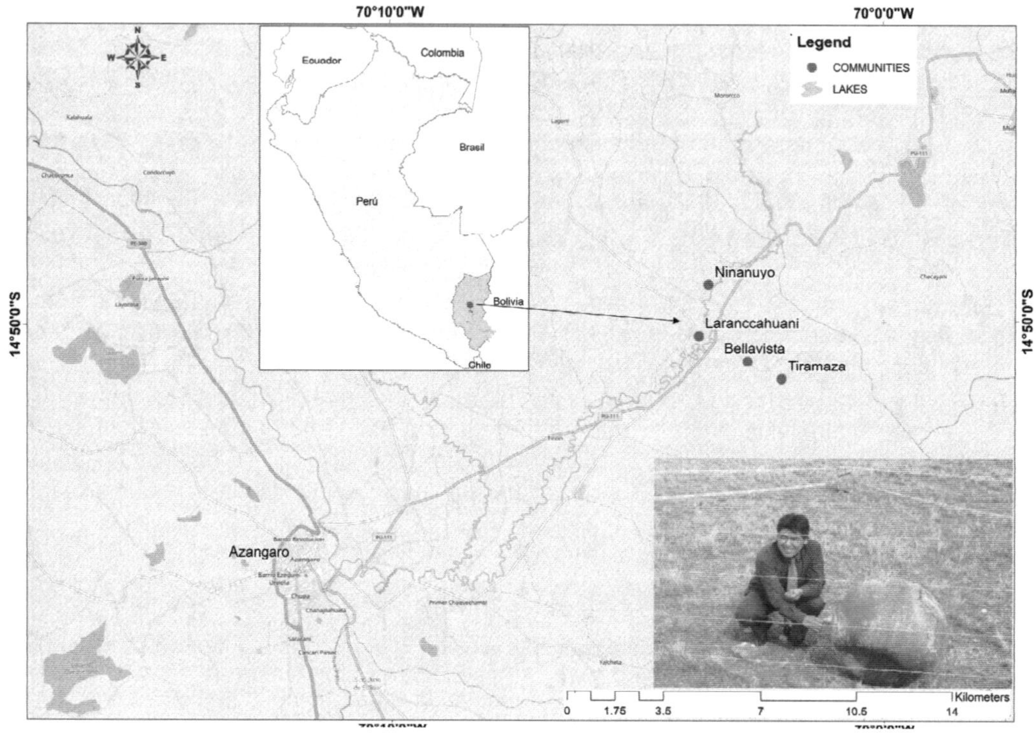

Figure 1. Location of the study area and location of falling objects.

2.3. Procedure

Before applying the factor analysis, the exploratory analysis is carried out, specifically the Barlett sphericity test is applied, a test that consists in verifying if the correlation matrix of the items results in an identity matrix, then the Kaiser-Meyer-Olkin (KMO) test should be used, which measures the suitability of the data to perform a factor analysis by comparing the values of the observed correlation coefficients with the partial correlation coefficients, the results of the factorial model will be acceptable if the KMO index is equal to or greater than 0.70.

2.4. Analysis of data

In the field of psychology, reliability is a psychometric property that refers to the absence of measurement errors; that is, the degree of consistency and stability of the scores obtained in the survey. An instrument that allows verifying the reliability is the Cronbach's alpha coefficient, this coefficient checks the internal consistency, it is based on the variance of each item and the total variance of the items (variables), the value of α varies between 0 and 1, a value of α greater than 0.6 means a good consistency of the results of the scores.

3. Results

The count on the perception of fear between men and women, using the chi-square test of independence, confirms independence at a significance level of 5%; that is, the intensity of fear is independent of gender, that is, there is no relationship between fear and gender (Table 1).

Table 1. Local perception of fear by gender.

Types of fear	Measurement scale	Gender		Total
		Men	Women	
	No fear	2	1	3
Fear of headaches and dizziness	Some fear	21	12	33
	Very afraid	10	3	13
	Total	33	16	49
		$chi^2(2) =$	0.7493	Pr = 0.688
	No fear	17	10	27
Fear that other spherical objects may fall	Some fear	15	6	21
	Very afraid	1	0	1
	Total	33	16	49
		$chi^2(2) =$	0.8799	Pr = 0.644
	No fear	13	10	23
Fear of anxiety	Some fear	18	6	24
	Very afraid	2	0	2
	Total	33	16	49
		$chi^2(2) =$	2.8345	Pr = 0.242
	No fear	16	8	24
Fear of staying overnight at night	Some fear	15	8	23
	Very afraid	2	0	2
	Total	33	16	49
		$chi^2(2) =$	1.0222	Pr = 0.600
	No fear	10	5	15
Fear of depression and panic	Some fear	22	11	33
	Very afraid	1	0	1
	Total	33	16	49
		$chi^2(2) =$	0.4949	Pr = 0.781
	No fear	1	0	1
Fear of the appearance of diseases	Some fear	17	12	29
	Very afraid	15	4	19
	Total	33	16	49
		$chi^2(2) =$	2.6517	Pr = 0.266

Likewise, the relationship between the perception of fear by the level of education, through the chi-square test, there is independence between the variables mentioned at a significance level of 5%, that is, fear is independent of the level of education (Table 2).

The results of the fear inventory offer internal consistency according to Cronbach's alpha reliability index ($\alpha = 0.78$), and there is evidence of positive correlation at a significance level of 5% and it is acceptable in terms of consistency, and subsequently, the principal component factor analysis was applied. The results show that the correlation matrix has a determinant close to zero (Table 3), confirming that there is a correlation between the types of fear; likewise, Bartlett's test of sphericity corroborates the existence of a joint correlation, i.e., the fear inventory is not correlated ($\chi^2 = 106.40$; df $= 15$). Therefore, principal component factor analysis is possible. From a statistical point of view, the eigenvalues of components 1 and 2 are 2.99 and 1.38, respectively, the first component explains 49.8% of the total variance, and the second component explains 23% of the variability, which means that the data can be condensed into 2 principal components.

IOP Conf. Series: Earth and Environmental Science 1091 (2022) 012015 doi:10.1088/1755-1315/1091/1/012015

Table 2. Local perception of fear by education level.

Types of fear	Measurement scale	Education level					Total
		No level	Incomplete primary school	Primary complete	Secondary incomplete	Secondary complete	
Fear of headaches and dizziness	No fear	0	0	3	0	0	3
	Some fear	0	1	9	10	13	33
	Very afraid	1	1	3	6	2	13
	Total	1	2	15	16	15	49
					$chi^2(8)$ =	12.9715	Pr = 0.113
Fear that other spherical objects may fall	No fear	0	1	8	8	10	27
	Some fear	1	1	6	8	5	21
	Very afraid	0	0	1	0	0	1
	Total	1	2	15	16	15	49
					$chi^2(8)$ =	4.5975	Pr = 0.800
Fear of anxiety	No fear	0	1	9	6	7	23
	Some fear	0	1	5	10	8	24
	Very afraid	1	0	1	0	0	2
	Total	1	2	15	16	15	49
					$chi^2(8)$ =	27.3509	Pr = 0.001
Fear of staying overnight at night	No fear	0	1	7	8	8	24
	Some fear	0	1	7	8	7	23
	Very afraid	1	0	1	0	0	2
	Total	1	2	15	16	15	49
					$chi^2(8)$ =	25.2072	Pr = 0.001
Fear of depression and panic	No fear	0	0	4	5	6	15
	Some fear	1	2	10	11	9	33
	Very afraid	0	0	1	0	0	1
	Total	1	2	15	16	15	49
					$chi^2(8)$ =	4.2962	Pr = 0.829
Fear of the appearance of diseases	No fear	0	0	1	0	0	1
	Some fear	0	1	10	9	9	29
	Very afraid	1	1	4	7	6	19
	Total	1	2	15	16	15	49
					$chi^2(8)$ =	4.7607	Pr = 0.783

The eigenvalues and the proportion of the variances of the principal components, where component 1 explains 50% of the total variance, and component 2 explains 23% of the variability, allow for the number of principal components as a function of the size of the eigenvalue according to the Kaiser criterion, i.e. the number of principal components is defined as a function of the number of eigenvalues greater than unity. In this case, the eigenvalues greater than unity are two ($\lambda_1 = 2.99$ and $\lambda_2 = 1.38$) first eigenvalues indicating the principal components (Table 4). Additionally, in Figure 2, the sedimentation figure is illustrated, where the first 2 components have eigenvalues greater than unity, which explains 73% of the total variance of the cumulative proportion.

Table 3. Correlation matrix of the fear inventory.

Variables	Fear of headaches and dizziness	Fear that other spherical objects may fall	Fear of anxiety	Fear of staying overnight at night	Fear of depression and panic	Fear of the appearance of diseases
Fear of headaches and dizziness	1.00					
Fear that other spherical objects may fall	0.3768* (0.0076)[a]	1.00				
Fear of anxiety	0.0860 (0.5567)	0.3885* (0.0058)	1.00			
Fear of staying overnight at night	0.1660 (0.2544)	0.6824* (0.000)	0.5337* (0.0001)	1.00		
Fear of depression and panic	0.2208 (0.1274)	0.5799* (0.000)	0.5052* (0.002)	0.5546* (0.000)	1.00	
Fear of the appearance of diseases	0.6091 (0.000)	0.4751* (0.0006)	0.0488 (0.73919	0.2098 (0.1480)	0.3269 (0.0219)	1.00

[a] Values in brackets are p-values.

Table 4. Eigenvalues and variance ratio of data.

Components	Eigenvalues (λ_i)	Variance ratio	Accumulated value
1	2.987	0.498	0.498
2	1.381	0.230	0.728
3	0.564	0.094	0.822
4	0.475	0.079	0.901
5	0.343	0.057	0.958
6	0.250	0.042	1.000

Likewise, the factorial analysis with varimax orthogonal rotation produced simpler factor solutions and theoretically more significant in their independence, where the axes remain at a 90° angle, which allows redistribution of the variance from the first factors to the last ones, to achieve a pattern of simpler factors (Table 5), the correlation between Factor 1 and fear that other spherical objects may fall is 0.855, which indicates that there is a positive relationship; Likewise, the correlation between Factor 1 and the fear of anxiety that the residents have perceived is 0.804, which means a degree of positive association, in addition, the correlation between Factor 1 and the fear of depression and panic of the residents is 0.777. While Factor 2 is positively correlated with fear of headaches and dizziness with a coefficient of 0.865, in the same way the correlation between Factor 2 and fear of the appearance of diseases is 0.887.

Table 5. Factorial loads between variables with varimax orthogonal rotation.

	Variables	Factor 1 (affective symptoms)	Factor 2 (physical symptoms)	Not explained
1	Fear of headaches and dizziness	0.070	0.865	0.248
2	Fear that other spherical objects may fall	0.717	0.478	0.257
3	Fear of anxiety	0.804	-0.112	0.341
4	Fear of staying overnight	0.855	0.125	0.254
5	Fear of depression and panic	0.777	0.241	0.338
6	Fear of the appearance of diseases	0.137	0.887	0.194

In figure 2, two factors are identified from the results of the factorial loads, in effect, the results show that Factor 1 is positively correlated with the fear that other spherical objects may fall, fear of anxiety, fear of staying overnight, fear of depression and panic, so this component measures the "affective symptoms" that impact a condition called panic disorder, which is essentially sudden feelings of terror with no apparent danger. Likewise, Factor 2 is positively associated with fear of headaches and dizziness and fear of disease onset, so this component mainly measures the "physical symptoms" that also contribute to the condition of panic disorder. The mental disorder is characterized by producing feelings of worry, anxiety, or fear, so strong that they interfere with the daily activities of the person suffering from them.

In addition, the fall of the satellite tank has caused feelings of fear and concern, it has even interfered in the development of the daily activities of the inhabitants. International authorities and the Peruvian Space Agency - CONIDA must anticipate the fall of debris and give alerts to avoid some potential damage, it is known that satellites at the end of their useful life are exorbitant (taken to a low orbit of decay) so that they re-enter the atmosphere under control and disintegrate for the most part in the oceans.

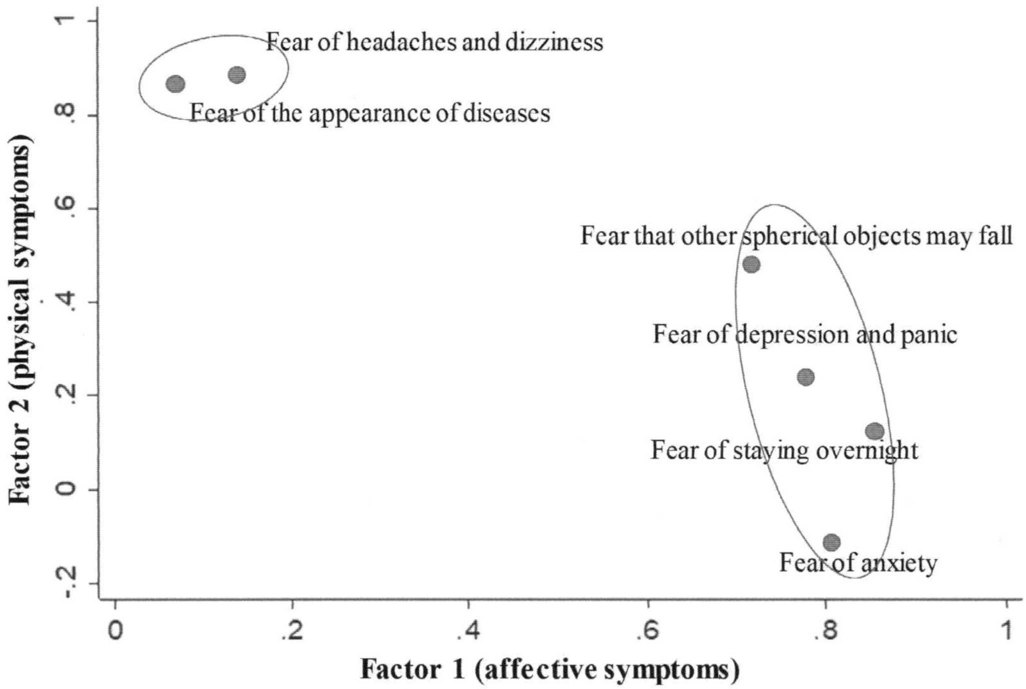

Figure 2. Clusters according to the fear inventory with varimax orthogonal rotation.

4. Discussion

The perception of fear between men and women has independence, however, other researchers state that depression is 5.8% in women and 2.5% in men [14], these differences between women and men regarding the prevalence of depression are consistently maintained regardless of the age group studied; on the other hand, the percentage of affected individuals increases with age in both genders [14]. But gender shows a clear influence on fear levels. Women reflect the highest levels in totals and factors, something expected according to the different parenting styles between men and women; women develop the affective component more [13]. Consequently, fear could affect emotional performance [13], the decrease in their labor productivity, a family dynamic with a greater probability of suffering a fatal outcome [14]. In addition, the relationship between the perception of fear and level of education, the variables are independent, since the prevalence increases with age and decreases with increasing schooling, the prevalence is higher in men in rural areas than in urban areas [14, 22].

The results of the fear inventory offer internal consistency according to Cronbach's alpha reliability index because there is evidence of a positive correlation between the types of fear, mental disorders generally have a strong impact on the lives of individuals, the family, and society as a whole [14], this guarantees the validity of the results in Table 2. Likewise, these results suggest an interrelation between stress, anxiety, and depression that should not be neglected [17]. According to Bedoya & García [15], Fear can be both real or imaginary, rational or irrational, objective and subjective, and is a function of the subject who suffers it, their perception, and the belief system and resources that can serve as a security system, which could be built by representations of previous experiences. This group of fears shows in two main components according to the size of the eigenvalues greater than the unit, which represent the less affective symptoms, the greater the physical symptoms. This is corroborated by the factor analysis that provides a more significant solution in its independence as a function of factor loads where the correlation between Factor 1 and fear that other spherical objects may fall is highly significant and similarly between Factor 1 and fear of the anxiety that the residents have perceived, in addition, the correlation between Factor 1 and the fear of depression and panic of the residents, since depression constitutes a serious public health problem that requires more and better attention [17], since fear affects both the worker and the organization, starting with the worker, who experiences this emotion and this generates psychological, physiological and behavioral reactions, which condition their performance and their way of relating in the organization [15].

Factor 2 is positively correlated with highly significant fear of headaches and dizziness, equally between Factor 2 and fear of disease, affecting a decrease in emotional intelligence as fear levels increase [13]. What generates Factor 1 are panic disorders that are sudden sensations due to falling spherical objects; However, other researchers indicate that they endanger human and animal life in various ways, affecting inhabited areas [18], while Factor 2 can interfere with the daily activities of the sufferer, basically due to the reaction with significant discomfort during the first days after trauma [23]. These behaviors of the affected inhabitants at an organizational or individual level in individuals have perceived a threat or a danger that can be real or imaginary, associated with work phenomena, and reacts to it psychologically, physiologically, and behaviorally, negatively influencing them and their performance, which affects organizational results and labor relations [15]. These traumatic experiences exceed the expectations of the person, leaving signs and symptoms at a cognitive, behavioral, affective, and physiological level, subsequently interfering in their family, social and work life [24].

The affected residents did not know the dangers and risks of falling spherical objects, once the impact occurred, noise and sound generated concern, despair, a lot of tension, fear, fatigue, and others, this is corroborated by Gonzáles-Forteza et al. [17], and these disorders are characterized by their impact on people's mood, generating alterations in other spheres of life, such as appetite, fatigue, sleep, cognitive difficulties, among others; and as time passes, the affected population leaves behind psychological stress and trauma, especially the children; this is a growing concern as stated by other researchers [18], likewise, fear affects the emotional performance of the resident, relapse into depression and this decreases their productivity, as confirmed by Pulido & Herrera [13].

Traumatic experiences have a clear impact on the psychosocial and physical health of the inhabitants, generating discomfort [24], creating insecurity and anguish daily [25], many times the residents feel displaced by the psychological trauma caused by the fall of the propellant tanks; that affects a higher level of perception of fear, therefore it would be more likely that the levels of satisfaction and performance are lower [15], altering the cognitive, behavioral, affective and physiological level, subsequently interfering in their family, social and work life [24].

The fall of the satellite tank has caused feelings of fear and concern, it has even interfered in the development of the daily activities of the inhabitants, as well as generating very serious emotional problems [26], therefore, the authorities must intervene in the control and the Peruvian Space Agency - CONIDA, must make more effective use of outer space as a geopolitical tool in two ways [27], since it has become a growing concern in recent years [18] and therefore the recent launch of space activities highlights the need to fully and effectively regulate [26, 28].

5. Conclusion

In the research, it was determined that affective symptoms have a positive correlation with the fear that other spherical objects may fall and as well as the fear of depression and panic of the inhabitants to the fall of objects, fear of anxiety, fear of spending the night, fear of depression. Physical symptoms have a positive association with the fear of suffering from headaches, dizziness, and fear of appearance of diseases, these mental disorders are characterized by provoking feelings of worry, anxiety, or fear, so strong that they can interfere with daily activities of those who suffer. In addition, gender and level of education are independent of the types of fear in the inventory of fears. However, in the residents of the study area, the fall of the satellite tanks has caused feelings of fear and concern, even to the point of interfering with the development of the daily activities of the residents.

Acknowledgment(s)

Our special thanks to the management of the Postgraduate School of the Universidad Nacional del Altiplano and the residents of the study area.

ORCID iDs

A P Calatayud-Mendoza https://orcid.org/0000-0002-1213-0035
H N Chui-Betancur https://orcid.org/0000-0001-8869-9423
G Belizario-Quispe https://orcid.org/0000-0001-9805-3714
E Huaquisto-Ramos https://orcid.org/0000-0003-1671-0747
R Alfaro-Alejo https://orcid.org/0000-0003-1672-9026

References

[1] Hall L 2014 The history of space debris *Space Traffic Management Conference, Embry-Riddle Aeronautical University - Daytona Beach*
[2] Debreceni M J, Kuo T K, Jaekle D E and North Andover M 1997 Development of atitanium propellant tank *Jt. Propuls. Conf. Exhib.* **20** 1–31
[3] López L I 2010 La seguridad aeroespacial en América del Norte *Norteamérica* **5** 173–219
[4] Tam W, Hersh M and Ballinger I 2003 Hybrid Propellant Tanks for Spacecraft and Launch Vehicles *Propuls. Conf.* **39** 1–16
[5] Tarn W H, Kawahara G H, Jaekle D E and Larsson L W 2000 Design and Manufacture of a Propellant Tank Assembly *Am. Inst. Aeronaut. Astronaut.* **36** 1–22
[6] Imburgia J S 2011 Space debris and its threat to national security: a proposal for a binding international agreement to clean up the junk *Vand. J. Transnat'l L.* **44** 589
[7] Bergamini E, Jacobone F, Morea D and Sciortino G P 2018 The Increasing Risk of Space Debris Impact on Earth: Case Studies, Potential Damages, International Liability Framework and Management Systems BT - Enhancing CBRNE Safety & Security: Proceedings of the SICC 2017 Conference ed A Malizia and M D'Arienzo (Cham: Springer International Publishing) pp 271–80
[8] Darya A M and Fernini I 2018 Analysis of space debris Re-entry over the Arabian Peninsula (2004 to 2018) *2018 International Conference on Signal Processing and Information Security (ICSPIS)* (IEEE) pp 1–4
[9] Rodríguez A 2004 *El miedo, el motor de la historia individual y colectiva* (Madrid: Eurotheo)
[10] Weiten W and Wight R D 1992 Portraits of a discipline: An examination of introductory psychology textbooks in America *Introductory Psychology Textbooks* pp 453–504
[11] Ruiz A L and Guerrero E A 2017 Afectaciones psicológicas en personal de primera respuesta: ¿Trastorno por estrés postraumático o estrés traumático secundario? *Rev. Puertorriquena Psicol.* **28** 252–65

[12] Alfie M 2017 Riesgo ambiental: La aportación de Ulrich Beck *Acta Sociológica* 171–94
[13] Pulido F and Herrera F 2015 Miedo e inteligencia emocional en el contexto pluricultural de Ceuta *Anu. Psicol.* **45** 249–63
[14] Kroenke K, Spitzer R L and Williams J B W 2001 The PHQ-9: Validity of a brief depression severity measure *J. Gen. Intern. Med.* **16** 606–13
[15] Bedoya C and García M 2016 Efectos del miedo en los trabajadores y la organización *Estud. Gerenciales* **32** 60–70
[16] Panta-Fiestas I 2018 Los objetos caídos del cielo en Puno, Perú *Cientificos.pe*
[17] González-Forteza C, Hermosillo A E, Vacio-Muro M de los Á, Peralta R and Wagner F A 2015 Depresión en adolescentes. Un problema oculto para la salud pública y la práctica clínica *Bol. Med. Hosp. Infant. Mex.* **72** 149–55
[18] Delgado-Martínez J G and Álvarez-León R 2018 Aspectos bioéticos relacionados con la basura espacial y sus efectos sobre la vida en la Tierra y la exploración aeroespacial *Pers. y Bioética* **22** 39–55
[19] Campo M 2013 The study of emotions in Vygotskij: The theory of emotions *Psychol. Sci. Educ.* **5** 323–37
[20] Hamm A O 2020 Fear, anxiety, and their disorders from the perspective of psychophysiology *Psychophysiology* **57(2)** 1–14
[21] Calleja N 2011 Inventario de Escalas Psicosociales en México 1984-2005 *Fac. Psicol.* 1–230
[22] Alfaro-Alejo R, Paredes-Mamani R P, Montenegro-Gambini J, Belizario-Quispe G and Flores-Condori E 2021 Geo-hydrological Risk Awareness and Disaster Preparedness in a Mountainous Area of Southern Peru Vulnerable to Disaster *IOP Conference Series: Earth and Environmental Science*
[23] Johnson L, Lodge C, Vollans S and Harwood P J 2019 Predictors of psychological distress following major trauma *Injury* **50** 1577–83
[24] Ohman A and Mineka S 2001 Fears, Phobias, and Preparedness: Toward an Evolved Module of Fear and Fear Learning *Psychol. Rev.* **108** 483–522
[25] Vance L and Mense A 2013 Value analysis for orbital debris removal *Adv. Sp. Res.* **52** 685–95
[26] Velázquez J C 2013 El derecho del espacio ultraterrestre en tiempos decisivos: ¿estatalidad, monopolización o universalidad? *Anu. Mex. Derecho Int.* **13** 583–638
[27] Gallego M J 2013 Hacia una geopolítica del cosmos: confrontación multipolar en la última frontera geográfica y tecnológica *Rev. Geopolítica* **4** 30–45
[28] Jakhu R S and Pelton J N 2017 *Global Space Governance: an international study* (Springer)

Using the TELEMAC model for analyzing the hydrodynamic regime in the Vam Nao River, Vietnam

Nga N Q Tra[*1,2], Kim T Tran[3], Nghia H Ho[3], Hoai C Huynh[1,2] and Bay T Nguyen[1,2]

[1] Ho Chi Minh City University of Technology (HCMUT), 268 Ly Thuong Kiet Street, 10 District, Ho Chi Minh City, Vietnam
[2] Vietnam National University Ho Chi Minh City, Linh Trung Ward, Thu Duc District, Ho Chi Minh City, Vietnam
[3] Ho Chi Minh University of Natural Resources and Environment, 236B Le Van Sy Street, Tan Binh District, Ho Chi Minh City, Vietnam

*Corresponding author: tnqnga@hcmut.edu.vn

Abstract. The Vam Nao River is located Vietnamese Mekong Delta's upstream, plays the role of transferring flow from the Tien River to the Hau River and balances the discharge for two of these main branches. Hence, the hydrodynamic regime is very complicated. In 2017, there was a severe riverbank failure, resulting in the loss of many houses and negative impacts on the local people. So, in this research, the TELEMAC model was applied to simulate flow and bed change, combined with field surveys, thereby analyzing the hydrodynamic regime of this river. The model was calibrated using the water level, discharge and suspended load data. The simulation results show that the flow has only one direction from the Tien River to Hau. The maximum velocity in the flood season is 1.66 m/s, the discharge is 8298 m^3/s (in 2017), the flow after the confluence appears the vortex. Moreover, the confluence area has a rather high erosion rate, up to 1.2 m/year. According to data collected nearly 30 years, the Vam Nao River's topographic data signify the riverbed is growing to expand and deepen, over 43 m after the confluence and distributed near the riverbank. With this development trend, it is necessary to continuously monitor to provide early warnings to minimize damages due to the risks of riverbank failure.

1. Introduction

In recent years, riverbank erosion in Vietnam is increasing, especially in the Vietnamese Mekong Delta (VMD), causing severe socio-economic impacts. VMD is one of the young deltas (formed about 7000 years ago), so it has weak geology. In addition, VMD is one of the largest deltas in the world, with a dense network of rivers, so it has a complex hydraulic regime. There are presently about 665 eroded sites along the rivers with a total length of riverbank erosion of 1048 km (Figure 1) [1].

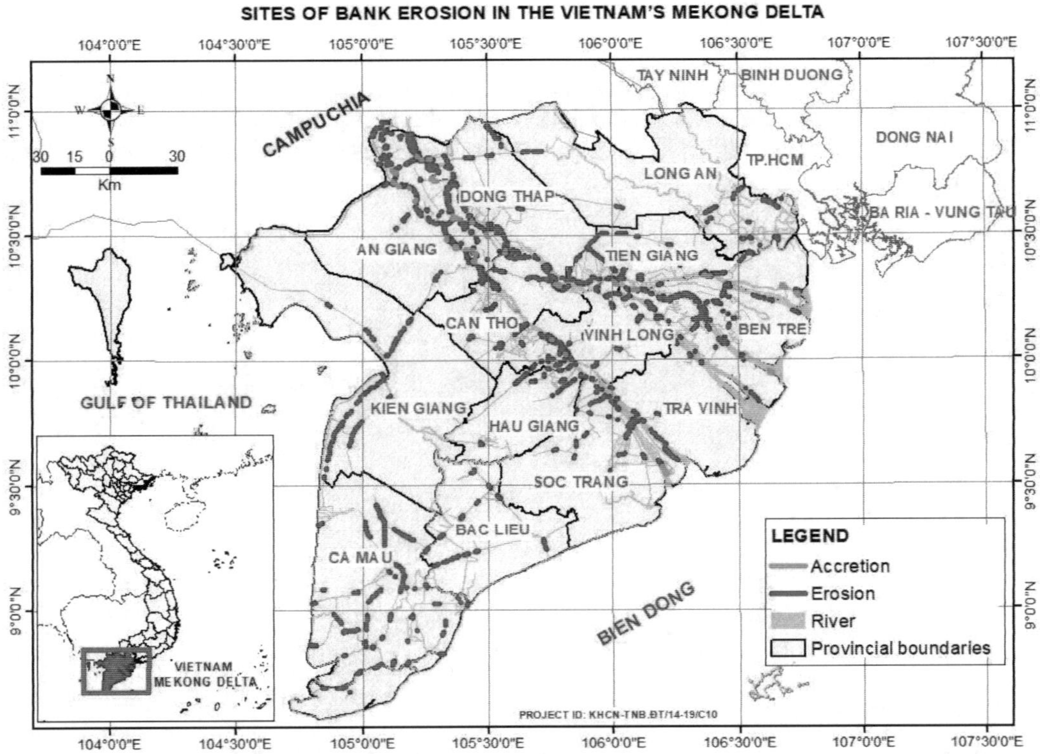

Figure 1. Sites of bank erosion in the Vietnam's Mekong Delta [1].

Vam Nao River is a tributary in the Mekong River system connecting the Tien River with the Hau River. Vam Nao River plays the role of transferring water from the Tien River to the Hau River, helping to balance the water volume between these two large tributaries, so the hydraulic and morphological characteristics are quite complicated. Therefore, the assessment of the bottom changes of the Vam Nao River is necessary, as a scientific basis, to help authorities come up with a plan to prevent and overcome, to minimize the impact on the economy and communities.

Commonly used methods to assess changes in hydrodynamic morphology include measurement and survey; physical model; modeling; remote sensing, and GIS. Hydraulics and its related fields such as the study of bedload transport or predicting the bed level change is one of the problems that has attracted more scientists since the late 19th century [2]. Based on the development of numerical simulation models and measured data, the researchers will more and more accurately predict the river bed morphology changes thanks to the alluvial-flow conditions [3]. Although, the 1D models only simulate the average cross-section but do not consider the horizontal flow as in the two-dimensional (2D) and three-dimensional (3D) models. So, the 1D model has not yet accurately reflected the flow hydrodynamics. Moreover, the hydraulic roughness coefficient applied in the 1D model usually does not change over a section (or the whole study area). Still, in reality, the hydraulic roughness coefficient varies on each section according to cross-section and along the flow direction [4]. Therefore, determining the hydrodynamics of the flow and the change of riverbed morphology is a complex process in the natural river system, requiring the support of appropriate mathematical models. The 2D and 3D models can simulate more detailed river morphology changes [5]. Due to the limitation of the data collected in the vertical direction (from the water surface to the river bottom), the 2D model is often accepted in the study instead of using the 3D model [6]. In Vietnam, many hydrodynamic models are commonly used such as: MIKE21[7-11], Delft3D Flow [12-14], CCHE2D [15, 16] and TELEMAC [17-19]. In this study, the TELEMAC model established by the National Hydraulics and Environment Laboratory of Electricité de France was employed to simulate and evaluate the change of the Vam Nao riverbed.

IOP Conf. Series: Earth and Environmental Science 1091 (2022) 012016 doi:10.1088/1755-1315/1091/1/012016

2. Methodologies

2.1. TELEMAC-2D for Hydrodynamics

The TELEMAC model is a numerical model system whose primary purpose is to simulate flow dynamics through shallow water equations integrating depth, using the finite element method. The governing equations considered are:

Continuity equation:

$$\frac{\partial h}{\partial t} + u.\vec{\nabla}(h) + h\mathrm{div}(u) = S_h \tag{1}$$

Momentum equation along x:

$$\frac{\partial u}{\partial t} + \vec{u}.\vec{\nabla}(u) = -g\frac{\partial Z}{\partial x} + S_x + \frac{1}{h}\mathrm{div}(h\nu_t\vec{\nabla}(u)) \tag{2}$$

Momentum equation along y:

$$\frac{\partial v}{\partial t} + \vec{u}.\vec{\nabla}(v) = -g\frac{\partial Z}{\partial y} + S_y + \frac{1}{h}\mathrm{div}(h\nu_t\vec{\nabla}(v)) \tag{3}$$

where h i the water depth (m); u, v are velocity components (m/s); g is the gravitational acceleration (m/s^2); z is the free surface elevation (m); t is time (s); x, y are horizontal space coordinates (m); S_h is the source or sink of fluid (m/s) and S_x, S_y are source or sink terms in dynamic equation (m/s^2), ν_t is momentum diffusion coefficient (m^2/s) and $\vec{\nabla}$ is divergence.

2.2. SISYPHE for Morphodynamics

The SISYPHE is the sediment transport module coupled with TELEMAC-2D to produce simulations of bed evolution. The total sediment transport is calculated based on the Soulsby-Van Rijn transport formula [20]. The transport rate formula is expressed:

$$q_t = A_s U\left(\sqrt{U^2 + \frac{0.018}{C_D}U_{rms}^2} - U_{cr}\right)^{2.4}(1 - 1.6\tan\beta) \tag{4}$$

The formula (4) applies to the total sediment transport rate per width of the flow. As representing the total bed load and suspended load transport, U is depth-averaged velocity, U_{rms} is root-mean-square wave orbital velocity, and C_D is drag coefficient due to the current alone. The threshold current velocity [21] is expressed:

$$U_{cr} = \begin{cases} 0.19(d_{50})^{0.1}\, log(4h/d_{90}), & d_{50} \le 0.5mm \\ 8.5(d_{50})^{0.6}\, log(4h/d_{90}), & d_{50} > 0.5mm \end{cases} \tag{5}$$

where d is mean grain diameter and h is water depth.

The SISYPHE simulates riverbed morphodynamics by calculating temporal changes in bed elevation Z_b using the Exner equation:

$$(1-n)\frac{\partial Z_b}{\partial t} + \nabla.Q_b = 0 \tag{6}$$

where n is the non-cohesive bed porosity, Z_b is the bed elevation (m), and Q_b is the solid volume transport (bedload) per unit width (m^2/s).

2.3. Study area

The study area is the upstream VMD from Tan Chau station (Tien River), Chau Doc station (Hau River) to My Thuan and Can Tho Bridge, limited with four boundaries as in Figure 2.

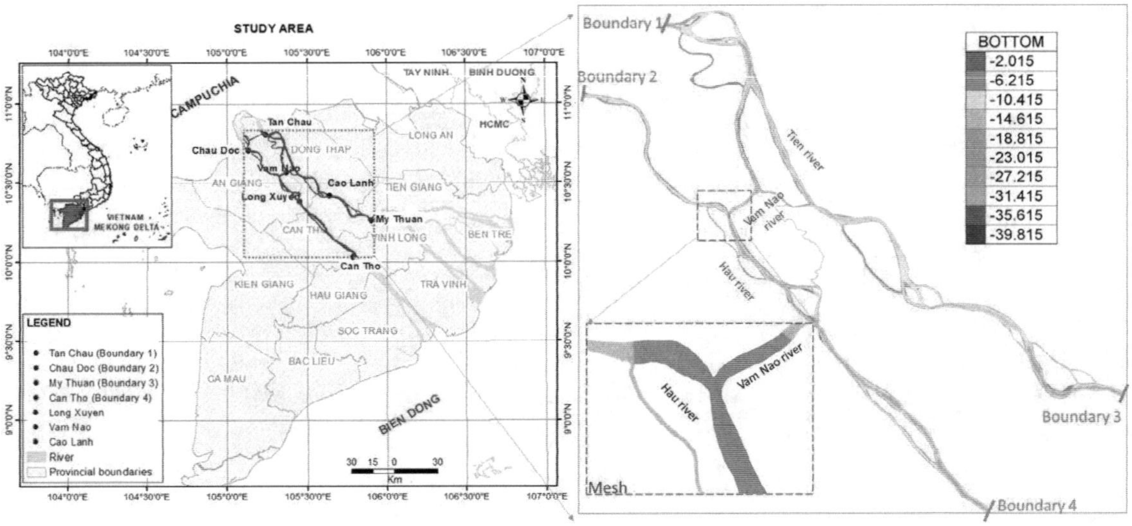

Figure 2. The study area.

The mesh is set up using BlueKenue. The mesh is divided into 2 areas: a fine mesh area with a distance between nodes of 20 m for the confluence of Hau River - Vam Nao River and the remaining areas with a distance mesh are over 100 m. The domain is an unstructured mesh with 62904 nodes and 117420 elements.

2.4. Data

2.4.1. Bathymetry data. The bathymetry data in 1991, 2003, 2013 and 2017 were collected from HCMC University of Technology and the Southern Institute of Water Resources Research, Vietnam. All bathymetric datasets (point data) were projected to WGS 1984 Universal Transverse Mercator, zone 48 N. These bathymetric data were used to evaluate the morphological changes. The Telematic model used bed elevation in 2017 to simulate the hydrodynamic regime of Vam Nao River.

2.4.2. Boundary conditions. The boundary conditions for hydrodynamic simulation were defined as discharges boundaries at the Tan Chau and Chau Doc station as water levels at My Thuan and Can Tho station. Measured data gave the suspended sediment concentration of four boundaries at the Tan Chau, Chau Doc, My Thuan and Can Tho stations for sediment simulation. These data in 2017 at all hydrological stations were collected from Hydro-Meteorological Data Center (HMDC).

2.5. Setup model

2.5.1. TELEMAC model's parameters. Calibration and validation are calculated for five days in April 2017 (dry season) and October 2017 (flood season), respectively. The initial condition is quiet water level at 0.49m. The bottom roughness uses the Strickler coefficient with a value of 45 for the whole region.

2.5.2. SISYPHE model's parameters. The particle size consists of three levels in mm of 0.005 (clay), 0.0275 (dust) and 0.15 (sand), respectively. The initial condition is average SSC in kg/m^3 by particle size of 0.013 (clay), 0.012 (dust), 0.010 (sand) corresponding. The settling velocity is ws_1 = 0.00145 m/s, ws_2 = 0.00399 m/s, ws_3 = 0.63164 m/s.

2.6. Model performance evaluation

The model performance is assessed based on the statistical analysis of the simulated and measured parameter agreement. To evaluate the performances of models, the three following statistical measures were employed: Nash - Sutcliffe efficiency index (NSE) [22], percentage of error (PBIAS) and the coefficients of determination (R^2). These coefficients are formulated as [23] and described as follows:

$$\mathrm{NSE} = 1 - \frac{\sum_{i=1}^{n}(P_{m,i}-P_{s,i})^2}{\sum_{i=1}^{n}(P_{m,i}-\overline{P_m})^2} \tag{7}$$

$$R^2 = \left[\frac{\sum(P_{m,i}-\overline{P_m})(P_{s,i}-\overline{P_s})}{\sqrt{\sum_{i=1}^{n}(P_{m,i}-\overline{P_m})^2\sum_{i=1}^{n}(P_{s,i}-\overline{P_s})^2}}\right]^2 \tag{8}$$

$$\mathrm{PBIAS} = \frac{\sum_{i=1}^{n}(P_{m,i}-P_{s,i})}{\sum_{i=1}^{n}P_{m,i}} \text{ x } 100\% \tag{9}$$

where, n is the number of data; $P_{m,i}$ $P_{s,i}$ are the measured and simulated of the ith value; $\overline{P_m}$, $\overline{P_s}$ are the average of measured and simulated parameter, respectively.

The standard for comparison between simulation and measurement proposed by Moriasi et al. (2007) [23] are shown in Table 1.

Table 1. The standard for evaluation [23].

Performance rating	Flow		Sediment
	NSE	R^2	PBIAS
Very good	NSE > 0.8	$R^2 > 0.85$	|PBIAS| < 10
Good	0.70 < NSE ≤ 0.80	$0.75 < R^2 \leq 0.85$	10 ≤ |PBIAS | < 15
Satisfactory	0.50 < NSE ≤0.70	$0.60 < R^2 \leq 0.75$	15 ≤ |PBIAS | < 20
Unsatisfactory	NSE ≤ 0.50	$R^2 \leq 0.60$	|PBIAS | ≥ 20

3. Results

3.1. Calibration and validation

Model calibration and validation were conducted for simulating flow and sediment concentration of dry season (April 2017) and flood season (October 2017). The discrepancy between observed and simulated results was quantified for each simulation, compared for water level at Vam Nao, Long Xuyen, Cao Lanh stations; discharge and sediment concentration using data at Vam Nao station.

Figure 3 to 5 present the plotted simulation and observation results for dry and flood seasons at the Vam Nao station. H represents water level, Q is discharge and SSC denotes suspended sediment concentration.

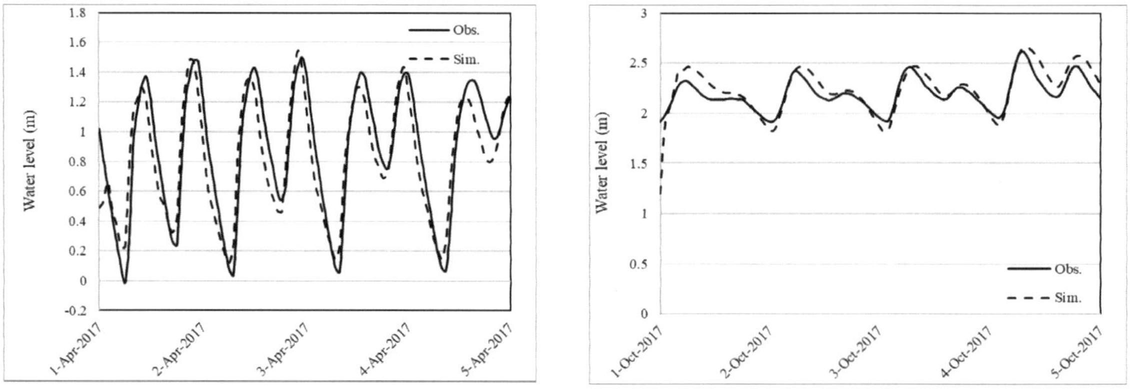

(a) (b)

Figure 3. Comparison of observed and simulated water levels during (a) dry and (b) flood seasons at the Vam Nao station.

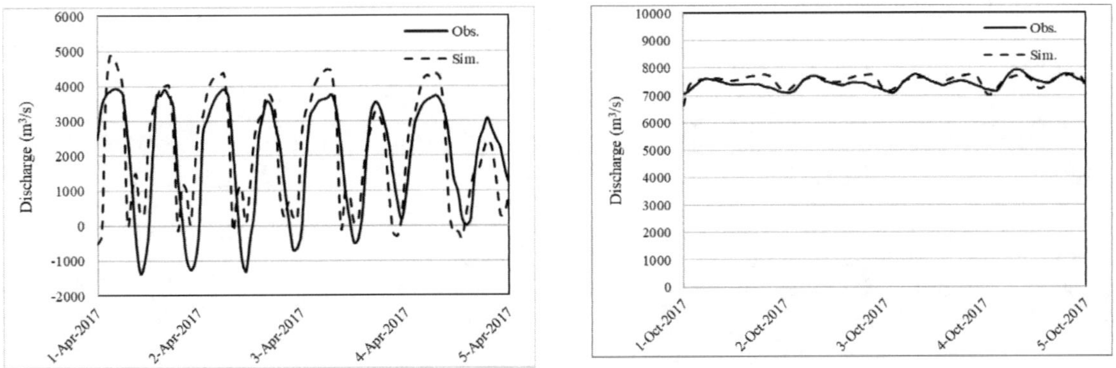

(a) (b)

Figure 4. Comparison of observed and simulated discharge during (a) dry and (b) flood seasons at the Vam Nao station.

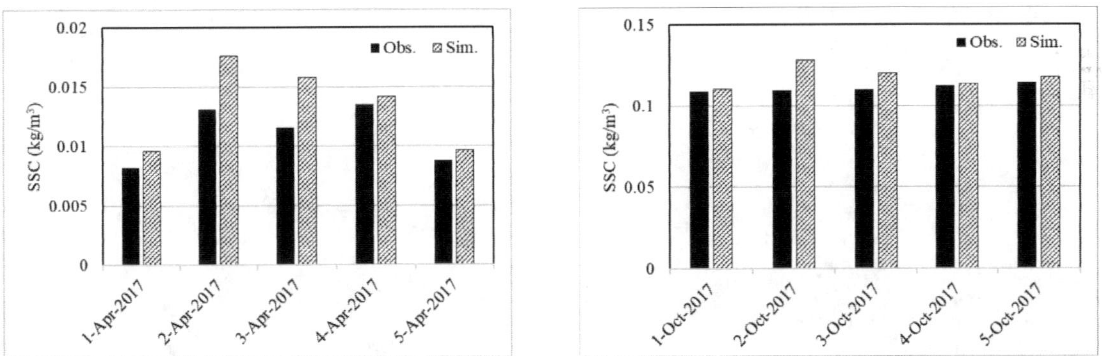

(a) (b)

Figure 5. Comparison of observed and simulated SSC during (a) dry and (b) flood seasons at the Vam Nao station.

The calibration and verification results of water level, discharges and total SSC presents in Table 2. Considering the goodness-of-fit statistics and comparison plots discussed above, it is generally concluded that this model can realistically simulate the discharge, water level, and suspended sediment concentration for the upstream VMD rivers. And the well-calibrated model was used to investigate the hydrodynamic, sediment transport, and accretion and erosion processes for the study area.

Table 2. The performance of the model for the simulation of water level (H), discharge (Q), and suspended sediment concentration (SSC) in the VMD.

Station	Dry season (April 2017)					Flood season (October 2017)				
	NSE		R^2		PBIAS	NSE		R^2		PBIAS
	H	Q	H	Q	SSC	H	Q	H	Q	SSC
Vam Nao	0.78	0.76	0.81	0.72	5.12	0.66	0.72	0.90	0.78	6.32
Long Xuyen	0.65	-	0.94	-	-	0.61	-	0.91	-	-
Cao Lanh	0.92	-	0.90	-	-	0.67	-	0.93	-	-

3.2. Spatial distribution of Velocity

After evaluating the model, the hydraulic and erosion regimes were simulated in two seasons of 2017. The dry and flood periods are from January 1 to June 30 and from July 1 to December 31, respectively.

Simulation results show that tides influence the flow in the dry season, but in the flood season, the flow only flows in one direction from upstream to the sea (Figure 6).

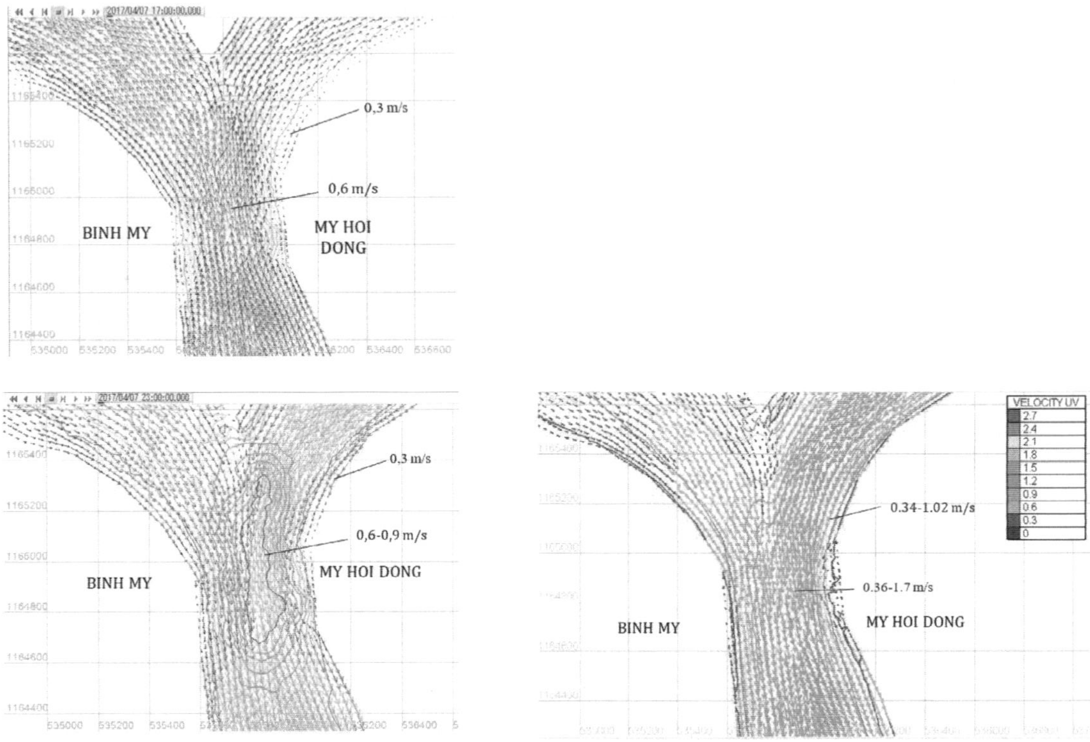

(a) (b)

Figure 6. The flow velocity during (a) dry and (b) flood seasons at the confluence of the Hau River and the Vam Nao River.

In the dry season, when the flow goes upstream, the tidal flow is affected by the river flow, so the velocities are about 0.3-0.6 m/s in the Vam Nao River. The current speeds nearly the bank are only about 0.3 m/s and range between 0.6 and 0.9 m/s in the mid-stream. When the flow flows from upstream to the sea, the speeds of the whole area are higher (about 0.3-1.2 m/s) with 0.4 m/s in the near bank area and reach 0.7-1.2 m/s in the mid-stream. The velocity distribution of the two rivers is significantly different (only 0.3-0.6 m/s in the Hau River and almost 0.6-0.9 m/s in Vam Nao River). Therefore, the flow in this area tends to push the flow to the left bank (from the upstream down) creating a vortex after the confluence. Beside that, the flow combined with the area's topography to form a scour hole near the riverbank, leading to an increase in riverbank slope, which is one of the causes of erosion. That is most evident in the riverbank of My Hoi Dong commune.

During the flood season, the speeds are about 0.34-1.02 m/s in the near bank area and increase to reach 1.02-1.7 m/s in the middle of the stream (sometimes up to 2.04- 2.38 m/s), flow velocities in flood season are stronger than in dry season.

3.3. Accretion and erosion processes

The results of riverbed change after six months of the dry season in 2017 (from January 1 to June 30) are shown on Figure 7(a). The alluvial area tends to focus on the area near the shore of Tan Hoa commune (approximately 0.18 m to 1.47 m). The erosion area is concentrated in the middle of the stream with a value of about -1.10 m to -2.09 m. Because the speed on Vam Nao River is higher than Hau River, so on Vam Nao River, there is a tendency of erosion, upstream of the confluence of Hau River has a tendency of accretion.

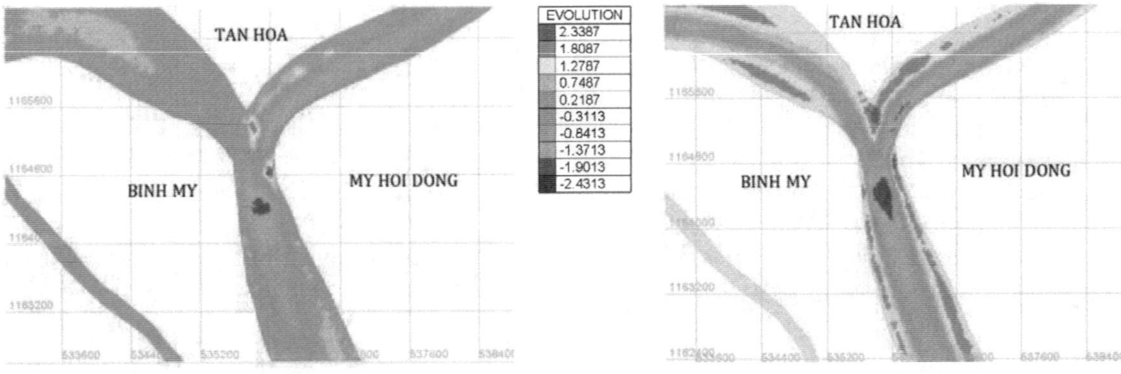

(a) (b)

Figure 7. Accretion and erosion processes during (a) dry and (b) flood seasons at the confluence of the Hau River and the Vam Nao River.

Figure 7(b) shows the results of the riverbed evolution after six months of the flood season in 2017 (from July 1 to December 31). The result indicates that the sedimentation trend is distributed along both sides of the river with an accretion level of about 0.36 m to 1.22 m. The erosion trend is distributed in the middle of the stream with values from -0.49 m to -2.51 m. On the Vam Nao River, the erosion trend is clearly in the middle of the river; there is an erosion area close to the bank of My Hoi Dong commune.

According to measured geometric data collected from 1991 to 2017 (Figure 8), it is found that the erosion hole after the confluence of the Hau River and the Vam Nao River tends to expand to the two banks, deepen and move towards the downstream side.

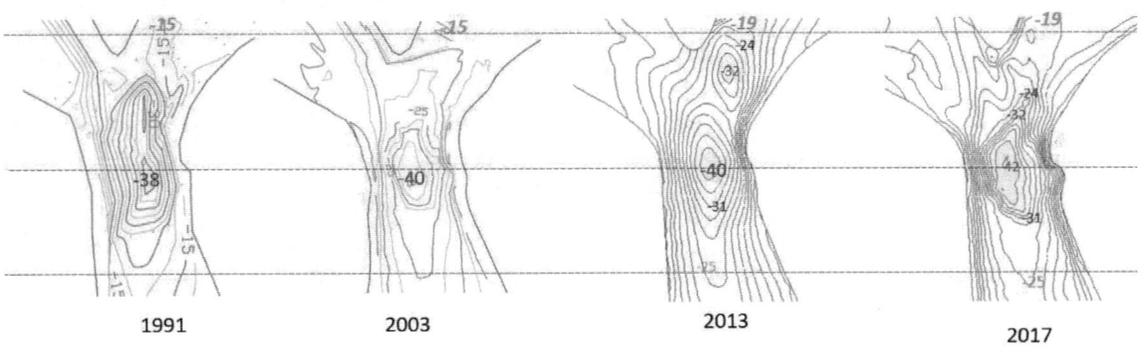

Figure 8. The Vam Nao River's bathymetry nearly 30 years is tending to expand and deepen.

4. Conclusions

The capabilities of TELEMAC model to simulate the hydrodynamic regimes and sediment transport at Vam Nao River were explored. The simulation results showed some similarities with the observed results. The Vam Nao River transfers water from Tien River to Hau River. After the confluence of the Hau River and Vam Nao River, the flow velocity distribution of the two rivers is significantly different. These create eddies that increase the erosion rate and the riverbank slope. So, the hydrodynamic regime strongly effects on morphology changes. In general, accretion and erosion processes in VMD change in time and space.

In fact, for nearly 30 years, the Vam Nao River's topographic data signify the riverbed is growing to expanding, deepening and distributed near the riverbank after the confluence. This is one of the main reasons leading to the risks of riverbank failure. Hence, it is necessary to continuously monitor and give early warnings to minimize damages.

In the future, simulations are needed to understand the combined effects, including hydropower development, sand mining, and climate change on flow, sediment transport, accretion and erosion processes to improve the sustainability of the VMD.

Acknowledgments

This research is funded by Ministry of Science and Technology of Vietnam under the grant number '08/2017/HĐ-KHCN-TNB.ĐT/14-19/C10'.
Besides, this study is a part of the first author's PhD work, which is funded by the Domestic Master/ PhD Scholarship Programme of Vingroup Innovation Foundation.

ORCID iDs

Nga N Q Tra https://orcid.org/0000-0002-0126-1398

References

[1] Hoai H C, Bay N T, Khoi D N and Nga T N Q 2019 Analyzing the causes producing the rapidity of riverbank erosion in Mekong delta *VN J. Hydrometeorol.* **703** pp 42-50
[2] Ancey C 2020 Bedload transport: a walk between randomness and determinism. Part 2. Challenges and prospects *J. Hydraul. Res.* **58** pp 18-33
[3] Zuo L Q, Lu Y J, Liu H X, Ren F F and Sun Y Y 2020 Responses of river bed evolution to flow-sediment process changes after three gorges project in middle Yangtze river: A case study of Yaojian reach *Water Sci. Eng.* **13** pp 124-35
[4] Wilson M and Atkinson P 2007 The use of remotely sensed land cover to derive floodplain friction coefficients for flood inundation modelling *Hydrol. Processes* **21** pp 3576-86
[5] Wu W 2007 Computational river dynamics *Book Computational river dynamics* vol (CRC Press)
[6] Jia Y and Wang S S 1999 Numerical model for channel flow and morphological change studies *J. Hydraul. Eng.* **125** pp 924-33
[7] Tran T H 2017 Application of MIKE 21 FM modelling to simulate the water quality at the coastal area Dinh Vu *Sci. Tech. Dev. J. - Natural Sci.* **1** pp 282-92
[8] Tran Anh D, Hoang L P, Bui M D and Rutschmann P 2018 Simulating future flows and salinity intrusion using combined one-and two-dimensional hydrodynamic modelling - the case of Hau River, Vietnamese Mekong delta *Water* **10**
[9] Tri D Q, Thai T H and Linh N T M 2018 Application of 2D modeling in simulation the erosion of dykes on Thach Han river basin in Vietnam *Lowland Tech. Int. J.* **20** pp 393-400
[10] Nguyen A T T, Tran H N, Vu H T and Nguyen T M 2020 Applying MIKE 21 FM (HD+ ST) model to evaluate the ability to regenerate sand and gravel for efficient management of river exploitation and protection: A case study at Red River crossing Phuc Tho District, Hanoi *JMES* **61** pp 86-94
[11] Van C T, Tuan L A and Tuan N C 2021 Application of two-dimensional hydrodynamic model (MIKE 21FM) to simulate the sediment regime on Hau river, piloted in Long Xuyen city-An Giang province *Sci. Tech. Dev. J. - Sci. Earth Environ.* **5** pp 20-33
[12] Vinh V D, Ouillon S, Thao N V and Tien N N 2016 Numerical simulations of suspended sediment dynamics due to seasonal forcing in the Mekong coastal area *Water* **8**
[13] Tu L X, Thanh V Q, Reyns J, Van S P, Anh D T, Dang T D and Roelvink D 2019 Sediment transport and morphodynamical modeling on the estuaries and coastal zone of the Vietnamese Mekong Delta *Cont. Shelf Res.* **186** pp 64-76
[14] Duc Anh N Q, Tanaka H, Tam H S, Tinh N X, Tung T T and Viet N T 2020 Comprehensive study of the sand spit evolution at tidal inlets in the Central Coast of Vietnam *J. Mar. Sci. Eng.* **8**
[15] Tan N P, Tri V P D and Thanh V Q 2014 The application of a two-dimensional hydraulic model to simulate the hydraulic properties and deposition and erosion patterns in the Dinh An estuary *Sci. J. of Can Tho University* **31** pp 8-17

[16] Lee S K, Dang T A and Le V T 2019 Assessment of river morphological change for Co Chien Estuary applying the CCHE2D model *J. Indian Soc. Remote Sens.* **47** pp 1623-32
[17] Vu T T, Nguyen P K, Chua L H and Law A W 2015 Two-dimensional hydrodynamic modelling of flood inundation for a part of the Mekong River with TELEMAC-2D *IJECC* **5** pp 162-75
[18] Binh D V, Kantoush S, Sumi T, Mai N P and Trung L V 2020 Dam-induced riverbed incision and saltwater intrusion in the Mekong Delta *River Flow 2020* (CRC Press) pp 2243-51
[19] Binh D V, Kantoush S A, Sumi T, Mai N P, Ngoc T A, Trung L V and An T D 2021 Effects of riverbed incision on the hydrology of the Vietnamese Mekong Delta *Hydrol. Processes* **35**
[20] Soulsby R L 1997 Dynamics of marine sands: a manual for practical applications *Book Dynamics of marine sands: a manual for practical applications* vol 9 (London: Telford)
[21] Rijn L C V 1984 Sediment transport, part II: suspended load transport *J. Hydraul. Eng.* **110** pp 1613-41
[22] Nash J E and Sutcliffe J V 1970 River flow forecasting through conceptual models part I - A discussion of principles *J. Hydrol.* **10** pp 282-90
[23] Moriasi D N, Arnold J G, Van Liew M W, Bingner R L, Harmel R D and Veith T L 2007 Model evaluation guidelines for systematic quantification of accuracy in watershed simulations *Transactions of the ASABE* **50** pp 885-900

Flash Flood Susceptibility Mapping in Sungai Pinang catchment using Weight of Evidence

A Saleh[1], N Sabtu[2] and M R Bunmi[2]

[1] Disaster Preparedness and Prevention Centre (DPPC), Malaysia-Japan International Institute of Technology (MJIIT), Universiti Teknologi Malaysia Kuala Lumpur, Jalan Sultan Yahya Petra, 54100 Kuala Lumpur, Malaysia
[2] School of Civil Engineering, Engineering Campus, Universiti Sains Malaysia, 14300 Nibong Tebal, Penang, Malaysia

*Corresponding author e-mail: azlansaleh10@gmail.com

Abstract. Flash floods are one type of natural disaster that frequently occurs in Malaysia, especially in urban areas. This paper's objective is to develop the flash flood susceptibility mapping (FFSM) using weight of evidence (WOE) in the Sungai Pinang catchment. The Sungai Pinang catchment is located in the highly urbanised areas of Georgetown. The FFSM was developed using six flash flood conditioning factors: elevation, slope, rainfall, land cover, distance from the river, and lithology. The study showed that the WOE method is capable of producing high accuracy, with a success rate of 0.839 and a validation rate of 0.923 for the receiver operating characteristic's area under the curve (AUROC).

1. Introduction

Floods are a form of natural disaster that occurs on a regular basis in Malaysia. A flood can result in fatalities, environmental devastation, economic upheaval, and property loss. According to the World Meteorological Organization (WMO), flooding remains the third most devastating natural disaster, taking numerous lives and destroying destruction on infrastructure [1]. Direct and indirect destruction can both caused by flooding. Floods have the potential to directly result in the deaths of humans and animals, and property destruction. Indirect damage includes ecological changes, disease spread, and other implications due to floods [2].

According to Singh [3], floods affected roughly 178 million people, and the number of floods observed has grown dramatically during the last two decades. This is due to unmanaged development, a lack of drainage infrastructure and stream overflow, and extreme rainfall, which can result in flash floods in low-lying regions [4]. According to Dewan [5], the Balu River is used as an illustration of how development or land opening in low-lying areas has a substantial impact on future floods. This unplanned expansion has the potential to significantly increase runoff. Furthermore, because the main of Dhaka's urban districts are low-lying plain, elevation plays a crucial effect in the urban's flood threat. [5]. Dewan [5] added that altitudes between 1 and 4 metres are classified as high-hazard, while altitudes higher than 9 metres are classified as low-hazard. Individuals who live in low-lying locations or near rivers, for example, are in danger of floods. This demonstrates that elevation is a critical factor in determining an urban area's susceptibility to flooding.

Thus, the primary purpose of this work is to build the FFSM using the WOE model in accordance with six confounding variables: elevation, slope, rainfall, land cover, distance from the river, and lithology. To develop the FFSM, the current study will focus on the following objectives: (i) selecting flash flood conditioning factors (elevation, slope, rainfall, land cover, distance from the river, and lithology); (ii) analysing the correlation between the conditioning factors using a multi-collinearity test; (iii) developing the FFSM using WOE models; and (iv) determining model performance using the area under the receiver operating characteristic (AUROC) curve.

2. Study Area

The Sungai Pinang runs through Georgetown, the capital city of Penang. It is a developed and densely populated area [6]. The catchment of Sungai Pinang is just 3.1 kilometres long and spans an area of 50.97 kilometres square [7]. The study location is restricted by the longitudes of 100° 14' 26" East and 100° 20' 48" East, and the latitudes of 5° 26' 50" North and 5° 21' 29" North. Sungai Pinang was chosen for developing a flash flood susceptibility map (FFSM) because it is a low-lying area and susceptible to flash flooding. This study area is also the largest river catchment area on Penang Island with the highest population density. On 4 November 2017, a flash flood struck Penang. According to Davies [8], Kepala Batas meteorological station in Penang reported 946 mm of rain between 4 and 6 November 2017, with 458 mm of that amount falling in only 24 hours on 6 November 2017. The location of the study area is presented in Figure 1.

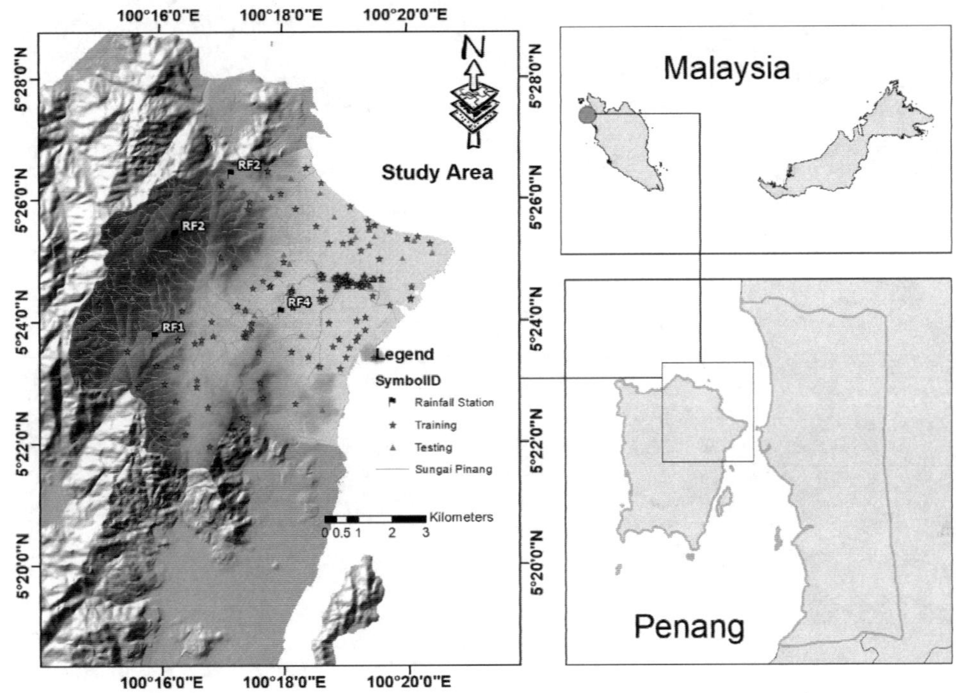

Figure 1. Location of study area Sungai Pinang catchment, Penang

3. Methodology

Firstly, a flash flood inventory will be created utilising data from the Penang GIS Centre (PEGIS) and the Majlis Bandaraya Pulau Pinang (MBPP) report. Based on the report, 110 flash flood points were recorded, which will be used in this study to develop FFSM. ArcGIS and Google Earth Pro were used to convert the report's inventory data to point data. Following data combination and validation, 70% of the entire data set was used for training and 30% for validation. This number is taken from past susceptibility mapping experiments, in which 70% of the time was spent on training and 30% on validation in the creation of FFSM [9–11].

Six conditioning factors were chosen for the analysis of FFSM. These conditioning factors are elevation, slope, rainfall, land cover, distance from the river, and lithology. *Elevation* is a very important factor in developing FFSM. Due to the movement of water from higher altitudes to lower altitudes areas, lower altitude or ground are prone to flooding. [12]. Flooding is more likely to occur in low-lying regions than in higher altitude places [13]. *Slope* is critical during a flood occurrence. When the angle of the slope increases, the infiltration time reduces, resulting in floods [14]. Furthermore, slope affects infiltration, drainage density, runoff, and erosion of the soil [15]. *Rainfall* is the essential component that contributes to the formation of a flood. Due to the common association of flash floods with heavy and rapid rainstorms, rainfall is critical in flash flood prediction [16]. The conditioning factor maps for this study are depicted in Figure 2.

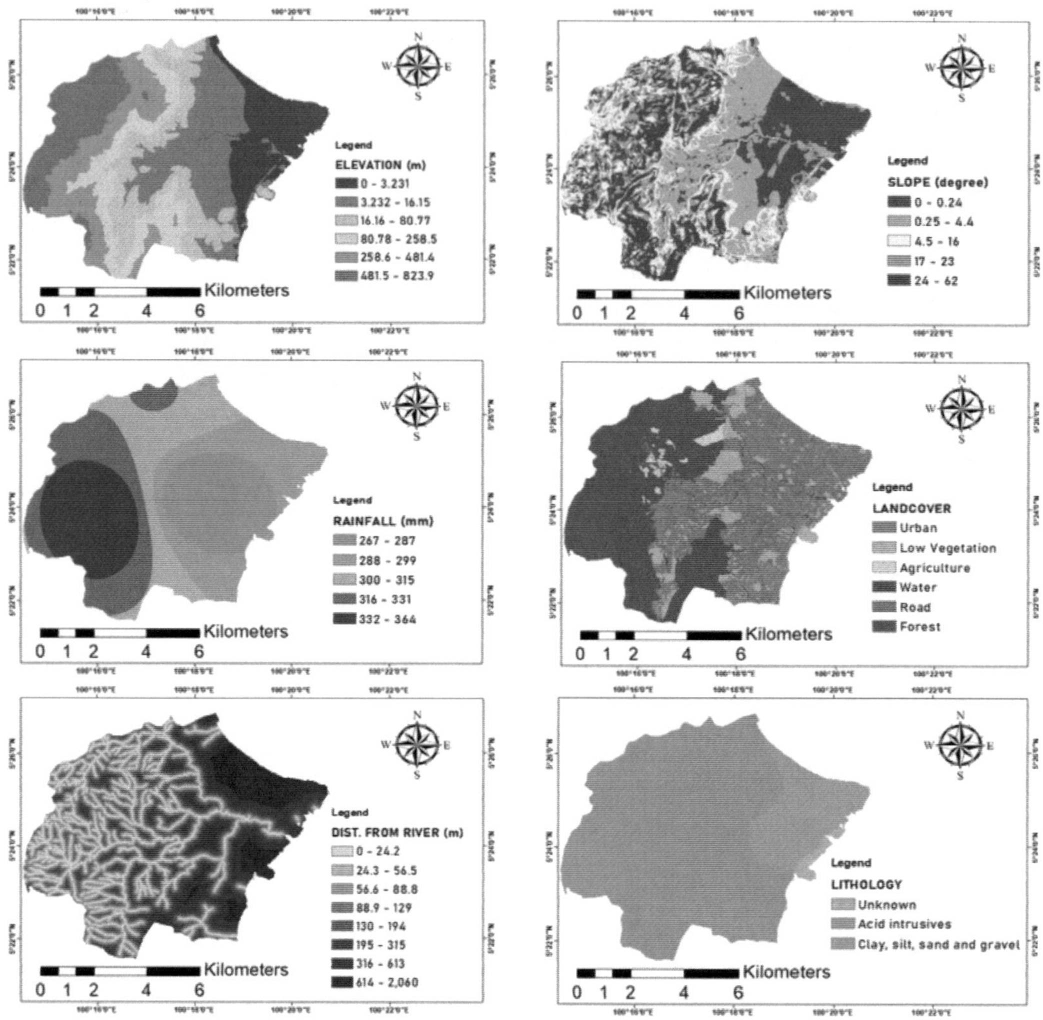

Figure 2. The conditioning factors of this study.

Land cover in FFSM is taken into account appropriately, since the dense vegetation and forest area coverage result in less surface runoff than developed regions. Land cover has a significant effect on runoff's velocity, interception, infiltration, and evapotranspiration [17,18]. *Distance from river* factor calculated in an ArcGIS environment using the Euclidean distance tool. It was revealed that decreasing the distance from the river component considerably increases the chances of flooding; increasing the distance significantly decreases the probability of flooding [19]. *Lithology*, the present study divided lithology into three groups by utilising a geological map obtained from PEGIS. (Unknown class, Acid Intrusive class, and Clay, Silt, and Gravel class).

Following the selection of the conditioning factors, multi-collinearity analysis was utilised to assess the correlation between them. This method was used by [10,20,21] to analyse the correlation between factors. Then, the FFSM will be developed after the analysis of the WOE method. The Wf's values of conditioning factors will be assigned to each raster layer using the raster calculator in ArcGIS to develop FFSM. Finally, the FFSM will be analysed using the area under the curve of receiver operating characteristic (AUROC) to determine the accuracy of the FFSM. Figure 3 shows the flowchart of the methodology.

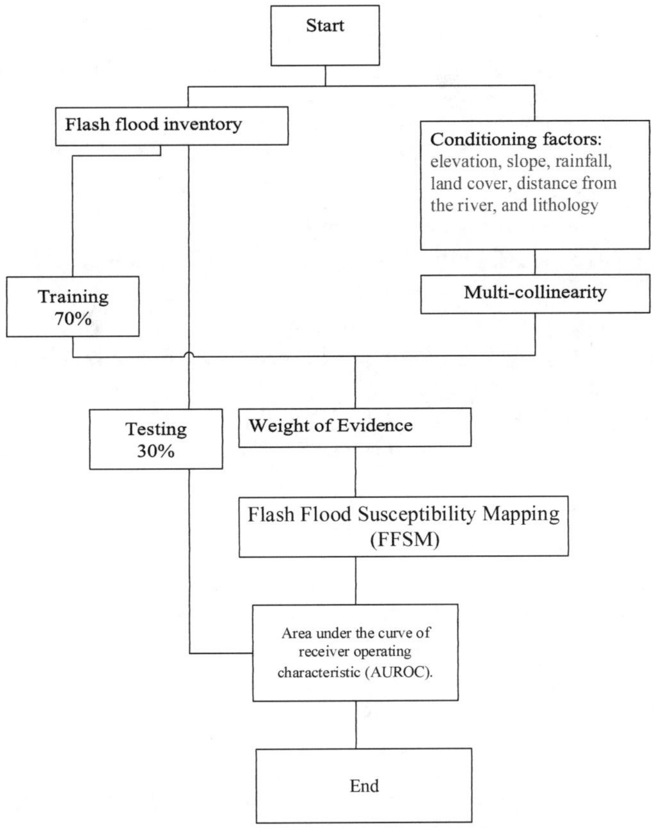

Figure 3. Flowchart of the methodology

3.1. Weight of Evidence (WOE)

In this study, the WOE method is used to compute the weights assigned to each variable category or class in relation to flash flood occurrences. As indicated previously by Costache et al. [22], in order to derive the WOE coefficients, the positive (W+) and negative (W-) weights must first be computed. Positive weights suggest a strong association between a variable category and flash flood spots or pixels, whereas negative weights suggest that this spatial relationship does not exist. Weights should be computed using equations (1) and (2) [23]:

$$W_i^+ = \ln \frac{P\{B|A\}}{P\{B|\bar{A}\}} \tag{1}$$

$$W_i^- = \ln \frac{P\{\bar{B}|A\}}{P\{\bar{B}|\bar{A}\}} \tag{2}$$

where *P* denotes the probability of a flash flood occurring and *ln* represents the natural log. *B* and $\bar{B}$ are denote the occurrence and non-occurrence of the variable, respectively. *A* and $\bar{A}$ represent the occurrence and non-occurrence of flash floods [24]. The final WOE (*Wf*) value given to every variable category or class can be obtained as equation (3) [25]:

$$Wf = W_{plus} + W_{mintotal} - W_{min} \quad (3)$$

where W_{plus} indicates the positive weight of a class variable, W_{min} represents the negative weight of a class independent variable, and $W_{mintotal}$ denotes the total of all negative weights in an independent variable.

3.2. Multi-collinearity Analysis

When more than two predictor variables in a multiple regression model have a significant linear correlation, it is referred to as multi-collinearity in statistics [26]. In this study, multi-collinearity was identified using TOL and VIF. SPSS software was used to perform the multi-collinearity analysis. Khosravi et al. [27] previously reported that VIF greater than 10 or tolerance below 0.1 shows that there is a problem with multi-collinearity among the variables. Equations (4) and (5) were applied to determine VIF and TOL:

$$TOL = 1 - R_v^2 \quad (4)$$

$$VIF = \frac{1}{TOL} \quad (5)$$

where R_v^2 represents the coefficient of determination of an explanatory variable's regression on all other explanatory variables [28]. Multi-collinearity analysis is performed using SPSS software.

4. Results and Discussion

4.1. Multi-collinearity

Multi-collinearity is often not an issue or problem when the TOL (tolerance) value exceeds 0.1 and the VIF (variance inflation factor) number is less than 10 [29]. The TOL and VIF studies of this data set revealed no indication of multi-collinearity. The slope factor had the lowest TOL value of 0.322, while the distance from river had the greatest TOL value, which was 0.910. The highest VIF value observed in this current study is slope (VIF=3.101). The findings suggest that all of the conditioning factors can be used to generate the FFSM since their values fall within the specified range (TOL > 0.1 and VIF 10). Table 1 presents the results of multi-collinearity analysis for this study.

Table 1. Multi-collinearity analysis

Conditioning factor	Collinearity Statistics	
	TOL	VIF
Elevation	0.452	2.213
Slope	0.322	3.101
Land cover	0.533	1.877
Rainfall	0.522	1.915
Distance from River	0.910	1.099
Lithology	0.723	1.384

4.2. Weight of Evidence

Table 2 shows the spatial relationship between flash flood occurrences and variables using the WOE method. The WOE method is used to determine the relationship between flash flood occurrences and classes of independent variables. Elevation factor values ranged from 1.40 in class 0-3.231 to 0.93 in class 3.231-16.155, 0.44 in class 481.431-823.925, 0.44 in class 80.777-258.486, 0.44 in class 258.486-

481.431, and -3.08 in class 16.155-80.777. It demonstrates that in the Sungai Pinang basin, the lowest elevation class has the largest effect on the occurrence of flash floods.

Based on Table 2, the greatest Wf value is 1.46 for the 0-1° slope class, followed by 0.96 for 0°, 14.20° (-0.66), 25-62° (-0.67), 1-5° (-0.98), and 5-14° (-1.58). The lowest Wf value is -2.96 for the 20–25° slope class. It demonstrates definitely that the low slope class is more vulnerable to flash floods. In terms of land cover, the road and urban classes have the highest weights of 0.97 and 0.88, respectively. Two components had a negative impact on flash floods: forest (-0.69) and agriculture (-0.14).

In terms of rainfall, the class 267–287 mm has the greatest Wf value (2.17), making it the sole class with a beneficial effect. The fourth class of this conditioning factor (316–330 mm) was discovered at a low Wf value of -3.84. The 88.9–129 m class had the greatest influence on the presence of flash floods, with a score of 1.028. There were just two classes where this conditioning factor was negative: 56.6-88.8 m and 88.9-129 m. The remaining classes had negative values, with the lowest being 130–194 m. The clay, silt, sand, and gravel classes all have a significant impact with 2.80 on flash floods in the Sungai Pinang basin. Acid Intrusive had a -1.85 value, whilst Unknown had a 0.24 value. Finally, a flash flood susceptibility map will be developed using ArcGIS software, using the WOE weights and equation 6:

$$FFSM_{WoE} = WoE_{elv} + WoE_{slope} + WoE_{landcover} + WoE_{rainfall} + WoE_{distriver} + WoE_{lithology} \quad (6)$$

Table 2. Weight of Evidence

Factor	Factor classes	W+	W-	Wf
Elevation	0 - 3.23	1.16	-0.51	1.40
	3.23 - 16.15	0.78	-0.42	0.93
	16.16 - 80.78	-2.62	0.18	-3.08
	80.78 - 258.49	0.00	0.16	-0.44
	258.49 - 481.43	0.00	0.16	-0.44
	481.43 - 823.93	0.00	0.16	-0.43
Slope	0	1.40	-0.07	0.96
	0 - 1	0.96	-1.02	1.46
	1 - 5	-0.40	0.06	-0.98
	5 - 14	-0.97	0.09	-1.58
	14 - 20	0.00	0.14	-0.66
	20 - 25	-2.32	0.13	-2.96
	25 - 62	0.00	0.15	-0.67
Land Cover	Urban	0.54	-0.46	0.88
	Low Vegetation	0.35	-0.05	0.27
	Agriculture	0.00	0.02	-0.14
	Water	0.14	0.00	0.01
	Road	0.91	-0.19	0.97
	Forest	0.00	0.56	-0.69
Rainfall	267 - 287	1.45	-1.56	2.17
	288 - 299	-0.46	0.09	-1.39
	300-315	0.00	0.21	-1.06
	316-330	-2.78	0.22	-3.84

	331-364	-2.02	0.19	-3.06
Distance from River	0–24.2	-0.12	0.02	-0.17
	24.3–56.5	-0.18	0.03	-0.24
	56.6-88.8	0.10	-0.02	0.09
	88.9-129	0.84	-0.22	1.03
	130-194	-0.19	0.02	-0.24
	195-315	-0.02	0.00	-0.06
	316-613	-0.10	0.01	-0.15
	614-2060	0.00	0.12	-0.15
Lithology	Unknown	-0.20	0.00	0.24
	Acid Intrusive	-0.93	1.36	-1.85
	Clay, Silt, Sand, and Gravel	1.43	-0.91	2.80

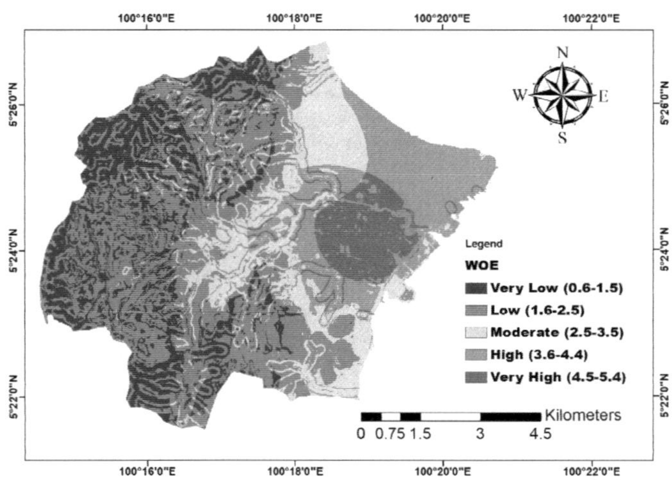

Figure 4. FFSM developed by WOE method

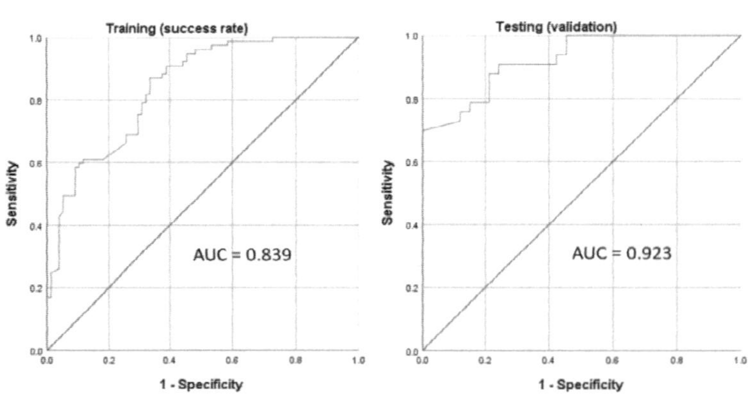

Figure 5. AUROC of WOE method

Following the development of the map using equation 6, it was divided into five susceptibility levels: very low, low, moderate, high, and very high. Figure 4 shows the FFSM using the WOE method. The success rate of the WOE method was 0.839, and the validation rate was 0.923. As Hong et al. [30] stated that the AUROC value can be divided into five classes, which are 0.5–0.6 (poor), 0.6–0.7 (average), 0.7–0.8 (good), 0.8–0.9 (very good), and 0.9–1 (excellent). Thus, an AUROC of 0.7 to 0.8 signifies a reasonable level of performance for the success rate. For the validation rate, it can be classified as excellent. The findings indicate that the WOE method can be used to create FFSM for this study area. Figure 5 shows the AUROC of the WOE method.

5. Conclusion

The WOE method was used in this work to create the FFSM for the Sungai Pinang catchment. The study considered six conditioning factors: elevation, slope, land cover, rainfall, distance from the river, and lithology. These components could be classified as standard or essential factors. The correlation between the conditioning factors was analysed using multi-collinearity analysis. These six conditioning factors demonstrate that there is no correlation and can be used to create an FFSM. Based on the FFSM, it indicates that the location near the major river of Sungai Pinang is the most vulnerable to flash floods. The AUROC of FFSM can be classified as good if it is greater than 0.7. The AUROC results show that the FFSM produced by the WOE model demonstrates very good performance for success rate and excellent validation rate. Thus, the FFSM of the WOE model can be used to help the local authorities manage and prevent flash floods in the study area. It is also important to give information about flash flood prone areas for future development and land use planning in the Sungai Pinang catchment areas.

Acknowledgments

The first author would like to thanks to his supervisor, Prof Ali Yuzir from Malaysia-Japan International Institute of Technology (MJIIT) for his advices that improved the quality of the manuscript. The first author also would like to acknowledge MJIIT and UTM for providing an Incentive Scheme to undertake his PhD program. The authors would also like to thank the MBPP, DID, JMG, PLANMalaysia, and PEGIS for providing the data for this study.

ORCID iDs

Azlan Saleh https://orcid.org/0000-0002-7457-3114
Nuridah Sabtu http://orcid.org/0000-0003-0564-1272
Mudashiru Rofiat Bunmi http://orcid.org/0000-0003-3655-0074

References

[1] DID 2017 Flood Preparedness Guidelines
[2] Koirala S 2011 Some flood damage estimate of the world *UNESCO-IHE Flood Manag. Educ. Platf.*
[3] Singh D 2012 Unplanned urbanization increasing flood impacts *UNDRR*
[4] Bhuiyan T R, Hasan Reza M I, Choy E A and Pereira J J 2018 Direct impact of flash floods in Kuala Lumpur City: Secondary data-based analysis *ASM Sci. J.* **11** 145–57
[5] Dewan A 2013 *Floods in a Megacity: Geospatial Techniques in Assessing Hazards, Risk and Vulnerability*
[6] Saad F N M, Rahman N N N A, Kadir M O A and Omar F M 2006 Identification of pollution sources within the Sungai Pinang River Basin *Proceeding Malaysian Res. Gr. Int. Conf.* 478–85
[7] Rahaman Z A, Farhana S, Rus C H E, Omar M A and Ismail W A N R 2016 Rivers and Lakes as Natural Heritage: Water Quality Status in the Northern States of Peninsular Malaysia **23** 109–28
[8] Davies R 2017 Malaysia – Severe Storm and Floods Leave 7 Dead, 10,000 Displaced

FloodList

[9] Saleh A, Yuzir A, Sabtu N, Abujayyab S K M, Mudashiru R B and Pham Q B 2022 Flash Flood Susceptibility Mapping in Urban Area using Genetic Algorithm and Ensemble Method *Geocarto Int.* **0** 1–27

[10] Al-Juaidi A E M M, Nassar A M and Al-Juaidi O E M M 2018 Evaluation of flood susceptibility mapping using logistic regression and GIS conditioning factors *Arab. J. Geosci.* **11** 765

[11] Costache R, Bao Q, Avand M, Thi N, Linh T, Lee S, Nguyen D, Thi P, Nhi T, Vojtek M, Vojtekov J, Pham Q B, Avand M, Thuy Linh N T, Vojtek M, Vojteková J, Lee S, Khoi D N, Thao Nhi P T, Dung T D, Bao Q, Avand M, Thi N, Linh T, Lee S, Nguyen D, Thi P, Nhi T, Vojtek M and Vojtekov J 2020 Novel hybrid models between bivariate statistics, artificial neural networks and boosting algorithms for flood susceptibility assessment *J. Environ. Manage.* **265**

[12] Khosravi K, Nohani E, Maroufinia E and Pourghasemi H R 2016 A GIS-based flood susceptibility assessment and its mapping in Iran: a comparison between frequency ratio and weights-of-evidence bivariate statistical models with multi-criteria decision-making technique *Nat. Hazards* **83** 947–87

[13] Tehrany M S, Pradhan B and Jebur M N 2013 Spatial prediction of flood susceptible areas using rule based decision tree (DT) and a novel ensemble bivariate and multivariate statistical models in GIS *J. Hydrol.* **504** 69–79

[14] Khosravi K, Pourghasemi H R, Chapi K and Bahri M 2016 Flash flood susceptibility analysis and its mapping using different bivariate models in Iran: a comparison between Shannon's entropy, statistical index, and weighting factor models *Environ. Monit. Assess.* **188**

[15] Conforti M, Aucelli P P C, Robustelli G and Scarciglia F 2011 Geomorphology and GIS analysis for mapping gully erosion susceptibility in the Turbolo stream catchment (Northern Calabria, Italy) *Nat. Hazards* **56** 881–898

[16] Borga M, Anagnostou E N, Blo G and Creutin J 2011 Flash flood forecasting , warning and risk management : the HYDRATE project **4**

[17] Choubin B, Moradi E, Golshan M, Adamowski J, Sajedi-Hosseini F and Mosavi A 2019 An ensemble prediction of flood susceptibility using multivariate discriminant analysis, classification and regression trees, and support vector machines *Sci. Total Environ.* **651** 2087–96

[18] Yalcin A, Reis S, Aydinoglu A C and Yomralioglu T 2011 A GIS-based comparative study of frequency ratio, analytical hierarchy process, bivariate statistics and logistics regression methods for landslide susceptibility mapping in Trabzon, NE Turkey *Catena* **85** 274–87

[19] Talukdar S, Ghose B, Shahfahad, Salam R, Mahato S, Pham Q B, Linh N T T, Costache R and Avand M 2020 Flood susceptibility modeling in Teesta River basin, Bangladesh using novel ensembles of bagging algorithms *Stoch. Environ. Res. Risk Assess.* **34** 2277–300

[20] Tehrany M S, Jones S and Shabani F 2019 Identifying the essential flood conditioning factors for flood prone area mapping using machine learning techniques *Catena* **175** 174–92

[21] Towfiqul Islam A R M, Talukdar S, Mahato S, Kundu S, Eibek K U, Pham Q B, Kuriqi A and Linh N T T 2021 Flood susceptibility modelling using advanced ensemble machine learning models *Geosci. Front.* **12** 101075

[22] Costache R, Ţîncu R, Elkhrachy I, Pham Q B, Popa M C, Diaconu D C, Avand M, Costache I, Arabameri A and Bui D T 2020 New neural fuzzy-based machine learning ensemble for enhancing the prediction accuracy of flood susceptibility mapping *Hydrol. Sci. J.* **65** 2816–37

[23] Van Westen C J, Rengers N and Soeters R 2003 Use of geomorphological information in indirect landslide susceptibility assessment *Nat. Hazards* **30** 399–419

[24] Tehrany M S, Pradhan B and Jebur M N 2014 Flood susceptibility mapping using a novel ensemble weights-of-evidence and support vector machine models in GIS *J. Hydrol.* **512** 332–43

[25] Costache R, Blaschke T, Pham Q B, Pandey M, Arabameri A, Blaschke T, Pham Q B, Pham B T, Pandey M, Arora A, Linh N T T and Costache I 2021 Flash-flood potential mapping using deep learning, alternating decision trees and data provided by remote sensing sensors *Sensors (Switzerland)* **21** 1–21

[26] Hong H, Liu J and Zhu A X 2020 Modeling landslide susceptibility using LogitBoost alternating decision trees and forest by penalizing attributes with the bagging ensemble *Sci. Total Environ.* **718** 137231

[27] Khosravi K, Pham B T, Chapi K, Shirzadi A, Shahabi H, Revhaug I, Prakash I and Tien Bui D 2018 A comparative assessment of decision trees algorithms for flash flood susceptibility modeling at Haraz watershed, northern Iran *Sci. Total Environ.* **627** 744–55

[28] Arora A, Pandey M, Siddiqui M A, Hong H and Mishra V N 2019 Spatial flood susceptibility prediction in Middle Ganga Plain: comparison of frequency ratio and Shannon's entropy models *Geocarto Int.* **0** 1–32

[29] Saha S, Arabameri A, Saha A, Blaschke T, Ngo P T T, Nhu V H, Band S S, Thao P, Ngo T, Ha V and Band S S 2020 Prediction of landslide susceptibility in Rudraprayag, India using novel ensemble of conditional probability and boosted regression tree-based on cross-validation method *Sci. Total Environ.* 142928

[30] Hong H, Duan H, Gigović L, Pamučar D, Wang Y, Chen W, Tien Bui D, Drobnjak S and Li S 2018 A Hybrid GIS Multi-Criteria Decision-Making Method for Flood Susceptibility Mapping at Shangyou, China *Remote Sens.* **11** 62

Alternative Design of Post-Earthquake Permanent Housing in Indonesia using Cold Formed Steel

R. Alghiffary[1], Data Iranata*[1], D. Irawan[1], B. Suswanto[1], H. Kristijanto[1]

[1] Department of Civil Engineering, Institut Teknologi Sepuluh Nopember, Surabaya, Indonesia

*Corresponding author e-mail: iranata80@gmail.com / data@ce.its.ac.id

Abstract. This article informs about permanent housing 2-story necessity for the Earthquake victim's refugees in Indonesia. Continuing about the temporary housing (Huntara 1.0) which has been studied previously. In this article, an alternative earthquake resistant design of permanent housing using Cold-Formed Steel is proposed. Same as previous research, the loading estimation used assumption that the buildings are located in Yogyakarta as one of the cities that has the highest value of PGA. Another assumption of loading is based on available Indonesia Building Code, including the calculation of structural ability and design. Stress ratio check is automatically done using AISI LRFD 96, then manually checked using Indonesia's code. The model also accommodates stiffness of masonry panel using Holmes equation. Overall, a new alternative design of post-earthquake permanent housing is presented in this article with relatively economic in budget and rapid construction method.

1. Introduction

Every country in an earthquake-prone area must have its mitigation actions to deal with natural disasters. Therefore, to avoid many casualties and unnecessary demolition of collapsed infrastructure due to disasters, disaster management strategies should be well planned and improved properly [1]. Indonesia is one of the countries that are prone to earthquakes, according to Indonesia's Meteorology, Climatology and Geophysics Agency, earthquakes occur almost every day [2]. The earthquake record in Indonesia presented in Figure 1 shows how vulnerable Indonesia is to earthquake attacks.

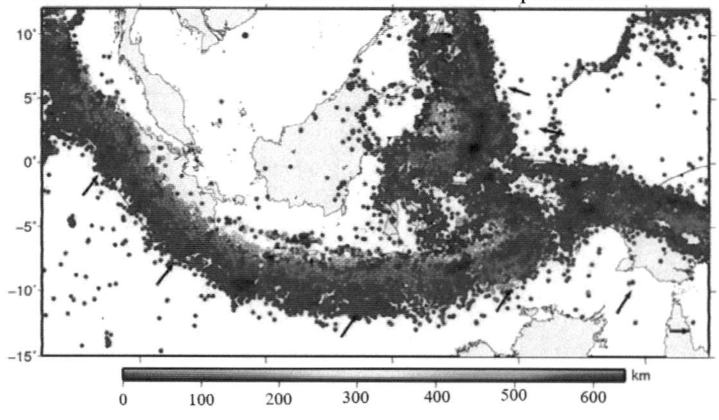

Figure 1. Earthquake record catalog 1907-2016

After an earthquake occurs, people who are victims of the earthquake disaster suffer in the economic, social, and physical integrity of life [3]. In a messy post-earthquake situation, house is a very important element not only to restore the life of the affected people, but also can prevent the increase of deaths and the spread of disease, once conditions of personal hygiene and protection against external factors are ensured [4]. So, there is an urgent necessity for protection and shelter. Moreover, earthquake-resistant permanent housing becomes the priority for the victims as a replacement for their collapsed houses in disaster. Permanent housing itself is defined as community-based housing without a designated length of stay. It means that permanent housing can return the daily activities such as school and work to normal. Permanent housing can take the form of a prefabricated home, a rented apartment, or a small shack. Building a permanent house that can withstand more earthquakes attack and meet the owner's need for safety feeling and a limited budget is a difficult job. Therefore, all those factors had created an idea about designing earthquake-resistant permanent housing.

Research on earthquake-resistant houses has been done before. Indonesia's Ministry of Public Works built an earthquake-resistant house called RISHA (Instant Simple and Healthy House), this product was first built after Aceh's disaster in 2004. RISHA is a house with the knockdown concept. Construction of RISHA uses precast panels that are connected by bolts and do not need cement and masonry blocks. With that method, the construction time of RISHA is faster than the conventional house [5]. Another development of RISHA was studied, RUSPIN (Instant Panel System Eminent Housing). The material used for RUSPIN is the same as RISHA, but it reduces the number of exterior displays of bolted connection room partition [6]. However, the use of precast concrete will experience problems in the material transportation process, if it will be constructed in a post-earthquake area. In 2018, Institut Teknologi Sepuluh Nopember launched a project called Huntara (Hunian Sementara / Temporary Housing), this Huntara was constructed from wood material and it has 34,56 m2 areas [7]. Although these materials are easily obtained from nature or the forest around the disaster areas, unfortunately, the quality and durability of these materials cannot be ensured.

Alternative construction material for housing nowadays was proposed. That material is cold-formed steel. The use of cold-formed steel (CFS) in the construction industry has become increasingly popular over the past decades. AISI has created its standard for cold-formed steel earthquake-resistant building design, so CFS can be widely used and developed [8]. Compared to hot-rolled steel, CFS members can lead to more economical and efficient design solutions due to their lightweight, easy and rapid construction, and greater flexibility in manufacturing [9].

Structural systems using Cold Formed Steel for simple structures can be considered for high-magnitude earthquakes [10]. Cold Formed Steel can achieve high strength by taking into account non-structural components of stiffness and lateral strength [11]. According to [12], with a cyclic test the structure of Cold Formed Steel can reach a drift ratio of more than 4%, which is very important for earthquake-resistant structures. Therefore, CFS was chosen in this paper as the main material in the design of earthquake-resistant permanent housing to meet the necessity of people affected by the earthquake disaster. However, the main column design is suggested from double lip channel cold formed with higher value of thickness due to limitation of cold-formed section dimension.

Previous research conducted by peterman about cold-formed steel structure and its seismic behaviour (including experimental test) has been done. Overall, the work demonstrates the good performance of these structures under seismic excitation and it should be underlined that the performance of the structure is related to the response of the overall structural system, not just the designated elements in the lateral-force-resisting system [13] and [14]. Another study conducted by McCrum investigates the performance of a single-story moment-resisting cold-formed steel (CFS) portal frame under seismic excitation through cyclic testing. The study concludes that not only perfect-fit tolerance bolt holes (PTBH) but also normal tolerance bolt holes (NTBH) connections had good hysteretic energy dissipation capacity, stable hysteresis, and good ductility [15]. Another alternative design of low-cost earthquake-resistant using CFS as the main material has been done, the house has 51.84 m2 areas and costs 36 million rupiahs [16].

2. Methodology

The methodology of this research consist of three different stages includes loading estimation, modelling, and data processing from the available model. Each stage refers to the proper literature review and will be described in the following explanation. All the stages used in this study must satisfy the main goal to get permanent housing with the following criteria: comfortable, easy to construct, and effective in cost.

2.1. Loading Estimation

The design proposed in this paper is expected to be widely used in Indonesia, especially for areas with high earthquake hazard potential. The potential for earthquake hazards in each region in Indonesia has been mapped in SNI 1726:2019 [17]. As can be seen in Figure 2, almost every area has a potential earthquake hazard. Therefore, areas with a high risk of earthquake hazards will appear with a higher number of PGA.

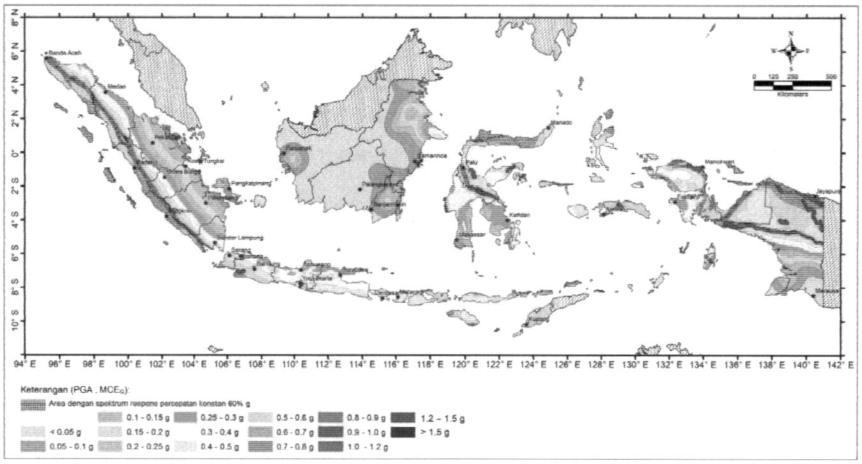

Figure 2. Earthquake Peak Ground Acceleration Maps

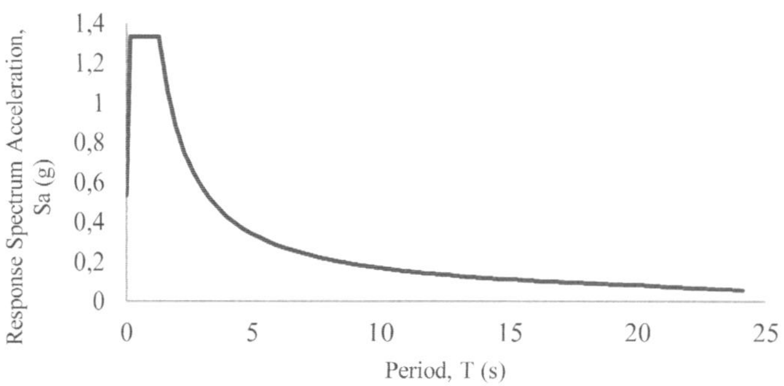

Figure 3. Response Spectrum Loading

Table 1. Seismic Parameter in Yogyakarata

Seismic Parameter	Values	Seismic Parameter	Values
PGA (g)	1.5	S_{M1}(g)	2.55
S_s(g)	2	S_{DS}(g)	1.333
S_1(g)	1.5	S_{D1}(g)	1.7
F_a	1	T_0(s)	0.157
F_v	1.7	T_s(s)	0.784
S_{MS}(g)	2	T_L(s)	20

Table 2. Seismic Parameter for Building in Yogyakarta

Seismic Parameter	Description
Risk Category	II
I	1
Seismic Design Category	D
R	4,5
Seismic Resistance System	Intermediate -Moment Resistance Frame

In this paper, the city used as the location model is Yogyakarta due to its high PGA value and other seismic parameter. Another reason is that the Yogyakarta earthquake disaster in 2006 with 5,9 SR caused 390,077 houses to collapse, this number is greater than other cities, so the seismic risk in this region needs to be considered. Yogyakarta itself has a 1.5g PGA value. Assumed that the locations of housing have soil category SD (medium soil).

Table 3. Design Load for Structure

Component	Load Values
Cold-Formed Steel	7850 kg/m^3
Lightweight Wall Panel	600 kg/m^2
Roof	10 kg/m^2
Wall Panel	10 kg/m^2
Rain Load	20 kg/m^2

For the seismic loading, some parameters to draw response spectrum function are listed in Tabel 1. SS and S1 values are obtained from the Seismic Code Maps that has the same configuration as PGA Maps in Figure 3 with different colour and value. After obtaining the value of SS and S1, we need soil category to get Fa and Fv values. Fa and Fv are defined as site classification. Using soil category SD (medium soil) as an assumption, the value of Fa and Fv can be obtained. Then the value of SMS and SM1 can be calculated by multiplying Ss with Fa and S1 with Fv, respectively. SDS and SD1 values are then calculated by multiplying two per three with SMS and SM1, respectively. Then with SDS and SD1 values are obtained before, calculate the value of T0, TS, and TL to make a response spectrum graph. The response spectrum graph resulting from available parameters for Yogyakarta is presented in Figure 3.

Another parameter to define seismic load is the building parameter. These building parameters are used to scale the response spectrum function loading, the building parameter include seismic primary factor (I) and coefficient of response modification (R). Those building parameters can be obtained from SNI 1726:2019. The Required parameters are listed in Table 2. Besides seismic loading, some loads need to define for this design process. loads are defined in accordance with Indonesia National Standard [18] and [19]. Other required loads that are used include: lightweight wall panel with 600 kg/m2, roof and wall panel with load maximum load values 10 kg/m2 and rain load used is 20 kg/m2 as seen in Table 3.

2.2. Design Modelling

Design modelling in this article is used finite element based structural analysis program. Firstly, the propose design from the architect are processed by using linear model to find optimal size of each structural element. The proposed housing architectural design in three-dimensional model and floor plan were captured in Figure 4. The design is chosen to accommodate enough room for earthquake victim's family needs. The models in structural analysis program are also modelled using three-dimensional model.

All the materials used in the proposed model are based on Cold formed steel catalogue from Ministry of Public Works [20] that also based on SNI 7971:2013 [21]. The model has been running automatically using AISI-LRFD96 design code [22] to check stress ratio before manually recalculated using

Indonesia's Code. Design of roof trusses frame used in this proposed construction model used previous research by Wuryanti using C75 profile [23], this frame has successfully resist 14083 N loads.

Half height of wall is designed using masonry panel. Masonry panel used in this proposed model are define as equivalent diagonal strut. For linear analysis in previous research found of Amalia [24] that as substitution of masonry panel Holmes equation [25] is the closest to the experimental test of Mehrabi [26]. Equivalent diagonal strut define in the structural analysis must have same thickness and elastic modulus with the masonry panel. Meanwhile, the calculation of equivalent width of strut (w) to be input can be calculated as 1/3 of diagonal length of infill panel.

Loading that used in this model are used previous assumption as described before. The response spectrum load function is entered as user defined function using .txt table from the formula specified before. Scale factor to be entered are derived from the multiply of I (seismic primary factor) and g (gravity acceleration of 9,81 m/s2) derived by R. Then, the scale factor calculated for this model is 2,1778 for mayor axis, and the thirty percent of that value in minor axis is 0,6533.

3. Result and Discussion

All presented results have been through a proper structural analysis process. From the modelling process, there are some results gathered, including three-dimensional models in structural analysis software.

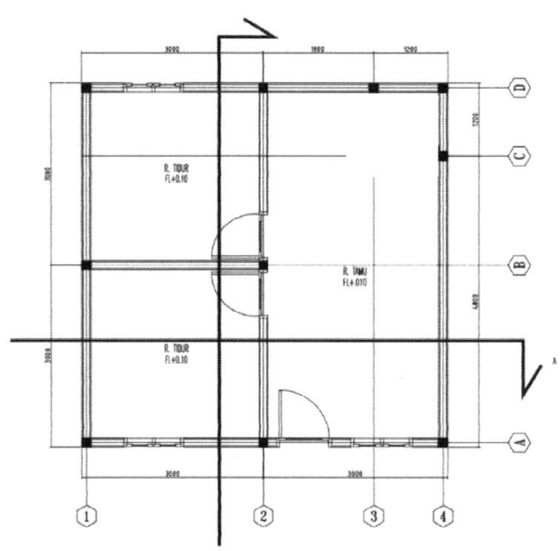

Figure 4. Floor Plan of The Permanent Housing

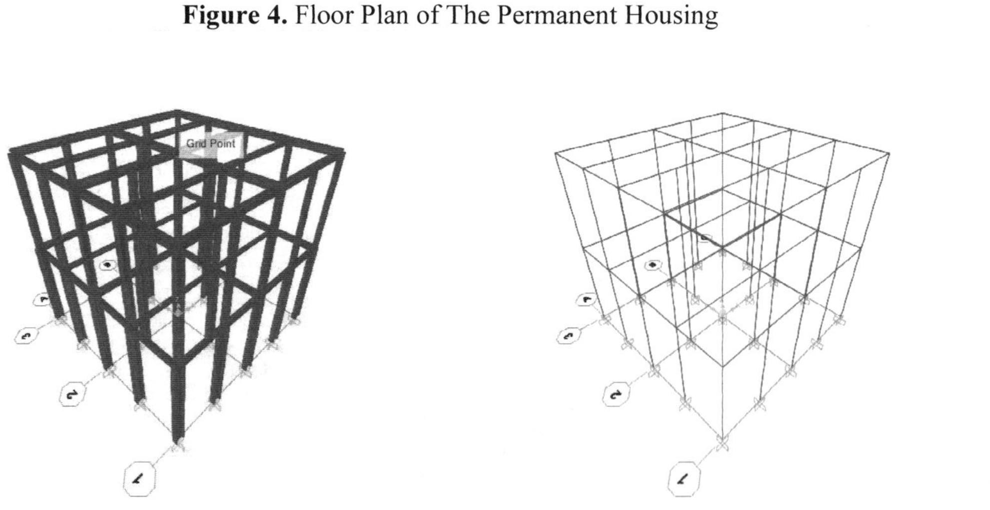

Figure 5. Three-dimensional view structural analysis model

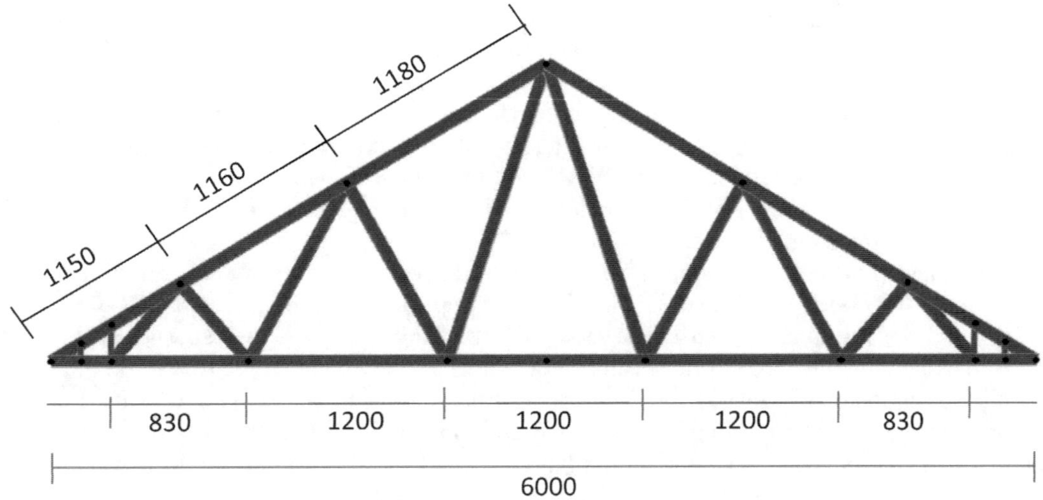

Figure 6. Trusses frame roof model configuration

Table 4. Design Load

Section	Total Weight
Roof Truss: Lip C75.35.15.1,5	196.54
Secondary Beam: Lip C150.65.20.4	236.4
Primary Beam: Lip C200.75.25.4,5	914.4
Column: Double Canal Lip C250.75.25.4,5	3754.8

First section of this result is directed to layout plan and three-dimensional view of this proposed building as presented in figure 4. There are 2-storiest in the permanent housing. The layout of this housing consists of two main bedroom and one functional room on every story. Bedrooms are separated using wall panel, 90 cm masonry wall height are designed around the building and in the required cover or separating panel. As the previous description the masonry wall are defined as equivalent diagonal strut using Holmes equation. The three-dimensional view of this proposed building designed to has simple view and enough ventilation. All this proposed designed has already been adjusted to available previous research.

The three-dimensional view of the structural analysis model can be seen in Figure 5. The left side of figure 5 is the extrusion mode and the right side is the normal mode of three-dimensional structural analysis model. The model use moment resistant frame system that connected every column using primary beam. As described in the previous chapter, the masonry wall in the modelling projected as equivalent diagonal strut using Holmes equation.

The roof trusses frame in this proposed design are using previous research by Wuryanti with lip C75.35.15.1,5 profile. This model are suggested to have moment resisting frame system that has beam and column as primary member to resist seismic loading. Column in this proposed design using C type profile but with higher thickness, double Lip C250.75.25.4,5. The primary beam used in the design model consist of C200.75.25.4,5 and the secondary beam use Lip C.150.65.20.4.

Material summary of this model are presented in table 4 below. From the cold formed material summary of this model 5102 kg of cold formed material need to be used for one housing. If we assume that price per kilograms of cold formed material is 17.000 rupiahs, so the total budget for material only is around 86.7 millions rupiahs. This price is still reasonable with the resistant of this proposed housing.

4. Conclusion

Natural disasters, especially earthquakes, will always occur in countries that are earthquake-prone areas. Each country must have mitigation actions to contain and cope with these natural disasters. There have been so many earthquakes that have occurred recently that have caused thousands of deaths and damaged houses. Communities in the disaster area suffer in various aspects. In such a situation, the house is not only a place to live to restore life, but the house can also reduce the death and spread of disease from the victims. So there is an urgent need for a place to live, namely a temporary house. Post-disaster management by preparing shelter plans for disaster victims is very important. Making a temporary house that is resistant to the impact of another earthquake and meets the owner's need for security and a limited budget is a tough job. By continuing previous research on designing temporary housing, all of these factors gave birth to ideas for designing permanent housing and studying the behaviour of structures against seismic loads.

The Indonesian Ministry of Public Works has research on earthquake-resistant buildings, in this case for housing. The products are named RISHA (Instant Simple and Healthy Home) and RUSPIN (Leading Housing with Instant Panel System). The material used for RUSPIN is the same as RISHA, which is made of precast concrete. However, because it still uses precast concrete, if it is to be built in a post-earthquake area, material transportation will be a problem as well as production limitations if it only relies on one type of material. In 2018, the Sepuluh Nopember Institute of Technology also launched Huntara 1.0 (Temporary Shelter), this shelter has an area of 34.56 m2 and is made of wood (ITS News, 2018). These materials are easily obtained from nature or forests around the affected areas, unfortunately the quality and durability of these materials cannot be ensured.

An alternative material for residential construction is currently proposed. It is a cold-formed steel (CFS) structural element that has undergone a dramatic expansion in various applications over the last decade. In this article 36 m2 permanent housing design using CFS material with 2-stories was proposed. The housing consists of two bedrooms and one functional room on every story. Model was assumed being built in Yogyakarta, as one of the regions that has highest value of PGA in Indonesia. Another assumption used in this model is roof and wall panel that has no more than 10 kg/m2 load, 20 kg/m2 rain loads and 40 kg/m2 wind load. Stress ratio check is automatically done using AISI LRFD 96, then manually checked using Indonesia's code. The model also accommodates stiffness of masonry panel using Holmes equation.

The roofing system used in this temporary housing using trusses frame, that using the same configuration as Wuryanti's research of trusses that consist of profile lip C75.35.15.1,5. Column in this proposed design using C type profile but with higher thickness, double Lip C250.75.25.4,5. The primary beam used in the design model consist of C200.75.25.4,5 and the secondary beam use Lip C.150.65.20.4 that stiffed in the middle joint using truss system. Overall design need 5102 kg of cold formed section and the material only price should be roughly 86.7 million rupiahs and this price are still relatively reasonable.

Near in the future, the prototype making, and experimental loading test are required to be done to ensure the safety of this proposed design to be used extensively. Join relation with a cold formed industry sector to create proper and easy to construct design are the next aim. That also leads to possibility of faster recovery in post-disaster area that support mitigation after earthquake.

Acknowledgement
This research has been funded by Institut Teknologi Sepuluh Nopember, Surabaya, Indonesia

References

[1] Parwanto N B 2014 Quantitative Study On Natural Disasters Risk Management Policy *National Graduate Institute for Policy Studies*

[2] Badan Meteorologi dan Geofisika 2019 Gempa Bumi Dirasakan https://www.bmkg.go.id/gempabumi/gempabumi-dirasakan.bmkg

[3] Amori K Y 2010 A Study of Life Recovery and Social Capital regarding Disaster Victims A Case Study of Indian Ocean Tsunami and Central Java Earthquake Recovery *Mayumi S AKAMOTO* 31(2) 13–20

[4] Félix D, Monteiro D, Branco J M, Bologna R, & Feio A 2015 The role of temporary accommodation buildings for post-disaster housing reconstruction *Journal of Housing and the Built Environment* 30(4) 683–699

[5] LITBANG PUPR 2019a RISHA http://eproduklitbang.pu.go.id/risha/
[6] LITBANG PUPR 2019b Rumah Unggul Sistem Panel Instan (RUSPIN) http://eproduklitbang.pu.go.id/ruspin/
[7] ITS News 2018 Huntara ITS 1.0 Ringankan Beban Korban Gempa https://www.its.ac.id/news/2018/09/03/huntara-its-1-0-ringankan-beban-korban-gempa/
[8] American Iron and Steel Institute AISI Manual Cold-Formed Steel Design - Vol.1 2013
[9] Mojtabaei S M, Kabir M Z, Hajirasouliha I, & Kargar M 2018 Analytical and experimental study on the seismic performance of cold-formed steel frames *Journal of Constructional Steel Research* 143, 18–31.
[10] Pourabdollah O, Farahbod F, & Rofooei F R 2017 The seismic performance of K-braced coldformed steel shear panels with improved connections *Journal of Constructional Steel Research* 135(April) 56–68
[11] Kechidi S, Bourahla N, & Castro J M 2017 Seismic design procedure for cold-formed steel sheathed shear wall frames: Proposal and evaluation *Journal of Constructional Steel Research* 128 219–232
[12] Pehlivan B M, Baran E, & Topkaya C 2020 An energy dissipating hold down device for cold formed steel structures *Journal of Constructional Steel Research* 166 105913
[13] Schafer B W, Ayhan D, Leng J, Liu P, Padilla-Llano D, Peterman K D, Yu C 2016 Seismic Response and Engineering of Cold-formed Steel Framed Buildings *Structures* 8 197–212.
[14] Peterman K D, Stehman M J J, Madsen R L, Buonopane S G, Nakata N, & Schafer B W 2016 Experimental Seismic Response of a Full-Scale Cold-Formed Steel-Framed Building II: Subsystem-Level Response *Journal of Structural Engineering* 142(12) 04016128
[15] McCrum D P, Simon J, Grimes M, Broderick B M, Lim J B P, & Wrzesien A M 2019 Experimental cyclic performance of cold-formed steel bolted moment resisting frames *Engineering Structures* 1811–14.
[16] Hardwani S M, & Patil A V 2013 Low cost Earthquake resistant house using Cold Formed Steel [CFS] member *A Case Study* 2(9) 2536–2540.
[17] Badan Standarisasi Nasional Tata cara perencanaan ketahanan gempa untuk struktur bangunan gedung dan non gedung *Pub. L. No. RSNI 3 03-1726-201x* 2019.
[18] Direktorat Penyelidikan Masalah Bangunan 1981 *Peraturan Pembebanan Indonesia untuk Gedung.*
[19] Badan Standarisasi Nasional 2013a *Beban minimum untuk perancangan bangunan gedung dan struktur lain.*
[20] Kementerian Pekerjaan Umum 2018 *Katalog Produk Baja Ringan Konstruksi. Jakarta: Direktorat Bina Kelembagaan dan Sumber Daya Jasa Konstruksi.*
[21] Badan Standarisasi Nasional Struktur Baja Canai Dingin *Pub. L. No. SNI 7971:2013* 2013.
[22] Brockenbrough R L *The 1996 AISI specification* 1996.
[23] Wuryanti W 2018 Experimental Study into The Capacity of Cold-Formed Roof Trusses with Various Geometric Model Under Static Loads *Jurnal Permukiman* 13(2) 90–103.
[24] Amalia A R, & Iranata D 2018 Comparative study on diagonal equivalent methods of masonry infill panel Comparative Study on Diagonal Equivalent Methods of Masonry Infill Panel *AIP Conference Proceeding* 1855(June 2017), 030011-1–11.
[25] Holmes 1961 Steel Frames With Brickwork and Concrete Infilling *Proceedings of the Institution of Civil Engineers* 473–478.
[26] Mehrabi A B, Shing P B, Schuller M P, & Noland, J 1996 Experimental Evaluation of Masonry-Infilled RC Frames *Journal of Structural Engineering ASCE* 122(3) 228–237.

Relationship of Building Form with Damage Pattern and Damage Level

Rita Laksmitasari Rahayu*[1], Sugeng Triyadi S[2], and Lily Tambunan[3]

[1] Architectural Doctoral Study Program, School of Architecture, Planning, and Policy Development (SAPPK), Bandung Institute of Technology, Indonesia
[2,3] Senior Lecturer, School of Architecture, Planning and Policy Development (SAPPK), Bandung Institute of Technology, Indonesia

*Corresponding author e-mail: ritalaxmi@gmail.com

Abstract. The level of damage to the building is influenced by the configuration of the building plan with the angle of re-entrant and the pattern of damage. Architects need to know and understand the pattern of building damage and the configuration of hospital plans with an architectural approach. A certain level of damage has an impact on the safety of the lives of patients and health workers. The method used is a quantitative method with 2 steps, namely literature study and correspondence. The technique used is a visual identification technique from photographs in 28 earthquake-related journals with 108 cases. There is a significant relationship between the shape of the re-entrant angled building and the level of damage caused by the earthquake. The configuration of the building plan with re-entrant angles U and L is closely related to the level of heavy structural damage (level 4). The pattern of structural damage in the configuration of the re-entrant corner plan U, L, H occurs in the re-entrant corner and the outside of the building. The hospital as a vital building must use a floor plan configuration without a re-entrant angle.

1. Introduction

The criteria for building resilience to earthquake damage are measured by the building's ability to save lives and the assets in it. The building's performance has an implicit earthquake code or standards. The structural system in the building determines the building's resistance to earthquakes. The structural system used is related to function, form, and its relation to other systems in the building [1]. Earthquake-resistant buildings consist of physical aspects of structural elements and non-structural elements including earthquake-resistant architectural factors.

The resilience of the building must continue to provide continuity of community life in the event of a disaster. The resilience or preparedness of vital buildings - such as hospitals, fire stations, and communication centers - in earthquake areas is a top priority. When an earthquake occurs, vital buildings are at great risk of damage and cannot function properly. Damage to vital buildings, especially hospitals, has an impact on the safety of the lives of patients and health workers. Vital buildings must not only remain standing but must also remain operational. Building resilience is vital in providing protection for the safety of the life of building users [2]. The building collapse resulting in the death of victims is proportional to the total number of buildings that completely collapsed [3].

Hospital buildings must be sufficiently resistant to the strength of an incoming earthquake. One of the factors that affect the resilience of the building is the calculated live load. The inertia force is proportional to the weight, if the building has a larger total weight, the greater the inertia force for a certain level of acceleration [4]. The strength of the earthquake causes swaying of the building structure. The shaking can cause damage to the building. The magnitude of the value of the mapped earthquake acceleration and the magnitude of the amplification value of the spectral acceleration on the ground surface [5] affect the calculation of the strength of the building against earthquakes. The number of floors is one of the variables that affect the vulnerability of buildings to earthquakes. The building attributes used to assess seismic susceptibility are construction period, number of floors, and roof shape [6]. The number of floors of the building or the height of the building greatly affects the length of the natural vibration period during an earthquake. [4].

Buildings that have a re-entrant angle can experience premature collapse due to an earthquake. The earthquake in the re-entrant angled building resulted in a sudden change in stiffness and an increase in strength in torsion [7; 2] Building torque occurs in structural elements that are not symmetrically located in the plan. The building experiences torsion because the center of rigidity or resistance and the center of mass for this configuration do not coincide [4; 2]. To overcome this condition, the buildings are tied together on the stress concentration line and locate the earthquake-resisting elements at the wing tips. Another solution is to separate the buildings into simple shapes. One example of an irregular building configuration is a re-entrant building plan. The performance of buildings with irregular configurations is not as good as buildings with regular configurations. An example of a building with a re-entrant angle is the characteristic shape of the floor plan of the letters L, H, T, X, and its derivatives [2]. In buildings with irregularity, the re-entrant angle has increased from 15% to 25% [8].

1.1 Architectural

Architects design hospital buildings with various alternative configurations or floor plans. The consideration of the process of designing a building plan consists of several aspects, namely the function of the building, the flow of user activities, the organization between spaces, the potential around the site, the shape of the site, and the area of the site. Alternative characteristics of the plan form are the results of the architectural analysis stage which have consequences for the strengthening of the building structure. Possible building configurations have the characteristics of regular and irregular floor plans. The meaning of order is that each part is placed correctly by paying attention to the other parts as a whole and according to its purpose. Regularity forms a harmonious whole and does not only look at geometric regularities [9].

Hospitals are designed with a particular floor plan configuration as a result of design thinking. The configuration or shape of the building plan comes from the main form. In order to facilitate the process of viewing and understanding a layout configuration, the layout configuration is simplified and made into an orderly form. Likewise, simple configurations or shapes can be reduced or increased so that the shape of the building becomes an irregular shape. Addition or subtraction of fields will result in irregular buildings [9].

One of the requirements for inpatient rooms in inpatient installations is to have easy reach of nurses to patient care rooms and an approach to reduce the spread of infection [10]. The priority of inpatient design lies in the circulation of users and medical equipment. It is important to provide easy access between spaces in providing medical services. The relationship between spaces should be made linearly [11].

The consequences of various forms of plans will affect the resistance of buildings to earthquakes. Hospital buildings as vital buildings need an understanding of the shape of the building plan in accordance with earthquake-resistant buildings. The strength of a certain earthquake will have a damaging impact with a certain level of damage to the building. One of the causes of damage to buildings with a heavy level of damage is the shape of the building. The relationship between the shape of the building plan and the level of damage needs to be known. The appropriate form of the building plan can minimize the level of damage to the hospital building.

1.2 Damage Level Building

The level of damage is so important to measure a building damage due to an earthquake. The damage measurement scale is in the form of a standard interpretation of building damage that is used to assess the level of damage [3]. Measurement of the level of damage using the European Macroseismic Scale

(EMS). In table 1, the level of damage to EMS consists of five levels, namely 1-5. The relationship between structural damage to buildings affects non-structural damage to buildings. At level 1 damage, the building did not suffer structural damage but the building suffered minor non-structural damage. Damage level 4 includes very heavy damage, where the building suffered heavy structural damage and the building suffered very heavy non-structural damage [12].

Tabel 1. Definition of damage level

Damage Level L_d	Definition	Notes
0	No damage	-
1	Slight damage	No SD, slight N-SD
2	Moderate damage	Slight SD, Moderate N-SD
3	Substantial to heavy damage	Moderate SD, heavy N-SD
4	Very heavy damage	Heavy SD, very heavy N-SD
5	Destruction	Very heavy SD

SD = Structural Damage; N-SD = Non-Structural Damage

2. Method and Research Area

2.1 Research Method

Quantitative methods were used in this study [13]. Correspondence relationship is sought between the shape of the building plan and the level of damage to the building. Keywords were obtained from the results of collecting field observation data and reviewing journal articles (literature studies). Some keywords are grouped with initial meaning and supported by theory. Some keywords are grouped into categories (content analysis). The initial meaning is carried out at this stage [14]. As part of content analysis, the open coding stage is carried out by identifying keywords from existing text data. Distribution analysis and correspondence analysis were carried out after the content analysis stage [15]. Data collection time is from December 2017 to January 2021. The literature study uses visual identification techniques from photographs in the journal. The interview technique was used in field observations. The correspondence between the variables of building form and the level of damage was used as a method in the second step. The relationship between the shape of the building and the degree of damage is the result of the second step.

2.2 Research Area and Sample Selection

This research is looking for the configuration of buildings that are resistant to earthquakes, for that we collected several cases of buildings that were affected by the earthquake. Earthquake-affected buildings were taken with several building configurations, various functions, and various heights. From various configurations, the pattern of building damage and the level of damage to the buildin can be determined. The pattern of damage and the level of damage to the building can be used as a reference in designing hospital inpatient buildings.

From 28 journals related to earthquakes in the world, there are 108 cases with 46% residential, 10% school, 11% hospital, 10% commercial, 7% mixed building, 7% office, and 7% other (figure 1). The strength of the earthquake ranged from 5.9 Mw to 9 Mw, mostly in the 7 Mw to 7.5 Mw range.

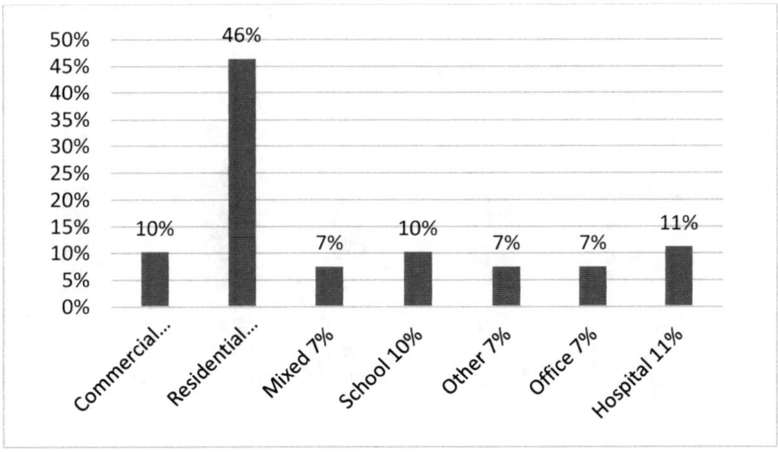

Figure 1. Percentage of building functions from observation cases.

3. Result and Discussion

3.1 result

The configuration of the building adapts to the characteristics of its function. Of the 108 cases there were 64 buildings or 59% of the building configurations in the form of "I". The square building configuration contains 18 buildings or 17% of the cases. The configuration of the "L" shaped building consists of 13 buildings or 12%. Other building configurations such as "H", "U", and irregular as many as 13 buildings or 12% (figure 2). Building configurations "I", "H", "L", and "U" can be assumed to have a row organization that has the same size and function. The spatial organization of the functions of hospitals, schools, offices, and flats (residential) form the "I", "H", "L", and "U" building configurations.

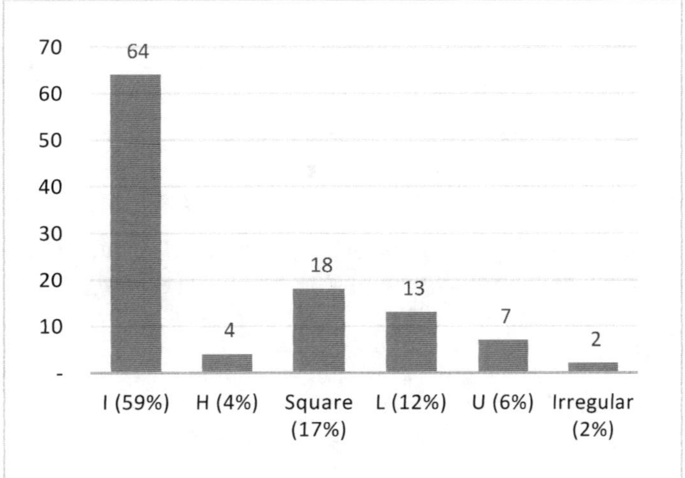

Figure 2. Building Configuration.

The building suffered structural damage with a level of heavy structural damage due to the earthquake as much as 29%. The level of very heavy structural damage is experienced by buildings as much as 26%. The level of damage is slight structural damage as much as 20% of buildings. In Figure 3 it looks almost the same from the various levels of damage to the building structure.

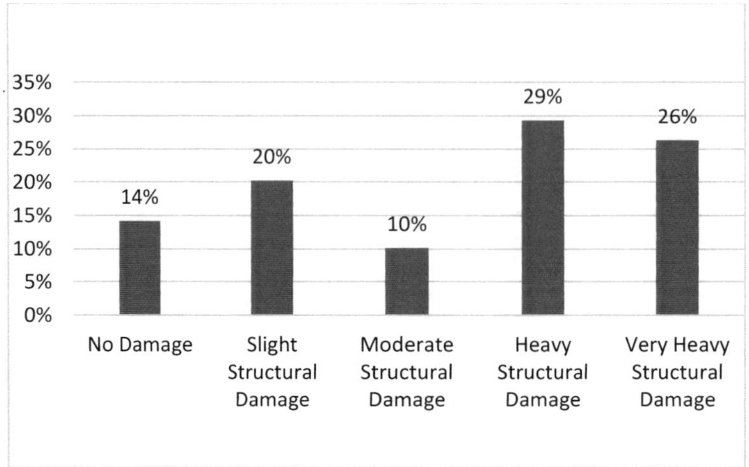

Figure 3. Structural Damage Level.

3.2 Discussion

In the "L" configuration the average damage is 3.9 or the level of damage 4. The "U" configuration building has an average damage of 3.8 or the level of damage 4. The "I" configuration building has an average damage at the level of damage 3. The three building configurations are susceptible to damage or are included in vulnerability cluster A. The square building configuration is not susceptible to damage (cluster B) and has an average damage level of 1.9 (cluster B). The "H" and irregular building configurations are included in 2 vulnerability clusters A and B, with an average damage level of 2.6 and 2.5 (table 2).

Table 2. Vulnerability Cluster. Cluster A untuk bangunan yang memiliki nilai kerentanan terhadap kerusakan lebih besar dibanding cluster B.

Level	Cluster		Mean
L	A		3,93
U	A		3,85
I	A		3,04
H	A	B	2,60
Irregular	A	B	2,50
Square		B	1,93

There is a significant relationship between the shape of the building and the level of damage to the building. Buildings that have a re-entrant angle will experience susceptibility to damage (cluster A) for "L", "U", "H", and irregular building configurations (figure 4). The "I" building configuration also has a vulnerability to building damage if the building is less than optimal in responding to shear and torsion forces. Despite the "H" and "Irregular" building configurations are included in cluster B, these configurations are also included in cluster A. The "H" and "Irregular" building configurations have damage levels of 2.6 and 2.5 or moderate damage.

The addition of stiffeners to the building is necessary to anticipate additional shear forces. The vulnerability is caused by the torsional behaviour of the building, and the additional shear force generated in the perpendicular direction [7]. The "I" building configuration will be stronger if the ratio between the width and length of the building is close to 1:1, or close to the square building configuration.

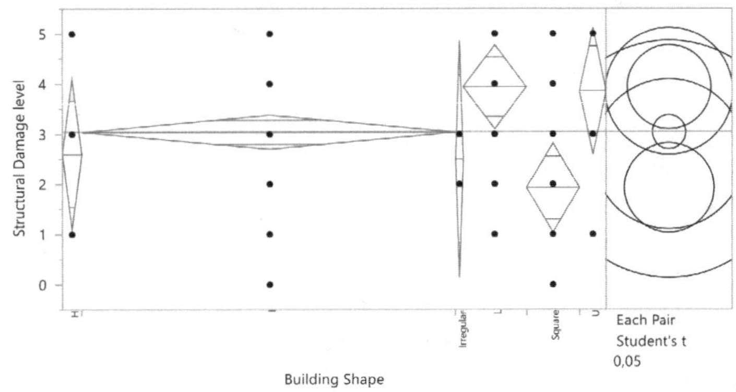

Figure 4. Relationship between building configuration and level of building damage.

In this study, 11% of the building functions as a hospital. Most of the hospitals have "I" building configurations as much as 63%, "U" configurations as much as 25%, and "H" building configurations as much as 13% (figure 5). Hospitals, especially inpatient installations, use a lot of linear space organization. Spaces that have the same function and repeat with the same space size form a linear spatial organization [9].

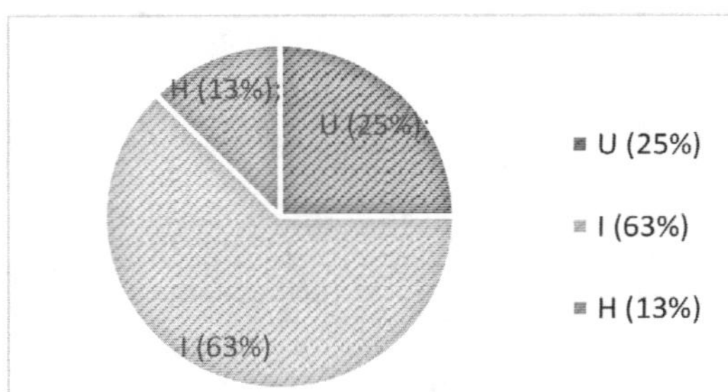

Figure 5. Hospital building configuration.

The patient room has a function as an inpatient room with the same size placed repeatedly in a row. The linear space organization will form the "I" building configuration. Access from the corridor connects between the inpatient rooms. Nurse stations, vertical circulation rooms, and service rooms are placed as connectors between rows of inpatient rooms (figure 6).

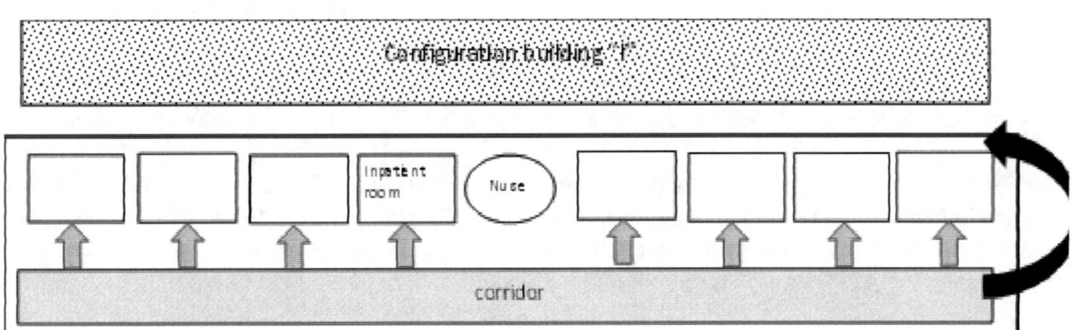

Figure 6. Functional relationships between rooms in inpatient installations. Functional relationships between rooms in inpatient installations. The location of the inpatient room is in a row with direct access to the corridor, forming an "I" building configuration.

The location of the damage often occurs on the edge of the 1st floor building (55%) and the edge of the 2nd floor building (25%) for all building configurations (figure 7). Damage to the edges of the 3rd and 4th floor buildings is 4%. Damage to the center of the building on the 1st floor is 6% and damage to the center of the building on the 2nd floor is 3%. Buildings with irregular configurations are more vulnerable than buildings with regular configurations.

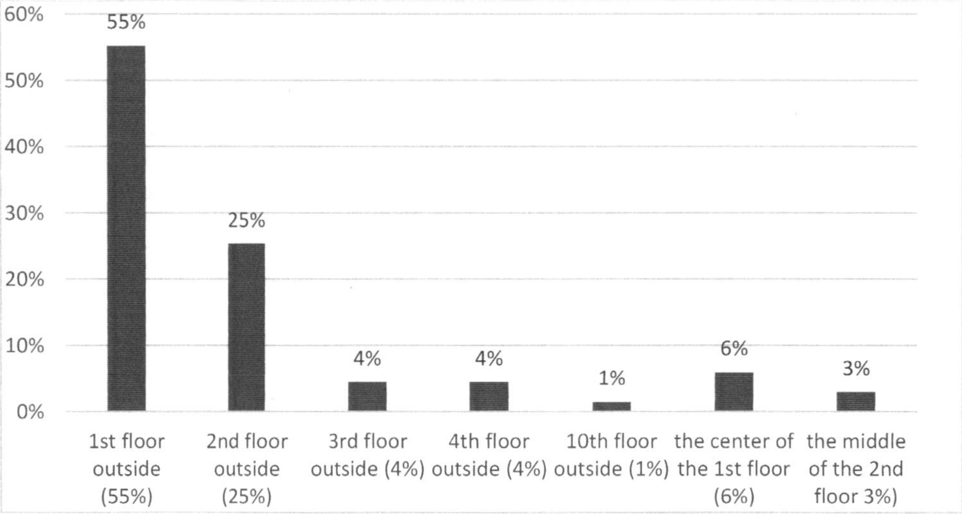

Figure 7. Graph of the location of damage to buildings in various building configurations.

Mostly adopted building configurations in hospitals are “I”, “H”, and “U” (figure 5). The configuration of the building "I" suffered damage to the building in the middle of the building (dotted line) as much as 3%. Damage to buildings located on the edge of the building (dots) is 97%. The "U" building configuration is damaged at 100% re-entrant angle. The configuration of the building "H" was damaged in the middle of the building (dotted line) as much as 33% and damage to the edges of the building (dots) as much as 67% (figure 8).

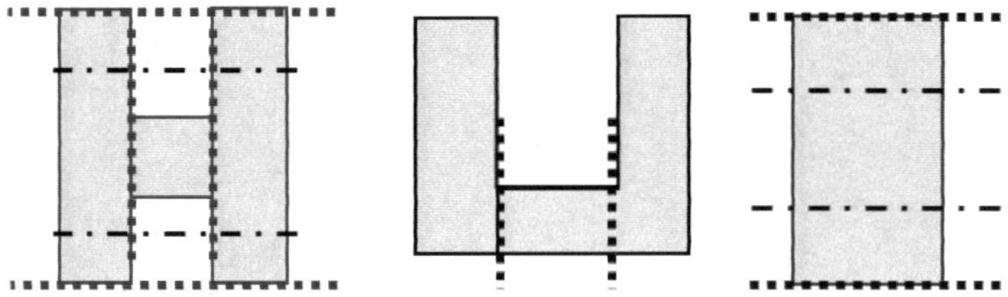

Figure 8. Location of damage to buildings in "H", "U", and "I" building configurations.

A series of repeating spaces with the same space size and function can be formed into a square configuration. The organization of the inpatient room uses an approach to the behaviour characteristics of room users, medical services, and medical support for patients, estimates of the number of rooms and room size, room layout, and room characteristics [10].

4. Conclusion

The inpatient installation as part of a vital health facility needs to pay attention to the configuration of the building. Most of the hospital building configurations are “I” (63%). The configuration of the “I” building corresponds to the organization of space in an inpatient installation but is prone to damage. The "I" building configuration has a damage level of 3 or substantial of heavy damage. The pattern of damage often occurs on the edge of the building with a building configuration of "I" (97%). As an anticipatory measure, the configuration of the "I" building needs special stiffeners so that the building can withstand sudden shear forces.

Hospital buildings also have "L" and "U" building configurations and both are also susceptible to damage, both have a high level of damage, namely 4 or very heavy damage. It is not recommended that inpatient buildings use the "L" and "U" building configurations. The pattern of damage often occurs at the corners of the re-entrants and the sides of the building edges. The "H" and "Irregular" building configurations have a low level of damage, namely 2.5 and 2.6. This building configuration often causes damage to the corner of the re-entrant and the edge of the building. The most earthquake-resistant building configuration is the square building configuration, with a level of damage 2 (figure 9).

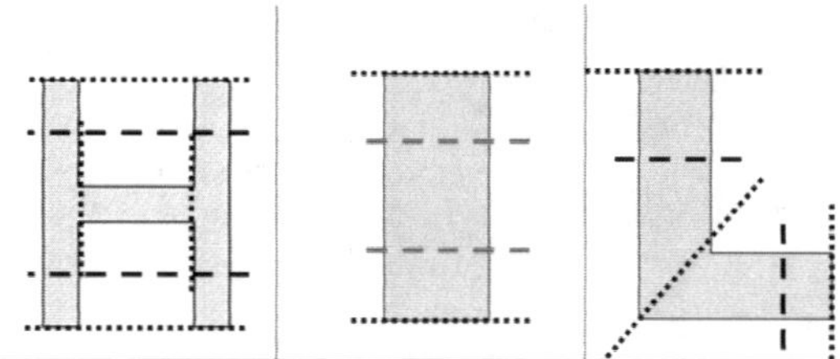

Cluster	A and B	A	A
Mean	2,6	3	4

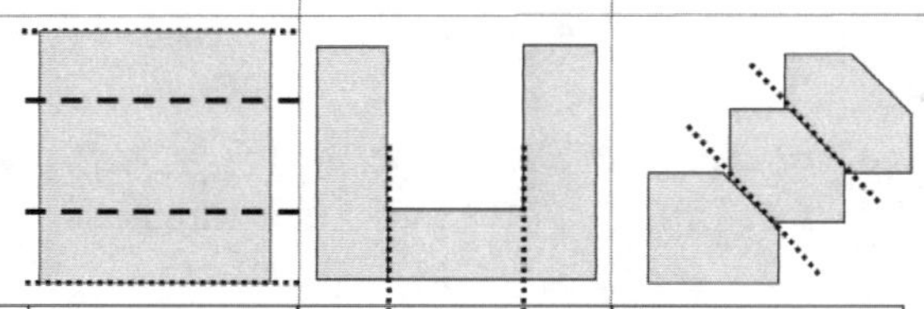

Cluster	B	A	B
Mean	2	4	2,5

Figure 9. Building configuration, damage pattern, and average level of building damage. Building configuration, damage patterns, and the average level of damage to buildings. Damage to the building occurs at the corner of the re-entrant and the edge of the building.

There is a significant relationship between the configuration of the building and the level of damage. Building configurations that have a re-entrant angle are susceptible to damage with a level above 2 (moderate damage). The inpatient building is quite safe by using a building configuration with no re-entrant angle. Hospitals, especially inpatient buildings that are resistant to earthquakes, can still use the "I" building configuration with building stiffeners so that the building does not experience torsion and is able to respond to shear forces. Damage patterns in certain building configurations can help design hospitals, especially inpatient buildings so that they can be considered for building reinforcement and reinforcement.

Acknowledgments
Thanks are conveyed to the Chair of Architecture Doctoral Program the Architecture, Planning, and Policy Development (SAPPK), Bandung Institute of Technology, who has provided facilities in carrying out the completion of this research. The Dean of FTIK Indraprasta University, PGRI and his staff who have provided support in completing this research. Thank you to all those who have provided input in the preparation of this manuscript.

Orcid iDs
Rita Laksmitasari Rahayu https://orcid.org/0000-0001-7539-9894
Sugeng Triyadi S https://orcid.org/0000-0002-4223-2294
Lily Tambunan https://orcid.org/0000-0002-1602-4301

References

[1] Noor Cholis Idham 2013 *Merancang Bangunan Gedung Bertingkat Rendah* (Yogyakarta: Graha Ilmu)

[2] Bungale. S. Taranath 2005 *Wind and Earthquake Resistant Buildings Structural Analysis and Design*, (New Jersey: Marcel Dekker) p 99

[3] Shigeyuki Okada and Nobuo Takai 2000 *Proc. Int. Conf. World Conference on Earthquake Engeineering* (Aukland) 12 (Upper Hutt, N.Z.: New Zealand Society for Earthquake Engineering) p 0705

[4] Andrew Charleson 2008 *Seismic Design for Architects Outwitting the Quake* 1st (MA: Elsevier)

[5] Sari Farlianti, and Sapta 2017 Perhitungan Respon Spektra Percepatan Gempa Kota Palembang berdasarkan SNI 1726; 2019 Sebagai Terhadap Revisi SNI 1726;2012 *Jurnal Ilmiah Teknika.* **6** p 167

[6] Lorenzo Diana, Julien Thiriot, Yves Reuland, Pierino Lestuzzi 2019 Application of Association Rules to Determine Building Typological Classes for Seismic Damage Predictions at Regional Scale: The Case Study of Basel Frontiers *In Built Environment* **5**

[7] Momen Mohamed Ahmed and Shehata E. Abdel Reheem 2016 Irregularity Effects on The Seismic Performance of L-Shaped Multi-Story Buildings *Journal of Engineering Sciences* **4** p 513

[8] Anis S. Shatnawi, Mazen Musmar, Laith I. Gharaibeh 2018 Evaluation of Seismic Analysis Procedures for Concrete Moment Resistant Frames with Horizontal Re-entrant Corners Irregularity *International Journal on Advanced Science Engineering Information Technology* **8** p 876

[9] Francis D.K. Ching 2015 *Architecture Form, Space, & Order* ed 4, (Inc New Jersey: John Wiley and Sons)

[10] Global Rancang Selaras 2010 *Arsitektur Rumah Sakit* (Yogyakarta: Global Rancang Selaras)

[11] Direktorat Bina Pelayanan Penunjang Medik dan Sarana Kesehatan, Direktorat Bina Upaya Kesehatan *Pedoman Teknis Bangunan Rumah Sakit Ruang Rawat Inap* 2012 (Jakarta: Kementerian Kesehatan Republik Indonesia)

[12] Mauro Dolce, Andreas Kappos, Angelo Masi, Gregory Penelis, Marco Vona 2006 Vulnerability Assessment and Earthquake Damage Scenarios of The Building Stock of Potenza (Southern Italy) Using Italian And Greek Methodologies. *Engineering Structures* **28** p 357-371

[13] John W. Creswell and J. David Creswell 2018 *Research Design: Qualitative, Quantitative, and Mixed Methods Approaches* fifth edition. (London: Sage Publications: International Education and Professional Publisher)

[14] H.E. Kusuma 2018 Analisis Isi, Analisis Data Teks secara Kualitatif dan Kuantitatif, Ikatan Peneliti Lingkungan Binaan Indonesia, Bandung

[15] Bunga Sakina, Hanson E Kusuma 2016 *Proc. Int. Symp. of the International Symposium on Architectural Interchanges in Asia* (Miyagi, Japan) 11 p 573

Generating City-Scale 3D Finite Element Models from GIS Features and Heuristic-based Framing Models using Feed-Forward Artificial Neural Networks

Jamie Alea B Ramos*, Pher Errol B Quinay and Karlo Daniel Q Colegio

University of the Philippines Institute of Civil Engineering, U.P. Campus, Diliman, Quezon City, 1101 Philippines

*E-mail: jamiealearamos1@gmail.com

Abstract. City simulation-based studies require numerous structural models of sufficient accuracy. In this study, a tool that automates the generation of three-dimensional (3D) finite element (FE) models given geographic information system (GIS) features is developed based on heuristic-based framing models. The developed tool utilizes feed-forward artificial neural networks (ANN), enhanced by topology-based partitioning and parallel computing by Message Passing Interface (MPI). Different sets of topologies are trained and tested using heuristic-based framing models to predict column locations and floor elevations given GIS features. Optimal topologies are then integrated into the frame model generating tool. With GIS features on building footprint and height as inputs, the developed tool performs several tasks: (1) pre-processing the given GIS features, (2) establishing the column locations, (3) generating the beam framing in the two-dimensional (2D) plane, (4) predicting the floor elevations, and (5) generating the 3D frame model. Using the developed tool, the 3D frame models of 8,753 and 7,400 target structures in two urban areas in Metro Manila, Philippines are created. With the generated 3D FE models, the application to city scenario seismic response simulation is demonstrated. Comparisons with a recent simulation-based study showed that the current ANN-based tool outputs more realistic framing configurations.

1. Introduction

Demand for virtual city models increases as their uses become extensive. Possible applications of city model generation include disaster management for hazards simulations, transportation improvements, environmental impact analysis, and urban planning [1]. However, in most cases, virtual city models are generated manually by creating the model of each structure in the target area [2]. This problem worsens for city response simulations because these are usually performed using the finite element method (FEM), which requires models to be of sufficient fidelity. As a result, numerous models with sufficient accuracy are needed to simulate the hazard response of a city. To overcome the time-consuming task of manually creating numerous finite element (FE) models, a program for an automated model generation must be developed [3].

Since city response simulations are sensitive to framing, utilizing machine learning to generate city-scale FE models is suggested. In particular, an artificial neural network (ANN) is used in this research because of its capability to solve problems with no definite solutions [4], making it a potential solution to the problem of automated frame model generation.

In general, ANN is composed of interconnected nodes in various layers that mimics the behavior of neurons. An ANN's layers include the input layer that contains the data from the surroundings, the hidden layer(s) that processes the data, and the output layer which contains the solution to the problem. Connection weight coefficients connect the nodes in different layers and can be adjusted based on the given input and expected output pairs [4]. In this paper, ANN topology refers to the number of nodes in each layer, and how these layers are connected.

Applications of ANN in automated model generation have already been investigated. Denipote and Assaf utilized ANN, image understanding (IU) techniques, and satellite images to generate the models of the structures in a target area. In the study, the ANN was trained to select a three-dimensional (3D) model from a set of limited pre-generated models [5]. Aside from satellite images, geographic information system (GIS) features such as building footprint and building height can be used to create the 3D shape of the structure [6]. Such GIS features are extracted from available remote sensing data at fine resolutions which are sufficient to provide information on a structure's exterior outline and height [7]. In this study, GIS features are utilized to create a structure's 3D model.

This paper outlines the development of the frame model generating tool that utilizes feed-forward ANN topologies which were trained and tested against heuristic-based framing models to predict the column locations and floor elevations given the GIS features on building footprint (or corner coordinates of the building) and building height. The developed tool's target application is the generation of models of low-rise reinforced concrete structures in two urban areas in Metro Manila, Philippines which will be referred to as City 1 and City 2. This paper will also present a potential application of the generated city-scale frame models, specifically city seismic response estimation.

2. Overview of Developed Frame Model Generating Tool

In this study, different sets of ANN topologies are trained and tested to predict column locations and floor elevations given GIS features on building footprint and height, respectively. A survey is conducted among local civil engineers to obtain heuristic-based framing and story design models for the training and test datasets. Based on the training and test results, the optimal ANN topologies for predicting the aforementioned structural components are identified and integrated into the frame model generating tool as ANN connection weights.

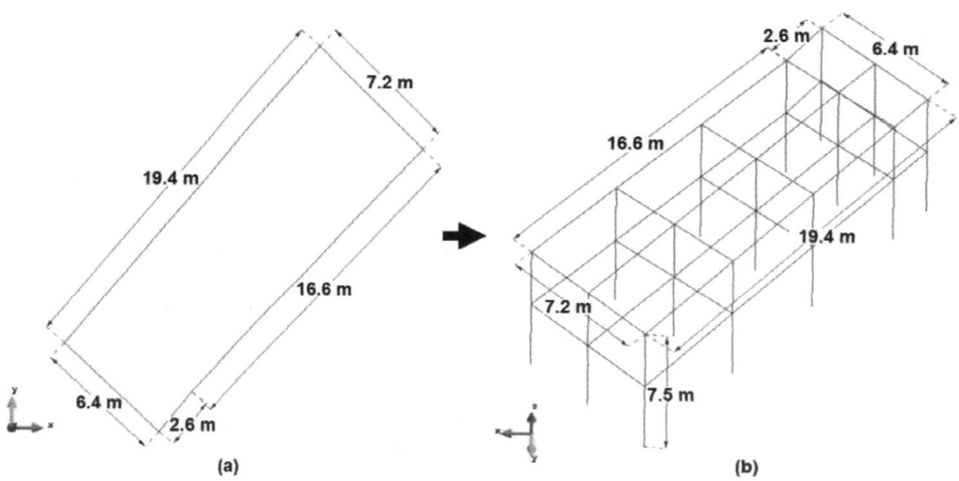

Figure 1. (a) GIS feature on building footprint, and **(b)** its corresponding 3D frame model generated using the developed tool.

The goal of the frame model generating tool is to transform the given GIS features on building footprint and height into a 3D frame model of the structure. Figure 1(a) depicts a sample feature that describes the building footprint, whereas figure 1(b) shows the corresponding 3D frame model generated for the given GIS features depicted in figure 1(a). Inputs to the developed tool include the GIS features on building footprint and height, particularly the 3D corner coordinates of the target structure. In this research, the features are derived from the high-resolution GIS dataset of the study areas produced by the Greater Metro Manila Area (GMMA) READY Project [8].

The automated frame model generating tool is developed using the C programming language. For each structure, the tool performs the following tasks: (1) pre-processing of given GIS features of the structure, (2) establishing column locations using the optimal ANN topologies for column location prediction, (3) generating beam framing in the two-dimensional (2D) plane based on the established column locations, (4) predicting floor elevations using the optimal ANN topology for floor elevation prediction, and (5) generating 3D frame model based on the generated beam framing in the 2D plane and the predicted floor elevations. The tool then executes the aforementioned tasks for all structures in the target area.

2.1. Pre-processing of GIS Features

After scanning the GIS features, the developed tool pre-processes the given 3D global coordinates of the structure to conform to the ANN input models. The 2D coordinates of the building footprint are first transformed into their corresponding local coordinates. The given building height, on the other hand, is normalized with respect to the maximum building height considered in this study (10 meters). Conversion factors for the transformation are stored in distinct variables.

2.2. Generation of Column Location

The generation of column locations is highly dependent on the configuration of the building footprint. The developed tool in this study covers four-sided, five-sided, and six-sided building footprints that are common among the structures in the target areas. Different ANN models for column location prediction are investigated for each n-sided footprint due to differences in the sizes of the input layers (determined by the number of sides of the footprint), and the quality of the available training and test datasets.

To generate the column locations given a four-sided building footprint, a numbered grid of possible column locations based on the footprint shape is first generated. ANN-predicted column locations can be obtained from the pre-generated locations using the feed-forward process implementation and the ANN connection weight matrices of the optimal topology for this application. In particular, the ANN model used consists of input nodes equal to the 2D corner coordinates of the footprint; and output nodes representing possible column locations with expected values equal to either 0 or 1. For the output nodes, a value of 1 indicates that the column exists in the pre-generated location while a value of 0 indicates otherwise. As a result, a value greater than 0.5 for an ANN-predicted output is established to indicate the presence of a column in the location that the particular ANN output node represents.

To overcome the difficulty of training and testing ANN topologies of larger sizes, another ANN model with fewer input and output nodes is used for five-sided and six-sided footprints. The model focuses on predicting the number of interior columns along both local axes of the building footprint with the input nodes representing the corner coordinates of the footprint, and each output node representing a number of interior columns. The expected value for the output node representing the number of interior columns along the axis is 1 while the expected values for the other output nodes are 0. Hence, the ANN output node with the value closest to 1 after the feed-forward process is identified. The corresponding number that this node represents is then established as the number of interior columns along the axis in consideration.

To represent the ANN-predicted number of interior columns along the local x and y axes of the footprint, vertical and horizontal lines are automated, respectively. The intersections of these lines are numbered and labelled as possible column locations.

Column locations are then determined by taking into account the position of each possible column location in relation to the building footprint as well as its proximity to other possible column locations. Using this method, the generated structural models are limited to column-to-column connections.

2.3. Generation of Beam Framing

For each n-sided building footprint, the beam framing in the 2D plane is generated using the numbered grid of possible column locations customized for the given footprint. The developed algorithm for beam framing generation identifies which two possible column locations can form a beam element. If both locations are identified as established column locations the program creates the beam element. If only one column is identified, the program stores the current column location and then looks for the next possible column location to which it can connect. This process is repeated until all possible side beam elements, horizontal beam elements, and vertical beam elements have been formed.

2.4. Prediction of Floor Elevation

After generating the beam framing in the 2D plane, the floor elevations are then predicted. the ANN model used for this application has an input layer with one node equal to the normalized building height, and an output layer with three nodes equal to the normalized floor elevations, with the first node always equal to the building height, and nodes that do not represent a floor equal to 0 during training and testing. Hence, after implementing the feed-forward process, the floor elevations normalized with respect to the maximum building height considered (10 meters) can be predicted using the connection weight matrices of the optimal ANN topology. The number of floors can also be identified indirectly by counting the number of floor elevations that are at least 2.7 meters high (minimum floor elevation based on the Revised Implementing Rules and Regulations of the National Building Code of the Philippines (IRR) 2005 [9]).

2.5. Generation of 3D Frame Model

Using the beam framing in the 2D plane and the predicted floor elevation, the nodes and elements of the 3D model are established. The x- and y-coordinates of the column locations are first transformed to global coordinates using the conversion factors from pre-processing. These 2D coordinates are then replicated across all identified floors by taking the predicted floor elevations as z-coordinates of each established column location. The elements are identified after the node coordinates have been determined. Except at elevation $z = 0$, the established beams in the 2D plane are duplicated on all floors. Columns, on the other hand, are generated by connecting each node to the node directly below it.

3. Algorithm of Implemented Feed-Forward ANN Code

For column location prediction and floor elevation prediction, different sets of ANN topologies are trained and tested using the ANN code implemented in the C programming language. The algorithm is based on the computational model of a feed-forward ANN. With a feed-forward ANN, the input layer is directly connected to the output layer if there are no hidden layers; otherwise, it is connected to the first hidden layer. The first hidden layer is then connected to other hidden layers if there are any, and the last hidden layer is connected to the output layer [10]. To increase the flexibility of the node that accepts the signal sources, all layers except the output layer have one bias node with a value always equal to 1 [11].

In this study, ANN topologies are represented as matrices or vectors in the implemented ANN code. Using matrix operations, the computational model of feed-forward ANN can be implemented in a less complicated manner [11]. The ANN code is programmed such that ANN layers can be represented as matrices with multiple rows. This implies that the ANN code can train and test two or more ANN topologies with the same number of nodes and layers using the same set of connection weights but different input and expected output pairs.

In general, the training process employs feed-forward and back-propagation processes to obtain the connection weight matrices. With these connection weight matrices, the feed-forward process is implemented using a dataset that the ANN never encountered during training. This process is called testing, and it is used to determine the ANN's generalization ability [12].

The developed ANN code automates the training and testing processes for different sizes of ANN topologies by performing the matrix operations with variables based on the size of the ANN topology.

Furthermore, topology-based partitioning and parallel computing by Message Passing Interface (MPI) are implemented to run continuously in 120 cores of high performance computing (HPC) resources of the Department of Science and Technology - Advanced Science and Technology Institute Computing and Archiving Research Environment (DOST-ASTI CoARE).

Based on the training and testing results, the optimal ANN topology for each application is identified. This topology is traditionally identified through trial and error and is defined as the smallest topology that can capture the relationship in the training data. Establishing an evaluation metric helps in the selection of the optimal topology because performance scores can be assigned to each topology [12]. For the ANN code implemented in this study, the performance score of the topology is identified as the maximum score among the scores calculated in all epochs. The topology with the highest performance score among the trained ANN topologies is then identified as the optimal topology.

The implemented ANN code is then customized for the target applications by adjusting the model for the input and output layers, the established evaluation metric, and the presentation of results. For each application, the number of input and output nodes is fixed as a consequence of the implemented ANN algorithm, while the number of hidden nodes and hidden layers is varied to train and test different ANN topologies. The established sets of ANN topologies in each application are also trained and tested thrice to yield three distinct sets of connection weights as a consequence of setting the initial connection weight coefficients randomly. Among the three performance scores calculated from the three runs, the highest score is chosen as the performance score of the ANN topology.

4. City-Scale 3D Model Generation

The developed frame model generating tool is used to create the line element meshed models of the target structures in two urban areas in Metro Manila, Philippines. For City 1, 8,753 structures out of 8,980 structures are modelled by the developed tool (97.472% efficiency). In total, the meshed model of City 1 is composed of 124,457 nodes and 150,382 elements. On the other hand, for City 2, 7,400 out of 8,662 structures are modelled by the developed tool (85.431% efficiency). The line element meshed models of the structures in City 2 consist of 134,443 nodes and 181,809 elements.

5. Application to City Seismic Response Estimation

Using the generated FE models, the city scenario seismic response of the study areas is estimated. The seismic response simulation is divided into three parts – pre-processing, analysis, and post-processing. During the pre-processing stage, the developed frame model generating tool exports the necessary input files needed for the analysis. In particular, the input files contain information on the line meshed model, the boundary conditions, the material properties, and the seismic load assignment. For the analysis and post-processing stages, the existing module developed by Quinay, Soliman, and Fader [7] is used. Given the input files exported by the pre-processing tool, the module performs linear static analysis to calculate the nodal displacements of the structures. The meshed model and seismic response estimation files of each structure are generated by the module for visualization. The seismic response of a structure or city can be visualized using the two files.

Three loading scenarios are considered to estimate the seismic response of the study areas. In the first scenario, the lateral force is applied to the nodes of the longest side (side 1), and the longer side adjacent to the longest side (side 2). In the second and third scenarios, the lateral force is applied to the nodes of side 1 and side 2, respectively.

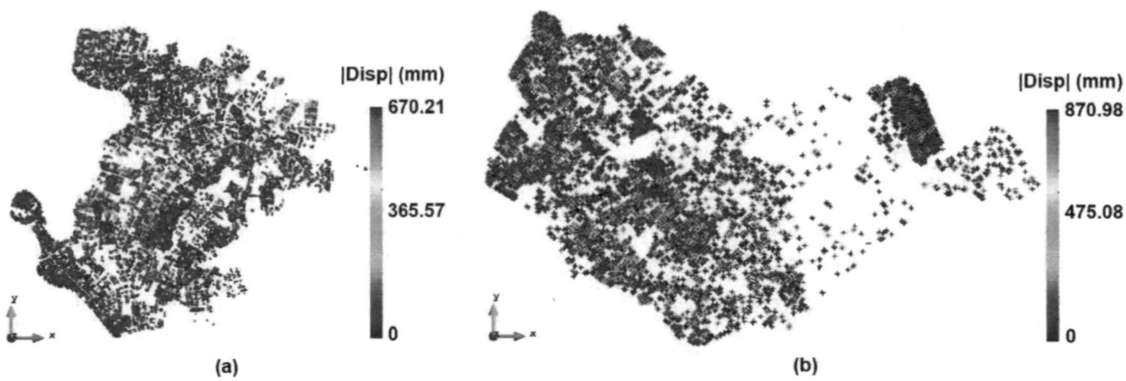

Figure 2. Nodal displacements model of **(a)** City 1, and **(b)** City 2.

For City 1, the maximum nodal displacement under the first, second, and third loading scenarios are 0.670 meters, 0.525 meters, and 0.388 meters, respectively. On the other hand, for the case of City 2, the maximum nodal displacement for the first, second, and third loading scenarios are 0.871 meters, 0.738 meters, and 0.554 meters, respectively. Figures 2(a) and 2(b) show the nodal displacements models of City 1, and City 2, respectively.

6. Comparison with a Simulation-based Study

Generated 3D frame models of selected structures and their estimated seismic response are compared to the results of the simulation-based study conducted by Quinay, Soliman, and Fader [7]. The tool developed in the said study generates models for structures with four-sided and six-sided building footprints based on maximum column-to-column spacing, beam spacing, and floor-to-floor elevation assumptions. In this paper, models generated for a structure with a four-sided building footprint and a structure with a six-sided building footprint are compared.

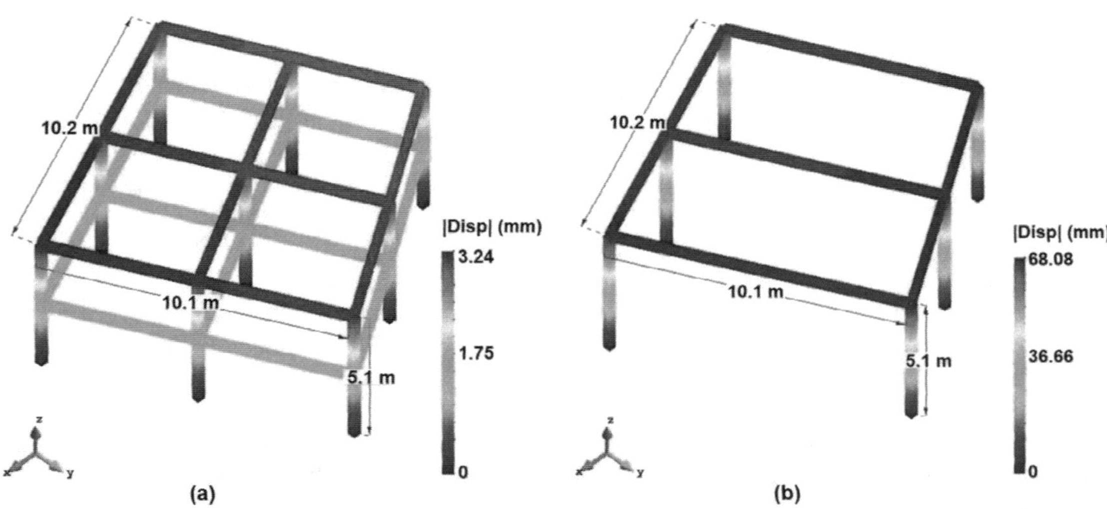

Figure 3. Seismic response of the models generated using **(a)** the tool developed by Quinay, Soliman, and Fader [7], and **(b)** the tool developed in this study for a structure with a four-sided footprint.

Figure 3 shows the generated models and estimated seismic response for the same GIS features on building footprint and height. In particular, figure 3(a) shows the seismic response of the 3D frame model created using the tool developed by Quinay, Soliman, and Fader [7], whereas figure 3(b) shows the seismic response of the 3D frame model created using the tool developed in this study.

When the 3D frame models in Figure 3 are compared, the model in figure 3(a) has more structural nodes and elements compared to the model in figure 3(b). This demonstrates the possibility of reducing computational costs during analysis by using the models generated by the tool developed in this study. In consequence, under similar loading conditions and assumptions on soil and material properties, the calculated nodal displacements of the model in figure 3(a) are smaller.

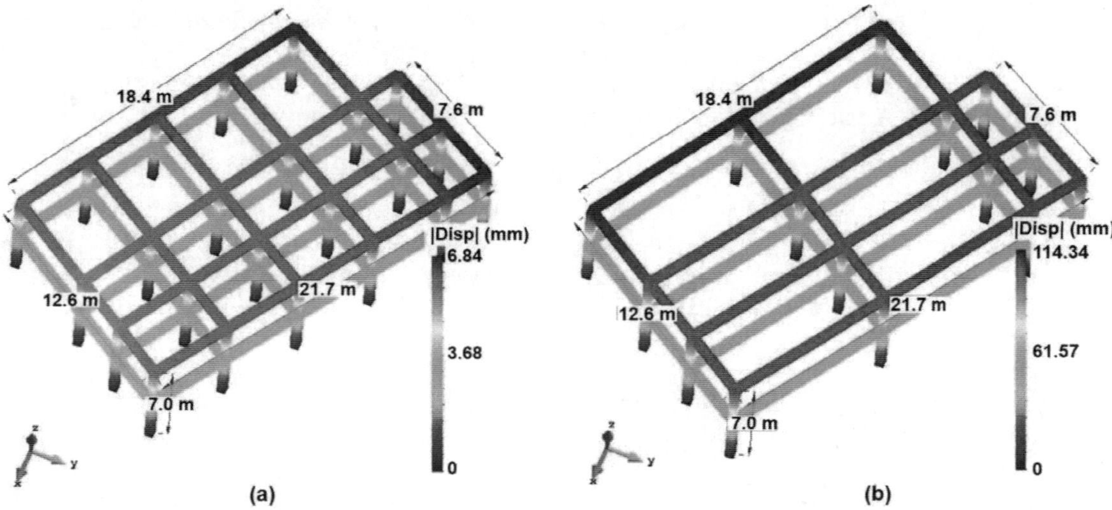

Figure 4. Seismic response of the models generated using **(a)** the tool developed by Quinay, Soliman, and Fader [7], and **(b)** tool developed in this study for a structure with a six-sided footprint.

Figure 4 depicts the generated models and estimated seismic response for another structure with a six-sided building footprint and a building height of 7.00 meters. As observed from the figure, the model in figure 4(a) has more structural nodes and elements, and consequentially, has smaller nodal displacements compared to the model in figure 4(b). These observations are consistent with the observations made for the models in figure 3.

More importantly, it can be noticed how the beams of the model in figure 4(a) are not parallel or perpendicular to other beams, which is a beam placement practice that is rarely observed in reality. The tool developed in this study attempts to fill this knowledge gap. In particular, the algorithm of the developed tool was designed such that beams are placed parallel or perpendicular to one side of the building footprint, as observed in the model in figure 4(b). This demonstrates that the tool developed in this study can generate models that are more similar to observed structural framings in practice.

7. Concluding Remarks

City-scale 3D FE models of the target urban areas in Metro Manila, Philippines are successfully generated from the GIS features on building footprint and height, and the heuristic-based framing and story design models. The line element meshed models are generated using the developed frame model generating tool that utilizes optimal feed-forward ANN topologies for column location and floor elevation prediction.

To better demonstrate the integration of ANN to automated frame model generation, it is highly suggested to extend the application of ANN to beam framing prediction. Nonetheless, optimal ANN topologies for column location and floor elevation prediction are successfully integrated into the developed tool in the form of ANN connection weights.

Furthermore, the possible application of city-scale FE model generation to city seismic response estimation is demonstrated in this paper. Given the generated city-scale FE models, the seismic response of two cities in Metro Manila, Philippines is estimated.

Lastly, the generated frame models and estimated seismic response of selected structures are compared to the results of the simulation-based study by Quinay, Soliman, and Fader [7]. When compared to the models generated by Quinay, Soliman, and Fader [7], the selected models generated using the tool developed in this study have fewer nodes and elements. With fewer degrees of freedom (DOF), the computational cost during the analysis of models created by the tool developed in this study may be reduced. Moreover, the tool developed in this study provides a possible solution to the knowledge gap in the column and beam placements identified in the previous study.

Aside from these comparisons, it is highly recommended that the frame models be validated using the actual framing of selected structures in the target urban areas.

Acknowledgments
The authors would like to acknowledge the civil engineers who participated in the conducted survey; the DOST-ASTI CoARE; and the Feature-based Earthquake Analysis Toolset for Urban Response Estimation (FEATURE) Project of the Department of Science and Technology – Philippine Institute of Volcanology and Seismology (DOST-PHIVOLCS).

References

[1] Santos B 2019 5 ways 3D city models plan a better tomorrow for urban communities *Microdesk* Available at https://www.microdesk.com/articles/5-ways-3d-city-models-plan-a-better-tomorrow-for-urban-communities/

[2] Kobayashi K and Terzidis K 2000 Extracting the geometry of buildings from satellite images using fuzzy multiple layer perceptrons *SIGraDi'2000 – Construindo (n)o Espacio Digital (Constructing the Digital Space)* pp 236–9

[3] Hori M, Ichimura T, Wijerathne L, Ohtani H, Chen J, Fujita K and Motoyama H 2018 Application of high performance computing to earthquake hazard and disaster estimation in urban area *Frontiers in Built Environment* **4** 103

[4] Lazarevska M, Knezevic M, Cvetkovska M and Trombeva-Gavrilooska A 2014 Application of artificial neural networks in civil engineering *Tehnicki Vjesnik-technical Gazette* **21** 1353–59

[5] Denipote J G and Assaf R 2005 Classification of elements in an 3D urban virtual environment using artificial neural nets *ResearchGate* Available at https://www.researchgate.net/publication/252063691_Classification_of_Elements_in_an_3D_Urban_Virtual_Environment_Using_Artificial_Neural_Nets

[6] Sugihara K and Hayashi Y 2006 Automatic generation of 3D building models from building polygons on GIS *Proc. of the 11th International Conference on Computing in Civil and Building Engineering, ICCCBE* pp 14–16

[7] Quinay P E, Soliman J M and Fader A R 2020 Development of simulation-based approach using frame models generated from GIS features and BIM data for application to city seismic response analysis of low- to mid-rise RC structures in Metro Manila *Journal of Earthquake and Tsunami* **14** 2050021-1–18

[8] GMMAREADY 2013 *GMMA READY Project 2013: Hazards Mapping and Assessment for Effective Community Based Disaster Risk Management,* Phase II, Project implemented by National Disaster Risk Reduction and Management Council-Office of Civil Defense, Collective Strengthening of Community Awareness on Natural Disasters Agencies (NDRRMC-CSCAND), administered by United Nations Development Programme (UNDP) and funded by Australian Government through the Australian Agency for International Development (AusAID) (Quezon City)

[9] National Building Code Review Committee (NBCRC) 2005 *Revised Implementing Rules and Regulations of the National Building Code of the Philippines (P.D. 1096)* (Manila: Professional Regulatory Board of Architecture (PRBoA))

[10] Panchal F and Panchal M 2014 Review on methods of selecting number of hidden nodes in artificial neural network *International Journal of Computer Science and Mobile Computing* **3** 455–64

[11] Thomas A 2017 *An introduction to neural networks for beginners* Available at https://www.adventuresinmachinelearning.com/wp-content/uploads/2020/02/A-beginners-introduction-to-neural-networks-V3.pdf

[12] Maier H R and Dandy G C 2000 Neural networks for the prediction and forecasting of water resources variables: a review of modelling issues and applications *Environmental Modelling & Software* **15** 101–24

IOP Conf. Series: Earth and Environmental Science 1091 (2022) 012021 doi:10.1088/1755-1315/1091/1/012021

Development of a k-Nearest Neighbor (kNN) Machine Learning Model to Estimate the SPT N-Values of Valenzuela City, Philippines

J Galupino*[1] and J Dungca[1]

[1] Department of Civil Engineering, De La Salle University, Manila, Philippines

*Corresponding author e-mail: joenel.galupino@dlsu.edu.ph

Abstract. Due to the intricate geological formation of geomaterials, they often exhibit a range of attributes at different sites on a given site. This has posed a problem for geotechnical engineers, as they require correct soil and rock information to plan and design geotechnical construction projects. Numerous efforts have been made at the local level to bridge this disparity by standardizing the quantification of soil parameters; nevertheless, these studies have limitations, and there are still areas with ambiguous or incorrect information. To help close this gap, this article will demonstrate an innovative technique for predicting representative Soil Penetration Test (SPT) N-values for each Valenzuela City Barangay/Zone using a k-Nearest Neighbor (k-NN) Machine Learning Model. Borehole data from the city of Valenzuela were collected and analyzed for this study. The k-NN Model's input parameters are the borehole's latitude, longitude, and depth, while the response parameters are the SPT N-Values. The k-NN technique was used to train the input data, identify patterns, and make classification decisions for the SPT N-Values per depth per location in Valenzuela City. The centroids of each Barangay/Zone in Valenzuela City were also extracted in Latitude and Longitude format and utilized to deploy the k-NN Model. To illustrate the findings, we presented subsurface information from Valenzuela City with SPT N-Values. For validation purposes, the accuracy rate of the machine learning model was obtained. The model's hyperparameter in terms of k was tuned to determine if the accuracy rate could be improved.

1. Introduction

Soils and rocks are natural geomaterials, and their properties are influenced by a range of spatially changing causes during their complex geological formation phase. As an outcome, geomaterials can exhibit a range of qualities depending on their location inside a specific site. This presents a problem for geotechnical engineers, as they require reliable soil and rock information to design and construct geotechnical projects [1]. Borehole logs, on the other hand, are often undertaken occasionally for a given project because of money and time constraints. As a result, geological and geotechnical data can be gathered exclusively at the project's location. Other areas' subsurface information must be extrapolated based on accessible data from historical site investigations, or, more commonly than not, empirical correlations are employed to obtain further information on a site to reduce cost, time spent, and effort [2]. Numerous efforts are being undertaken to close this gap by merging the quantification of soil parameters, the mapping of soil properties, and the mapping of hazard assessment in the Philippines [3-8]. Moreover, many empirical models were developed using traditional numerical and statistical techniques [9-22], however, there are still limitations, there are still areas of uncertainty or unreliability in the data, and the majority of them are not publicly available. By constructing models that create soil parameters for unknown sites, geotechnical engineers may reliably anticipate the soil parameters they will encounter at their target locations.

Furthermore, in the dawn of new technology, data are being captured, stored, manipulated, analyzed, managed, and presented. Soft computing techniques such as Machine Learning have effectively been applied to eliminate the previous information requirement about the interactions among inputs, parameters, and outputs in soil explorations. Machine Learning learns from the soil data, recognizes patterns, and makes predictions. Machine Learning models that have been successfully utilized [23-25] in estimating geotechnical parameters in the Philippines.

In this study, a k-Nearest Neighbor (kNN) Machine Learning Model was developed to estimate the SPT N-Values of Valenzuela City, Philippines. By creating these models that estimate soil parameters for unknown sites, geotechnical engineers would accurately predict the soil parameters they will encounter at their target locations. The developed model will assist engineers in determining the appropriate form of foundation for a specific site and able to be used for Disaster Risk Reduction (DRR) planning to minimize ground-related disasters; government zoning and policymakers for land use purposes, and the real estate industry as an initial guide before investing. To validate the result, the accuracy rate of the kNN model was determined together with the root mean square error (RMSE). Furthermore, the hyperparameter, in terms of k (number of neighbors), of the model was tuned to check if the accuracy rate may be improved.

2. Methodology

2.1. Data

The borehole data used were obtained from previous research or construction companies conducting borehole tests in the chosen locale, which was used to assess the soil properties. Borehole data were also collected in areas surrounding the locale, to ensure that the data are well sparse. The independent and dependent variables are shown in Table 1, as patterned in previous studies [6]. The output soil data is a numerical parameter, which is very suitable for a Machine Learning Model.

Table 1. Dependent and Independent Variables of the Study

Independent Variable(s)	Dependent Variable(s)
1. Longitude 2. Latitude 3. Elevation	1. SPT N-Value

2.2. Research Locale

The location of the study covers the area of Valenzuela City in the Philippines, it has an area of 47.02km^2. It is known as the "Northern Gateway to Metropolitan Manila," which comprises 32 Barangays. Valenzuela is one of the most urbanized and fastest-growing cities in Metro Manila. Since it is crossed by two major highways – the MacArthur Highway and the North Luzon Expressway – it is dubbed the "Northern Gateway to Metropolitan Manila." In addition, it is also due to its proximity to the cities of Manila, Quezon, and Caloocan, this formerly picturesque village has been converted into a bustling, highly urbanized metropolis. The area is home to both small and large-scale enterprises, including food goods, metals, plastics, and textiles, all of which contribute to the town's rapid economic expansion. The city's ongoing physical expansion earned it additional recognition and money, earning it the moniker "economic miracle of the decade." Additionally, this industrial expansion provided a plentiful supply of work for the rising labor force, thereby enhancing the citizens' economic standing. As a result, structures are being built to house these developments, and as previously mentioned, soil explorations are a prerequisite to these projects. However, these subsurface explorations, which often involve borings, are costly.

To be consistent with the density of 1 borehole per square kilometer, since some of the Barangays are more than 1 km^2, these barangays were divided into zones, shown in Figure 1.

Figure 1. Map of Valenzuela City with Divided Barangay/Zones

2.3. Data Processing

Borehole data will be extracted in a format compatible with Microsoft Excel. Matlab will be used for programming. These parameters are then processed using Machine Learning. The model will be implemented on a Barangay basis level. In the Philippines, the common practice in determining the geotechnical properties of a specific site is using the Standard Penetration Test (SPT). The extracted data from the borehole logs will represent its location and removed redundant borehole locations will not be used and will not affect the modeling. Since Latitude and Longitude is a more reliable methods of positioning. The Latitude and Longitude of the borehole exploration will be used to define and identify its position.

2.4. Machine Learning Model

Machine learning automated soil data processing models by learning from the soil data, recognizing its patterns, and making decisions with minimal human intervention. A k-Nearest Neighbor (kNN) Machine Learning Model was developed to estimate the SPT N-Values of Valenzuela City, Philippines.

k-Nearest Neighbor Modelling is a non-parametric classification technique that can be applied to both classification and regression. In both instances, the input consists of the data set's k nearest training examples. The performance is determined by whether the k-NN algorithm is used for classification or regression. For k-NN, most of its neighbors vote to classify an object, with the object being assigned to the most prevalent class among its k nearest neighbors. If k equals 1, the object is simply assigned to the class of the object's closest neighbor [26].

k is a user-defined constant in the classification process. An unlabeled vector (a query or test point) is categorized by assigning the mark that occurs most frequently among the k training samples closest to the query point. Euclidean distance is a frequently used distance metric for continuous variables. For discrete variables, such as those used in text classification, another metric, such as the overlap metric such as Hamming Distance, may be used. k-NN may be used in conjunction with correlation coefficients such as Pearson and Spearman. The Euclidean distance between two points in Euclidean space is the length of a line segment between the two points [27].

2.5. Data Split Ratio, Accuracy Rates, Hyperparameter Tuning, and Deployment

Some studies provide a detailed analysis of the training and testing process in machine learning [28]. Meaningful data has become more challenging as a result of the proliferation of data. To make sense of the data, a field called data mining has developed. Data mining is a collection of procedures that are used to extract meaning from data [29]. To detect the actions of a machine learning model, we must use observations that were not used during the training phase. Otherwise, the model's assessment will be skewed. The datasets are (1) Training Dataset, (2) Validation Dataset, and (3) Test Dataset.

The Training Dataset is the real dataset on which the model is trained. The model observes and gains knowledge from this data. The Validation Dataset is used to validate a given model, but it should be evaluated frequently. This information is used to adjust the model's hyperparameters. The Test Dataset data sample was used to conduct an unbiased assessment of the final model fit on the training dataset. The Test dataset serves as the gold standard against which the model is evaluated. Generally, the test set is used to compare competing models. The Dataset Split Ratio is 70% Training Data, 15% Validation Data, and 15% Testing Data.

The accuracy rates for the machine learning model were gathered for validation. Accuracy is the ratio of the number of correct predictions to the total number of input samples, shown in Equation 1:

$$Accuracy\ (\%) = \frac{No.of\ Correct\ Predictions}{Total\ No.of\ Predictions} \tag{1}$$

The hyperparameter, in terms of k, of the model was tuned to check if the accuracy rate may be improved. Hyperparameter tuning is the problem of selecting the appropriate collection of hyperparameters for a machine learning system [30].

To be consistent with the density of 1 borehole per square kilometer, since some of the Barangays are more than 1 km2, these barangays were divided into zones, shown in Figure 1. The centroids, in Latitude and Longitude format, of each Barangay was determined through Geographic Information System (GIS).

3. Results and Discussions

3.1. Data Points

There are twenty-four (24) Borehole logs used for data processing, shown in Figure 2, accumulating to 243 data layers.

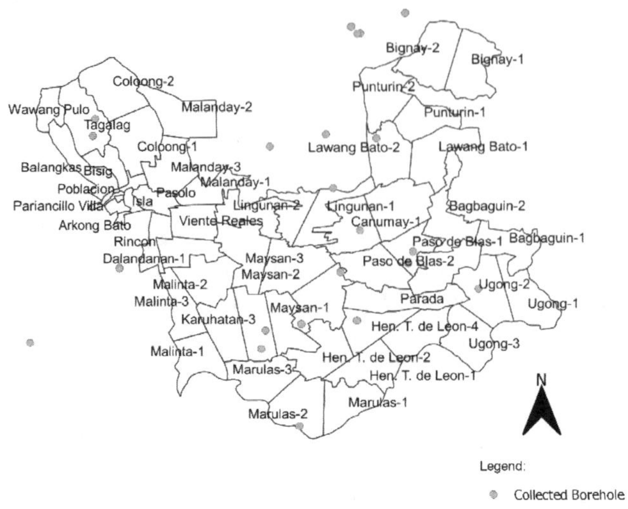

Figure 2. Collected borehole data for Valenzuela City

Using k-NN utilizing its neighbors vote to classify, with the object being assigned to the most prevalent class among its k nearest neighbors, a total of 1,912 data layers were created to estimate the subsurface layer of each Valenzuela Barangay/Zones in terms of SPT N-values, these data layers were embedded in the various points shown in Figure 1.

3.2. Accuracy Rate Validation and Hyperparameter Tuning

Accuracy is the ratio of the number of correct predictions to the total number of input samples, it is computed using Equation 1. k-Nearest Neighbor Model was used to train the data and generated accuracy rate. It was further tuned to verify if the accuracy rate will increase if the number of neighbors (k) is also increased, the results are shown in Figure 3.

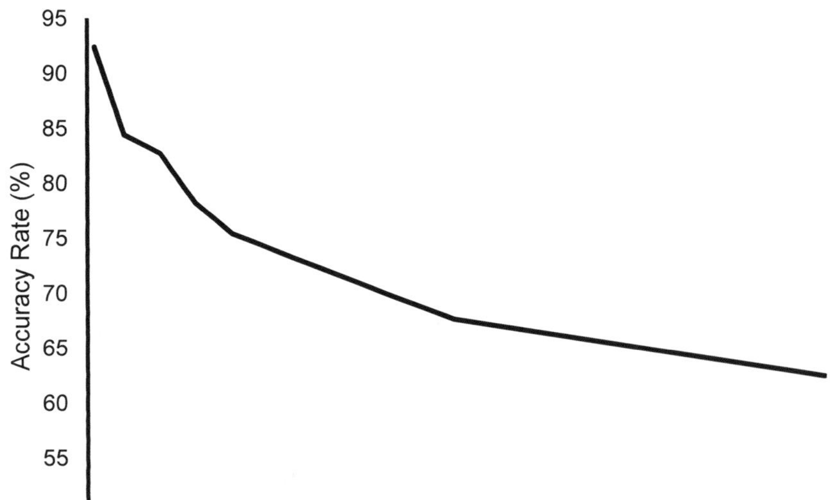

Figure 3. Hyperparameter Tuning of the Model

When the value of 'k' is reduced, the anticipated value becomes less stable. In contrast, increasing the 'k' value stabilizes the projected value while increasing the inaccuracy. As a result, the 'k' value in the middle should be preferred, and it was also suggested that an odd number should be used as a tie-breaker. As a result, a 'k' of five (5) was used for deployment in the study. An accuracy rate of 84.4% was garnered in k=5. k-NN used distance to classify an unknown parameter. The k-NN algorithm assumes that similar things exist nearby. In other words, similar things are near to each other [31].

3.3. Elevation
On-site measurement gathered the highest elevation point of Valenzuela is 38 meters (125 ft) above mean sea level. Having a surface gradient of 0.55% and a gentle slope, hilly landscape. The average elevation point is 2 meters (6.6 ft) above mean sea level.

k-NN Model provided a maximum elevation of 34 meters above mean sea level and an average elevation point of 2.4 meters above mean sea level, their comparison is shown in Table 2.

Table 2. Comparison of Old and k-NN Generated Parameters

Parameters	Measured (m)	k-NN (m)	% Diff.
Maximum Elevation	38	34	10%
Average Elevation	2.0	2.4	16%

The percentage difference for the Maximum Elevation in terms of the Old Data compared to the k-NN generated data is 10%, while for Average Elevation, it has 16%, which still has a strong association.

3.4. SPT N-Values
SPT N-value is the number of blows or penetration resistance (N) value is registered for each penetration interval at a given depth. Additionally, if each of the intervals requires more than 50 blows or if the total number of blows required reaches 100, the test is terminated. Locally, all boreholes should be drilled to a depth of at least five (5) meters into hard strata or until a suitable bearing layer is reached, unless otherwise indicated by the consulting Geotechnical Engineer. The Old SPT N-Value for each borehole is shown in Figure 4, while the k-NN Generated SPT N-values are shown in Figure 5.

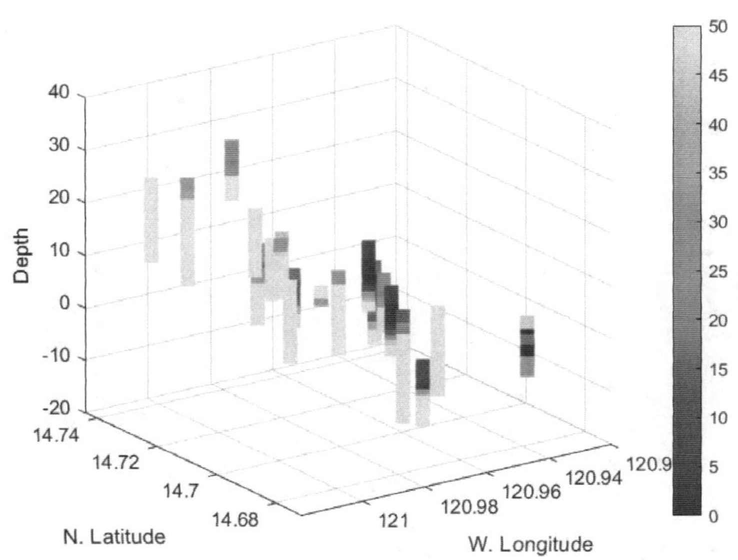

Figure 4. SPT N-values of the BH Logs

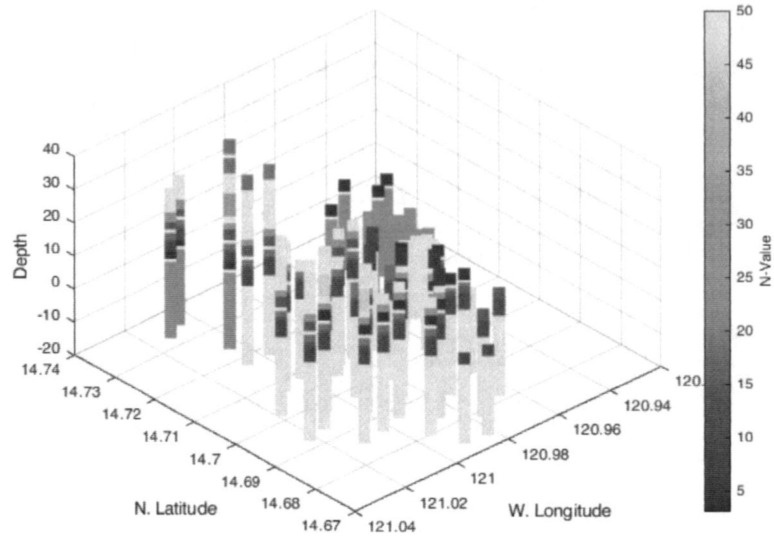

Figure 5. SPT N-values generated by k-NN Model

It can be noticed that Valenzuela City is dominated by hard strata with a 50 SPT N-value, shown on the Histogram in Figure 6, that is very suitable to support small and large-scale enterprises.

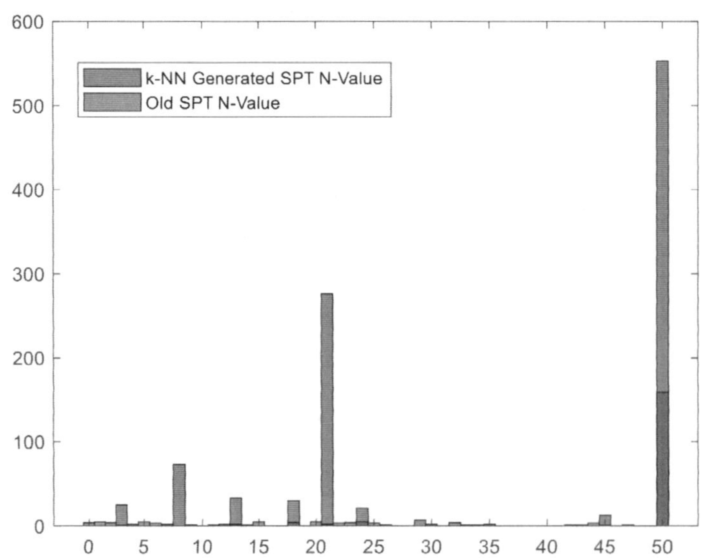

Figure 6. Histogram of SPT N-Values

Due to the geological properties of the area, it is common to find 50 SPT N-Values. In Valenzuela, a mixture of sands, clays, and gravels remained the predominant soil type. However, in the k-NN generated data, the northwestern part of Valenzuela is dominated by lower SPT N-values; this may be a result of the k-NN algorithm, in which the layer is classified by the majority of its neighbors, with the classification assigned to the most prevalent class among its k nearest neighbors. The layers were influenced by Malabon City soil data, which is characterized by low N-values.

In Figure 7, the histogram of SPT N-Values is shown. SPT N-value of 50 may be encountered mostly in 23 Valenzuela Barangays, namely:

- Arkong Bato
- Bagbaguin
- Bignay
- Canumay
- Coloong
- Dalandanan
- Hen. T. de Leon
- Isla
- Karuhatan
- Lawang Bato
- Lingunan
- Malinta
- Mapulang Lupa
- Marulas
- Palasan
- Parada
- Paso de Blas
- Pasolo
- Punturin
- Rincon
- Tagalag
- Ugong
- Wawang Pulo

These Barangays may support shallow foundations, however, must be approved and recommended by the consulting Geotechnical Engineer.

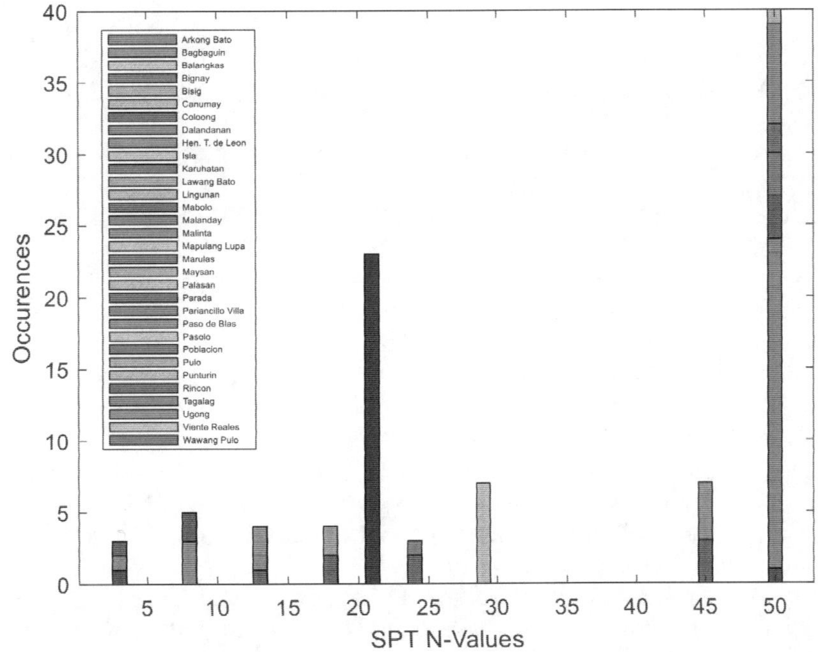

Figure 7. Histogram of SPT N-Values per Barangay

4. Conclusion

Soils and rocks are natural geomaterials, and their properties are influenced by a range of spatially changing causes during their complex geological creation phase. As a result, geomaterials can exhibit a range of qualities depending on their location inside a specific site. This study utilized Machine Learning, specifically the k-Nearest Neighbor (k-NN) model, to estimate representative Soil Penetration Test (SPT) N-values for each Valenzuela City Barangay. This provided engineers with a reference soil profile to help them determine what to expect at a particular project site.

Twenty-four (24) borehole logs were used to process the data, and a total of 1,912 data layers were constructed to estimate the subsurface layer of each Valenzuela Barangay in terms of SPT N-values. The accuracy rate of 84.4% was garnered in k=5. The percentage difference between the Old Data and the k-NN generated data for Maximum Elevation is 10%, while 16% for Average Elevation indicates a strong relationship.

It can be seen that Valenzuela City is dominated by hard strata with a 50 SPT N-value; however, the k-NN generated data shows that the northwestern section of Valenzuela is dominated by lower SPT N-values; this may be due to the soil data from Malabon City, which is dominated by lower N-values.

5. References

[1] Hu Y, Zhao T, and Wang Y 2020 Bayesian Learning of Site-Specific Spatial Variability Using Sparse Geotechnical Data pp 553–558

[2] Anbazhagan P, Uday A, Moustafa S and Al-Arifi N 2016 Correlation of densities with shear wave velocities and SPT N values *Journal of Geophysics and Engineering* **13** pp 320–341

[3] Dungca J and Chua R 2016 Development of a Probabilistic Liquefaction Potential Map of Metro Manila *International Journal of GEOMATE* 2016(April) pp 1804–1809

[4] Dungca J 2020 A reference for the allowable soil bearing capacities in Quezon city, Philippines *International Journal of GEOMATE* **19** pp 42–47

[5] Galupino J, Garciano L, Paringit M and Dungca J 2017 Location-based prioritization of Surigao municipalities using probabilistic seismic hazard analysis (PSHA) and geographic information systems (GIS) Paper presented at the HNICEM 2017 - 9th International Conference on Humanoid, Nanotechnology, Information Technology, Communication and Control, Environment and Management, 2018-January 1-7

[6] Galupino J and Dungca J 2019 Quezon City soil profile reference *International Journal of GEOMATE* **16**

[7] Dungca J, Concepcion I, Limyuen M, See T and Vicencio M 2017 Soil bearing capacity reference for Metro Manila, Philippines *International Journal of GEOMATE* **12** pp 5-11

[8] Uy E, Paringit M, Cutora M, Galupino J, Garciano L and Dungca J 2020. Characterization of Cebu province municipalities using probabilistic seismic hazard assessment (PSHA) and geographic information system (GIS) Paper presented at the IOP Conference Series: Earth and Environmental Science **479**

[9] Galupino J and Dungca J 2015 Permeability characteristics of soil-fly ash mix *ARPN Journal of Engineering and Applied Sciences* **10** pp 6440-6447

[10] Dungca J and Galupino J 2016 Modeling of permeability characteristics of a soil-fly ash-bentonite cut-off wall using response surface method *International Journal of GEOMATE* **10** pp 2018-2024

[11] Dungca J, Galupino J, Sy C and Chiu, A 2018. Linear optimization of soil mixes in the design of vertical cut-off walls *International Journal of GEOMATE* **14** pp159-165

[12] Dungca J, Galupino J, Alday J, Barretto M, Bauzon M and Tolentino A 2018 Hydraulic conductivity characteristics of road base materials blended with fly ash and bottom ash. *International Journal of GEOMATE* **14** pp 121-127

[13] Adajar M, Galupino J, Frianeza C, Aguilon J, Sy J and Tan P 2020 Compressive strength and durability of concrete with coconut shell ash as cement replacement *International Journal of GEOMATE* **17** 183-190

[14] Dungca J and Dychangco L 2016 Strength properties of road base materials blended with waste limestones *International Journal of GEOMATE* **11** pp 2493-2498

[15] Adajar M, Aquino C, dela Cruz J, Martin C and Urieta D 2019 Investigating the effectiveness of rice husk ash as a stabilizing agent of expansive soil *International Journal of GEOMATE* **16** (58)

[16] Adajar M and Cutora M 2018 The effect of void ratio, moisture content, and vertical pressure on the hydrocompression settlement of copper mine tailing *International Journal of GEOMATE* **14** pp 82-89

[17] Adajar M, de Guzman E, Ho R, Palma C and Sindico D 2017 Utilization of aggregate quarry waste in the construction industry *International Journal of GEOMATE* **12** pp 16-22

[18] Uy E and Adajar M 2017 Assessment of critical-state shear strength properties of copper tailings *International Journal of GEOMATE* **12** pp 12-18

[19] Uy E and Dungca J 2018. A comparative settlement prediction of limestone blended materials using Asaoka and hyperbolic method *International Journal of GEOMATE* **14** 63-69

[20] Ubay I Alfaro M, Alfaro M and Blatz J 2020 Stability assessment of an aging earth-fill dam considering the anisotropic behavior of clay *International Journal of GEOMATE* **18** pp 84-91

[21] Adajar J, Ubay I, Alfaro M and Chen Y 2020. Discrete element model parameters to simulate slope movements *International Journal of GEOMATE* **18** pp 192-199

[22] Adajar J, Ubay I, Alfaro M and Chen Y 2020 Discrete element modeling of undrained consolidated triaxial test on cohesive soils Paper presented at the Geotechnical Special Publication 2020-February (GSP 317) pp 172-182

[23] Elevado K, Galupino J and Gallardo R 2019 Compressive strength optimization of concrete mixed with waste ceramics and fly ash *International Journal of GEOMATE* **16** pp 135-140

[24] Elevado K, Galupino J and Gallardo R 2018 Artificial neural network (ANN) modeling of concrete mixed with waste ceramic tiles and fly ash *International Journal of GEOMATE* **15** pp 154-159

[25] Dungca J and Galupino J 2017 Artificial neural network permeability modeling of soil blended with fly ash *International Journal of GEOMATE* **12** pp 76-83

[26] Zhang Z 2016 Introduction to machine learning: K-nearest neighbors *Annals of Translational Medicine* **4**

[27] Lubis A, Lubis M and Al-Khowarizmi 2020 Optimization of distance formula in the k-nearest neighbor method *Bulletin of Electrical Engineering and Informatics* 9 pp 326–338

[28] Uçar M, Nour M, Sindi H and Polat K 2020 The Effect of Training and Testing Process on Machine Learning in Biomedical Datasets *Mathematical Problems in Engineering* 2020

[29] Witten I, Frank E and Hall M 2013 Data Mining-Practical Machine Learning Tools, and Techniques, Morgan Kaufmann, Burlington, MA, USA

[30] Ghawi R and Pfeffer J 2019 Efficient Hyperparameter Tuning with Grid Search for Text Categorization using kNN Approach with BM25 Similarity *Open Computer Science* **9** pp 160–180

[31] Harrison O 2018 Machine Learning Basics with the K-Nearest Neighbors Algorithm

IOP Conf. Series: Earth and Environmental Science 1091 (2022) 012022 doi:10.1088/1755-1315/1091/1/012022

Attitude and flood preparedness of households in Pasir Mas, Kelantan, Malaysia

F A Noor Diyana
HADR Research Centre, National Defence University of Malaysia, Kem Sungai Besi, 57000 Kuala Lumpur, Malaysia

Abstract. Kelantan is a state that is often impacted by the yearly monsoonal floods resulting in the displacement of a large number of households as well as damages to crops, livestock and properties. The flood is a yearly occurrence that necessitates the community to prevent loss of lives and extensive property damage; however, previous studies have found that communities are unprepared to face flood disasters, especially in flood-prone areas. Thus this study aims to determine the relationship between attitude towards flood preparedness and the level of flood preparedness of households in Pasir Mas, Kelantan. A total of 220 respondents from 10 villages were selected with the assistance of the District Office to ensure the selection of villages with experience of flood or is flood-prone. Descriptive analysis was conducted to determine the level of attitude and flood preparedness. Results from the analysis found that the attitude and preparedness level was moderate. Therefore, it is suggested that the community's attitude and preparedness level can be further enhanced with the assistance of the responding agencies and local authorities to empower a more resilient community.

1. Introduction

Over the last 20 years, Malaysia has experienced extreme weather events such as La Nina in 2011 and 2012, the 2014 Flood, and most recently in the late 2020 and early 2021, which affected 27 districts from several states, namely Johor, Pahang, Kelantan, Selangor and Perak. The flood caused an estimated of 11,973 families to evacuate to 397 evacuation centres, with eight death reported [1]. The state of Kelantan in the East Coast of Peninsular Malaysia is one of the most flood-prone states due to the annual northeast monsoon flood causing considerable and significant damages. Previously in the 2014 flood, Kelantan was one of the worst affected states in Peninsular Malaysia as 10,000 people were displaced. Furthermore, between 2016 to 2018, RM8.41 million of financial aid from the government was distributed to the victims in Kelantan, which was the most significant amount among the states in Peninsular Malaysia [2]. Despite mitigation efforts such as high concrete levees and pumps placed along the Kelantan River [3] and the flood warning and forecasting system in place [4], the community still faces regular flooding, which affects their daily activities and livelihood. Since the area is prone to regular flood events, it is essential to identify the community's attitude towards flood preparedness and its level.

2. Literature Review

A community that experiences frequent floods is expected to be more resilient against disasters is one of the strategies for disaster risk reduction as identified by the United Nations International Strategy for Disaster Risk Reduction (UNISDR) through the enhancement of community-based disaster risk management [5]. This approach implements top-down and bottom-up approaches, which comprises activities and significant role-playing between the residents, local authorities, and the responding agencies. These activities include instilling awareness across the community through increasing their knowledge of flood risk, safe routes and familiarisation with the type of flood warning messages and emergency responses. The community-based disaster risk management concept is advocated as a priority and as part of the Sendai Framework for Disaster Risk Reduction guiding principle, which strives towards risk reduction at all levels.

Since the community needs to be well-prepared since Kelantan is frequently affected by floods, especially during the northeast monsoon, their attitude towards flood disasters reflects the community's preparedness. Previous studies have found that attitude was a key determinant of disaster preparedness [6]. Attitude is a potential habitual tendency towards a psychological matter with favourableness or unfavourableness [7]. The Theory of Reasoned Action (TRA) explains the process of belief, which changes over time and how belief influences the attitude through the Fishbein expectancy-value model. Thus, the attitude towards flood in this study can be described as the potential tendency to take preparedness measures for flood disasters.

Subsequently, preparedness is being in a state of readiness to respond to a disaster or emergency [8]. It focuses on sustaining the viability of the community by protecting lives and properties. An individual must be self-reliant and self-sufficient during the first 72 hours of the disaster event to save themselves [9]. Self-preparedness includes various self-protective behaviours and planning activities such as packed clothing and essential items for evacuation, plans to arrange furniture to minimise damage, identifying safe evacuation routes and evacuation centres. These preparedness measures could reduce personal injury, loss of life and damage to properties during disasters [10]. Past studies have found that people exposed to hazards remain unprepared despite countless efforts [11]. Therefore, this study aims to identify the level of attitude towards flood disasters and the community's preparedness level in the district of Pasir Mas in Kelantan.

3. Method

The study employed a multi-stage sampling method to select respondents through flood evacuees data between 2012 to 2017 obtained from NADMA. Through the data obtained, one state from each of the four zones (central, north, south and east coast) in Peninsular Malaysia was selected based on the highest number of evacuees within the zone. Kelantan was selected to represent the east coast zone. Subsequently, the district of Pasir Mas was then selected as it recorded one of the highest numbers of evacuees in the state between the period of 2012 to 2017, with a total of 18,753 evacuees [12]. The selection of the villages was made with the recommendation of the Pasir Mas Land and District Office to select the villages (Table 1) that experienced evacuation during a flood.

Table 1. List of villages selected.

District	Population (N)	Villages
Pasir Mas	13,260	Kubang Sepat
		Kankong
		Chetok
		Bandar Pasir Mas
		Rantaun Panjang
		Gual Periuk
		Lemal
		Meranti
		Alor Mas
		Gelang Mas

Next, the households were selected with assistance from the head of the village. The head of the household was selected as the respondent; however, if the head of household was not present or unavailable, the spouse was selected. A total of 220 respondents were obtained during the data collection process. The data collection was carried out during March 2018 using self-administered questionnaires distributed to the respondents. Three sections were included in the questionnaire consisting of the demographic background of the respondents, attitude towards flood and flood preparedness.

Items in the demographic background include age, gender, education background, income level and working status. The attitude variable was adopted from one author and consisted of eight statements [13] using a Likert-type scale ranging from 1(strongly disagree) to 5(strongly agree). The measurement was selected as it represented the attitude of the respondent towards preparing for a flood disaster. Items in the measurement consisted of positive (No.2, No.3, No.5 and No.6) and negative statements (No.1, No.4, No.7 and No.8) as tabulated in Table 2. Prior to the analysis, the data was recoded using Statistical Package for Social Sciences (SPSS) in order to prepare the data for analysis and report an accurate level of attitude.

Table 2. Items to measure attitude

No.	Items
1.	Preparing for Tsunami is inconvenient for me.
2.	Preparing for tsunami will significantly reduce damage to my home.
3.	Preparing for tsunami will improve my everyday living conditions.
4.	It is unnecessary to prepare for tsunami as assistance will be provided by
5.	Preparing for tsunamis will improve my ability to deal with disruption to
6.	Preparing for tsunami will save lives.
7.	I don't know how I can prepare for a tsunami.
8.	It's difficult to prepare for a tsunami.

As for flood preparedness, a total of 23 statements were adopted [6] were listed down using 'Yes' or 'No' statements. The items consists of statements related to preparation measures such as placing latches on cabinets or wardrobes, financial preparedness, identifying safe routes and evacuation centres. All the items used to measure preparedness were adapted from various sources in order to ensure suitability for this study and a sample of the statements are tabulated in Table 3.

Table 3. Sample of statements to measure preparedness

No.	Statements
1.	Stockpile drinking water for a three day supply
2.	I have an emergency savings of at least 3 months of my salary
3.	Have a first aid kit
5.	I have gathered important documents (e.g.insurance policies, deeds etc.) in one place that is easy to retrieve
6.	I have purchased additional insurance that cover damages caused by disaster
7.	I have planned to bring spare clothes if I need to evacuate
8.	I know where the nearest evacuation centre to my house
9.	I have identified a safe place to keep my car in case of flood

Before the survey, a pre-test was conducted among communities in flood-prone areas. A total of 20 respondents were selected to participate in the pre-test survey to test the reliability of the attitude variable. The attitude towards flood was found to have an alpha score higher than 0.6. The self-administered questionnaire was then distributed to the respondents. However, an interview mode was used for the respondents that unable to read or had difficulty reading. A token of appreciation was given to the respondents when the completed questionnaire was returned. Statistical Package for Social Science (SPSS) was used to analyse the descriptive data, determine the levels of attitude and preparedness as well as to identify the relationship between attitude and preparedness.

4. Results and Discussion

The descriptive analysis of the socio-demographic shows that more than half of the respondents were male (69%) with an average age of 50 years old. At least half of the respondents reported secondary school as the highest education level (62%). They had an average household income of RM1,067.00, much lower than Kelantan's average household income (RM4,214) [14]. Most of the respondents' livelihood depended on subsistence work (47%) and operating their own business (21%).

Table 4. Socio-demographic characteristics of Pasir Mas respondents.

Background	N=220	Average	Percentage (%)
Gender			
Male	152		69
Female	68		31
Age		50	
Highest education level			
Did not attend school	27		12
Primary School	34		16
Secondary School	135		62
Tertiary level	23		10
Household income		RM 1,067.00	
Working status			
Government Sector	6		3
Private Sector	9		4
Entrepreneur	47		21
Subsistence work	103		47
Not working	35		16
Retired	20		9

Results from the analysis depicted in Table 5 show that the respondents reported a moderate attitude towards flood (51%), and only a handful reported a high level of attitude (20%). The item with the highest mean was that respondents believe flood preparedness saves lives, reduces property damage, and improves their daily lives. These results suggest that the respondents believe that taking precautionary measures can reduce the impact of the flood. However, the average mean was below 4, demonstrating room for improvement in the community's attitude towards floods as they often experience floods during the monsoon season.

Table 5. Attitude level.

Level of attitude	Frequency (N)	Percentage (%)
Low	64	29
Moderate	112	51
High	44	20

Next, the level of preparedness obtained from the analysis, as illustrated in Table 6, found that the respondents had a moderate level of flood preparedness (44%). Not many had a high level of preparedness (23%). Among the preparedness measures taken by the respondents was safekeeping of essential documents, planning the placement of their furniture to avoid damage and identifying the nearest evacuation centres. However, in terms of financial preparedness, more than half of the respondents were unable to save at least three months of their income, purchase life insurance or additional car and house insurance policy to cover damages caused by disasters. Additionally, at least half of the respondents had no first aid kit and no knowledge of conducting Cardiopulmonary Resuscitation (CPR). It is suggested that the respondents had difficulty purchasing insurance or could have at least three months of savings due to their average household income reported in this survey which is below the average household income in Kelantan.

Table 6. Flood preparedness level.

Level of preparedness	Frequency (N)	Percentage (%)
Low	73	33
Moderate	96	44
High	50	23

5. Conclusion

Results obtained from this study indicate that the community's attitude in Pasir Mas can be further improved as they experience frequent floods, especially during the northeast monsoon season. The community believes that preparing can save lives and reduce the damage to their property if they take the necessary precautionary measures; however, the results suggest they also depend on the government agencies for assistance. Subsequently, the results also indicate that the community's level of preparedness was at a moderate level, which can be further enhanced with the help of the local authority and responding agencies. The community could participate in disaster risk management activities with the aid of the local agencies as ineffective programs are caused by the lack of community participation [15]. Moreover, financial assistance from the government through an affordable insurance plan could reduce the community's financial burden within the lower-income group who are more vulnerable to disasters [16]. Awareness programs that promote community-based disaster preparedness can be further developed with assistance from the local government and related agencies to cultivate a more resilient community towards flood disasters. Hence, a collaborative effort from all stakeholders such as the local community, local government and local responding agencies, specifically in an area susceptible to frequent floods, is imperative to strengthen their resilience towards floods.

Acknowledgement

We want to express our gratitude to UK Space Agency for providing the funds to conduct this project and thank NADMA for their support. We would also like to acknowledge all individuals and organisations who have contributed to the successful completion of this project either directly or indirectly.

6. References

[1] International Federation of Red Cross and Red Crescent Societies (IFRC) 2020 *Operation Update Report Malaysia: Floods*

[2] National Disaster Management Agency of Malaysia (NADMA) 2018. *Annual Report* Putrajaya, Malaysia

[3] Tuan Pah Rokiah S H, Abd Rahim M N, and Hamadi I 2014. The level of satisfaction towards flood management system in Kelantan, Malaysia *Pertanika Journal of Social Science and Humanities* **22**(1) 257-269

[4] Mohit M A and Sellu G M 2013. Mitigation of climate change effects through non-structural flood disaster management in Pekan Town, Malaysia *Procedia-Social and Behavioral Sciences* **85** 564-573.

[5] Shaw R 2015. Recovery from the Indian Ocean Tsunami *A Ten Year Journey*

[6] Frandsen M, 2012 *Promoting community bushfire preparedness: Bridging the theory–practice divide* (Doctoral dissertation, University of Tasmania)

[7] Fishbein M and Ajzen I 2011 Predicting and changing behaviour: The reasoned action approach. *Psychology Press*

[8] Haddow G D, Bullock J A and Coppola D P, 2014 The disciplines of emergency management: mitigation. *Introduction to Emergency Management* 71-100

[9] Kohn S, Eaton J L, Feroz S, Bainbridge A A, Hoolachan J, and Barnett D J 2012 Personal disaster preparedness: an integrative review of the literature *Disaster medicine and public health preparedness* **6**(3) 217-231

[10] Lindell M K, Prater C S, Gregg C E, Apatu E J, Huang S K, and Wu H C 2015 Households' immediate responses to the 2009 American Samoa Earthquake and Tsunami *International Journal of Disaster Risk Reduction* **12** 328–340

[11] Becker J S, Paton D, Johnston D M, and Ronan K R 2013 Salient beliefs about earthquake hazards and household preparedness *Risk Analysis* **33**(9) 1710–1727

[12] National Disaster Management Agency of Malaysia (NADMA) 2018 Statistik Mangsa Banjir 2012-2017 Unpublished manuscript (Putrajaya)

[13] McIvor D and Paton D 2007 Preparing for natural hazards: Normative and attitudinal influences. *Disaster Prevention and Management* **16**(1) 79–88

[14] Department of Statistics Malaysia 2020 *Household Income and Basic Amenities Survey Report by State and Administrative District* (Putrajaya)

[15] United Nations Office for Disaster Risk Reduction (UNISDR) 2015 *Sendai Framework for Disaster Risk Reduction* 2015-2030

[16] Chan N W 2012. Impacts of disasters and disasters risk management in Malaysia: The case of floods, in Sawada Y, and S Oum (eds.) *Economic and Welfare Impacts of Disasters in East Asia and Policy Responses* ERIA Research Project Report 2011-8 (Jakarta: ERIA) 503-551

Finding an Optimal Use of Tides for Leaching of Pond Waters in Channel Networks with Various Size and Simultaneous Opening Number of Gates, Case Study of Sei Teras Fishpond Irrigation Area, Central Kalimantan

R Ciptadi[1], A P Rahardjo*[1], B Kamulyan[1]

[1] Civil and Environmental Engineering Department, Universitas Gadjah Mada, Yogyakarta 55281, Indonesia
*Corresponding author e-mail: rahardjo.adam@ugm.ac.id

Abstract. Sei Teras Village has been developing aquaculture since 2006 and has fresh and saltwater sources. Recently, there has been a plan for upgrading the traditional extensive ponds to intensive ones. It is expected to cultivate vannamei shrimp, which delivers a better return. An average of 5-15% of the total water volume in the ponds is recommended to be replaced every day to keep the water quality within the range of the requirements. This study aims to find an alternative design and operation of pond gates for better use of tides for fishpond water leaching. Measurements were carried out for topography, tides, and current boundary conditions for calibration. The flow simulations on the fishpond irrigation network are modeled with two scenarios of pond water gate size and variations in the number of fishpond gates operated simultaneously using HEC-RAS software. Flow simulation is carried out to determine the process of entering and leaving water into ponds. The results of the simulations are evaluated based on their water replacement capacity and changes of discharge and water level at the control point. The initial results show that wider gate opening increases the capacity of water replacement and simultaneous opening operation gives the maximum result.

1. Introduction

In responding to the issue of the world food crisis issued by the FAO, the Indonesian government decided to develop a fishpond irrigation area in the wetland of Central Kalimantan in a national food estate development project. Wetlands, with very high biological productivity, tremendous resource potential, and environmental function, are a unique ecosystem in the earth and a gathering area of biological diversity and threatened species [1][2]. Wetlands provide food and habitat that supports plants, animals, and people. Urbanization, industrialization, and population growth in coastal areas put tremendous pressure on many coastal wetlands. Brackish water ponds are often built-in coastal wetlands for aquaculture cultivation.

One of the traditional aquaculture areas is the Sei Teras. Based on interviews with local communities, aquaculture in the Sei Teras area has been developed since 2006 [3]. The development of aquaculture in this area has been slow. Early development of ponds in this area was used to catch shrimp and fish carried away by tidal floods. Several ponds have been used extensively for polyculture ponds of shrimp, crabs, and milkfish with low productivity. The ponds in this area are of traditional extensive ponds with more than 2 ha for each pond area. The ponds are equipped with 40-70 cm deep channels. Some ponds have no maintenance, and some others are added with chemicals such as fertilizers and pesticides. In general, farmers' gross earning per month is less than 9 million rupiahs per pond.

Therefore, to improve farmer income and increase the national shrimp exports, it is necessary to introduce cultivation technology and bring the aquaculture in this area to an intensive level.

Intensive shrimp farming technology is a cultivation technique with a high stocking rate and optimal management. In general, intensive shrimp cultivation is carried out in ponds with a size of 0.5-1 ha with a water depth of 1-1.2 m [4] with ideal salinity conditions of 15-25 ppt [5][6][7]. The stocking density of vannamei shrimp at an intensive level between 400.000-1.000.000 org per ha [8]. Stocking density affects the amount of feed used; these two are the primary sources of waste in ponds. Shrimp pond waste contains organic material consisting of shrimp feces, feed residue, shrimp body remains, and dead microorganisms (plankton) aggregates that are often found settled at the pond's bottom. [9]. In intensive cultivation without recirculation or water replacement, the organism waste, food, dead microorganisms, toxic metabolites, such as ammonia and nitrite, can be disastrous for pond farmers because they degrade water quality quickly[10].

Many would argue that the best management strategy for aquaculture ponds is increased flow rates and flushing [11]. When used in an intensive system, water recirculation or exchange may increase viability by reducing the significant release of nutrients associated with intensive cultivation[12] [13]. Water exchange can effectively remove this waste and excess nutrients from the cultivation system [14]. Water exchange benefits maintaining and improving water quality, as well as increasing the growth of aquaculture organisms in the Sei Teras Area. Therefore, water exchange using sluice gates is necessary and possible at the beginning of the development of the intensive pond in this area. This study uses variations in the width and height of the gates following the Irrigation Planning Standard [15], while the variations in the number of simultaneous gate openings used were 0, 8,20, 30, and 62 with evaluation based on water discharge and volume. In addition, a simulation effect of the number of gates openings on the flow and volume of water at the control points on the network is also carried out. This study aims to find an alternative design and operation of pond gates for better use of tidal waters for fishpond water leaching by exchanging and replacing water.

2. Method

2.1. Research Area

Sei Teras aquaculture area is located in Kuala Kapuas Regency, Central Kalimantan Province, Indonesia (3°25'48.11 "S, 114°20'14.15"E). This area consists of three fishpond irrigation areas (FIA), Sei Teras FIA, Pelampai FIA, and Cemara Labat FIA, with a total area of approximately 3,278 ha. It can be seen that the shape of the pond in this area is mostly rectangular, and some others have an irregular shape. The square-shaped ponds are developed through the government, while the irregular ones are developed independently by the local community. The area of each pond plot ranges from 2 to 25 ha, and it is equipped with wooden sluice gates. The water depth in the inner pond perimeter (caren) channels ranges from 0.8 to1 m, and that is in the middle area of the pond ranges from 0.4 to 0.5 m. The average pond's bottom elevation is -0.2 m.

Some farmers maintain traditional cultivation in this area by only harvesting brackish water shrimps, fishes and crabs that naturally enters the pond. However, some others have started the green mangrove crab enlargement and milkfish cultivation. The three fishpond irrigation areas are in the same water system. The pond water system is influenced by tides from three channel ends at the coastline that carry saltwater (Number 1 to 3 in lower part of **Figure 1**) and one source of fresh water from the upstream of Sei Teras river.

IOP Conf. Series: Earth and Environmental Science 1091 (2022) 012023 doi:10.1088/1755-1315/1091/1/012023

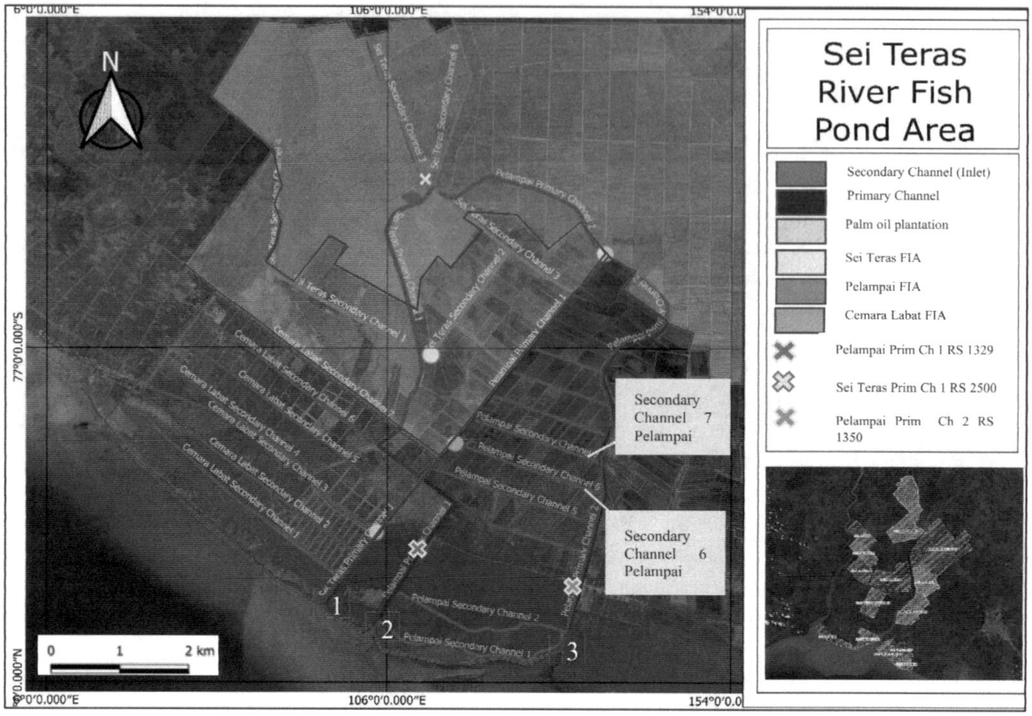

Figure 1. Sei Teras Fishpond Irrigation Area (Modified from [16]).

2.2. Data Collection

Primary data were collected through interviews, discussions, and observations topography with the native. Tidal data was collected by installing a Peilschall and IP Camera at RS 1329 Pelampai primary channel 1 near the channel coast end number 2 (3°26'28,50"S, 114°19'31,06"E) in ***Kesalahan! Sumber referensi tidak ditemukan.*** that monitored every hour for 17 days. While secondary data in the form of channel measurement data and topography were obtained from the local water resources authority, I

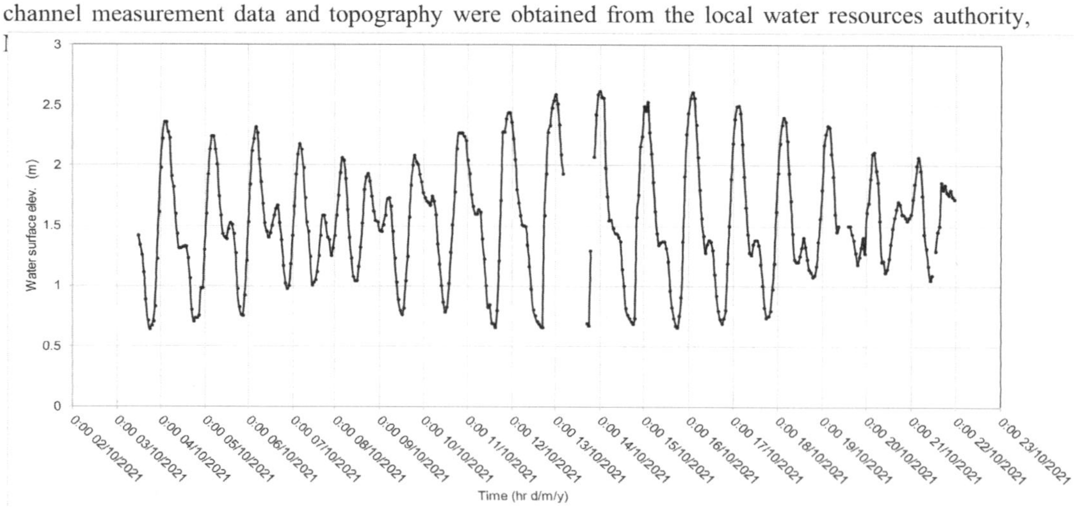

Figure 2. Tidal observation data at RS 1329 Primary channel 1 (station no. 2).

2.3. Experimental Design

In the existing condition, Pelampai secondary channels 6 and 7 connect Pelampai primary channels 1 and 2, for more details can be seen in **Figure** 3. In the development of intensive cultivation, it is recommended to equip ponds with dedicated supply and drain channels. Therefore, in the simulation model, Pelampai secondary channels 6 and 7 are used as the supply channels. In the simulation, channels are specifically modeled to only be connected to the Pelampai primary 1. For drainage needs, in this model, a new drain channel is added between Secondary channels 6 and 7. This drain channel is only connected to Primary channel 2.

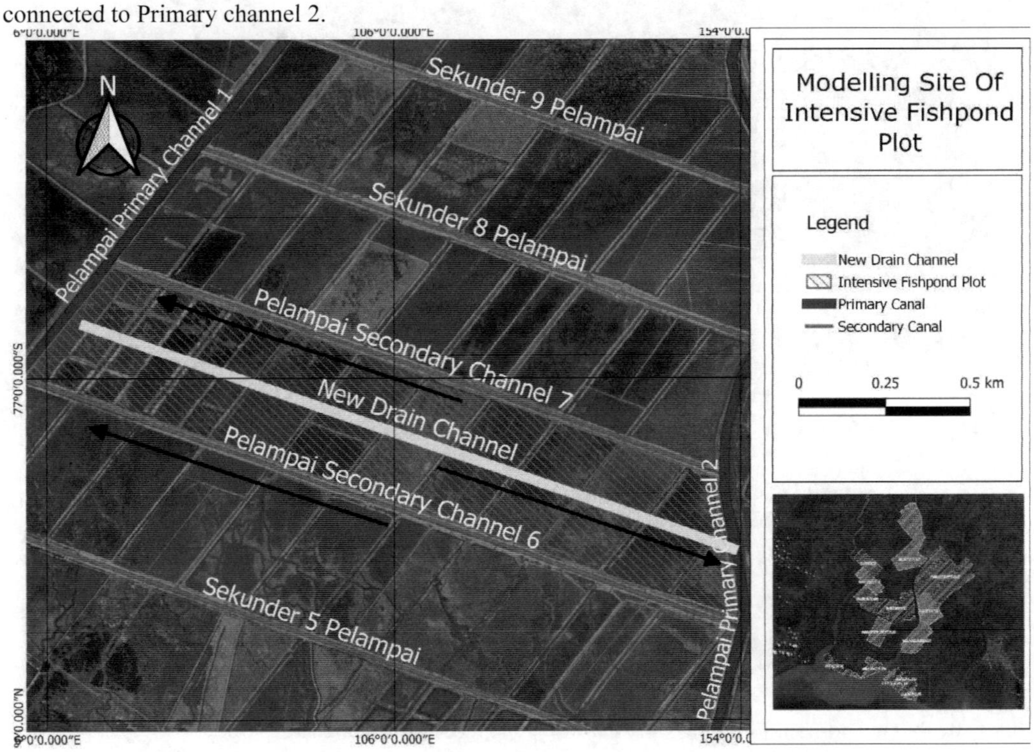

Figure 3. Modelling site of intensive pond plot in Sei Teras aquaculture area (Modified from [16]).

2.4. Boundary Condition Model

The hydraulic simulation was made using HEC-RAS 5.0.3 application software. Downstream boundary conditions use a stage hydrograph, while flow hydrograph is used for upstream boundary conditions with a constant flow rate of 0 m^3/s. The geometry of the model is based on measurement data in the Block D Food Estate Planning Report with adjustments based on field observations. The size of the ponds in the existing conditions is too wide and exceeds the size criteria for intensive cultivation ponds, which is around 0.5-1 ha. At the modeling location, the size of the pond is about 8 ha, so in the model, the pond size is adjusted to a smaller size of about 0.9 ha per pond. In the HEC-RAS program, ponds can be modeled using a storage area (SA) module, but in this study, ponds are modeled as channels. This is intended to support the simulation of water quality in further research and can adjust pond's slope. Pond's slope is set at 0.2% towards the outlet gate. The elevation at the inlet gate is -0.2 m and at the outlet gate -0.6 m. The new drain channel has a length of 2100 m with a top width of 20 m and a bottom width of 17 m and has a depth of 2.15 m.

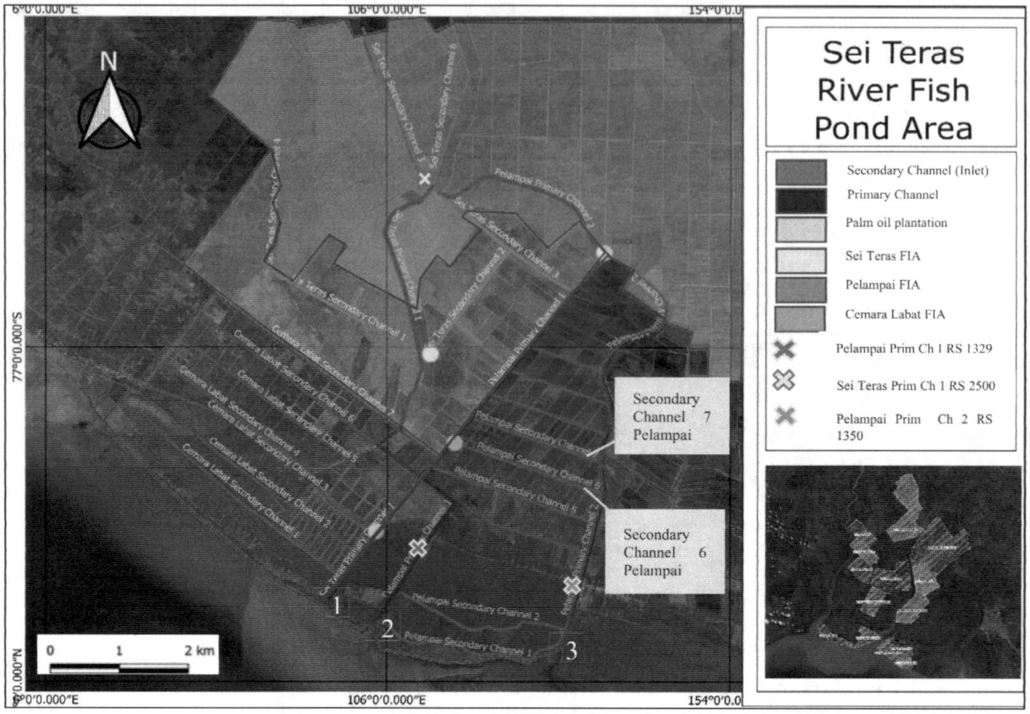

Figure 1. Sei Teras Fishpond Irrigation Area (Modified from [16]).

2.2. Data Collection

Primary data were collected through interviews, discussions, and observations topography with the native. Tidal data was collected by installing a Peilschall and IP Camera at RS 1329 Pelampai primary channel 1 near the channel coast end number 2 (3°26'28,50"S, 114°19'31,06"E) in ***Kesalahan! Sumber referensi tidak ditemukan.*** that monitored every hour for 17 days. While secondary data in the form of channel measurement data and topography were obtained from the local water resources authority, Ministry of Public Works, and Public Housing of the Republic of Indonesia.

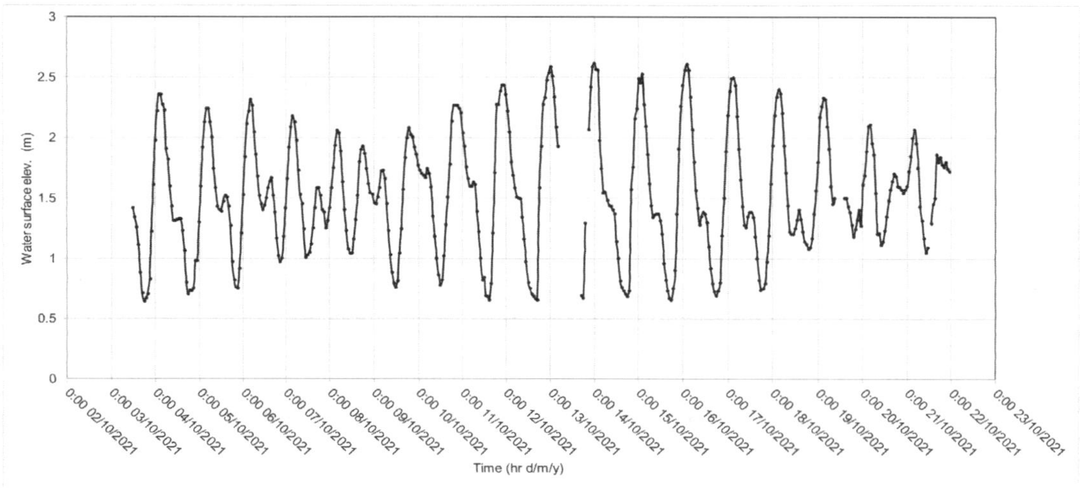

Figure 2. Tidal observation data at RS 1329 Primary channel 1 (station no. 2).

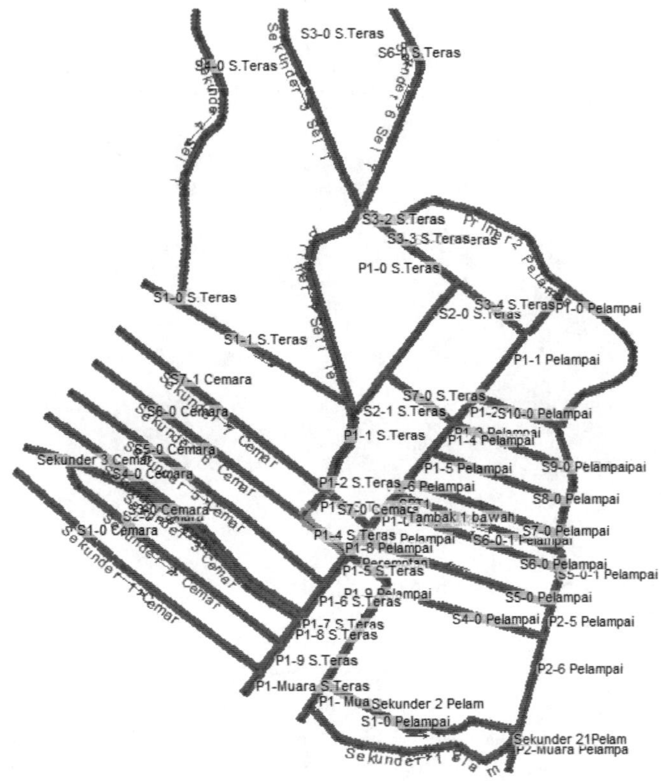

Figure 5. The existing geometry of Sei Teras fishpond irrigation area.

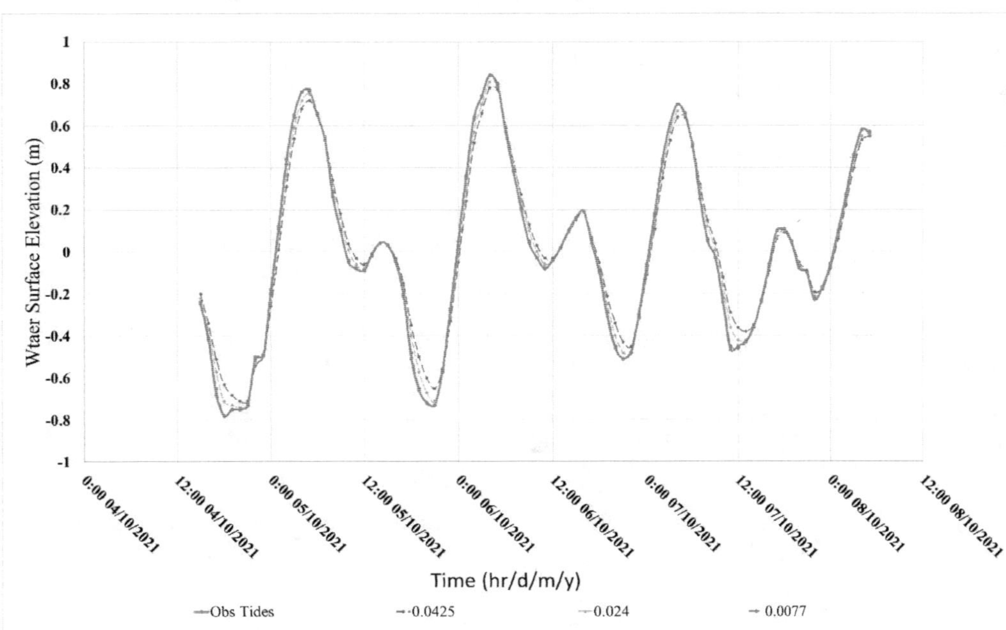

Figure 6. Comparison of tidal observation data and model hydrograph with variation of n Manning values.

In **Figure 6**, it can be seen the comparison of model results with n Manning values of 0.0425, 0.024, and 0.0077. The coefficient of n Manning 0.0077 gives model result values closest to the measured data. The n Manning coefficient below 0.0077 causes an error in the simulation model run.

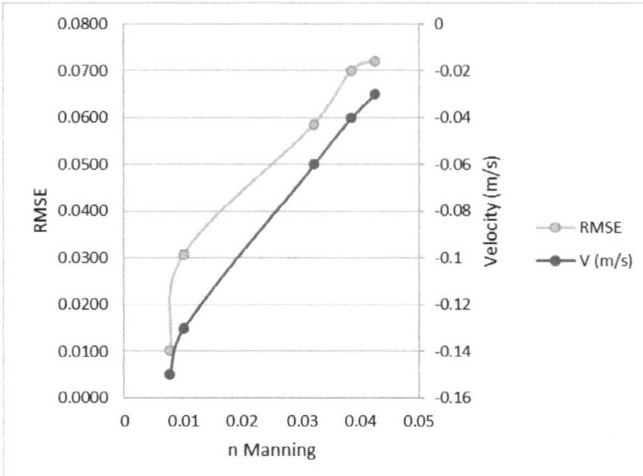

n Manning	V (m/s)	RMSE
0.0077	-0.15	1.03%
0.01	-0.13	3.07%
0.032	-0.06	5.86%
0.0385	-0.04	7.02%
0.0425	-0.03	7.21%

Figure 7. Root mean square error and velocity calibration of exciting geometry Sei Teras Fishpond Area 2021.

Analysis of the Root Mean Square Error (RMSE) was carried out between the measurement tides and the model at the observation point by comparing the Flow Velocity of the model. The flow was measured on October 7 at 23:30, measuring of the velocity is 0.1495 m/s at low tide (towards the sea). The modelling was carried out from 04 October 21– 08 October 21 with flow velocity results closest to the field measurements and the lowest error rate at the n Manning coefficient of 0.0077. The results of the RMSE analysis show an error rate of 1.03%, and the speed of the model is at -0.15 m/s.

3.2. Selection of Gates Size Variations (Scenario 1)

Several simulations were carried out with variations in the size of the gates in one fishpond plot to see the gates' capacity. The outlet gates were simulated in a pond full of water to determine how to rinse water if it experienced flooding or rain. The upstream and downstream boundary conditions in Scenario 1 were the same as existing conditions simulation but with additional opening gates. The opening gates were regulated by using the boundary conditions of the time series gate opening. Gates were simulated to be drained and then refilled. Drain simulation was carried out for the period of 17:00-22:00 on October 4, 2021, while for refilling was carried out for the period of 02:00-07:00 on October 5, 2021.

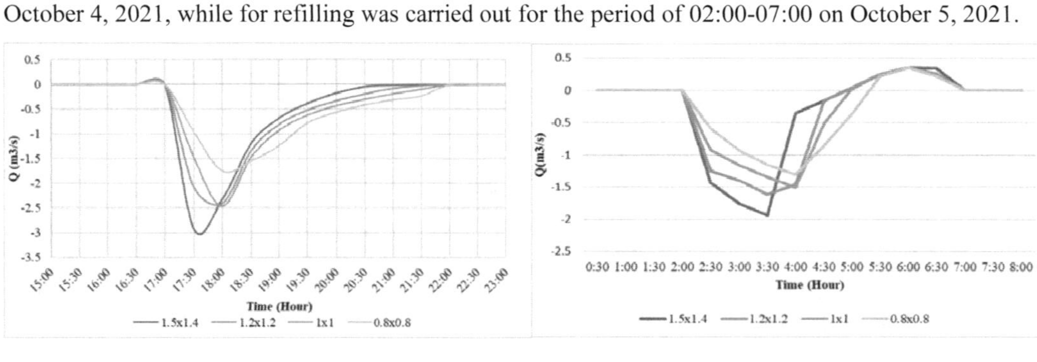

Figure 8. Comparative graph of the relationship between discharge and time based on variations in the size of the sluice gate (left outlet gate graph-right inlet gate graph).

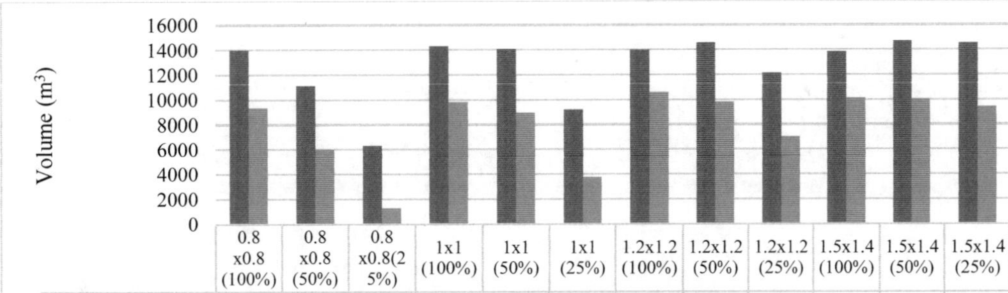

	0.8 x0.8 (100%)	0.8 x0.8 (50%)	0.8 x0.8(25%)	1x1 (100%)	1x1 (50%)	1x1 (25%)	1.2x1.2 (100%)	1.2x1.2 (50%)	1.2x1.2 (25%)	1.5x1.4 (100%)	1.5x1.4 (50%)	1.5x1.4 (25%)
■ Volume Outlet Gate	14004	11160	6336	14328	14058	9216	13986	14580	12114	13878	14706	14544
■ Volume Inlet Gate	9360	5994	1242	9792	8964	3744	10602	9846	7002	10170	10026	9450

Figure 9. Comparison of the volume for each gate opening variation.

It shows on **Figure 8** and **Figure 9**, that the largest fastest flushing and filling of water were obtained at a gate opening size of 1.5m width and 1.4 m height. The gate setting can rinse as much as 13,878 m^3 of water and can dry up ponds. The volume that can enter the pool is of 10,170 m^3, achieved in 2.5 hours of gate opening. Replacing 10% of pond water with a depth of 0.8-1.2 m requires a volume of water entering and leaving the pond of approximately 950 m^3 per day. Based on the simulation results above, a 5-15% water change can be replaced with a tidal peak of more than +0.6 m

3.3. *Variation in the Number of Ponds Drained and Filled (Scenario 2)*

Simulation of Scenario 2 uses the same upstream and downstream boundary conditions as the previous. In this scenario, the gate size and boundary conditions of the last simulation results were applied to 62 fishponds. Changes of discharge and water level due to variations in gate opening simultaneously at the control point of Pelampai primary channel 1 RS 1329 can be seen in **Figure 10**. There is a reduction in the discharge for the simulation of Plans 62 and 30 when filling the pond. This is due to the pond's water condition not being able to drain properly due to the volume of water being drained or discharged that exceeds the capacity of the drainage channel to drain water into the Primary 2 Pelampai channel.

In **Figure 11**, the water level of the pond can be compared at the end of the drain for Plan 62 and 8. The water level in the pond on Plan 62 is still high so that the water in each pond is wasted into the channel when filling, this causes a discharge of 19.38 m^3/s. The effect of simultaneously operating pond gates also affects the condition of the control point RS 2500 upstream and RS 1350 downstream.

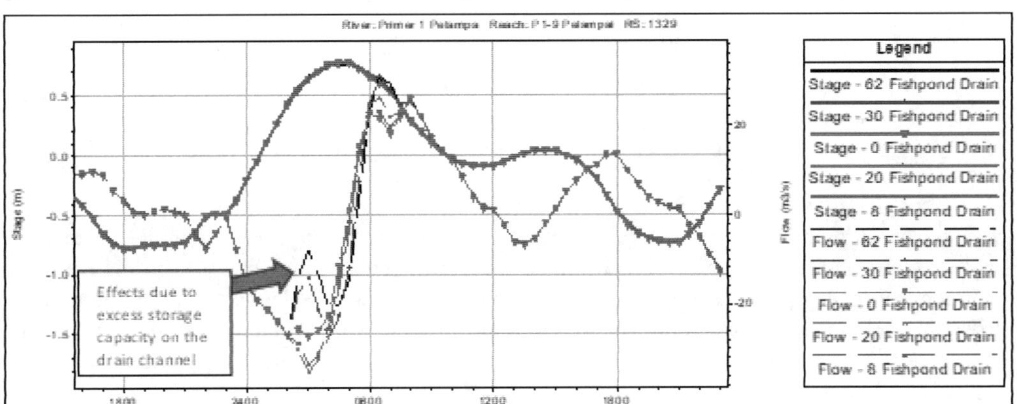

Figure 10. Comparison of discharge and water level due to variations of the number of fishpond gates that are operated simultaneously in Pelampai primary channel 1 RS 1329

IOP Conf. Series: Earth and Environmental Science 1091 (2022) 012023 doi:10.1088/1755-1315/1091/1/012023

In Plans 62 and 30 (**Figure 12**), the water level changes about 3-5 cm during draining and filling, while in the other plans it does not change. Filling the ponds in Plans 62 and 30 accelerated the peak discharge, while the in other plans it caused a shift of the peaks of the flood tides, and caused the low tides to retard.

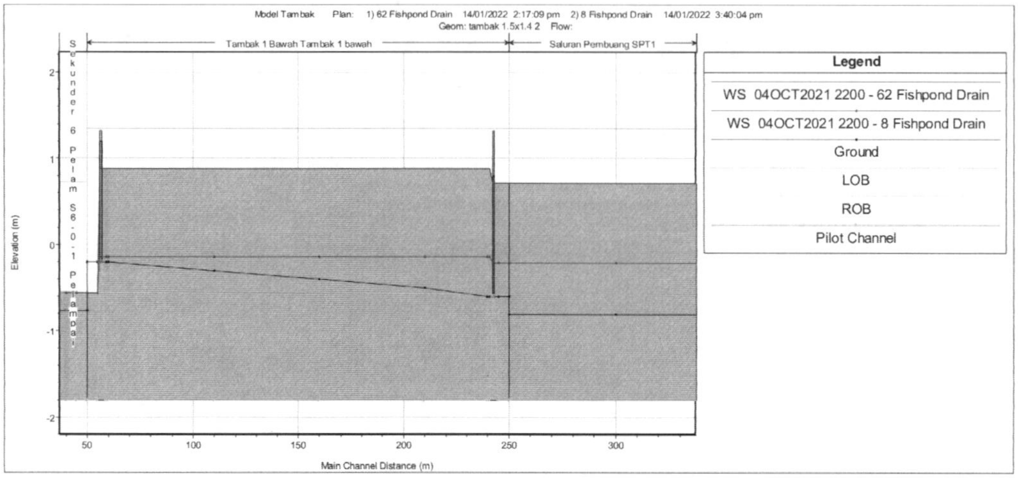

Figure 11 Comparison of water level in ponds due to draining 62 fishponds and 8 fishponds

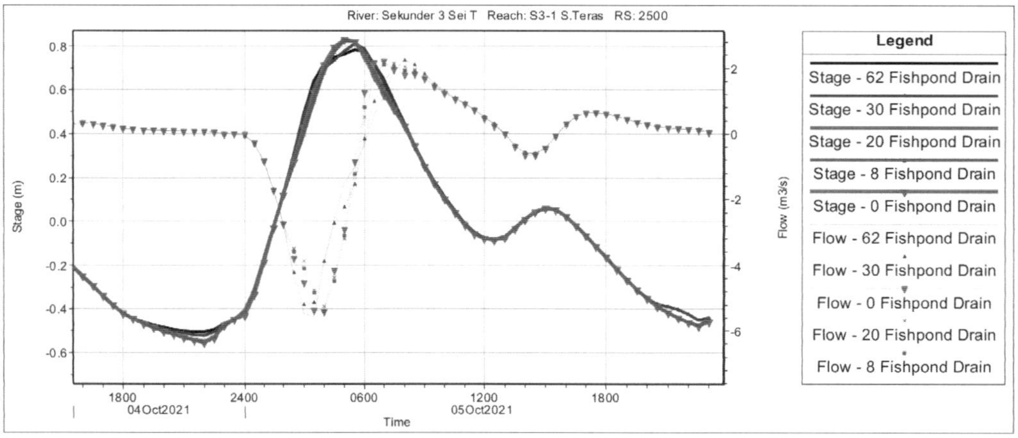

Figure 12. Comparison of discharge and water level due to variations of the number of fishpond gates that are operated simultaneously in Sei Teras secondary channel 3 RS 2500.

At the control point of Pelampai primary channel 2 RS 1350 (**Figure 13**) there was an increase in discharge when draining due to applying separate irrigation and drainage channels, the Primary 2 Pelampai channel indirectly becomes the main channel of the pond irrigation network. In **Figure 13**, it can be seen that the more gates are opened simultaneously, the higher discharge during draining at the control point until it reaches a maximum at the simultaneous opening of 30 and 62 gates. The maximum drain discharge achieved is 12.09 m^3/s, while at the same time, the discharge under normal conditions is 2.44 m^3/s or 9.65 m^3/s higher.

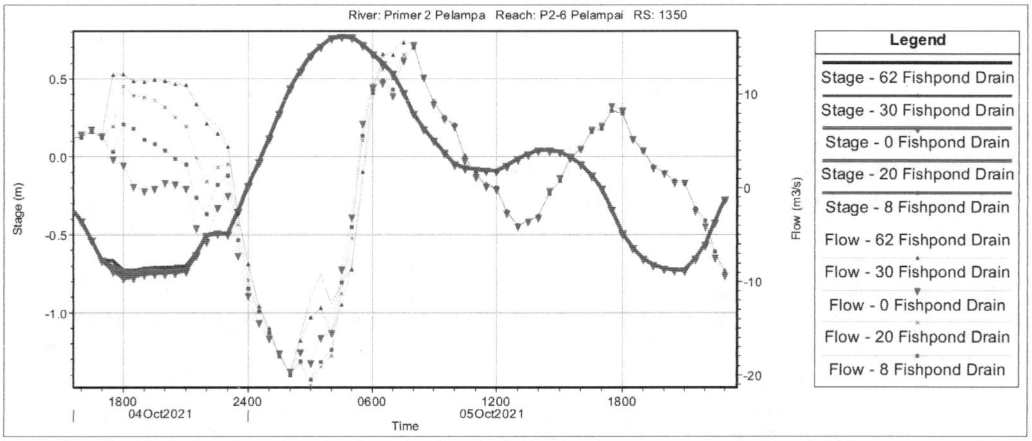

Figure 13 Comparison of discharge and water level due to variations of the number of fishpond gates that are operated simultaneously in Pelampai primary channel 2 RS 1350

4. Conclusion

Sei Teras aquaculture is an extensive traditional aquaculture area with a good water supply volume and is planned to develop into an intensive vannamei shrimp pond. Intensive systems apply high stocking densities in ponds with strict management. The existing pond area is divided into smaller sizes. The waste from intensive shrimp ponds needs to be appropriately managed. The way for maintaining water quality is to replace the water in ponds in every tidal cycle. The waste that becomes pond bottom sediment requires a large discharge to dispose.

Calibration of the existing model geometry has been carried out, which the result will be used as the basis for the design simulations. Calibration was conducted by concerning two parameters namely water level and velocity where both parameters were measured in the field. The n Manning roughness coefficient was the adjusting parameter that used to fit the field condition. The value of 0.0077 was the best fit one, with RMSE of 1.03%.

Design of gate openings for rinsing the pond water has shown possible settings to be applied in Sei Teras aquaculture areas. Variations in the size of the width and height gate openings indicate that the larger the gate, the greater the volume of water that can be discharged. Using a gate opening of 1.5 m width x 1.4 m height makes it possible to dispose of all the water in the pond within 2.5 hours. The water volume that can be removed from the gate is 13,878 m^3. The daily water change of 10% (950 m^3) can be removed and replaced in one tidal cycle. Based on the simulation results, the more gates that are operated properly, the higher the discharge during draining and filling downstream of the network.

Incomplete draining causes the discharge to be reduced, so it is necessary to pay attention to the pond drains' capacity to carry out gate operations. In order to maximize the ability of pond flushing and drains, it is necessary to consider the size of the drainage channel and the number of pond gate operations in one flushing together. Simultaneous operation of 20 pond gates provides maximum discharge effect.

Acknowlegments

The authors would like to thanks the Ministry of Public Works and Housing which give full support of this research. The assistance from Mr. Parlinggoman, and Mr. Yakubson during the survey period is highly appreciated. Finally, many thanks for Prof. Budi Wignyosukarto who has shared his experiences that were very beneficial during this research.

ORCID iDs

R. Ciptadi https://orcid.org/0000-0002-6471-1472
A.P. Rahardjo https://orcid.org/0000-0002-9023-6089

References

[1] Thakur D and Gill G S 2013 Some Geo-Environmental Hazards in Dharamshala Area of Himachal Pradesh and Their Mitigation *Res. Artic. Int. J. Adv. Earth Environ. Sci.* **1** 41–53
[2] Hongyu L 1998 Conservation of wetlands especially as waterfowl habitat in northeast China *Chinese Geogr. Sci.* **8** 281–8
[3] Ciptadi R, Rahardjo A P and Kamulyan B 2021 Evaluation and Enhancement of Sustainable Organic Fishpond Farming in the Sei Teras Fishpond Irigation Area, Central Kalimantan *IOP Conf. Ser. Earth Environ. Sci.* **930**
[4] Amri, khairul & Kanna I 2008 *Budi Daya Udang Vaname - Google Play* (PT.Gramedia Pustaka Utama)
[5] Bett C and Vinatea L 2009 Combined effect of body weight, temperature and salinity on shrimp Litopenaeus vannamei oxygen consumption rate *Brazilian J. Oceanogr.* **57** 305–14
[6] Rakhfid A, Baya N, Bakri M and Fendi F 2017 Pertumbuhan dan kelangsungan hidup udang vaname (Litopenaeus vannamei) pada padat tebar berbeda Growth and survival rate of white shrimp (Litopenaeus vannamei) at different density **1** 1–6
[7] Rahman F, Putra I and Rusliadi R 2016 Growth And Survival Rate Of Western White Prawns (Litopaneaus vannamei) On Different Salinity *J. Online Mhs. Bid. Perikan. dan Ilmu Kelaut.* **3**
[8] M.Ghufran H Kordi K. 2010 *Budi Daya Udang Laut* (Paccinongang, Gowa, Sulawesi Selatan: Lily Publisher)
[9] Nanga K O M P P 2019 Perencanaan Instalasi Pengolahan Air Limbah **6** 5–8
[10] Esparza-Leal H M, Cardozo A P and Wasielesky W 2015 Performance of Litopenaeus vannamei postlarvae reared in indoor nursery tanks at high stocking density in clear-water versus biofloc system *Aquac. Eng.* **68** 28–34
[11] Rogers G L 1989 Aeration and circulation for effective aquaculture pond management *Aquac. Eng.* **8** 349–55
[12] Reid B and Arnold C R 1992 The Intensive Culture of the Penaeid Shrimp Penaeus vannamei Boone in a Recirculating Raceway System *J. World Aquac. Soc.* **23** 146–53
[13] Mishra J K, Samocha T M, Patnaik S, Speed M, Gandy R L and Ali A M 2008 Performance of an intensive nursery system for the Pacific white shrimp, Litopenaeus vannamei, under limited discharge condition *Aquac. Eng.* **38** 2–15
[14] Boyd C E, Queiroz J, Lee J, Rowan M, Whitis G N and Gross A 2000 Environmental assessment of channel catfish Ictalurus punctatus farming in Alabama *J. World Aquac. Soc.* **31** 511–44
[15] Direktorat Irigasi dan Rawa 2013 Standar Perencanaan irigasi. Kriteria Perencanaan Bagian Standar Pintu Pengatur Irigasi: Perencanaan, Pemasangan, Operasi dan Pemeliharaan KP - 08 63
[16] ArcGis World Imagery

Application of Delft3D-WAVE in the investigation of the typhoon-induced high wave in Central Vietnam

L T Anh*[1], D D T Trang[1], N D Thao[1] and N Kiet[1]

[1] Department of Port and Coastal Engineering, Ho Chi Minh City University of Technology (HCMUT), Vietnam

*Corresponding author e-mail: tuananh131188@hcmut.edu.vn

Abstract. With a long coastline of around 3260 km, tropical typhoons induce the deadliest disasters in Central Vietnam's highly populated coastal area. The focus of this paper is two-fold. The first objective is to investigate the feasibility of utilizing Delft3D-WAVE in forecasting typhoon-induced high wave events in Central Vietnam. By running a WRF model simultaneously in multiple domains with different grid resolutions, wind field distribution data during typhoon events were reproduced. Using the generated wind field data, typhoon Kaemi (2000) was employed as the case study to calibrate and validate the wave models. Comparisons between simulated results and observed data showed a good agreement, especially during peak wave height. The combined wind-wave model was applied to estimate the temporal and spatial impacts of high waves caused by the typhoon Ketsana (2009). The second objective is to assess the potetial risk from typhoon-induced wave in Central Vietnam based on the models output. From these results, this study recommends the application of Delft3D-WAVE to investigate high waves induced by typhoon events in Vietnam's coastal areas.

1. Introduction

Disasters induced by typhoons are the most rampant in Vietnam. The Ministry of Natural Resources and Environment of Vietnam (MONRE) pointed out that the number of high-intensity typhoons affecting the country had increased in the recent years due to global climate change. While researches on typhoon-induced disasters in Vietnam typically focus on storm surge, disasters brought about by the high waves from typhoon events are often overlooked. With a 3260-km long coastline, an extremely high coastal disaster risks threatens the coastal communities of Vietnam.

Studies about typhoon-induced high waves have been investigated by many researchers worldwide [1]. Cardone et al. (1996) [2] confirmed the advantage of third-generation wave model over the first and second-generation in high waves simulation during extratropical storm cases. Nowadays, third-generation wave models such as the Wave Action Model (WAM) [1]; WAVEWATCH III (WWIII) [3] [4] are applied to study the mechanism of high waves caused by typhoons. WAM and WWIII show the advantages when reproducing deepwater wave [5] while SWAN, which considers the effects of bottom roughness, is more suitable for coastal/nearshore modelling [6].

In Vietnam, many researchers have investigated the mechanism of typhoon and typhoon-related disasters. Chien et.al. 2014 [7] investigated the storm surge and high wave during typhoon Xangsena (2006) along the coast of Central Vietnam by applying the SWAN wave model, which was forced by the wind field simulated by the Fujita typhoon model. However, the typhoon models also have their own limitation. According to Jun T. and Takeshi I., 2013 [8], the typhoon model often fails at predicting wind fields outside the typhoon region. They also mentioned that the accuracy of the wave simulation depends on the extensive domain wind field. Thus, the underestimations from typhoon models will affect the result and the applicability of the wave model.

This study focuses on the application of a wind-wave model in Vietnam and has two objectives. First, it intends to evaluate the performance of Delft3D-WAVE model (mainly based on SWAN) using the wind field simulated by the meteorological model WRF (The Weather Research and Forecasting) v4.0 in reproducing the wave field during significant extreme weather events. Second, it aims to investigate the potential risks from typhoon-induced high wave that the coastal communities in Central Vietnam may potentially face.

2. Methodology

The detail steps implementation to achieve the first objective are described as follow. After determining of the research area, the data related to significant typhoon events within the area including paths, maximum wind speed, coordinates of the typhoon centre will be collected. The meteorological model WRF will be implemented beforehand. Then, the output wind field from this model will be applied as the external force for the Delft3D-WAVE. The authors will be looking at the performance of these combined models in reproducing wave field during the typhoon events. For the second objective, based on the simulated results of wave field, the qualitative evaluation could be conducted to determine the most vulnerable place where suffered the highest wave height within the research area.

2.1. Research area

The authors focus on the area around DaNang, including the north and the south neighboring provinces Hue and Quang Nam (Figure 1). According to the statistical data from the Vietnam Maritime Search and Rescue Coordination Center (VIETNAM MRCC) (http://vmrcc.gov.vn/), 459 typhoon events made landfall in Vietnam in the last 70 years. The area of DaNang experienced the most number of typhoon landfalls with 41 events for 45 years, or an average of nearly one event per year. This place is considered vulnerable to typhoon disasters, as well as the typhoon-induced high wave.

Figure 1. The research area.

2.2. WRF model

2.2.1. Domain. The Weather Research and Forecasting v4.0 (WRF), a widely-used mesoscale meteorological model, was used to reproduce the wind field, including wind speed and central pressure. It was developed by The National Center for Atmospheric Research (NCAR) and the National Center for Environmental Prediction (NCEP). In this study, two domains were utilized in the WRF simulation run: the ~30 km resolution outer domain (d01) (73x73 grid points) covering the entire area of the East Sea, and the inner domain (d02) with a resolution of 10km x 10km (73x73 grid points) focusing on the offshore area in DaNang (Figure 2). The inputted uniform time step was 30 s for both domain simulations. For the lateral boundary condition of the outer domain, the NCEP Final (FNL) Operational Global reanalysis dataset with 1° × 1° horizontal resolution and a 6 h interval was applied.

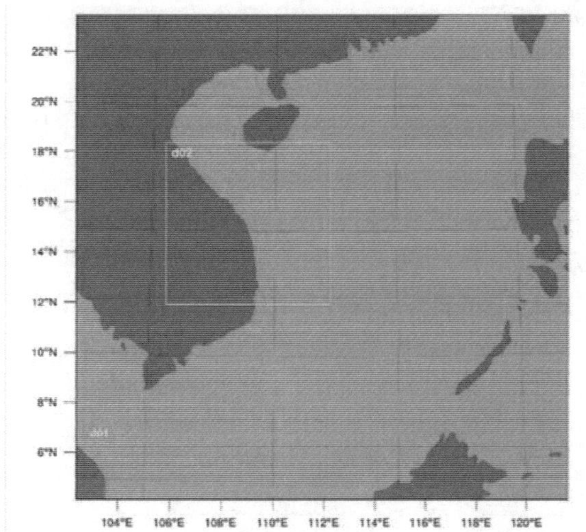

Figure 2. WRF domains.

2.2.2. Configuration. The forecast results of the mesoscale models usually come with errors caused by the uncertainties of the physical representations of the mesoscale model. To overcome this problem, in the WRF setting, several physical schemes need to be parameterized, and the most significant difficulty is how to choose a suitable set of representative schemes. The WRF model provides several parameterization schemes to be selected for the description of each physics option, including Cumulus parameterization (CU), Planetary boundary layer (PBL), Microphysics parameterizations (MP), Land-Surface Models (LSMs), Short Wave Radiation (SWR) and Long Wave Radiation (LWR).

Islam et al. [9] used the case of super typhoon "Haiyan" to access the sensitivity of the MP, CU, and PBL physics schemes and pointed out how CU scheme influenced the typhoon intensity from WRF estimation. Srinivas et al. [10] chose 13 scheme combinations, including CU, MP, and PBL schemes to assess the most suitable WRF combination simulations for 21 typhoons. Based on these literatures, the effect of several schemes on the simulations in terms of typhoon track/ intensity was assessed while the other physics schemes were fixed.

In this study, the configuration of the WRF model was referred from Radhika D. et.al, 2015 [11] that investigated the effect of physical parameterization schemes on track and intensity of cyclone Laila (2010) in the Bay of Bengal. The study suggests to apply the BMJ for CU physical scheme due to its more intense storm generation than the others. For the PLB physic scheme, the study pointed out that the intensity simulated by MYJ is close to the observed data. The WSM6 was chosen for the MP physical scheme because it shows the ability to reproduce a more reasonable feature of the cyclones than the other options. The other schemes were kept as default. The configuration of the WRF model is described in Table 1 below.

IOP Conf. Series: Earth and Environmental Science 1091 (2022) 012024 doi:10.1088/1755-1315/1091/1/012024

Table 1. Configuration of WRF model

Simulation time (all domains)	
Typhoon Kaemi (2000)	0:00 UTC 17/08/2000 – 0:00 UTC 24/08/2000
Typhoon Ketsana (2009)	0:00 UTC 25/09/2009 – 0:00 UTC 01/10/2009
Grid resolutions	
(d01; d02)	(30.0; 10.0 km)
Pressure top of the model	50 hPa
Number of vertical layers	27
Nesting scheme	2-way nesting (2 domains)
Micro physics (MP)	WSM 6-class graupel scheme
The cumulus parameterization (CU)	Betts-Miller-Janjic scheme
Planetary boundary condition (PLB)	Mellor-Yamada-Janjic (Eta) TKE scheme;
Map projection	Lambert conformal
Bathymetry data	USGS

2.3. Delft-3D WAVE model

This study employed the Delft3D-WAVE model, which is mainly based on the SWAN spectral wave model. The third-generation wave model SWAN is suitable for simulating random, short-crested, and wind-generated waves in coastal regions and inland waters. The governing equation of the SWAN model is based on the action balance equation with sources and sinks.

2.3.1. Domain. Two computational domains were used in Delft3D-WAVE (Figure 3). The large domain (DW1) with spatial resolutions of 0.05° × 0.05° consisting of 120 x 120 grids points were considered for the East Sea that encompasses Central Vietnam. The small domain (DW2) over 0.01° × 0.01° grid resolution (250 x250 grid points) was nested into and used the the boundary condition generated from the large domain (DW1).

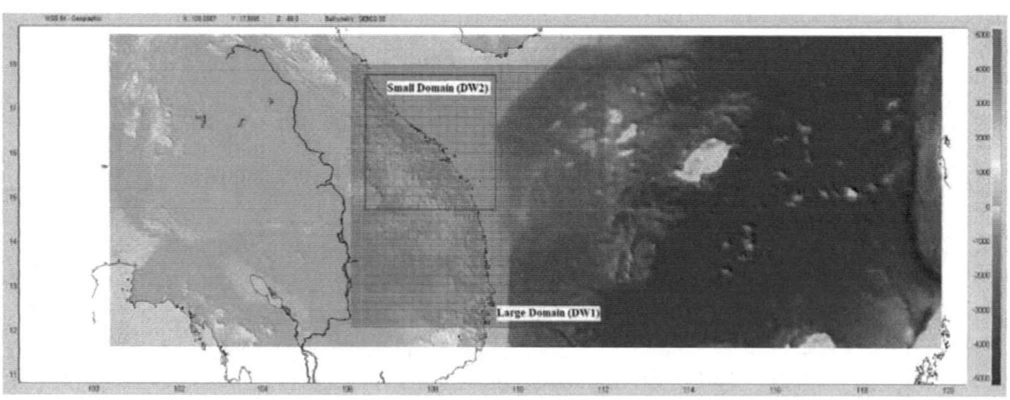

Figure 3. The nested domains in Delft3D-WAVE model.

2.3.2. Configuration. The simulated period utilized for the wave model was the same as the calculation period used for the WRF model (Table 1). The Delft3D-WAVE model simulations used the bathymetric data retrieved from the GEBCO 08 and were forced by the input wind field variables generated from the WRF simulation results such as eastward/northward wind components.

For the wave simulation in Delft3D-WAVE, physical process including depth-induced wave breaking, effect from the bottom friction, wind-wave growth, white-capping effect, non-linear triad, and quadruplet interactions were assigned. The wave spectral parameters were set as default, with the frequencies ranging between 0.05 Hz and 1 Hz and divided into 24 bins, while the wave directions were divided into 36 bins separated by 10° for each bin. The computational time step was set up as 60 minutes.

2.4. Data used for model calibration and validation

The observed wind speed and wave height data during typhoon events is limited in Vietnam. Thus, the actual data used to verify the accuracy of models were referred to published papers. In this study, two significant events of typhoon Kaemi (2000) and typhoon Ketsana (2009) (Figure 4), of which the measured wind speed and wave height are available, were chosen as the case studies.

Measured data for the case of typhoon Kaemi (2000) were published by Nguyen Ngoc Tran et. al., 2003 [12]. In this paper, the authors measured the wind speed at the Tien Sa lighthouse in the Son Tra peninsula and the wave height at the station named Wave Observation with coordinates E 108°15'34'' N 16°11'27.7'' as shown in Figure 4. For the case of typhoon Ketsana, the observed wind speed was collected by Thai T H et. al., 2018 [13] in LySon island, which is located along the typhoon's path. Wave data, however, is unavailable. The abovementioned data presented from these two publications were incorporated in this study. The maximum wind speed for the two events were used for calibration and validation of the WRF model. The performance of the WRF model was also evaluated by comparing with the extracted data from Reanalysis Data from ECMWF (ERA-5 dataset). The accuracy of the wave model was verified through comparisons with observed wave height data during typhoon Kaemi (2000), after which the model was applied to hindcast the wave field generated by tyhoon Ketsana (2009).

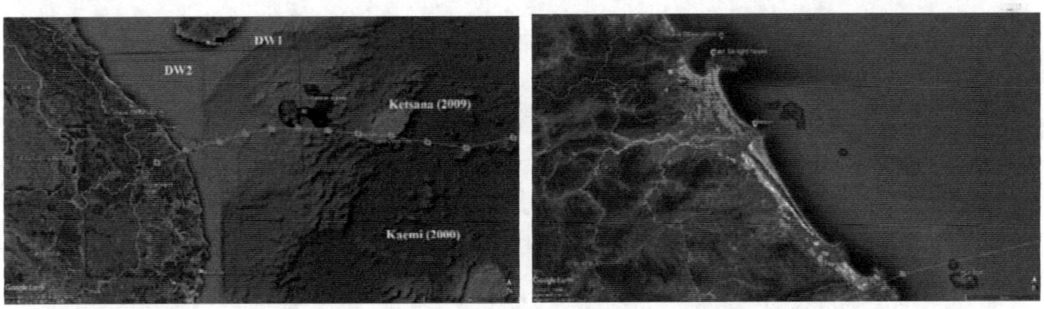

(a) (b)

Figure 4. (a) Typhoon Kaemi (2000) and Ketsana (2009) tracks.
(b) Wave observation and wind observation points.

3. Results and Discussion

Calibration of the WRF model produced results that are in good agreement with the observed data derived from Nguyen Ngoc Tran et. al., 2003 [12]. Figure 5 illustrates that the maximum 10-minute average wind speed during Typhoon Kaemi (2000) was estimated at 21m/s, while the highest instantaneous wind speed reached 26 m/s measured at Tien Sa Lighthouse. The WRF model-generated data also shows discrepancy with the Reanalysis Data (ERA-5 dataset), illustrating the significantly underestimated wind speed from the ERA-5 dataset compared to the observed data. These results prove the suitability of the selected physical schemes for WRF model.

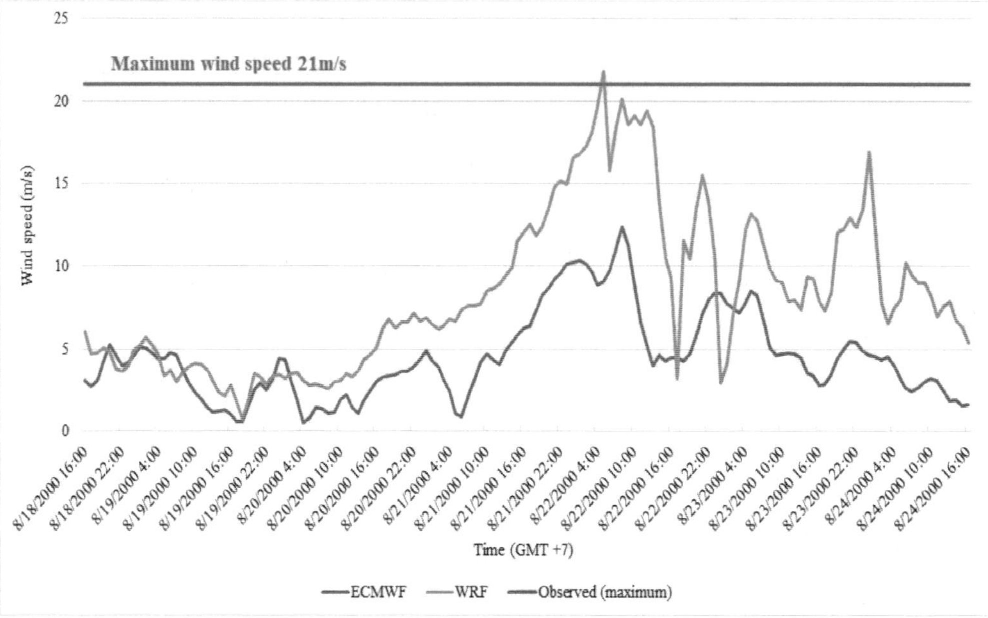

Figure 5. Wind speed at DaNang during typhoon Kaemi (2000) from WRF model.

The wave model Delft3D-WAVE was calibrated using the observed wave data during typhoon Kaemi at a wave observation station. The bottom friction was adjusted to correct the reliability of the model, and the bottom friction from Madsen [14] was used with the coefficient of 0.05. The time series of the simulated wave heights compared with the records are shown in Figure 6. The explanation for the the underestimation of calculation results after occurrence of the peak wave is the limitation of the wind model after the typhoon crossed the land. The high correlation coefficient between the estimated value from the numerical model and the observed data (Figure 7) shows good performance of the wave model in reproducing the wave height variation tendency during the typhoon period. Since predicting the maximum wave height is crucial for disaster mitigation planning, the bias and RMSE values in Table 2 demonstrate that the wave model is a reliable tool at estimating the peak wave height value of 5m. Figure 8 shows the wave height distribution causing by typhoon Kaemi when it approached land.

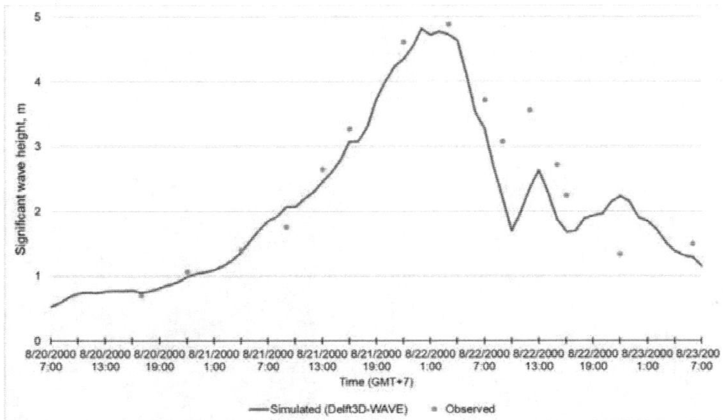

Figure 6. Significant wave height during typhoon Kaemi (2000) at a wave observation station.

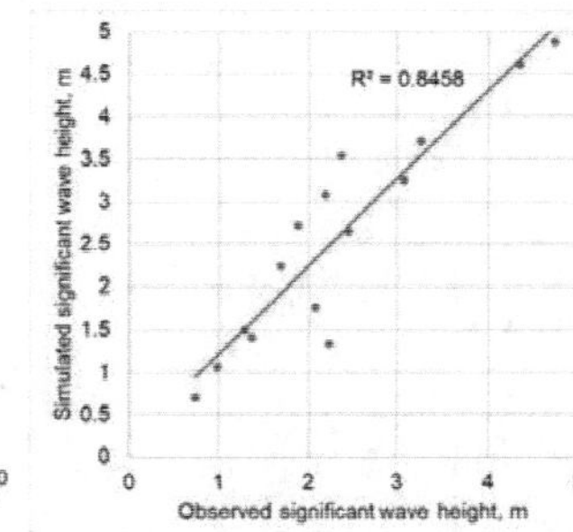

Figure 7. The correlation coefficient between simulated and observed wave height during typhoon Kaemi.

Table 2. Bias and RMSE between simulated and observed wave height during typhoon Kaemi

	Full simulation time	Before and during landing time
Bias	-0.25233	-0.18891
RMSE	0.550378	0.354452

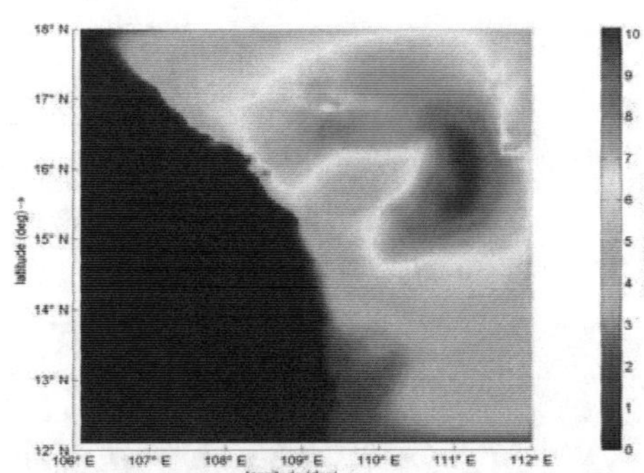

Figure 8. Significant wave height distribution offshore during typhoon Kaemi (2000).

Regarding typhoon Ketsana, the reproduced time series wind speed from WRF models also matched well with observed data described by Thai T H et. al., 2018 [13], in which the maximum wind speed that occurred at LySon island reached nearly 30 m/s (Figure 9). Furthermore, the result from WRF model also shows a good agreement with the wind speed extracted from Reanalysis dataset ERA5. As shown in Table 3, difference of bias and RMSE values between simulated WRF data and reanalysis data is insignificant, demonstrating the acceptable accuracy of the two datasets and thereby validating the WRF

IOP Conf. Series: Earth and Environmental Science 1091 (2022) 012024 doi:10.1088/1755-1315/1091/1/012024

model. Since observed wave data is unavailable during the typhoon Ketsana event, the wave field was hindcasted by applying the same parameters utilized for the case of typhoon Kaemi. The wave height distribution when typhoon approached the land is depicted in Figure 10.

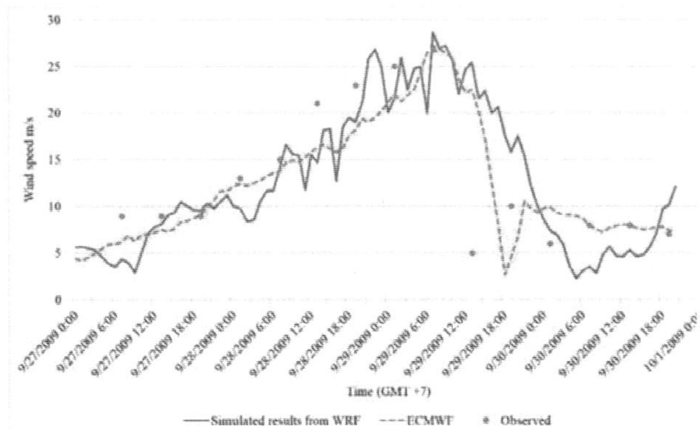

Figure 9. Wind speed at LySon during typhoon Ketsana (2009).

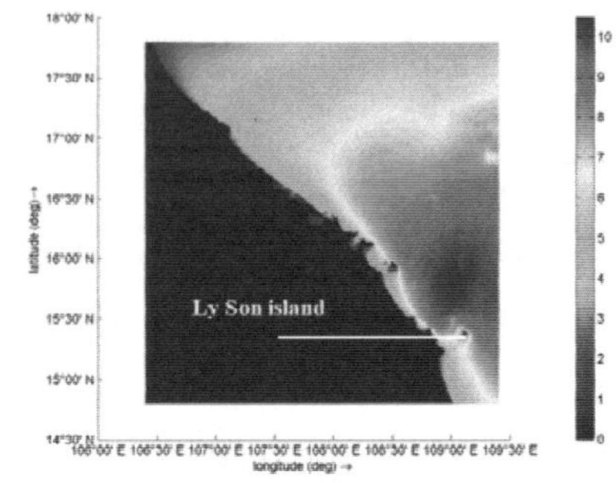

Figure 10. Wave distribution during typhoon Ketsana (2009).

Table 3. Bias and RMSE between simulated and observed wind speed during typhoon Ketsana

	WRF	ECMWF
BIAS	0.1538	-0.178
RMSE	6.3042	5.2907

For an accurate determination of the level of risk from typhoon-induced wave where observed data is not available, the wave height at several locations (Figure 11) within the research area were extracted from the numerical model (Table 4). The qualitative evaluation is implemented based on the value of the wave height at each location, the higher wave height the higher level of vulnerability. According to the highest value in Table 4 (in red colour) the most vulnerable place is LySon island. Another island located in the north of the research area named ConCo is at the second most vulnerable. DaNang, the largest city in Central Vietnam and blocked from typhoon-induced high wave generated by the Son Tra peninsula, only experienced a wave height of less than 3m, leading to a lower threat level.

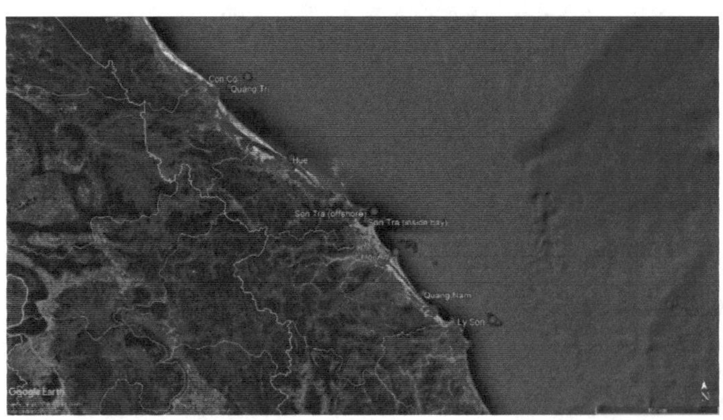

Figure 11. Vulnerable locations in Central Vietnam.

Table 4. Estimated wave height at vulnerable places in Central Vietnam

Locations		ConCo	Quang Tri	Hue	Son Tra (offshore)	Son Tra (inside bay)	HoiAn	Quang Nam	LySon
Coordinates		N 17.17 E107.33	N 17.1 E107.2	N 16.58 E107.65	N16.19 E108.25	N 16.13 E108.2	N 15.9 E108.4	N 15.6 E108.6	N 15.4 E109.1
Wave height, m	**Kaemi**	5.30	5.39	4.40	5.15	2.30	3.24	4.08	6.02
	Ketsana	5.10	5.16	4.42	5.94	2.78	4.15	5.70	8.10

The estimated high wave generated by the typhoon events raised the threat in Central Vietnam, especially on the remote islands such as LySon, ConCo. According to Anh et. al. 2019 [15] the local people on the remote island in South Vietnam are often quite worrisome towards the impacts of typhoon events. Compared to the people residing inland, people from island feel isolated whenever a typhoon approaches as they have no information where to evacuate. It can be noticed that the number of typhoons that made landfall in Central Vietnam is significantly higher than in southern part, which means that the level of threat from the typhoon in Central Vietnam is also higher. However, until now, there is not any investigation regarding how local people on the island in Central Vietnam prepare to face with typhoon disasters. According to Figure 10 the wave height of around 7-8 m occurred along the large area in Central Vietnam during typhoon Ketsana. According to the statistical data, such a strong typhoon accompanied by a high wave caused 169 people die and caused massive property loss both for inland and the islands.

4. Conclusion

This study investigated the applicability of WRF and Delft3D-WAVE as assessment tools to evaluate the potential threat from typhoon-induced high wave for the coastal area. A WRF model using a set of suggested physical schemes shows reasonable accuracy in reproducing the wind field during typhoon Kaemi (2000) and Ketsana (2009). The simulated significant wave height during typhoon Kaemi (2000) from Delft3D-WAVE forced by the WRF model output also agreed well with the observed value. The extreme high wave distribution during typhoon Ketsana (2009) was reproduced by the combined wind-wave models. The qualitative assessment based on the typhoon-induced wave results demonstrated that Central Vietnam are very vulnerable to typhoons and typhoon-induced waves, especially the remote islands. Further site investigations need to be conducted to determine a more comprehensive level of risk each location in terms of physical damage caused by typhoons, as well as the social issues including people awareness and preparedness against extreme weather events.

Acknowledgments
This research was supported by The Murata Science Foundation.

ORCID iDs
L T Anh https://orcid.org/0000-0002-1027-1360
N D Thao https://orcid.org/0000-0001-9097-2863

References

[1] WAMDI Group 1988 The WAM model - A third generation ocean wave prediction model *J. Phys. Oceanogr* **18** 1775–1810

[2] Cardone V J, Jensen R E, Resio D T, Swail V R and Cox A T 1996 Evaluation of contemporary ocean wave models in rare extreme events: The "halloween storm" of October 1991 and the" storm of the century" of March 1993 *Journal of Atmospheric and Oceanic Technology* **13** (1) 198–230

[3] Tolman H L 1989 The numerical model WAVEWATCH: a third-generation model for the hindcasting of wind waves on tides in shelf seas *Communications on Hydraulic and Geotechnical Engineering* **89**-2 (Delft University of Technology) ISSN 0169–6548 72 pp

[4] Tolman H L 1991 A third-generation model for wind waves on slowly varying, unsteady, and inhomogeneous depths and currents *J. Phys. Oceanogr.* **21** 782-97

[5] Samiksha S V, Polnikov V G, Vethamony P, Rashmi R, Pogarskii F and Sudheesh K 2015 Verification of model wave heights with long-term moored buoy data: Application to wave field over the Indian Ocean *Ocean Engg.* **104** 469–79

[6] Parvathy K G, Umesh P A and Prasad K Bhaskaran 2017 Inter-seasonal variability of wind waves and their attenuation characteristics by mangroves in a reversing wind system *International Journal of Climatology* **37**(15) 5089–106

[7] Chien D D, Thuy N B, Sao N T, Thai T H and Kim S 2014 Investigation of the interaction between wave and storm surge using numerical model *Scientific and Technical Hydro Meteorological Journal* **647** (in Vietnamese)

[8] Jun T, Takeshi I 2013 Prediction of tropical cyclone induced wind field by using mesoscale model and JMA best track *Proc. The Eighth Asia-Pacific Conference on Wind Engineering (Chennai, India)*

[9] Islam T, Srivastava P K, Rico-Ramirez M A, Dai Q, Gupta M and Singh S K 2015 Tracking a tropical cyclone throughWRF–ARW simulation and sensitivity of model physics *Nat. Hazards* **76** 1473 –95

[10] Srinivas C V, Rao D B, Yesubabu V, Baskaran R and Venkatraman B 2013 Tropical cyclone predictions over the Bay of Bengal using the high-resolution Advanced Research Weather Research and Forecasting (ARW) model *Q. J. Roy. Meteor. Soc.* **139** 1810–25

[11] Radhika D Kanase and P S Salvekar 2015 Effect of Physical Parameterization Schemes on Track and Intensity of Cyclone LAILA using WRF Model *Asia-Pac. J. Atmos. Sci.* **51**(3) 205–27

[12] Tran N N, Nagai K, Kubota H A, Hue N N and Quang D X 2003 Statistical characteristics of unusual waves observed at DaNang, Vietnam *Proceedings of the 2nd International Conference: Asian and Pacific Coasts 2003 (Makuhari, Japan)*

[13] Thai T H, Tri D Q and Hoang D V 2018 Study on the impact of wave and storm surge in coastal area in Central Vietnam *Journal of Hydrometeorology* 3/2018 (in Vietnamese)

[14] Madsen O, Poon Y K and Graber H 1988 Spectral wave attenuation by bottom friction: Theory *Proceedings 21st International Conference Coastal Engineering, ASCE* pages 492-504

[15] Anh L, Takagi H and Thao N 2019 Storm Surge and High Waves due to 1997 Typhoon Linda: Uninvestigated Worst Storm Event in Southern Vietnam *Journal of Japan Society of Civil Engineers*, Ser. B3 (Ocean Engineering) **75**(2)

IOP Conf. Series: Earth and Environmental Science 1091 (2022) 012025 doi:10.1088/1755-1315/1091/1/012025

The condition of upper soil layer that causes severe damage to land due to liquefaction

H Muhanifah[1,2], A D Adi[1] and F Faris[1]

[1] Department of Civil and Environmental Engineering, Faculty of Engineering, Universitas Gadjah Mada, Yogyakarta 55281, Indonesia
[2] Directorate General of Housing, Ministry of Public Works and Housing, South Jakarta 12110, Indonesia

agusdadi1@ugm.ac.id

Abstract.

An evaluation using the liquefaction potential index (LPI) was still applicable to the site conditions in Central Sulawesi in 2018's earthquake with Mw 7.5. In some areas, severe liquefaction occurred and destroyed many buildings. This study was conducted to determine the depth of the liquefaction layer in any soil conditions that may cause severe damage to the land. This is intended to obtain guidance to select an area for housing, especially for low-budget housing. The possibility of land severely damages due to liquefaction occurred when the LPI value was more than 15. The analysis was conducted by creating various models of soil lithologies using the simplified procedure based on the standard penetration test (SPT) data. The upper layer was designed to have liquefaction potential, while the lower layer had not. The results were presented in graphs of correlation of the upper layer depths with the LPI greater than 15 using the N-SPT data ranging from 5 to 20 with a fines content (FC) of 5%, 15%, and 35%. Based on these results, selecting a safe location for housing could be done easily to prevent a massive disaster in the future.

1. Introduction

Earthquake is one of the natural phenomena that become a concern due to their effects. In an earthquake, major destruction may occur due to liquefaction. Liquefaction is a phenomenon of soil losing its strength [1]. One of the severe liquefaction disasters occurred at Central Sulawesi on September 28, 2018, due to the earthquake with a moment of magnitude Mw 7.5, triggering flow-slides in Balaroa, Petobo, Jono Oge, and Sibalaya [2,3], as shown in figure 1. Based on data from the National Disaster Management Authority (BNPB), as many as 4028 buildings were damaged [4], most of them were residential housings. Figure 2 shows the damage condition due to flow-slide.

IOP Conf. Series: Earth and Environmental Science 1091 (2022) 012025 doi:10.1088/1755-1315/1091/1/012025

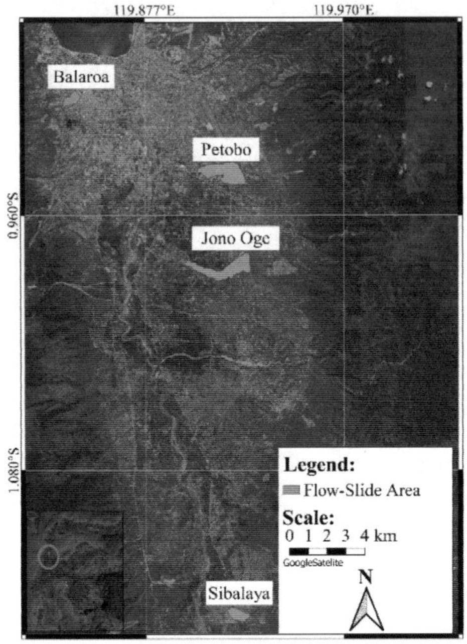

Figure 1. Location of flow-slide due to the Central Sulawesi earthquake on September 28.

Figure 2. Damage condition: (a) Balaroa; (b) Petobo; (c) Jono Oge; (d) Sibalaya (Ministry Public Works and Housing documentation, 2018).

Several researchers have analysed the potential for liquefaction in the flow-slide area after the September 28, 2018, Central Sulawesi earthquake [5–8]. Muhanifah *et al.* [6] conducted further analysis of the factor of safety (SF) value obtained from the analysis using the simplified procedure based on the Standard Penetration Test (SPT) data developed by Idriss and Boulanger [9] into the liquefaction potential index (LPI) by Iwasaki *et al.* [10] to determine the level of potential liquefaction in the liquefied and non-liquefied areas in Balaroa, Central Sulawesi. The result showed that the liquefaction potential level obtained from the LPI interpretation matched the fact in the site where at a very high liquefaction potential level (LPI > 15), severe damage occurred due to liquefaction [6].

Soil characteristics are the main factors affecting liquefaction, where sites with high soil strength resistance have a relatively low liquefaction susceptibility, while sites with low soil strength resistance have more susceptibility [11]. The contribution of the upper soil layer to liquefaction plays a dominant role in the occurrence of severe liquefaction [6]. In addition, according to Ishihara [12], the thickness of the unliquefied soils overlaying the liquefied layer and the thickness of the liquefied layer affect the damage that occurs on the ground surface due to liquefaction.

From previous studies, the upper soil layer condition such as the N-SPT value and its depth that causes severe damage to the land due to liquefaction was not observed and appropriately analysed. The study was conducted to determine the condition of the upper soil layer that may cause severe damage to the land due to liquefaction by making several soil layer models. The result is intended to be used as a guideline in selecting a safe area for housing, especially for 1-2 stories of simple structures and low-budget housing where it is impossible to undertake any complex liquefaction mitigation efforts. By selecting a safe area, massive disasters due to liquefaction can be avoided.

2. Research Method

Several stages were arranged to conduct the study, including setting the conditions on the model and analysing the liquefaction potential. The analysis was carried out to determine the depth of the liquefaction layer in any soil conditions that may cause severe damage to the land.

2.1. Model Condition

The analysis was carried out by modelling five soil lithologies, i.e., gravel, well-graded sand and gravel, coarse or medium sand, well-graded sand, and fine or silty sand with fines content (FC) were 5%, 15%, and 35%. The SPT for each soil was carried out at 1 m depth intervals and started from 1.5 m to 19.5 m. The soil structure was made of 2 layers where the upper soil layer was a liquefied layer with a thickness of H, and the layer below was unliquefied, as shown in figure 3. Groundwater level (GWL) was assumed at the ground surface; hence the soil was fully saturated. This condition represents the worst condition for the liquefaction potential.

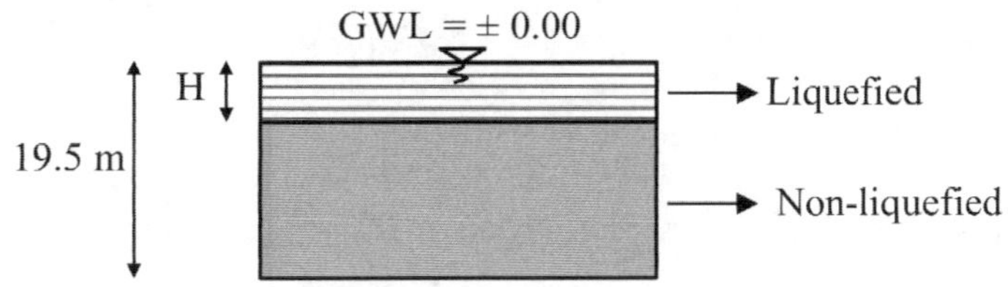

Figure 3. Model of soil layers.

The model was started with soil with a liquefied layer thickness of 1.5 m and continued for each depth until a very high liquefaction potential level was obtained. The liquefied layer was soil with a relatively loose density with an N-SPT value of 5 to 20. Table 1 shows the unit weight obtained from the correlation between soil lithology and relative density based on Australian Standard AS 4678-2002 Earth-Retaining Structures [13].

Table 1. Unit weight of soils. [13]

No.	Lithology	γ_m, Moist Unit Weight (kN/m^2)		γ_s, Saturated Unit Weight (kN/m^2)	
		Loose	Dense	Loose	Dense
1	Gravel	16.0	18.0	20.0	21.0
2	Well-graded sand and gravel	19.0	21.0	21.5	23.0
3	Coarse or medium sand	16.5	18.5	20.0	21.5
4	Well-graded sand	18.0	21.0	20.5	22.5
5	Fine or silty sand	17.0	19.0	20.0	21.5

2.2. Analysis of liquefaction potential

The liquefaction potential of the soil layer was analysed using a simplified procedure using the Idriss and Boulanger method [9] to obtain soil conditions that matched the desired model. This method was the closest in estimating the suitability of liquefaction potential and occurrence in the field [14]. The analysis produced a factor of safety (FS) that described the presence or absence of liquefaction potential in each soil layer shaken by an earthquake. The earthquake used in this study was similar to the earthquake that caused a flow-slide in Balaroa with the moment magnitude of Mw 7.5 and the peak ground acceleration of 0.33 g [6]. Various depths of the liquefied layer were modelled, and then LPI analysis using the method by Iwasaki *et al.* [10] was carried out to determine the level of potential liquefaction.

2.2.1. Simplified procedure. FS is the ratio of the cyclic resistance ratio (CRR) to the cyclic stress ratio (CSR) as in equation (1). When the FS < 1, the soil layer has liquefaction potential, whereas when the FS ≥ 1, the soil layer has no liquefaction potential.

$$FS = \frac{CRR}{CSR} \tag{1}$$

CSR due to the earthquake was calculated using equation (2) Seed and Idriss [15].

$$CSR = 0.65\left(\frac{\sigma_{vc}}{\sigma'_{vc}}\right)\left(\frac{a_{max}}{g}\right)r_d \tag{2}$$

Where σ_{vc} and σ'_{vc} are the total vertical stress and effective vertical stress respectively in kN/m^2, a_{max} is the maximum ground surface acceleration in g, and r_d is the stress reduction coefficient calculated using equation (3) to (5) Idriss and Boulanger [9].

$$r_d = exp[\alpha(z) + \beta(z).M] \tag{3}$$

$$\alpha(z) = -1{,}012 - 1{,}126\sin\left(\frac{z}{11{,}73} + 5{,}133\right) \tag{4}$$

$$\beta(z) = 0{,}106 + 0{,}118\sin\left(\frac{z}{11{,}28} + 5{,}142\right) \tag{5}$$

Where z is the soil depth in analysis in meter and M is the moment magnitude of the earthquake in Mw.

CRR represents the liquefaction resistance of the soil in situ, and the most common method to determine liquefaction resistance is using the SPT data. The field SPT data needed some correction for field testing procedures and overburden pressure ($(N_1)_{60}$). The higher the N-SPT value will give the higher liquefaction resistance [16]. Based on the field performance and the $(N_1)_{60}$, Seed *et al.* [17] suggested an approximate range of potential damage that can be identified, as shown in table 2.

Table 2. Potential damage ranges based on SPT dan field performance data. [17]

$(N_1)_{60}$	Potential Damage
0 – 20	High
20 – 30	Intermediate
> 30	No significant damage

$(N_1)_{60}$ is the multiplication of the N-SPT value corrected with field testing procedures (N_{60}) and the overburden pressure factor (C_N) as shown in equation (6).

$$(N_1)_{60} = N_{60}\, C_N \tag{6}$$

Where the N_{60} was obtained from equation (7) by Idriss and Boulanger [9], which is the multiplication of the correction value which can be seen in table 3, and the field N-SPT value (N_m). Meanwhile, the C_N was found using equation (8) with the atmospheric pressure (P_a) which is equal to 101 kPa.

$$N_{60} = C_E C_B C_R C_S N_m \tag{7}$$

$$C_N = \left(\frac{P_a}{\sigma'_{vc}}\right)^{0.784-0.0768\sqrt{(N_1)_{60}}} \leq 1.7 \tag{8}$$

Table 3. Factors of field-testing procedure correction for the N-SPT value.

Factor	Description
Energy ratio (C_E)	Donut hammer United States, $C_E = 0.75$ [17].
Borehole diameter (C_B)	Borehole diameter 65 – 115 mm, $C_B = 1.0$ [18].
Rod length (C_R)	Rod lengths are assumed to be equal to depth plus 1.5 m (for the above-ground extension). The correction factor for rod length [19]: Rod length < 3 m, $C_R = 0.75$ Rod length 3 – 4 m, $C_R = 0.80$ Rod length 4 – 6 m, $C_R = 0.85$ Rod length 6 – 10 m, $C_R = 0.95$ Rod length 10 – 30 m, $C_R = 1.00$
Sampler (C_S)	Standard sampler, $C_S = 1.0$ [18].

With the same $(N_1)_{60}$ value, sand with a higher FC will be more resistant to liquefaction than clean sand (FC ≤ 5%) [17], hence a correction due to FC was required. The correction of the value $(N_1)_{60}$ to the equivalent clean-sand $(N_1)_{60cs}$ was calculated using equation (9) and (10) [9].

$$(N_1)_{60cs} = (N_1)_{60} + \Delta(N_1)_{60} \tag{9}$$

$$\Delta(N_1)_{60} = exp\left(1.63 + \frac{9.7}{FC + 0.01} - \left(\frac{15.7}{FC + 0.01}\right)^2\right) \tag{10}$$

After the above values were obtained, the CRR was calculated using equations (11) to (15) [9].

$$CRR_{M,\sigma'_{vc}} = CRR_{M=7.5,\sigma'_{vc}=1} \,.MSF \,.K_\sigma \tag{11}$$

$$CRR_{M=7.5,\sigma'_{vc}=1} = exp\left(\frac{(N_1)_{60cs}}{14.1} + \left(\frac{(N_1)_{60cs}}{126}\right)^2 - \left(\frac{(N_1)_{60cs}}{23.6}\right)^3 + \left(\frac{(N_1)_{60cs}}{25.4}\right)^4 - 2.8\right) \tag{12}$$

$$MSF = 6.9\, exp\left(\frac{-M}{4}\right) - 0.058 \leq 1.8 \tag{13}$$

$$K_\sigma = 1 - C_\sigma ln\left(\frac{\sigma'_v}{P_a}\right) \leq 1.1 \tag{14}$$

$$C_\sigma = \frac{1}{18.9 - 2.55\sqrt{(N_1)_{60cs}}} \leq 0.3 \tag{15}$$

Where $CRR_{M=7.5,\sigma'_{vc}=1}$ is the cyclic resistance ratio for earthquake magnitude of 7.5 and effective vertical stress of 1 atm. $CRR_{M,\sigma'_{vc}}$ is the cyclic resistance ratio for other values of earthquake magnitude and effective vertical stress, MSF is a magnitude scaling factor used to adjust the CRR to a common value of M, K_σ is overburden correction factor, and C_σ is the relative soil density factor.

2.2.2. *Liquefaction potential index.* The FSs obtained from the liquefaction potential analysis using a simplified procedure were used to find the LPI based on the formula developed by Iwasaki *et al.* [10] using equations (16) to (19). The LPI classified the level of potential liquefaction and provided an overview of follow-up suggestions as proposed by Iwasaki *et al.* [10], which can be seen in table 4. By taking the reference if the LPI value is greater than 15, the soils could be severely damaged due to liquefaction [6,10], the maximum depth of the liquefied soil layer could be defined.

$$LPI = \int_0^{20\,m} F(z).w(z).dz \tag{16}$$

$$F(z) = 1 - FS \text{ for } FS < 1 \tag{17}$$

$$F(z) = 0 \text{ for } FS \geq 1 \tag{18}$$

$$w(z) = 10 - 0{,}5z \tag{19}$$

Table 4. LPI-based classification of potential liquefaction level. [10]

LPI	Potential Liquefaction Level	Suggestions
LPI = 0	Very Low	Detail investigations on soil liquefaction are not needed in general
0 < LPI ≤ 5	Low	Detail investigations on soil liquefaction are needed only for especially important structures
5 < LPI ≤ 15	Rather High	Detail investigations on soil liquefaction are needed for important structures, and countermeasures of soil liquefaction are needed in general
LPI > 15	Very High	Detail investigations and countermeasures on liquefaction are needed

3. Result and Discussion

The analysis based on SPT data using the Idriss and Boulanger [9] method resulted in FSs from the gravel, well-graded sand and gravel, coarse or medium sand, well-graded sand, and fine or silty sand soil model in various FC and N-SPT values. Figure 4 shows graphs of the liquefaction potential in fine or silty sand with a thickness of the liquefied soil layer (H) 3.5 m, the field N-SPT value on the liquefied soil layer 5, and the FC of 15%.

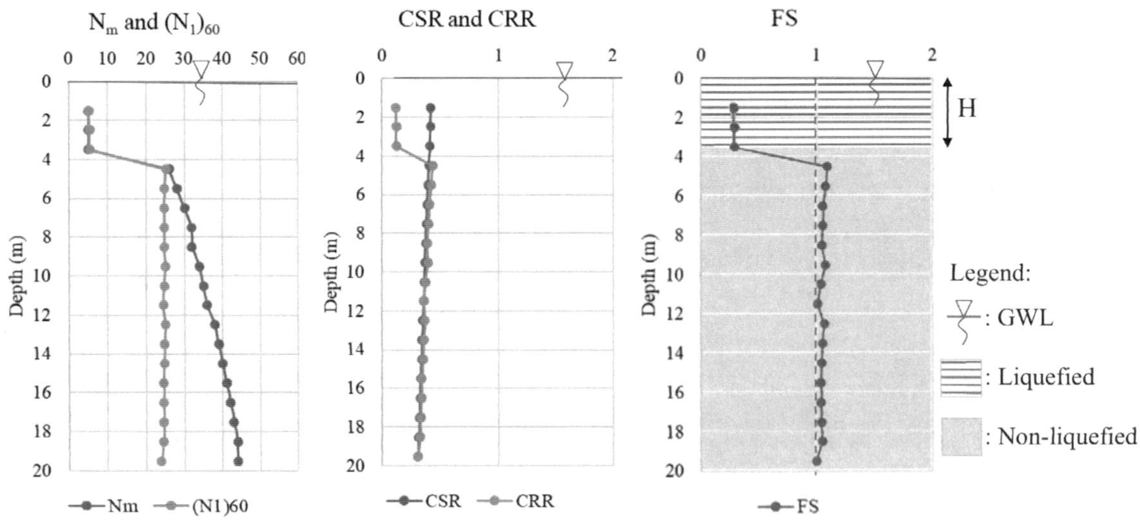

Figure 4. Example curve of N_m, $(N_1)_{60}$, CSR, CRR, and FS at fine or silty sand with H = 3.5 m, N_m on the liquefied soil layer = 5, and FC = 15%.

Figure 4 shows that fine or silty sand with a relatively loose density with a field N-SPT value of 5 at a depth of 1.5 m to 3.5 m and FC 15%, when shaken by an earthquake with Mw 7.5 and maximum ground surface acceleration of 0.33 g will have a smaller the CRR than the CSR, so that, at those depths, it produces FS less than 1. This result indicates that the soil layer up to a thickness of 3.5 m from the soil

IOP Conf. Series: Earth and Environmental Science 1091 (2022) 012025 doi:10.1088/1755-1315/1091/1/012025

surface has liquefaction potential. Analysis using the FS values found an LPI of 20.399, which could be classified as a very high liquefaction potential level.

Figure 5 shows graphs of the relationship between H and LPI on fine or silty sand with FC values of 5%, 15%, and 35%.

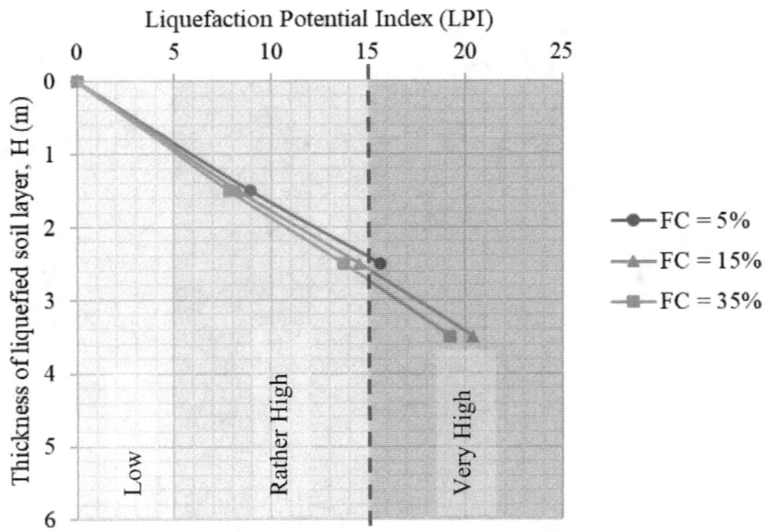

Figure 5. LPI for each thickness of liquefied soil layer at fine or silty sand with N_i on the liquefied soil layer = 5.

The thickness limit of the liquefied soil layer has the potential for severe damage due to liquefaction in fine or silty sand which in the liquefied soil layer has a field N-SPT value of 5 can be seen in figure 5. In fine or silty sand with FC 5%, the potential for severe damage is due to liquefaction occurs when the thickness of the liquefied soil layer is more than 2.4 m. In fine or silty sand with FC 15%, the potential for severe damage due to liquefaction occurs when the thickness of the liquefied soil layer is more than 2.6 m. Meanwhile, in fine or silty sand with FC 35%, the potential for severe damage due to liquefaction occurs when the thickness of the liquefied soil layer is more than 2.7 m. The results show that the thickness limit of the liquefied soil layer increases with increasing the FC.

The results of the liquefied soil layer thickness having severe potential damage to the land due to liquefaction in several soil conditions when shaken by an earthquake are shown in figure 6.

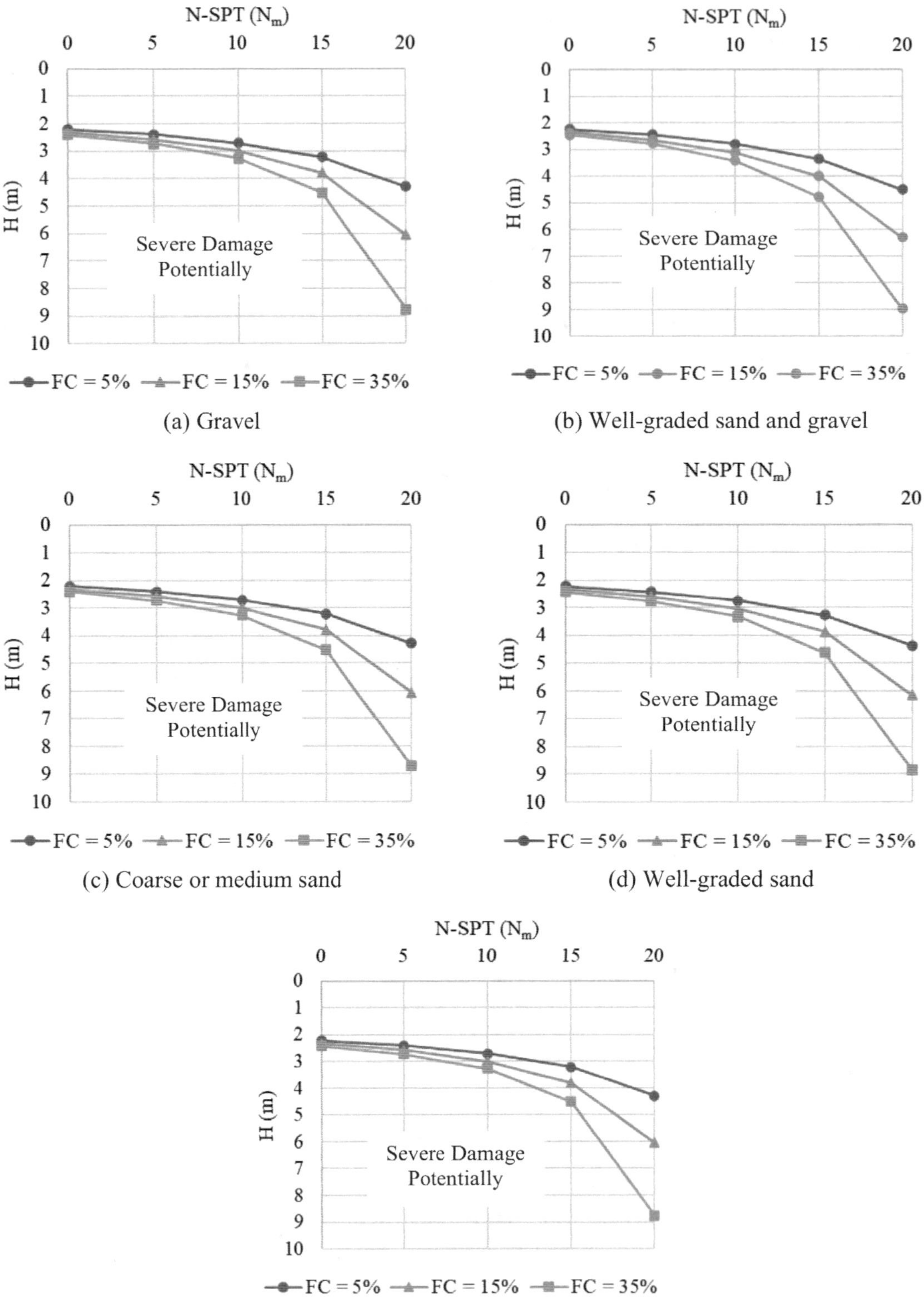

Figure 6. Relationship between the limit of the thickness of the liquefied upper soil layer (H) and the potential for severe damage to land due to liquefaction.

Figure 6 shows the thickness of the liquefied upper soil layer against the potential for severe damage to land due to liquefaction in several lithologies, i.e., gravel, well-graded sand and gravel, coarse or medium sand, well-graded sand, and fine or silty sand. The thickness limit is determined based on the field N-SPT value in the liquefied soil layer and the FC. At the thickness of the soil layer below the limit of the thickness line, severe damage to land due to liquefaction has the potential to occur. With the same FC, the higher the N-SPT value in the field, the deeper the thickness limit of the liquefied upper soil layer. Similarly, with the same N-SPT value in the field, the higher the FC, the deeper the thickness of the liquefied upper soil layer. In addition, with the same field N-SPT and FC values, the higher the unit weight, the deeper the thickness limit of the liquefied upper soil layer.

Based on figure 6, soil with low field N-SPT values, FC, and unit weight in the upper soil layer has a high potential for severe damage to land due to liquefaction when shaken by an earthquake. This study gives suggestions in selecting a safe area where housing can be built and the areas that need to be avoided for housing purposes. In the unsuggested area, housing development needs special treatment for the soil or building structures to mitigate the liquefaction hazard. For simple and low-cost buildings such as public housing with a structure of 1-2 stories, this will require a high cost and can not be recommended.

4. Conclusion

Disaster due to liquefaction causes severe damage in several locations where the thickness of the liquefied upper soil layer influences the severity. By modelling various soil conditions and analysing their liquefaction potential, a graph of the thickness limit of the liquefied upper soil layer against the potential for severe damage to land due to liquefaction was generated. The lower the value of N-SPT in the field, FC, and unit weight, the thickness of the liquefied upper soil layer is shallower than the soil surface to produce severe damage to the land. With a minimum depth of the liquefied upper soil layer of about 2 m, when an earthquake shakes the ground with a moment magnitude of 7.5 and a maximum ground surface acceleration of 0.33 g, there is the potential for severe damage to the land due to liquefaction. This study can be used in selecting a safe area where simple housing can be built with a low budget to mitigate liquefaction disasters. Further studies are needed to provide complete information for soil conditions with various soil layers and earthquake magnitude models. In addition, it is also necessary to verify the model by conducting laboratory tests or using data on liquefaction that has occurred.

Acknowledgments

The authors would like to thank The Indonesian Ministry of Public Works and Housing and Japan International Cooperation Agency (JICA) Project for AUN/SEED-Net for supporting and funding this research and publication.

References

[1] Das B M and Ramana G V. 2011 *Principles of Soil Dynamics* (Stamford: Cengage Learning)
[2] Kiyota T, Furuichi H, Hidayat R F, Tada N and Nawir H 2020 Overview of long-distance flow-slide caused by the 2018 Sulawesi earthquake, Indonesia *Soils Found.* **60** 722–35
[3] Kusumawardani R, Chang M, Upomo T C, Huang R-C, Fansuri M H and Prayitno G A 2021 Understanding of Petobo liquefaction flowslide by 2018.09.28 Palu-Donggala Indonesia earthquake based on site reconnaissance *Landslides* **18** 3163–82
[4] PuSGeN 2018 *Kajian Gempa Palu Provinsi Sulawesi Tengah 28 September 2018 (M7.4)* ed M Irsyam, N R Hanifa, D Djarwadi and D A Sarsito (Bandung: Research and Development Agency of Ministry of Public Works and Housing)
[5] Andiny A N, Faris F and Adi A D 2021 Re-liquefaction hazard evaluation in flow-slide affected area of Jono Oge , Central Sulawesi , Indonesia Re-liquefaction hazard evaluation in flow-slide affected area of Jono Oge , Central Sulawesi , Indonesia *IOP Conf. Ser. Earth Environ. Sci.* **861**
[6] Muhanifah H, Adi A D and Faris F 2021 Liquefaction investigation of Balaroa , Central Sulawesi

on liquefied and non-liquefied areas Liquefaction investigation of Balaroa , Central Sulawesi on liquefied and non-liquefied areas *IOP Conf. Ser. Earth Environ. Sci.* **861**

[7] Hazarika H, Rohit D, Pasha S M K, Maeda T, Masyhur I, Arsyad A and Nurdin S 2021 Large distance flow-slide at Jono-Oge due to the 2018 Sulawesi Earthquake , Indonesia *Soils Found.* **61** 239–55

[8] Pratama A, Fathani T F and Satyarno I 2021 Liquefaction potential analysis on Gumbasa Irrigation Area in Central Sulawesi Province after 2018 earthquake *IOP Conf. Ser. Earth Environ. Sci.* **930**

[9] Idriss I M and Boulanger R W 2008 Soil Liquefaction during Earthquake *Earthquake Engineering Research Institute MNO-12* (United States of America: EERI Publicatiom)

[10] Iwasaki T, Tokida K and Tatsuoka F 1981 Soil Liquefaction Potential Evaluation with Use of the Simplified Procedure *First International Conference on Recent Advances in Geotechnical Earthquake Engineering and Soil Dynamics* (St. Louis, Missouri) pp 209–14

[11] Mase L Z, Farid M, Sugianto N and Agustina S 2020 The Implementation of Ground Response Analysis to Quantify Liquefaction Potential Index (LPI) in Bengkulu City, Indonesia *J. Civ. Eng. Forum* **6** 319

[12] Ishihara K 1985 Stability of Natural Deposits During Earthquakes *Proc. Elev. Int. Conf. soil Mech. Found. Eng.* **1** 321–76

[13] Standards Australia 2002 *Australian Standard AS 4678-2002 earth-retaining structures* (Sydney: Standards Australia)

[14] Mase L Z 2018 Studi Kehandalan Metode Analisis Likuifaksi Menggunakan SPT Akibat Gempa 8,6 Mw, 12 September 2007 di Area Pesisir Kota Bengkulu *J. Tek. Sipil* **25** 53–60

[15] Seed H B and Idriss I M 1970 *A Simplified Procedure for Evaluating Soil Liquefaction Potential* (Berkeley, California)

[16] Day R W 2012 *Geotechnical Earthquake Engineering Handbook with The 2012 International Building Code* (New York: McGraw-Hill)

[17] Seed H B, Tokimatsu K, Harder L F and Chung R M 1985 Influence of spt procedures in soil liquefaction resistance evaluations *J. Geotech. Eng.* **111** 1425–45

[18] Skempton A W 1986 Standard penetration test procedures and the effects in sands of overburden pressure, relative density, particle size, ageing and overconsolidation *Geotechnique* **36** 425–47

[19] Youd T L, Idriss I M, Andrus R D, Arango I, Castro G, Christian J T, Dobry R, Finn W D L, Harder Jr. L F ., Hynes M E, Ishihara K, Koester J P, Liao S S C, Marcuson III W F, Martin G R, Mitchell J K, Moriwaki Y, Power M S, Robertson P K, Seed R B and Stokoe II K H 2001 Liquefaction resistance of soils: summary report from the 1996 NCEER and 1998 NCEER/NSF workshops on evaluation of liquefaction resistance of soil *J. Geotech. Geoenvironmental Eng.* **127** 817–33

IOP Conf. Series: Earth and Environmental Science 1091 (2022) 012026 doi:10.1088/1755-1315/1091/1/012026

Slope stability analysis during an earthquake in flow-slide affected area of Jono Oge

A N Andiny[1,2], F Faris[1,3] and A D Adi[1]

[1] Department of Civil and Environment Engineering, Faculty of Engineering, Universitas Gadjah Mada, Yogyakarta 55281, Indonesia
[2] Directorate General of Highways, The Ministry of Public Works and Housing, South Jakarta 12110, Indonesia
[3] Centre of Excellence of Technological Innovation for Disaster Mitigation (GAMA-InaTEK) Universitas Gadjah Mada, Yogyakarta 55281, Indonesia

Corresponding author: fikri.faris@ugm.ac.id

Abstract. On September 28th, 2018, the earthquake with a moment magnitude (Mw) 7.5 due to strike-slip Palu-Koro Fault in Central Sulawesi caused a major geological failure. In the Jono Oge area, the earthquake-induced liquefaction and flow-slide occurred on relatively flat ground with an average slope of 2%. In this paper, the safety factor for slope stability analysis was determined by shear strength reduction caused by the earthquake at the potentially liquefied soil layer, varying peak ground acceleration values, and groundwater level conditions. This analysis evaluated the potential of ground movement occurrence in Jono Oge with the condition of soil layer pre and post the Palu-Donggala earthquake that happened in 2018 using GeoStudio Slope/W geotechnical software. The analysis was conducted on the cross-section of the soil layer based on three boreholes incorporating standard penetration test (SPT) data on the study area. The analysis results showed ground movement potential because of the seismic loads and groundwater level conditions. The lower the safety factor of slope stability, the higher the ground movement potential in flow-slide affected area. This ground movement potential could trigger the occurrence of flow-slide in Jono Oge area.

1. Introduction

On September 28th, 2018, The United States Geological Survey (USGS) reported an earthquake struck on Sulawesi, Indonesia with a moment magnitude (Mw) 7.5. The rupture occurred on the left-lateral, north-south striking fault with the epicenter of the earthquake was approximately at 10 km depth and 72 km north of Palu City [1,2]. The 2018 Palu-Donggala Earthquake caused major geological failure in several Central Sulawesi areas, i.e., tsunami, liquefaction, landslide, and structural damages. Figure 1 shows the areas where the impact of earthquake-induced liquefaction and flow-slide were most visible occurred in Balaroa, Petobo, Jono Oge, and Sibalaya.

In the preliminary study, many researchers analysed the flow-slide occurrence in Central Sulawesi several areas [3–7] and evaluated the re-liquefaction in flow-slide affected area of Jono Oge [8]. Soil aging and cementation are among the main factors that govern liquefaction because compared to the older soil deposits, newly deposited soil tends to be more susceptible to liquefaction [9]. Palu Valley has the alluvial plain at the center with young alluvial fans and the older fan surfaces at the east and west edges that the alluvial layer thins toward the mountain [5,10]. Furthermore, the head scarp of Jono Oge flow-slide area in Figure 2 falls close to the edge of young alluvial fan that meets the older alluvial fan.

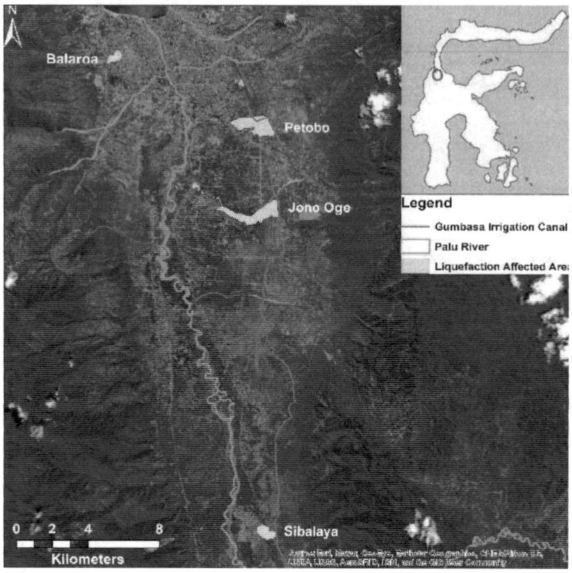

Figure 1. Liquefaction and flow-slide affected area in Balaroa, Petobo, Jono Oge, and Sibalaya.

Figure 2. The head scarp of Jono Oge flow-slide near the irrigation canal (Source: The Ministry of Public Works and Housing Documentation, 2018).

Most of the infrastructure damages were caused by the liquefaction-induced flow-slide that occurred in Jono Oge. Mason et al. [5] explained the geological failure in which the earthquake-induced liquefaction and the unlined irrigation canal played a significant role in initiating the slope failure. The existence of the irrigation canal affected the shallowness of the groundwater table which could increase the potential of liquefaction. This flow-slide was the seismic liquefaction-induced landslide that failed along very gentle slopes with an average slope of 1% in Jono Oge [7,11]. Hazarika et al. [6] explained the occurrence of Jono Oge post-earthquake displacements in 4 zones. The breach of the channel and the destruction of the water gate occurred near the irrigation canal, the flow-slide began at the west side of irrigation canal with the length of displacement was about 2 – 50 m, the deposits of the flow-slide area with 300 – 1,300 m length of displacement were collected at the west side of the rural road, and the flood water flowed until it joined the stream to Palu River.

Based on the re-liquefaction analysis by Andiny et al. [8], the flow-slide affected area of Jono Oge has high to very high liquefaction potential in the west side of the irrigation canal to the former rural road and low liquefaction potential at the irrigation canal area. This study aims to evaluate the potential of the liquefaction-induced ground movement in Jono Oge based on the safety factor of slope stability. The analysis of this study was determined by shear strength reduction caused by the earthquake at the potentially liquefied soil layers, varying peak ground acceleration values, and groundwater level conditions. The evaluation of this analysis could be used to identify the liquefaction mitigation requirements of the infrastructure in the study area especially the irrigation canal and road reconstruction planning projects.

2. Methods

The slope stability analysis in this study was located at the flow-slide affected area of Jono Oge, Central Sulawesi. The analysis was observed on three boreholes in flow-slide affected area of Jono Oge, as shown in Figure 3. The soil borehole data, including the SPT and laboratory data was carried out from geotechnical report of Japan International Cooperation Agency (JICA) Team in Central Sulawesi and Central Sulawesi National Road Implementation Agency.

Figure 3. The location of boreholes and the cross-section of slope modelling in Jono Oge flow-slide affected area.

The study area of this analysis focused on the cross-section of the slope at the flow-slide area of Jono Oge (Figure 3). The slope modelling was constructed based on the slope contour data of pre and post Palu-Donggala earthquake that happened in 2018. The soil parameter correlations of the slope stability analysis were the SPT and laboratory data. This research analysed the slope modelling in several scenarios which had various PGA values and groundwater level conditions with the shear strength reduction at the potentially liquefied soil layer. Furthermore, the slope stability analysis by GeoStudio Slope/W was used to determine the safety factor and the illustrated slip surface of the slope.

2.1. *Liquefaction Potential Analysis*

The earthquake-induced liquefaction played a significant role in initiating the slope failure in Jono Oge area. Andiny et al. [8] evaluated the liquefaction potential in Jono Oge area using the simplified procedure SPT-based by Idriss-Boulanger method [12]. The SPT is a test process in the bottom of the borehole where a split-barrel sampler is driven a given distance after seating using a hammer weighting for each blow. In this method, the two controlling factors of liquefaction were cyclic resistance ratio (CRR) and cyclic stress ratio (CSR). The ratio of CRR and CSR values resulted the safety factor against liquefaction (FS_{Liq}) that was used to evaluate the liquefaction potential at a certain depth of soil. In addition, this analysis evaluated the liquefaction occurrences probability. The probability value was calculated by Juang et al. [13] probability approach at a certain depth of soil.

The potentially liquefied soil layer in the analysis result for BH-4 was 1.5 m thick of each loose to medium dense silty sand layer at 7.5 m and 13.5 m depth. In the BH-6 area, the potentially liquefied soil layer was 1 m thick of each soil layer at 4.5 m, 5.5 m, 8.5 m, 10.5 m, 11.5 m, 18.5 m and 19.5 m depth that consisted of sand and gravelly sand layer. The BH-5 area had loose to medium dense sand soil layer at 2.5 m, 4.5 m, 5.5 m, 9.5 m and 10.5 m but there was no potentially liquefied soil layer because the groundwater table was at the 14.08 m depth below the ground surface.

2.2. *Slope Stability Analysis*

The analysis of slope modelling in Jono Oge used GeoStudio Slope/W (R2 2018) geotechnical software. The study area with an average slope of 2% was modelled in two soil layer conditions which are the result of pre and post Palu-Donggala earthquake. The first soil profile was the condition before the ground movement caused by the 2018 earthquake. The analysis used the BH-1 soil investigation data to represent the first soil profile condition. The BH-1 soil investigation data had less damage or was not affected by the ground movement after the 2018 earthquake. The second soil profile was the condition after the ground movement caused by the 2018 earthquake that modelled based on the slope cross-section, as shown in Figure 3.

The cohesionless soils are the soil types that susceptible to liquefaction. The least to most resistance to liquefaction cohesionless soils listing is clean sands, nonplastic silty sand, nonplastic silt, and gravel [9]. In this study area, the earthquake-induce liquefaction had triggered the long-distance flow-slide occurrence. Hence, this analysis evaluates the liquefaction-induced ground movement by assuming the liquefaction occurs at the most susceptible soil layer to liquefy. The area of BH-1, BH-2, and BH-3 have loose to medium dense sandy soil layers that are considered potentially liquefied.

The slope stability analysis used the limit equilibrium method. One of the methods to conduct the slope stability analysis was Morgenstern-Price method which satisfied both moment and force equilibrium [14]. The ground movement occurs when the static driving forces exceed the value of soil shear strength along the slip surface, and then the safety factor becomes less than 1 ($FS < 1.0$) [9]. The Slope/W generated several trial and error of slip surfaces to find the most probable minimum safety factor value of slip surface [15]. The slope stability analysis evaluated the safety factor in several scenarios based on the soil parameter, seismic load, and pore water pressure.

The liquefaction-induced ground movement simulation by Slope/W used shear strength reduction which reduced the friction angle of the potentially liquefied soil layer. Gradually reducing the shear strength parameter from this method is an effective approach for searching the most critical slip surface [16]. The soil friction angle as the shear strength parameter was gradually reduced 10%, 25%, 50%, and 75%. One of the advantages of this method was that the critical slip surface could be found from the shear strain arising of the gravity loads application and the reduction of shear strength [14].

To sum up, this analysis on the first soil profile evaluated the ground movement occurrence in Jono Oge caused by the earthquake-induced liquefaction in 2018. The analysis indicated the ground movement potential in varying scenarios in the second soil profile. The potentially liquefied soil layer was determined based on the liquefaction potential analysis of BH-4, BH-5, and BH-6 in the previous study by Andiny et al. [8].

2.3. Soil Parameter Correlation

Mohr-Coulomb described the strength of the materials in this stability analysis. The analysis required several soil parameters such as cohesion, friction angle, and unit weight of soil. The unit weight of soil for this analysis was using laboratory testing of field samples data in BH-1, BH-2, and BH-3. The friction angle value was determined by the correlation between N_{60} and the friction angle of soils. In this analysis, the sand and gravelly sand soil layer was considered as soil layer with the cohesion value equal to zero (0). The soil parameter correlation for the silt soil layer in this analysis used the Australian Standards, as shown in Table 1.

Table 1. The values of c' and φ' of typical soil [17].

Soil group	Typical soils in group	Soil parameters	
		c' (kPa)	φ' (degrees)
Poor	Soft and firm clay of medium to high plasticity, silty clays, loose variable clayey fill, loose sandy silts	0 – 5	17 – 25
Average	Stiff sandy clays, gravelly clays, compact clayey sands and sandy silts, compacted clay fill (Class II)	0 – 10	26 – 32
Good	Gravelly sands, compacted sands, controlled crushed sandstones and gravel fills (Class I), dense well-graded sands	0 – 5	32 – 37
Very good	Weak weathered rock, Controlled fills (Class I) of roadbase, gravel and recycled concrete	0 – 25	36 – 43

The friction angle for BH-1, BH-2, and BH-3 were determined based on the correlation between N_{60} values and friction angle [18]. The correlation regression of friction angle and the N_{60} value are calculated by the equation (1) and (2), respectively. The N_m in equation (2) is an N-value in which the blow count represented the penetration resistance of soil [19]. The N-values are obtained from the standard penetration test (SPT) carried out in the study area.

$$\varphi = 0.70N_{60} + 18.0 \quad (1)$$

$$N_{60} = C_E \cdot C_B \cdot C_R \cdot C_S \cdot N_m \quad (2)$$

Whereas N_{60} is the corrected N-value, N_m is the measured blow count or N-value, C_E is the correction factor of energy ratio, C_B is borehole diameter correction factor, C_R is the correction factor for rod length, and C_S is sampler correction factor [12,20].

2.4. Peak Ground Acceleration (PGA)

The slope stability analysis evaluated the slope safety factor in varying seismic loads represented by the maximum ground acceleration at ground surface. In the Slope/W analysis, the pseudostatic analysis represented the effect of earthquake shaking by acceleration that created inertia forces. The horizontal pseudostatic force acts through the centroid of sliding mass, acting in an out-of-slope direction [9]. The vertical force generated by the vertical pseudostatic force is ignored because it usually has much less effect on the stability of a slope.

The seismic load for this analysis used the PGA value from USGS data 3.2 m/s^2 or 0.33g. This USGS data was derived from the 2018 Palu-Donggala earthquake. The other PGA value for this study was determined by the regression of attenuation relation by Kanno et al. [21] that only used two parameters, moment magnitude (M_w) and source distance (X). Equation (3) was for shallow events where the focal depth $D \leq 30$ km.

$$log\ pre = a_1 M_w + b_1 X - \log(X + d_1 \cdot 10^{e_1 M_w}) + c_1 + \varepsilon_1 \quad (3)$$

Where, as *pre* is the predicted PGA (cm/sec^2), a_1, b_1, c_1, and d_1 are the regression coefficients, coefficient e_1 for this equation is 0.5, and ε_1 is error between observed and predicted values. The previous study explained that the PGA values from attenuation relation by Kanno et al. [21] was similar with the recorded PGA and proposed ground motion prediction equation (GMPE) with the distance up to 200 km [22].

The PGA value 0.33g from USGS report was used for the analysis. In addition, this analysis used two other PGA values that were calculated by equation (3). The source distance for this equation was 81.1 km that is measured by the length from the epicenter of the 2018 Palu-Donggala earthquake to the flow-slide affected area of Jono Oge. The moment magnitude values were 6.8 M_w and 7.5 M_w based on the Earthquake Sources Map in Indonesia for Palu and USGS report, respectively [23].

2.5. Groundwater Table Conditions

The initial groundwater table affected the decrease in slope stability safety factor [24]. The preliminary study explained that the unlined irrigation canal played a significant role in slope failure [5]. The groundwater table depth that close to the ground surface is one of the main factors that caused earthquake-induced liquefaction at the loose sandy soil layer. Sassa et al. [25] concluded that if the groundwater table was near a critical level to cause a landslide, even a minor earthquake could trigger a large-scale landslide. Other than the earthquake, rainfall could trigger the slope ground movement. The rainfall wetting affected the initial groundwater table condition. In the preliminary study, Ariesta [24] explained that the initial groundwater table affected the decrease of safety factor while wetting process by the rainfall did not cause a significant decrease of the slope stability safety factor.

In this slope stability analysis, the ground movement of the slope was determined at the various groundwater level. The groundwater level condition in Jono Oge was affected by the irrigation canal included the susceptibility of liquefaction and slope failure [26,27]. The groundwater level for the first soil layer condition was at -3.5 m depth based on the soil investigation data in the previous study when the irrigation canal was on the operation condition [27]. The second soil layer condition used the groundwater level data based on the SPT data at site ranged between 0.5 – 3.2 m below the ground surface at the flow-slide affected area caused by the 2018 earthquake and at depth -14.08 m for the site near the irrigation canal.

3. Results and Discussion

The ground movement analysis was observed on the slope cross-section based on the BH-1, BH-2, and BH-3 data in flow-slide affected area of Jono Oge. The soil profile in this slope modelling was divided into three types of the soil layer, i.e., sand soil layer, silt soil layer, and gravelly sand soil layer. Table 2 shows the soil parameters in this study. The PGA values for the slope stability analysis with 6.8 M_w was 0.17g and 7.5 M_w earthquakes were 0.33g and 0.34g. The evaluation applied the shear strength reduction methods at each scenario of slope stability analysis.

The first soil profile expressed the ground movement due to the 2018 Palu-Donggala earthquake triggered by the liquefied soil layer under the ground surface. The slope stability analysis of the second soil layer condition expressed the potential of ground movement with the three different PGA values and two groundwater level conditions.

Table 2. Soil parameter values for each soil layer before and after the 2018 Palu-Donggala earthquake

Soil parameter	Soil layer before the 2018 Palu-Donggala earthquake						Soil layer after the 2018 Palu-Donggala earthquake											
	Silt			Sand			Silt				Sand			Gravelly sand				
	I	II	III	I	II	III	I	II	III	IV	I	II	III	I	II	III	IV	V
c' (kPa)	5	10	5	0	0	0	5	10	5	10	0	0	0	0	0	0	0	0
φ' (degrees)	25	32	25	19.6	26.9	48.8	25	32	25	32	20.9	27.7	46.6	43.6	24.7	41.6	47.6	49.5
Unit weight (kN/m^3)	15.5	19.5	17.5	19.5	19.5	19.5	18.1	18.6	17.5	20.4	20	21.7	19.6	21.7	21	19.5	20.8	21

3.1. The First Soil Profile

The slope stability analysis evaluated the first soil profile with a scenario in which the PGA value was 0.33g, and the groundwater level was at 3.5 m depth below the ground surface. In this slope modelling, the Sand II layer was modelled to be liquefied due to earthquake. The safety factor value initial condition of slope modelling was 6.588 which indicated the slope was stable ($FS \geq 1.0$). The slip surface occurred at the slope with the groundwater level at -3.5 m depth, PGA value 0.33g, and the liquefied soil layer with 10%, 25%, and 50% shear strength reduction. The slope stability safety factor decreased when the seismic load was applied, as shown in Figure 4. The slip surface in Figure 5 shows that the sliding surfaces begin at about 140 m from the BH-1 location. The head scarp of sliding surface of this analysis was similar to the crest of the ground movement in Jono Oge flow-slide affected area in 2018 that near the irrigation canal where the young alluvial fan meet the old alluvial fan [6].

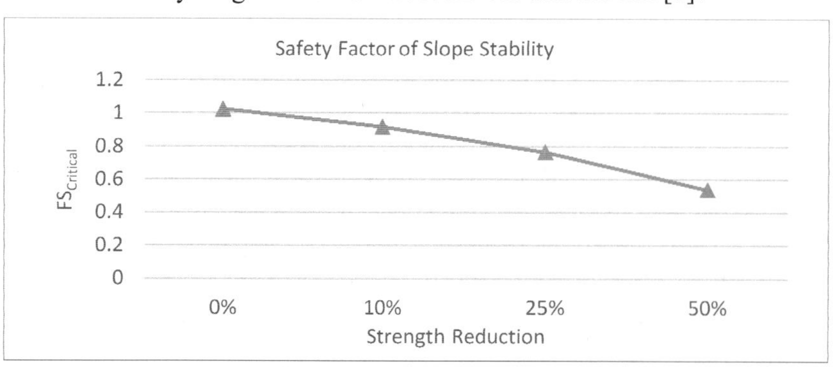

Figure 4. The safety factor value of slope modelling before the 2018 Palu-Donggala earthquake.

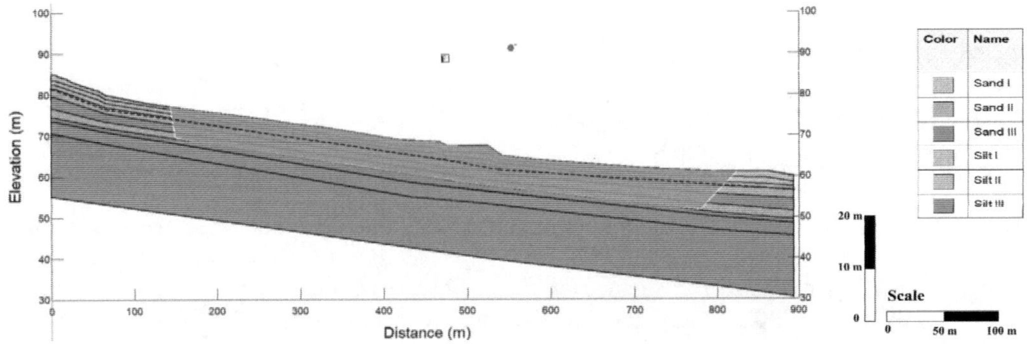

Figure 5. The critical slip surface on the first soil layer condition with 50% shear strength reduction and the PGA value 0.33g.

3.2. The Second Soil Profile

The slope stability analysis of the second soil profile expressed the potential of ground movement with six scenarios of different PGA values and groundwater table conditions. The PGA values for this analysis were 0.17g, 0.33g, and 0.34g. For the groundwater level condition, this slope stability analysed the groundwater level based on the borehole data, as shown in Figure 6(a) and at a depth of -3.5m as the assumption that the unlined irrigation canal was in the operating condition, as shown in Figure 6(b). In this slope modelling, the Sand I layer and Sand II layer were modelled to be liquefied due to earthquake.

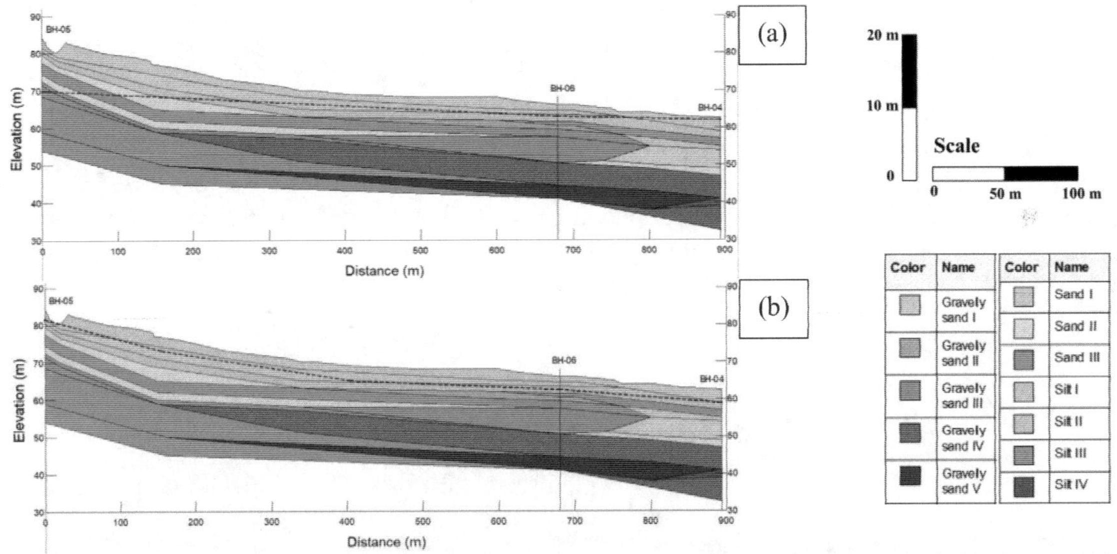

Figure 6. Slope modelling with soil profile post-earthquake; (a) Soil profile with with the irrigation canal off operating condition; and (b) Soil profile with the irrigation canal on operating condition.

The slope stability safety factors at initial condition for the irrigation canal on and off operating conditions are valued more than 1 ($FS \geq 1.0$), which means the slope was stable. As the shear strength reduction increased due to the earthquake, the slope stability safety factor decreased. The safety factor plotted in Figure 7 shows that the slope failure is occurred when the slope modelling generates the 25%, 50%, and 75% shear strength reduction. The assumption of irrigation canal was in operating condition affected the safety factor of slope stability. The slope failure and liquefaction potential could be triggered by the irrigation operating condition and the construction condition [26,27].

Figure 7(a) and Figure 7(b) are the curve of slope stability safety factor with the irrigation canal off and on operating conditions, respectively. The safety factor of slope stability decreased along with the increasing of shear strength reduction at each scenario of slope stability analysis [14]. For the irrigation canal off operating condition, the slope started to fail when shear strength reduction was at 50% except for the seismic load 0.17g scenario. The analysis of PGA 0.17g scenario with 75% shear strength reduction resulted a slip surface with 600 m length, as shown in Figure 8(a).

The other scenario was the irrigation canal on operating condition that has the groundwater level at -3.5 m depth. The slope modelling with PGA values 0.33g failed and had a long-distance slip surface when liquefied soil layer had 50% and 75% shear strength reduction. The different cases occurred at the slope modelling with PGA 0.17g which failed at 50% and 75% shear strength reduction. The scenario of PGA 0.17g resulted the 200 m length slip surfaces with the head scarp lied after the irrigation canal. The slope stability analysis with PGA 0.34g showed that the slope began to slide when the ground received the seismic loads, as shown in Figure 8(b).

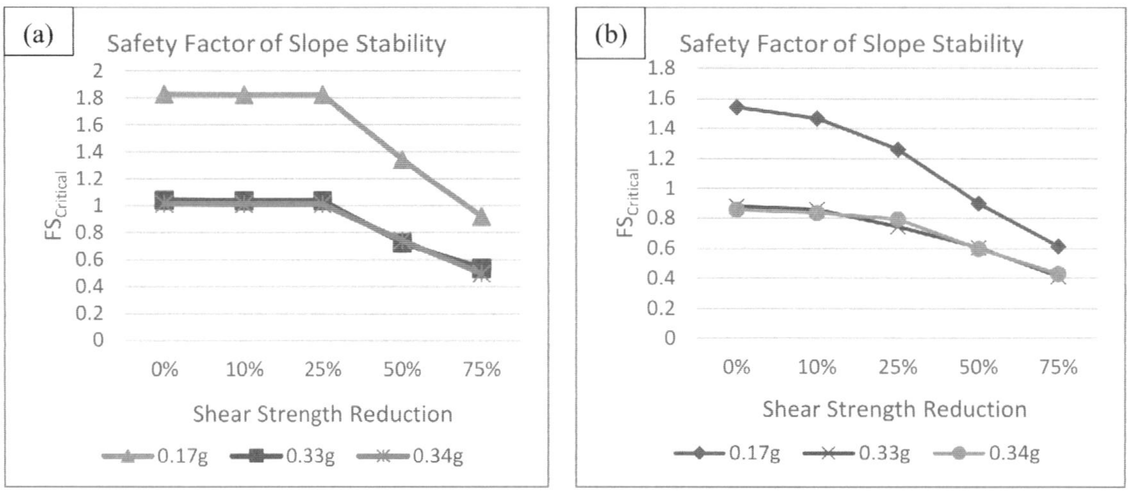

Figure 7. The safety factor values of slope modelling post the Palu-Donggala earthquake; (a) Slope stability analysis with the irrigation canal off operating groundwater level condition; and (b) Slope stability analysis with the irrigation canal on operating groundwater level condition.

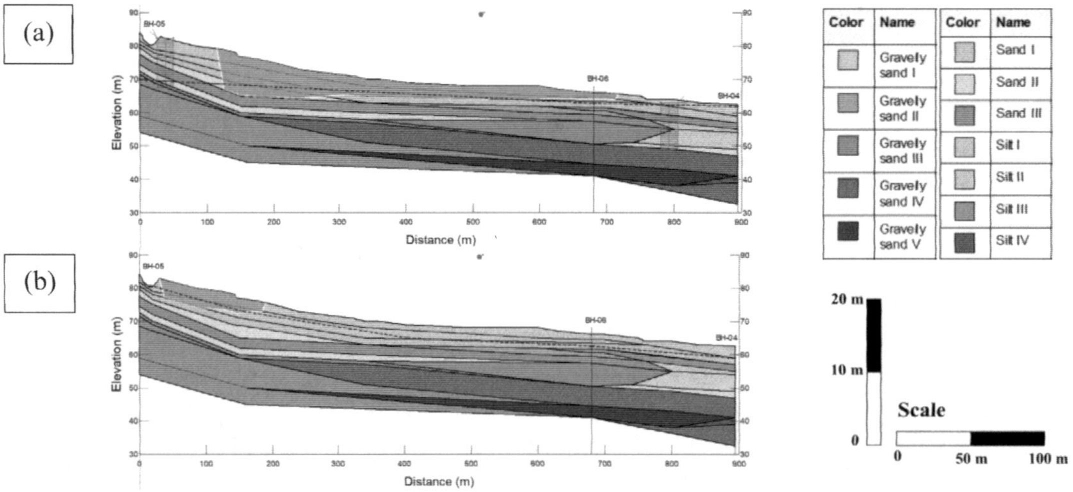

Figure 8. The slip surfaces in varying PGA values, groundwater level conditions, and shear strength reduction; (a) Slip surface with irrigation canal off operating condition, 75% shear strength reduced, and PGA 0.17g; (b) Slip surface with shallow groundwater table, no shear strength reduction, and PGA 0.34g.

4. Conclusion

The analyses included the slope failure in Jono Oge pre and post the 2018 Palu-Donggala earthquake incorporating soil profile data both in liquefied and non-liquefied area. The PGA value of 0.33g scenario at pre-earthquake condition analysed the slope failure that yield safety factor (*FS*) < 1.0. The slip surface began to slide at the west side of irrigation canal which was similar to the head scarp of the flow-slide occurred in 2018. The slope began to fail when the liquefied soil layer shear strength was 10% reduced.

The analysis PGA 0.33g and 0.34g and based on post-earthquake soil profile with the irrigation canal off operating condition, the analysis resulted in the slope failing at the 50% and 75% shear strength reduction. The length of slip surface was more than 600 m, and the head scarp was about 100 m from the irrigation canal. The PGA 0.17g earthquake scenario analysis resulted in the slope failing at the 75% shear strength reduction.

The analysis using PGA 0.33g and 0.34g with the irrigation canal on operating condition resulted in the slope failing when the seismic load was applied. The shear strength reduction at the liquefied soil layer decreased the safety factor values. The length of long-distance slip surface was more than 640 m, and the head scarp was at the west side of irrigation canal. The analysis of PGA 0.17g earthquake scenario resulted in the slope failing at the 50% shear strength reduction. The head scarp was after the side of irrigation canal which could cause damage to the irrigation canal construction.

The analysis of this study described a similar slope failure that occurred due to the 2018 Palu-Donggala earthquake. With the same method, slope stability analysis generated various scenarios. The analysis resulted that the safety factor of slope stability decreasing because of the liquefied soil layer assumption. This slope failure possibly turned into a long-distance ground movement due to the earthquake and shallow groundwater level conditions. The lower the safety factor of slope stability, the higher the ground movement potential in Jono Oge flow-slide affected area. Moreover, this ground movement potential could trigger the occurrence of flow-slide in Jono Oge area. However, this evaluation in Jono Oge flow-slide affected area required more advanced study and laboratory tests to determine the mechanism of the flow-slide and the numeric analysis for the flow-slide probability.

Acknowledgments

This work was funded by the Japan International Cooperation Agency (JICA). The authors would like to express their gratitude for the support given by Central Sulawesi National Road Implementation Agency, the Indonesian Ministry of Public Works and Housing.

References

[1] Valkaniotis S, Ganas A, Tsironi V and Barberopoulou A 2018 *A preliminary report on the M7.5 Palu earthquake co-seismic ruptures and landslides using image correlation techniques on optical satellite data*

[2] Mason H B, Gallant A P, Hutabarat D, Montgomery J, Reed A N, Wartman J, Irsyam M, Prakoso W, Djarwadi D, Harnanto D, Alatas I, Rahardjo P, Simatupang P, Kawanda A and Hanifa R 2019 *Geotechnical reconnaissance: The 28 September 2018 M 7.5 Palu-Donggala, Indonesia Earthquake*

[3] Miyajima M, Setiawan H, Yoshida M, Ono Y, Kosa K and Oktaviana I S 2019 Geotechnical damage in the 2018 Sulawesi earthquake, Indonesia *Geoenvironmental Disasters* **6** 1–8

[4] Faris F, Fathani T F and Wang F 2019 Report on the UNESCO Chair workshop on geoenvironmental disaster reduction 28th April - 1st May, 2019, Palu - Jakarta, Indonesia *Geoenvironmental Disasters* **6**

[5] Mason H B, Montgomery J, Gallant A P, Hutabarat D, Reed A N, Wartman J, Irsyam M, Simatupang P T, Alatas I M, Prakoso W A, Djarwadi D, Hanifa R, Rahardjo P, Harnanto D S, Kawanda A, Himawan A and Yasin W 2021 East Palu Valley flowslides induced by the 2018 Mw 7.5 Palu-Donggala earthquake **373**

[6] Hazarika H, Rohit D, Pasha S M K, Maeda T, Masyhur I, Arsyad A and Nurdin S 2021 Large distance flow-slide at Jono-Oge due to the 2018 Sulawesi Earthquake *Soils Found.* **61** 239–55

[7] Zhao B 2021 Landslides triggered by the 2018 Mw 7.5 Palu supershear earthquake in Indonesia *Eng. Geol.* **294** 106406

[8] Andiny A N, Faris F and Adi A D 2021 Re-liquefaction hazard evaluation in flow-slide affected area of Jono Oge, Central Sulawesi, Indonesia *IOP Conference Series: Earth and Environmental Science* vol 861 p 052030

[9] Day R W 2012 *Geotechnical Earthquake Engineering Handbook: With the 2012 International Building Code* (McGraw-Hill)

[10] Thein P S, Pramumijoyo S, Brotopuspito K S, Kiyono J, Wilopo W, Furukawa A, Setianto A and Putra R R 2015 Estimation of S-wave velocity structure for sedimentary layered media using Microtremor array measurements in Palu City , Indonesia *Procedia Environ. Sci.* **28** 595–605

[11] Kiyota T, Furuichi H, Hidayat R F, Tada N and Nawir H 2020 Overview of long-distance flow-slide caused by the 2018 Sulawesi earthquake, Indonesia *Soils Found.* **60** 722–35

[12] Idriss I M and Boulanger R W 2008 *Soil Liquefaction During Earthquakes* (Oakland: Earthquake Engineering Research Institute (EERI))

[13] Juang C H, Fang Y F and Li D K 2008 Reliability Analysis of Liquefaction Potential of Soils Using Standard Penetration Test *Reliability-Based Design in Geotechnical Engineering* ed K K Phoon (Taylor & Francis) pp 497–526

[14] Cheng Y M, Lansivaara T and Wei W B 2007 Two-dimensional slope stability analysis by limit equilibrium and strength reduction methods *Comput. Geotech.* **34** 137–50

[15] Salmasi F and Jafari F 2019 A Simple Direct Method for Prediction of Safety Factor of Homogeneous Finite Slopes *Geotech. Geol. Eng.* **37** 3949–59

[16] Ma J Z, Zhang J, Huang H W, Zhang L L and Huang J S 2017 Identification of representative slip surfaces for reliability analysis of soil slopes based on shear strength reduction *Comput. Geotech.* **85** 199–206

[17] Standards Australia 2002 *Australian Standard AS 4678-2002 Earth-retaining structures*

[18] Mujtaba H, Farooq K, Sivakugan N and Das B M 2018 Evaluation of relative density and friction angle based on SPT-N values *KSCE J. Civ. Eng.* **22** 572–81

[19] ASTM International 2011 *D1586 - 11 Standard Test Method for Standard Penetration Test (SPT) and Split-Barrel Sampling of Soils*

[20] Seed H B, Tokimatsu K, Harder L F and Chung R M 1985 The influence of SPT procedures in soil liquefaction resistance evaluations: Berkeley, University of California **I** 15

[21] Kanno T, Narita A, Morikawa N, Fujiwara H and Fukushima Y 2006 A new attenuation relation for strong ground motion in Japan based on recorded data *Bull. Seismol. Soc. Am.* **96** 879–97

[22] Anbazhagan P, Kumar A and Sitharam T G 2013 Ground motion prediction equation considering combined dataset of recorded and simulated ground motions *Soil Dyn. Earthq. Eng.* **53** 92–108

[23] Masyhur I, Widiyantoro S, Natawidjaja D H, Meilano I, Rudyanto A, Hidayati S, Triyoso W, Hanifa N R, Djarwadi D, Faizal L and Sunarjito 2017 *Peta Sumber dan Bahaya Gempa Indonesia Tahun 2017* (Badan Penelitian dan Pengembangan, Kementerian Pekerjaan Umum dan Perumahan Rakyat)

[24] Ariesta D 2019 The Effect of Initial Groundwater Table and Rainfall Wetting Towards Slope Stability (Case Study of Landslide in Tangkil Hamlet, Banaran Village, Pulung Subdistrict, Ponorogo Regency) *J. Civ. Eng. Forum* **5** 149

[25] Sassa K, Fukuoka H, Wang F and Wang G 2007 Landslide Induced by a Combined Effect of Earthquake and Rainfall *Progress in Landslide Science* pp 193–207

[26] Watkinson I M and Hall R 2019 Impact of communal irrigation on the 2018 Palu earthquake-triggered landslides *Nat. Geosci.* **12** 940–5

[27] Pratama A, Fathani T F and Satyarno I 2021 Liquefaction potential analysis on Gumbasa Irrigation Area in Central Sulawesi Province after 2018 earthquake *Earth Environ. Sci.* **930**

Mitigation plan against liquefaction potential on Gumbasa Irrigation Canal around Jono Oge area after 2018 Central Sulawesi Earthquake

A Pratama[1,2], T F Fathani[1,3,*], I Satyarno[1,3]

[1] Department of Civil and Environmental Engineering, Faculty of Engineering, Universitas Gadjah Mada, Yogyakarta 55281, Indonesia
[2] Directorate Irrigation and Lowland, Directorate General Water Resources, Ministry of Public Works and Housing, Republic of Indonesia
[3] Center for Disaster Mitigation and Technological Innovation (GAMA-InaTEK), Universitas Gadjah Mada, Yogyakarta 55281, Indonesia

*Corresponding author: tfathani@ugm.ac.id

Abstract. Gumbasa Irrigation Area was one of the public infrastructures affected by the liquefaction and flow-slide. Restoring it to operate as before required a reconstruction plan with consideration on potential damage caused by the earthquake and liquefaction. The objective of this study was to assess the performance of the mitigation plan applied on Gumbasa Irrigation Area (canal and rice field). Two mitigation plans to be applied in the Gumbasa Irrigation Canal are evaluated, i.e., (1) implementing impermeable lining and geogrid; (2) applying impermeable lining and steel-sheet pile with drain. Each mitigation plan was compared using the finite element method to examine the stability and displacement mechanism. Based on the result, the first mitigation plan was identified as more effective to reduce displacement caused by earthquake and liquefaction potential. The lowering groundwater method was applied on Jono Oge rice fields area by maintaining the recommended groundwater level obtained from the simulation of groundwater level using the simplified method. We concluded that mitigation plans on irrigation canal by applying impermeable lining and geogrid and by lowering groundwater on the rice fields area were the best mitigation plan for Gumbasa Irrigation Area.

1. Introduction

The Gumbasa Irrigation Area in Central Sulawesi Province suffered the worst damage due to the 7.5 magnitude earthquake on September 28, 2018. The earthquake triggered a tsunami and liquefaction followed by flow-slide [1], [2]. The liquefaction followed by flow-slide affected four locations on Gumbasa Irrigation Area: Petobo, Lolu, Jono Oge, and Sibalaya. Approximately 95 km of the Gumbasa Irrigation Canal was damaged, causing irrigation failure to 8,180 hectares of irrigation area.

There have been several studies related to the effect of Gumbasa Irrigation on the liquefaction phenomenon followed by flow-slide. Lateral spreading and landslide occurred only in areas exceeding the critical density of flood-irrigated land [3]. Shallow groundwater level was due to water seepage in the Gumbasa Irrigation Canals that triggered liquefaction events followed by flow-slide during the 7.5 M_w earthquake [4]. Regarding these studies, a study on the mitigation plan to be applied in the reconstruction of Gumbasa Irrigation is required to reduce the liquefaction potential.

Restoring the Gumbasa Irrigation system to be back in operation is urgent. However, the reconstruction plan must consider the potentials of reoccurring liquefaction events. The mitigation plans should reduce the vulnerability and damage if earthquake and liquefaction events reoccur in the future. The objective of this study was to compare the most optimal mitigation plan to be applied, especially for the reconstruction plan of the Gumbasa Irrigation Area.

2. Methods

This study was carried out in Gumbasa Irrigation Area, especially in the Jono Oge liquefaction area, as shown in Figure 1. In the Jono Oge liquefaction area, about 181 ha of irrigable farmlands were lost due to liquefaction. The data used in this study is referred to the geotechnical investigation conducted by JICA in 2019 after the 7.5 M_w earthquake [5], [6]. The topography and groundwater monitoring data were referred to the survey results conducted by the Ministry of Public Works and Housing in 2021 [7].

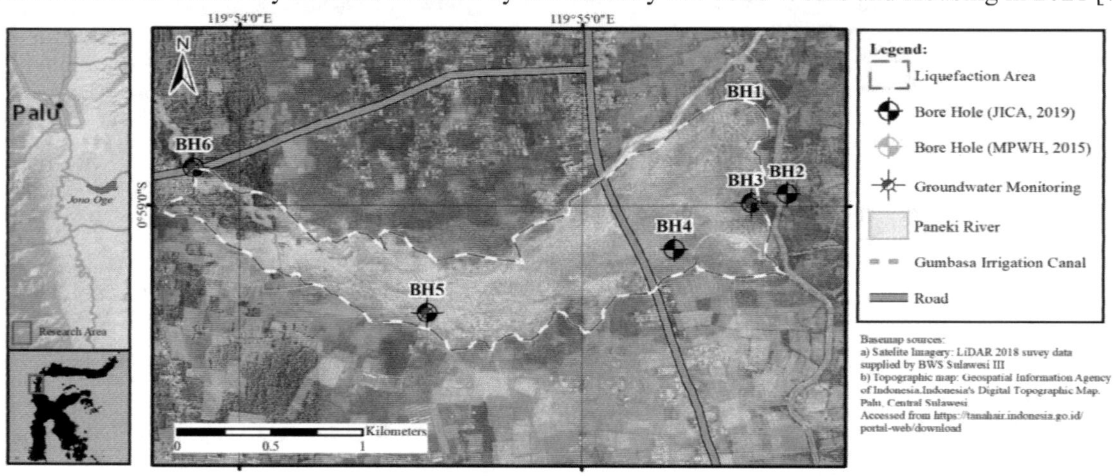

Figure 1. Jono Oge liquefaction area on base map sources (1) 2018 LiDAR survey data supplied by BWSS III [7] and (2) Indonesia's digital topographic map from Geospatial Information Agency of Indonesia [8])

The boundary area to be analyzed is the damaged Gumbasa Irrigation Canal along 1300 m. In this area, there are six boreholes and three groundwater monitoring. There is a breached irrigation canal in the Jono Oge area at the southeastern corner of the flow-slide [9]. The cross-section of breached Gumbasa Irrigation Canal is presented in Figure 2. This cross section was used for two-dimensional finite element method.

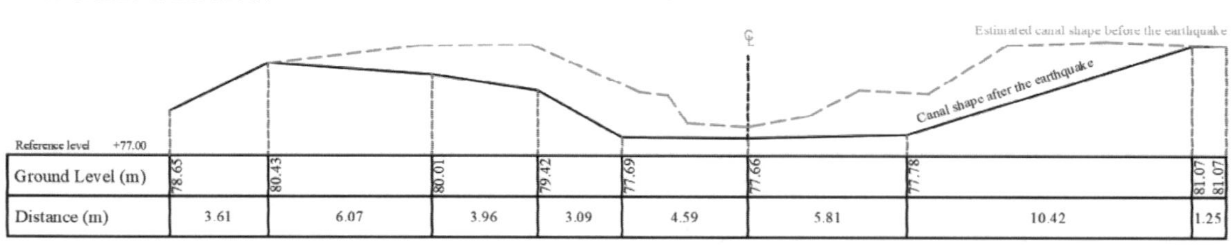

Figure 2. Cross-section of damage point 1 modified from MPWH, 2021 [7].

Figure 2 showed a difference in ground level between the conditions before and after the earthquake. The existing cross-section (after the earthquake) was referred to MPWH, 2021 [7], and the estimated elevation (before the earthquake) referred to MPWH, 2015 [10]. The earth lining canal dominated along the Gumbasa Irrigation Canal in the Jono Oge area, so that there was a high seepage. The seepage process triggers the soil condition around the canal bed, and the canal bank becomes saturated. It was confirmed by the recorded groundwater elevation of 3.5 m on BH1, stated on MPWH, 2015 [10].

The followings are the steps to reduce the effects of liquefaction: (1) assessing the liquefaction potential; (2) evaluating the effects of liquefaction; (3) determining the remedial against liquefaction; and (4) evaluating the remedial effectiveness [11]. The assessment of the liquefaction potential referred to previous study in 2021 [12]. The evaluation of the liquefaction effect was obtained from a technical assessment of the Jono Oge area to inventory the damage.

The liquefaction mitigation plan was selected from several studies. The Japanese Geotechnical Society classified the countermeasures to mitigate damaging liquefaction effects into two categories: (1) remediating the liquefiable soil through ground improvement; and (2) relieving the effect of liquefaction by structures strengthening [11]. The countermeasures to mitigate liquefaction in this study applied by strengthening structures.

The effectiveness of the mitigation plans on the Gumbasa Irrigation Canal was compared by using the two-dimensional finite element method of Plaxis 2D v8.6. Each mitigation plan is modeled by a plane strain model and uses a pseudo-static method to determine the earthquake load behavior. The Plaxis 2D v8.6 results were the numerical simulation to examine the displacement on each mitigation plan. The mitigation plans on Gumbasa Canal were performed under the following scenario, i.e. (1) without implementing impermeable lining; (2) implementing impermeable lining and geogrid; (3) applying impermeable lining and steel-sheet pile with drain.

The effectiveness of the mitigation plan on the rice field area was evaluated by calculating the liquefaction potential analysis using the Idris Boulanger 2008 method [13]. This method compares the Cyclic Stress Ratio (*CSR*) and Cyclic Resistance Ratio (*CRR*).

$$CSR_{M=7.5,\sigma'_{vc}=1} = 0.65\frac{a_{\max}}{g}\frac{\sigma_{vc}}{\sigma'_{vc}}r_d\frac{1}{MSF}\frac{1}{K_\sigma} \tag{1}$$

where $a_{\max}$ is site modified peak ground acceleration (m/s^2), g is the gravitational acceleration (m/s^2), σ_{vc} is total vertical stress (kN/m^2), σ'_{vc} is effective vertical stress (kN/m^2), r_d is coefficient of shear stress reduction, MSF is the factor of magnitude scaling, and K_σ is the factor of overburden correction. The *CRR* equation using N-SPT is expressed in equation 2

$$CRR_{M=7.5,\sigma'_{vc}=1} = \exp\left(\frac{(N_1)_{60cs}}{14.1} + \left(\frac{(N_1)_{60cs}}{126}\right)^2 - \left(\frac{(N_1)_{60cs}}{23.6}\right)^3 + \left(\frac{(N_1)_{60cs}}{25.4}\right)^4 - 2.8\right) \tag{2}$$

where $(N_1)_{60cs}$ is the corrected penetration resistance. The Safety Factor against liquefaction (*FS*) is computed by using the following equation:

$$FS = \frac{CRR}{CSR} \tag{3}$$

The results of the calculation *FS* indicate that safety factor against liquefaction where ($FS < 1$ indicates liquefaction potentially to occur, and $FS \geq 1$ indicates no potential for liquefaction to occur. Using the Sonmez and Gokceoglu method [14], the Safety Factor against liquefaction (*FS*) was converted to Liquefaction Severity Index (*LSI*).

$$L_S = \int_0^{20} P_L(z).W(z).dz \tag{4}$$

where $P_L(z)$ is liquefaction probability value related to depth function.

3. Results and Discussion

This study used the assessment of liquefaction potential from the previous study. The role of groundwater level in Gumbasa Irrigation Area was related to liquefaction potential [12]. One of the scenarios was by simulating the on-operating condition given with 7.9 M_w earthquake. As presented in Figure 3, the results indicated potential reoccurring liquefaction on Jono Oge area. The potential ranged from very low to moderate on the Gumbasa Irrigation Canal and rice field. The technical assessment results used to evaluate the effects of liquefaction, such as damages in Gumbasa Irrigation Area in the Jono Oge segment, are presented in Figure 4.

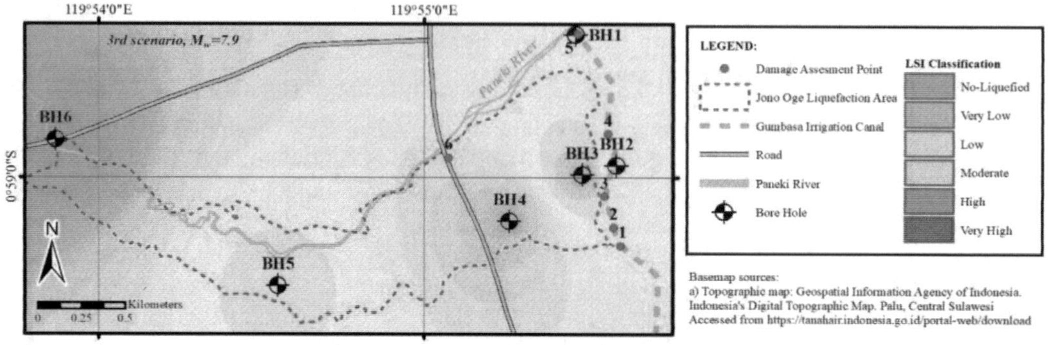

Figure 3. Areas with liquefaction susceptibility of 3rd scenario (M_w=7.9) based on Liquefaction Severity Index (*LSI*) modified from [12].

(a) (b) (c)

Figure 4. Field survey documentations: (a) aerial documentation [15], (b) and (c) technical assessment.

The damages occurred along the canals, buildings, and irrigation areas. The worst damaged canal is presented in Figure 4a. At this point, the left bank canal was breached, causing water in the canal to flow into the liquefaction and flow-slide area. The damage to the irrigation canal embankment was caused by flow-slide movement in the liquefaction and flow-slide area [16]. The irrigation structure in that segment experienced structural failure, as shown in Figure 4b. The canal damage was dominated by cracks and deformation of the canal bed and the lateral spreading of the canal bank. Figure 4c presents the formerly rice fields area damage that indicated irregular land contours. The formerly rice field area beside the Paneki River was filled with sediment carried by the river during the flood. Within 1300 m of damaged irrigation canals in the Jono Oge area, attention should be paid to the segments point 1 to point 3 (300 m).

Based on liquefaction susceptibility (Figure 3) and verified by technical assessment (Figure 4), the reconstruction plan of the Gumbasa Irrigation Area must be accompanied by a mitigation plan against liquefaction. It should be applied on irrigation canals and areas affected by liquefaction followed by flow-slide.

3.1. Mitigation plan on the irrigation canal

The mitigation plans for irrigation canals should consider the damage that occurred. The earth lining irrigation canal in the Jono Oge area experienced a large deformation and lateral spreading when the earthquake hit on September 28, 2018. Other studies concluded that water flow on the Gumbasa Irrigation Canal also accelerated the emergence of mudflows through the ground fractures [17]. In addition to understanding the geotechnical conditions of the soil around the irrigation canal is needed. Interpretation is carried out based on the results of drill tests on longitudinal sections BH1 to BH2 and cross-sections BH1 to BH6. The interpretation was conducted by connecting the soil layers resulting from the drill log at each drill point, as shown in Figure 5.

IOP Conf. Series: Earth and Environmental Science 1091 (2022) 012027 doi:10.1088/1755-1315/1091/1/012027

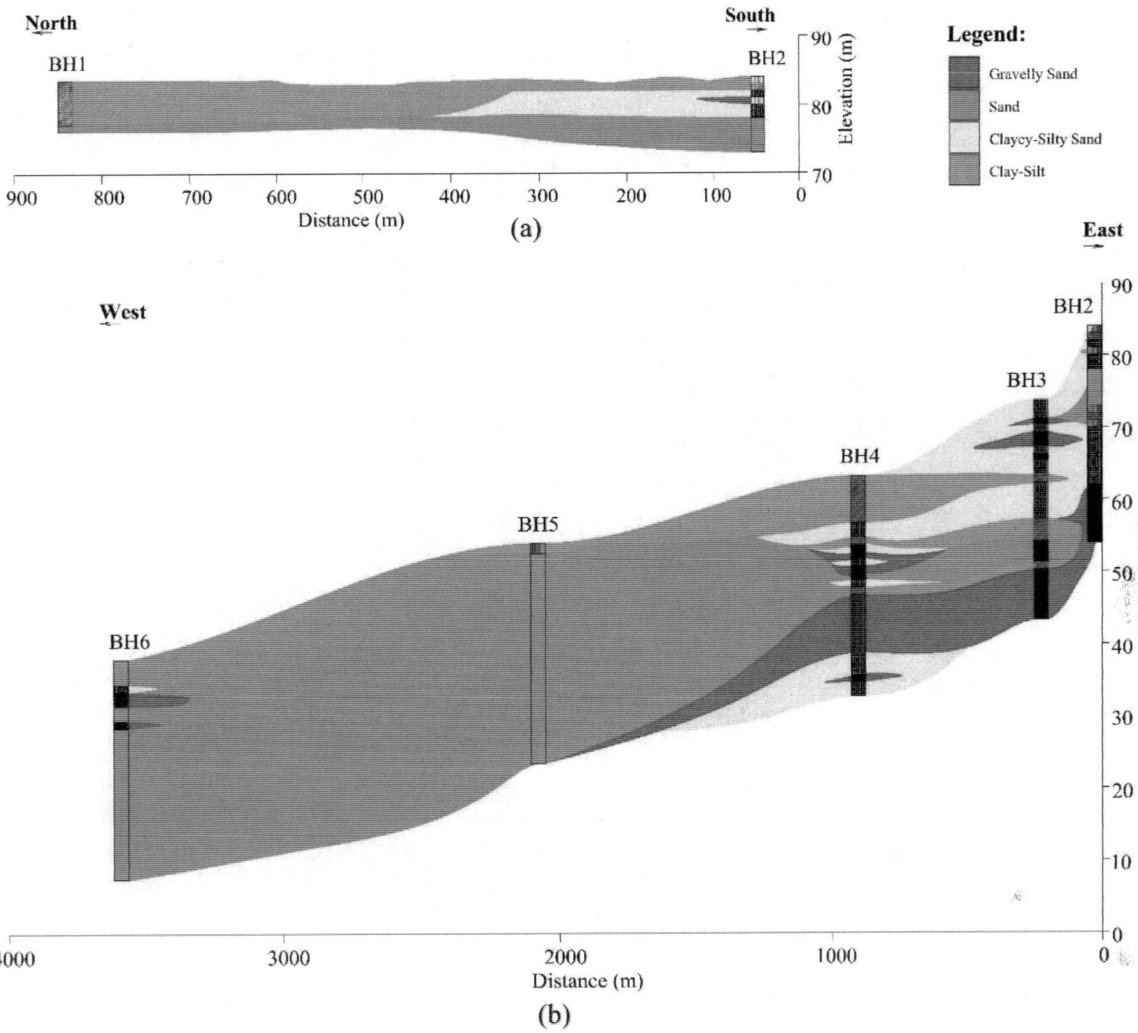

Figure 5. Geotechnical condition after the earthquake, (a) long section profile of BH1 to BH2; and (b) cross section profile of BH2 to BH6.

The long section of the irrigation canal BH1 to BH2 (Figure 5a) was dominated by a layer of clayey-silty sand. Figure 5b shows the cross-section of BH2 to BH6 dominated by sand and clay-silt sand. Based on long section and cross section, the irrigation canal in the Jono Oge segment was dominated by a layer of clayey-silty sand which had liquefaction potential [18]. The operational goal of an irrigation area depends on the certain water table elevation to enable water enter the fields as needed. Thus, the new Gumbasa Irrigation Canal must be able to minimize the deformation and lateral spreading either in normal operation or during an earthquake.

The first proposed mitigation plan was the use of impermeable lining to intercept water seepage in Gumbasa Irrigation Canal. Replacing the flood-irrigation and earth irrigation canals with more efficient irrigation practices that can significantly reduce the probability of future landslide events [19]. The impermeable lining canal is made by combining geomembrane layers and then covering them with precast concrete. Applying the geogrid to the irrigation canals embankment aims to increase the strength to withstand earthquake loads.

The second proposed mitigation plan was the use of steel-sheet pile for the embankment. One of the studies related to mitigation methods in irrigation canals is using sheet-pile with drain [20]. This method effectively reduces deformation of the flume structure (i.e., uplift/sinking and inclination). The steel-sheet pile with drain method will be applied on the side of impermeable lining for new fill embankment. The illustration of the first damage point applied with the mitigation plan is presented in Figure 6.

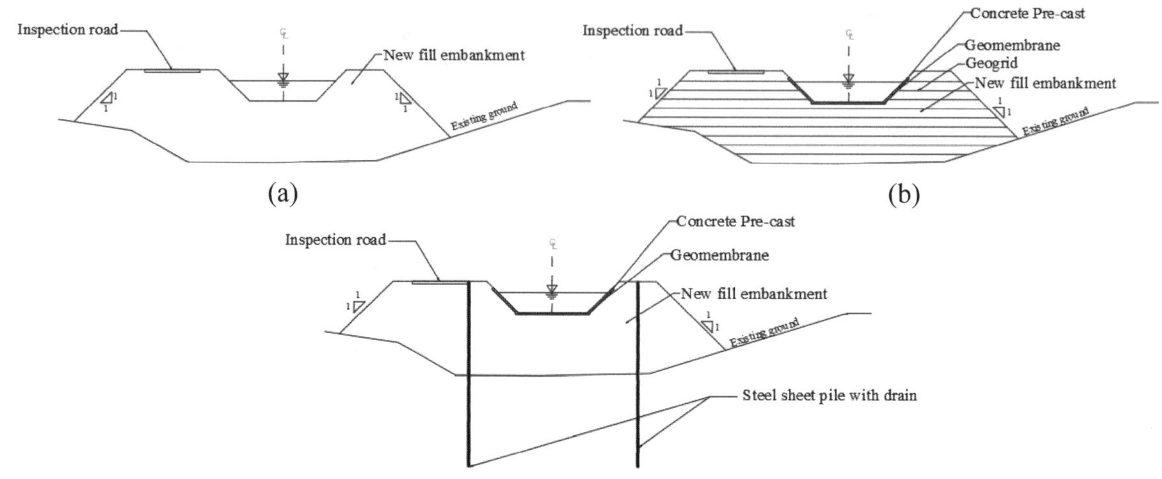

Figure 6. Mitigation plan for 1st damage point, (a) 1st scenario without implementing impermeable lining; (b) 2nd scenario implementing impermeable lining and geogrid; (c) 3rd scenario with applying impermeable lining and steel-sheet pile with drain.

The new irrigation canal elevation in Figure 6 is referred to the results of MPWH hydraulics calculations in 2021. The canal shape must be in a trapezoid with an upper width of 7.1 meters, and a canal bed width is 3.70 meters, while the lowest elevation of the canal must be at +81.05. A new embankment is required to obtain the elevation of the canal bed. It required up to 3.38 meters embankment at the first damage point cross-section. Each layer of the fill embankment consists of 25 cm soil compacted to reach 95 percent of standard proctor maximum density. The geogrid layers are installed on the embankment layer every 50 cm depth. The embankment analyzed in the finite element method has a slope ratio of 1:1 (vertical to horizontal). The load to be input is 10 kN/m^2, located on the inspection road (the traffic load) and canal bed (the irrigation water load).

Moreover, to determine the effectiveness of the liquefaction mitigation plan, the finite element method was carried out with the help of the Plaxis 2D v8.6. Figure 6a showed the 1st scenario without implementing impermeable lining as a comparison with the other scenario. The Plaxis 2D v8.6 output aims to determine the displacement. Soil data obtained from the N-SPT correlation value of several references was used as input in the material model [21], [22]. The material model used Mohr-Coulomb with drained material type. The input parameter for concrete precast, steel-sheet pile, and geogrid is presented in Table 1. Therefore, the input parameters of existing soil and embankment soil are presented in Table 2. For example, the 2nd mitigation scenario model is seen in Figure 7.

Table 1. Structural parameter input in modeling

Name	EA (kN/m)	EI (kNm^2/m)	w (kN/m/m)	d (m)	ν
Concrete Precast	6.91E+03	5.18E+01	7.2	0.300	0.150
Steel-sheet pile	4.73E+06	1.02E+05	1.860	0.509	0.200
Geogrid	1.00E+05				

Table 2. Soil parameter input in modeling

Layer	Material model	Type of behavior	γ_{unsat} kN/m^3	γ_{sat} kN/m^3	E_{ref} kN/m^2	ν -	c_{ref} kN/m^2	ϕ °	ψ °	*N-SPT*
Silt 1	MC	Drained	17	20	15000	0.350	20.000	25.000	0	12
Clay 1	MC	Drained	19	19	50000	0.400	13.000	28.000	0	19
Silty Sand 1	MC	Drained	17	20	10000	0.250	20.000	30.000	0	4
Silt 2	MC	Drained	17	17	13000	0.325	20.000	25.000	0	10
Silty Sand 2	MC	Drained	17	20	20000	0.300	20.000	34.000	0	18
Sand 1	MC	Drained	19	21.5	80000	0.400	1.000	37.000	0	>50
Silt 3	MC	Drained	20	20	20000	0.350	20.000	25.000	0	>50
Clay 2	MC	Drained	20	20	50000	0.400	13.000	28.000	0	>50
Silt 4	MC	Drained	20	20	20000	0.350	20.000	25.000	0	>50
Silty Sand 3	MC	Drained	19	21.5	20000	0.400	20.000	34.000	0	>50
Gravelly Sand 1	MC	Drained	21	23	100000	0.450	1.000	38.000	0	>50
Fill	MC	Drained	19	21.5	20000	0.400	50.000	37.000	0	-
Silt 1	MC	Drained	17	20	15000	0.350	20.000	25.000	0	12

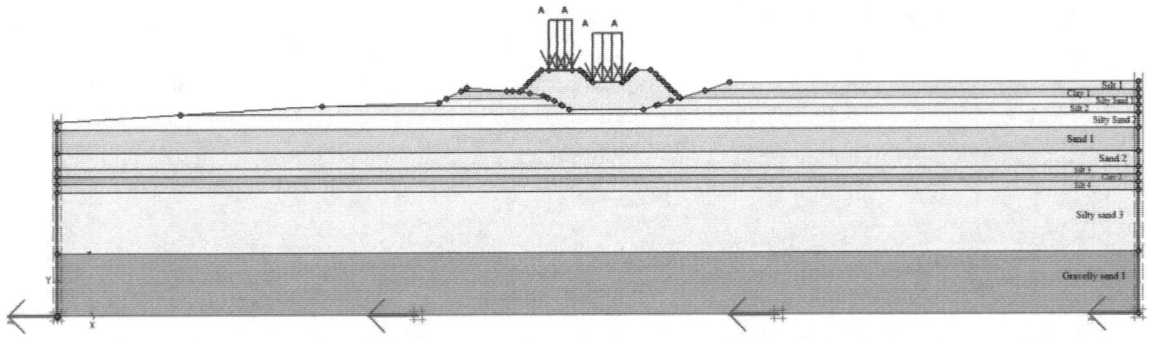

Figure 7. Soil, plate, and load modeling input for 2nd scenario.

The pseudo-static method was used to determine the behavior during the earthquake. Each mitigation plan is simulated by using PGA_M. It was inputted to Plaxis 2D v8.6 by inserting x-acceleration and y-acceleration. The PGA_M value was 0.418 g for the maximum considered earthquake. This PGA_M value was obtained from *Desain Spektra Indonesia*'s webpage, 2021 [23] and multiplied by site coefficient [24]. The Plaxis 2D v8.6 simulation results of each scenario for $PGA_M = 0.418$ g are presented in Figure 8.

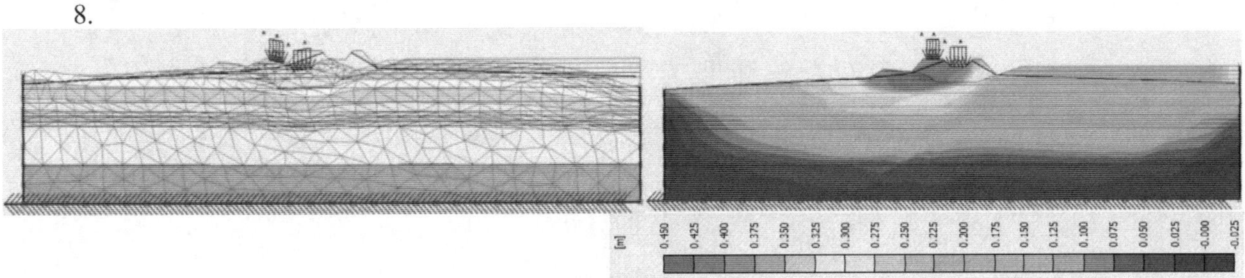

(a)

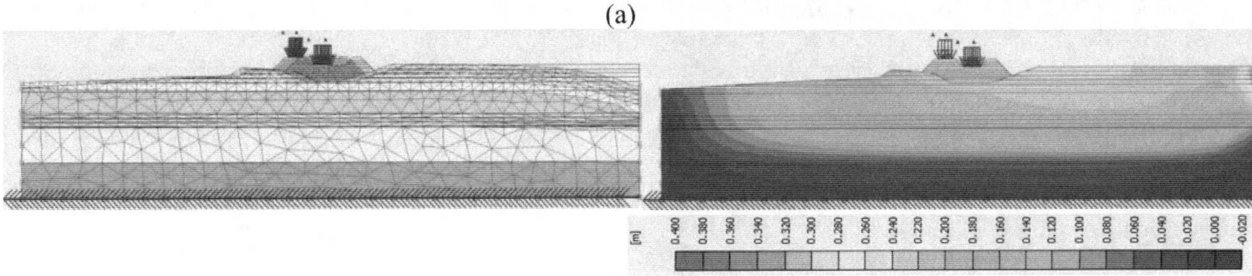

(b)

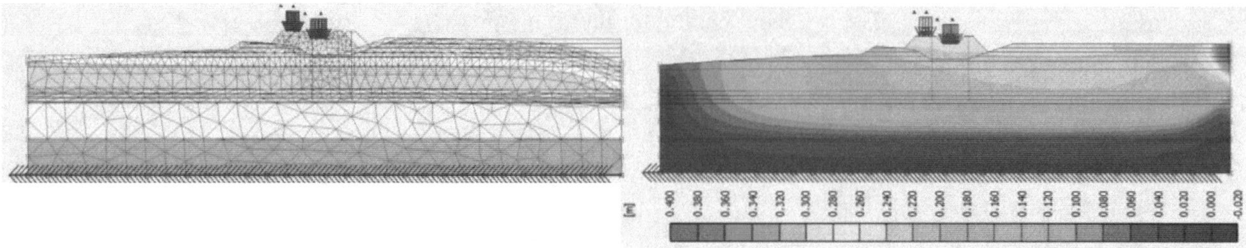

(c)

Figure 8. Result of Finite element modeling for $PGA_M = 0.418$ g, (a) 1st scenario without implementing impermeable lining; (b) 2nd scenario implementing impermeable lining and geogrid; (c) 3rd scenario with applying impermeable lining and steel-sheet pile with drain.

Figure 8a shows that based on finite element modeling for $PGA_M = 0.418$ g, the new irrigation canal without impermeable lining resulted in soil body collapse. In this figure, the finite element modeling stops at $PGA_M = 0.312$ g. Figure 8b the new irrigation canal with impermeable lining resulted in 0.220-0.260 m displacement. This result is similar to Figure 8c (the new irrigation canal applied with impermeable lining and steel-sheet pile with drain).

The displacement value that occurs above the left embankment of the channel used as reference. Then the displacement value is plotted in Figure 9 to determine its relationship with the PGA_M value.

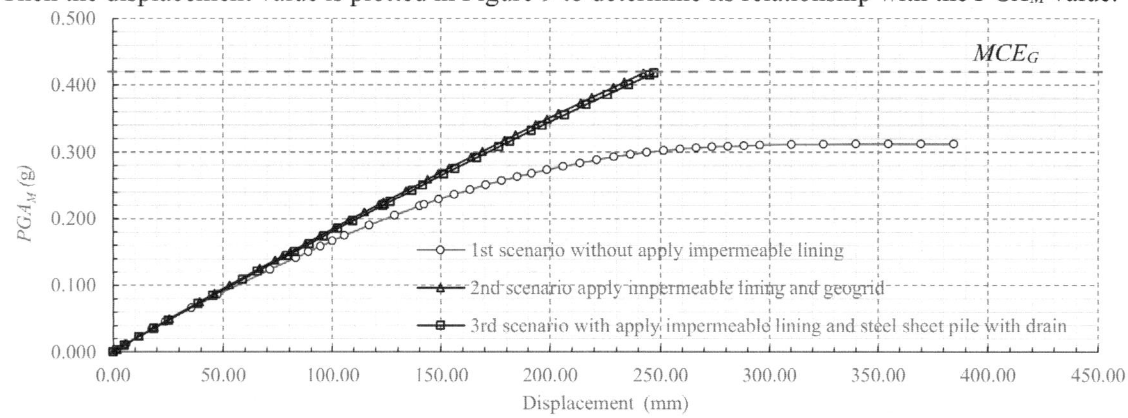

Figure 9. Displacement graph on the bottom of new canal embankment (point B).

Figure 9 showed that the most significant displacement for PGA_M=0.418 g is the third scenario, about 247 mm. The smallest displacement was the second scenario, about 242 mm. It proved that maintaining the embankment and the bed irrigation canal in unsaturated conditions could increase the shear strength to resist the earthquake.

3.2. Mitigation plan on areas affected by liquefaction followed by flow-slide

As an area affected by liquefaction followed by a flow-slide, the most effective mitigation approach for Jono Oge was by lowering the groundwater level because such mitigation plan is able to accommodate wide area. Lowering the groundwater level in a wide existing area to be appropriate to prevent damage due to liquefaction [25]. The optimum mitigation plan (lowering the groundwater level) is obtained by determining the groundwater elevations threshold against liquefaction potential. The groundwater elevation threshold was determined by calculating the liquefaction potential analysis using the Idris Boulanger 2008 method. The resulted safety factor was converted to *LSI* values using the Sonmez and Gokceoglu method. The groundwater table was simulated at various levels to obtain the low to non-liquefied *LSI* values. The liquefaction potential analysis was calculated using post-earthquake geotechnical data, and PGA_M value 0.418 g for maximum considered earthquake $M_w = 7.9$. The results of groundwater level recommendations are presented in Table 3.

Table 3. Groundwater level recommendation

Borehole	Shallowest groundwater level (m)	Recommended groundwater level (m)	Average groundwater level (m)	*LSI* values	
				before mitigation	after mitigation plan
BH3	-5.96	-7.50	-7.62	moderate	low
BH4	-0.81	-2.50	-6.41	very low	low
BH5	-0.70	-3.50	-2.33	moderate	low
BH6	-1.23	-1.50	-2.79	non-liquefied	Non-liquefied

In the Table 3, the liquefaction severity index can be reduced from moderate to non-liquefied become low to non-liquefied level by maintaining the groundwater level as recommended. If the groundwater level threshold is compared with the average groundwater level recording data, the BH5 needs to lower the groundwater level about 1.5 m.

4. Conclusions

The results of soil profile interpretation showed that non-plastic (cohesionless) soils are dominated in Gumbasa Irrigation Area on Jono Oge. This soil type criteria was susceptible to liquefaction potential, so the mitigation plan must accompany the reconstruction plan.

Various simulations used Plaxis 2D v8.6 to analyze the performance of the mitigation plan applied in the Gumbasa Irrigation Canal. Under the first scenario, the displacement for PGA_M= 0.418 g resulted in soil body collapse. It caused the soil embankment that was saturated and affected by the irrigation water seepage.

Under the second scenario, the displacement for PGA_M= 0.418 g resulted in 242 mm displacement. It proved that maintaining the embankment and the bed irrigation canal in unsaturated conditions could increase the shear strength to resist the earthquake.

Under the third scenario, the displacement for PGA_M= 0.418 g resulted in 247 mm displacement. There is a small difference in displacement between the second and the third scenarios. The advantage of the second scenario is a simple and low-cost method, but the disadvantage is the inability to drain the pore water pressure increases during the earthquake. The advantage of the third scenario is the drain ability installed on the steel-sheet pile. It was helpful to reduce the excess pore water pressure when the pressure increases during the earthquake. The reduction of excess pore water pressure can reduce the probability of liquefaction. The second scenario (applying impermeable lining geogrid) was the best alternative to mitigate the liquefaction on Gumbasa Irrigation Canal by considering the simple and low-cost method. This mitigation plan applying impermeable lining geogrid to withstand the peak ground acceleration during the maximum considered earthquake. In the future, each of the above mitigation plans needs to be tested using dynamic methods to compare with the pseudo-static simulation results.

The mitigation plan on liquefaction and flow-slide area, the lowering groundwater method can reduce the liquefaction potential. This mitigation plan is applied by keeping the groundwater elevations threshold. The liquefaction severity index can be reduced to a low-non liquefiable level by maintaining the groundwater level as recommended. However, it is necessary to study the groundwater modeling further to test the number of production wells to reach and maintain the recommended groundwater level.

5. Acknowledgement

The authors would like to express their gratitude for the support given by the Directorate Irrigation and Lowland and *Balai Wilayah Sungai Sulawesi III*, Directorate General of Water Resources, Ministry of Public Works, and Public Housing, Republic of Indonesia.

6. References

[1] Bao H, Ampuero J P, Meng L, Fielding E J, Liang C, Milliner C W D, Feng T and Huang H 2019 Early and persistent supershear rupture of the 2018 magnitude 7.5 Palu earthquake *Nat. Geosci.* **12** 200–5

[2] Socquet A, Hollingsworth J, Pathier E and Bouchon M 2019 Evidence of supershear during the 2018 magnitude 7.5 Palu earthquake from space geodesy *Nat. Geosci.* **12** 192–9
[3] Bradley K, Mallick R, Andikagumi H, Hubbard J, Meilianda E, Switzer A, Du N, Brocard G, Alfian D, Benazir B, Feng G, Yun S H, Majewski J, Wei S and Hill E M 2019 Earthquake-triggered 2018 Palu Valley landslides enabled by wet rice cultivation *Nat. Geosci.* **12** 935–9
[4] Watkinson I M and Hall R 2019 Impact of communal irrigation on the 2018 Palu earthquake-triggered landslides *Nat. Geosci.* **12** 940–5
[5] Japan International Cooperation Agency (JICA) 2019 *Boring survey for basic response for Central Sulawesi earthquake (Phase I)*
[6] Japan International Cooperation Agency (JICA) 2019 *Geotechnical Survey - Geotechnical Investigation of Landslide Palu Area*
[7] Ministry of Public Work and Housing 2021 *Gumbasa Irrigation Drawing - Package 2*
[8] Badan Informasi Geospasial 2021 Central Sulawesi Province Map *[online] Available at: https://tanahair.indonesia.go.id/portal-web/download/perwilayah*
[9] Mason B, Hutabarat D and Prakoso W 2019 Geotechnical Reconnaissance: The 28 September 2018 M7.5 Palu-Donggala, Indonesia Earthquake PacTrans-SSI Bridge Project View project Aceh-Pidie Mw 6.5 Earthquake Aftershock 2016 Monitoring View project *Rep. Geotech. Extrem. Events Reconnaissance, Geotech. Extrem. Events Reconnaiss. Assoc.* 1–86
[10] Ministry of Public Work and Housing 2015 *Laporan Akhir Pekerjaan SID Rehabilitasi Dareah Irigasi Gumbasa Tahap II Kabupaten Sigi (4073 Ha)*
[11] Japanese Geotechnical Society 1998 *Remedial Measures Against Soil Liquefaction: From Investigation and Design to Implementation* vol 148 (CRC Press)
[12] Pratama A, Fathani T F and Satyarno I 2021 Liquefaction potential analysis on Gumbasa Irrigation Area in Central Sulawesi Province after 2018 earthquake *IOP Conf. Ser. Earth Environ. Sci.* **930** 012093
[13] Idriss I M and Boulanger. R W 2008 Soil liquefaction during earthquakes *Earthq. Eng. Res. Inst.* **160** 43
[14] Sonmez H and Gokceoglu C 2005 A liquefaction severity index suggested for engineering practice *Environ. Geol.* **48** 81–91
[15] Ministry of Public Work and Housing 2019 *Paparan Penanganan Daerah Irigasi Gumbasa Pasca Gempa Bumi dan Likuifaksi*
[16] Kiyota T, Furuichi H, Hidayat R F, Tada N and Nawir H 2020 Overview of long-distance flow-slide caused by the 2018 Sulawesi earthquake, Indonesia *Soils Found.* **60** 722–35
[17] Jalil A, Fathani T F, Satyarno I and Wilopo W 2021 Liquefaction in Palu: the cause of massive mudflows *Geoenvironmental Disasters* **8**
[18] Aydan Ö, Ulusay R and Atak V O 2008 Evaluation of ground deformations induced by the 1999 Kocaeli earthquake (Turkey) at selected sites on shorelines *Environ. Geol.* **54** 165–82
[19] Garcia-Chevesich P, Wei X, Ticona J, Martínez G, Zea J, García V, Alejo F, Zhang Y, Flamme H, Graber A, Santi P, McCray J, Gonzáles E and Krahenbuhl R 2021 The impact of agricultural irrigation on landslide triggering: A review from chinese, english, and Spanish literature *Water (Switzerland)* **13** 1–17
[20] Otsushi K, Kato T, Hara T, Yashima A, Otake Y, Sakanashi K and Honda A 2010 Study on a liquefaction countermeasure for flume structure by sheet-pile with drain *Geotech. Soc. Singapore - Int. Symp. Gr. Improv. Technol. Case Hist. ISGI'09* 437–43
[21] Bowles J E 1997 *Foundation Analysis and Design* (Singapore: The McGraw-Hill Companies)
[22] Lindeburg M 2011 *Civil Engineering Reference Manual for the PE exam* (www. ppi2pass. com)
[23] Badan Standardisasi Nasional 2019 *SNI 1726-2019 Tata Cara Perencanaan Ketahanan Gempa Untuk Struktur Bangunan Gedung dan Non Gedung*
[24] PuSGeN Desain Spektra Indonesia *[online] Available at: http://rsa.ciptakarya.pu.go.id/2021/*
[25] Yasuda S and Hashimoto T 2016 New project to prevent liquefaction-induced damage in a wide existing residential area by lowering the ground water table *Soil Dyn. Earthq. Eng.* **91** 246–59

IOP Conf. Series: Earth and Environmental Science 1091 (2022) 012028 doi:10.1088/1755-1315/1091/1/012028

Isolation and characterization of filamentous biological soil crust forming algae from Malaysia

P Chheang[1], F T Y Min[1], J Md Khudzari[1], K Iwamoto*[1]

[1] Malaysia-Japan International Institute of Technology, Universiti Teknologi Malaysia, 54100 Kuala Lumpur, Malaysia

*Corresponding author e-mail: k.iwamoto@utm.my

Abstract. The biological soil crust method (BSC Method), which was established in Japan, may be a viable strategy for addressing the erosion and sedimentation problems in Malaysia. However, because this strategy has never been tested in a tropical environment, there is concern that it will be washed away by heavy rains and biological disturbance. Thus, identification and characterization of indigenous BSC-forming algae should be prioritized to ensure the effectiveness of the technology. This study was to confirm the presence of biocrust filamentous algae in Malaysia, followed by the collection and isolation of the biocrusts and the identification of their morphological properties. Algal growth and soil-fixing ability were also determined. In this work, we have successfully isolated indigenous filamentous algae, which can be potential candidates for the BSC Method in Malaysia. Results have shown that cultivation with aeration increased the biomass by 1.9-fold after one-week of cultivation. Based on soil erosion tests, it was demostrated that the presence of filamentous alga helped increase the soil surface resistance. Our findings suggest by implementing the BSC Method on selected slopes, soil erosion in Malaysia can be mitigated.

1. Introduction

Malaysia is located near the equator and experiences a tropical climate with two distinctive monsoon seasons, the Southwest Monsoon (April to September) and the Northeast Monsoon (October to March) [1]. Malaysia receives rainfall of up to 700 mm per month during the Northeast Monsoon. Several states, particularly Kelantan, Terengganu, Pahang, and east Johor in Peninsular Malaysia and Sarawak, often suffer from severe flooding due to heavy precipitation during this season [Saimi, 2020 #56;Lan, 2012 #4]. Apart from flooding, the heavy downpour also induces other natural disasters like landslides and mudslides [3]. Due to land scarcity, housing and infrastructure development have been expanded to hilly terrain areas [4], which triggers the naturally-occurring landslides and mudslides in Malaysia [5]. The combination of slope instability caused by development activity and intense rainfall in a tropical climate increases the risk of slope failures. Thus, mitigation methods are urgently needed as development activity is continuously to grow.

The current control method for slope collapses focuses on enhancing slope stability to reduce slope failures. The slope stability could be improved by lowering slope angles, terracing, branching, enhancing slope drainage, utilizing rock bolts and retaining walls, and incorporating wire cables and wire fences to prevent rock falls [6]. Concrete structures are frequently used in Malaysia due to their effectiveness and readily-available high-quality raw materials [7]. However, their greyish appearance is unappealing to the public and restrains the growth of plants, resulting in shallow ecological values [8]. Given the fact that current measures solely rely on material structures, which are costly and less environmentally friendly, periodic repair and maintenance are necessary to maintain the protectability [9]. On the other hand, geotextiles could provide immediate erosion control and slope stability, and natural materials such as jute, straw, and coir, as well as synthetic products such as polypropylene and polyester materials, are generally used [6]. Although it has been confirmed to effectively transfer the excessive shear stress from soil to geotextile materials [10]; it cannot be applied on rocky slopes and some synthetic products may result in runoff. Mulching is an effective agronomic method to reduce erosion by rainfall, runoff, and wind. However, mulching is a temporary solution that may also be washed away by heavy rain [11]. Therefore, cost-effective and environmentally friendly control measures are urgently needed.

Biological soil crust (BSC) or biocrust is a group of microorganisms that live within a few millimeters of the topsoil layer and is primarily composed of green algae, cyanobacteria, lichens, and mosses [12]. This crust cover serves as the earth's living skin in the form of vegetative cover, particularly in a water-scarce environment such as deserts and arid regions.[13]. Besides being the vegetation covers, BSCs also play an essential role in ecology by improving soil nutrients and soil aggregation stability by the secretion of EPS [14]. Despite the benefits of biocrust, naturally occurring biocrust takes years to develop. Therefore, research on developing the artificial BSC has been of interest to many researchers, including a Japanese company called NIPPON KOEI Co., Ltd. This company has developed the BSC Method that involves spraying BSC-1 (filamentous green alga indigenous to Japan) onto target areas. This method has been successfully implemented in Japan (e.g., Okinawa, Hokkaido, and Tokyo-Miyake Jima) and Nepal's environment [15]. While no evidence of vegetation recovery was observed, it could be seen that BSC could reduce the susceptibility of the slope against a very intense rainfall of 300 mm per 48 hours [16].

Based on previous findings, it was hypothesized that the BSC Method may help mitigate soil erosion problems in tropical regions like Malaysia. However, the use of the BSC Method in Malaysia's environment with a foreign algal strain has yet to be validated. Thus, it would be preferable to use indigenous species over imported strains. In this work, we checked the BSC formation in Malaysia, collected several biocrust filamentous algae from the BSC site, isolated them, and identified their morphological characteristics. In addition, the algal growth and soil-fixing ability were also determined.

2. Materials and Methods

2.1. *Biocrust sampling*

Biocrust algae were collected from bare and disturbed soil surfaces at (i) Malaysia-Japan International Institute of Technology (MJIIT), Universiti Teknologi Malaysia, Kuala Lumpur, (ii) Hutan Bandar (Recreational Forest), Cyberjaya, Selangor, and (iii) R&R Seri Kembangan North Bound, Seri Kembangan, Selangor (Figure 1). These areas represent locations in on-campus areas, construction sites, as well as highways.

Figure 1. Sampling points. (i) MJIIT, Kuala Lumpur (ii) Hutan Bandar, Cyberjaya, Selangor (iii) R&R Seri Kembangan North Bound, Seri Kembangan, Selangor

2.2. *Isolation and identification of biocrust-forming algae*
The biocrust samples were air-dried at room temperature for one night to remove unwanted species and allow easy separation of soil and algae. After wetting the dried samples with distilled water, the single string of algae was picked up using tweezers and a needle under the stereoscopic microscope (G1200, Mustool, China). The algal strain was washed with distilled water several times on a triple well microscope cavity slide before being inoculated with AF6 medium [17] into a 12-well culture plate (Nunclon™ Delta Surface, USA). The culture was kept for a week under constant white LED light (60 μmol photons $m^{-2} s^{-1}$). The algal strings were then transferred to a new, fresh medium. This procedure was repeated until the unialgal filaments were obtained. Once the unialgal status was confirmed under the stereoscopic microscope, the algal was preserved in a culturing vessel. Finally, stereoscopic and inverted microscopes (Eclipse TS100, Nikon, Japan) were used to examine the algal morphology.

2.3. *Physiological properties assessment*

2.3.1. Algal culture. Algae were grown in AF6 medium, aerated with either air or 2% CO_2, kept at 25–30 °C irradiated under 60 μmol photons $m^{-2} s^{-1}$ of white LED light. Prior to cultivation, the algal cell was separated into smaller pieces in two different ways: by manually tearing and by whisking with an electric whisker (China). Table 2 summarizes the preculture treatment and cultivation conditions.

Table 1. Experimental design.

Samples*	Sample conditions	Aeration conditions
WA	Whisking	Air
TA	Tearing	Air
WC	Whisking	2% CO_2
TC	Tearing	2% CO_2

*Description:
WA: the alga was whisked before inoculation and aerated during cultivation;
TA: the alga was torn into smaller pieces before inoculation and aerated during cultivation;
WC: the alga was whisked before inoculation and aerated with 2% CO_2 during cultivation;
TC: the alga was torn into smaller pieces before inoculation and aerated with 2% CO_2 during cultivation.

2.3.2. Biomass yield. After one week of culture, the algal cells were harvested on filter paper (Whatman™ GF/F, 47 mm, Cytiva, USA) and dried for 12 hours at 45°C. The increase in biomass was calculated by subtracting the weight of filter paper. Biomass yield was used to determine the best culture conditions for filamentous alga.

*2.3.3. Soil-*fixing ability. The soil erosion test was conducted following the methods described in Kuwabara *et al.* [18]. Red soil was collected from a construction area in Cyberjaya, Selangor. The soil was dried naturally for a few days and before being sieved to eliminate debris and gravel before testing. Then, the soil was soaked with distilled water, and the slurry was transferred to a rectangular container (12 cm × 5 cm). The slurry was dried in an oven for 24 hours at 70°C. The algal mixture was then sprayed onto the half-rectangular container and incubated for five days under the white LED light. During cultivation, water was sprayed once daily. On the fifth day, the water was sprayed vigorously and evenly across the soil surface, simulating a scenario of heavy rains. Finally, the impact of BSC cover on water erosion was evaluated.

3. Results and Discussion

3.1. *Existence of natural biocrust in Malaysia*
Natural biocrusts were discovered in Malaysia's tropical climate. The biocrust shown in Figure 2 was found on different soil surfaces, including inside planters and disturbed soil surfaces near construction sites.

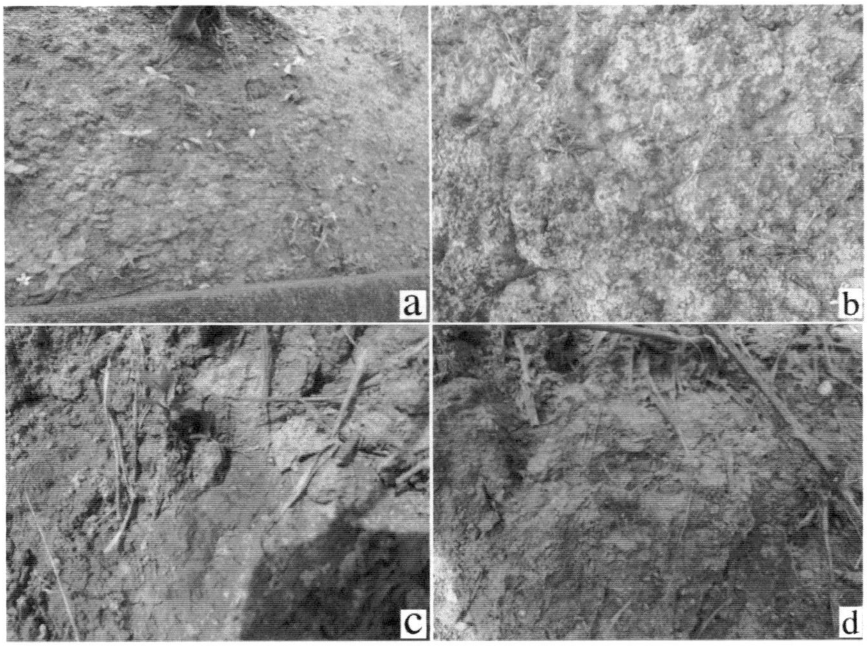

Figure 2. The natural biocrust found in Malaysia's environment. Samples were collected from (a) the planter around the campus, (b) slopes at the recreational forest, (c and d) highway slopes.

It is worth noting that biocrusts appear to occur at various successional stages. The biocrust that we discovered on the campus areas and highways were in the early stage of succession, as shown in Figure 2 (a, c, and d), as evidenced by their simple nature structure and light green color. This biocrust type quickly forms on the soil surface, yet it is susceptible to strong sunlight, drought, and other field constraints. This algal crust also exhibits low protectability against water and wind erosion [19]. The dominant community is cyanobacteria and/or green algae, namely *Microcoleus* sp. and *Nostoc* sp. (cyanobacteria), *Klebsormidium* sp., and *Zygogonium* sp. (green algae) taxa [20]. On the other hand, the biocrust discovered (Figure 3. b) near an active construction site in Selangor appears to be in the later successional stage, as indicated by the brown color. According to Lan *et al.* [21], this later BSC that is dark-colored and colonized by lichens or mosses presenting a better protective ability compared to light-colored biocrust.

3.2. *Isolation and Identification of biocrust-forming algae*

As illustrated in Figure 3, the algae from natural biocrust were successfully isolated. Following confirmation of morphology using a microscope, strain B4 was chosen was chosen.

The morphology of strain B4 was revealed under the microscope (Figure 3). The B4 appeared to adhere to one another in a batch, and the structure was identical to that of algae filaments. The B4 consists of a long filament, non-branching cell of 5-8 μm width. The cell length is relatively shorter than the cell width. The description is similar to the characteristics of *M. vaginatus* described by Giraldo-Silva *et al.* [22]. It could be concluded that the B4 strain could be the non-branching bundle-forming filamentous cyanobacteria *M. vaginatus*. This strain is known as cosmopolitan species and tiare the major contributor to the BSC community in arid and semi-arid environments [23]. This cyanobacterium is considered the primo-colonizer thanks to its filaments and EPS secretion. The secretion provides a more favorable environment for the colonization of subsequent successional species [24]. This pioneer strain is primarily used as an inoculum to initiate the formation of BSC cover and restore degraded soil conditions in dryland ecosystems [24]. They are frequently studied due to their high resistance to environmental constraints imposed by human activity [20] and their ability to grow and colonize on soil surfaces quickly. This potential strain was identified and isolated successfully in this study, indicating a greater chance of using the BSC Method to prevent soil erosion in Malaysia.

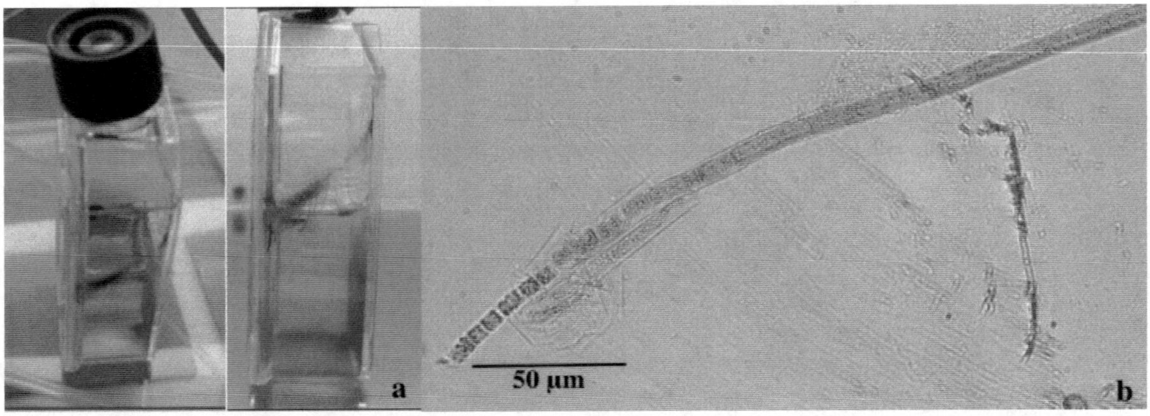

Figure 3. (a) Filamentous algae in AF6 medium and (b) image of filamentous algae under x40 magnificent.

3.3. *Physiological assessment*

3.3.1. Biomass yield. Table 2 shows the growth of algal biomass one week after cultivation. The filamentous algae grew well under air and 2% CO_2 supply. The results have shown that the final biomass increased by twice as much as the initial biomass in almost all cases. According to the ratio, the highest yield was found in WA and TA, increasing 1.89 and 1.90-fold, respectively. Although CO_2 is often used to boost the growth rate and biomass productivity [25], the yield under CO_2 was relatively lower compared to under air. The biomass changed between 1.43–1.58 times under TC and WC, orderly. In other words, using air could promote biomass yield compared to under CO_2. This finding suggests that CO_2 is not necessarily needed to boost the growth and biomass yield of filamentous algae. *[Couradeau, 2019 #13]*

Table 2. The increase of biomass under different culture conditions.

Sample	Initial (mg)	Final (mg)	Ratio
WA	1.71	3.23±0.12	1.89±0.04
TA	1.14	2.17±0.06	1.90±0.03
WC	1.72	2.70±0.10	1.58±0.06
TC	1.65	2.37±0.12	1.43±0.04

3.3.2. Soil-fixing ability. After one week of incubation, the inoculum was successfully colonized on the soil surface, as evident by the appearance of alga on the BSC area compared to the control plot (without BSC). According to Wang *et al.* [24], the cyanobacterium could quickly colonize nine days after cultivation. Figure 8 shows the soil surface conditions after a heavy rainfall simulation. The section with BSC was more resistant to water spraying compared to the section without BSC. The resistance of section with BSC was likely due to the presence of B4 alga. The alga grows and secretes EPS which binds soil particles and form the sheet-like biofilm [21], providing a protective barraier against erosive agents. Additionally, the same result was previously reported by Kuwabara *et al.* [18]. This result demonstrated that the presence of biocrust on specific areas could help reduce the susceptibility of the soil surface to water erosion.

Figure 4. The soil surface with and without algal inoculation after rainfall simulation test.

4. Conclusions and Recommendations

In this study, an indigenous filamentous biocrust-forming alga was successfully isolated and the morphology was examined under an inverted microscope. Notably, the natural biocrusts are abundantly available in Malaysia, particularly on bare and disturbed soil surfaces. According to the soil conditions, some biocrusts are still at the early stage of succession, and some are in the later stage. Algal cultivation without CO_2 aeration showed higher yields than those with CO_2. The size of alga had insignificant effect on the biomass. As demonstrated by the soil-fixing ability test, soil with filamentous alga was more resistant to water erosion. The finding could serve as a preliminary assessment of the ability of indigenous filamentous algae to fix the soil. Further study should focus on the identification of B4 alga by molecular approach. Importantly, the functional feasibility of B4 strain in fixing soil needs to be confirmed in laboratory and outdoor conditions.

Acknowledgments

This work is funded by the AUN/SEED-Net [Grant Number R.K130000.7343.4B403] under the Collaborative Education Program for Sustainable Environmental Engineering Network (CEP-SEEN). We would also like to thank the Malaysian Ministry of Higher Education and Universiti Teknologi Malaysia for their support of this work through the FRGS (PY/2019/03135) and Professional Development Research University (PY/2020/04020) grants, respectively.

References

[1] Ministry of Natural Resources and Environment 2015 First Biennial Update Report to the UNFCC. Available from: https://unfccc.int/sites/default/files/resource/MALBUR1.pdf (Accessed on: March 1, 2022)

[2] Saimi F M, Hamzah F M, Toriman M E, Jaafar O and Tajudin H 2020 Trend and linearity analysis of meteorological parameters in Peninsular Malaysia *Sustainability* **12** 9533

[3] Majid N A, Ruslan R, Wan I and Wan M 2018 Spatial modeling various types of slope failure using artificial neural network (Ann) in Pulau Pinang, Malaysia *Jurnal Teknologi* **80** 135–46

[4] Ramli Zamri Bin 2019 Hazard mapping in landslides and earthquakes in Malaysia. Available from: http://ancst.org/wp-content/uploads/2019/10/LS_EQ-Mapping_JMG_Zamri_Final.pdf (Accessed on: March 1, 2022)

[5] Majid N 2020 Historical landslide events in Malaysia 1993-2019 *Indian Journal of Science and Technology* **13** 3387–99

[6] Dorairaj D and Osman N 2021 Present practices and emerging opportunities in bioengineering for slope stabilization in Malaysia: an overview *PeerJ* **9** 1–34

[7] Mizal-Azzmi N., Mohd-Noor N and Jamaludin N 2011 Geotechnical approaches for slope stabilization in residential area *Procedia Engineering* **20** 474–82

[8] Leung F T Y, Yan W M, Hau B C H and Tham L G 2015 Root systems of native shrubs and trees in Hong Kong and their effects on enhancing slope stability *CATENA* **125** 102–10

[9] Aimee H and Normaniza O 2014 Physiological responses of Melastoma malabathricum at different slope orientations *Joural of Tropical Plant Physiology* **6** 10–22
[10] Niroumand H, Kassim K, R Ghafooripour A and Nazir 2012 The role of geosynthetics in slope stability *Electronic Journal of Geotechnical Engineering* **17** 2739–46
[11] Jordán Antonio, Zavala L M and Gil J 2010 Effects of mulching on soil physical properties and runoff under semi-arid conditions in southern Spain *CATENA* **81** 77–85
[12] Büdel Burkhard, Dulić Tamara, Darienko Tatyana, Rybalka N and Friedl T 2016 Cyanobacteria and algae of biological soil crusts *Biological Soil Crusts: An Organizing Principle in Drylands* 226 ed B Weber, B Büdel and J Belnap (Cham: Springer International Publishing) chapter 4 pp 55–80
[13] Belnap J 2003 Comparative Structure of Physical and Biological Soil Crusts *Biological Soil Crusts: Structure, Function, and Management* 150 ed J Belnalp and OL Lange (Berlin: Springer Berlin Heidelberg) chapter 15 pp 177–91
[14] Chamizo S, Cantón Y, Emilio R C and Domingo F 2015 Biocrusts positively affect the soil water balance in semiarid ecosystems *Ecohydrology* **9** 1208–21
[15] Ministry of Economy Trade, and Industry 2020 Climate change adaptation good practices by Japanese private sector. Available from: https://www.meti.go.jp/policy/energy_environment/global_warming/adaptation_goodpractice_FY2020ENG.pdf (Accessed on: June 25, 2021)
[16] NIPPON KOEI 2021 BSC method introduction. Available from: http://www.asiapacificadapt.net/adaptationforum2020/wp-content/uploads/2021/04/bsc-method_introduction.pdf (Accessed on: January 25, 2022)
[17] Ashyikin N, Suhaimi N, Kaha M, Hara H, Zakaria Z, Sugiura N, Othman N A and Iwamoto K 2018 Lipid production enhancement in tropically isolated microalgae by azide and its effect on fatty acid composition *Journal of Applied Phycology* **30** 3063–73
[18] Kuwabara T, Iwamoto K, Hara H, Yamaguchi T, Mohamad S E, Abdullah N, Md Khudzari J and Othman F-S 2021 Prevention of soil erosion using microalgae in Malaysia *IOP Conference Series: Materials Science and Engineering* **1051** 012047
[19] Lan S, Hu C, Rao B, Wu L, Zhang D and Liu Y 2010 Non-rainfall water sources in the topsoil and their changes during formation of man-made algal crusts at the eastern edge of Qubqi desert, inner Mongolia *Science China Life Science* **53** 1135–41
[20] Szyja M, Büdel B and Colesie C 2018 Ecophysiological characterization of early successional biological soil crusts in heavily human-impacted areas *Biogeosciences* **15** 1919–31
[21] Lan S, Wu L, Zhang D and Hu C 2013 Assessing level of development and successional stages in biological soil crusts with biological indicators *Microbial Ecology* **66** 394–403
[22] Giraldo-Silva A, Nelson C, Barger N and Garcia-Pichel F 2019 Nursing biocrusts: isolation, cultivation, and fitness test of indigenous cyanobacteria *Restoration Ecology* **27** 793–803
[23] Muñoz-Martín M, Becerra-Absalón I, Perona E, Fernández-Valbuena L, Garcia-Pichel F. and Mateo P 2019 Cyanobacterial biocrust diversity in Mediterranean ecosystems along a latitudinal and climatic gradient *New Phytologist* **221** 123–41
[24] Wang W, Liu Y, Li D, Hu C and Rao B 2009 Feasibility of cyanobacterial inoculation for biological soil crusts formation in desert area *Soil Biology and Biochemistry* **41** 926–9
[25] Singh S and Singh P 2014 Effect of CO_2 concentration on algal growth: a review *Renewable and Sustainable Energy Reviews* **38** 172–9

Liquefaction Potential Analysis and Mitigation at The Sediment Control Structures on Bangga River, Central Sulawesi

Fauzan[1], A Rifa'i[*1], S Ismanti[1]

[1]Department of Civil and Environmental Engineering, Faculty of Engineering, Gadjah Mada University, Yogyakarta 55281, Indonesia

*Corresponding Author. Email: ahmad.rifai@ugm.ac.id

Abstract. The Palu Earthquake in 2018 has triggered a debris flow incident in Bangga River. As a form of protection for similar disasters, it is planned to construct Sediment Control Structures in the Bangga River. The Bangga River area also has liquefaction potential that needs to be considered to ensure the structure's stability. This study aims to perform liquefaction analysis at the Sediment Control Structure area. Liquefaction potential is calculated using the Simplified Method and Liquefaction Potential Index Method based on 15 bore-log data that includes SPT. Further analysis is conducted by calculating lateral spreading and settlement potential. Liquefaction potential is found in the Consolidation Dam area, leading to lateral spreading and settlement at the footing of the structure. The lateral spreading potential is found from a depth of 11 to 13 m, with the potential movement of 0.886 m and located only 1 m below the Consolidation Dam foundation. Mitigation planning is proposed by compaction method to increase the liquefaction resistance. The safety against liquefaction is achieved by increasing NSPT value of the vulnerable soil to be greater than 36. This increasing soil strength can reduce lateral spreading potential up to 95% and the settlement potential up to 89%.

Keywords: Liquefaction potential, Sediment Control Structure, Bangga River, compaction method.

1. Introduction

The Bangga River is located on the west side of the Palu valley, Central Sulawesi, near the Palukoro fault zone Figure 1[1,2]. After the September 2018 Palu and Donggala Earthquake, many slope failures and large-scale collapses have occurred in the mountainous area in the Bangga basin. Due to torrential rain from February 2019, the collapsed sediment flowed downstream as sediment flow [1]. The sediment that flowed out deposited in Bangga village was estimated to be about 1,000,000 m^3, and the sediment depth reached up to 3m [1]. Considering the level of vulnerability to a debris flow in the Bangga river, the Ministry of Public Works and Public Housing plans to build a sediment control system in the downstream area of the Bangga river.

Located in a zone with high earthquake susceptibility, the downstream area of the Bangga river is also vulnerable to another earthquake-triggered disaster such as landslides and liquefaction [3,4]. Liquefaction is interpeted as the transformation of soil layer from solid state to a liquefied state as a result of the increased pore-water pressure and decreased effective stress due to cyclic loading [4]. The potential for liquefaction in the Bangga river is influenced by the relatively shallow groundwater level and the soil layer consisting of sandy soil [5]. Based on the geological map of Pasangkayu, the downstream area of the Bangga river mainly consist of alluvium rock formation [6]. The alluvium formation is classified as young sediment consisting of silt, sand, clay, and gravel [6]. The soil conditions in this formation are classified as non-well compacted soil[7].

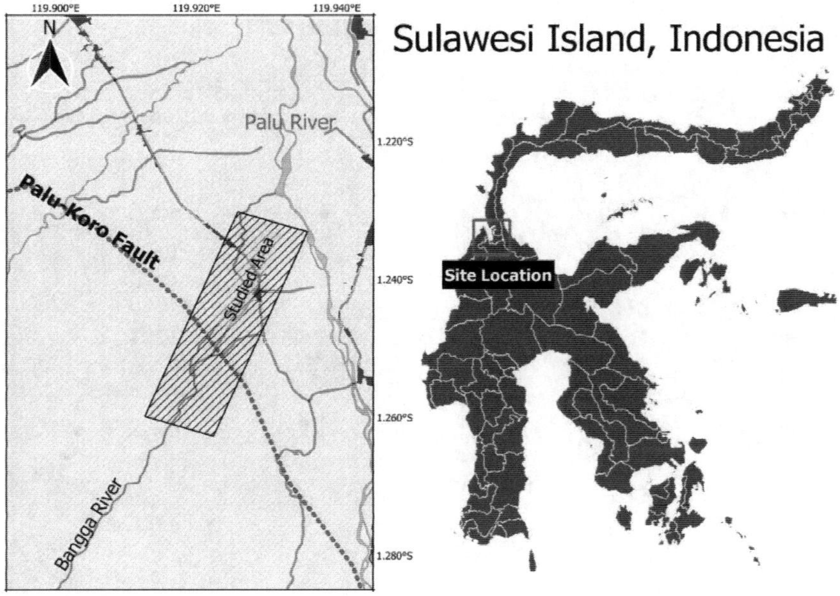

Figure 1. Location of the studied area ([1,2] with modification)

Considering the importance of the building and the potential risks, this study aims to analyze the potential for liquefaction in the construction area of the Sediment Control Structures on the Bangga River and mitigation plans that can be conducted to minimize the risk of the disaster.

2. Research Method

2.1 Research Area

This research focuses on the construction site of the Sediment Control Structures on the Bangga River, which consists of 3 Sabo buildings and 3 Consolidation Dam buildings. The liquefaction potential analysis was carried out based on bore-log data for 15 SPT points conducted by JICA at the construction site of the Sediment Control Structures. Bore-hole points are scattered at each location of the Sediment Control Structures, as can be seen in Figure 2. From Figure 2, it can be seen that the Palu-koro fault cuts the pound river area near the location of Sabo Dam 2. The downstream area of Bangga Rivers consists of 3 (three) rock formations, which is intrusive rock formations (Tmpi($_{d,g}$)), pakuli formations (Qp), and alluvium formations (Qa) [6]. Drilling test results show that most of the downstream area of the Bangga river consists of sandy soils with moderate to dense relative densities. The bore-log data shows a relatively shallow groundwater level at a depth of fewer than 5m.

The NSPT points used in this study are spread over 6 sediment control buildings. One point is BH-01 at the Sabo Dam 1. Five points, namely BH-02 to BH-06 are at the Sabo Dam 2. The next five points, namely BH-07 to BH-11, are at the Sabo Dam 3. In the Consolidation Dam 1 building there is one SPT point, namely BH-12, then two SPT points on the Consolidation Dam 2 building, namely BH-13 and BH-14 and one SPT point at the Consolidation Dam 3 building location, namely BH-15. The average depth of each borehole is 20 meters. The penetration test was carried by dropping a hammer weighing 63.5 Kg (140 lbs) and falling freelyover a constant height of 76 cm (30") along the drill pipe.

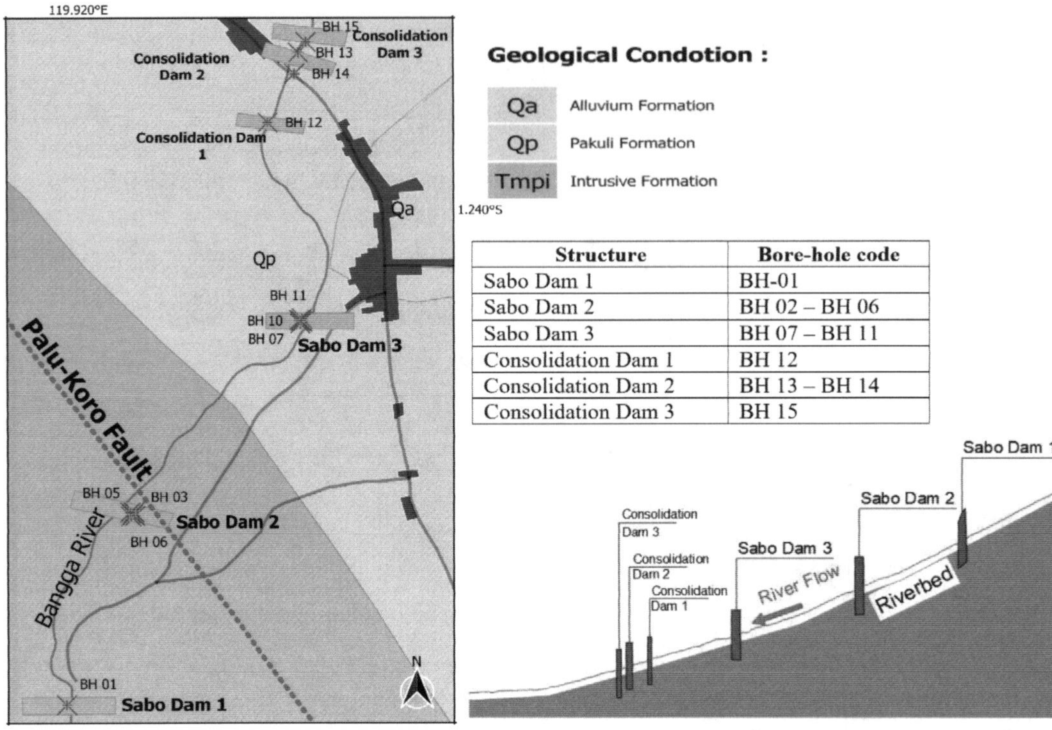

Structure	Bore-hole code
Sabo Dam 1	BH-01
Sabo Dam 2	BH 02 – BH 06
Sabo Dam 3	BH 07 – BH 11
Consolidation Dam 1	BH 12
Consolidation Dam 2	BH 13 – BH 14
Consolidation Dam 3	BH 15

Figure 2. Bore-hole Position layout and the geological structure at the lower reaches of Bangga River ([1,6] with modification)

2.2 Liquefaction Analysis Methods

The calculation of the liquefaction potential is carried out using a simplified method [8]. The simplified method is developed by Idriss and Boulanger based on the Seed and Idriss equation using NSPT data [9]. Seed and Idriss demonstrated that sufficiently significant shear stress in the sandy soil might be treated as a rigid body [10]. The most significant stress created by the earthquake (τ_{max}) was estimated using the time history's shear stress, which was transformed to cyclic stress using the maximum ground acceleration and adjustment for the earthquake's stress factor [9]. Seed and Idriss represent earthquake-induced cyclic stresses (*CSR*) as a function of maximum cyclic shear stress divided by the vertical effective stress (σ'_{vc}) [9]. In 2008, Idriss and Boulanger added a correction factor for the static load into the Seed and Idriss's equation, which is described in equation 1 [11].

$$CSR = 0.65\,\frac{\tau_{max}}{\sigma'_{vc}}\,\frac{1}{K_\alpha} = 0.65\,\frac{\sigma_{vc}}{\sigma'_{vc}}\,\frac{a_{max}}{g}\,r_d\,\frac{1}{K_\alpha} \qquad (1)$$

where a_{max} is the peak ground acceleration, σ_{vc} is total vertical stress, r_d is shear stress reduction coefficient, and Kα is the static load correction factor.

Idriss and Boulanger then modeling the function of soil resistance against liquefaction (CRR) in the form of equations 2-3 by referring to the value of N_{60} data corrected for various factors ($Ni_{(60)}$) in the equation [11].

$$CRR = \exp\left(\frac{(N_1)_{60cs}}{14{,}1} + \left(\frac{(N_1)_{60cs}}{126}\right)^2 - \left(\frac{(N_1)_{60cs}}{23{,}6}\right)^3 + \left(\frac{(N_1)_{60cs}}{25{,}4}\right)^4 - 2{,}8\right) .\ MSF \qquad (2)$$

$$MSF = 6{,}9\ EXP\left(\frac{-Mw}{4}\right) - 0{,}058 \qquad (3)$$

Where $(N_1)_{60cs}$ is corrected SPT blow count, *MSF* is the magnitude scaling factor, and *Mw* is earthquake magnitude.

The safety factor of liquefaction (FS) then defined as the value of Cyclic Resistance Ratio (CRR) divided with the value of Cyclic Stress Ratio (CSR) as seen in equation 4. Where the soil layer with a FS value less than 1 (one) is classified as susceptible to liquefaction.

$$FS = \frac{Cyclic\ Resistance\ Ratio}{Cyclic\ Stress\ Ratio} \quad (4)$$

The Liquefaction Potential Index method is used to measure the level of liquefaction potential that may occur. Iwasaki et.al. developed the Liquefaction Potential Index (LPI) methods in 1981 [12]. In 2003, Sonmez modified the Iwasaki equation by considering the uncertainties parameters [13,14]. Sonmez formulates the value of the potential level of liquefaction in equation (5-10) and sets a safe number for liquefaction at 1.2, meaning that the soil is considered to have no liquefaction potential if the FS value is greater than 1.2. Sonmez divides the liquefaction probability category into 5 (five) categories which are Non-liquefiable, Low, Moderate, High, and Very High.

$$LPI = \int_{0}^{20} F_{(z)} \cdot W_{(z)}\, dz \quad (5)$$

$$F_{(z)} = 1 - F_L \quad for \quad 0.95 > F_L \quad (6)$$

$$F_{(z)} = 2\,x\,10^6\, e^{-18,427 F_L} \quad for \quad 0.95 < F_L < 1.2 \quad (7)$$

$$F_{(z)} = 0 \quad for \quad F_L \geq 1.2 \quad (8)$$

$$W_{(z)} = 10 - 0.5\,z \quad for \quad 20\,m > z \quad (9)$$

While F_L is liquefaction safety facor value, and z is the depth of the soil layer in meter.

2.3 Post Liquefaction Analysis Method

Liquefaction can have a wide variety of implications, depending on the site circumstances, seismic loading characteristics, and the type of any constructions on the site [11]. These consequences include deformation or instability of the soil mass ranging from mildly sloping ground to embankment slopes, increased lateral pressures on retaining structures, loss of bearing support for shallow or deep foundations, loss of lateral support for embedded structures or piles, and settlement caused by reconsolidation of the liquefied soils [11,15]. Lateral spreading and settlement are consequences of post-liquefaction that illustrate the range of often involved concerns [11].

The lateral spreading displacement is derived by integrating maximum shear stresses versus depth, which introduced as lateral displacement index" (LDI) [16]. The Lateral displacement index is computed using equation (11-14).

$$LDI = \int_{0}^{z_{max}} \gamma_{max} \cdot dz \quad (11)$$

$$\gamma_{max} = min\left(\gamma_{lim}\,, 0{,}035\left(2 - FS_{liq}\right)\left(\frac{1 - F_\alpha}{FS_{liq} - F_\alpha}\right)\right) \quad (12)$$

$$\gamma_{lim} = 1{,}1859\left(1{,}1 - \sqrt{\frac{(N_1)_{60cs}}{46}}\right) \geq 0 \quad (13)$$

$$F_\alpha = 0{,}032 + 0{,}69\sqrt{(N_1)_{60cs}} - 0{,}13\,(N_1)_{60cs} \quad (14)$$

Where z_{max} is maximum depth, γ_{max} is maximum shear strain, dz is the thickness of the soil layer, γ_{lim} is limiting value of shear strain and FS_{liq} is factor of safety of liquefaction.

To calculate the potential of settlement caused by the liquefaction effect (S_{v-1D}), is proposed as equation (15-16) [16].

$$S_{v-1D} = \int_0^{z_{max}} \varepsilon_v . dz \tag{15}$$

$$\varepsilon_v = 1.5 \exp(-0.369\sqrt{(N_1)_{60cs}}) . \min(0.008, \gamma_{max}) \tag{16}$$

Where S_{v-1D} is ground surface settlement for one-dimensional consolidation and ε_v is volumetric strain.

2.4 Mitigation Planning

Mitigation planning for liquefaction potential in the Sediment Control Structures area in the Bangga River is carried out by considering several aspects, namely the characteristics of the soil in the area, liquefaction conditions that may occur, the depth and thickness of the soil layer that is susceptible to liquefaction, ease of installation and suitability to the site location and cost factor. Table 1 shows the summary of various mitigation methods for liquefaction used as a primary reference in selecting appropriate mitigation methods in this research [17].

Table 1. Summary of Soil Liquefaction Mitigation Methods [17]

Methods	Parameters to Modify	Methods	Advantages	Limitations
Densification	• Can increase the shear strength and stiffness of the material	• Vibro-compaction • Dynamic-compaction • Blast-densification	• Inexpensive • It can be used to structures nearby	• Accelerate ground settlement • It cannot be used on materials with finer grain size, such as silts.
Reinforcement	• Increases shear strength and stiffness • Shortens the drainage route • Reduce pore water pressure	• Stone-columns • Jet grouting	• Strengthens the soil • Increases the earth's lateral pressure	• It cannot be utilized on pre-existing structures. • Expensive
Drainage	• Reduce excess pore water pressure	• Gravel-drains • Vertical-drains	• Decrease the likelihood of excess pore water pressure accumulating	• It is not suitable for installation beneath existing structures. • Is incapable of minimizing seismic settlement
Solidification	• Stabilizes and aggregates the soil	• Grouting • Soil mixing	• Low viscosity chemicals may be utilized	• Expensive • Chemicals have the potential to be hazardous.

3. Result and Discussion

3.1 Liquefaction Potential Analysis at Sediment Control Structure in Bangga River

The liquefaction potential test in the Sediment Control Structures area on the Bangga river was carried out against a planned earthquake of 7.9 Mw with the Peak Ground Acceleration (*PGA)* value based on the spectrum response design issued by the Ministry of Public Works, and Public Housing of Indonesia and the Book of Map and Sources of Indonesia Earthquake Hazards in 2017 [2]. The analysis results show that several bore-holes are vulnerable to liquefaction (having a safety value below 1), mainly at the Consolidation Dam area. The depth of the soil layer that is susceptible to liquefaction varies from the deep of 3m and 10 to 13m. Figure 3 shows the analysis results for each bore-hole's safety number (FS.liq). When compared with the NPST value from the bore-log data, it can be seen that soil layers with a relatively low NSPT value (below 30) tend to be vulnerable to liquefaction hazards.

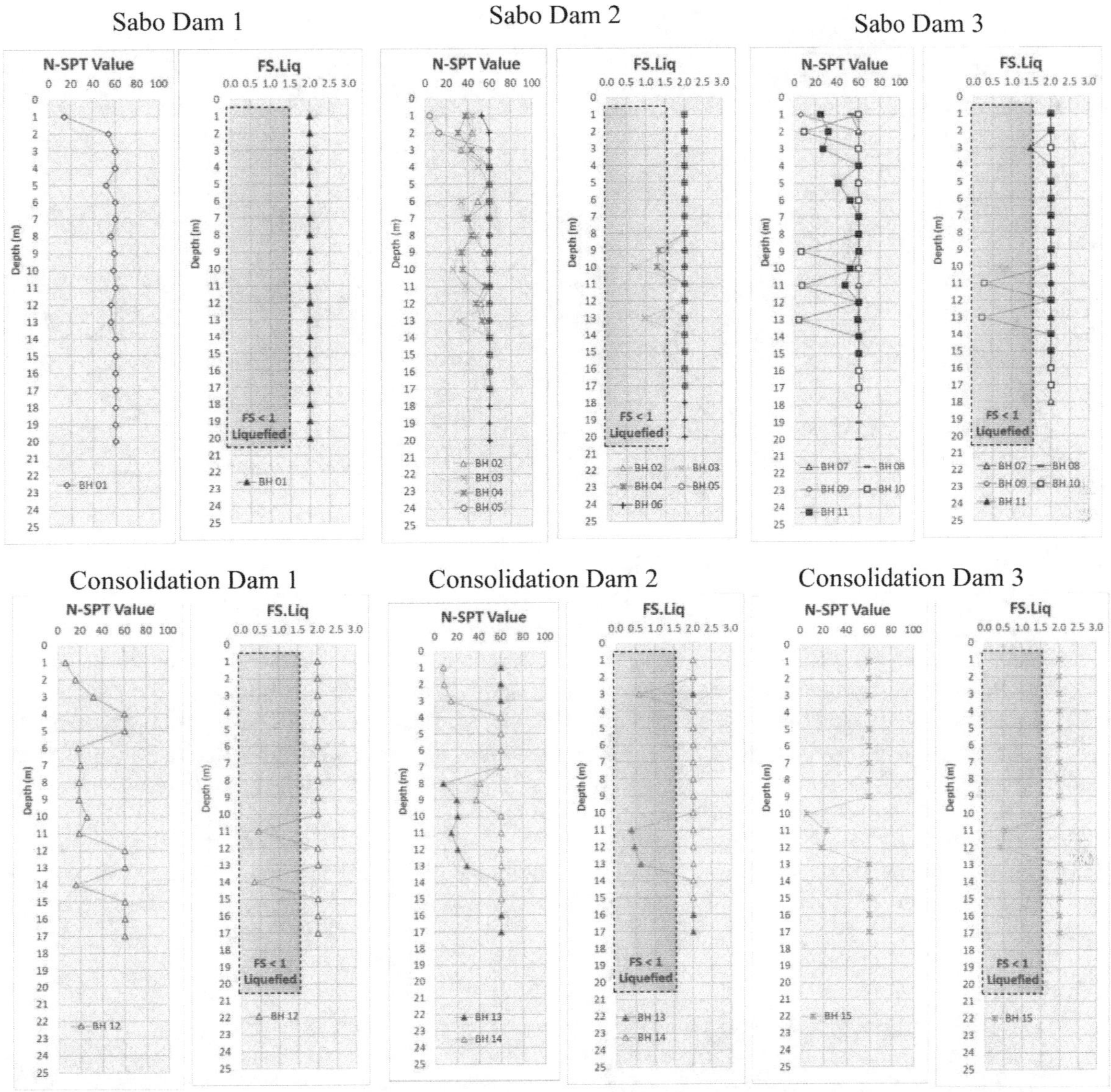

Figure 3. Value of N-SPT and Factor of Safety to Liquefaction

The safety factor (FS. Liq) value is then used as input to categorize the levels of potential liquefaction using the Liquefaction Potential Index (LPI) method. The LPI method can represent the potential liquefaction levels for each point based on the value of the safety factor and the depth function. The results are presented in a vulnerability map displayed in Figure 4. It can be seen from the map that the point with the highest level of vulnerability is in the Consolidation Dam 2 zone with a high liquefaction potential vulnerability scale. The bore-hole points of BH 13 and BH 14, which represent the construction site of Consolidation Dam 2, show susceptibility to liquefaction at a depth of 11m-13m. The Sabo dam 1, Sabo dam 2, and Sabo dam 3 areas are considered as non-liquefiable to low liquefaction potential zone. Meanwhile, Consolidation dams 1 and 3 are classified as moderate-level potential zone areas.

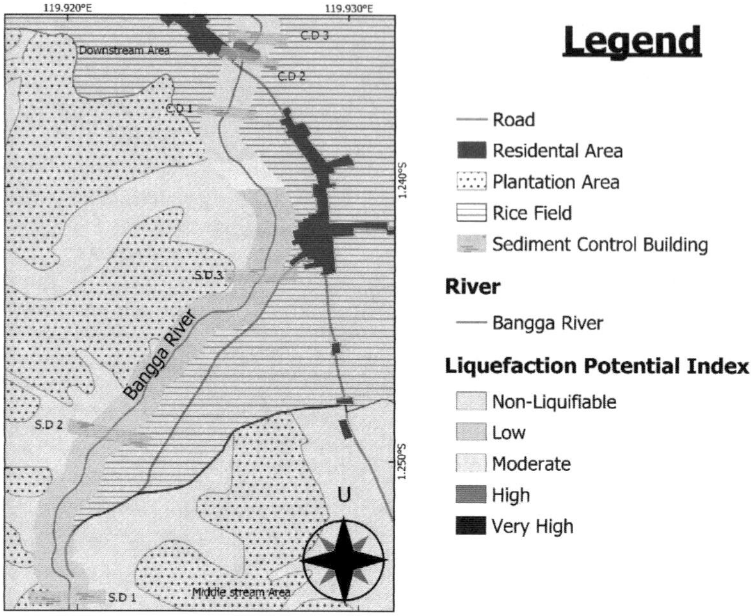

Figure 4. Liquefaction Potential Index Mapping of Sediment Control Structure Area in Bangga River

3.2 Post-liquefaction Analysis

Analysis of the post-liquefaction effect was carried out at the most critical point based on the potential mapping results, namely at the Consolidation Dam 2 construction site. The maximum potential for lateral spreading is estimated to be 0.45m at a depth of 11m, with total potential are calculated to be about 0.886m. The total potential for settlement due to liquefaction is 0.088m at a depth of 11m to 13m (Figure 5). The footing of the Consolidation Dam 2 building is at a depth of 10m, so during the construction process, soil cutting was carried out to a depth of 10m. The footing of the Consolidation Dam is only one meter above the potentially deformed soil layer (Figure 5). So, it is necessary to propose a mitigation approach to maintain the structure's safety against the potential danger of liquefaction.

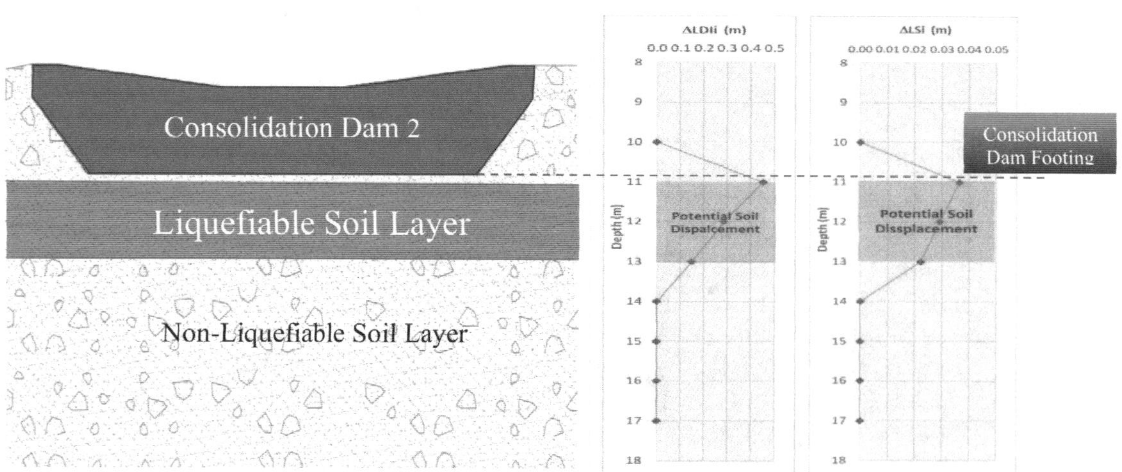

5.a. Cross Section of Consolidation Dam 2 Structure 5.b. Lateral Spreading and Settlement Potential

Figure 5. Potential Displacement of Soil Layer under The Consolidation Dam Structure

3.3 Mitigation Planning

The location of Consolidation Dam 2, which has a high-scale liquefaction potential, is the most critical point selected in the mitigation analysis. Considering the conditions where the potential liquefaction found is in a shallow layer and close to the soil surface and the geological structure that is mostly sandy soil, the authors choose the densification method as a countermeasure to overcome the danger of liquefaction in the study area. Another consideration in determining this method is seeing the condition of the site, which is quite far away with limited transportation access, and considering the cost factor, the densification technique is considered as a practical choice.

The compaction method aims to increase the shear strength of the soil so that it has sufficient resistance to liquefaction. To determine the level of strength that is safe against liquefaction potential, a back analysis is carried out on the Simplified Method equation from Idriss and Boulanger. The safety factor value is set at 1.2, based on Sonmez's study in 2005 [18]. The analysis results are shown in Table 2 which summarizes the NSPT values that need to be achieved to achieve safety against liquefaction at bore-hole 13.

Table 2. NSPT value required to reach a safe point against liquefaction

Depth	Initial NSPT Value	FS. Liq Initially	Required NSPT Value	Desired FS. Liq
11	15	0.39	36	1.2
12	21	0.47	38	1.2
13	29	0.62	39	1.2

The results show that to achieve a level of safety against liquefaction (safety factor equals to 1.2), the strength of the soil in the liquefied layer needs to be increased to have an NSPT value above 36. Using Bowles's correlation for the relationship between NSPT and relative density (D_r), it was concluded that the soil density needed to be increased by an average of 42% to achieve the desired bearing capacity value.

One of the compaction methods that can be used is dynamic compaction to get the required soil strength. To obtain an overview of the required compaction technique, the authors use a mathematical approach to simulate the behavior of granular soils under dynamic compaction [19]. This equation shows the relationship between the NSPT value of the soil after compaction with the weight factor of the load, the height of the pendulum falling during compaction, and the depth of influence of the compaction carried out as shown in Figure 6 and equation 17-18.

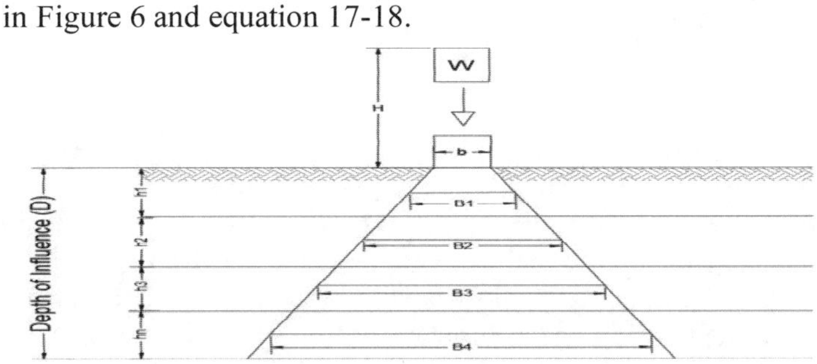

Figure 6. Model of Dynamic Compaction [20]

$$N_{n+1} = N_n + \frac{\Delta n}{D}(145 - N_n) \tag{17}$$

$$\Delta n = \sqrt{\frac{8.W.H.D}{3\,N_n^{\;2.5}\,(D + 2b^2)}} \tag{18}$$

Where N_{n+1} is NSPT value after compaction, N_n is NSPT value before compaction, W is tamper weight, H is dropping height, and D is the depth of influence.

The compaction scenario is carried out using a tamper weight of 10 t, with dimensions of 1.8m x 1.8m and a falling height of 5 meters. The compaction is carried out at a depth of 10 meters which is the top layer of the modeling soil. The analysis results show that to achieve the desired NSPT value, it is necessary to compact with a total of 6 strokes for each point. The estimated change in NSPT value after compaction and its effect on the liquefaction potential is shown in Figure 7.

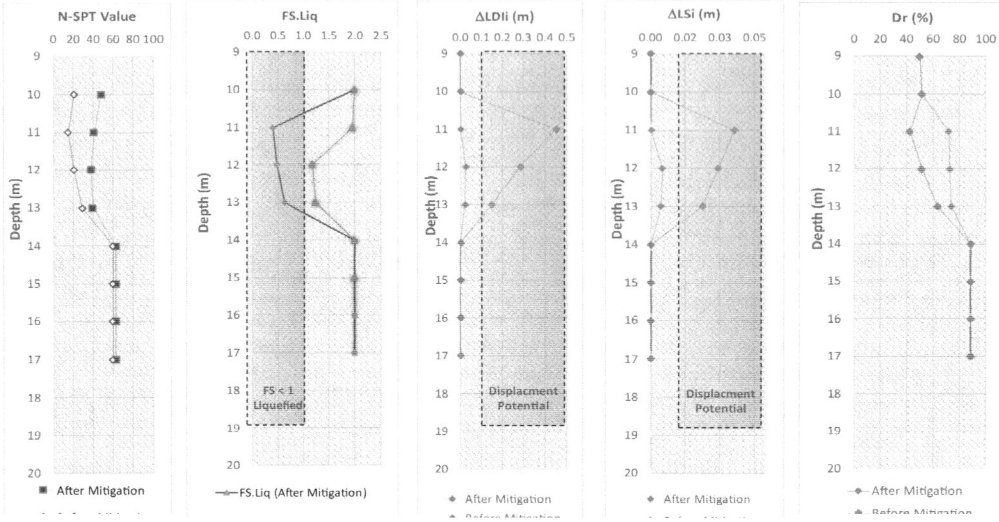

Figure 7. Comparison of Post-Liquefaction effect before and after soil reinforcement

Increasing the soil bearing capacity strength in layers that are susceptible to liquefaction can minimize the potential for lateral spreading and settlement, as shown in table 2. The total lateral spreading potential is reduced to the value of 0.046m (95%). Meanwhile, the total potential for settlement due to liquefaction is reduced to the value of 0.015m from initially 0.009m (89%).

Table 3. Comparison of Total Displacement Potential, Before and After Soil Improvement

Total of displacement potential	Before soil improvement	After soil improvement
Lateral spreading	0.886 m	0.046 m
Settlement	0.088 m	0.009 m

4. Conclusion

The potential liquefaction analysis using the Simplified Method for a design earthquake of 7.9 Mw at the Sediment Control Structure in the Bangga River shows that there is potential for liquefaction in some areas. The liquefaction potential is found in several layers of soil in the Consolidation Dam construction area, specifically at the downstream area of the Bangga river. Liquefaction Potential Index analysis depicts the level of potential hazards in each area. The results of this analysis are presented in a liquefaction disaster potential map. The mapping results show that the most critical zone is in the Consolidation Dam 2 construction area, classified as a high potential area for liquefaction.

Based on the post-liquefaction effect analysis in the Consolidation Dam 2 area, the liquefaction potential can cause deformation of the soil layer in the form of lateral spreading and settlement. The potential for lateral spreading is found from a depth of 11 to 13 m, with a total movement of 0.886 m. Meanwhile, settlement potential due to liquefaction is also found at a depth of 11 to 13 m with a total movement of 0.088 m. The susceptible soil layer to deformation is only 1 m below the Consolidation Dam foundation. One of the mitigation methods for liquefaction disasters that can be conducted is by improving the soil bearing capacity. Compaction work was proposed as the soil improvement. The analysis results show that the safety against liquefaction is achieved if the NSPT value of the vulnerable soil layer is more than 36. This increasing soil strength leads to reducing the lateral spreading potential up to 95% and the settlement potential due to liquefaction up to 89%.

This study was only analyzed based on the bore-log and geological data. To obtain more detailed results, further research is required to be conducted by considering the dynamic properties of soil. Mitigation methods that can be applied are not limited to the compaction method. Further research by evaluating the effectiveness of other mitigation methods can be comprehensive research.

References

[1] Japan International Cooperation Agency 2019 *The Project for Development of Regional Disaster Risk Resilience Plan in Central Sulawesi-Detailed Design on Additional Servis for Irrigation and River Sector After The 2018 Earthquake at Central Sulawesi*

[2] Tim Pusat Studi Gempa Nasional 2017 *Peta Sumber Dan Bahaya Gempa Indonesia Tahun 2017* (Bandung: Pusat Penelitian dan Pengembangan Perumahan Pemukiman, Badan Penelitian dan Pengembangan Kementerian Pekerjaan Umum)

[3] Badan Geologi og Kementerian ESDM Republik Indonesia 2018 Peta Potensi Likuifaksi Berdasarkan Kerusakan Geologi Permukaan *Hazard Map* **2018** 2018

[4] Marcuson W F 1978 Definition of Term Related to Liquefaction *J. Geotech. Eng.* 104

[5] Fauzan, Rifa'i A og Ismanti S 2021 Study of Liquefaction Potential at Sabo dam Construction on Poi and Bangga River, Sigi Regency, Central Sulawesi *IOP Conf. Ser. Earth Environ. Sci.* **930** 012083

[6] Sukido, Sukarna D og Sutisna K 1993 *Peta Geologi Lembar Pasangkayu, Sulawesi* (Bandung)

[7] Widyaningrum R 2012 Penyelidikan geologi teknik potensi liquifaksi daerah Palu , Provinsi Sulawesi Tengah *Kementrian Energi dan Sumber Daya Miner.* 43

[8] Boulanger R og Idriss I M 2014 *CPT and SPT Based Liquefaction Triggering Procedures* (Davis, California)

[9] Seed H B og Idriss I M 1971 *A Simplified Procedure For Evaluating Soil Liquefaction Potential* (California)

[10] Mase L Z 2014 Analisis Pendahuluan Potensi Likuifaksi di Kali Opak Imogiri Daerah Istimewa Yogyakarta *Semin. Nas. Geotek. di Yogyakarta* 10–1

[11] Idriss I M og Boulanger R . 2008 *Soil Liquefaction During Earthquakes* (Oakland: Earthquake Engineering Research Institute)

[12] Iwasaki T, Tokida K og Tatsuoka F 1981 Soil Liquefaction Potential Evaluation with Use of the Simplified Soil Liquefaction Potential Evaluation with Use of the Si *Int. Conf. Recent Adv. Geotech. Earthq. Eng. Soil Dyn.* **12** 209–14

[13] Sonmez B, Ulusay R og Sonmez H 2008 A study on the identification of liquefaction-induced failures on ground surface based on the data from the 1999 Kocaeli and Chi-Chi earthquakes *Eng. Geol.* **97** 112–25

[14] Sonmez H 2003 Modification of the liquefaction potential index and liquefaction susceptibility mapping for a liquefaction-prone area (Inegol, Turkey) *Environ. Geol.* **44** 862–71

[15] Youd T L, Idriss I M, Andrus R D, Arango I, Castro G, Christian J T, Dobry R, Finn W D L, Jr L F H, Hynes M E, Ishihara K, Koester J P, Liao S S C, Iii W F M, Martin G R, Mitchell J K, Moriwaki Y, Power M S, Robertson P K, Seed R B og Ii K H S 2001 Liquefaction Resistance of Soils : Summary Report From the 1996 NCEER and 1998 NCEER/ NSF Workshops on Evaluation Liqeufaction Resistance of Soils *J. Geotech. Geoenvironmental Eng.* **127** 297–313

[16] Zhang G, Robertson ; P K, Asce M og Brachman R W I Estimating Liquefaction-Induced Lateral Displacements Using the Standard Penetration Test or Cone Penetration Test

[17] Muttaqa U Z, Kassim K A og Mohammed A S 2018 Bio-desaturation and bio-sealing techniques for mitigation of soil liquefaction: A review *MATEC Web of Conferences* b 250 (EDP Sciences)

[18] Sonmez H og Gokceoglu C 2005 A liquefaction severity index suggested for engineering practice *Environ. Geol.* **48** 81–91

[19] Ebid A M 2015 Simple Mathematical Approach to Simulate Granular Fill Behavior under Dynamic Compaction *Int. J. Eng. Innov. Technol.* **5** 24–31

Hydrological Drought Responding to Meteorological Drought in Prek Thnot River Basin

Chhordaneath Hen[1], Ty Sok[1,2]*, Sophal Try[3], Ratboren Chan[1,2], Ilan Ich[1], Kimleang Chum[1], Chantha Oeurng[1]

[1] Faculty of Hydrology and Water Resources Engineering, Institute of Technology of Cambodia, Cambodia
[2] Research and Innovation Center, Institute of Technology of Cambodia, Cambodia
[3] Disaster Prevention Research Institute, Kyoto University

*Corresponding author e-mail: sokty@itc.edu.kh

Abstract. Drought is an inevitable consequence of climate variability and is pervasive across many regions all over the globe. This study aims to explore the relationship of drought and to evaluate the adaptivity and sensitivity in the Prek Thnot River Basin, Cambodia. The drought indicators were used between the occurrence of droughts based on Standardized Precipitation Index (SPI) and Standardized Runoff Index (SRI) to assess the impact in 8 stations in its base months (June, August, and November) for each time scale (1-month, 3-month, and 9-month). The results indicated that 1998 and 2005 were the most critical years. Furthermore, the dry event could be seen in each basin mostly in June and August as a state in drought frequency as in all stations. The outcomes indicated that most stations experienced drought conditions. The small amount of rainfall occurs over these stations during the year indicated, and they had a high frequency of drought. The drought conditions during the period could have many impacts on population living livelihood, especially on sanitation, economy while the most damaging negative impact is on crop production in the regions, given Cambodians' dependence on agricultural resources.

1. Introduction

Drought is one of the natural hazards or disasters of varying degrees affecting some parts of the world every year and almost every season (1). The impacts are related to hydrological drought (HD), such as drought in rivers, reservoirs, lakes, and groundwater (2), while the meteorological drought (MD) causes the shortage of precipitation over some period of timescale (3). Consequently, hydrological drought follows meteorological drought, principally due to the continuation of the MD (4). Hence, there is a close relationship between the Hydrological and Meteorological Drought.
To determine and assess the drought characteristics: the Standardized Precipitation Index (SPI) (5) is an enormously utilized index to evaluate MD in the world which represents abnormal wetness and dryness (6). The correlation coefficient is extensively applied to analyze the linear correlation of HD and MD by using the different timescale between HD and MD in the same method to determine the response time from the MD to the HD (7).
Therefore, to further enhance the understanding of the relationship between HD and MD, our study aims to present a useful methodology for determining the influence of the rainfall and flow during a drought period on the multi-scale correlations between HD and MD at the basin level. The main objectives of this study are (1) to assess the relationship between the occurrence of drought based on the application of a drought index SPI; (2) to determine the correlation between hydrological and meteorological

drought and comparing its relation. The result of this study could improve our decision-making ability under highly uncertain HD conditions and provide useful information on optimizing water resources allocation during drought.

2. Methodology

2.1.Study area

The Prek Thnot river basin has the drainage area of about 6,666 km^2, and its main river is 232 km in length (8). It lies between latitudes 11°00' and 12°10' N and longitudes 103°80' and 105°00' E (Figure 1) (9). The highest elevation is 1,543 m above mean sea level. The boundary of Prek Thnot River is covered by several provinces namely Koh Kong, Kampong Speu, Kampot, Takeo, Kandal, and Phnom Penh Capital City. The location of the eight selected rainfall stations in this study is shown in Figure 1. The area receives about 1,500 mm of rainfall per year, and the average runoff is estimated to be 6,042 m^3/s (10). The forest area in Prek Thnot River was greatly decreased by 56.5% in 2018, which was converted to arable land 22.7%, permanent pasture 8.5% and other 11.4% in 2018 (11).

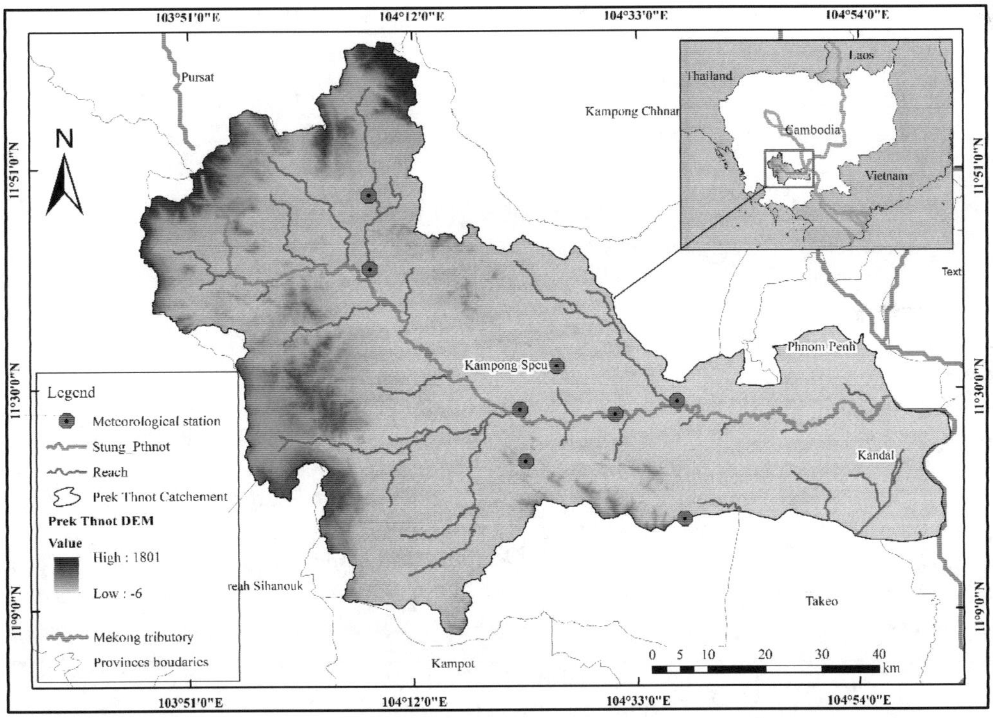

Figure 1. The study area of Prek Thnot River Basin and meteorological stations

2.2.Standardized Precipitation Index (SPI) and Standardized Runoff Index (SRI)

The Standardized Precipitation Index method found by MacKee (5) allows determining the wet and drought meteorological conditions location by using precipitation as input information. SPI effectively provides early drought warning and drought damage control (12). The short-term time scale, such as a 3-month comparison to seasonal precipitation and long-term time scale, reflects long-term precipitation patterns (24- 48-month). The comparison between the short-term and long-term time scales of SPI is deemed necessary; the short-term time scale could occur in occurrences of drought or flood events in the middle of long-term time scale that would only be visible over a long period (12).

Table 1. The categorization of SPI value

Categories	SPI & SRI
Extremely Wet	≥ 2.0
Very Wet	1.5 to 1.99
Moderately Wet	1.0 to 1.49
Near Normal	-.99 to .99
Moderately Dry	-1.0 to -1.49
Severely Dry	-1.5 to -1.99
Extremely Dry	≤ -2

Source: The effect of the length of record on the standardized precipitation index calculation (13)

SPI is calculated by dividing the difference between normalized seasonal precipitation which is determined by SPI tool and its long-term seasonal mean by the standard deviation.

(1)

where:
X = normalized monthly rainfall amount
$\overline{X}$ = mean of rainfall calculated from time series of monthly value
S = standard deviation of rainfall calculated from time series of monthly value

3. Results and Discussion

3.1.SPI-Three-month timescale for meteorological stations
Eight meteorological stations show unstable events, mainly in June and November. Noticeably, Kong Pisey experienced severe dry periods continuously from 2001 to 2007 in its base month in June and August. In 1997, Oral and Trapeang Chor went under extremely dry conditions in June and August. In 2002, Kompong Speu, Oral and Prey Pdao met extreme drought (August, June, November, respectively). In the next two years, Phnom Srouch and Peam Khley suffered from severe drought in June and November. Assumingly, there is not much difference in drought conditions in all eight stations, and they similarly underwent the same drought condition in almost the same year.

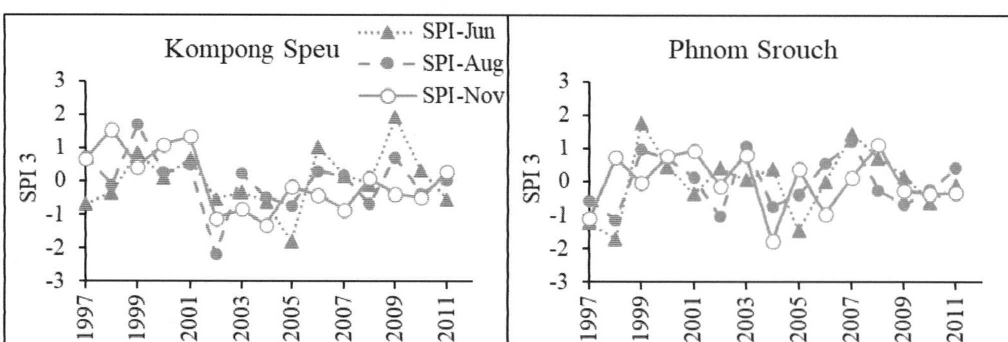

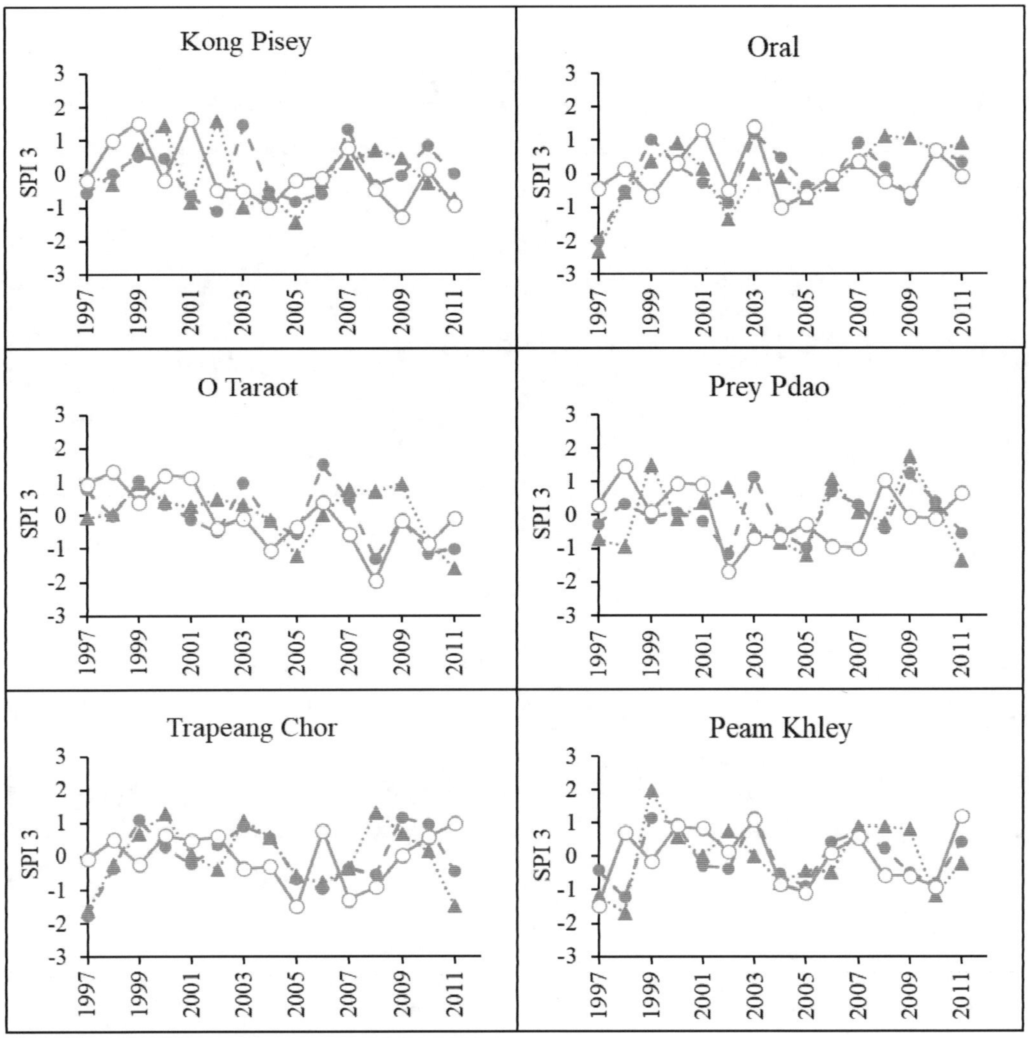

Figure 2. Three-months timescale SPI values for June, August, and November of eight stations

3.2.Six-months timescale

Short time scales of drought index SPI were mainly found in 1998, 2004, and 2005 in Phnom Srouch, Kong Pisey, Prey Pdao, and Peam Khley in its different base months (Figure 3). However, 2005 was the dry year that happened in each station (August and June). O Taraot was the one that met severe drought

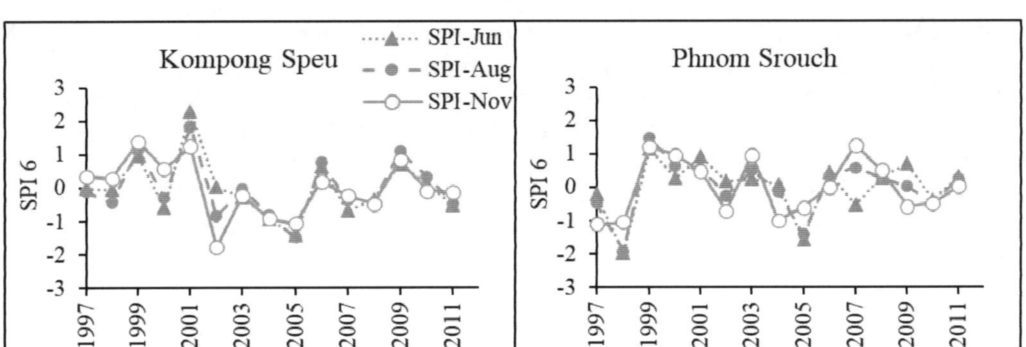

in 2008. According to the six-month timescale, it can be seen that droughts changed less slowly, lasted longer, and had less variability compared to those in 1-month and 3-month timescales.

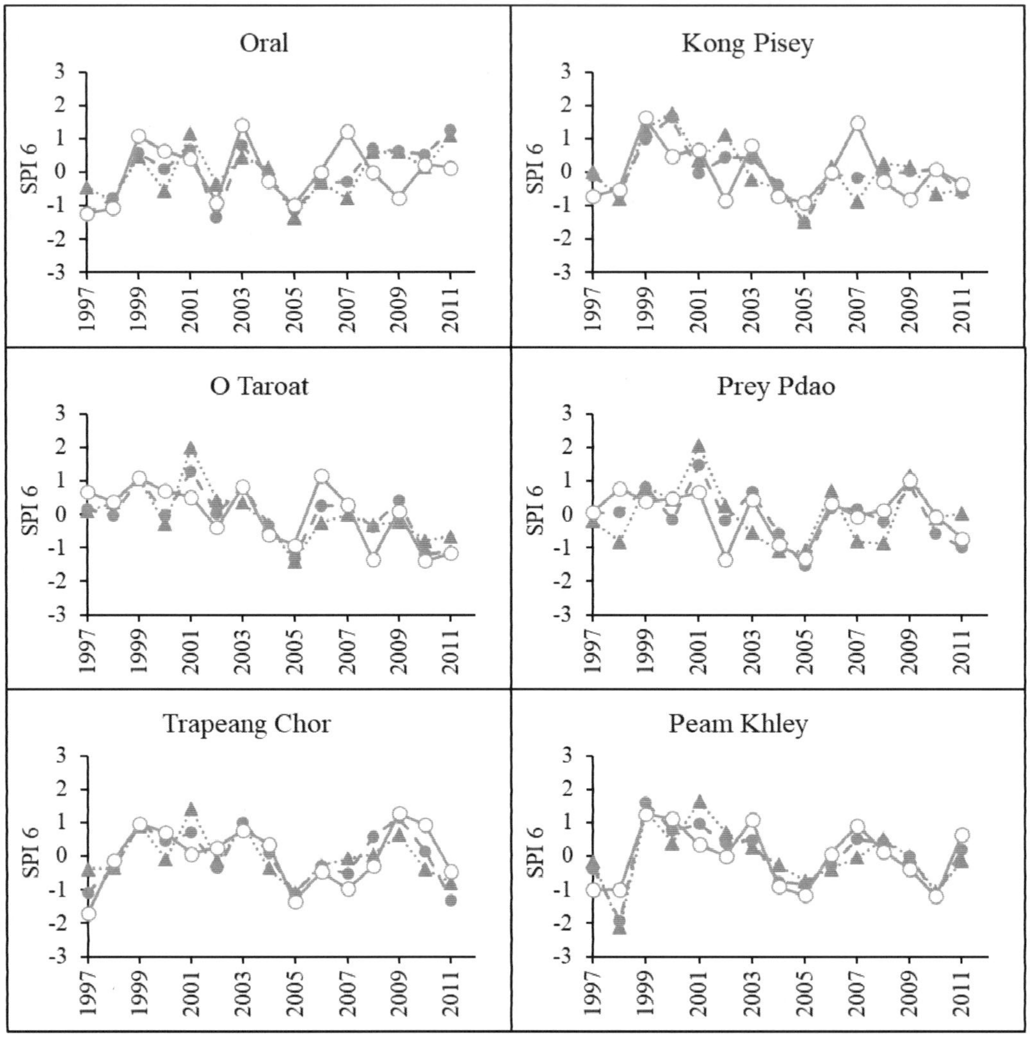

Figure 3. Six-months timescale SPI values for June, August, and November of eight stations

3.3. Correlation analysis

According to Table 2(a), the calculation presents the correlation values of SPI with the 1-, 3-, and 6-month in terms of meteorological and hydrological droughts of the period from 1997 to 2011. The Peam Khley HD had a stronger correlation with Peam Khley MD in a more extended accumulation period. For example, the correlation coefficient of SPI value in Peam Khley HD versus Peam Khley MD from 0.5024 for the 6-months accumulation period to 0.3712 and 0.2048 for the 3- and 1-month accumulation period, respectively. As true as other stations in Table 2 (b), (c), (d), they all moved gradually to positive near one correlation value.

IOP Conf. Series: Earth and Environmental Science 1091 (2022) 012030 doi:10.1088/1755-1315/1091/1/012030

Table 2. R^2 between the SPI accumulation in the hydrological versus meteorological droughts among the observed stations

Peam Khley HD versus Peam Khley MD (a)			
	SPI 1	SPI 3	SPI 6
SRI 1	**0.2048**	0.2760	0.1825
SRI 3	0.0813	**0.3712**	0.3632
SRI 6	0.0685	0.2771	**0.5024**
Peam Khley HD versus Phnom Srouch MD (b)			
	SPI 1	SPI 3	SPI 6
SRI 1	**0.1513**	0.2423	0.2874
SRI 3	0.0296	**0.2426**	0.4205
Tasal HD versus Oral MD (c)			
	SPI 1	SPI 3	SPI 6
SRI 1	**0.3806**	0.3659	0.2269
SRI 3	0.0994	**0.4832**	0.4300
SRI 6	0.0320	0.1778	**0.5218**
Kompong Toul HD versus Prey Pdao MD (d)			
	SPI 1	SPI 3	SPI 6
SRI 1	**0.1050**	0.2056	0.2337
SRI 3	0.0314	**0.2338**	0.3609
SRI 6	0.0230	0.1456	**0.3805**

3.4. Three-months timescale

The drought frequency in three types of drought conditions is shown in Figure 4 for three-month time scale. Prey Pdao and Peam Khley stations had the highest moderate dry period in June. These stations had the same amount of percentage of moderate drought in November compared to Kong Pisey station where it is higher. The two stations rarely had a problem with the extreme drought severity as it showed in the figure. The extreme drought only happened in June and August for these three stations. Oral rainfall station faced approximately 5% of drought occurrence. The extreme dry event had the same values in all three-month scale. August and November had less severe conditions compared to June and the absence of moderate drought in June. Last but not least, O Taraot was in moderately dry condition throughout the base months and the other drought conditions had almost the same drought occurrences.

IOP Conf. Series: Earth and Environmental Science 1091 (2022) 012030 doi:10.1088/1755-1315/1091/1/012030

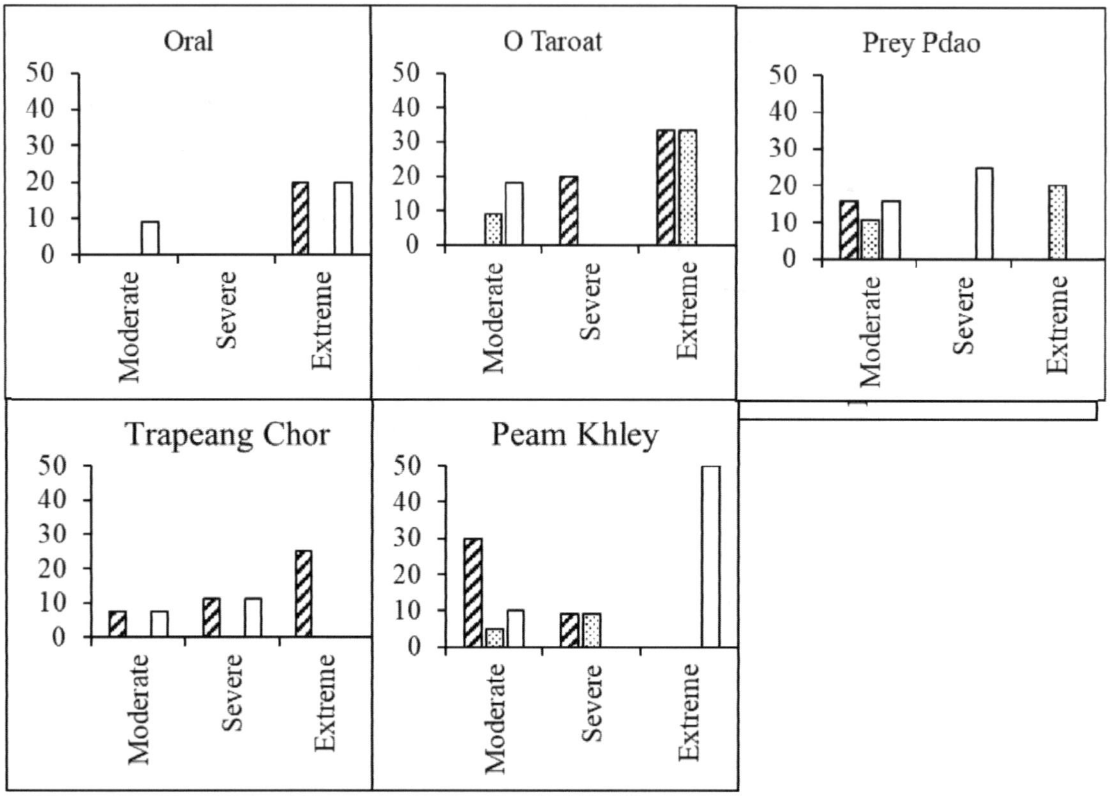

Figure 4. Drought frequency (%) base on a three-month timescale in eight stations

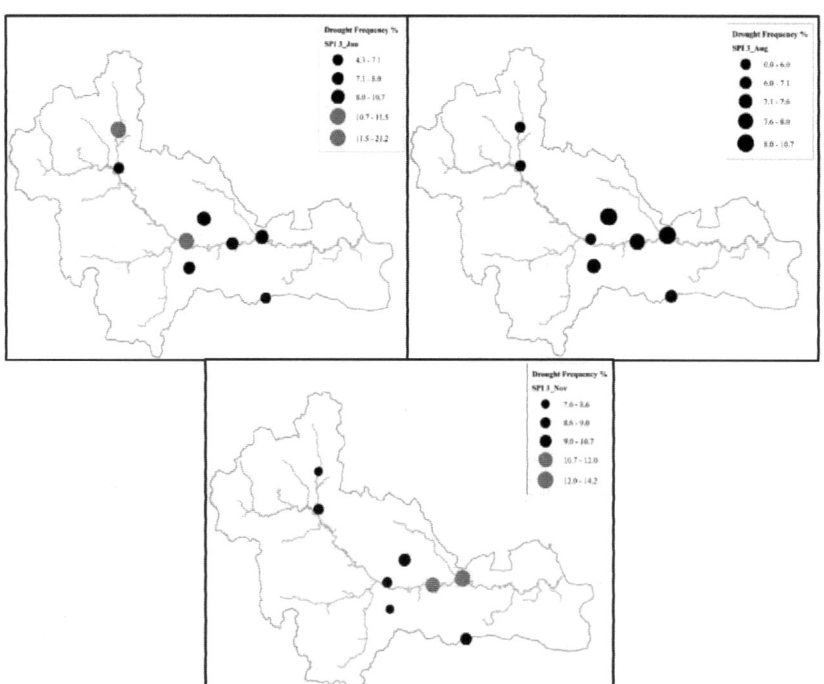

Figure 5. Drought Frequency occurrences based on SPI 3

3.5.Six-months timescale

In Figure 6, the number of moderate drought occurrences underwent more than severe and extreme events in most stations which were alike to 1-month and 3-month timescales. All these stations were coming across the same event except Prey Pdao and Peam Khley that were noticeably high in drought frequency moderate according to its base months (June, August, and November). Presumably, Prey Pdao, Kong Pisey, and Peam Khley were likely to have drought severity issues several times in June, August, and November. However, Prey Pdao station rarely experienced extreme drought. To sum up, the 6-month scale did not change as rapidly as the two-time scales as mentioned in the sections above.

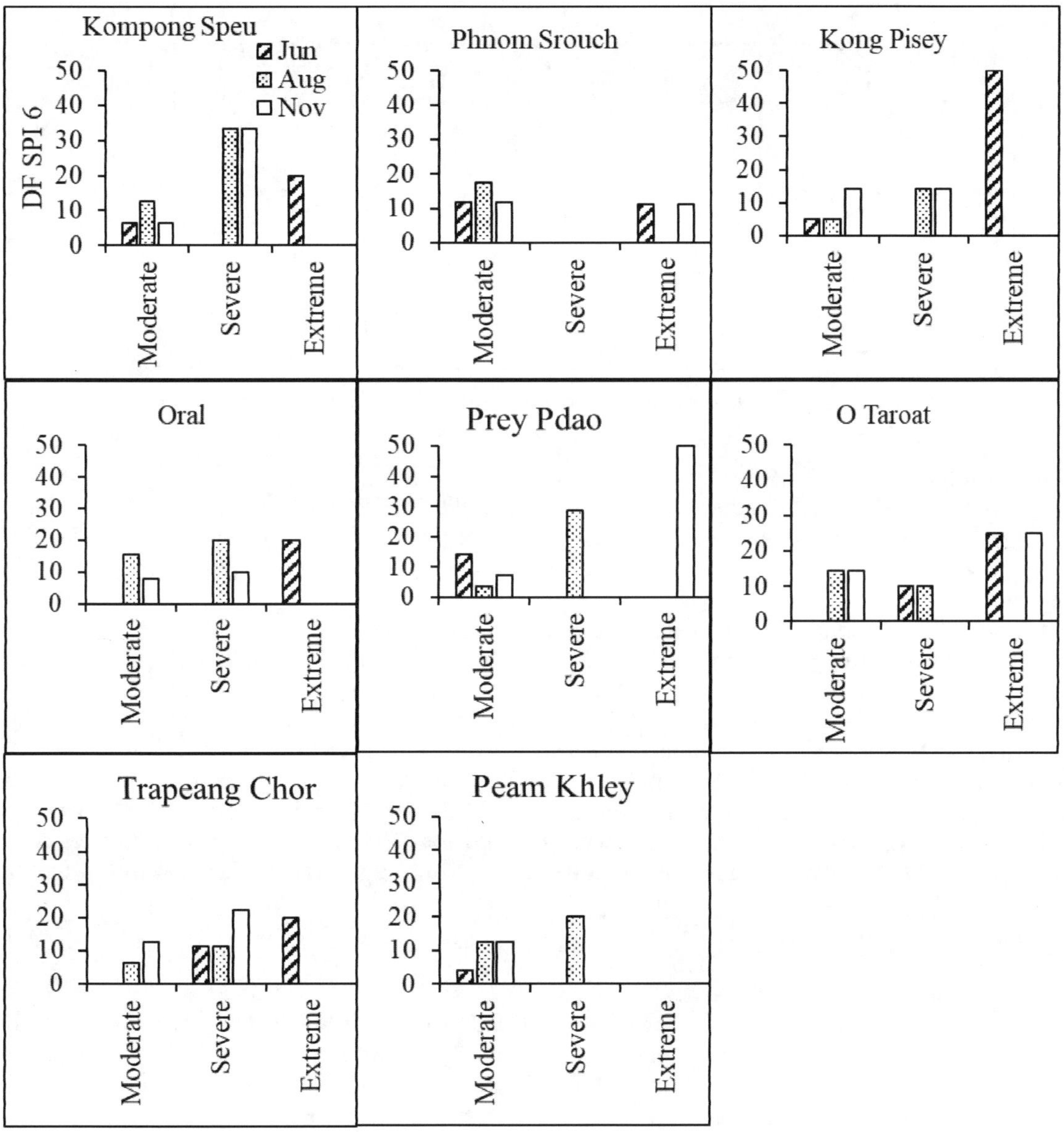

Figure 6. Drought frequency (%) base on a one-six timescale in eight stations

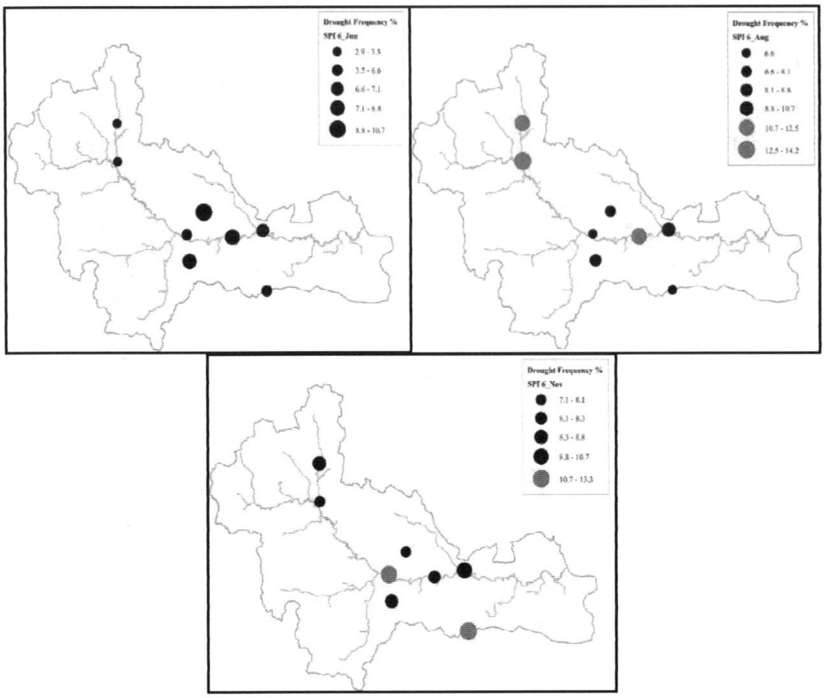

Figure 7. Drought Frequency occurrences based on SPI 6

4. Conclusions

This study presented a useful framework for examining the influence of rainfall and flow during a drought on the multi-scale correlation between HD and MD using the drought index Standardized Precipitation Index (SPI) and Correlation analysis comparing the two different types of droughts in Prek Thnot River Basin.

Based on the drought index SPI method, the results showed that 1998 and 2005 were the most critical years identified with drought occurrence. Hence, each different rainfall station showed different drought conditions. In general, the dry event could be seen in each basin mostly in June and August as shown in the drought frequency for all stations. Moreover, Phnom Srouch, Oral, Peam Khley, and Prey Pdao had the most experienced drought conditions, which means that the small amount of rainfall over these stations during the year. The drought condition might have negative impact on crop production in the regions during the period. Thus, this phenomenon should be taken into consideration seriously to prevent the damage that can happen and in turn impact crop production. Therefore, future infrastructure should be designed and constructed to store the rainwater for agriculture to give sufficient water in the dry season.

The relationship between MD and MD had a significant influence on each other, which 2005 and 2002 turned correlated. Plus, the correlation values of both years were inclined to 1, indicating a strong correlation in longer duration timescales.

The presence of this study enables us to gain a better understanding of the evaluation of HD and MD under the drought effect condition within multi-timescale and correlation of its intensity of human activities, especially of the reservoir operation. The future research should consider to conduct drought forecasting in the Prek Thnot River Basin in order to mitigate drought impacts and reduce its damages. Plus, there is a need to make drought research suggestions to evaluate a longer duration timescale for a better perspective between the MD and HD.

Acknowledgments
The authors would like to acknowledge that the work was funded by Cambodia Higher Education Improvement Project (Credit No. 6221-KH). The authors also thank the Ministry of Water Resources and Meteorology (MOWRAM) of Cambodia for providing the data.

References

[1] Wilhite DA, Buchanan-Smith MJD, water crises: Science t, issues m. *Drought as hazard: understanding the natural and social context.* 2005;3:29.

[2] Van Loon AF, Gleeson T, Clark J, Van Dijk AI, Stahl K, Hannaford J, et al. *Drought in the Anthropocene*. 2016;9(2):89-91.

[3] Quiring SMJGC. *Monitoring drought: an evaluation of meteorological drought indices*. 2009;3(1):64-88.

[4] Lorenzo-Lacruz J, Vicente-Serrano SM, González-Hidalgo JC, López-Moreno JI, Cortesi NJCR. *Hydrological drought response to meteorological drought in the Iberian Peninsula*. 2013;58(2):117-31.

[5] McKee TB, Doesken NJ, Kleist J, editors. *The relationship of drought frequency and duration to time scales. Proceedings of the 8th Conference on Applied Climatology*; 1993: California.

[6] Du J, Fang J, Xu W, Shi PJSer, assessment r. *Analysis of dry/wet conditions using the standardized precipitation index and its potential usefulness for drought/flood monitoring in Hunan Province, China.* 2013;27(2):377-87.

[7] Huang S, Li P, Huang Q, Leng G, Hou B, Ma LJJoH. *The propagation from meteorological to hydrological drought and its potential influence factors*. 2017;547:184-95.

[8] Khorn N, Ismail MH, Kamarudin N, Nurhidayu SJPJoS, *Technology. Land Use Change using Geospatial Techniques in Upper Prek Thnot Watershed in Cambodia*. 2020;28(3).

[9] Nuorteva P, Keskinen M, Varis OJJoW, Change C. *Water, livelihoods and climate change adaptation in the Tonle Sap Lake area, Cambodia: learning from the past to understand the future*. 2010;1(1):87-101.

[10] Water Environment Partnership in Asia and M.o.t.E.o. Japan. *State of water environmental issues.* . 2004.

[11] CIA. Cambodia. 2021.

[12] Liu L, Hong Y, Bednarczyk CN, Yong B, Shafer MA, Riley R, et al. *Hydro-climatological drought analyses and projections using meteorological and hydrological drought indices: a case study in Blue River Basin, Oklahoma*. 2012;26(10):2761-79.

[13] Wu H, Hayes MJ, Wilhite DA, Svoboda MDJIJoCAJotRMS. *The effect of the length of record on the standardized precipitation index calculation.* 2005;25(4):505-20.

Assessing Flood Hazard Index using Analytical Hierarchy Process (AHP) and Geographical Information System (GIS) in Stung Sen River Basin

Chanseyma Khoeun[1], Ty Sok[1,2*], Ratboren Chan[1,2], Sotheanea Khe[1], Ilan Ich[1], Kimsan Chan[1], Chantha Oeurng[1]

[1] Faculty of Hydrology of Water Resources Engineering, Institute of Technology of Cambodia, Russian Federation Blvd., P.O. Box 86, Phnom Penh, Cambodia.
[2] Research and Innovation Canter, Institute of Technology of Cambodia, Phnom Penh, Cambodia

*Corresponding author e-mail: sokty@itc.edu.kh

Abstract: Flood is a water-related disaster that causes negative impacts on Human and societies. In Cambodia, flood has become more severe and frequent in most regions, especially, in the floodplain areas. This study aims to simulate the flood hazard index and vulnerability located in the Sen River Basin using the Analytical Hierarchy Process (AHP) method and Geographical Information System (GIS). GIS was applied for the estimated Flood Hazard Index (FHI) in which seven parameters were selected: Flow accumulation, Elevation, Density Drainage, Rainfall Intensity, Slope, Soil Type, and Land Use. Additionally, the relative importance of each parameter of physical factors has been compared in the pairwise matrix 7×7 to gain the weight value by AHP Method. The Flood Hazard Zones was classified into five classes from Very Low to Very High and were observed as Very low 6175 Km^2, Low 4094 Km^2, Medium 2002 Km^2, High 818 Km^2, and Very High 859 Km^2. The FHI was verified with the flood hazard map from global surface water and Flood history from Google Earth Engine. Consequently, the results were found to be reasonable and adequate, revealing a realistic representation of hazards on the corresponding flood hazard map.

1. Introduction

Floods have been defined as one of the most terrifying natural disasters and induces substantial negative impacts on humans and societies as a whole. In recent decades, the increasing trend in frequency and magnitude of flood events has been recorded which mainly resulting of the ongoing climate changes, land-use change as well as land-use changes driven by human activities. As a result, from social and economic growth, the potential for flood damage in many rivers is projected to rise [1]. Especially, the occurrence of flood events has increased significantly all worldwide over the last three decades [2].

Cambodia is considered as one of the most disaster-prone countries, where there is abundant water during the rainy season that leads to flooding. Obviously, flood events deteriorate households, rice production in many areas along floodplain, property, crops, loss of livestock, and deterioration of health conditions owing to waterborne diseases. The 2011 flood event, which is the worst flooding in terms of economic devastation, is a particular flood which has been extensively studied in the hydrology domain. The flood hazard has been able to provide much of the required information for mapping the extent of a flood, assessing the damage, and evaluating the vulnerability within the flooded areas [3].

To identify the flood-prone areas, several approaches have been developed and utilized in different geographic regions. Numerous researchers have used Multi-Criteria Decision Analysis (MCDA) like the Analytical Hierarchy Process (AHP) to assess the flood hazard/risk potential [4]. The AHP approach is known as one of the most powerful approaches of MCDA in the field of flood risk management, due to a host of benefits including its direct involvement in decision-making, a comprehensive geographic information system (GIS), and criteria evaluation consistency [5]. However, this approach is not widely used in Cambodia as well as in the river basin within. Additionally, Cambodia has very limited researches on flood hazard mapping with an inadequate comprehensive flood hazard program in many areas, including the Tonle Sap region, due to data limitation in the study area. To the best of our knowledge, there has been no studies conducted to assess flood hazard zone by using AHP method in the Sen River basin which is the Tonle Sap's major tributary and commonly regarded as the flood-prone basin.

Therefore, this study aims to simulate the flood hazard index along with the Sen River Basin using AHP. The main objectives are to classify parameters of the flood hazard map as well as estimate the weight of the AHP method then overlays the parameters and validation to the sensitive region of the flood hazard index on the map.

2. Materials and Methods

2.1. Sen River Basin

The Sen River Basin, a largest sub-basin in the Tonle Sap Lake Basin with a total area of about 16,000 km^2. The Sen River originates from the Dangrek Mountains in Preah Vihear Province, the border of Cambodia and Thailand, and drains about 500 km across Kampong Thom into the Tonle Sap Lake as shown in Figure 1 [6]. The basin's climate is dominated by tropical monsoon with two distinct seasons, wet season and dry season. The annual rainfall in the basin approximately ranges from 1,100 to 1,800 mm with the average of 1,400 mm [7]. The average temperature is about 27.5 ºC with the minimum and maximum temperatures of 20ºC and 35ºC, respectively [8]. The peak discharge at the basin's outlet, which was measured from 1982 to 2000 at the Kampong Thom Station, was 702 m^3/s [9].

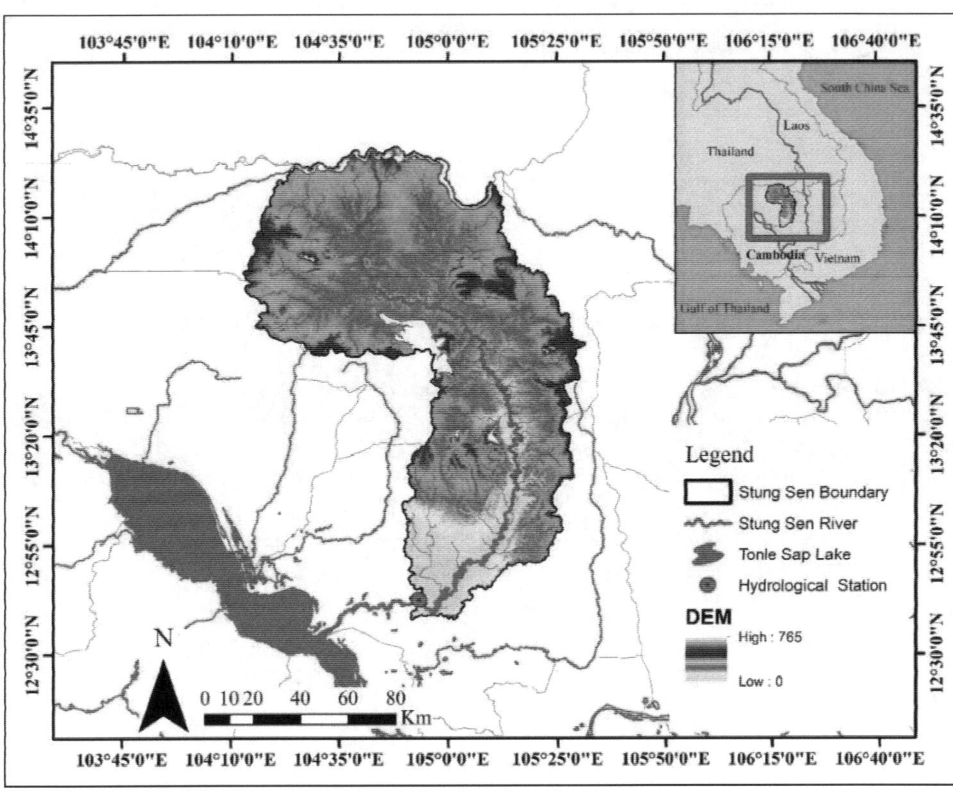

Figure 1. Location of Sen River Basin and Tributary.

2.2. The Framework of Study
The overall procedure of the methodology was presented as a flowchart to achieve the specific objectives as stated in this study (Figure 2). Accordingly, an index model has been developed in a GIS environment aiming to define flood hazard areas with a regional focus. The developed model performs a multi-criteria analysis incorporating with Flood Hazard Index (FHI). The FHI was used to assist the identification of hotspots related to flood risk and allow a comparative analysis between different basins [10]. The AHP element was computed to define weight of each indicator based on different criteria. In overall, the comparison of the two indices and the corresponding flood hazard maps supports the identification of prone areas, while records of historical flood events verify the accuracy of the methodology.

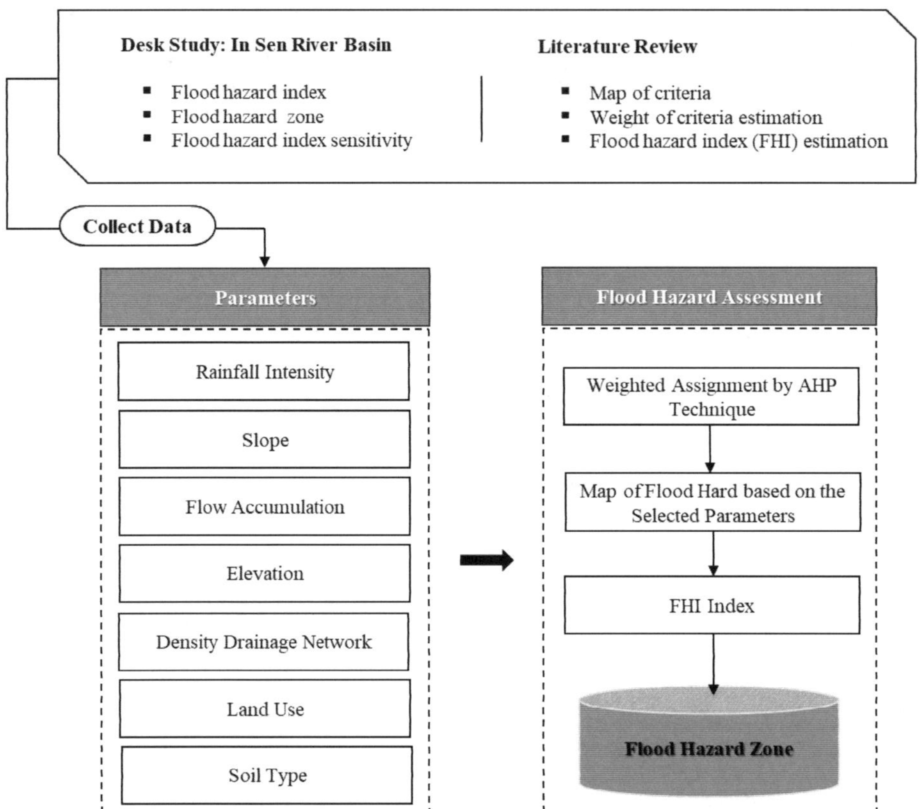

Figure 2. Overall flowchart of Multi-criteria of AHP and FHI.

2.3. Data Acquisition
The data of shapefile were acquired from Mekong River Committee (MRC). The topographic factors (flow accumulation, elevation, and surface slope) can be derived from the DEM with a spatial resolution of 30 m. Land use and soil type with resolution 250m × 250m were given from a high-accuracy land use and soil map. Rainfall intensity was calculated using the annual rainfall from rainfall data obtained from 3-gauge stations. These parameters include river network were integrated in with the GIS environment to obtain thematic maps showing the research results relative to flood hazard. Natural hazards such as floods are multidimensional, and hence, any analysis and mapping of these events require a comprehensive database [11].

2.4. Determining criteria weight
Analytical Hierarchy Process (AHP) is a method of organizing and analysing complex decisions and decision-making that developed since 1980 by Thomas Saaty [12]. It can be considered as an effective approach that can be used for assessing and intergrading the impacts of various factors, and all factors/parameters (criteria and sub-criteria) that matched each other, are normally defined by using

pairwise comparisons [13]. In general, the AHP is comprised of five steps: (1) break down a decision problem into component factors, [14] arrange these factors in a hierarchical order, (3) assign numerical values to determine the relative importance of each factor based on subjective relevance, (4) set up a comparison matrix, and (5) compute the normalized principal eigenvector, which gives the weight of each factor [15].

The analytical hierarchy process (AHP) was selected in this study for comparing the relative importance and defining the weight values of each factor through a pairwise comparison matrix A = [a_{ij}], a 7 × 7 matrix shown in Table 1, where a_{ij} is the elements of the matrix for $i, j = 1, 2, ..., n$, $a_{ij} = 1$ for $i = j$ and $a_{ij} = 1/a_{ij}$ for $i \neq j$, while a_{ji} is the reciprocal value of a_{ij}.

Table 1. Parameters of flood hazard of Analytical Hierarchy Process (AHP).

Parameters	Flow	Elevation	Density	Rainfall	Land use	Slope	Soil
Flow	1	2	3	4	3	6	7
Elevation	0.5	1	2	4	2	3	3
Density	0.25	0.5	1	3	1	2	2
Rainfall	0.5	0.25	0.33	1	2	2	3
Land use	0.33	0.5	1	0.5	1	3	2
Slope	0.17	0.33	0.5	0.5	0.33	1	1
Soil	0.14	0.33	0.5	0.33	0.5	1	1

The weight of the criteria was classified regarding their relative importance to the cause of flood (Table 1). The important scale varies from 1 to 9, indicating less important to much more important criteria, respectively (Table 2). The weights of each factor (w_i) was estimated by using Saaty's Method through normalizing pairwise comparison matrix A, that matrix A is transformed into matrix B = [b_{ij}], by calculating the arithmetic averages from the row of the normalized comparison matrix B according to the following equation:

$$w_i = \frac{\sum_{j=1}^{n} b_{ij}}{n} \quad (1)$$

Where, b_{ij} is the element of matrix B after normalizing pairwise comparison matrix A, defined by equation [14] and n is number of elements in the row.

$$b_{ij} = \frac{a_{ij}}{\sum_{i=1}^{a_{ij}} a_{ij}} \quad (2)$$

To ensure the differences between the pairwise comparisons and the reliability weights that were calculated, then the consistency ratio was computed to check the consistency. The consistency of the judgement or the comparison matrix can be checked by the CR that allows the judgement or the preference to be consistent and accepted if the value of CR is less than 0.1. The judgement and weight of each criterion need to be check and calculate again if CR is excess 0.1 [12]. The consistency ratio (CR) can be defined using the following equation:

$$CR = \frac{CI}{RI} \quad (3)$$

$$CI = \frac{\lambda_{max} - n}{n - 1} \quad (4)$$

$$\lambda_{max} = \frac{1}{n} \sum_{i=1}^{n} \frac{(Aw)_i}{w_i} \quad (5)$$

where, CI is the consistency index calculated by equation (4), and RI is the random index whose value is defined by the number of criteria (n), shown in Table 3. λ_{max} is the largest eigenvalue of the comparison matrix and n is the number of criteria. A is referred to matrix A, w is the eigenvector of matrix A or the weight of each criterion ($w = [w_i]$; $i = 1, 2, ..., n$), and n is the total number of eigenvector's elements.

Table 2. Scale of decision comparison matrix of conditioning factor.

Scale	Definition
9	Extreme importance
8	Very Strong to extreme importance
7	Very strong importance
6	Strong to very strong importance
5	Strong importance
4	Moderate to strong the importance
3	Moderate importance
2	Equal to moderate importance
1	Equal importance

2.5. Analytical Hierarchy Process (AHP) Evaluation Model

After the weight was defined, the seven parameters (Flow accumulation, Rainfall intensity, Soil type, Land use, Slope, Elevation, and Density from the drainage network) were classified into classes (51 sub-categories). Then, each sub-category was specified for a suitability rating value (Table 5) and reclassified in order to get the final flood hazard map. However, before getting the final map, all shapefile maps of each factor are required to convert to raster maps to active the reclassification of sub-categories based on the new rating.

Finally, the weighted sum overlay in the GIS environment was applied based on the equation (6) to estimate the flood hazard index of Sen River Basin.

$$FHI = \sum_{i=1}^{n} r_i . w_i \quad (6)$$

Where, r_i is the rating of the parameter in each point, w_i is the weight of each parameter, and n is the number of the criteria.

Table 3. Random Index (R.I.) for n=1, 2 … 8 [16].

Number of criteria (n)	1	2	3	4	5	6	7	8
Random Index (RI)	0	0	0.58	0.9	1.12	1.24	1.32	1.41

3. Results and Discussion

3.1. Normalized Weights and Rating of the Factors

Table 4. illustrates that the flow accumulation has the most weight with the value of 0.346, while the soil type weighs 0.051 which is the lowest. The rating follows in order such as Flow accumulation, Elevation, Density, Rainfall Land use, slope and soil type whose the rating are 1,2,3,4,5,6, and 7. The weight of each parameter is defined following the Analytical Hierarchy Process (AHP) with the Consistency Ratio (CR) = 0.046.

AHP is a structured technique used for analyzing complex problems, where a large number of interrelated objectives or criteria are involved [15]. The weights of these criteria are defined after they are ranked according to their relative importance [11]. The first stage of the AHP technique consists to define of the decision problem. Preparing a pairwise comparison matrix constitutes the second stage of the technique. An $n \times n$ dimensional pairwise comparison matrix of the conditioning factors is prepared at this stage. The importance value scale proposed by Saaty in 1980 is shown in Table 2.

Table 4. Normalized comparison matrix for evaluation of flood hazard of Analytical Hierarchy Process (AHP).

Parameters	Flow	Elevation	Density	Rainfall	Land use	Slope	Soil	Weight
Flow	0.346	0.407	0.36	0.30	0.305	0.333	0.368	0.346
Elevation	0.173	0.204	0.24	0.30	0.203	0.167	0.158	0.206
Density	0.087	0.102	0.12	0.225	0.102	0.111	0.105	0.122
Rainfall	0.173	0.051	0.04	0.075	0.203	0.111	0.158	0.116
Land use	0.114	0.102	0.12	0.038	0.102	0.167	0.105	0.107
Slope	0.059	0.067	0.06	0.038	0.034	0.056	0.053	0.052
Soil	0.048	0.067	0.06	0.025	0.051	0.056	0.053	0.051

3.2. Multi-influencing factors of Flood Hazard Index

Flow Accumulation. The Flow accumulation was derived from the flow direction raster. The highest flow accumulation is 14,582,000 pixels. High flow accumulation values indicate that water flow is more concentrated, possibly leading to a higher flood risk. In this sense, an increase in flow accumulation should reflect an increase in flood susceptibility. Accumulation flow sums the water flowing indicates a region of concentration flow and consequently highest flood hazard [19]. The class of flow accumulation raster was defined and classified into five classes as shown in Figure 3.

3.2.1. Elevation. The altitude has a signification impact on the propagation of flood Hazard. Flat regions are in low elevation are inundated quicker than areas at higher altitudes. Naturally, low elevations have been assigned the highest ratio as a prone region, which the highest is 765m from the sea level. The digital elevation model was interpolated from contour lines and elevation points in the topographic map. Elevation mostly inversely affects flooding. Flood frequency increases with decreasing elevation. The original hypsometric intervals were reclassified into five classes of susceptibility according to elevation levels as shown in Figure 3.

3.2.2. Slope. The Slope has been calculated from DEM and it is considered as the important parameter from the weight score of AHP. A slope degree map was generated from DEM. Moreover, slope classes used for reclassification were defined for the natural Breaks (Jenks) in five classes as shown in Figure 3. Slope weigh was 5.2% according to the AHP calculator by according to Table 4.

3.2.3. Rainfall Intensity. Averages of annual rainfall were calculated according to monthly rainfall data from three station rainfall station for 10 years of data from 2011 to 2020, which the mean annual precipitation is around 16000 mm. Subsequently, a mean annual rainfall was generated by employing interpolation techniques by using ArcGIS software as displayed in Figure 3. Rainfall intensity was, use past tense to describe methodology, present tense to describe facts and numbers that are always true in all studies performed considering the allocation of stations in the studied area. Rainfall intensity was divided into five classes from 1,325.9mm to 1,931.4mm (Table 5).

3.2.4. Density Drainage Network. Drainage density is an important element of control measures which expresses the length of the river per unit of area (Km/Km^2). As the river network density, higher density generally means increased surface runoff and increased probability of flooding. The highest drainage density is 1.2 km/km^2. Density drainage network map was produced employing by ArcGIS then reclassified into five classes from low to high according to Figure 3.

3.2.5. Land Use. Land use influences infiltration rate the interrelationship between surface and groundwater, and debris flow. Thus, while forest and lush vegetation favour infiltration, urban and pasture area support the overland flow of water. The group of land-use regions is shown in Figure 3.

3.2.6. Soil Type. The texture and structure of each soil type can influence significantly their permeability and consequently their capacity of water storage. Soil Type of Flood Hazard is an important criterion because it may extenuate or aggravate the magnitude a flood event. Soil type was divided into eight classes from low to high (Figure 3) and the weight of AHP is 0.051 as shown in Table 5.

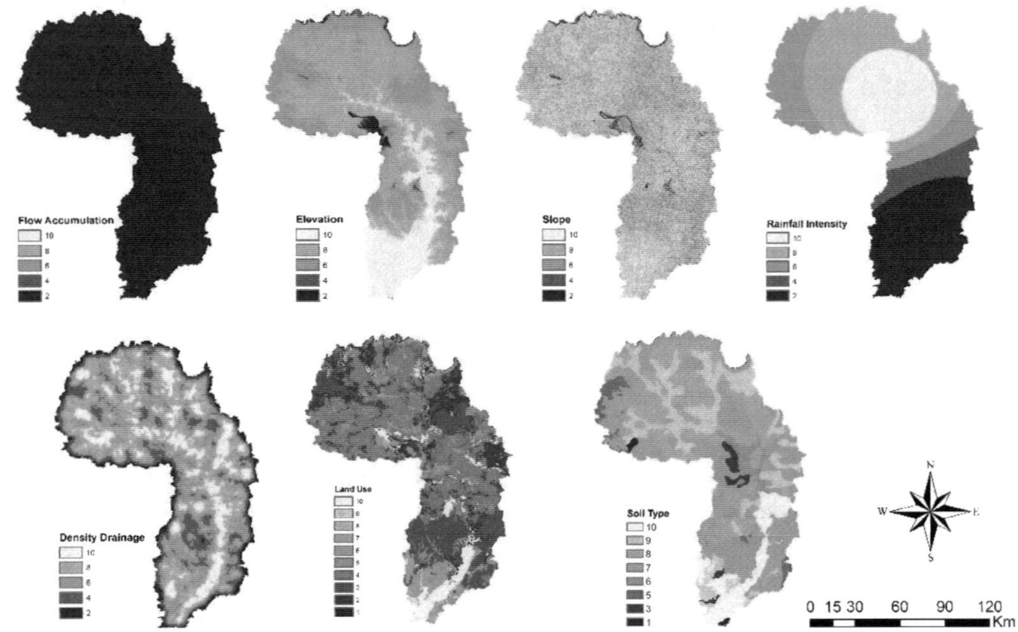

Figure 3. The map of multi-Criteria (7 parameters).

According to Table 5, there were 7 criteria that detail the reclassification of each parameter. The rating was given by each class which was the maximum 10 and minimum of 1. The highest weight of AHP is 0.346 (Flow accumulation) and the minimum of AHP is 0.051 (Soil Type).

Table 5. Class of parameters, rating, and weight.

Parameters	Class	Rating	Weight
Flow accumulation (pixels)	11, 952,000 – 1,450,200	10	0.346
	8,177,300 - 11, 952,000	8	
	2,516,200 - 8,177,300	6	
	629,030 - 2,516,200	4	
	0 - 629,020	2	
Rainfall (mm/day)	1,572 - 1,631	10	0.116
	1,514 - 1,572	8	
	1,442 - 1,571	6	
	1,375 - 1,442	4	
	1,325 - 1,375	2	
Slope (degree)	0 - 3	10	0.052
	3- 7	8	
	7 - 11	6	
	12 - 21	4	
	22 - 62	2	

Elevation (m)	0 - 54	10	0.206
	54 - 98	8	
	98 - 188	6	
	188 - 364	4	
	364 - 765	2	
Density drainage (Km/Km^2)	0.96 - 1.2	10	0.122
	0.83 - 0.95	8	
	0.7 - 0.82	6	
	0.51 - 0.69	4	
	0.049 - 0.5	2	
Land use	Agriculture and Paddy	10	0.107
	Water and urban	9	
	Cropping	8	
	Grarigue	7	
	wood shrubland	6	
	Regrowth	5	
	Deciduous	4	
	Evergreen	3	
	Mix evergreen	2	
	Forest land	1	
Soil Type	Gleyic Acrisol	10	0.051
	Gleyic-plinthic	9	
	Ferric Acrisol	8	
	Haplic Acrisol	7	
	Eutric Leptosol	6	
	Dystric	5	
	Dystric Leptosol	3	
	Rock out crop	1	

3.3. Assessing of the Flood Hazard Potential Indices

Flood Hazard Index map based on AHP as shown in Figure 4 was produced in GIS environment. The proposed methodology linearly combined the selected parameter, taking into account the relative weights. After applying the weighting sum of all conditioning factors, the map of flood hazard showed a pattern of flood influenced by the flow parameter due to its high weight of 0.346 of AHP. There were five classes of Flood Hazard Index such as Very Low, Low, Medium, High, and Very High as shown in flood hazard map (Figure 4.a).

Based on the flood hazard map, the flood hazard index of High and Very Low class covered 5.87% and 44.26 % of the entire basin. The Medium covered 14.35 %, while the Low recorded 29.35% of the area. The very high flood hazard covered 6.16% of the basin as shown in Table 6.

Table 6. Area coverage of flood Hazard Index classes.

Flood Hazard Index	Area (km^2)	Area (%)
Very Low	6175.179	44.265
Low	4094.524	29.35
Medium	2002.486	14.354
High	818.753	5.869
Very High	859.568	6.162

The effectiveness of the AHP methodology could be confirmed by very agreement between flood hazard zone that displayed in flood hazard map and flood historical map, possible limitation, and source of uncertainties should be discussed. The revised weight factors and the corresponding maps were compared with those obtained in the initial hypothesis. First, the number of flood conditioning factors as well as the question of which factor should be chosen in order to characterize the physical characteristic that determines the propensity of the study area to flooding. In this study, seven conditioning factors were selected similarly as in another flood hazard index.

FHI and Flood history maps are represented in Figure 4. The Flood map was downloaded from NASA which represents the maximum flood extent during 1984 – 2020. The flood history map derived from Google Earth Engine was divided into resolution 30m and the thresholds - 15 which display more flood areas than the map from global surface water [17]. Based on FHI and the flood history map, it can be seen that the AHP method applied to FHI was acceptable by comparing to both of the flood history maps.

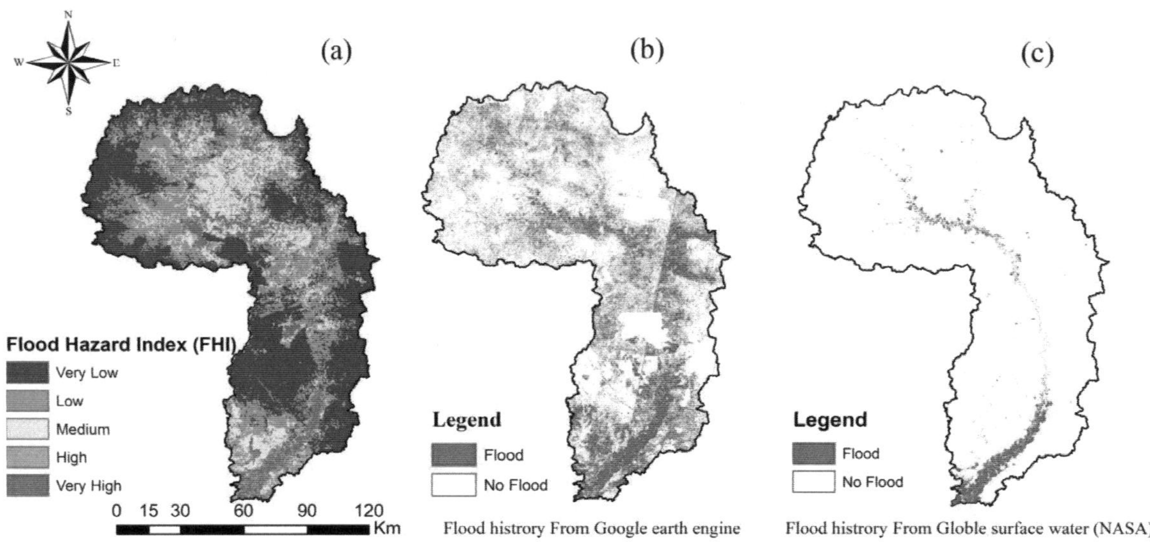

Figure 4. Flood Hazard: (a) Flood hazard Index map from AHP methodology, (b) Flood history map from google earth engine and (c) Flood Hazard history map from Global surface water.

During the monsoon season, Tonle Sap Lake expands as far as about 10 km west of Kampong Thom [6] which leads to why most floods occur at the downstream most floods occur at the downstream region and riverbank. The uplands are <100 m in the uppermost part of the lower reach, and they gradually descend in elevation to less than 20 m around Kampong Thom [14]. Residual hills of sandstone are distributed in a line above the upland southeast of Kampong Thom. Thus, there are less flood area at upstream of the basin. Through the validation with the flood history maps, this can be predicted to the future flood hazard area where the very high risk remains in the same area as presented in the FHI map. This study is the first attempt to assess the inland storm flood risk by GIS-AHP method and the first production of rainstorm flood risk zoning map. However, FHI mapping is particularly on a national or larger scale, requires the resolution of the problem of finding the optimal balance between the complexity of the flood hazard model and the accuracy and quantity of the input data. The result of the flood vulnerability maps could be further compared and validate through with the national data base and satellite image for some of the historical severe flood events to reinforce the validity. Future work will address the challenges associated with data consolidation, in terms of the original scale, resolution, or generalization, as well as their integration.

4. Conclusion

Integrating AHP with the GIS environment for evaluating FHI is applicable in Stung Sen River Basin. Flow accumulation was found be main factor that contributes to the flood hazard followed by elevation. Consequently, flood hazard often occurs at the downstream region, Stung Sen Town. The comparison of the flood hazard maps obtained from the FHI and Flood hazard history from NASA and manages to describe better high- and very high-risk areas.

In addition, FHI was compared with the flood historical point showing very good agreement between the FHI zone and historical flood events of which 6.162% were coincident with very high Flood Hazards.

At the same time, the simulate flood inundation areas were compared with flood maps derived from NASA imagery to check accuracy. As the result, it was confirmed that these results were reasonable and adequate. Furthermore, FHI will be included with vulnerability to predict the human vulnerability and also apply with the region vulnerable.

Acknowledgments
The study was supported and funded by SUMERNET 4 All Joint Action Project for strengthening flood risk management caused by climate change in Stung Sen River Basin, Cambodia (FloodCam). In addition, the authors would like to thank the high participation to JICA Project for AUN/SEED-Net for fully covering the authors to join this regional conference.

References

[1] Schanze, J., E. Zeman, and J. Marsalek, Flood risk management: hazards, vulnerability and mitigation measures. Vol. 67. 2007: Springer Science & Business Media.
[2] Kourgialas, N.N. and G.P.J.H.S.J.J.d.S.H. Karatzas, Flood management and a GIS modelling method to assess flood-hazard areas—a case study. 2011. 56(2): p. 212-225.
[3] Klemas, V.J.J.o.C.R., Remote sensing of floods and flood-prone areas: an overview. 2015. 31(4): p. 1005-1013.
[4] Meyer, V., Scheuer, S., & Haase, D., A multicriteria approach for flood risk mapping exemplified at the Mulde river, Germany. 2009. 48(1): p. 17-39.
[5] Luu, C., Von Meding, J., & Kanjanabootra, S., Assessing flood hazard using flood marks and analytic hierarchy process approach: a case study for the 2013 flood event in Quang Nam, Vietnam. 2018. 90(3): p. 1031-1050.
[6] Nagumo, N., T. Sugai, and S.J.G. Kubo, Late Quaternary floodplain development along the Stung Sen River in the Lower Mekong Basin, Cambodia. 2013. 198: p. 84-95.
[7] Oeurng, C., et al., Assessing climate change impacts on river flows in the Tonle Sap Lake Basin, Cambodia. 2019. 11(3): p. 618.
[8] Theara, T., et al., Integrated modeling to assess flow changes due to future dam development and operation in Stung Sen River of Tonle Sap Lake Basin, Cambodia. 2020. 11(4): p. 1123-1133.
[9] Nagumo, N., T. Sugai, and S.J.G.a. Kubo, Location of a pre-Angkor capital city in relation to geomorphological features of lower reach of the Stung Sen River, central Cambodia. 2010. 23(5-6): p. 255-266.
[10] Kazakis, N., I. Kougias, and T.J.S.o.t.T.E. Patsialis, Assessment of flood hazard areas at a regional scale using an index-based approach and Analytical Hierarchy Process: Application in Rhodope–Evros region, Greece. 2015. 538: p. 555-563.
[11] Chakraborty, S. and S.J.N.H. Mukhopadhyay, Assessing flood risk using analytical hierarchy process (AHP) and geographical information system (GIS): application in Coochbehar district of West Bengal, India. 2019. 99(1): p. 247-274.
[12] Mu, E. and M. Pereyra-Rojas, Practical decision making: an introduction to the Analytic Hierarchy Process (AHP) using super decisions V2. 2016: Springer.
[13] Thomas, P.G. and P.C. Doherty. The Analytic Hierarchy. in Process: Planning, Priority Setting, Resource Allocation, McGraw-Hill. 1980. Citeseer.
[14] Lin, Z., et al., Defect engineering of two-dimensional transition metal dichalcogenides. 2016. 3(2): p. 022002.
[15] Saaty, T.L., Fundamentals of the analytic hierarchy process, in The analytic hierarchy process in natural resource and environmental decision making. 2001, Springer. p. 15-35.
[16] Siddayao, G.P., et al., Analytic hierarchy process (AHP) in spatial modeling for floodplain risk assessment. 2014. 4(5): p. 450.
[17] Dalin, C., Konar, M., Hanasaki, N., Rinaldo, A., & Rodriguez-Iturbe, I., Evolution of the global virtual water trade network. 2012. 109(16): p. 5989-5994.

River Water Quality Implication after the Earthquake

Leangkim Kong*[1], Aznah Nor[1], Mohd Hakim[1], NF. Said[2], J. Asanuma[3]

[1] **Department of Chemical and Environmental Engineering, Malaysia-Japan International Institute Technology, Universiti Teknologi Malaysia, Kuala Lumpur, Malaysia.**

[2] **Water Quality, Laboratory, and Material Testing Branch, Sabah State Water Department, Malaysia.**

[3] **Faculty of Life and Environmental Science, Tsukuba University, Japan.**

***Corresponding author email:** kongleangkim@gmail.com

Abstract. The earthquake of 6.0 Magnitude occurred in the Ranua district of Sabah lasted for 30 seconds on 5th June 2015. This study investigated the various water quality parameters such as turbidity, color, dissolved oxygen, pH, electric conductivity, total dissolved solids, and so on from two monitoring stations (Bambangan and Kimolohing). The secondary data was requested from the Water Department of Sabah, where the Liwagu river is the downstream system of Mesilau. Through the temporal trace of each parameter, the water quality parameters showed an abrupt change one day before the earthquake. Consequently, around two weeks after the quake, it reached its initial state rapidly. After the earthquake, the mean turbidity is 436.25 NTU with a standard deviation of 1416.86 NTU. Its peak was on 17th June with 5550 NTU, total dissolved solids were 58.68 mg/l, and the electrical conductivity was 122.98 μs/cm at the Bambangan station, while it was 673.00 NTU with a standard deviation of 1869.90 NTU and it peaked on 17th June with 3750 NTU in the Kimolohing. The dissolved oxygen hit the lowest level of 3.27 mg/l on 17th June when 4.19 mg/l. The nitrate concentration increased to the maximum of 0.18 mg/l and 0.15 mg/l in the Bambangan, and the Kimolohing station, respectively. The metal concentrations also increased after the earthquake at both stations. However, alkalinity and hardness had decreased since 30th April before the earthquake and continued to decline until 14th August, when it reached the initial level. The value of pH and chloride did not seem to be affected by the earthquake, as the concentration maintained the average level.

1. Introduction

Water quality change in the vulnerable area, or the location where an earthquake occurred, might significantly disrupt the environmental system and society. Over decades, water quality changed due to earthquakes in numerous regions [1-11]. The shacking impaired the sewage system, infrastructure, and treatment facility. It causes micro-organisms and pathogen indicators with sediment discharge from the raw sewage into the river. That increases the potential risk to human health for those involved in recreational and work-related activities in the river environment. In addition, the impact of secondary hazards such as landslides, debris flow, and sediment liquefaction might cause a disturbance in water quality and take a long time to recover.

Moreover, changing a dominant parameter in water can destroy living organisms and the ecosystem in the catchment. The landslide triggered by the quake brings a hefty amount of sediment into the river deposits in the stream, making the channel shallower and increasing sediment concentration, turbidity, and the flood plain [12]. In addition, it also destroys the landscape vegetation, which boosts the surface runoff and flood velocity [13], the source of nutrient input in the river [14]. The reach-in sediment concentration in water also intrudes the absorbance of oxygen into the water, which is vital for lives in water and might cause the algae to bloom. Further, the high risk of water-borne diseases in water and the limitation of drinking water sources due to pollution concern the public's health [15]. The earthquake resulted from physical changes (drying up of springs, temporary damming of streams, and increase in erosion) and ecosystem services destruction [16]. After the earthquake, water quality changes, such as in groundwater, spring, and surface water, have been well established and documented [1-4, 6-11, 17].

On the other hand, the research paper on the influence of various sources of pollution, such as human activity [18-19], and other natural phenomena, such as climate change and floods, are well produced [20]. However, the research paper on surface water quality degradation caused by earthquakes seemed significantly limited. Moreover, the threat to surface water quality from human activity, earthquakes, or other natural disaster might temporarily or permanently affect living organisms, ecosystems, and humans in catchment areas where surface water is the primary source. Therefore, it is necessary to address this issue by producing more research papers.

Ranau experienced an earthquake of 6.0 Mw on 5th June 2015. This earthquake lasted about 30 seconds. The shaking from the mainshocks and aftershocks triggered the extensive landslide on the very steep slope of Mount Kinabalu, stripping about 1500 hectares (15 Km^2) of vegetarian cover and producing a high amount of loose material. The debris flow wipes away the living organism, and its deposit influences water quality downstream. In addition, it results in increased turbidity, which is unusable for human use for several weeks. To protect the system, the authority needs to close down the water intake for drinking water treatment.

Since the region lies on the blind fault area, the earthquake will possibly trigger again in the future. Therefore, the study aims to evaluate water quality parameters in the Liwagu river before and after the earthquake on 5th June 2015 in Sabah. Hence, this study will provide valuable and additional information for the decision-maker and relevant stakeholders to implement the strategy and follow the river's status to recover and reduce the damage.

2. Material and Methods

2.1. Study area and data collection

This study was conducted at the Ligawu river in Ranau district, Sabah, Malaysia. Sabah had been hit by 65 earthquakes since 1973, with magnitudes ranging from 3.3 to 6.5 on a Richter scale [21], located around 1000 Km from the Philippine Trench and 1000 km from the Eurasian-Philippine plate boundary, sitting on the semi-stable South China Sea region. Therefore, the region faults of Sabah are closely related to the active part in the Sulu Sea and Celebs Sea region. The latest earthquake occurred in 2015. The 2015 Mount Kinabalu Earthquake, with a magnitude of 6.0 and a focal depth of 10 km located 6.014°N 116.563°E in the Kinabalu Mountain, Sabah in Borneo, came as follows: a surprise and lasted for about 30 seconds claiming 18 lives. Three aftershocks occurred in the earthquake's immediate aftermath, with the first and second measuring 4.3 magnitudes, while a third was 2.8 Mw. A total of 33 aftershocks have been reported at Ranau, ranging in magnitude from 1.6 to 4.5, recorded by the Malaysian Meteorological Department. Two significant earthquakes occurred earlier at Ranau, with the first recorded in 1966 at magnitude 5.3 and the second at magnitude 5.2 in 1991. Active tectonics of the converging and diverging plate margins are responsible for the occurrence of earthquakes. However, the absence of such an active plate margin in the earthquake location is uncommon [22].

In 2015 (**Table 1**), the secondary water quality data were from the Sabah water department; each parameter is monitored and analyzed following the American Public health Association (APHA) standard. There are two monitoring stations along the Liwagu river, 200km in length, where the river is a downstream system of the Mesilua river. The distance between stations is 1.6 Km, whereas Bamangan and Kimolohing are water intake points for drinking water treatment (**Figure 1**). The river flows eastwards off the southern slope of Mount Kinabalu and then into the Labuk River. The variation pattern of each parameter gives information on water quality variation before and after the earthquake Liwagu river is the primary water source for people in the Ranau district. This river suffered much damage from the earthquake in 2015.

Table 1: List of parameters

Parameters	Units	Limitation[a]
Turbidity	NTU	1000
Color	HU	300
pH	-	5.5-9.0
Electrical Conductivity	µs/cm	-
Total Dissolved solid	mg/l	1500
Dissolved Oxygen	mg/l	-
Ammonia	mg/l	1.5
Nitrate	mg/l	10
Iron	mg/l	1
Manganese	mg/l	0.2
Aluminum	mg/l	-
Alkalinity	mg/l	-
Total-Hardness	mg/l	500
Chloride	mg/l	250
Sulphate	mg/l	-

Limitations following the National Water Quality Standard for Malaysia

Figure 1: Location of water intake stations (Bambangan and Kimolohing)

3. Result and discussion

The water quality data were observed daily from January to December, covering before and after the earthquake on 5th June 2015. The data was divided into the phase before and after the mean and standard deviation earthquake to compare the data variation. **Table 2** and **Table 3** differentiate the mean value of each water quality parameter at Bambangan and Kimolohing stations before and after the earthquake and portray the dispersion of the dataset. The turbidity value was 436.25 NTU ± 37.92 NTU than before the quake, 36.10 NTU ± 1416.86 NTU (**Table 2**). This number indicates the rising of turbidity after the shock.

Similarly, at the Kimolohing station, as shown in **Table 3**, the turbidity level increased after the quake to 309.61 NTU ± 954.49 NTU. From **Figure 2** (a) and (b) Bambangan station and Kimolohing station, turbidity increased quickly from 4th June, the day before the earthquake on 5th June, until around two weeks after the earthquake on 17th June. Thus, the turbidity level remains higher than usual after the aftermath. The color level rose quickly from 4th June to 17th June and decreased to 14th August. The heavy rain causes this increase in turbidity and color after the incident, producing debris flow and mudflow in the stream [22]. Color in water is known to indicate water clarity. It was high compared to before the earthquake in both stations because of the turbidity related to suspended solids in water. Hence, the high level of the suspended substance in water causes the color of the water to be cloudy, affecting the river's living organism if this high level remains for a long time. As a result of reducing the light penetrating water, it also slows down the photosynthesis process. Prompt the eutrophication to release a harmful toxic substance to humans and living organisms.

The total dissolved solids and electrical conductivity decreased by comparing both stations' mean and standard deviation. **Figure 4** (a); (b), the total dissolved solids and electric conductivity concentration dropped on the same day, from 30th April to 21st May, rose on 4th June, and fell again on 17th June. These two parameters are reported as significant changes related to seismic activity in the groundwater system [6, 23]. Additionally, total dissolved solids and electric conductivity have a substantial relationship. Therefore, the increase of total dissolved solids also increases the concentration of EC, and diluted rainwater volume into the stream is probably a reason.

Dissolved oxygen, the most vital element to support the living organism, dropped slightly in both stations despite the mean value. According to **Figure 3** (c) and (d), the dissolved oxygen concentration in the Bambangan station gradually decreased from 21st May until 17th June. Unlikely, dissolved oxygen in the Kimolohing station rapidly reduced from 4th June until 17th June. Plus, a high turbidity level might intrude the absorbance of oxygen from the atmosphere into water. However, there is an agreement of the lowest concentration of dissolved oxygen on the same day (17th June) in both stations, which indicates the change in dissolved oxygen level. The pH level remains at the mean value of around 7.0 before and after the earthquake. The pH level remains exact before and after the quake (**Figure 3**). Therefore, no significant change could indicate the event's impact.

The mean concentration of ammonia was 0.03 mg/l at Bambangan, before the earthquake, 0.06 mg/l. In comparison, in the Kimolohing station, it is 0.03 mg/l less than before the earthquake, 0.04 mg/l, and the limit of its concentration in water is 1.5 mg/l. Therefore, it is still at the safe level. **Figure 5** (a); (b) demonstrate the conflict of the variation pattern of ammonia between the stations of concentration changing related to the earthquake's impact. The nitrate concentration in Bambangan station **Figure 5** (c) increased from 4th June until 17th June. The nutrient concentration in the river usually contains fertilizer from agriculture fields or soap and detergent from the household. Other nitrogen sources have in the soil, discharge into the river, surface runoff, and soil erosion.

In the Bambangan station (**Table 2**), the mean concentration of nitrate is 0.04 mg/l ± 0.02mg/l (**Table 3**), while in Kimolohing is 0.05 mg/l ± 0.02 mg/l before the earthquake. After the earthquake, its concentration was 0.05 mg/l ± 0.04 mg/l. In Kimolohing **Figure 5** (d), nitrate concentration rapidly increased from 4th June until 17th June. A previous study by Sato *et al.* [24] found the change in the nitrate concentration due to the rainfall after the incident, which transported the nitrate into the river's catchment. Therefore, the transportation of sediment and material from the upstream and nearby catchment triggered by an earthquake might be the factor that makes it increases.

The metal concentration of aluminum, iron, and manganese increases according to the mean value and standard deviation except for aluminum in Kimolohing. Before the earthquake, the mean concentration is 0.05 mg/l; after the quake, is 0.04 mg/l. Therefore, aluminum concentration quickly increased from 4th June until 17th June and after the concentration of aluminum reached the initial state as shown in **Figure 6** (a); (b). Likewise, the concentration of iron rapidly increased from 4th June until 17th June and decreased until 14th August **Figure 6** (c); (d). The increase in manganese concentration started on 4th June until 17th June and dropped again on 11th September **Figure 7** (a); (b). A previous study [25] stated that the overall random increasing or decreasing trends in primary heavy metal contents after the earthquake portrayed that this imbalance might be due to mingling of the soil of multifarious composition caused by land sliding. The alkalinity level decreased after the quake at 60.97 mg/l $\pm$ 17.82 mg/l, 57.93 mg/l $\pm$ 23.24 mg/l before and after the earthquake. However, the average hardness level increased in both stations.

The alkalinity concentration in the Bambangan station **Figure 8** (a) decreased from 30th April until 14th August and rose again on 25th September. The Kimolohing **Figure 8** (b) showed a similar decreasing pattern except on 17th June, when it increased, even though it remained at an equivalent level of the Bambangan station. The hardness level in each station was decreased from 14th May until 4th June and increased again until 17th June. A report of rock falling from Mount Kinabalu during the shaking, where the small gravel and stone were transported into the river [22]. A similar study by Qian et al. [9] also found that the Wenchuan earthquake reduced the decrease in hardness in China. Besides, the chloride and sulfate concentration remains the same in both stations.

At the Bambangan station **Figure 7** (c), the sulfate concentration increased from 4th June until 17th June, increased again until 14th August, and decreased until 11th September. However, there is a disagreement between both stations; in the Kimolohing station **Figure 7** (d), there was no peak concentration before and after the earthquake. Therefore, the level of chloride change is minimal (**Figure 9**), which means the quake perhaps does not have the majority impact on its concentration in water [26].

The Liwagu river falls into Class I - Class II based on Water Quality Standard for Malaysia (NWQSM) and Water Quality Index [27]. Therefore, the longest variation time is around three months before the earthquake, and the shortest period is one day before the earthquake, which can considerably change rapidly. The recovery time is around two weeks after the quake until it returns to its initial condition. The 120 aftershocks were reported three months after the earthquake, and only five were considered significant [22]. Parameters such as sulfate, alkalinity, hardness, chloride, total dissolved solids, and electric conductivity fluctuated during the aftershock period. Nevertheless, the lack of information to verify if this aftershock has an impact. Conversely, a report of electric conductivity and total dissolved solids in underground water fluctuated during the aftershock in another region [6].

Moreover, heavy rain was recorded on 17th June. Thus, it might make sense that parameters such as turbidity, color, and metal concentration increase at the mentioned time in the river. The review of the water quality status in Malaysia by Garba *et al.* [28] stated that the primary sources of pollution in Sabah are logging and deforestation. In another aspect of the earthquake indicator, the temporal and spatial variation of water quality and human activity must be the factor to be considered. It must reduce the impact on the dataset, such as the location of the monitoring point, to eliminate the unrelated pollution of nearby activity of humans and distance from the epicenter [29]. The interaction of groundwater and surface water is a phenomenon to deliberate that the discharge from groundwater during seismic activity may cause a change in surface water such as dissolved matter, and salinity. Indeed, the series of earthquake events may provide more evidence and information on impending earthquakes' influence on river water quality and groundwater because groundwater is known to interact with the ground movement of seismic activity [30-31]. Thus, any assumption is not strongly justified by just one earthquake episode.

Table 2: Mean and standard deviation of parameters in Bambanganstation

Parameters	Before earthquake		After earthquake	
	Mean	Standard d.	Mean	Standard d.
Turbidity	36.10	37.92	436.25	1416.86
Color	91.54	79.72	760.00	2283.69
pH	7.55	0.40	7.53	0.12
EC	130.72	33.33	122.98	21.86
TDS	62.23	16.05	58.68	10.49
DO	8.28	0.58	7.84	1.48
NH3	0.06	0.07	0.03	0.03
NO4	0.04	0.02	0.05	0.04
Fe	0.25	0.23	1.13	1.11
Mn	0.03	0.02	0.14	0.15
Al	0.03	0.01	0.06	0.05
Alkalinity	60.97	17.82	57.93	23.24
Hardness	62.14	19.37	64.00	15.53
Cl	8.15	7.85	7.93	5.27
SO4	4.92	2.72	5.33	3.50

Table 3: Mean and standard deviation of parameters in Kimolohingstation

Parameters	Before earthquake		After earthquake	
	Mean	Standard d.	Mean	Standard d.
Turbidity	17.28	19.23	309.61	954.49
Color	58.93	67.77	673.00	1896.90
pH	7.53	0.23	7.55	0.10
EC	124.89	34.99	118.63	22.11
TDS	59.42	16.86	55.87	11.08
DO	8.25	0.58	8.11	1.55
NH3	0.04	0.04	0.03	0.04
NO4	0.05	0.02	0.05	0.04
Fe	0.16	0.11	1.07	0.54
Mn	0.03	0.02	0.15	0.20
Al	0.05	0.07	0.04	0.04
Alkalinity	59.51	18.20	54.79	23.02
Hardness	55.59	18.52	59.83	12.50
Cl	7.00	3.57	7.07	3.59
SO4	5.00	2.77	4.73	4.20

IOP Conf. Series: Earth and Environmental Science 1091 (2022) 012032 doi:10.1088/1755-1315/1091/1/012032

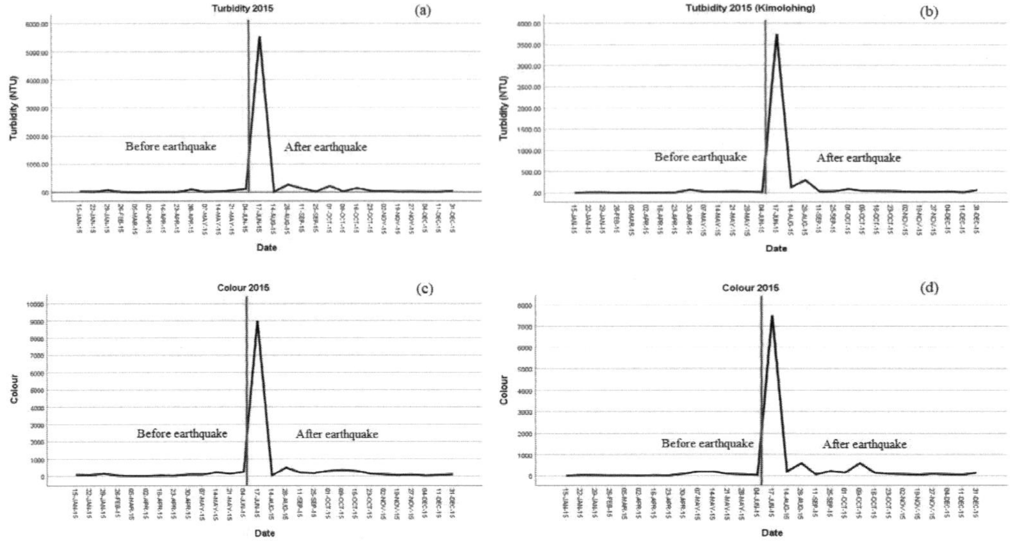

Figure 2: Turbidity and Colour (a), (c) Bambangan; (b), (d) Kimolohing

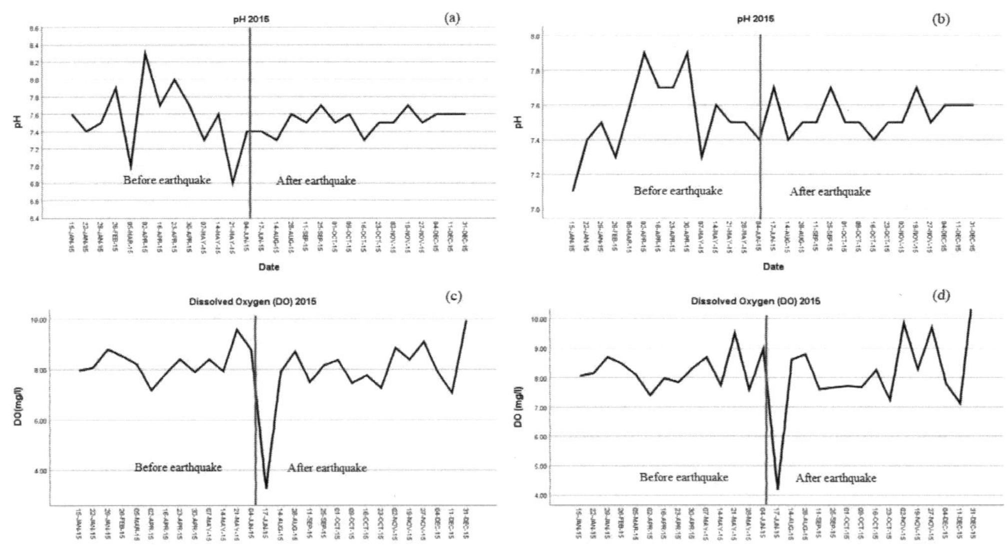

Figure 3: pH and DO (a), (c) Bambangan; (b), (d) Kimolohing

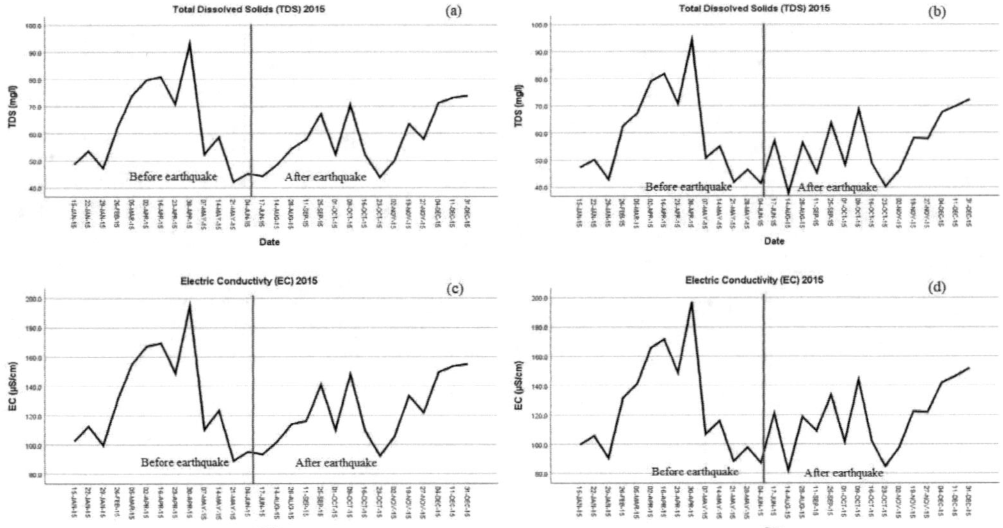

Figure 4: TDS and EC (a), (c) Bambangan; (b), (d) Kimolohing

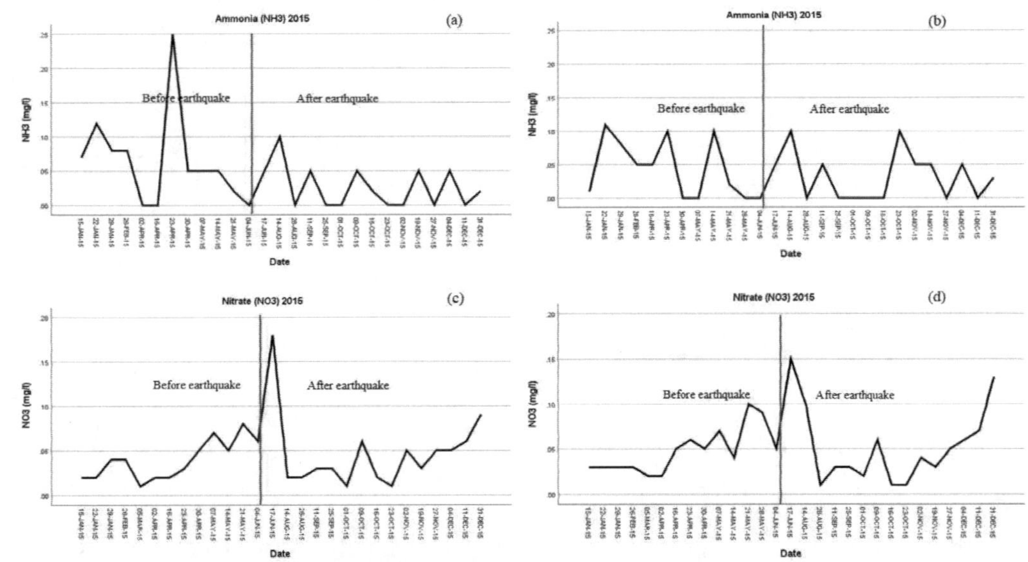

Figure 5: NH3 and NO3 (a), (c) Bambangan; (b), (d) Kimolohing

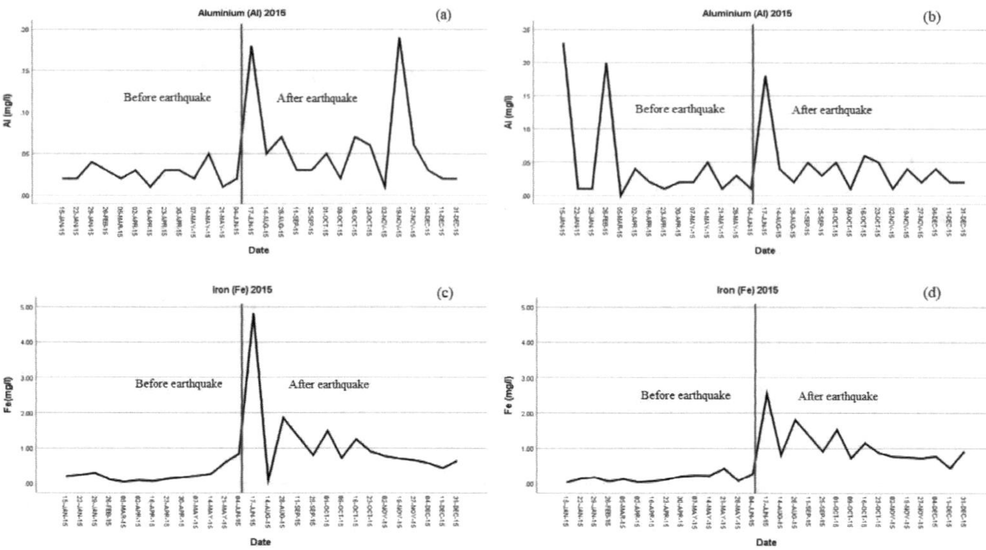

Figure 6: Al and Fe (a), (c) Bambangan, (b), (d) Kimolohing

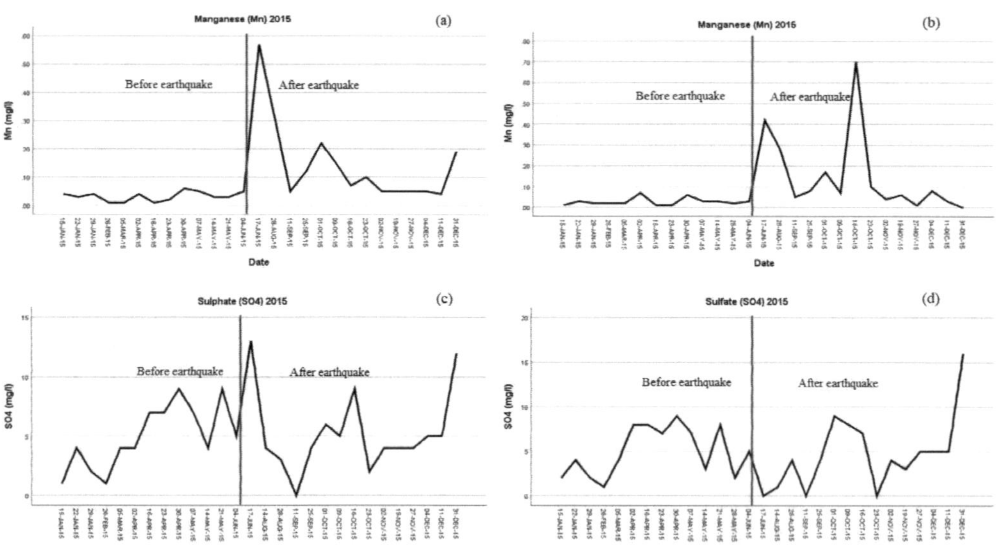

Figure 7: Mn and SO4 (a), (c) Bambangan; (b), (d) Kimolohing

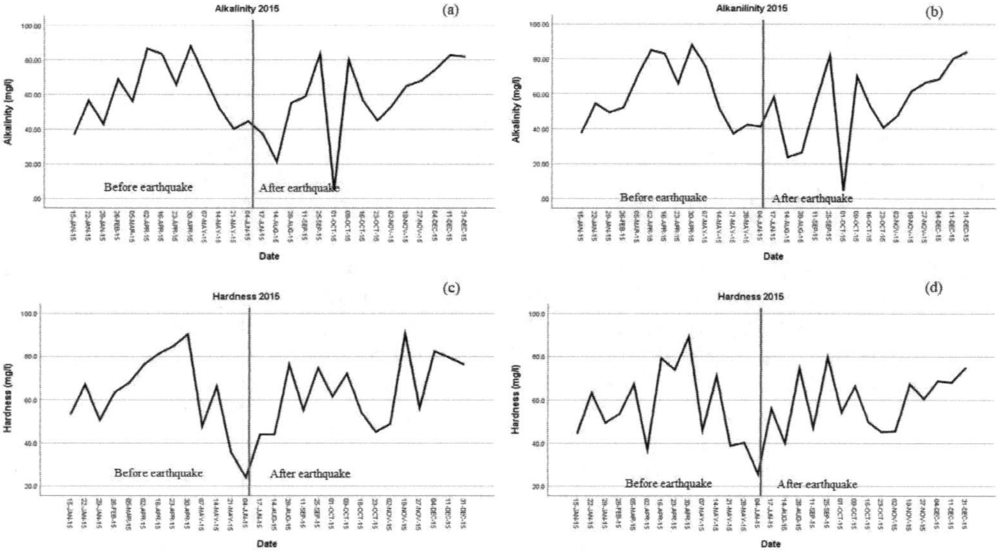

Figure 8: Alkalinity and Hardness (a), (c) Bambangan; (b), (d) Kimolohing

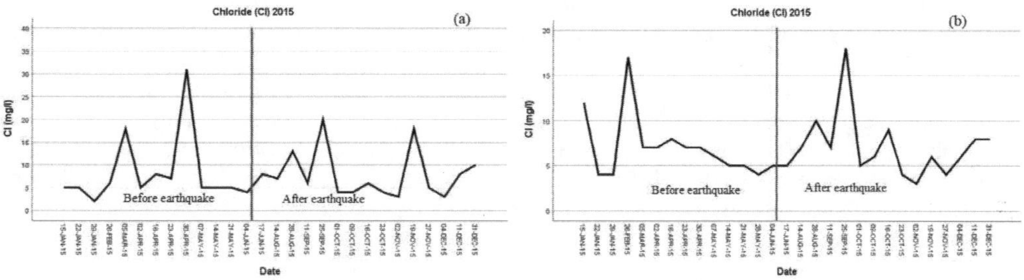

Figure 9: Cl (a) Bambangan; (b) Kimolohing

4. Conclusion and Recommendation

In conclusion, the earthquake has not only brought fatalities and damaged infrastructure but also has a significant disturbance to water quality on the surface and underground. The excessive total suspended solids cause damage to drinking water treatment and limit the accessibility of safe and clean water. The 5th June 2015 earthquake had a remarkable influence on the water quality of the Liwagu river. Through observing water quality in 2015, several river water quality parameters such as turbidity, color, total dissolved solids, electric conductivity, nitrogen content, and metal concentration were changed from one day to three months before the earthquake and still around two weeks later. Even though the fluctuation was detected, according to the national standard, it still follows the safe level, excluding turbidity and color showing higher values levels during and after the earthquake. These two parameters are the most vulnerable and highly affected; the debris and mudflow triggered by the earthquake might be the reason for the high turbidity and color level. It is advised to study a series of earthquake events to investigate the behavior of river water quality response to the earthquake process. The monitoring location is also a factor for future work to reduce the influence of point source and non-point source pollution. Plus, the relationship analyses between the parameters are recommended further to investigate the effect between the parameters. Water quality indexes are suggested to classify the status and impact in future circumstances.

Acknowledgments
This work is funded by the AUN/SEED-Net Grant no: R.K130000.7343.4B403 under Collaborative Education Program for Sustainable Environmental Engineering Network (CEP-SEEN).

References:

[1] Chen L, Wang G Hydrochemical changes of a spring due to the May 30, 2014 Ms 6.1 Yingjiang earthquake, southwest China 2021 *Environmental Pollution* **284** 117125

[2] Chien S-HJ, Chi W-C, Ke C-C Precursory and coseismic groundwater temperature perturbation: An example from Taiwan 2020 *Journal of Hydrology* **582** 124457

[3] Jakovljević D, Lozanov-Crvenković Z Water quality changes after Kraljevo earthquake in 2010 2015 *Natural Hazards* **79** 2033–53

[4] Kagabu M, Ide K, Hosono T, Nakagawa K, Shimada J Describing coseismic groundwater level rise using tank model in volcanic aquifers, Kumamoto, southern Japan 2020 *Journal of Hydrology* **582** 124464

[5] Lee S-H, Lee JM, Moon S-H, Ha K, Kim Y, Jeong DB, Kim Y Seismically induced changes in groundwater levels and temperatures following the ML 5.8 (ML 5.1) Gyeongju earthquake in South Korea 2021 *Hydrogeology Journal* **29** 1679–89

[6] Malakootian M, Nouri J Chemical variations of ground water affected by the earthquake in bam region Malakootian, M 2010 International Journal of Environmental Research **4** 443–54

[7] Morimura S, Zeng X, Noboru N, Hosono T Changes to the microbial communities within groundwater in response to a large crustal earthquake in Kumamoto, southern Japan 2020 Journal of Hydrology **581** 124341

[8] Pant BR, Rawal DS, Thapa K, Shrestha SM, Koju R, Pandey DR Post-earthquake water quality in Bhaktapur district, Nepal 2019 *Journal of water security* 1–7

[9] Qian F, Hu B, Liu JJ, Lin W, Xiong MB, editors. *Analysis of the Relationship between River Flow* and Water Quality before and after the Wenchuan Earthquake. Advanced Materials Research; 2013: Trans Tech Publ.

[10] Xiang Y, Sun X, Gao X Different coseismic groundwater level changes in two adjacent wells in a fault-intersected aquifer system 2019 *Journal of Hydrology* **578** 124123

[11] Yan R, Wang G, Ma Y, Shi Z, Song J Local groundwater and tidal changes induced by large earthquakes in the Taiyuan Basin, North China from well monitoring 2020 *Journal of Hydrology* **582** 124479

[12] Meena SR, Bhuyan K, Chauhan A, Singh RP Changes in the flood plains and water quality along the Himalayan rivers after the chamoli disaster of 7 February 2021 2021 *International Journal* of Remote Sensing **42** 6984–7001

[13] Zhang G, Cui P, Jin W, Zhang Z, Wang H, Bazai NA, Li Y, Liu D, Pasuto A Changes in hydrological behaviours triggered by earthquake disturbance in a mountainous watershed 2021 *Science of the Total Environment* **760** 143349

[14] Li L, Chen X, Zhang M, Zhang W, Wang D, Wang H The spatial variations of water quality and effects of water landscape in Baiyangdian Lake, North China 2022 *Environmental Science and* Pollution Research **29** 16716–26

[15] Ito Y, Shrestha Malla S, Bhattarai AP, Haramoto E, Shindo J, Nishida K Waterborne diarrhoeal infection risk from multiple water sources and the impact of an earthquake 2020 *Journal of* Water and Health **18** 464–76

[16] Chaudhry MN, Hussain S, Baloch IH, Dawood H Impact of 8th October 2005 Earthquake Associated with Kashmir Boundary Thrust (KBT), Pakistan 2021

[17] Lee SH, Ha K, Hamm SY, Ko KS Groundwater responses to the 2011 Tohoku earthquake on Jeju Island, Korea 2013 *Hydrological Processes* **27** 1147-57

[18] Mishra BK, Kumar P, Saraswat C, Chakraborty S, Gautam A Water security in a changing environment: Concept, challenges and solutions 2021 *Water* **13** 490
[19] Giri S Water quality prospective in Twenty First Century: Status of water quality in major river basins, contemporary strategies and impediments: A review 2021 *Environmental Pollution* **271** 116332
[20] Nazari-Sharabian M, Taheriyoun M Climate change impact on water quality in the integrated Mahabad Dam watershed-reservoir system 2022 *Journal of Hydro-environment Research* **40** 28–37
[21] Mohamed A, editor Monitoring active faults in Ranau, Sabah Using GPS. 19th United Nations regional cartographic conference for Asia and the Pacific Bangkok, Thailand; 2012; Bankok, Thailand.
[22] Tongkul F The 2015 Ranau Earthquake: Cause and Impact 2016 *Sabah Society Journal* **32** 1-–28
[23] Barberio MD, Barbieri M, Billi A, Doglioni C, Petitta M Hydrogeochemical changes before and during the 2016 Amatrice-Norcia seismic sequence (central Italy) 2017 *Scientific Reports* **7** 11735
[24] Sato T, Takahashi HA, Kawabata K, Takahashi M, Inamura A, Handa H Changes in the nitrate concentration of spring water after the 2016 Kumamoto earthquake 2020 *Journal of Hydrology* **580** 124310
[25] Mirza MA, Khuhawar M, Arain R, Choudhary MA, Mahar KP Post-Earthquake Impact on Quality of Spring Water in Northern Pakistan 2015 *Asian Journal of Chemistry* **27** 1658
[26] Isidore F, Han LK, Bidin K Water quality status of Liwagu River, Tambunan, Sabah, Malaysia 2013 *Journal of Tropical Biology & Conservation (JTBC)*
[27] Cleophas N, Isidore F, Lee K, Han K, Bidin K Water quality status of Liwagu River, Tambunan, Sabah, Malaysia 2013 Journal of tropical biology and conservation **10** 67-73
[28] Garba TE, Richard RL, Thani NEA, Majid MAA, Lawal M Geological Effects on Water Quality: A Review of Issues and Challenges in Malaysia 2021 *Sains Malaysiana* **50** 1857–70
[29] King C-Y, Zhang W, Zhang Z Earthquake-induced groundwater and gas changes 2006 *Pure and Applied Geophysics* **163** 633–45
[30] Barbieri M, Franchini S, Barberio MD, Billi A, Boschetti T, Giansante L, Gori F, Jónsson S, Petitta M, Skelton A Changes in groundwater trace element concentrations before seismic and volcanic activities in Iceland during 2010–2018 2021 *Science of the Total Environment* **793** 148635
[31] Hosono T, Saltalippi C, Jean J-S Coseismic hydro-environmental changes: insights from recent earthquakes 2020 *Journal of Hydrology* **585** 124799

IOP Conf. Series: Earth and Environmental Science 1091 (2022) 012033 doi:10.1088/1755-1315/1091/1/012033

BSC (biological soil crust) Method for soil erosion control and vegetation recovery: Case studies in Japan

T C T Nguyen[1], J Md Khudzari[1], F Imamura[2], S Sethi[2] and K Iwamoto*[1]

[1] Malaysia-Japan International Institute of Technology (MJIIT), Universiti Teknologi Malaysia, Jalan Sultan Yahya Petra, 54100 Kuala Lumpur, Malaysia
[2] Research & Development Center, Nippon Koei Co., Ltd, Tokyo 102-8539, Japan

* Corresponding author e-mail: k.iwamoto@utm.my

Abstract. The biological soil crust (BSC) is a naturally occurring vegetation layer in the topsoil that is composed of a diverse array of microscopic living organisms such as cyanobacteria, green algae, lichen, liverworts, moss, micro-fungi, and other bacteria. BSC cover is well recognized for mitigating soil erosion caused by wind forces and overland flow. By forming a sheet-like soil microbe community, BSC cover can bind loose soil particles, stabilize and protect soil surfaces. This article reviews the applications of human-induced BSC, namely BSC Method, that was developed by the Japanese company Nippon Koei Co. Ltd. The BSC Method utilizes an algal species called BSC-1. We also discussed the BSC projects undertaken by Nippon Koei Co. Ltd between 2016 and 2021. BSC Method has been demonstrated to be effective in preventing and controlling soil erosion and regenerating vegetation; this method is simple, eco-friendly, and inexpensive.

1. Introduction

Soil erosion control practices can be categorized into three namely, agronomics, agrostological, and mechanical [1]. The agronomics practices include tillage operation, intercropping, cover crop, mulching, and organic matter. The agrostological practices include the cultivation of grass. The mechanical practices include contour farming, bund, micro basin tillage, contour terracing, and geo-textile. It is revealed that there are 11 major strategies for erosion control used in Asia's agricultural land between 2014 and 2020 (Figure 1). The most popular strategy is tillage operation followed by mulching practice, cover crop, and grass cultivation. This study is reported that China had 23 case studies on control procedures; in comparison, South Korea and India each had six, Indonesia had two, Iran had one, and Israel had one. It demonstrates that soil conservation is a significant environmental challenge that requires additional research and investigation.

Biological soil crust (BSC) is one major element in soil composition, which has a crucial role in the ecological ecosystem by improving soil stability and enhancing erosion prevention effects. Moreover, BSC enables to supply shelter for a variety of taxonomic groups, soil microbes (soil algae, filamentous bacterium, fungus, lichen, moss), and vascular vegetative communities [2]. The whole biological soil crust (BSC) society comprises hundreds of species that mainly contain the primary producer group and provide an important food source for higher-target consumers at the trophic level [3–5]. Another evidence for biocrust contribution is that North America, represented as one dryland region, has a less variety of biocrust than the vascular plant (tracheophytes); however, biocrust is reported to support up to 20 different species in group taxa [3,6].

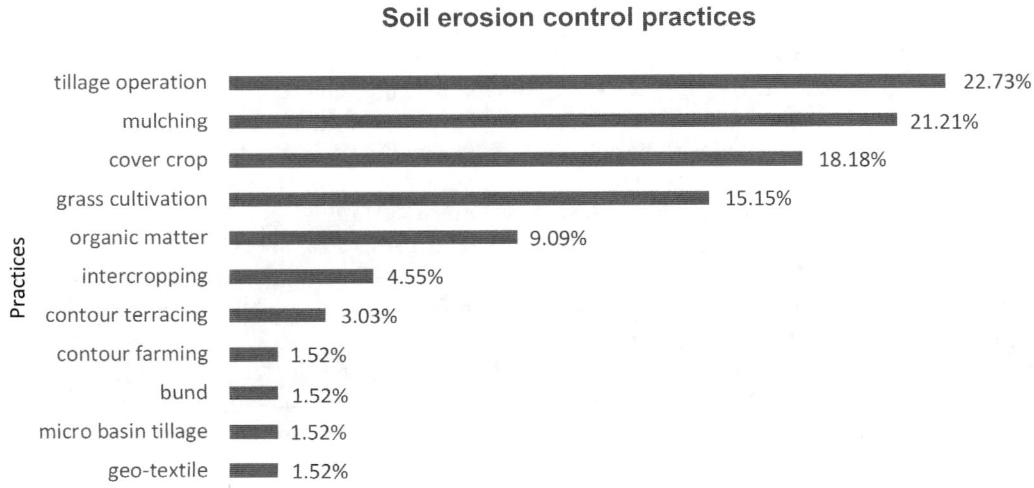

Figure 1. Research on soil erosion control practices in Asia's agricultural land. Modified from Nasir Ahmad et al.[1].

BSC has a critical role in ecological succession in several regions with specific crust successional dynamics existing in each climate region or terrestrial condition [7]. BSC keeps the soil moisture content and contributes nitrogen together with organic carbon sources. The accumulation of these contents are expected to respond rapidly and vary based on stressor factors [3]. It is asserted that BSC is well-known as a highly sensitive community for disturbance indication [8–10], regarding grazing or livestock trampling [11]; human-induced disturbance [12]; water infiltration of crust or water runoff [12,13].

Soil components are devastated by calamities predominantly caused by the increasing environmental changes, leading to soil deterioration as one typical example of ecological issues [2]. It asserts that BSC being demolished increases the possibility of desertification, soil destruction, and natural disaster occurrences; however, BSC could obtain an opportunity to recover natural vegetation even though this procedure takes time [2]. Therefore, it is recommended for both dry and desert-like locations. The priority concern is to boost BSC growth on the outermost layer of the soil surface in both quantitative and qualitative terms. At the same time, in the arid and semi-arid regions, the utilization of the microalgae inoculant is a promising application to promote BSC sheet-like cover and enhance vegetation restoration.

Numerous research has been conducted and concentrated on BSC restoration-based progress, involving strain isolation and development, inoculation, biomass production, soil rehabilitation, and ecological succession management [2,3]. For instance, many studies reveal that the harsh environmental condition hinders and slow down the growth rate. It has been shown that biocrusts actively develop and promote succession in nature [14]. One moss-dominated soil crust, named *Syntrichia caninervis*, a typical desiccation-tolerant moss, has managed two-month biomass production when cultured on a growth chamber, equivalent to ten-year obtainment in the field [15]. Moreover, in such disturbed locations where the soil quality is not well, the speed of biocrust is estimated at a slow rate with long-time rehabilitation [16]. In some cases, it could take centuries to recover from susceptible land [17–19]; or in the semi-arid region, it is reported to show vegetation get revived around 10 – 20 years [20]. To compare BSC with other conventional methods in erosion control, BSC has various outstanding outcomes. BSC can be applied to protected areas and requires a lower cost than other methods. There is no slope sharping or mesh installation needed for BSC method. Conventional controls (check dam, slope drain, ground cover) are supposed to be more expensive, not eco-friendly and temporary, contributing to ecological imbalance and climate change effects. Meanwhile, BSC sometimes can be encouraged to combine with other solutions (mixed with seed spray or sheet covering) without impact on the environment. Therefore, at present, enhancing the BSC mechanism in topsoil communities is one promising target for soil conservation measures that has drawn scientists' attention, is an ideal sustainable and environment-friendly solution. It is also highlighted that several case studies are presented in this review paper to demonstrate the usefulness of the BSC Method both within Japan and in one foreign country (Nepal) under a variety of climate conditions.

2. Technical background and features of BSC Method in Japan

In Japan, according to Nippon Koei Co., Ltd (NK), BSC is considered a 'sheet-like soil microbes' colony that is an eco-friendly biotechnology tool and a good candidate for reconstructing soil ecosystem function. It works in enhancing the soil aggregation and porosity and then obtaining the excellent capture retention on soil water content; thus, BSC has been shown to reduce physical crusting impact, thereby boosting the hydrological functioning task in the biocrusts surface. Furthermore, BSC is also a soil substrate of microorganism colonies occupied from soil algae, filamentous fungus, lichens, and moss in the upper soil layers. These communities accelerate vegetation recolonization, give a boost to natural recovery, and progressively stabilize the soil surface.

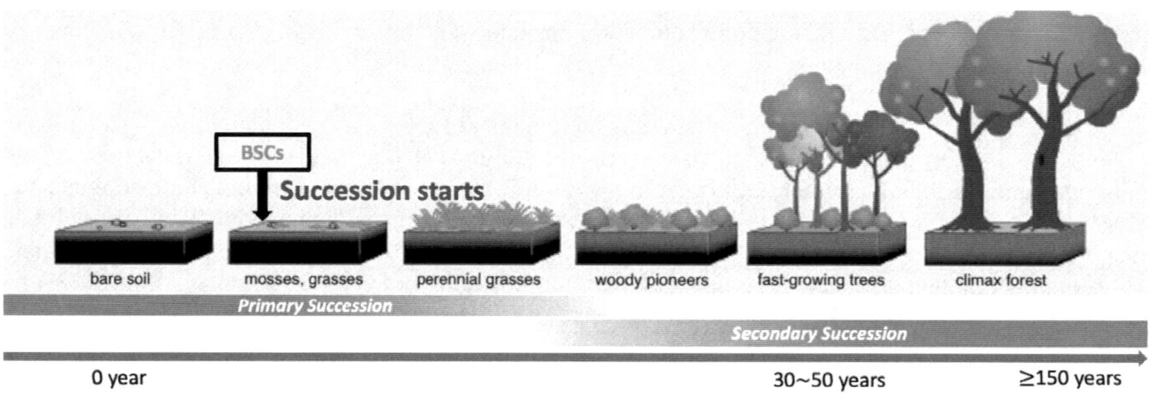

Figure 2. BSC is a natural phenomenon found at the first stage of vegetation succession.

In terms of ecological function, together with mosses, grasses and lichens, BSCs are expected to appear naturally at the beginning of vegetation succession, as shown in Figure 2. The BSC formation mainly comes from soil algae which are the pioneer contributor; then it proceeds the invasion of higher-level plant communities such as grasses, shrubs, and trees. There are four key roles of BSC in ecological succession which are: soil being fixed due to the filamentous structure and mucilage secretion; moisture retention in soil; supply of ammonium or nitrate compounds by the nitrogen-fixing blue-green algae; and production of organic carbon sources through photosynthesis mechanism. Figure 2 illustrates the time duration indicative of the sequence of the xerarch succession series that could occur in nearly 50 years or even more than 150 years. This is known as the estimated period, which can be ranged based on various internal and external agents (soil type and condition, climate region, the existence of soil-organism communities). Based on NK's principle, when BSC formation occurs, the natural vegetation succession could start and develop. Another convincing statement is BSC would be ideal to be formulated at the first stage due to the potential of surface soil erosion prevention. BSC formation would protect the surface, enhance soil stability, and act like a natural measure to prevent soil erosion. Thenceforth, one emerging approach is established in the collaboration between NK and Public Works Research Institute (PWRI), namely BSC methodology or BSC Method. The aim is to deal with the intense precipitation, to protect the soil surface against soil erosion and slope collapse. In 2009, the BSC method obtained a Japanese patent, certifying it as a Soil Erosion Prevention Method (Patent No. 3718203 and 4412628). The methodology is to spray the materialized cosmopolitan soil algae onto the top bareland. This method, once adapted, can fulfill the requirement of four sustainable goals, namely

SDG 6, SDG 11, SDG 13, and SDG 15 [21]. Therefore, BSC Method is one of the preferred feasible solutions compared to conventional methods due to the outstanding advantage – environmentally-friendly technology utilizing soil organisms.

In a primary term, the BSC technical method takes advantage of biological soil crust forming, reducing soil erosion and surface runoff. As a result, the vegetative communities gain a chance to recolonize, maintain and develop at the soil base. Finally, it can facilitate vegetation restoration in the biocrusted soil. Besides, the BSC Method also benefits from utilizing BSCs and the surrounding sites; the principle is to proceed with the natural plant invasion artificially and quickly. In outdoor condition, there must be several existing soil microorganisms available at the surface, even the buried or non-buried seeds distributed by flying birds or wind dispersion. With the support of BSC, these potential livings could enhance the root settlement and plant development shown at the soil surface and facilitate vegetation gradually invading, more vascular plants formation, or provide compatible surrounding shelters for many insects and wild animals. Meanwhile, the soil without BSC might easily record the sediment runoff by wind force or the high intense precipitation. It not only wipes out the soil particles at the surface, but also slips seeds off somewhere without germination. Hence, the restoration might take a longer time to recolonize and recover ecosystem processes in such disturbed areas.

BSC-1 is highly recommended for BSC-Method which has functions in controlling soil erosion and accelerating natural vegetation recovery. In the prime of BSC-1 formation, NK partnered with Nikken Sohonsha Co. Ltd – the Japanese expert in algae cultivation technology for over 30 years. Thus, the newly outstanding technique for cultivating soil algal species in water was established. Figure 3 shows as the morphotypes of BSC-1 under the microscope. This filamentous microalgae used for the BSC method has a global habitat, distributed in the wide range and found in most parts of the world under various ecological conditions. Being a hermaphrodite, the reproductive system is known as clonal proliferation, not sexual reproduction; thus, there is no risk of genetic pollution and almost no adverse environmental impacts on native species.

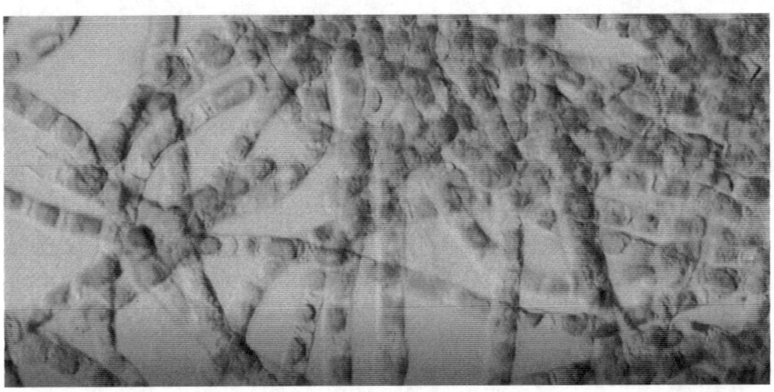

Figure 3. Morphotypes of BSC-1, captured from the corporate NK video about the BSC Method [22]

NK then points out the significant advantages while applying BSC-Method with BSC-1. Most of all, the operation cost is expected to be affordable than the conventional method. BSC-1 is prepared by adding water and mixing it with fiber agents and fertilizers (mainly NPK). This makes it easily attached to the surface when spraying onto the target sites. In a large area such as construction sites or hillslope, it is greatly recommended to use a hydroseeding machine, where these ingredients are constantly stirred in a mixing tank, and the seeding truck is installed with hose and pump for spraying. The construction sites are usually collapsed area of surface soil, the devastated land by construction, constructed slope surface. In some cases, the hoses can extend to 100 meters, depending on the pump capacity and type of equipment used. Therefore, it is possible to spray on various topography from the flat landscape to the high-position slope. Another plus point is no need for slope shaping/leveling or lath/ mesh installation compared to the seed-spraying technique. It makes BSC-Method to meet both the demand for temporary construction reduction and the risk associated with it.

Moreover, it is a claim that the vegetation restores faster with BSC without seed or plant implementation during the whole progress. One most considerable advantage of BSC-Method is its eco-friendliness. In the areas that need conservation (historical sites, national parks, restricted or protected areas), BSC Method does not affect the natural vegetation and the local ecology. Lastly, this methodology can be combined with other conventional measures. Some cases are reported on bringing back the natural vegetation when BSC artificially covers the land with a vegetation sheet.

3. Application of BSC Method

As a technology of environment–friendly slope restoration with soil algae, a brief on case study review is listed out, divided into two mains: domestic usage and international usage example.

3.1. Domestic usage example:

There are several individual cases for trials in Japan, which can be classified based on purposed outcomes such as general usage, test site at protected areas, applying location with sand pump, test among various soil conditions and construction site. Table 1 summarizes the significant domestic cases of BSC-Method in terms of slope restoration.

3.1.1. Cases of general usage. Three typical examples in total exemplify the general use of BSC-Method. Represented as the cold area, Northern Hokkaido is chosen for testing for a span of 24 months, proving that BSC can accelerate green recovery at a slope. It further suggests the BSC formation before snowing time. Another testing location is in Okinawa Prefecture, where typhoons frequently occur. After 4-day spraying, a typhoon attacked this area; however, there was no damage and no erosion effects. The vegetation got recovered and grew well after nine months. At a mountain road in Akita Prefecture, BSC was sprayed and monitored within six months. The result displays a slope restoration at the site; however, no other methods worked due to rocky terrain.

3.1.2. Cases in the protected area. Due to no impact on natural vegetation, BSC-Method is a proper solution for those places where environmental conservation needs to be considered, especially at National Park. In most nature parks, the intrusion of invasive species is in high concern and much restricted. In addition, conventional methods such as seed spray are not allowed to apply in a protected area because it might contain the potential risk of contamination from invasive seed species. The Fuji-Izu-Hakone National Park at Miyake Island witnesses a remarkable slope restoration during 8-month testing, where the vegetation was restored and the roots developed in a short span of time. Yanbaru National Park at Okinawa Island, one of the restricted areas, is executed in collaboration with the Okinawa Prefectural Government. It is proved that the slope surface without BSC becomes eroded again after a few months (Figure 4); meanwhile, the plant gets a quick recovery effect at topsoil (Figure 5). One test for checking the effect is conducted and shows progressive improvement from time to time. Another example is in Sanriku Fukkoh National Park at Miyagi Prefecture; even BSC indicates the development after only one month, then vegetation is constantly recovered.

Figure 4. Site monitoring at Yanbaru National Park at Okinawa Island, (a) at the beginning and (b) slope restoration after 1 year 8 months [21].

Figure 5. Status of 3.5 years later from BSC spray, Yanbaru National Park at Okinawa Island, (a) undeveloped vegetation after two years of collapse and (b) vegetation recovered at the same level as neighboring area [23].

3.1.3. Cases with sand pump. There is one case about spraying BSC-1 using the general sand pump. This area is located in Southern Kantoh with an existing slope with other method installation. This location is known as an unsuccessful current construction area, having a bad vegetation recovery condition. By applying BSC-1 with a sand pump, after two months, plant species quickly took roots and developed at high efficiency in vegetation coverage.

3.1.4. Cases at various soil conditions. Southern Kyushu is an area with white pumiceous volcanic soil texture, which is expected to be quickly eroded and less organic content in the soil. There is a record that heavy rain (approximately 1,358mm) followed by BSC spray, but no impact on erosion effects is figured out, and vegetation restoration is facilitated to develop. Without BSC, this area is eroded within a month (eroded by 380mm rainfall in September), displaying the rill erosion at a slope. Another testing is in Central Hokkaido where contains volcanic ash soil. Also, it is recorded that BSC is highly formulated on the area with no occurrence in soil erosion. The third case is the soft rock area in Aichi Prefecture. The original condition is less or no existence of vegetation. However, within one month, BSC is formulated in the spaces among rock, spreading steadily onto the surface. The last application is from Northern Kyushu Area, where exists the seed spray approach in both areas. It is expected to illustrate for BSC case combined with one conventional method. After monitoring in three months, the BSC area could mitigate the seed runoff, and then vegetation coverage capacity is accelerated at a higher speed than the control area.

IOP Conf. Series: Earth and Environmental Science 1091 (2022) 012033 doi:10.1088/1755-1315/1091/1/012033

Table 1. Example cases of BSC-Method in Japan in slope restoration with soil algae, extracted from NK [21].

Start date	Duration (months)	Location	Description/ Site condition	Execution method
10/2016	42	Okinawa	Yanbaru National Park. Exotic species are prohibited.	Slope spray
07/2017	24	Northern Hokkaido	Cold area. Grasses cannot grow well	Slope spray
09/2017	9	Okinawa	Area with typhoons attack frequently. There is a collapsed slope of mountain road without sharpening.	Slope spray
11/2017	8	N/A*	Additional fertilization is required caused by heavy rain	Groundwater seepage spray
03/2018	2	Okinawa	Site is in poor results of sheet covering.	Sheet-covering spray with seed
05/2018	2	Southern Kantoh	Site is one unsuccessful existing construction area, slope with other methods (not good vegetation recovery).	Slope spray
07/2018	8	Miyake	Fuji-Izu-Hakone National Park. Exotic species are prohibited.	Slope spray
09/2018	1	Southern Kyushu	White pumiceous volcanic soil. Area is eroded easily and less organic, pyroclastic-flow deposits in low erosion resistance.	Slope spray
09/2018	3	Northern Kyushu	There is seed spray onto the slope surface.	Slope spray
09/2018	12	Central Hokkaido	Volcanic ash soil. Less or no vegetation.	Slope spray
03/2019	6	Akita	Mountain road. Slope includes many stones.	Slope spray
05/2019	5	N/A	Area is a nearly bare slope, with no or less vegetation.	Sheet-covering spray
01/2020	9	Kagoshima	Site is in poor results of sheet covering.	Sheet-covering spray
05/2020	1	Miyagi	Sanriku Fukkoh National Park. Surface is covered with degraded base material.	Degraded-base spray
06/2020	0.5	Akita	Site is in poor results of sheet covering.	Sheet-covering spray with seed
11/2020	1	Aichi	Soft rock. Less or no vegetation.	Slope spray
N/A	1.5	Hokkaido	Slope surface contains gravels.	Slope spray
N/A	3	Okinawa	Main Island. Site is a collapsed slope of mountain stream	Slope spray
N/A	4.5	Okinawa	Main Island. Completed bridge construction before testing.	Slope spray
N/A	9	Okinawa	Island area. High risk of slope erosion.	Slope spray

* Data not available

3.1.5. Cases at the construction site. Two testing sites, which are Akita Prefecture and Okinawa Island, both maintain the sheet covering installation. The BSC-Method is applied with seed supplementation, showing vegetation recolonized and established within two weeks in Akita Prefecture and two months in the latter location. In Kagoshima Prefecture, during more than one year, the vegetative transformation is not recorded due to a poor quality of sheet covering. However, with the support of BSC-1 material, there is a significant change in the way plant species get adapted and occupy the target surface after nine months. After another 5-month testing onto the bare slope, it is found that BSC communities developed in the sheet covering, then spreading to surrounding areas to establish the complete plant recovery.

3.1.6. Other specific cases. To obtain the Best Award for 2015 Subtropical Greening Case, NK conducted several short-term cases among Okinawa Main Island [24] and Hokkaido wherein, both locations are related to the construction area. One 91-day experiment is monitored by spraying onto the collapsed slope of a mountain stream as an application example for red soil discharge prevention. Another location is the finished bridge construction. It works as demonstrating for the early recovery of vegetation for 4.5 months, especially one indigenous species (Japanese pampas grass) tends to recolonize after spraying BSC. One Hokkaido case is conducted in 1.5 months. BSC sprayed onto the target surface containing gravels to testify the effect on preventing gravel falls due to slope erosion. Finally, the positive achievement gains in treated-BSC slope compared with the untreated area. Another testing is carried out at the slope with a high risk of erosion in Okinawa Island within nine months. By formulating BSC, the erosion prevention effect was visible, followed by the promotion of vegetation intrusion. The green status then develops, gradually covering as similar as the surrounding area. There is one study of BSC application onto the minor groundwater seepage. Due to the heavy rainfall, supplementation fertilizers are required to constrain the loss of chemical fertilizers. Around eight months, the vegetation invasion gradually started in the rainy season in the BSC treated area while the seed spray area occurred with tiny rills, gully, or ravines.

3.2. International usage example:
In a mission to broadcast BSC technology outside Japan territory, Japan International Cooperation Agency (JICA), in cooperation with two private companies (NK Co., Ltd and Nikken Sohonsha Corporation), promoted a project as one of JICA outputs. It was under a contract of "Collaboration Program with the Private Sector for Disseminating Japanese Technology for Environmental – Friendly Slope Restoration with Soil Algae in Nepal". Meanwhile, the coordinator was Ministry of Transport of Nepal in 2019. This is one typical example conducted overseas and applied sand pump spray, commencing in February 2019 and terminating in March 2021. The application is carried out in the Sindhuli Road in Nepal, one of the mountain roads with slope failures [25]. To be a part of this program, a workshop is organized about "Biological Soil Crust Method for Slope Protection", introducing the BSC-Method. With the sand pump spray technique, BSC from the cosmopolitan soil algae has roles in promoting vegetative succession with no effects on the traffic flow, thus alleviating slope failures nearby Sindhuli Road. Figure 6 describes the changes after one month of spraying BSC onto the slope surface. A high-intensity rainfall occurred, it is recorded that the BSC area enables to confront heavy intensity rainfall, approximately 300mm within 48 hours. It displays that the control area without BSC supply, this area has suffered more soil runoff on the slope surface than the BSC area with shallower hollows at the cover. After rainfall, small eroded areas are merged into one larger one, and control area is recorded as more susceptible on slope surface than in BSC area. To compare the erosion effect, the area and volume of eroded soil are investigated. It is shown that in BSC area, the area and volume with the eroded status increase to approximately twice after 6 months (from 41.64 $m^2/100m^2$ to 79.14 $m^2/100m^2$; and from 6.29 $m^3/1000m^2$ to 13.64 $m^3/1000m^2$). Meanwhile, the bare land (control area) has a higher increase in both area and volume of eroded soil, showing more soil being damaged than before (area increased by 3.6 times from 62.37 $m^2/100m^2$ to 230.07 $m^2/100m^2$ and volume increase by more than 12 times from 9.34 $m^3/1000m^2$ to 117.76 $m^3/1000m^2$). BSC applied in this study is also indicated that the vegetation

IOP Conf. Series: Earth and Environmental Science 1091 (2022) 012033 doi:10.1088/1755-1315/1091/1/012033

could be recovered and developed among rill erosion, plant invasions with some new vegetation existing on the surface.

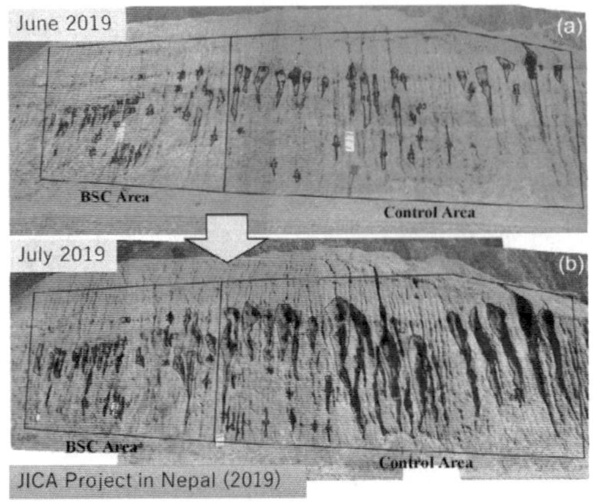

Figure 6. JICA Project in Nepal (2019) on the effects of erosion control and vegetation recovery by BSC-Method, (a) at the beginning and (b) high rainfall after one month [25].

4. BSC Method for soil erosion prevention: At a glance

4.1. Laboratory data
Before applying at the outdoor testing site, lab-scale experiments are conducted to assess BSC potentials [24]. One test is waterway simulation on the erosion effects (Figure 7a), there are 6 different collecting times (0; 2; 4; 6; 8 and 10-minute period) in the group without BSC (Figure 7b); meanwhile, there are 5 time-durations (2; 4; 6; 8 and 10-minute period) in the group with BSC (Figure 7c). The effluent from the soil with BSC appeared to be less muddy compared to the soil without BSC. Figure 7 illustrates that soil enables to against erosion from water, maintaining at least 10 minutes in the BSC treatment. In the control group, muddy water is collected in yellow/ orange tone color from the beginning (0 min), immediately when the first water flows. This indicates surface soil being runoff quickly and less resistance capacity compared to BSCs treatment. The muddier the water sample shows, the more erosion exists.

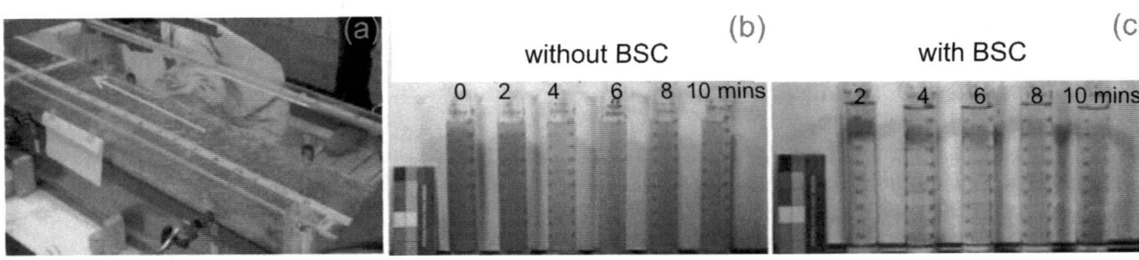

Figure 7. Testing waterway erosion, (a) experimental setup, (b) water runoff without BSCs and (c) water runoff with BSCs [24].

Figure 8a describes the experiment of rotating flow erosion with the continuous agitation by a mixer with a continuous water inlet in both with and without BSCs. Treatment without BSC was applied by 4 rotation speeds (50 rpm; 100 rpm; 150 rpm; 200 rpm) corresponding to flow velocity (38 cm/s; 76 cm/s; 114 cm/s; 152 cm/s) (Figure 8b). While 7 rotation speeds (50 rpm; 75 rpm, 100 rpm, 125 rpm, 150 rpm, 200 rpm, and 350 rpm) were applied in treatment with BSC (Figure 8c). The data is collected based on the amount of soil eluted from the system. It shows that without BSC, the first cylinder (50 rpm with 38 cm/s) was the least muddy among four; the others were shown in yellow/ orange water output. Meanwhile, treatment with BSC has a significantly positive effect on soil erosion up to the flow speed of 350 rpm, equivalently the velocity of 196 cm/s. With supplying BSC, it records less soil is eroded when the water keeps continuously mixing. There is an evident increase in soil aggregation and strength, reducing movement of adjacent soil sediments near the surface, and re-establishing soil stability. There is a certain difference in the soil texture between the two treatments. In the aspect of BSCs, the soil seems much denser and contains BSC coverage in green visualization.

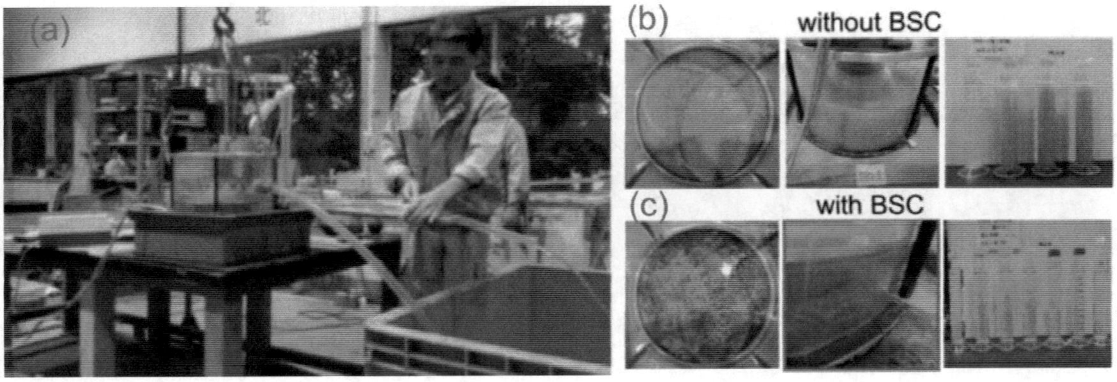

Figure 8. Simulation for rotating flow erosion, (a) experimental setup, (b) results without BSCs and (c) results with BSCs [24].

4.2. Significant BSC milestones

In 2009, NK collaborated with PWRI on a series of research projects controlling soil sediment, focusing on red soil around rivers and coastlines in the Southwest Islands, Japan. It is based on two main findings: slope site with existing soil algae enables to maintain much soil sediment; field area covered by soil algae points out the sediment yield dropped from 10 to 20 times compared with control field. Thus, one Japanese patent was issued BSC technology to prevent soil erosion (No. 3718203 and 4412628) [27]. Furthermore, in partnership with Nikken Sohonsha, NK continuously invented a materialized cosmopolitan soil algae: the BSC-1 product. Following that, NK won the Best Award for 2015 Subtropical Greening Case Presentation: Okinawa.

After several successful projects, the Ministry of Land, Infrastructure, Transport, and Tourism recognized the BSC Method and registered it in its New Technology Information System (NETIS). Cultivating soil algae in the water is a noteworthy achievement, leading to further innovation in BSC-1 development. BSC Method has also been tested outside Japan through JICA Projects. The project was along Sindhuli Road in Nepal from February 2019 until March 2021.

5. Conclusions

BSC-Method is one promising technology in preventing soil erosion and restoring vegetation. This environmentally-friendly technology offers a lower operating cost than other erosion control strategies. This technology can also be combined with other methods to maximize the benefits. The case of Nepal demonstrates the practical usefulness of BSC technology outside of Japan. Given the potential of BSC technology, further studies are needed, especially on scaling up the production of BSC-1 materials and expanding the technology for use in other countries.

Acknowledgments

The authors would like to thank Malaysia – Japan International Institute of Technology (MJIIT) Universiti Teknologi Malaysia (UTM). This work is funded by the AUN/SEED-Net grant (R.K130000.7343.4B403) under the Collaborative Education Program for Sustainable Environmental Engineering Network (CEP-SEEN). We also thank the Ministry of Higher Education Malaysia for funding this study under the grant FRGS/1/2019/TK10/UTM/02/14. The authors are also thankful to Universiti Teknologi Malaysia for the Professional Development Research University grant (05E05).

JMK is a Researcher of UTM under the Postdoctoral Fellowship Scheme for the project: Improvement of biological soil crust forming algae to prevent soil erosion.

References

[1] Nasir A N S B, Mustafa F B, Muhammad Y S Y and Didams G 2020 A systematic review of soil erosion control practices on the agricultural land in Asia *Int. Soil Water Conserv. Res.* **8** 103–15
[2] Labor A 2016 Potentials of the microalgae inoculant in restoration of biological soil crusts to combat desertification *Int. J. Environ. Sci. Technol.* **13** 2521–32
[3] Bowker M A, Maestre F T, Eldridge D, Belnap J, Castillo-Monroy A, Escolar C, Soliveres S 2014 Biological soil crusts (biocrusts) as a model system in community, landscape and ecosystem ecology *Biodivers. Conserv.* **23** 1619–37
[4] Bowker M A, Maestre F T and Escolar C 2010 Biological crusts as a model system for examining the biodiversity-ecosystem function relationship in soils *Soil Biol. Biochem.* **42** 405–17
[5] Bowker M A, Soliveres S and Maestre F T 2010 Competition increases with abiotic stress and regulates the diversity of biological soil crusts *J. Ecology* **98** 551–60
[6] Garcia-Pichel F, Loza V, Marusenko Y, Mateo P and Potrafka R M 2013 Temperature drives the continental-scale distribution of key microbes in topsoil communities *Science* **340** 1574–77
[7] Lan S, Wu L, Zhang D and Hu C 2015 Analysis of environmental factors determining development and succession in biological soil crusts *Sci. Total Environ.* **538** 492–99
[8] Belnap J and Eldridge D 2001 Disturbance and Recovery of Biological Soil Crusts *Ecol. Studies* **50** 363–83
[9] Cardoso E J B N, Vasconcellos R L F, Bini D, Miyauchi M Y H, Santos C A, Alves P R L, Paula A M, Nakatani A S, Pereira J M and Nogueira M A 2013 Soil health: Looking for suitable indicators. What should be considered to assess the effects of use and management on soil health? *Sci. Agric.* **70** 274–89
[10] Chamizo S, Cantón Y, Lázaro R, Solé-Benet A and Domingo F 2021 Crust composition and disturbance drive infiltration through biological soil crusts in semiarid ecosystems *Ecosyst.* **15** 148–61
[11] Concostrina-Zubiri L, Huber-Sannwald E, Martínez I, Flores Flores J L, Reyes-Agüero J A, Escudero A, Belnap J 2014 Biological soil crusts across disturbance-recovery scenarios: Effect of grazing regime on community dynamics *Ecol. Appl.* **24** 1863–77
[12] Chamizo S, Stevens A, Cantón Y, Miralles I, Domingo F and Wesemael B 2012 Discriminating soil crust type, development stage and degree of disturbance in semiarid environments from their spectral characteristics *Eur. J. Soil Sci.* **63** 42–53
[13] Chamizo S, Cantón Y, Afana A, Lázaro R, Domingo F and Solé-Benet A 2020 How development and disturbance of biological soil crust do affect runoff and erosion in drylands? *Adv. Studies Desert.* **41** 203 – 6
[14] Belnap J 2006 The potential roles of biological soil crusts in dryland hydrologic cycles *Hydrol. Process.* **20** 3159–78
[15] Xu S, Yin C, He M and Wang Y 2008 A technology for rapid reconstruction of moss-dominated soil crusts *Environ. Eng. Sci.* **25** 1129–38
[16] Chock T, Antoninka AJ, Faist AM, Bowker MA, Belnap J and Barger NN 2019 Responses of biological soil crusts to rehabilitation strategies *J. Arid Environ.* **163** 77–85
[17] Belnap J and Warren S D 2002 Patton's tracks in the Mojave Desert, USA: An ecological legacy *Arid. Land Res. and Manag.* **16** 245–58
[18] Doherty K D, Antoninka A J, Bowker M A, Ayuso S and Johnson N C 2015 A novel approach to cultivate biocrusts for restoration and experimentation *Ecol. Restor.* **33** 13–6

[19] Weber B, Bowker M, Zhang Y and Belnap J 2016 Natural recovery of biological soil crusts after disturbance *Biological Soil Crusts: An Organizing Principle in Drylands. Ecological Studies* ed. B Weber, B Büdel, J Belnap (Springer, Cham) pp 479–98

[20] Lázaro R, Cantón Y, Solé-Benet A, Bevan J, Alexander R, Sancho LG, Puigdefabregas 2008 The influence of competition between lichen colonization and erosion on the evolution of soil surfaces in the Tabernas badlands (SE Spain) and its landscape effects *Geomorphology* **102** 252–66

[21] United Nations *THE 17 GOALS - Sustainable Development Goals.* Available from: https://sdgs.un.org/goals (Accessed December 9, 2021)

[22] Nippon-Koei 2021 BSC method to protect slope against soil erosion and accelerate vegetation recovery [Video]. Available from: https://www.youtube.com/watch?v=5wfWbY5Jfzg&t=24s (Accessed December 7, 2021)

[23] Nippon-Koei 2021 Examples of BSC Method in Japan *7th Asia-Pacific Climate Change Adaptation Forum: Enabling Resilience for All: The Critical Decade to Scale-up Action*

[24] Nippon-Koei 2021 BSC method PR brochure *7th Asia-Pacific Climate Change Adaptation Forum: Enabling Resilience for All: The Critical Decade to Scale-up Action*

[25]. Nippon-Koei 2021 Environmentally-friendly Technology for Preventing Surface Soil Erosion Utilizing Soil Organisms *7th Asia-Pacific Climate Change Adaptation Forum: Enabling Resilience for All: The Critical Decade to Scale-up Action*

[26] Nippon-Koei 2021 Environment-friendly Slope Restoration with Soil Algae (BSC Method) *7th Asia-Pacific Climate Change Adaptation Forum: Enabling Resilience for All: The Critical Decade to Scale-up Action*

[27] Ministry of Economy, Trade, and Industry (METI) 2020 Methodology for prevention of soil surface erosion with soil algae（BSC methodology） *Climate Change Adaptation Good Practices by Japanese Private Sector*

Assessment of Reservoir and Canals Revitalization to Improve The Reclamation Process at Unit Tamban Tidal Irrigation Area, Central Kalimantan

S S Rahajeng[1,2], R Jayadi[1,*], and A Maas[3]

[1] Department of Civil and Environmental Engineering, Faculty of Engineering, Universitas Gadjah Mada, Yogyakarta
[2] Directorate Irrigation and Lowland, Directorate General Water Resources, Indonesian Ministry of Public Works and Housing, Jakarta
[3] Department of Soil Science, Faculty of Agriculture, Universitas Gadjah Mada, Yogyakarta

* Rachmad Jayadi: rjayadi@ugm.ac.id

Abstract. The Unit Tamban is one of the tidal irrigation areas in the Central Kalimantan Mega Rice Project (MRP). The Unit Tamban is an irrigation system consisting of three primary canals with a settling pond or reservoir at each end. After being developed in 1970 by the Tidal Rice Field Development Project (P4S), the three reservoirs no longer function because they are full of sediment. Sedimentation also occurs in primary and secondary canals, impacting tidal attenuation and decreasing canal capacity. The reservoir at the end of the primary canal assists water regulation for irrigation purposes and leaching acid sulfate water. Three scenarios can be used to increase the capacity of the irrigation system, namely reservoir revitalization, secondary canals normalization, and combining two previous scenarios. The performance of both options was assessed by hydraulic simulation modeling using the HEC-RAS software. The hydraulic simulations show that reservoir revitalization can increase the volume of incoming water during high tide by about 90% more than without a reservoir. If the secondary canal normalization is carried out, the fresh water volume entering the secondary canal exceeds 19% before normalization.

Keywords: tidal irrigation area, reservoir, normalization, hydraulic simulation

1. Introduction

The potential of swampland in Indonesia is enormous, with a total area of 33,393,570 ha; 60% consists of tidal swamps and 40% of *lebak* swamps [1]. With this potential, swampland for agriculture can support the food self-sufficiency program realization in Indonesia. The most significant project initiated in 1995 is the one million hectares Mega Rice Project (MRP) in Central Kalimantan Province. The government of Indonesia initiated the MRP to convert one million hectares of peat and lowland swamps into rice cultivation areas, consisting of five working areas: block A, block B, block C, block D and block E [2]. It is bounded by the Sebangau River (west), the Barito River (east), the Java Sea (south), while its northern border roughly follows the Palangka Raya.

One of the tidal irrigation areas in the MRP is the Unit Tamban. This irrigation area has been developed since 1970 with *Proyek Pembukaan Persawahan Pasang Surut* (Tidal Rice Fields Opening Project). At first, the locals cultivated using the traditional system by digging a canal perpendicular to the river for tidal land reclamation. The canal is called *handil*, which removes excess water in the land for agricultural activities. After more than 50 years of development, the reclamation process at the Unit Tamban has not yet been fully completed. Better water management is needed to support the reclamation process to increase agricultural productivity. Water management in swamplands must fulfill three objectives [3]. The first is to provide water for washing and diluting acid water. The second is a drainage system to remove excess water during the rainy season. And the third is to keep the potential acid sulfate in a reduced condition so that oxidation does not occur.

There is much sedimentation in both the primary and secondary canals currently at Unit Tamban due to poor maintenance and no rehabilitation for a long time. The reservoirs at the end of the primary canal are also no longer functioning because they are full of sediment and shrubs. The reservoir was developed in the 70s as a reservoir for soil leaching material that did not get out because it was pushed back by the tide and as a flow regulator in the water system [4]. Sedimentation in canals and reservoirs causes the water holding capacity to decrease and affects the leaching process of the soil. These problems impact the performance of the Unit Tamban, as evidenced by the presence of pH < 4 in water and soil on agricultural land. The average rice productivity is 3.5 tons/ha, with the majority applying the cropping pattern once a year.

Acid sulfate soils develop from accumulating parent material rich in pyrite (FeS_2) [5]. Pyrite accumulates in stagnant soils rich in organic matter and is rinsed with sulfate solutions. This acidification can cause agronomy, engineering, and environmental problems. For example, in Western Finland, acidification compounds from acid sulfate soil drainage will be pushed into rivers. It will affect fish and underwater life due to the low pH of the water and the high concentration of Al [6]. Therefore, a solution must be found for this problem; more fresh water is needed to enter the canal so that the water circulation is better because good water circulation will help the process of acid diluting [7].

This paper aims to model and simulate three revitalization scenarios to increase the fresh water supply. The scenarios are reservoir revitalization, secondary canals normalization, and combining the two previous scenarios. It is hoped that the volume of fresh water can help the diluting process of the acid to improve the water quality, soil quality and agricultural productivity. In Tuong (1993) [8], Konsten et al. argue that within 200 days, the water used for leaching is > 1000 cm, which means that it takes 5 cm in a day. The result of this study is a comparison of the volume of fresh water that can enter the canal in these three scenarios.

2. Methodology

The Unit Tamban is one of the irrigation areas under the central government's authority located in Kapuas Regency, Central Kalimantan Province, Indonesia. The existing water management in Unit Tamban is a fork system with the Kapuas River as its water source. Universitas Gadjah Mada introduced the Fork system and designed it to overcome acidity problems. Many of these systems have a large pond (reservoir) at the end of each secondary canal [5]. The water system network consists of three primary canals; the main primary canal is divided into the left pond primary canal, the middle pond primary canal, and the right pond primary canal with a total functional area of 3,506 ha. Each primary canal has a reservoir at each end. Agricultural land covers seven villages, namely Sei Jangkit, Tamban Luar, Bangun Harjo, Warnasari, Sidomulyo, Sidorejo, and Bandaraya. The irrigation scheme of the Unit Tamban can be seen in Figure 1 [9]. This study will focus on the middle pond primary canal that irrigates the secondary canal in Bangun Harjo Village and Sidomulyo Village.

In addition to supporting agricultural activities in the Unit Tamban, it is generally done by improving soil quality by adding ameliorant or lime and water management. This study will focus on water management by modeling and simulating three revitalization scenarios to increase the fresh water supply and improve water and soil quality. The water management in tidal irrigation areas mainly utilizes tidal energy. High tide pushes fresh water into the canal, while low tide removes acid water. Considering the hydro-topography condition, agricultural land in the Unit Tamban is classified into categories A, B and C, as seen in Figure 2 [10]. Category A was 22.8% area, then category B 39.8% area and 37.4% area in category C. The hydro-topography characteristics category is described in Table 1 [11].

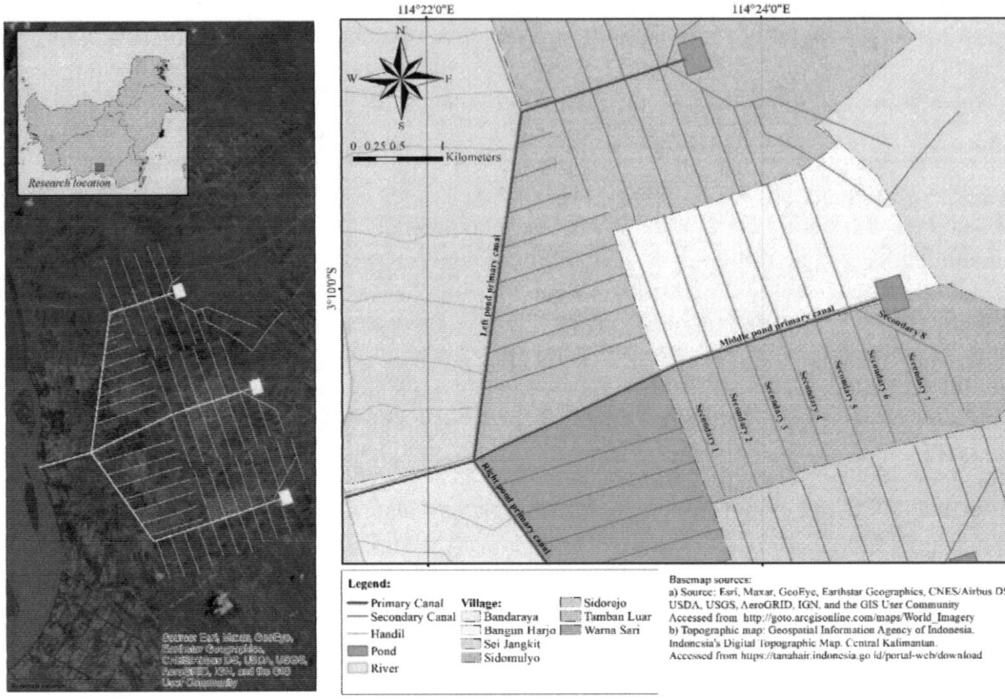

Figure 1. Location spot image and irrigation canal network of the Unit Tamban.

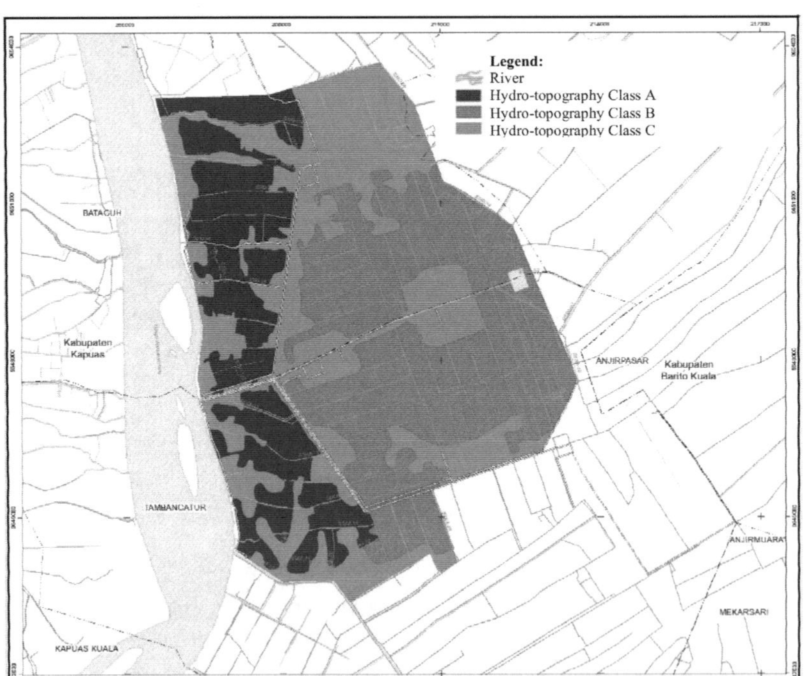

Figure 2. The Hydro-topography of the Unit Tamban.

Table 1. Hydro-topography category of tidal lowland

Hydrotopography	Characteristics
Category A	Land overflowed at least 4-5 times per spring tide cycle at wet and dry season
Category B	Land overflowed at least 4-5 times per spring tide cycle at wet season only
Category C	Land is not overflowed by the tide, but it affects the groundwater level
Category D	Land cannot be overflowed, and the groundwater level does not influence

In tidal irrigation, canal flow is determined by changes in water level elevation due to tidal movements in the sea and upstream discharge [4]. Furthermore, one of the mathematical models that can represent the unsteady flow phenomenon by describing the conservation and momentum equations is the HEC RAS software. This study's hydraulic simulation will be run using the existing conditions and several scenarios. The first scenario is to reactivate the reservoir at the end of the primary canal without changing the dimensions of the existing canal. The second scenario is the normalization of the secondary canal. Changes are made to the canal's existing dimensions; the canal's bottom will be widened but not deepened to prevent pyrite oxidation and land subsidence. Then the third scenario will combine the two previous scenarios by adding a reservoir at the end of the primary canal and normalizing the secondary canal.

Before the three scenarios simulation, the model was calibrated with several different manning's coefficients. The manning's coefficient was continuously changed until the variations between the observed and simulated water levels were within acceptable limits [12]. The calibration procedure gives the correct manning's coefficient of the river reach. After that, the hydraulic simulation will be continued by modeling the existing conditions. After getting the simulation results of the existing conditions, three scenarios are modeled as described in the previous paragraph. The data used in this study is secondary data from the geometry of the Unit Tamban and tidal data in the Kapuas River as the downstream boundary condition. All data were obtained from *Balai Wilayah Sungai* (River Basin Organization) Kalimantan II, Ministry of Public Works and Housing, Indonesia. The model for this hydraulic simulation consists of 279 canal cross-sections, including 65 primary canal cross-sections and 214 secondary canal cross-sections. The primary canal's wide, varying from 30 m to 70 m and the secondary canal is about 20 m.

From the hydraulic simulation results of the three scenarios, a comparison of the increase in the fresh water volume entering the primary and secondary canals was carried out. Another research also conducted a hydraulic simulation using HEC RAS to evaluate the fresh water supply with several water management scenarios for irrigation and acid dilution. The increase of fresh water volume is seen by calculating the incoming fresh water at high tide and low tide to determine the water volume to be discharged or taken out [13].

3. Results and Discussion

3.1 Calibration

Kapuas River tidal data at the mouth of the Unit Tamban was used as a downstream boundary condition obtained from secondary data. With 48 hours of tidal data, the highest tide is at 0.821 m, and the lowest tide is at -1.369 m. Calibration was calculated using two different manning values, $n = 0.03$ and $n = 0.04$. This manning value was chosen because it is closest to the primary condition of the canal in the form of a natural canal [14]. This value is used for all primary and secondary canals. After comparing with the measurement data in Warnasari Village, a manning value of 0.04 was chosen, which is more representative, as shown in Figure 3. The calibration accuracy is dependent on the quality of measurement data. Because the data is an hourly measurement, the actual tide peak cannot be observed. The analysis of Root Mean Squared Error (RMSE) indicators was carried out to see the calibration accuracy. The analysis results obtained that the RMSE value is 0.11, showing that manning values can be used.

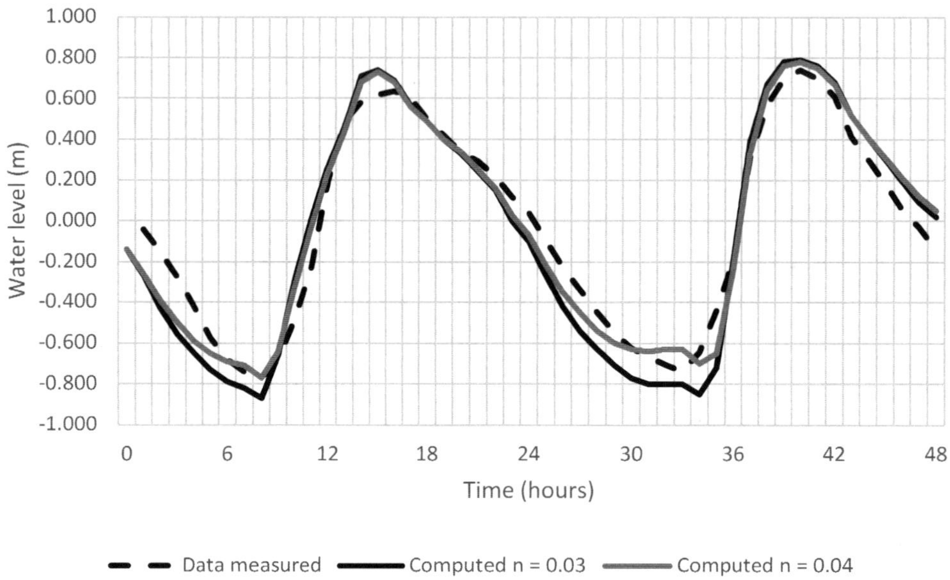

Figure 3. Calibration result at Warnasari Village.

3.2 Simulation Scenarios

3.2.1 Existing Condition. The hydraulic simulation in this study focuses on the middle primary canal with a length of ± 6 km. The middle primary canal covers two villages for the irrigation service area, namely Bangun Harjo Village and Sidomulyo Village. In Bangun Harjo Village, there are six secondary canals with each length 1.5 km. Meanwhile, Sidomulyo Village consists of eight secondary canals with an average length of 1.4 km.

3.2.2 Reservoir/ Settling Pond. At the Unit Tamban, three reservoirs are currently not functioning because they have been covered with sediment and shrubs, as shown in Figure 4. Due to the non-functioning reservoirs as a flow regulator, the sedimentation in the primary canal is getting worse. Furthermore, this also obstructed fresh water in and out the secondary canal for the irrigation supply. This study tries to model the primary canal and the reservoir as the former design (400 m x 400 m) with a depth of 1 m.

3.2.3 Secondary Canal Normalization. The sedimentation in the Unit Tamban secondary canal caused the reduction of the water holding capacity. Normalization that has not been done for a long time and poor maintenance are also the cause of sedimentation. Canal normalization activities have not been carried out for a long time to return the canal to its initial dimensions. Efforts to excavate the canal are also constrained by the exposure of acid sulfate soil under the canal. Therefore, this study's normalization scenario was carried out only by widening the canal bottom but not deepening it. The canal bottom width was 1 to 1,5 m and became 3 m after the normalization scenario.

Figure 4. Reservoir/Settling Pond condition at Unit Tamban.

3.3 Hydraulic Simulation Results

As previously explained, three different scenarios were simulated in this study. The hydraulic simulation results of the three scenarios are compared with the existing simulation conditions. The point to be reviewed is in the middle of the primary canal, approximately 4.5 km from the mouth. The tidal curve comparison of the existing condition and the three scenarios can be seen in Figure 5.

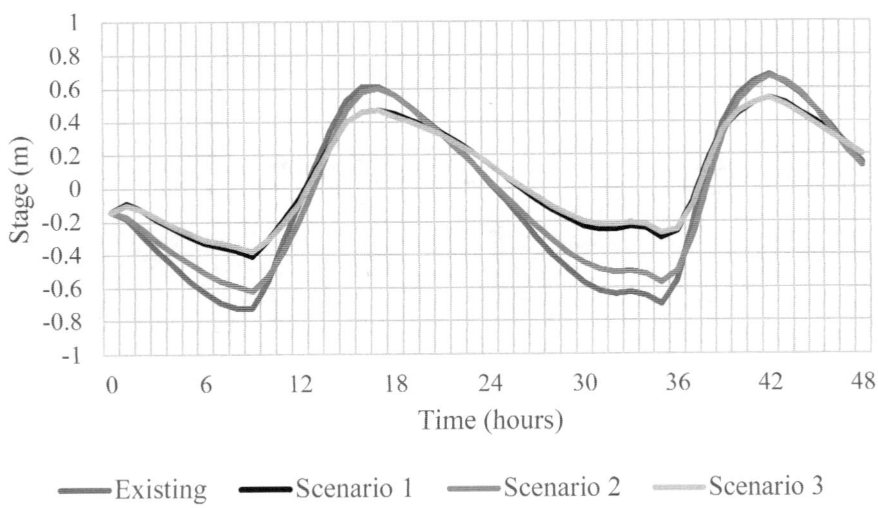

Figure 5. The tidal curve at the middle primary canal (4.5 km from the mouth).

Based on Figure 5, a large reservoir can reduce tides, as seen in scenario 1 and scenario 3. The water volume in the reservoir from the primary canal can only raise the water level slightly and makes the difference in water levels upstream and downstream becomes larger. The differences of water levels upstream and downstream affect the flow energy, or called the energy grade line. The energy grade line can be seen in Table 2 and Table 3. A more significant energy grade line is found in scenario 1 and scenario 3, which have a reservoir as a flow regulator. The difference in water level elevation upstream and downstream becomes more significant. The water level differences make the flow velocity higher.

Table 2. The energy grade line at high tide in the existing condition and three scenarios.

Distance (m)	Energy Grade Line			
	Existing	Reservoir (Scenario 1)	Normalization (Scenario 2)	Combine (Scenario 3)
0 5785	0.00005013	0.00013137	0.00006223	0.00013483

Table 3. The energy grade line at low tide in the existing condition and three scenarios.

Distance (m)	Energy Grade Line			
	Existing	Reservoir (Scenario 1)	Normalization (Scenario 2)	Combine (Scenario 3)
0 5785	-0.00012619	-0.00020743	-0.00014866	-0.00020916

The hydraulic simulation results also compare inflow and outflow from the existing condition and three scenarios. The canal discharge becomes higher with a reservoir, both at high and low tide, as shown in Figure 6 due to the higher energy grade line.

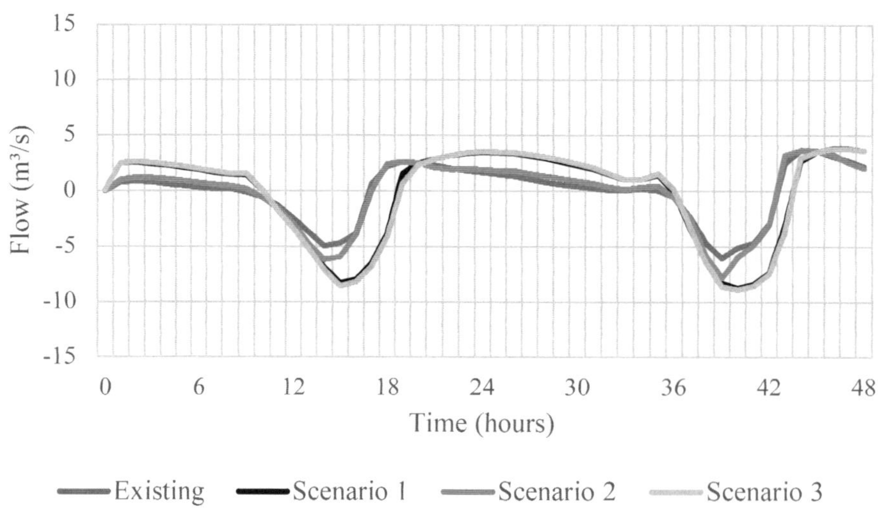

Figure 6. The curve of inflow (-) and outflow (+) at the middle primary canal in the existing condition and three scenarios (4.5 km from the mouth).

Figure 6 also shows the drainage time of each scenario. Drainage time is the time during low tide or what can be considered as outflow, where the water is out from the system into the river. Although the drainage time is not much different, the velocity of each scenario will be different. Scenario 1 and 3 with the reservoir have a higher average velocity than the scenario without a reservoir, as shown in Table 4. The difference in velocity affects the removal speed of acid leaching substances into the river. In scenario 1 and scenario 3, with a drainage time of 18 hours, the acid leaching substances can be removed as far as 11-12 km. In the existing condition with a drainage time of 18 hours, it can only remove the acid leaching substances as far as 2 km. The acid leaching substances will be pushed back into the canal by the high tide before reaching the river because the length of the primary canal is 6 km.

Table 4. The velocity during low tide in the existing condition and three scenarios.

Scenarios	Drainage Time (hours)	Average Velocity (m/s)	Distance (m)
Existing	18	0.033	2,124
Reservoir (Scenario 1)	18	0.186	12,024
Normalization (Scenario 2)	19	0.052	3,528
Combine (Scenario 3)	18	0.182	11,808

The maximum cumulative fresh water volume with the existing simulation is 79,128 m^3. With a more significant discharge due to the reservoir (scenario 1), the cumulative fresh water volume will increase to 150,144 m^3. The maximum cumulative fresh water volume with secondary canal normalization simulation (scenario 2) is 94,140 m^3. Moreover, in scenario 3, with combining reservoir and normalization, the maximum cumulative fresh water is 151,092 m^3. An explanation regarding the fresh water volume comparison can be seen in Figure 7 and Table 5. Figure 7 shows the cumulative comparison of fresh water volume at the middle primary canal. The cumulative amount is calculated by adding up the water that enters at high tide in one tidal cycle.

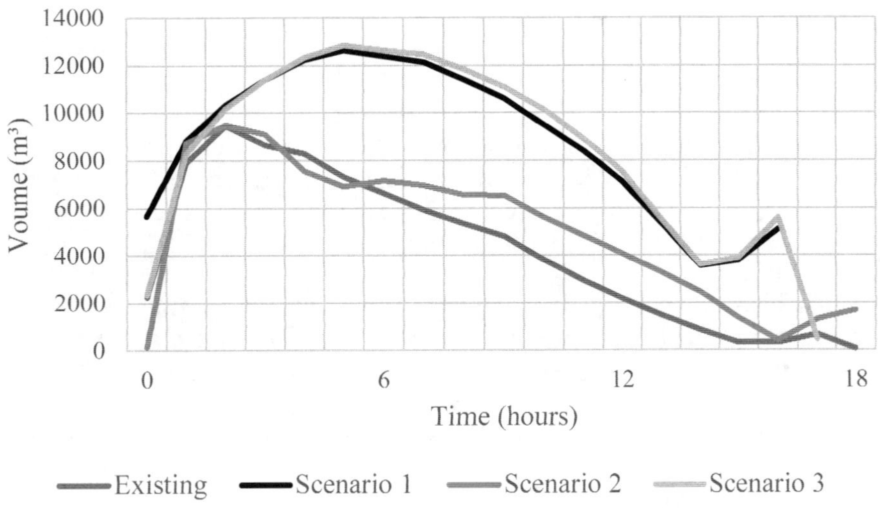

Figure 7. The cumulative fresh water volume at the middle primary canal in one tidal cycle (4.5 km from the mouth).

Table 5 shows the most increase of fresh water volume is in scenario 3 by combining the reactivation reservoir and secondary canal normalization as much as 90.9% compared to the existing condition. Meanwhile, if only reactivation reservoir (scenario 1), the fresh water volume will increase by 90.1%. The increase of fresh water volume in scenario 1 and scenario 3 is not much different. This is probably caused by secondary canal normalization which has a more negligible effect on increasing fresh water volume in the primary canal. This is also evidenced by the hydraulic simulation results in scenario 2, with the result that the fresh water volume is only 19% more compared to the existing condition.

Table 5. The cumulative fresh water volume at the middle primary canal in one tidal cycle.

Scenario Geometries	Maximum Volume (m^3)	Volume increase (%)
Existing	79,128	
Scenario 1	150,444	90.1
Scenario 2	94,140	19.0
Scenario 3	151,092	90.9

4. Conclusions
This study aims to improve water and soil quality to support agricultural productivity using tidal energy. The hydraulic simulation was run using HEC RAS software with the existing conditions and several scenarios. Scenario 1 revitalizes the reservoir, and scenario 2 is the secondary canal normalization. Furthermore, scenario 3 combines the two previous scenarios.

The removal speed of acid leaching substances increases with the reservoir scenarios. The drainage time in scenario 1 and scenario 3 is 18 hours. With this time, the distance that can be achieved to remove the acid leaching is approximately 12 km. The acid leaching substances will reach the river before being pushed back by the high tide. Meanwhile, a drainage time of 18 hours in the existing condition can only bring water as far as 2.1 km, so the acid leaching substances will be pushed back into the canal by the high tide before reaching the river.

From the hydraulic simulation, scenario 3 can increase the volume of incoming water by 90.9% compared to the existing condition, or equivalent to 151,092 m^3. From the energy grade line curve, scenario 3 also has the most significant results with a value of 0.00013483 at high tide and -0.00020916 at low tide. Reservoir revitalization has a significant role in increasing the fresh water volume entering the canal because it functions as a flow regulator. However, the effort to revitalize the reservoir is not easy because the reservoir is a location of acid leaching deposition that has accumulated for decades.

5. Acknowledgment
The authors would like to express their gratitude for the data and financial support given by the Directorate Irrigation and Lowland and *Balai Wilayah Sungai Kalimantan II*, Directorate General of Water Resources, Ministry of Public Works and Public Housing, Republic of Indonesia. This research is also under supervision of Prof. Dr. Budi Santoso Wignyosukarto, Dipl.HE. as Hydraulic Lecture from Department of Civil and Environmental Engineering, Faculty of Engineering, Universitas Gadjah Mada, Yogyakarta.

ORCID iDs
SS Rahajeng https://orcid.org/0000-0003-1544-0933
R Jayadi https://orcid.org/0000-0003-1664-8103

References

[1] Direktorat Irigasi dan Rawa 2010 *Audit Proyek Pengembangan Daerah Rawa* (*P2DR*) (Jakarta: Ministry of Public Works The Republic of Indonesia)

[2] Euroconsult Mott MacDonald and Deltares 2008 *Master Plan for the Rehabilitation and Revitalisation of the Ex-Mega Rice Project Area in Central Kalimantan* The Governments of Indonesia and the Netherlands Palangkaraya

[3] Wignyosukarto B S 2013 Leaching and Flushing of Acidity in The Reclamation of Acid Sulphate Soil, Kalimantan, Indonesia. (Irrigation and Drainage 62 vol 1) Wiley Online Library p 75-81

[4] Wignyosukarto B S 2005 *Reliabilitas Rehabilitasi Kolam Pasang pada Jaringan Irigasi Pasang Surut Unit Tabung Anen, Kalimantan Selatan* (Forum Teknik vol 29) no.1 p 1-12

[5] Suryadi FX 1996 *Soil and Water Management Strategies for Tidal Lowland in Indonesia* Dissertation (Rotterdam: IHE Delft)

[6] Sutela T and Vehanen T 2017 *The Effects of Acidity and Aluminium Leached from Acid Sulphate Soils on Riverine Fish Assemblages* Boreal Environment Research Helsinki

[7] Imanudin M S, Bakri B and Armanto M 2018 *Land And Water Management Option of Tidal Lowland Reclamation Area to Support Rice Production (A Case Study in Delta Sugihan Kanan of South Sumatra Indonesia)* (Journal of Wetlands Environmental Management vol 6) no.2

[8] Tuong TP 1993 *An Overview of Water Management of Acid Sulphate Soils* Ho Chi Minh City Symp. on Acid Sulphate Soils (The Netherland: ILRI-Wageningen) Publication 53 pp 265-279

[9] Badan Informasi Geospasial 2021 Central Sulawesi Province Map *[online] Available at: https://tanahair.indonesia.go.id/portal-web/download/perwilayah*

[10] Yodya Karya, Wahana Krida Konsulindo, and Rancangtama Bangunmandiri Indracipta 2020 *Survey Investigasi dan Desain (SID) Rehabilitasi dan Peningkatan Jaringan Irigasi Rawa Wilayah Kerja Blok D (Paket 4)* Balai Wilayah Sungai Kalimantan II Palangkaraya

[11] Direktorat Jenderal Sumber Daya Air 2017 *Surat Edaran nomor 19/SE/D/2017 tentang Pedoman Peningkatan Jaringan Irigasi Rawa Pasang Surut* Ministry of Public Works and Housing Jakarta

[12] Abbas S A, Al-Aboodi A H, and Ibrahim H T 2020 *Identification of Manning's Coefficient Using HEC-RAS Model: Upstream Al-Amarah Barrage* (Journal of Engineering vol. 2020 Hindawi)

[13] Fauzan A K, Wignyosukarto B S and Jayadi R 2021 *Water Management Evaluation for Upgrading Tidal Irrigation System, Katingan, Kalimantan* IOP Conference Series: Earth and Environmental Science 794

[14] Chow V T, Maidment D R and Mays L W 1988 *Applied Hydrology* (New York: McGraw-Hill Inc United States of America) p 33-35

Behavior of Flow of Channel Network of Dadahup Swampy Irrigation Area

M Ihsan Arif[1], Djoko Legono*[1], Djoko Luknanto[1], Budi Wignyosukarto[1], and Maruddin F Marpaung[2]

[1]Civil and Environmental Engineering Department, Gadjah Mada University, Jl. Grafika No.2, Yogyakarta, 55281, Indonesia
[2]Technical Implementation Unit for Lowlands, Directorate of Irrigation and Lowlands, Directorate General of Water Resources, Ministry of Public Works and Housing of Republic Indonesia, Indonesia

Email: *djokolegono@ugm.ac.id

Abstract. Dadahup is a swampy irrigation area that belongs to the Indonesian national development program, namely the food estate project. The food estate project is a revitalization program for the former peatland development to increase national food security. Peatlands located in lowlands (less than +2.00 m above mean sea level) were a significant problem in their land development activities; besides, the land is generally infertile. The peatland development has been discontinuing, and farmers shifted the jobs to another job for a living. In terms of the water-related problems, The Dadahup swampy irrigation area development faces various challenges, some of which are flood inundation during the rainy season and difficulty in irrigating the land during the dry season. This paper evaluates the behavior flow of the channel network using the HEC-RAS software. The irrigation area has a channel network comprising a primary channel, a perimeter channel, and several secondary channels. It also has connections with three main rivers, i.e., the Barito, Mangkatip, and Kapuas Murung Rivers. The simulation results with several scenarios have contributed better understanding of the flow behavior. Further discussion has promoted a proper water management implementation strategy to overcome the inundation issue.

Keywords: Swampy area, channel network, flood, the behavior of the flow.

1. Introduction

Indonesia is an archipelagic country with more than 17,000 islands traversed by the equator. Several large islands are Java, Sumatra, Kalimantan, Sulawesi, and Papua. The Dadahup swamp irrigation area (further called DIR Dadahup) is located on the island of Kalimantan, precisely in Kapuas District, Central Kalimantan Province (see Figure 1). Most of the area on the island of Kalimantan is a forest whose management is regulated by the Indonesian government.[1] According to their designation, almost all forests in Indonesia are state-owned and mapped quite accurately by the government. The Ministry of Forestry is responsible for land with permanent forest status that has been allocated for use as conservation forest, protection forest, limited production forest, or production forest. There are many mangroves and peat swamp forests in Kalimantan. Peat swamp forests in Kalimantan can be vulnerable to forest fires when the top layer of peat dries out and begins to smolder. Such fires produce high emissions of particulate matter, CO, and other compounds from incomplete combustion, making them very dangerous for the respiratory system. These fires can last for several months and cause the characteristic smog often seen in the dry season.

Until now, rice cultivation has been practiced by most of the local people in Kalimantan, including the Dayak people.[2] In some Kalimantan languages, the word food is translated as rice. To grow this sacred crop, the Dayaks used (and still do today) a shifting or shifting farming system ("slash and burn"). This agricultural approach reflects the vision of Dayak prosperity, which holds that rivers, land, and forests are all critical to Dayak identity. The same image is reflected in the changing mosaic land-use patterns created by the Dayaks in the forest ecosystems in which they live. In a typical Dayak land-use mosaic, there are patches of natural forest, cultivated forest, shifting/fallow fields, and permanent areas compatible with the ecological conditions of the mountains, wetlands, and river valleys in specific community areas.

Permanent wet rice fields are the only non-forest areas. It is sometimes said that shifting cultivation has significant negative consequences for natural forests. At the same time, there is no doubt that the changing nature of traditional agriculture has, to some extent, an impact on rainforests. Research shows that this impact is limited compared to industrial deforestation.[3,4] In Dayak culture, rice is a sacred plant, and rice cultivation is considered a religious and economic activity. It is the link between humans and the supernatural world. In traditional belief systems and knowledge, rice and its cultivation are conveyed to humans by the gods and through rituals can bring spiritual blessings to this world for the benefit of humans. Therefore, the Dayaks believe that, except for processing, the rice should not be beaten, abused, or thrown away. If tampered with, a ritual of apology and forgiveness must be performed.

DIR Dadahup is a part of the Indonesian former peatland development project called Ex-Mega Rice Project (EMRP) with an area of ± 21,226 ha.[5,6] The current condition of DIR Dadahup is that this irrigated area experiences flooding in the rainy season and drought in the dry season. For a long time in the rainy season (several months), some part of the DIR Dadahup land is inundated, whereas the land becomes dry.[7,8] The inundation mentioned above may be caused by seawater tides, excess rainwater, and or overflow of the river. However, such inundation may also be caused by the absence of regulatory structures and protective systems such as embankments [9,10] Understanding the flow behavior of the DIR Dadahup channel network is considered necessary.

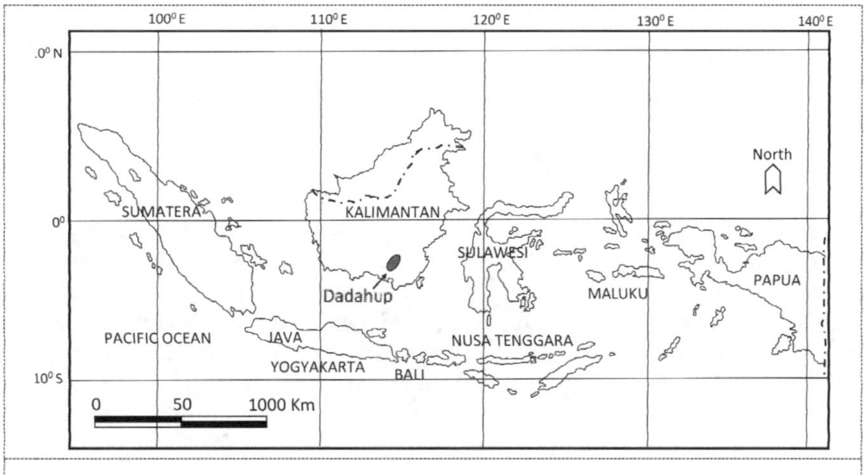

Figure 1. Location of Dadahup Swamp Irrigation Area.

2. Research Methods

2.1 General

In almost every case of identifying the physical behavior of the flow in a channel network, understanding the channel network's functional design is required. Figure 2 shows the flow chart of the study implementation. Understanding the functional design of the DIR Dadahup is initiated by reviewing the physical land condition of Kalimantan in general and the history of the swamp area irrigation development of Kalimantan. The complete process of understanding the functional design of the channel network is then provided with familiarizing the hydraulic concept of the DIR Dadahup. Further implementation involves the application of the HEC-RAS software to examine the behavior or the performance of the channel network.

The HEC-RAS (Hydrologic Engineering Center's River Analysis System) allows users to perform one-dimensional steady, one- and two-dimensional unsteady flow hydraulics, sediment transport/mobile bed computations, water temperature modeling, and generalized water quality modeling.[11] The simulation may accommodate various obstructions such as bridges, culverts, weirs, regulatory gates, and other structures. However, no obstacle is considered for this study on the flow behavior of the DIR Dadahup channel network. This study mainly used the one-dimensional unsteady flow simulation by adopting several scenarios as a function of the boundary conditions' types and magnitudes and the channel roughness coefficient. Examination of the hydraulic logic is introduced to assess the response of the hydraulics performance of the DIR Dadahup channel network, including the required running time, the stability, and the sensitivity of simulation results.

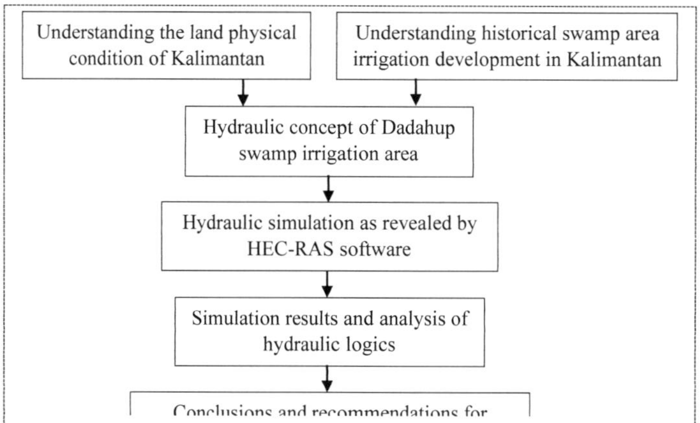

Figure 2. Flow chart of study implementation.

2.2 Land and Water Source Condition of DIR Dadahup

2.2.1 Land Condition. Land in Central Kalimantan Province generally consists of non-tidal land and tidal land in swampy areas with low elevation. This wide and flat swamp area is prone to puddles of water from high rainfall or tidal river estuaries. Tidal land has a non-uniform topography with a lot of biological variation. Swamp areas are also dominated by peat soil types which are very susceptible to changes in physical properties (especially the drainage capacity) due to land management.[5,12] Management of peatlands, including peatland development programs (PLG) that are accompanied by logging and construction of canal networks, can cause the drainage ability of this land to change. Subsequent changes in drainage ability can cause the ground to become very dry and difficult to return to its previous state. The continuation of the peatland development program called the Ex-Mega Rice Project (EMRP) includes the rehabilitation and revitalization of DIR Dadahup.[13] Figure 3 shows several blocks as parts of the EMRP where Block A consists of DIR Jenamas in the north and DIR Dadahup in the south. The tidal river of Barito has been selected as the primary source of water for DIR Dadahup development, provided with two tributaries of Mengkatip and Kapuas Murung River in the westside and south-east side, respectively. The area of Block A lies in the two districts of Barito Selatan and Kapuas. The area south part of Block A is more developed than that at the north region.

2.2.2 Water Source Condition.

The hydrological condition of DIR Dadahup is generally similar to that of the situation in Kalimantan Island. Previous research mentioned that surveys in 364 villages in Kalimantan Island show that flood frequency is associated with land use in catchment areas, including forest cover and condition.[14,15] The probability of floods becoming more frequent over the past three decades was higher for villages closer to mines but lower in watersheds with the greater cover of selectively logged forest. To some extent, the Indonesian peatland development changes the forest cover condition and increases potential flooding or inundation. Some situations would, in turn, provide constraints in further development of the land and the water management in peat/swampy land areas, including DIR Dadahup. Two alternatives may be possible to get the water sources for the DIR Dadahup development. The first alternative is water obtained from the surrounding rivers, considering the tide and water quality characteristics, mainly the acidity (pH) and the salinity. The second alternative is water obtained from the rain.

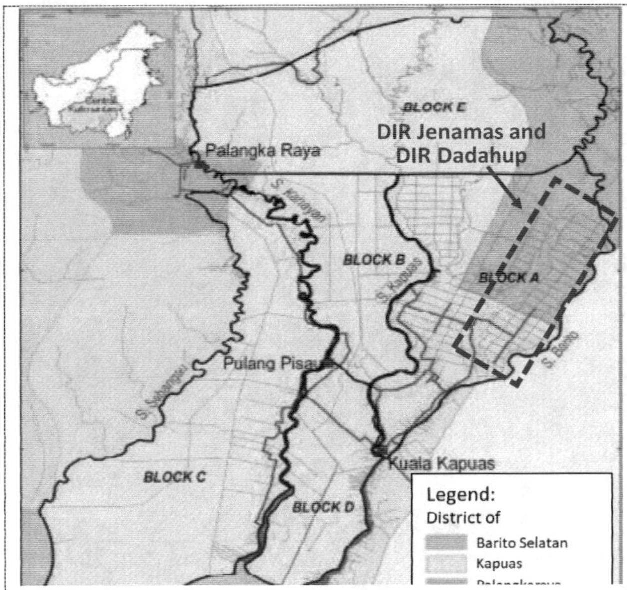

Figure 3. Position of DIR Dadahup on Ex-Mega Rice Project.

3. Results and Discussion

3. 1 DIR Dadahup Channel Network Geometrical Characteristics

The main water source to serve the agricultural development of DIR Dadahup is obtained from the Barito River, where the main primary canal or main primary channel (hereinafter referred to as SPU) is built with a length of about 58.00 km. This SPU has several additional primary channels or *Saluran Primer Pembantu* (also called SPP). The SPP water is then distributed over several secondary channels or *Saluran Sekunder* (further called SS). Supplying water from the SS into the rice field is made possible through the tertiary channels or *Saluran Tersier* (further called ST), whereas the drainage mechanism is carried out by utilizing the collector channel (further called SK). In addition to the Barito River, two rivers serve as water supply to paddy fields and drainage out of the land through the collector drainage, namely the Mengkatip River and the Kapuas Murung River. Figure 4 shows the schematization of the DIR Dadahup channel network that consists of one channel system SPU, three-channel systems of SPP, two-channel systems of SK, several numbers of channel systems of SS and ST.[6] To avoid the complexities, the simulation of the channel network of DIR Dadahup does not include channel systems of ST. No obstruction (e.g., gate, bridge, weir, or any other) is introduced in the channel systems, except embankments along SPU, SPP, SS, and SK. The more visible channel network of the DIR Dadahup is presented in Figure 5, whereas the channel description is shown in Table 1. The n-Manning roughness coefficient of 0.035 is used for the entire channel cross-sections over the whole system. Such assumption of the roughness coefficient for the channel systems may be too low for the lower channel orders or maybe too high for the main primary channel or SPU. A structured related hydrometry measurement is highly advisable to obtain a more appropriate value of such roughness coefficient. Several boundary conditions have been introduced in the simulation. This includes the discharge hydrograph at the upper river and stage hydrograph at the lower reaches. Some evaluations regarding the agreement of the simulation results are also carried out by utilizing data obtained from the real-time monitoring system. Understanding the behavior of the flow of the DIR Dadahup is also provided by comparing the simulation results to the real-time data at a location in SPU (near SS Q – Q2).

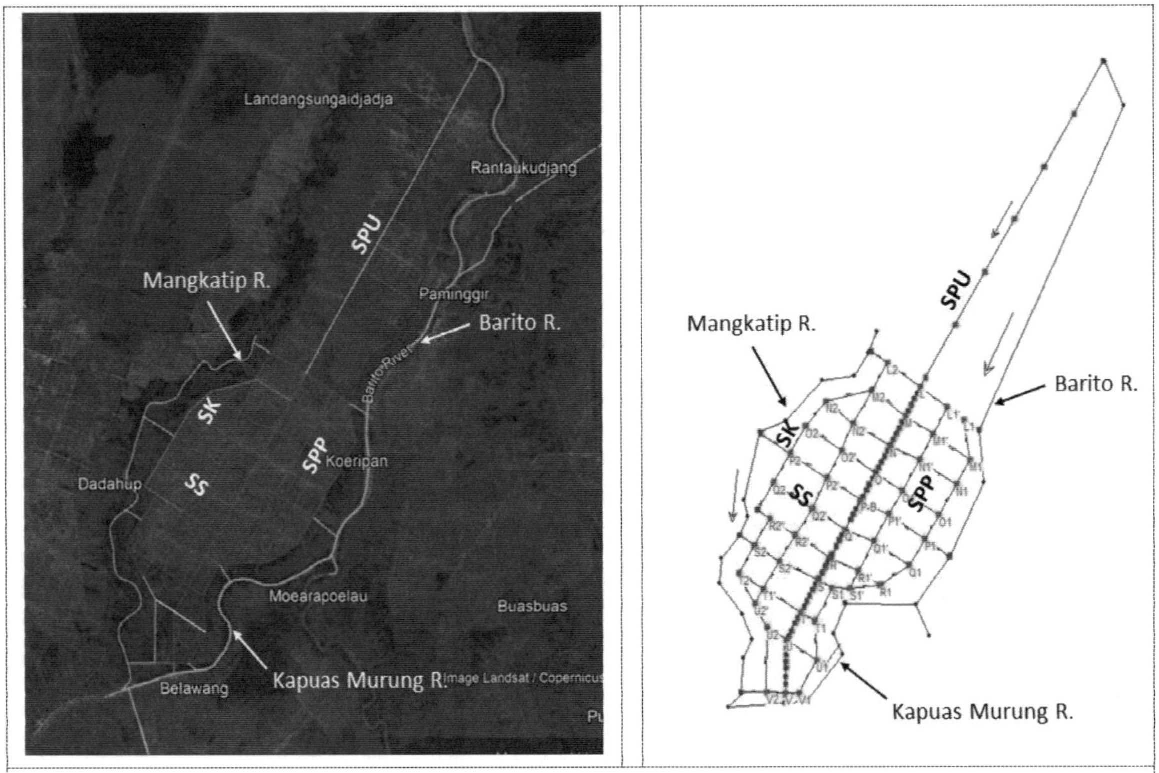

Figure 4. The schematization of the DIR Dadahup channel network. (a) On map. (b) In sketch.

The number of simulated segments for the SPU, SPP, SS, and SK are 13, 2, 20, and 2, respectively. The total length of all segments is 255.80 km, with the SPU being the longest to approximately 59.00 km. Assessment of flow behavior includes examining the longitudinal profile of maximum water level along with selected channel systems or segments. This includes the SPU, SS Q – Q1, and SS Q – Q2. The simulation assumes that no overflow may occur since all channel systems are provided with embankments that perform well. Besides, some control gates have been built in the DIR Dadahup. Such assumptions may cause significant discrepancies between the simulation results and real behavior.

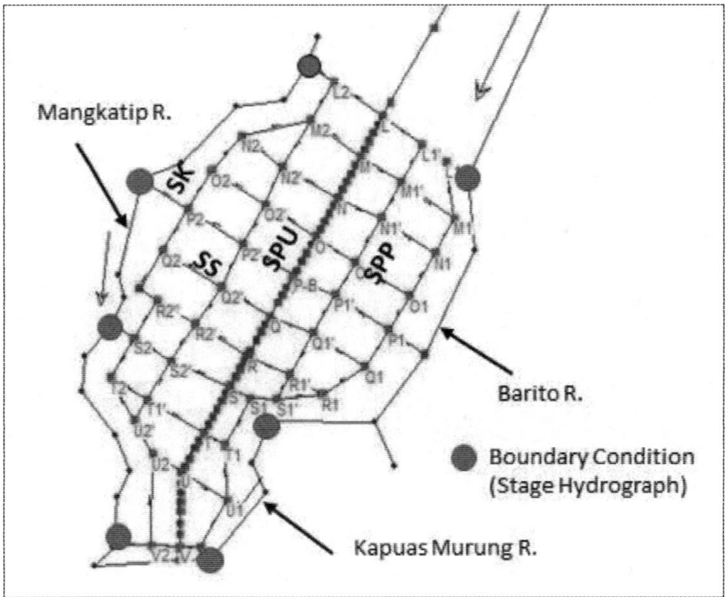

Figure 5. DIR Dadahup channel network and boundary condition locations.

Table 1. Channel network description of DIR Dadahup.

Name of Channel	Segment	Total length (Km)
Main Primary Channel (SPU)	Barito – L	59
	L – M	
	SPU L – L1	
	SPU L – L2	
	M – N	
	N – O	
	O – P	
	P – Q	
	Q – R	
	R – S	
	S – T	
	T – U	
	U – V	
Additional Primary Channel (SPP)	West SPP	35.52
	East SPP	
Secondary Channel (SS)	SS M – M1	96.31
	SS M – M2	
	SS N – N1	
	SS N – N2	
	SS O – O1	
	SS O – O2	
	SS P – P1	
	SS P – P2	

IOP Conf. Series: Earth and Environmental Science 1091 (2022) 012035 doi:10.1088/1755-1315/1091/1/012035

Table 1. Channel network description of DIR Dadahup (continued).

Name of Channel	Segment	Total length (Km)
	SS Q – Q1	
	SS Q – Q2	
	SS R – R1	
	SS R – R2	
	SS S – S1	
	SS S – S2	
	SS T – T1	
	SS T – T2	
	SS U – U1	
	SS U – U2	
	SS V – V1	
	SS V – V2	
Collector Channel (SK)	West SK	65.06
	East SK	
The total length of the channel is simulated		**255.8**

3.2 Boundary Conditions for HEC-RAS Simulation

Two scenarios have been introduced in the simulations. The first scenario utilized the discharge hydrograph at most upstream SPU (its meeting point with Barito River), and the second scenario utilized the stage hydrograph instead of the discharge hydrograph. Both simulation scenarios used eight boundary conditions in the form of stage hydrographs. Figure 6 shows the stage hydrographs at nine locations of the boundary conditions.

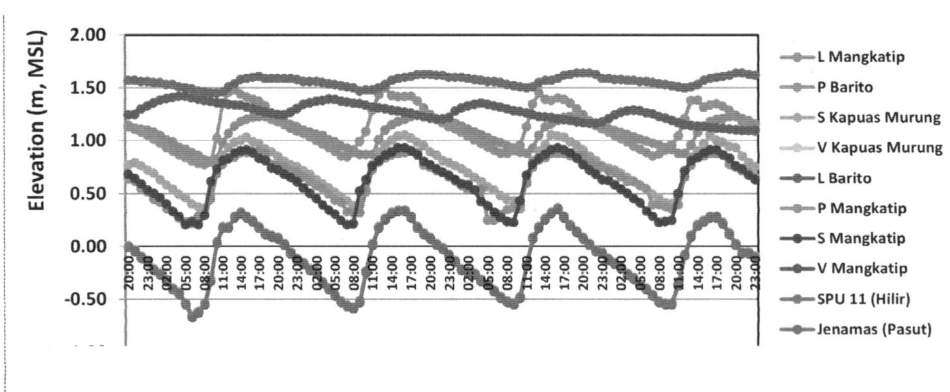

Figure 6. Boundary conditions (on 30 May - 2 June 2021).

3.3 Water Surface Profiles

The water surface profile along the SPU at maximum condition shows that the existing embankment of SPU is generally high and no overflowing takes place (see Figure 7). The water surface profile at cross-section or river station at secondary channel L point L1' or La2 which is 3.3 km downstream from the flow direction scenario near the Barito River shows that water flows alternately towards the east and west (see Figure 8). Figure 9 shows a water telemetry device installed on the main primary channel point Q which is 14 km downstream. The use of this tool is a collaboration between River Basin Organization II Palangka Raya and Gadjah Mada University. Data obtained from the use of this tool, one of which is the water level on October 24 to 27, 2021, is used as a comparison with the results of the HEC-RAS modeling from May 30 to June 2, 2021 (see Figure 10).

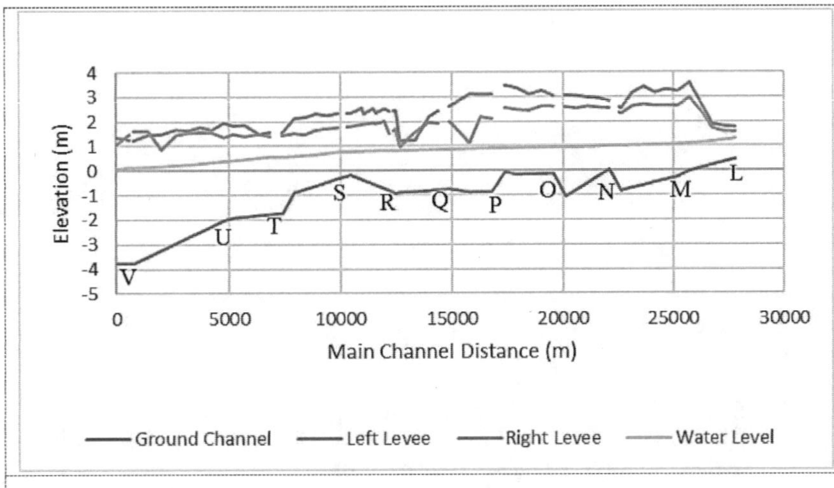

Figure 7. Water surface profile along with main primary channel on 31 May 2021 at 08.00 p.m.

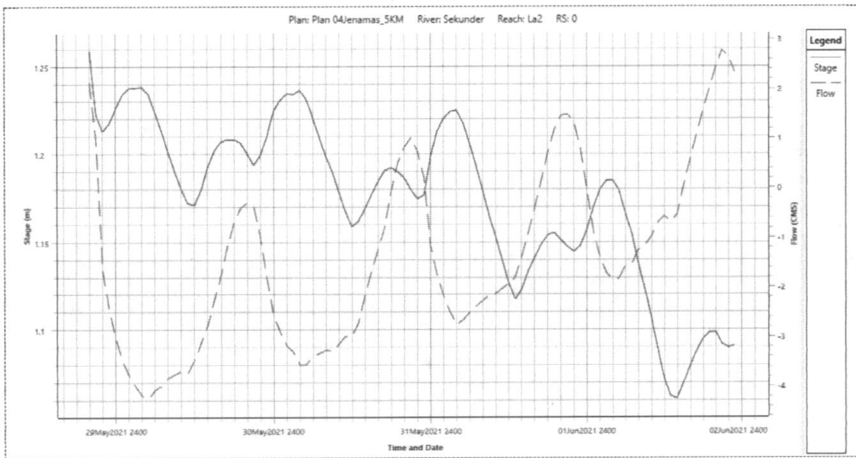

Figure 8. Water surface profile along with SPU L – L1.

Figure 9. Water telemetry on the main primary channel point Q. (a) The building where the water telemetry device is placed. (b) Water telemetry device.

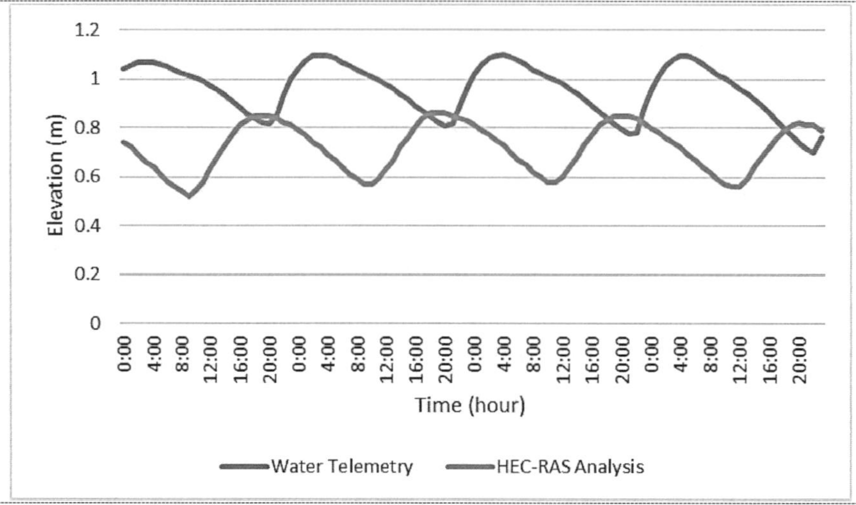

Figure 10. Water surface profile along with main primary channel between HEC-RAS results from the analysis on May 30 - June 2 2021 and observation data from water telemetry on October 24 to 27, 2021.

4. Conclusions

The following are the conclusions related to the behavior of the DIR Dadahup. The hydraulic simulation of DIR Dadahup using HEC-RAS with two different upstream boundary conditions does not show a significant difference in the simulation results. This is due to the equivalence of the discharge hydrograph with the stage hydrograph being applied in the simulation. The Barito River seems to contribute more water supply into the SPU than that of the Mengkatip River. Whereas Kapuas Murung appears to behave as the terminal of flow from SPU. The simulation results also indicate that the land with an elevation of less than 1.00 MSL may be prone to inundation. Unless some polder system is introduced, such land may not be productive for agriculture. Further simulation adopting several scenarios with obstructions may benefit getting proper operation or management of the channel networks.

Acknowledgments

The authors would like to express their profound gratitude to the Directorate General of Water Resources, Ministry of Public Works and Housing, the Government of the Republic of Indonesia for valuable financial support of this research, to all colleagues in the Department of Civil and Environmental Engineering, Universitas Gadjah Mada, for continuous encouragement related to sustainability of this study. The authors also thank JICA through the MJIIT Project for the author's chance to participate in the 9th-RCND Conference.

References

[1] Ratner M, Hardiono M, Alfred RJ *A Report: Borneo-Treasure Island at Risk* [Internet] (Switzerland: WWF) 2005 [cited 2021 Dec 10] Available from: https://wwf.panda.org/wwf_news/?21037/Report-Borneo-Treasure-Island-at-Risk/

[2] Arisanty D, Hastuti K P, Angriani P, Rajiani I 2019 Indigenous Knowledge of Banjarese Farmers to Predict the Season in Agriculture Area Swampland *Proceeding of ADVED – 5th International Conference of Advances in Education and Social Sciences* Prof. Dr. Ferit Uslu (Istanbul: International Organization Center of Academic Research (Ocerint)) pp 66-74

[3] Santosa PB, Yuwati TW, Rachmanadi D, Rusmana, Graham LLB 2014 Response of Peat Swamp Forest Species Seedlings to Flooding *Technical Papers*: *Tropical Peat Swamp Forest*

Silviculture in Central Kalimantan (Jakarta: Indonesia-Australia Forest Carbon Partnership) pp 37-45

[4] Zulkifley MTM, Fatt NT, Konjing ZZ, Ashraf MA 2016 Development of Tropical Lowland Peat Forest Phasic Community Zonations in the Kota Samarahan-Asajaya Area, West Sarawak, Malaysia *Earth Sci. Res. J.* **20** (1) pp 1-10 http://dx.doi.org/10.15446/esrj.v20n1.53670

[5] Euroconsult Mott MacDonald and Deltares 2008 Technical Report No. 4 Land and Water Management in the Ex-Mega Rice Project Area in Central Kalimantan *Master Plan for the Rehabilitation and Revitalisation of the Ex-Mega Rice Project Area in Central Kalimantan* (Palangka Raya: Delft Hydraulics in association with DHV, Wageningen UR, Witteveen+Bos, PT MLD and PT INDEC) pp 12-27

[6] PT. Virama Karya Persero (KSO) Cabang Semarang – PT. Wiratman – PT. Archiegama Bangun Cipta Pratama 2020 *Laporan Akhir Survey Investigasi Desain (SID) Rehabilitasi dan Peningkatan Jaringan Irigasi Rawa Wilayah Kerja Blok A (Paket 1) Kalimantan Tengah* (Palangka Raya: Balai Wilayah Sungai Kalimantan II) pp 4-2 – 4-7

[7] Hasyim S, Marpaung MF, Dhiaksa A 2020 Studi Investigasi Lapangan dalam Rapid Desain Rehabilitasi dan Peningkatan DIR Dadahup Provinsi Kalimantan Tengah *Proceeding Pertemuan Ilmiah Tahunan (PIT) Himpunan Ahli Teknik Hidraulik Indonesia (HATHI) ke-37* (Palembang: Himpunan Ahli Teknik Hidraulik Indonesia (HATHI)) pp 1015-1026

[8] Istianto H, Indrawan D, Suryadi FX 2020 Analisis Sistim Tata Air di DIR Kalimantan Tengah dengan Data Terbatas *Proceeding Pertemuan Ilmiah Tahunan (PIT) Himpunan Ahli Teknik Hidraulik Indonesia (HATHI) ke-37* (Palembang: Himpunan Ahli Teknik Hidraulik Indonesia (HATHI)) pp 1038-1047

[9] Wignyosukarto B 2000 Review konsep pengembangan pola tata air Lamunti, Dadahup dan Palangkau, Proyek Pengembangan Lahan Gambut, Kalimantan Tengah *Forum Teknik* **24** (3) (Yogyakarta: Universitas Gadjah Mada) pp 342-357

[10] Wignyosukarto BS 2013 Leaching and Flushing of Acidity in the Reclamation of Acid Sulphate Soil, Kalimantan, Indonesia *Irrig. and Drain* **62** (1) pp 75-81 https://doi.org/10.1002/ird.1777

[11] US Army Corps of Engineers 2021 *HEC-RAS River Analysis System: User's Manual Version 6.0* (United States: Institutes for Water Resources Hydrologic Engineering Center) pp 3-1 – 3-17

[12] Hirani A, Noor M 2021 Water management for increase rice production in the tidal swampland of Kalimantan, Indonesia: Constraints, limitedness, and opportunities *Proceeding of IOP Conf. Ser.: Earth Environ. Sci.* **724** 012021 https://doi.org/10.1088/1755-1315/724/1/012021

[13] Suriadikarta DA 2009 Pembelajaran dari kegagalan Penanganan Kawasan PLG sejuta hektar menuju pengelolaan lahan gambut berkelanjutan *Pengembangan Inovasi Pertanian* **2** (4) (Bogor: Pusat Perpustakaan dan Penyebaran Teknologi Pertanian) pp 229-242

[14] Wells JA, Wilson KA, Abram NK, Nunn M, Gaveao DLA, Runting RK, Tarniati N, Mengersen KL, Meijaard E 2016 Rising floodwaters: mapping impacts and perceptions of flooding in Indonesian Borneo *Proceeding of IOP Conf. Ser.: Environ. Res. Lett.* **11** 064016 https://doi.org/10.1088/1748-9326/11/6/064016

[15] Herawati HJ, Soeryamassoeka SB 2017 Drought mitigation participatory at Pinang Luar lowland, West Borneo (Indonesia) *Proceeding of MATEC Web of Conferences* **101** 04022 https://doi.org/10.1051/matecconf/201710104022

IOP Conf. Series: Earth and Environmental Science 1091 (2022) 012036 doi:10.1088/1755-1315/1091/1/012036

Climate change adaptation measures in the agricultural sector in Southeast Asia: A mini-review

S Sha'arani[1*], NSA Sabri[1], NA Hamdi[1], FA Riyadi[1]

[1] Malaysia-Japan International Institute of Technology, Universiti Teknologi Malaysia, Jalan Sultan Yahya Petra, 54100, Kuala Lumpur, Malaysia

*Corresponding author: wana.shazwana@gmail.com

Abstract. By 2050, the human population will have risen massively, so the demand for food and agricultural field will be the greatest global challenge in the future. However, biodiversity is declining as it experiences the impacts of deforestation, river pollution, the release of greenhouse gases, and many more detrimental effects that include global warming - the long-term increase in the planet's temperature. The agricultural field is primarily dependent on climate and thus climate change could affect agriculture in various ways. The situation demands alternative approaches to overcome and concurrently maintain food security. Hence, this paper presents a review of the adaptation measures employed by the agricultural sector to address climate change. The focus is Southeast Asia as climate change has been progressively more threatening in this region. Several adaptive measures and gaps were discovered, as presented in this paper. The present review will be a viable resource for communities, especially local farming communities in Southeast Asia, which are facing a future of climate change.

1. Introduction

People have been depending on plants as a source of food since ancient times. Since then, agriculture initiated a significant change when it was developed as a human food source. Feeding an expanding global population in a changing climate poses a major challenge for humankind, as the weather and climate significantly impact agricultural production. Southeast Asia has become the global region most vulnerable to climate change [1-3]. Climate change can be defined as the shift in climate patterns that is driven primarily by greenhouse gas emissions from natural processes and human actions [4,5]. Climate change is expected to influence agriculture and food security considerably, although the impact will vary by region and crop [6]. Agriculture and climate change are intrinsically linked in numerous ways since the latter is the primary cause of biotic and abiotic stress that has a detrimental effect on the agriculture sector [7]. Changes in climatic conditions, like an increase in abnormally hot or cold days, are expected to affect the optimal locations for diverse crop cultivation. It means crop management will need to keep pace with these transitions [8,9]. Henceforth, climate change will affect not only this region but also the whole world.

A considerable volume of research has been conducted on the adaptation measures to address the ways changing climatic patterns have impacted agricultural productivity, yet studies have rarely focused on Southeast Asia. The Food and Agriculture Organization revealed that the growing frequency and intensity of floods and droughts have negatively affected food security in most global regions [3]. Drought is agriculture's most devastating opponent, costing USD 37 billion in agricultural and livestock production losses from 2008 to 2018 [3]. A study by Sekhar, 2018 [2] discovered that shifting rainfall patterns could have a detrimental impact on 50% of the rice cultivation in Southeast Asia [2]. Hence, agriculture sustainability depends on adapting to climate change and acquiring precise ecological and biological information. Those are vital for forecasting future challenges and the implementation of appropriate countermeasures.

2. Overview of agriculture sector in Southeast Asia in facing climate change

A commitment was made to achieve sustainable development by 2030 through the Sustainable Development Goals (SDGs) aimed at 17 objectives with 169 targets. There are four goals in SDGs that have an interlinked connection with food and agriculture in the SDGs goal – indicators under Food and Agriculture Organization of the United Nations (FAO) (SDG 2, zero hunger; SDGs 6, clean water and sanitation; SDG 14, life below water; SDG 15, life on land) [10]. The key to achieving those entire set of SDGs is through food and agriculture [11]. Hence, agriculture plays one of the pivotal roles in the SDGs not only generate a substantial benefit for food security but in another context will be a driver for economic growth in facing climate change in the future.

Southeast Asia, as shown in Figure 1, is composed of Brunei, Myanmar, Cambodia, Indonesia, Laos, Malaysia, Philippines, Singapore, Thailand, Vietnam, and Timor-Leste. Rice, maize, and coconuts are cultivated across 115 million hectares (ha) of land in Southeast Asia (ASEA), while crops supply food for 45% of the population [1]. A substantial proportion of climate change research, especially in Malaysia, is related to agriculture [12]. This is due to its significance as a major national contributor to the Gross Domestic Product (GDP) [13]. As reported by the Department of Statistics (2020), in 2019, the agriculture sector contributed 7.1% (RM101.5 billion) to the GDP. Furthermore, exports from the agriculture sector in 2019 amounted to RM115.5 billion, an increase of 0.9% compared to 2018 (RM114.5 billion).

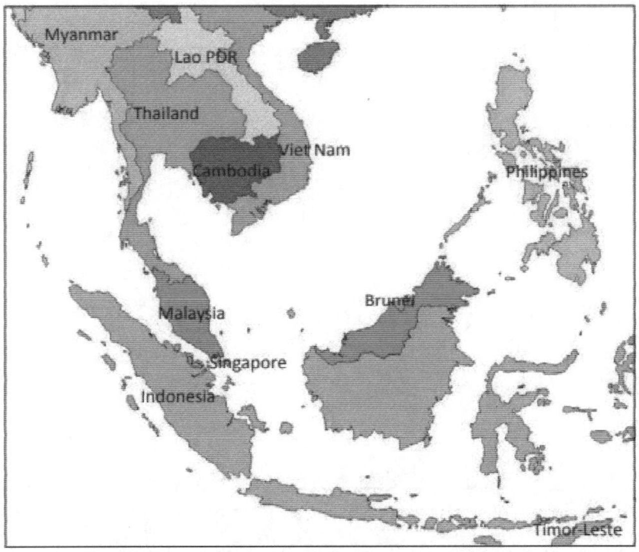

Figure 1. Map of Southeast Asia [14]

Three-quarters of the underprivileged in South Asia live in rural areas, and agriculture is the primary source of their income [15]. Climate change affects crop yield in a variety of ways. Temperature, soil moisture, radiation, carbon dioxide (CO_2) concentration, and rainfall are variables that significantly influence agricultural productivity, and their linkages are not simply linear [15]. Additionally, water stress currently exists in Southeast Asia, and future climate change is anticipated to result in more variable rainfall patterns, further affecting agriculture and therefore decreasing yield. As a result, this can greatly impact low-income rural populations who rely on traditional agricultural systems, such as those in Thailand's Khon Kaen Province, whereby rice production was reduced due to drought [16]. Meanwhile, agriculture in Cambodia is tightly linked to the rain-fed system; as a result, any variation in climate will impact agricultural output and, eventually, the country's economy [17]. Hence, adaptation to climate change is considered a fundamental aspect in combating climate change as well as mitigation measures across the world.

3. Soil management

World soil resources are scarce, unequally dispersed across varied locations, fragile, and prone to degradation due to land exploitation and soil mismanagement, as well as catastrophic occurrences linked to climate change [18]. Climate change has a significant direct and indirect impact on soil functions [19]. Temperature, precipitation, and moisture regime changes are all direct consequences of climate change, while their indirect effects encompass irrigation, crop rotation changes, and tillage practices. Figure 2 illustrates the effects of climate change and how it alters soil properties. However, it must be remembered that climate change may have both positive and negative consequences in the agricultural sector, which need to be uncovered [18]. Without healthier and sustainable soil management, the climate crisis cannot be addressed, nor can we produce enough food or adapt to a changing climate.

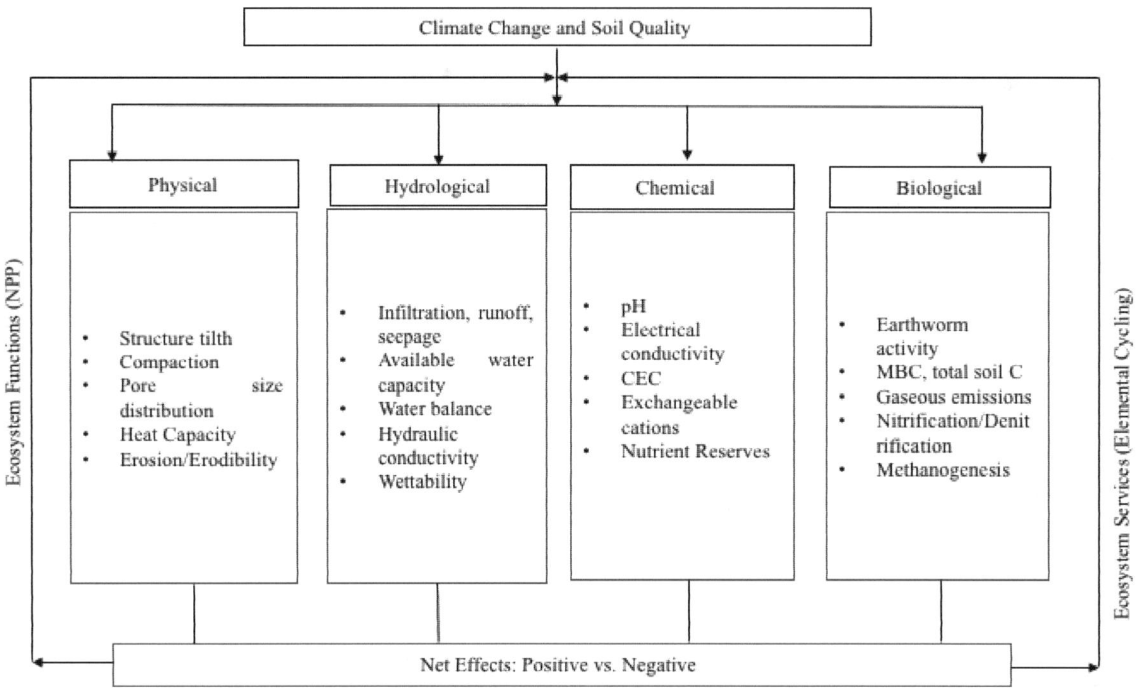

Figure 2. Impact of climate change on changes in soil quality and processes (both positive and negative) [18]

Some of the adaptation measures that have been applied for soil management include no-tillage and sequestering carbon in SOC. No-tillage (NT), often known as zero tillage or direct seeding, is a cropping method whereby seeds are directly deposited into untilled soil by drilling (with minimal soil disturbance), which maintains the previous crop residue [20]. In comparison to traditional production systems, it was discovered by Sapkota *et al.* 2015 [21] that changes in tillage practices moderate the effects of high temperature (the canopy temperature reduced by 1–4 °C and increase irrigation water productivity by 66–100%, thereby enabling the soil to adapt to water and heat stress situations. In addition, sequestering carbon in SOC is viewed as one strategy to reduce atmospheric carbon dioxide, thereby capable of mitigating climate change and improving soil quality [22].

4. Cropping system

From the perspective of food security and farming livelihoods in Southeast Asia, the sustainability of the rice–wheat system is a major challenge [23]. Different agricultural management practices can be employed through adaptation methods in Southeast Asia in response to climate change. Some farmers in Kandal Province, Cambodia, have divided their rice fields into two. They use different management practices to overcome rainfall unpredictability whereby i) half of the rice plot is planted with conventional wet-paddy rice that can withstand heavy rains and ii) the other half employs a method of rice intensification, which is drought-tolerant and requires less water [14][24]. Besides that, an alternative approach is to change crop varieties in response to climatic stresses and changes. As such, the seasonal and semi-permanent adoption of climate-resilient rice varieties is practiced on commercial farms in Vietnam [25]. In addition, climate-smart agriculture has emerged as a major topic for research and application in the face of climate change. By employing the modern agricultural technique, this new area of research and technology has addressed the deficiencies in agricultural practices and tools available to farmers [17]. A range of modern technologies can be incredibly beneficial to assist farmers (such as: power generators, solar panels, monitoring systems). Table 1 demonstrates several modern agricultural methods that have been implemented to improve crop yield.

Table 1. Implementation of several modern methods/techniques to improve agricultural yield

Method/Technique	Crop	References
Proposed a partially closed solar regenerated liquid desiccant aided evaporative cooling system	Lettuce	[26]
Leafy vegetables land use model (LYLY) model	Leafy vegetables	[27]
Proposed solar cooling system	Lettuce, Tomato, Cucumber	[28]

Likewise, climate-smart agriculture (CSA) is a technique of assisting those in charge of agricultural systems in efficiently responding to climate change [29]. It targets to tackle three main purposes: sustainably increasing agricultural productivity; adapting and building resilience to climate change; mitigation of greenhouse gas emissions (GHG) emission, whenever possible [29].

5. Water and land management

Water is a key component in agricultural productivity as it aids food security, while maintaining the soil and water quality are imperative for agriculture productivity. Competition for water resources is predicted to increase due to population expansion, urbanization, and climate change, with more direct investments being made in agriculture [30]. Improved irrigation is one approach to preparing the local agricultural systems for climate change [31-33]. The 'bucket and drip' irrigation system is a device that might prove beneficial in some regions of Southeast Asia. Water is managed using a gradual release system and supplied by a raised bucket to pipes with emitters throughout all plots [14].

As for land management, the most visible landscape marks demonstrating traditional knowledge are agricultural terraces. The Ifugao Rice Terraces in the Philippines and the Bali Rice Terraces in Indonesia are some of the most well-known examples of ancient Asian terraces that have been continuously cultivated [37]. Despite being highly labor-intensive, agricultural terracing is one of the world's most common and long-standing upland farming methods, with evidence of terracing dating from approximately 4000 to 6000 years ago [34-37]. Besides that, shifting land-use practices can mitigate the effects of climate change on agricultural productivity by i) rotating or shifting the production between livestock and crops, ii) shifting production out of marginal areas, and iii) evaluating different types of capital and labor.

6. Pest and disease control

In the twentieth century, increasing nitrogen fertilizer usage helped to maintain global population growth. Chemical fertilizer causes significant environmental damage, so a current worldwide target is to limit the usage of chemical pesticides. Global demand for nitrogen, phosphorus, and potassium for fertilizer use was expected to rise between 2016 and 2022 [38], as shown in Table 2. Hence, new sustainable agronomic approaches and new cultivars must be developed to prevent future environmental damage. Furthermore, optimal agronomic practices have to be developed to attain sustainable and cost-effective production. When crop biomass reduces owing to water and temperature stress over time as a result of climate change, the fertilizer application rate and timing should be modified to fulfill the nutritional demands of plant development and minimize nutrient loss. Adjustments to management techniques such as fertilization optimization play a substantial role in closing the yield gap, improving nutrient use efficiency (NUE), and reducing environmental concerns in the context of climate change. Moreover, crop rotation diversity improves soil physical traits, nutrient uptake, and soil microbial diversity, resulting in higher crop yields and reduced environmental risk. Most soils in indigenous crop production systems are depleted and low in nutrients. Legumes as green manures, cultivated in intercropping systems, as part of a crop rotation scheme, can help to alleviate this problem to some extent. These potential adaptation strategies could all be implemented in Southeast Asia.

Table 2. World demand for nitrogen, phosphorus and potassium for fertilizer use (thousand tonnes), 2016-2022 [38]

Year	2016	2017	2018	2019	2020	2021	2022
Nitrogen, N	105,148	105,050	105, 893	107,424	108,744	110,193	111,591
Phosphorus, as P_2O_5	44,481	45,152	45,902	46,587	47,402	48,264	49,096
Potassium, as K_2O	35,434	36,349	37,171	37,971	38,711	39,473	40,232
Total (N+ P_2O_5 + K_2O)	185,063	186,551	188, 966	191,981	194,857	197,930	200,919

7. Crops genetic modification

Since they were first commercialized in 1996, the growing area of genetically modified (GM) crops has significantly expanded [39]. Generally, technological advancement has aided agriculture productivity, particularly in the field of genetic crop modification. Genetically modified seeds represent an advanced step in agricultural production. One such exploration was the utilization of CRISPR/Cas9 to knockout OsRR22, a gene linked with salt susceptibility in rice [40]. Researchers discovered that, compared to wild-type plants, edited plants had much less severe biomass losses due to salt exposure [40]. Gene editing is a significant and emerging strategy in plants and animals that are being used in response to climate change. While it is important to emphasize that gene editing is not the only possible agricultural development strategy, its potential for addressing climate change in agriculture cannot be underestimated.

8. Conclusion

Maintaining food security despite the escalating human population requires an expansion in agricultural production. Hunger, poverty, and climate change need to be addressed collectively. One important finding is that agricultural strategies that aid adaptions to climate change are accessible, yet the institutional framework to integrate and disseminate such technology needs to be reinforced. Hence, researchers, professionals, farmers, governments, and NGOs must cooperate to identify and characterize the current state of the policies, measures, and strategies aimed to reduce the burden of climate change impacts. These actions are essential to evaluate the effectiveness of adaptation support, inform governance at various levels about adaptation needs, and justify funding allocations. A systematic adaptation plan at the cropping system scale should be implemented to fully leverage the positive impacts of climate change on agricultural yields. Whilst developments in technology and research are being made; it is critical to have reliable seasonal forecasting to provide farmers with early notice of changing climate patterns. That allows them sufficient time to devise a plan for their farming systems.

Acknowledgments

The authors would like to express their appreciation for the support of the Ministry of Higher Education Malaysia (MOHE) and Universiti Teknologi Malaysia (UTM) under the Fundamental Research Grant Scheme (FRGS) (R.K130000.7843.4F991). We thank Assoc. Prof. Dr. Hirofumi Hara and Assoc. Prof. Dr. Nor'Azizi Othman for fruitful discussions regarding the agricultural sector in Southeast Asia.

References

[1] Dedicatoria R M M and Diomampo C B 2019 Status of climate change adaptation in Southeast Asia region. In Alam M, Lee J and Sawhney P (eds) *Status of climate change adaptation in Asia and the Pacific* pp 153-182

[2] Sekhar C S C 2018 Climate change and rice economy in Asia: Implications for trade policy. The State of Agricultural Commodity Markets (SOCO) (Rome, FAO) p 62

[3] FAO 2021 The impact of disasters and crises on agriculture and food security: 2021. The Impact of Disasters and Crises on Agriculture and Food Security: 2021. FAO https://doi.org/10.4060/cb3673en.

[4] Hamann E, Blevins C, Franks S J, Jameel M I and Anderson J T 2021 Climate change alters plant-herbivore interactions *New Phytol.* **229** 1894-1910

[5] IPCC 2007 Climate Change 2007: Synthesis Report. Bernstein L, Bosch P, Canziani Chen Z, Christ R, Davidson O, Hare W, Huq S, Karoly D, Katt sov V, Kundzewicz Z, Liu J, Lohmann U, Manning M, Matsuno T, Menne B, Metz B, Mirza M, Nicholls N, Nurse L, Pachauri R, Paluti kof J, Parry M, Qin D, Ravindranath N, Reisinger A, Ren J, Riahi K, Rosenzweig C, Rusti cucci M, Schneider S, Sokona Y, Solomon S, Stott P, Stouff er R, Sugiyama T, Swart R, Tirpak D, Vogel C, Yohe G, Cambridge University Press, Cambridge.

[6] Anderson R, Bayer P E and Edwards D 2020 Climate change and the need for agricultural adaptation *Curr. Opin Plant Biol.* **56** 197-202

[7] Mendelsohn R 2014 The impact of climate change on agriculture in Asia *J. Integr. Agric.* **13** 660-665

[8] Brooks J, Oxley D, Vedlitz A, Zahran S and Lindsey C 2014 Abnormal Daily Temperature and Concern about Climate Change Across the United States *Rev. Policy Res.* **31** 199-217

[9] Dubois G, Ceron J P, Gössling S and Hall C M 2016 Weather preferences of French tourists: lessons for climate change impact assessment *Clim. Change* **136** 339-351

[10] FAO 2019. Tracking progress on food and agriculture-related SDG indicators. A report on the indicators under FAO custodianship. FAO publication. www. fao.org.

[11] FAO 2016. Food and agriculture. Key to achieving the 2030 agenda for sustainable development. FAO publication. www. fao.org.

[12] Tang K H D 2019 Climate change in Malaysia: Trends, contributors, impacts, mitigation and adaptations *Science of the Total Environment* **650** 1858-1871

[13] Department of Statistics Malaysia Official Portal, Selected Agricultural Indicators, Malaysia 2020 https://www.dosm.gov.my/v1/index.php?r=column/cthemeByCat&cat=72&bul_id=RXVKUVJ5TitHM0cwYWxlOHcxU3dKdz09&menu_id=Z0VT

[14] Lasco R D, Habito C M D, Delfino R J P, Pulhin F B and Concepcion R N 2011 Climate change adaptation for smallholder farmers in Southeast Asia. World Agroforestry Centre Philippines p 65

[15] Anbumozhi V, Breiling M, Pathmarajah S and Reddy V R (Eds.) 2012 *Climate change in Asia and the Pacific: how can countries adapt?* SAGE Publications India p 388

[16] Raghavan S V, Ze J, Hur J, Jiandong L and Ngoc N 2019. ASEAN Food Security under the 2 C-4 C Global Warming Climate Change Scenarios *Towards a Resilient ASEAN* **1** 37-52

[17] Shrestha R P, Raut N, Swe L M M and Tieng T 2018 Climate change adaptation strategies in agriculture: Cases from southeast Asia *Sustain. Agric. Res.* **7** 39-51

[18] Lal R 2014 Climate strategic soil management *Challenges* **5** 43-74

[19] Hamidov A *et al.* 2018 Impacts of climate change adaptation options on soil functions: A review of European case-studies *Land Degrad. Dev.* **29** 2378-2389

[20] Powlson D S, Stirling C M, Jat M L, Gerard B G, Palm C A, Sanchez P A and Cassman K G 2014 Limited potential of no-till agriculture for climate change mitigation *Nat. Clim. Change* **4** 678-683

[21] Sapkota T B, Jat M, Aryal J P, Jat R and Khatri-Chhetri A 2015 Climate change adaptation, greenhouse gas mitigation and economic profitability of conservation agriculture: Some examples from cereal systems of Indo-Gangetic Plains *J. Integr. Agric.* **14** 1524–1533

[22] Aryal J P, Sapkota T B, Khurana R, Khatri-Chhetri A, Rahut D B and Jat M L 2020 Climate change and agriculture in South Asia: Adaptation options in smallholder production systems *Environ. Dev. Sustain.* **22** 5045-5075
[23] Jalota S K, Kaur H, Ray S S, Tripathi R, Vashisht B B and Bal S K 2012 Mitigating future climate change effects by shifting planting dates of crops in rice–wheat cropping system *Reg. Environ. Change* **12** 913-922
[24] Resurreccion B P, Sajor E E and Fajber E 2008 Climate adaptation in Asia: Knowledge gaps and research issues in South East Asia; full report of the South East Asia team p 76
[25] Snidvongs A 2006 Vulnerability to climate change related water resource changes and extreme hydrological events in Southeast Asia *AIACC Final Reports Project No. AS, 07* p 115
[26] Ghosh A and Ganguly A 2017 Performance analysis of a partially closed solar regenerated desiccant assisted cooling system for greenhouse lettuce cultivation *Sol. Energy* **158** 644-653
[27] Mawois M, Le Bail M, Navarrete M and Aubry C 2012 Modelling spatial extension of vegetable land use in urban farms *Agron. Sustain. Dev.* **32** 911-924
[28] Davies P A 2005 A solar cooling system for greenhouse food production in hot climates *Sol. Energy* **79** 661-668
[29] FAO 2021 Climate-Smart Agriculture https://www.fao.org/climate-smart-agriculture/overview/en/
[30] The World Bank 2020 https://www.worldbank.org/en/topic/water-in-agriculture#1
[31] Rosenzweig C and Parry M L 1994 Potential impact of climate change on world food supply *Nature* **367** 133–138
[32] Elliott J. *et al.* 2014 Constraints and potentials of future irrigation water availability on agricultural production under climate change *Proc. Natl Acad. Sci.* **111** 3239–3244
[33] Licker R, Johnston M, Foley J A, Barford C, Kucharik C J, Monfreda C and Ramankutty N 2010 Mind the gap: how do climate and agricultural management explain the 'yield gap' of croplands around the world? *Global Ecol. Biogeogr.* **19** 769–782
[34] Dotterweich M 2013 The history of human-induced soil erosion: Geomorphic legacies, early descriptions and research, and the development of soil conservation—A global synopsis *Geomorphology* **201** 1-34
[35] Sandor J A and Homburg J A 2017 Anthropogenic soil change in ancient and traditional agricultural fields in arid to semiarid regions of the Americas *J. Ethnobiol.* **37** 196-217
[36] Spencer J E and Hale G A 1961 The origin, nature, and distribution of agricultural terracing *Pacific viewpoint* **21** 1-40
[37] Wei W, Chen D, Wang L, Daryanto S, Chen L, Yu Y, Lu Y, Sun G and Feng T 2016 Global synthesis of the classifications, distributions, benefits and issues of terracing *Earth-Science Reviews* **159** 388-403
[38] FAO 2021 Crops and livestock products. http://www.fao.org/faostat/en/#data/QCL
[39] Khan F and Hakeem K R 2015 Genetic modification of crop plants: issues and challenges *Crop Prod. Glob. Environ. Issues* pp 369-384
[40] Zhang Y. *et al.* 2019 CRISPR/Cas9-mediated sheep MSTN gene knockout and promote SSMSCs differentiation *J. Cell. Biochem.* **120** 1794–1806

IOP Conf. Series: Earth and Environmental Science 1091 (2022) 012037 doi:10.1088/1755-1315/1091/1/012037

Utilizing Google Street View for Rapid Seismic Vulnerability Assessment: Case Study in the City of Manila, Philippines

M K Evangelista*, D H Aquino and J A Ramos

University of the Philippines Institute of Civil Engineering, U.P. Campus, Diliman, Quezon City, 1101 Philippines

*E-mail: mjevangelista1@up.edu.ph

Abstract. Earthquake Impact Reduction Study for Metro Manila (MMEIRS) estimated that around 40% of the total number of residential buildings within Metro Manila will either collapse or be affected by the M7.2 generated by the West Valley Fault. Need arises to evaluate the seismic vulnerability of structures in the area to enhance the city's resilience to seismic hazards. Rapid seismic vulnerability assessments are typically conducted by means of sidewalk surveys. However, advances in digital technologies such as Google Street View (GSV) provide the potential to do remote assessments, particularly amid mobility restrictions brought about by the COVID-19 pandemic. This paper aims to demonstrate the usefulness of GSV in collecting data needed for rapid seismic vulnerability assessments through the case of buildings in the City of Manila. Six 300 m x 300 m blocks were evaluated using GSV for identifying seismic-related building parameters. Results show the ease of use of GSV in data collection on areas encompassing commercial and residential zones within the city and poses difficulty for blocks dominated by informal settlements. Among the challenges observed in formal zones include blockages in views due to fences, trees, and/or vehicles parked in front. For informal settlements, much of the buildings are not visible in GSV for evaluation thereby necessitating supplemental data collection. Overall, GSV demonstrates usefulness, and has the potential to speed up seismic vulnerability assessments in urban areas in conjunction with existing in situ assessments currently conducted.

1. Introduction

According to the Earthquake Impact Reduction Study for Metro Manila (MMEIRS), around 40% of residential buildings within Metro Manila will be damaged by M7.2 generated by the West Valley Fault [1]. Hence, seismic vulnerability assessments are crucial for the development of the rapidly urbanizing metropolitan. Evaluating the seismic vulnerability of the structures, especially non-engineered ones, will provide the governing body an opportunity to implement mitigation measures before an earthquake occurs. Once implemented, the risk of everyone to earthquakes is reduced, and the overall resilience of the cities is increased.

Rapid seismic vulnerability assessments are typically conducted by means of field survey or on-site investigation. FEMA P-154 [2], one of the most used rapid visual methods for evaluating the seismic vulnerability of structures, rates buildings based on information obtained through sidewalk surveys. Buildings are typically assessed based on their physical attributes which includes vertical and plan irregularities, height, structural material, and the lateral force-resisting system. On the other hand, the %NBS Rating System in New Zealand evaluates the performance of a building compared with a similar new one in terms of protecting life in the occurrence of an earthquake [3]. In the Philippines, the

Philippine Institute of Volcanology and Seismology (PHIVOLCS) created a self-check questionnaire for the evaluation of one to two-story concrete hollow block (CHB) houses in the Philippines [4]. The aforementioned tools assess the seismic vulnerability of a building through site inspection.

In the field of remote sensing, satellite imagery was used for visual-based seismic vulnerability assessments. A high-resolution optical satellite imagery in Germany allowed the extraction of building shape, position, and height to some extent [5]. In Yancheng, China, a mid-resolution satellite imagery in the study areas provided seismic vulnerability estimation while field investigation calibrated the assessment [6]. Given the limitations of these methods, further inquiry in the potential of remote sensing can provide a convenient, but not ultimate, assessment of seismic vulnerability of buildings.

In situ assessments have recently been challenging to implement due to mobility restrictions brought about by the COVID-19 pandemic. In particular, the City of Manila imposed hard lockdowns and curfews to curb the spread of the virus [7]. Thus, alternative means to collect building parameters necessary for the seismic vulnerability assessment of structures need to be explored.

Google Street View (GSV), launched in 2007, is a technology embedded in Google Maps and Google Earth which allows the view of buildings from a street perspective. It seeks to provide captured and processed street-level imagery that can provide an immersive experience to the users [8]. Initially, GSV only covered a handful of cities, but the platform has immensely expanded to include urban and rural areas globally. Such a rich dataset has the potential to be tapped for visual-based assessments. In a similar study, GSV was used for the automatic identification of soft-story buildings at a city scale using deep learning [9]. In the field of health science, it was highlighted that using GSV was feasible for auditing neighborhood environments albeit limited by temporal elements [10]. As for urban planning, GSV was deemed suited for assessing urban street greenery in cities [11]. The aforementioned studies highlighted the advantage of GSV in simplifying the data collection procedure.

This paper investigated the utility of GSV in conducting virtual structural surveys for potential usage to rapid seismic vulnerability assessments in an urban area in a developing country. Specifically, we test the usefulness of GSV in the data collection needed for vulnerability assessments through the case of buildings in selected districts in the City of Manila, Philippines. This was established by extracting the houses accessible through GSV and identifying two key parameters for seismic vulnerability — the building type and the number of floors. Different areas in the city were evaluated based on these criteria. Furthermore, this study was limited to data acquisition for seismic vulnerability assessment through remote means in the absence of physical site inspection.

2. Research Method

Case studies are utilized for both quantitative and qualitative research, and a single-case design is used for testing a particular theory [12]. As a methodology, it drives the validation of findings through analyzing context-dependent structures [13].

Testing the applicability of GSV for seismic vulnerability assessments was conducted by looking at the case of buildings in the City of Manila — the capital of the Philippines. Having 409,987 occupied housing units [14], it is the center of commerce, socio-political activities, and industrial development in the country. However, the city faces urbanization problems such as informal settlements thriving in the city — owing to the accessibility of employment [15]. Testing GSV through a case study in the City of Manila yields insights on its application in urban areas. Moreover, it would permit further inquiry on the limitation of this method at the city level.

2.1. Identifying the Sample Blocks

To investigate the utility of GSV for data collection needed for the conduct of seismic vulnerability assessment in the City of Manila, sample blocks were derived from the predominant land-use zones in the city. These were identified to be high density residential/mixed-use zones (R-3/MXD), medium intensity commercial/mixed-use zones (C-2/MXD), and high intensity commercial/mixed-use zones (C-3/MXD) based on the Official Zoning Maps 2005 - 2020 [16]. The three major zones encompass the majority of residential settlements in the city. Afterwards, a stratified sampling was conducted.

Using ArcGIS Pro, the "Create Fishnet" function under "Geoprocessing" was used to define cells of 300 meters by 300 meters grid, and the City of Manila Boundary was used for the "Template Extent" in the program. The created grid was superimposed to the Official Zoning Map and two blocks per zone type were randomly identified. Figure 1 shows the sample blocks within the City of Manila in Google Earth layout, with the following areas: Baseco (R-3/MXD), Sampaloc (R-3/MXD); Tondo (C-2/MXD), Paco (C-2/MXD); Binondo (C-3/MXD), and Sampaloc (C-3/MXD).

Figure 1. Sample blocks (300m x 300m) in the City of Manila for virtual structural survey in GSV.

2.2. *Virtual Structural Survey through GSV*

Seismic vulnerability assessments require data on the building characteristics. Using GSV, these were extracted remotely in the City of Manila sample blocks by identifying the building type and number of floors of each building. The key building types in Metro Manila that were established by the University of the Philippines Diliman - Institute of Civil Engineering (UPD-ICE) [17] were used as reference for this study. If both information were identified using GSV, the structure was then labeled as "Classified." For buildings with limited view using GSV for which the building type and/or number of floors cannot be determined, these were labeled as "Unclassified."

To supplement the data gathering and account for the buildings that are not accessible in GSV, the Google Earth Satellite View was utilized. The building footprints within each block were identified by manually tracing the roof of a building, and if visible, the walls that were projected from the view. As a rule, buildings partially situated within a sample block were included in the count.

3. Results and Discussion

Virtual structural survey in the City of Manila through GSV allowed the extraction of building parameters essential for seismic vulnerability assessments. The investigation was conducted in each sample block as shown in figures 2, 3, and 4. The building footprints were marked blue, pertaining to the buildings visible in GSV and the red marks, otherwise. The white border indicates the boundary of the sample block.

From figure 2(a), there is only one main road in Baseco accessible using GSV as observed from the blue outlines in the area. Majority of the houses in the block are inaccessible. Notably, most of the buildings have makeshift building types, making identification from both GSV and satellite view difficult.

On the other hand, figure 2(b) shows the extent of the GSV in the Sampaloc area. It was observed that most building footprints are in blue as most buildings are accessible using GSV. Despite the grid configuration, there are still some roads in the area accessible to motorized vehicles yet have no available GSV data. This explains the red marks in between the blue marks that are apparent in the figure.

(a)

(b)

Figure 2. Satellite view of study block in R-3/MXD zones in **(a)** Baseco and **(b)** Sampaloc.

The block in figure 3(a) is an area in Tondo. It can be observed that there are evident clumps of red-marked buildings on the east as a consequence of having informal settlements in the area. There are also some roads with no access to GSV, which are apparent on the southwestern and northern areas of the block. The blue-marked buildings are those that are adjacent to the roads.

For the block shown in figure 3(b), there are some houses in Paco that are inaccessible in GSV. Commonly, these areas have a small opening or "esquinita" which serves as access points for the residents. As motorized traffic is not allowed, there is limited GSV in such areas.

(a)

(b)

Figure 3. Satellite view of study block in C-2/MXD zones in **(a)** Tondo and **(b)** Paco.

The highly commercial areas of Binondo and Sampaloc in figure 4 show that the majority of the areas are visible in GSV. Hence, there is an ease in assessing the buildings. Though the east part of Sampaloc has an area which lapsed GSV, most of its roadways have visibility in the virtual structural survey. Notably, the buildings in Binondo have relatively larger footprints, with condominiums and malls evident in the satellite view. However, there are evident obstructions that block the view to low-rise buildings, such as tents of street vendors.

(a)

(b)

Figure 4. Satellite view of study block in C-3/MXD zones in **(a)** Binondo and **(b)** Sampaloc.

Figure 5 shows the distribution of buildings within each block in terms of its accessibility in GSV. Notably, R-3/MXD areas have high percentages of inaccessibility, partly due to the high population density of the area, as well as the proliferation of informal settlements. Tondo also has more than half of the total surveyed with inaccessible buildings, particularly due to the limited GSV in the block. On the other hand, areas in C-3/MXD have high accessibility rates, marking the wide extent of GSV in these areas.

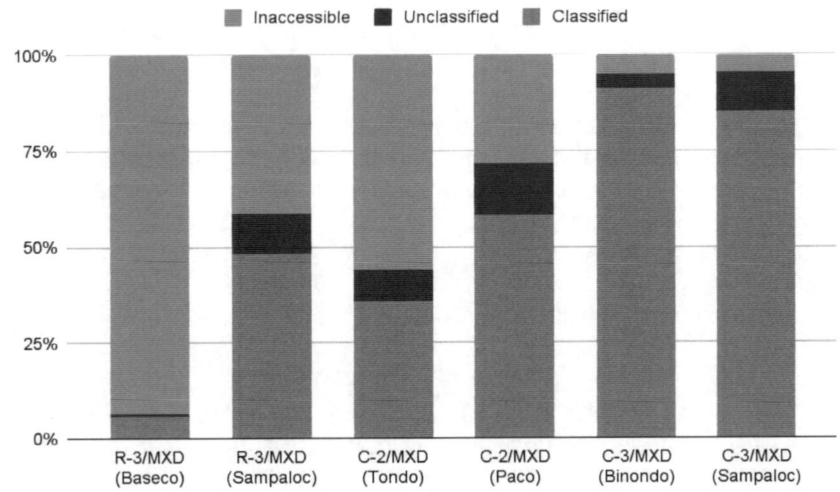

Figure 5. Building distribution in terms of accessibility based on conducted virtual structural surveys through GSV in selected areas in the City of Manila.

Several challenges were identified during the virtual structural survey. One of these is the limited perspective that is available in GSV. In particular, the building characteristics were easier to identify for buildings located at corner streets as there were multiple views available. It was also observed from the virtual structural survey that there were cases wherein roads can accommodate vehicular traffic, yet do not have access to GSV.

Another challenge encountered was obstructions such as tents, vendors, trees, and fences, that block the view to the buildings. Such obstructions constitute for the majority of the low-rise buildings labelled as "Unclassified". Tents, especially if the GSV is located underneath it, completely block the view to a building. There are areas, particularly for the case of buildings in Binondo, that were difficult to assess as these were located within small alleyways or "esquinita." Therefore, the limited angles in GSV hinder the identification of building parameters.

In assessing commercial areas, there was a need for multiple views to identify the building type properly, especially for buildings that have larger building footprints such as malls and condominiums. Projection was also an issue especially in high-density commercial areas. Condominium towers and apartments generally have a projected view in Google Earth satellite view. As such, adjacent low-rise buildings were seldomly partitioned and, in some cases, invisible from the satellite view. Hence, GSV was needed to supplement the identification of the building footprint.

Informal settlements in high density residential areas also exhibited low accessibility to GSV. Hence, incomprehensive data collection through GSV merits the need for further investigation on the site.

Data derived from GSV may be supplemented by data gathered through stakeholder interviews and through previously published reports.

4. Conclusion and Recommendation

Site limitation hinders physical investigation for seismic vulnerability assessment, making remote assessments a crucial endeavor in the time of COVID-19 pandemic. This paper investigated the utility of GSV for the conduct of data collection for seismic vulnerability assessments. The building characteristics, namely building type and number of floors were identified for buildings in the selected areas in the City of Manila using GSV. Results showed that high intensity commercial zones, and areas with regular spatial patterns exhibited high accessibility in GSV. Therefore, GSV has the potential to simplify the data collection in seismic vulnerability assessments.

However, remote assessments like the conducted virtual structural survey using GSV have its own limitations. In this study, obstructions, such as fences and trees, and limited angles provided by GSV hinder the identification of the building characteristics. Moreover, informal settlements and areas with organic spatial configurations lack accessibility to GSV, making the process of data collection inconsistent. As such, secondary sources of data may be utilized to accommodate this limitation. It is highly recommended to conduct onsite validation of the data acquired through remote means for a comprehensive seismic vulnerability assessment.

Further advancements in GSV and satellite imagery may improve the coherence of the data collection process. For future research, automatic tracing of the building footprint, especially for high-density residential areas, may simplify the virtual structural survey process. Automated detection of building attributes may also be facilitated by deep learning techniques.

Acknowledgements
The authors would like to thank Iana Angela Fajardo for her contributions in the virtual structural survey.

References

[1] Metropolitan Manila Earthquake Impact Reduction Study (MMEIRS) *2004 Earthquake impact reduction study for Metropolitan Manila, Republic of the Philippines final report* Japan International Cooperation Agency (City of Makati, Philippines)

[2] Applied Technology Council 2015 *Rapid visual screening of structures for potential seismic hazards: A handbook 3rd edition FEMA P-154* (Redwood, California, USA)

[3] Innovation and Employment Ministry of Business 2018 *Seismic assessment of existing buildings* Available at https://www.building.govt.nz/building-code-compliance/b-stability/b1-structure/seismic-assessment-existing-buildings/

[4] Philippine Institute of Volcanology and Seismology (PHIVOLCS) 2014 *How safe is my house? Self-check for earthquake safety of Concrete Hollow Block (CHB) houses in the Philippines* (Quezon City, Philippines)

[5] Mueller M, Segl K, Heiden U and Kaufmann H 2006 Potential of high-resolution satellite data in the context of vulnerability of buildings *Natural Hazards* **38** 247-258

[6] Fan X, Nie G, Xia C and Zhou J 2021 Estimation of pixel-level seismic vulnerability of the building environment based on mid-resolution optical remote sensing images *International Journal of Applied Earth Observation and Geoinformation* **101** 102339

[7] Calimbahin C 2021 The City of Manila under COVID-19: Projecting mayoral performance amid crisis *Contemporary Southeast Asia: A Journal of International and Strategic Affairs* **43** 70-76

[8] Anguelov D, Dulong C, Filip D, Frueh C, Lafon S, Lyon R, Ogale A, Vincent L, and Weaver J 2010 Google Street View: Capturing the world at street level *Computer* **43** 32-38

[9] Kalfarisi R, Hmosze M and Wu Z 2022 Detecting and geolocating city-scale soft-story buildings by deep machine learning for urban seismic resilience *Natural Hazards Review* **23**

[10] Rundel A, Bader M, Richards C, Neckerman K and Teitler J 2011 Using Google Street View to audit neighborhood environments *Am J Prev Med* **40** 94-100

[11] Li X, Zhang C, Li W, Ricard R, Meng Q and Zhang W 2015 Assessing street-level urban greenery using Google Street View and a modified green view index *Urban Forestry & Urban Greening* **14** 675-685

[12] Yin R 1981 The case study as a serious research strategy *Knowledge* **3** 97-114

[13] Case J and Light G 2011 Emerging methodologies in engineering education research *Journal of Engineering Education* **100** 186-210

[14] Philippine Statistics Authority 2017 *2015 Census of population, special report on housing characteristics* (Quezon City, Philippines)

[15] Connell J 1999 Beyond Manila: walls, malls, and private spaces *Environment and Planning A: Economy and Space* **31** 417-439

[16] City of Manila City Planning and Development Office 2006 *Manila Comprehensive Land Use Plan and Zoning Ordinance 2005-2020* (City of Manila, Philippines)

[17] University of the Philippines Diliman Institute of Civil Engineering (UPD-ICE) 2011 Development of vulnerability curves of key building types in the Philippines *2019 Pacific Conference on Earthquake Engineering* University of the Philippines Diliman, Quezon City

The Effect of Various Peak Ground Acceleration Level in Petobo Irrigation Canal Area Post 2018's Liquefaction Disaster

S Sauri[1,2], A Rifa'i*[1], H C Hardiyatmo[1]

[1]Department of Civil and Environmental Engineering, Faculty of Engineering, Universitas Gadjah Mada, Yogyakarta 55281, Indonesia
[2]Directorate General of Human Settlements, Ministry of Public Works, and Housing Indonesia

*Corresponding Author. Email: ahmad.rifai@ugm.ac.id; shpnsauri@rocketmail.com

Abstract. The soil liquefaction disaster in Petobo, Central Sulawesi, caused massive terrain damage. With a total displacement of up to 800 meters and a total sliding area of 1.43 km^2, the soil structure was deformed. The 7.5 Mw earthquake with the maximum PGA of 0.45 was causing flow-induced liquefaction with a distance from the crown to the foot of the landslide as far as 2 km. The research aims to find out how well the post-earthquake soil can withstand the potential to liquefaction if it is hit by various earthquake levels. This research will conduct several simulations to determine the soil performance against earthquakes. The empirical calculation method is carried out by referring to Idris-Boulanger to determine the liquefaction safety factor. Based on several scenarios that have been carried out, post-earthquake land can be re-liquefy when shaken by an earthquake with a certain level below the magnitude of the 7.5 Mw 2018's Palu earthquake. There are a total of six different scenarios of earthquakes with various level of PGA. The soils still have the possibility to liquefy with the PGA of 0.20 and can be considered safe as the PGA value of 0.13. Mitigation efforts, including soil remediation, must be carried out before reconstructing phase of the Petobo irrigation canal starts.

Keywords: Soil Liquefaction, Irrigation Canal, Safety Factor, Petobo, PGA, Earthquake.

1. Introduction

In 2018, Palu was hit by several strong earthquakes for about 7 hours. 7.5 moment magnitude was the mainshock that affected a wide area for almost 100 km from the epicenter. The earthquake was considered a shallow earthquake with a depth of 10 km below the surfaces [1].

In terms of the earthquake, Palu is has a relatively high level of seismicity since the location was close to the Palu-Koro fault. 160 km surface rupture was created by the 2018 earthquake, stretching from the south of the epicenter to the west coast of Donggala, Palu – Central Sulawesi [2][3][4].

Soil liquefaction and massive landslides have occurred at various locations, as illustrated in figure 1, namely Petobo, Jono Oge, Sibalaya, Lolu, and Balaroa [5][6][7]. Liquefaction is defined as a condition that soils deposits lose their bearing capacity due to a strong earthquake and increasing pore water pressure [8]. When the earthquakes occur, causing dynamics loads, the condition of the soil starts to liquefy for a moment, while its shear strength will decrease drastically to near zero.[9]. This phenomenon also affects the plasticity and stiffness of a soil deposit, which gradually changes when subjected to cyclic loads or earthquakes [10].

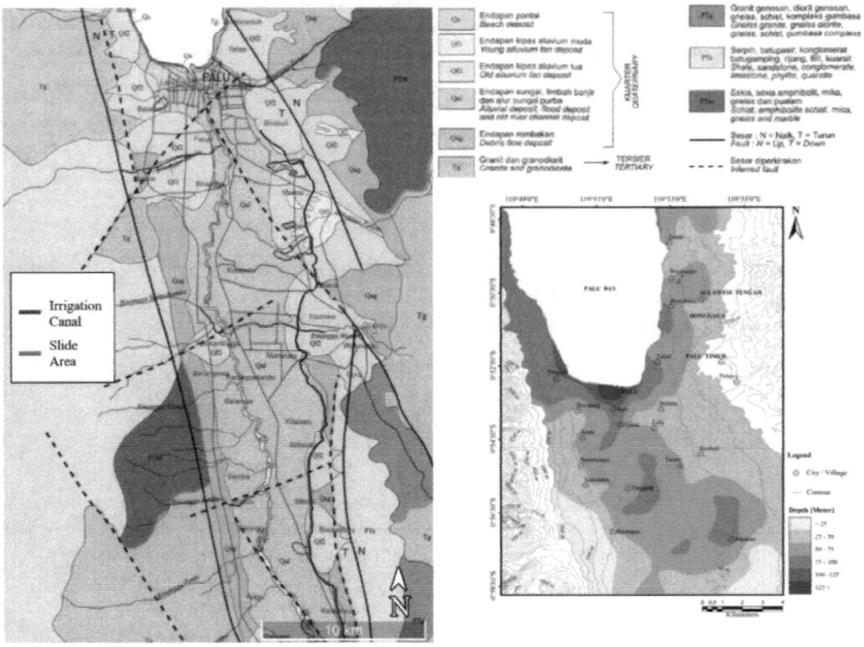

Figure 1. Geology Condition and Sediment Thickness in Palu [1]

This paper will calculate several earthquakes from various PGA levels to determine the critical limit that causes soil liquefaction. The locus of the study will focus on irrigation canals located in the liquefaction area in Petobo. The Gumbasa Irrigation Canal is estimated to influence the potential for liquefaction due to the high groundwater table and leaning slopes [11][12].

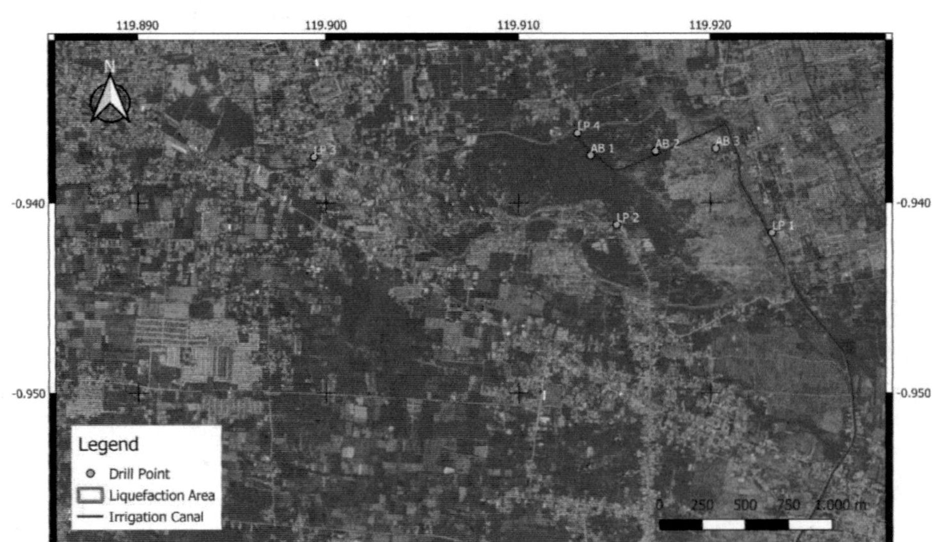

Figure 2. Post-earthquake imagery of Petobo liquefaction and landslides[13]

There are three boreholes data from SPT were taken out to determine the soil condition post-2018 earthquakes, namely AB 1, AB 2, and AB 3. These boreholes location was exactly the Petobo irrigation canal which is going to be reconstructed soon. AB 3 was located on the upper side of the canal while AB 1 is on the down side. AB 2 was located in between as can be seen at figure 2.

2. Research Methodology

This empirical PGA calculation formula that has been used in this paper was based on Ground-Motion Attenuation that was published in 2006 [14]. The justification using that formula was conducted by matching the similarity of the input data among other formulas such as the value of moment magnitude, depth of epicenter, distance of the hypocenter, and even the geological aspect since the data set used in formulating the equation based on Japan's earthquake data set which is similar to Palu earthquake in the Ring of Fire region.

Hence the earthquake level could be determined by many variables, in this paper, the PGA would be the independent variable while a constant Mw of 7.5 is the dependent variable. The study of potential liquefaction vulnerability in this paper was conducted based on several SPT data post-2018 Palu earthquakes. The liquefaction vulnerability test will be determined quantitatively using the Idriss-Boulanger 2008 method [8].

2.1 Peak Ground Acceleration Calculation Formula

The ground motion model in the formula is divided into two sections based on the depth of the epicenter of 30 km. In this paper, the 7.5 Mw Palu 2018 earthquake would be the base data determining the Peak Ground Acceleration, PGA. While other factors would be the independent variables, this research calculated many PGA values from several different earthquake scenarios. Those scenarios were simulated several earthquakes from various locations but with the exact value of 7.5 Mw. Since the location was disparate, the PGA may be different. The PGA calculation formula from Kanno[14] can be seen below.

$$\log \text{pre} = \alpha_1 M_w + b_1 X - \log(X + d_1 10^{0.5 M_w}) + C_1 \qquad (1)$$

Where pre is in cm/s^2, $\alpha_1 = 0.56$; $b_1 = -0.0031$; $c_1 = 0.26$; $d_1 = 0.0055$. Since the PGA value given is in bedrocks, SNI 1726:2019 have site classification for adjustment based on soil density as can be seen at the table 1 below. The soil density was determined using SPT data available.

Table 1. Site Coefficient FPGA[15]

Site Class	PGA ≤ 1	*PGA = 0.2*	*PGA = 0.3*	*PGA = 0.4*	*PGA ≥ 0.5*	*PGA ≥ 0.6*
SA	0.8	0.8	0.8	0.8	0.8	0.8
SB	0.9	0.9	0.9	0.9	0.9	0.9
SC	1.3	1.2	1.2	1.2	1.2	1.2
SD	1.6	1.4	1.3	1.2	1.1	1.1
SE	2.4	1.9	1.6	1.4	1.2	1.1

2.2 Liquefaction Vulnerability Analysis

The liquefaction vulnerability analysis was determined using the Idriss-Boulanger method (2008) [8] to obtain the factor of safety (FS). If the value of FS < 1 is obtained, it can be said to have a vulnerability to liquefaction. On the other hand, FS > 1 indicates not being susceptible to liquefaction. The equation regarding the safety factor can be seen in the following equation (2) to (9).

$$\mathrm{FS} = \frac{CRR}{CSR} \tag{2}$$

$$CSR = 0.65\, r_d \frac{a_{max}}{g} \frac{\sigma_v}{\sigma_v'} \frac{1}{MSF} \frac{1}{K_\sigma} \tag{3}$$

$$r_d = \exp\left(\alpha(z) + \beta(z) M\right) \tag{4}$$

$$\alpha(z) = -1.102 - 1.126 \sin\left(\frac{z}{11.73} + 5.133\right) \tag{5}$$

$$\beta(z) = 0.106 + 0.118\, sin\left(\frac{z}{11.28} + 5.142\right) M_w \tag{6}$$

$$MSF = 6.9 \exp\left(\frac{-M_w}{4}\right) - 0.058 \tag{7}$$

$$K_\sigma = 1 - C_\sigma ln\left(\frac{\sigma_{vc}}{P_a}\right) \le 1.1 \tag{8}$$

$$C_\sigma = \frac{1}{18.9 - 2.55\sqrt{(N_1)_{60}}} \le 0.3 \tag{9}$$

Where,

CRR	= Cyclic Resistance Ratio
CSR	= Cyclic Stress Ratio
r_d	= Stress Reduction
z	= Depth of soil layer
Mw	= Moment Magnitude scale
MSF	= Magnitude Scaling Factor
K_σ	= Correction factor of overburdened stress
C_σ	= Overburden corrected penetration resistances

Cyclic Resistance Ratio (CRR) is the ratio of cyclic resistance, a parameter of the resistance of a soil deposit to liquefaction. If the CSR value of any soil deposit is greater than its CRR value, the soil is vulnerable to liquefaction. To get the value of CRR, the equation (10) to (13) is used.

$$CRR = exp\left(\frac{(N_1)_{60cs}}{14.1} + \left(\frac{(N_1)_{60cs}}{126}\right)^2 - \left(\frac{(N_1)_{60cs}}{23.6}\right)^3 + \left(\frac{(N_1)_{60cs}}{25.4}\right)^4 - 2.8\right) \tag{10}$$

$$(N_1)_{60} = N_m.C_N.C_E.C_B.C_R.C_s \tag{11}$$

$$(N_1)_{60cs} = (N_1)_{60} + \Delta\,(N_1)_{60} \tag{12}$$

$$\Delta\,(N_1)_{60} = \exp\left(1.63 + \frac{9.7}{FC + 0.01} - \left(\frac{15.7}{FC + 0.01}\right)^2\right) \tag{13}$$

Where,

$(N_1)_{60}$ = SPT blow count corrected to ER = 60%
C_N = SPT overburden correction factor
C_E = SPT correction factor for energy ratio
C_B = SPT correction factor for borehole diameter
C_R = SPT correction factor for rod length
C_s = SPT correction factor for omitting sampler liners
$(N_1)_{60cs}$ = equivalent clean sand $(N_1)_{60}$

3. Result and Discussion

The calculation to determine several PGA values was conducted using the Kanno formula. Since the formula data sets are in a different location from the study area, some local adjustments has been made. SNI 1726:2019 [15] and Indonesia Earthquake Hazard Map as the major reference for the adjustment. There is a total of 6 scenarios that were done in this paper between 0.45 g until 0.13 g of PGA value with 7.5 Mw as the dependent variable. Those earthquake scenarios were given out to 6 different locations and depths of the epicenter. The details about the earthquake level scenarios can be seen in the table below.

Table 2. Earthquake Level Scenarios

PGA Value	Epicentrum (km)	Hypocenter (km)	*FPGA*
0.45 g	10	75	1.2
0.30 g	20	125	1..7
0.25 g	20	138	1.7
0.20 g	40	155	1.7
0.15 g	85	200	2.4
0.13 g	90	210	2.4

As can be seen above, different PGA value was carried out based on independent variables, the depths of the epicenter, and the distances of the hypocenter. PGA values were needed as one of the major factors to determine the liquefaction vulnerability of soils deposit. Using Idriss-Boulanger's frameworks, the calculation was done to find the critical value of the earthquake level that can be said as the safe level. The details about liquefaction safety factors can be seen in the table below.

Table 3. Liquefaction Vulnerability Safety Factor in Various PGA Level

PGA Value	Liquefaction Vulnerability	Borehole Points AB 1	Borehole Points AB 2	Borehole Points AB 3	Status
0.45 g	0.11	0-16 m	0-12 m	0-14 m	Unsafe
0.30 g	0.43	0-14 m	0-6 m	0-11 m	Unsafe
0.25 g	0.51	0-14 m	0-5 m	0-10 m	Unsafe
0.20 g	0.72	0-7 m	n/a	n/a	Unsafe

PGA Value	Liquefaction Vulnerability	Borehole Points AB 1	Borehole Points AB 2	Borehole Points AB 3	Status
0.15 g	0.85	0-4 m	n/a	n/a	Unsafe
0.13 g	1.00	n/a	n/a	n/a	Safe

2018 Palu's Earthquake with the PGA of 0.45 g based on the Kanno formula resulted from the soil deposit in the Petobo irrigation canal being unsafe condition. Liquefaction vulnerability analysis was re-conducted using several PGA levels reduction. There is a total of 6 simulations has been carried out until the PGA level shows the safe condition to liquefy. The PGA level of 0.13 g resulted from the non-liquefy soils deposit in both of three boreholes. While the other simulation using higher PGA tends to result in higher vulnerability soils to liquefy in the various depth between 0-14 m.

4. Conclusion and Recommendation

The simulation of several earthquake levels was conducted based on PGA values. Several adjustments has been made that refers to Indonesia National Standard of SNI 1726:2019 and Indonesia Earthquake Hazard Map 2017. The liquefaction vulnerability soils still can be found within the lower PGA of the 2018 Palu's Earthquake. With the constant variable of 7.5 Mw earthquake, the soils still have the possibility to liquefy with the PGA of 0.20 and can be considered safe as the PGA value of 0.13 with the critical safety factors of 1.00 at the AB 1. Due to the result, mitigation efforts, including soil remediation, must be carried out before reconstructing phase of the Petobo irrigation canal starts.

5. Acknowledgment

The authors would like to express their sincere gratitude to Balai Wilayah Sungai Sulawesi III, Directorate General of Water Resources, Ministry of Public Works, and Public Housing of Indonesia for the support and suggestion provided during the field investigation.

References

[1] Joint Research Centre 2018 Mw 7.5 Earthquake in Indonesia, 28 Sep 2018 *JRC Emerg. Reporting-Activation* 21

[2] Tim Pusgen 2018 *Kajian Gempa Palu Provinsi Sulawesi Tengah 28 September 2018 (M7.4)* vol 1

[3] Irsyam M, Cummins P R, Asrurifak M, Faizal L, Natawidjaja D H, Widiyantoro S, Meilano I, Triyoso W, Rudiyanto A, Hidayati S, Ridwan M, Hanifa N R and Syahbana A J 2020 Development of the 2017 national seismic hazard maps of Indonesia *Earthq. Spectra* **36** 112–36

[4] Jaya A, Nishikawa O and Jumadil S 2019 Distribution and morphology of the surface ruptures of the 2018 Donggala–Palu earthquake, Central Sulawesi, Indonesia *Earth, Planets Sp.* **71**

[5] Kiyota T, Furuichi H, Hidayat R F, Tada N and Nawir H 2020 Overview of long-distance flow-slide caused by the 2018 Sulawesi earthquake, Indonesia *Soils Found.* **60** 722–35

[6] Mason H B, Montgomery J, Gallant A P, Hutabarat D, Reed A N, Wartman J, Irsyam M, Simatupang P T, Alatas I M, Prakoso W A, Djarwadi D, Hanifa R, Rahardjo P, Faizal L, Harnanto D S, Kawanda A, Himawan A and Yasin W 2021 East Palu Valley flowslides induced by the 2018 MW 7.5 Palu-Donggala earthquake *Geomorphology* **373**

[7] Gallant A P, Montgomery J, Mason H B, Hutabarat D, Reed A N, Wartman J, Irsyam M, Simatupang P T, Alatas I M, Prakoso W A, Djarwadi D, Hanifa R, Rahardjo P, Faizal L, Harnanto D S, Kawanda A, Himawan A and Yasin W 2020 The Sibalaya flowslide initiated

by the 28 September 2018 MW 7.5 Palu-Donggala, Indonesia earthquake *Landslides* **17** 1925–34

[8] Idriss I M and Boulanger R W 2008 *Soil Liquefaction During Earthquakes*

[9] Robert W. Day 2012 *Geotechnical Earthquake Engineering Handbook*

[10] Mase L Z, Farid M, Sugianto N and Agustina S 2020 The Implementation of Ground Response Analysis to Quantify Liquefaction Potential Index (LPI) in Bengkulu City, Indonesia *J. Civ. Eng. Forum* **6** 319–29

[11] Watkinson I M and Hall R 2018 Impact of communal irrigation on the 2018 Palu earthquake-triggered landslides *Nat. Geosci.*

[12] Hazarika H, Rohit D, Pasha S M K, Maeda T, Masyhur I, Arsyad A and Nurdin S 2021 Large distance flow-slide at Jono-Oge due to the 2018 Sulawesi Earthquake, Indonesia *Soils Found.* **61** 239–55

[13] Sauri S, Rifa'i A and Christady Hardiyatmo H 2021 Liquefaction Vulnerability Analysis using N-SPT Value and Grain Size Analysis on Gumbasa Irrigation Canal in the Post- Disaster Petobo Area , Sulawesi *ICWRDEP Conf.*

[14] Kanno T, Narita A, Morikawa N, Fujiwara H and Fukushima Y 2006 A new attenuation relation for strong ground motion in Japan based on recorded data *Bull. Seismol. Soc. Am.* **96** 879–97

[15] Badan Standardisasi Nasional 2019 Indonesia National Standard: SNI 1726-2019 Tata Cara Perencanaan Ketahanan Gempa Untuk Struktur Bangunan Gedung dan Non Gedung *SNI* 254

Analysis of the Effective Length of Handil Channel for Tidal Irrigation at Palingkau Irrigation Area in Central Kalimantan Province for National Food Estate Program

E.E. Ramadhani[1,3], J. Sujono[1], and Taryono[2]
[1] Department of Civil and Environmental Engineering, Faculty of Engineering, Universitas Gadjah Mada, Yogyakarta
[2] Agrotechnology Innovattion Centre and Faculty of Agriculture, Universitas Gadjah Mada, Yogyakarta
[3] Directorate General of Water Resources, Ministry of Public Works and Housing

*Corresponding author e-mail: elvira.e.r@mail.ugm.ac.id

Abstract. The handil irrigation system is a traditional water management system developed by local farmers and the government to irrigate rice fields using tidal power for supply and drainage. The canal with a width of 25 meters and a length of about 2 kilometres was developed to irrigate 1,397 ha of irrigated agricultural land in the Palingkau Irrigation Area. At high tide, freshwater enters the canal from the Kapuas Murung River, and drains into the river at low tide. After a long period of development of the area, sediment deposition and changes in channel cross-section occur so that the water system does not function optimally. For the sake of rehabilitation, an analysis was carried out to determine the effective canal length of the canal that is influenced by tidal water, considering channel capacity, tides and land elevation. Hydraulic analysis was performed by using HEC-RAS mathematical modelling. The results showed that tidal fluctuations could not irrigate two handil's rice fields out of 8 existing handils. Canal rehabilitation in the form of increasing channel capacity is carried out by considering the length of the canal and the irrigation area irrigated, including its effect on improving water quality.

Keywords: irrigation, tidal river system, hydraulic analysis, agricultural

1. Introduction

The Indonesian government has programmed to build an irrigation network covering an area of 500.000 hectares and rehabilitate an irrigation area of 2.5 million hectares to support national food security. One of the programs is national food by building an integrated food estate area in Central Kalimantan Province with an area of 165.000 hectares. One of the areas included in the food estate is the Handil Rakyat Palingkau Irrigation Area, with an area of 1397 hectares. Based on construction history, Handil is a traditional water management system applied to irrigate rice fields using a tidal river system. Canals with a width of 25 meters and a length of about 2 kilometres were built to irrigate agricultural land. Freshwater enters the canal from the Kapuas Murung River at high tides, whereas low-quality water flows back into the river at low tides. After a long period of development of Handil Rakyat Palingkau Irrigation Area, sediment deposition and changes in channel cross-section occurred so that the water system did not function optimally. The conditions of overflow and puddle vary significantly from one area to another. From the conditions, an analysis was carried out to determine the effective length affected by the tides, taking into account the canal capacity, the tides, and the elevation of the irrigation area.

2. Material and Method

2.1. *Description of Area*

Administratively, the Handil Rakyat Palingkau Irrigation Area is located in the Kapuas Regency, Central Kalimantan Province. The Canal systems consists of traditional Handil canals namely Handil Mampai, Lumbah, Jaharap, Mangkatip, Madang, and Bangkinang with a note there are 2 branching handils. In some of these canals, people still use them for water transportation, such as traditional boats. Handil Rakyat Palingkau Irrigation Area can be seen in Figure 1.

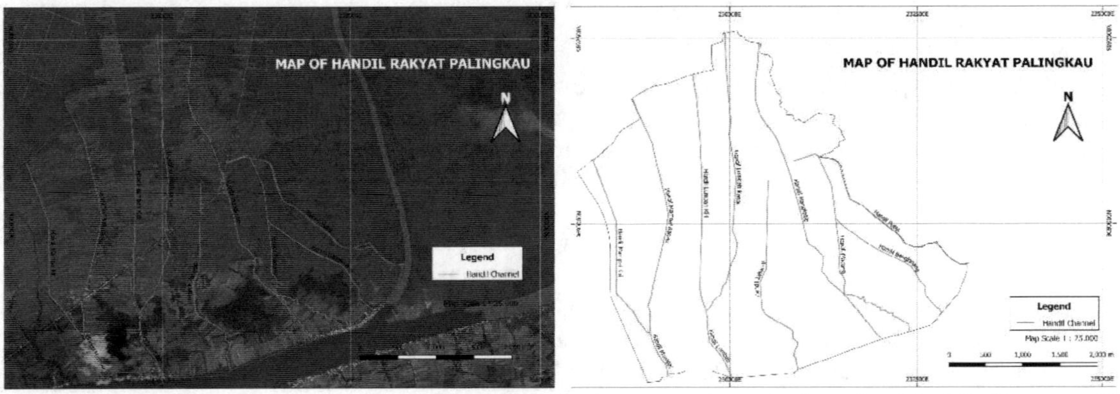

Figure 1. Location of study area in the Central Kalimantan Province, Indonesia (BWS Kalimantan II)

2.2. *Topography dan Hydro topography condition*

Topography is the shape of the earth's surface and other objects, while hydro topography describes the relative elevation of land to water level elevation which serves as a reference for water level elevation. The need for the management of swamp irrigation networks is determined by the hydro topography of the land. This is very important in assessing the potential for agricultural land development. Water will flow faster on media with a greater slope than on media with a lower slope. Likewise with river discharge, the slope will affect the water flow rate. The greater the slope of a river, the greater the river discharge (assuming the same cross-sectional area) [2]. Identifying of hydro-topography categories is the first step in delineating a water management zone. Hydro-topography can be defined as the elevation of the field in relation to the water level of a river or a canal at the nearest open water or intake point, and is expressed by the number of tidal irrigation during the rainy and dry s analyze how high the water level is needed to reach the irrigation land. Hydro-topography is identified into four categories, namely category A which is tidal irrigated areas in both rainy and dry seasons, category B which is tidal irrigated areas only in the rainy season, category C which is areas just above < 0.50 m the average high water level in the rainy season and category D as high areas which are water level is free from the influence of tidal. Based on the elevation of the topography map and the hydro topography of Handil Rakyat Palingkau Irrigation Area are mostly in Category B, where the area is suitable for rice cultivation. However, there are several handil channels that fall into Category C. Topography Map and Hydro-topography can be seen in Figure 2.

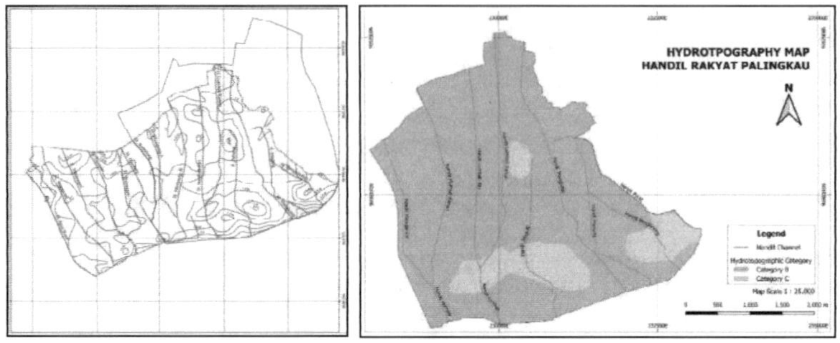

Figure 2. Topography Map and Hydro-topography Map (QGIS Mapping)

2.3. Hydraulic Simulation

One of the essential keys in managing lowland swamps is the problem of river hydraulic behaviour. The hydraulic effect of the channel is important because it will have a lot to do with extensive use and management of the environment. Channel hydraulic behaviour related to swamp management. [2] In this hydraulic simulation, the water level elevation on the handil needs to be analyzed to determine whether the water can reach the irrigation land elevation. In analyzing the condition of the river, the HEC-RAS 5.0.3 program can be used. The HEC-RAS 5.0.3 program was developed by The Hydrologic Engineer Centre (HEC) part of the U.S Army Corps of Engineers. The HEC-RAS 5.0.3 program uses data settings where different flow data calculations can be performed with the same geometry data. Geometry data consists of modelling layouts with cross-sections for the modelled channels. [4] Flow data is placed separately from geometry data. The data flow that can be selected is unsteady with a hypothetical discharge. Hypothetical discharge data is determined at 800 m^3/sec taking into account the cross-sectional area of the river with velocity is 0.5 m/sec. Each of these data flows must have the appropriate boundary conditions and initial conditions to run the modelling. [5-6] Furthermore, calculations can be done by creating a simulation scenario. In the HEC-RAS software, the condition of the reliable channel water elevation in the influence of its hydraulics and further handling as needed. [7]

3. Result and Discussion

Handils schemes and geometric data for hydraulic simulation using HEC-RAS are given in Figure 3.

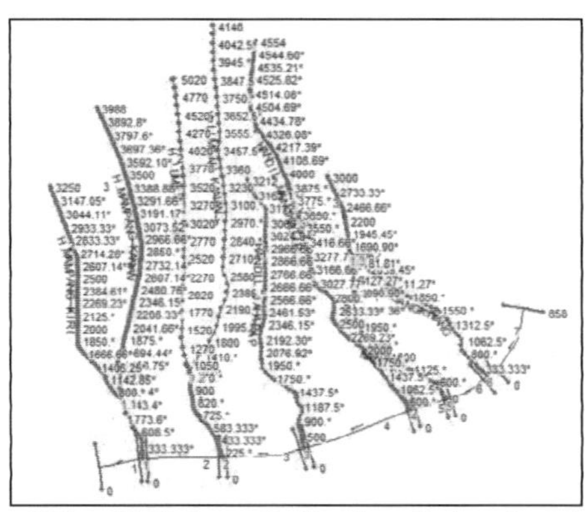

Figure 3. Geometric data of Handil Rakyat Palingkau Irrigation Area (HEC-RAS Analysis)

3.1. Hydraulic simulation

Based on the results of the HEC-RAS modelling analysis (Figure 4), it can be seen that Handil Mampang can still function as tidal irrigation for agricultural purposes. The water level is at an elevation of 1.5 to 2 meters while the land is at an elevation of 1.2 meters. This result is the same as the field conditions as shown in Figure 5, where the water can still reach the rice fields.

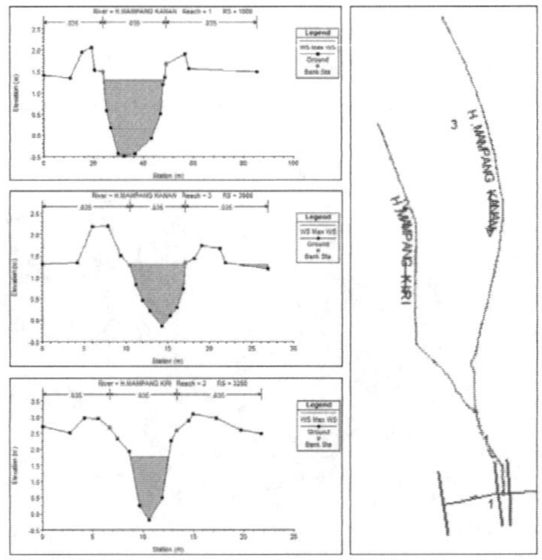

Figure 4. Cross section of Handil Mampang (HEC-RAS Analysis)

Figure 5. Condition of Handil Mampang Kiri and Handil Mampang Kanan

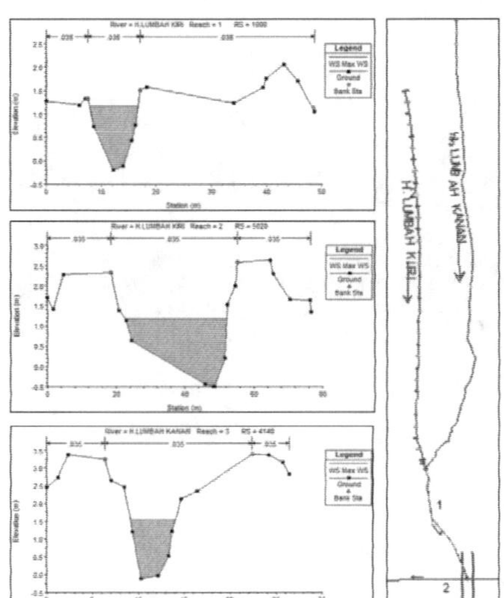

Figure 6. Cross section of Handil Lumbah (HEC-RAS Analysis)

Based on the results of the HEC-RAS modelling analysis (Figure 6), it can be seen in the cross-section of Handil Lumbah and compared with a topographic hydro map has a similar result as Handil Mampai that the water at this elevation can still be used as tidal irrigation for agricultural purposes. The water level is at an elevation of 1.2 to 1.6 meters while the irrigation land is at an elevation of 1.2 meters. This result is similar to the Handil and irrigation land condition as depicted in Figure 7, where the water can still reach the rice field.

Figure 7. Condion of Handil Lumbah Kiri and Handil Lumbah Kanan

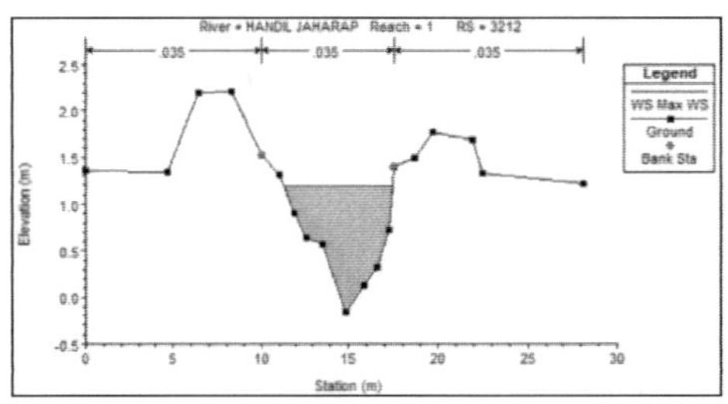

Figure 8. Cross section of Handil Jaharap (HEC-RAS Analysis)

The water level of Handil Jaharap is at an elevation of 1.3 meters while the land is at an elevation of 1.2m (Figure 8). It is proven by looking at the documentation that the water can still reach the land to irrigate the rice field (Figure 9).

Figure 9. Condition of Handil Jaharap

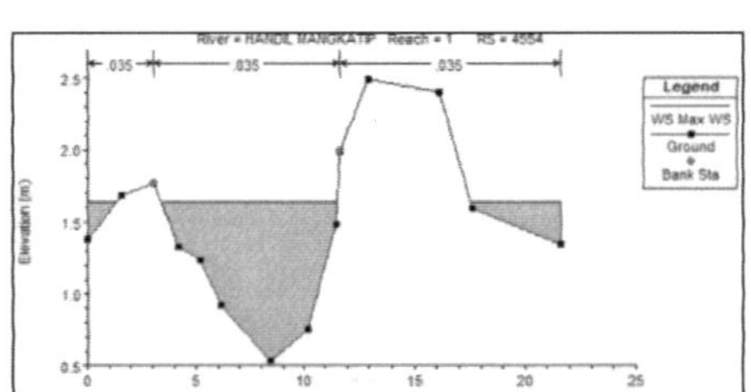

Figure 10. Cross section of Handil Mangkatip (HEC-RAS Analysis)

The water level of Handil Mangkatip is at an elevation of 1.7 meters while the land is at an elevation of 1.2m (Figure 10). It is proven by looking at the documentation that water can still reach the land to irrigate the rice field (Figure 11).

Figure 11. Documentation of Handil Mangkatip

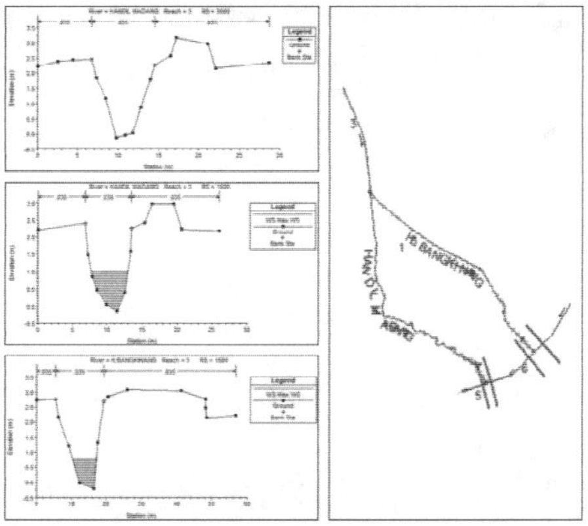

Figure 12. Cross section of Handil Madang and Handil Bangkinang (HEC-RAS Analysis)

Based on the results of the HEC-RAS modelling analysis (Figure 12), it can be seen that the cross-section of Handil Madang and Handil Bangkinang are compared with a topographic hydro map shows that the water elevation cannot reach the tip of the handil channel so cannot be used as tidal irrigation for agricultural purposes. The water level is at an elevation of 0.8 to 1 meter while the land is at an elevation of 1.5 meters. It is proven by looking at the documentation (Figure 13) that water cannot reach the land to irrigate the rice field and bushes starts to grow and covers the channel.

Figure 13. Documentation of Handil Madang (left) and Handil Bangkinang (right)

3.2. *Analysis of the Effective Length*

Analysis of the length of the canal is needed to find out how far the water can reach the end of the channel for irrigation purposes. In this analysis, topography and hydro-topography play an important role. It is also essential to take the tides into account, considering channel capacity, tides and land elevation.

Table 1. Data of handil channels and distances that can be reached by tidal waters based on the average velocity in the channel section

Handil Channel	Width (m)	Velocity (m/s)	Length (m)	Distance from mouth (m)	Distance (m)
		0.190	250	250	5472
Mampang	25.320	0.150	500	500	4320
		0.110	1000	1000	3168
		0.100	500	1500	2880
Mampang Kiri	5.730	0.030	1000	2000	864
		0.010	2000	3000	288
		0.100	500	1500	2880
Mampang Kanan	11.470	0.090	1000	2000	2592
		0.060	2000	3000	1728
		0.170	250	250	4896
Lumbah	9.560	0.130	500	500	3744
		0.100	1000	1000	2880
		0.100	500	1500	2880
Lumbah Kanan	9.820	0.080	1000	2000	2304
		0.070	2100	3100	2016
		0.050	500	1500	1440
Lumbah Kiri	6.910	0.020	1000	2000	576
		0.010	2020	3020	288
		0.190	1000	1000	5472
Jaharap	7.330	0.080	2000	2000	2304
		0.020	3000	3000	576
		0.130	1000	1000	3744
Mangkatip	15.070	0.110	2000	2000	3168
		0.060	3000	3000	1728
		0.100	500	500	2880
Madang	7.400	0.050	1000	1000	1440
		0.000	2200	2200	0
		0.000	50	2250	0
Madang Branch	7.780	0.000	160	2360	0
		0.000	800	3000	0
		0.110	500	500	3168
Bangkinang	19.180	0.050	1000	1000	1440
		0.000	2062	2062	0

Based on Table 1, it can be seen that the farther toward the end of the handil, the velocity decreases. It can also be seen in this analysis that the wider the channel, the velocity also increases. It can occur in the case of high tides with a significant difference in energy level at high tide (high water level) and at low tide (water level is almost the same at all points). There is a difference in velocity during high tide and stagnant water. It can be assumed that a cross-section with two different types of width, namely wide and narrow, with the same energy difference will flow water in a wider cross-section. The speed is determined by the high energy or the difference in water level upstream and downstream.

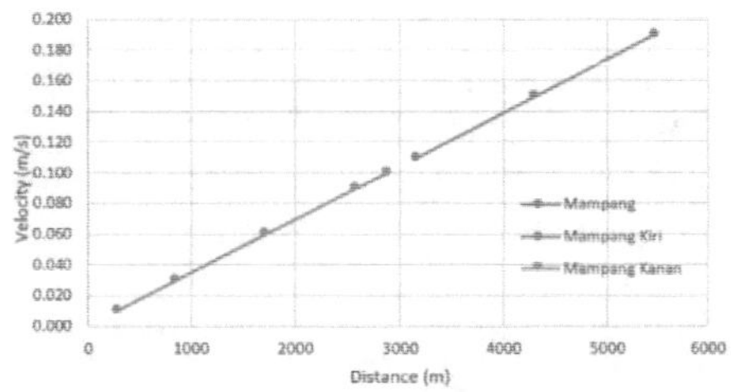

Figure 14. Relationship between velocity and the length of the handil channel to determine the effective length of the Handil Mampang

From the graph of the relation between velocity and the length of the handil channel, we can determine the effective length of the Handil Mampang can be determined. With a channel length of 3250 meters (Figure 3) for Mampang Kiri and Mampang Kanan, the minimum water velocity according to Figure 14 is 0.12 m/s. Comparring with the existing conditions, the current velocity is still sufficient.

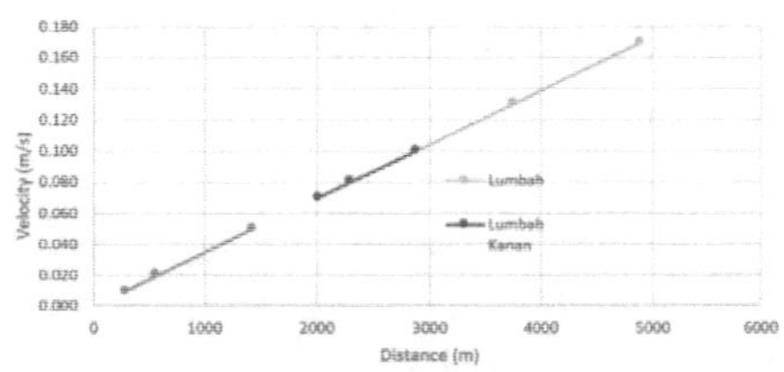

Figure 15. Graph of the relationship between velocity and the length of the handil channel to determine the effective length of the Handil Lumbah

From Figure 15, we can determine the effective length of the Handil Lumbah Kiri and Lumbah Kanan. The channel length of 3988 meters (Figure 3), the minimum velocity is 0.14 m/s. This velocity can be reached with the existing conditions.

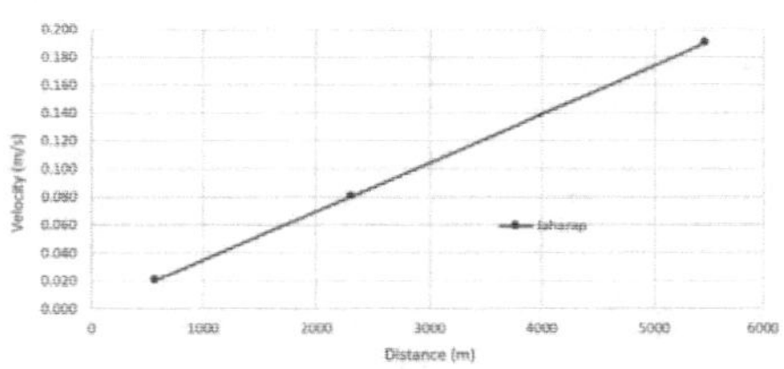

Figure 16. Graph of the relationship between velocity and the length of the handil channel to determine the effective length of the Handil Jaharap

For Handil Jaharap (Figure 16), we can determine from the graph for the effective length of the Handil Jaharap with a channel length of 3212 meters (Figure 4), the speed that can be planned is 0.12 m/s. With existing conditions, the current velocity is still sufficient.

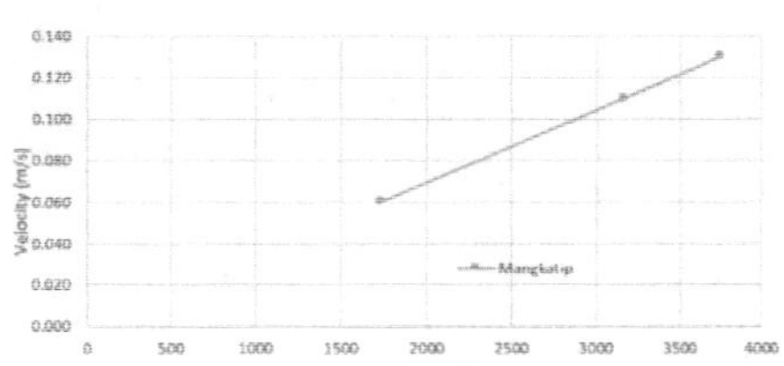

Figure 17. Graph the relationship between velocity and the length of the handil channel to determine the effective length of the Handil Mangkatip

From the graph (Figure 17) of the relation between velocity and the length of the handil channel, we can determine the effective length of the Handil Mampai. With a channel length of 4554 meters (Figure 4) for Handil Mangkatip, the speed that can be planned is 0.16 m/s. With existing conditions, the current velocity is still sufficient.

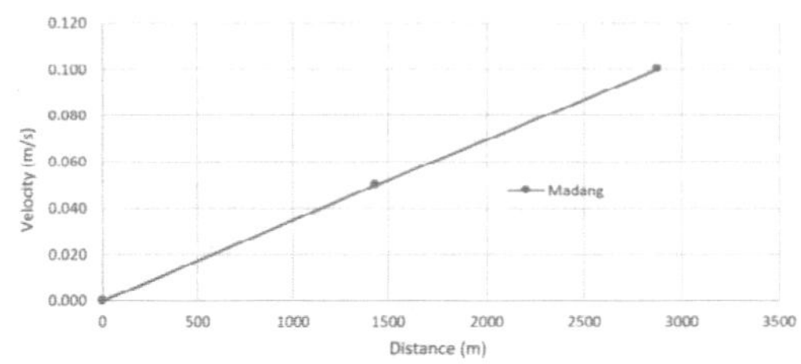

Figure 18. Graph of the relationship between velocity and the length of the handil channel to determine the effective length of the Handil Madang.

Handil Madang has a fairly steep topography, so discharge with low velocity cannot reach the tip of the handil channel to irrigate rice fields. Handil Madang has an initial speed of 0.1 m/s and then drops to 0 (Figure 18).

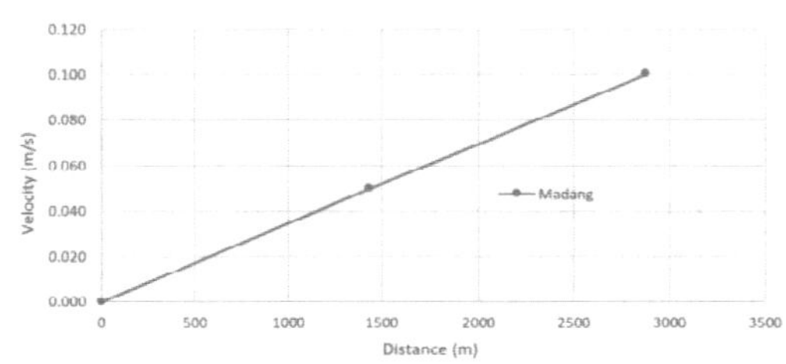

Figure 19. Graph of the relationship between velocity and the length of the handil channel to determine the effective length of the Handil Bangkinang

Similar with Handil Madang, Handil Bangkinang has steep topography, discharge with low velocity cannot reach the tip of the handil channel to irrigate rice fields. Handil Bangkinang has an initial speed of 0.11 m/s and then drops to 0 (Figure 19).

3.3. Relation between channel length and velocity

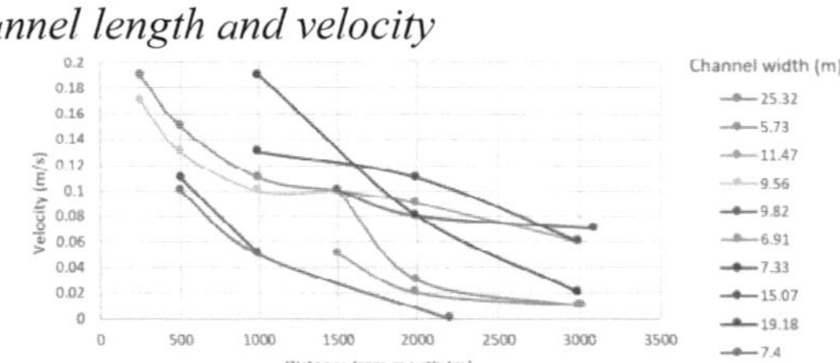

Figure 20. Graph of the relationship between velocity and distance from mouth depends on each channel width

From the graph (Figure 20) of the relation between velocity and distance from the mouth, we can determine the effective length from each channel width. From the graph, we can see that more wide the channel, the velocity also increases.

3.4. Calibration
Based on the survey documentation (Figure 5, 7, 9, 11, and 13) and hydraulic simulation (Figure 4, 6, 8, 10 and 12), the result of water level was quite similar. From modelling analysis, there are two handils do not function optimally. At the time of the survey, it was also found that there were two handils did not function optimally as irrigation channels due to steep topography and low velocity.

4. Conclusion
Hydraulic analysis on Handils was carried out using HEC-RAS mathematical modelling. The results showed that among 8 Handils, 2 Handils could not function properly, since the tidal water elevation at the Handils is lower than the irrigation land. Based on the survey, the two reliable channels cannot be drained and do not function optimally. Effective length analysis shows that the further toward the end of the handil, the velocity decreases. The wider the channel, the velocity also increasing. It can occur in the case of high tides with a significant difference in energy level at high tide (high water level) and at low tide (water level is almost the same at all points). There is a difference in velocity during high tide and stagnant water. It can be assumed that a cross-section with two different types of width, namely wide and narrow, with the same energy difference will flow water in a wider cross-section. The speed is determined by the high energy or the difference in water level upstream and downstream. At the same high tide level, a larger wide channel with the same change in energy height will have a larger cross-section than a narrow cross-section so that the discharge will be greater because the velocity is the same but the appearance is different. It is necessary to rehabilitate or normalize the cross-section of Handils to accelerate the flow to reach the required distance.

Acknowledgement
This research was conducted in one of the swamp schemes under BWS Kalimantan II, Ministry of Public Works and Housing. This research is also under supervision of Prof. Dr. Budi Santoso Wignyosukarto, Dipl.HE. as Hydraulic Lecture from Department of Civil and Environmental Engineering, Faculty of Engineering, Universitas Gadjah Mada, Yogyakarta and Parlinggoman Simanungkalit as Head of Sub Directorate Irrigation and Lowland, Directorate General of Water Resources, Ministry of Public Works and Housing Indonesia.

ORCID IDs
EE Ramadhani https://orcid.org/0000-0003-3544-7147
joko sujono https://orcid.org/0000-0002-3500-9246
Taryono https://orcid.org/ 0000-0003-1425-969X

Reference
[1] Widjaja-Adhi, I P.G., K. Nugroho, D.A. Suriadikarta, and A.S. Karama. 1992. Pengembangan Pertanian Lahan Rawa Pasang Surut dan Lebak. (in Bahasa) Pusat Penelitian dan Pengembangan Tanaman Pangan.
[2] Human Resources Development Agency. 2016. Kesesuaian Lahan Rawa. (in Bahasa) Ministry of Public Works and Housing.
[3] Suryadi, F.X. 1996. Soil and Water Management Strategies for Tidal Lowlands in Indonesia. Proefschrift Technische Universiteit Delft.
[4] HEC-RAS software, USACE, Hydrologic Engineering Center DAVIS, CA 9561-4687, USA.
[5] Fauzan, A.K., Wignyosukarto, B.S., Jayadi. R. 2020. Water Management Evaluation for Upgrading Tidal Irrigation System, Katingan, Kalimantan. Gadjah Mada University.
[6] Wignyosukarto, B.S. 2013. Leaching and Flushing of Acidity in the Reclamation of Acid Sulphate Soil, Kalimantan, Indonesia.
[7] Noor, H., Suhardjono, Prayogo, T.B. 2018. Evaluasi dan Pengembangan Jaringan Irigasi Rawa Pasang Surut Terhadap Pola Operasi Pintu Air D.I.R Pematang Limau Kabupaten Seruyan. (in Bahasa). Brawijaya University.

Review of flood prediction hybrid machine learning models using datasets

Ainaa Hanis Zuhairi[1], Fitri Yakub[1], Sheikh Ahmad Zaki[1], Mohamed Sukri Mat Ali[1]

[1] iKohza Wind Engineering for (Urban, Artificial, Man-made) Environment Laboratory, Level 10, Malaysia-Japan International Institute of Technology, Universiti Teknologi Malaysia.

*Corresponding author e-mail: ainaahanis@graduate.utm.my, mfitri.kl@utm.my

Abstract. Floods are among the most destructive natural disasters, and they are extremely difficult to model. Over the last two decades, machine learning (ML) methods have made significant contributions to the advancement of prediction systems that provide better performance and cost-effective solutions by mimicking the complex mathematical expressions of physical flood processes. Because of the numerous benefits and potential of ML, its popularity has skyrocketed. Researchers hope to discover more accurate and efficient prediction models by introducing novel ML methods and hybridising existing ones. The main focus of this paper is to show the state of the art of hybridising ML models in flood prediction. The most effective strategies for improving ML methods are hybridization, data decomposition, algorithm ensemble, and model optimization.

1. Introduction
Extreme weather occurrences have become more common in many regions of the world, possibly as a result of a shift in the climatic scenario [1]. Identification of natural catastrophes in earlier stage like floods, can considerably aid us in limiting the degree of damage caused by these events. This leads to requirement for efficient flood management system, effective and accurate forecasting system is one of the most important elements of any flood management programme. Machine learning is a problem-solving strategy based on bionics that replicates the operation of the human brain, it is often composed of a sequence of mathematical operations and transformations carried out by computer programmes linked to current electronic processors. It has the potential of becoming a system that is economical and dependable in flood prediction.

2. ML Methods in Flood Prediction
Historical flood event records, as well as cumulative data in real time from variety of sensors as well as rain gauges over varied intervals, are frequently utilised to create the ML prediction model. The dataset's conventional sources are usually rainfall as well as water level, which are monitored with rain gauges on land or by more recent remote sensing methods using satellites technology, multi sensor systems, and or using radars [2].

2.1 Short Term and Long-Term Flood Forecasting Methods

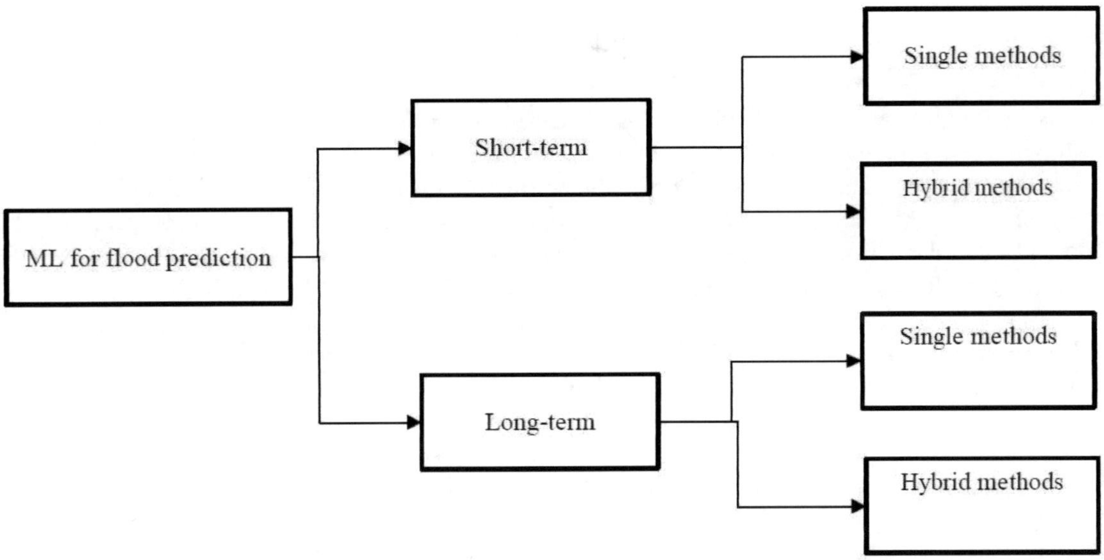

Figure 1. Machine Learning prediction methods [3].

ML has demonstrated considerable success in developing models that identifies patten in fields ranging from computer vision to recognising speech, understanding text and game Artificial Intelligence or AI [4 - 7]. Aside from these classical or traditional fields, ML, particularly deep learning, is becoming increasingly relevant and successful in engineering as well as sciences [5 – 7]. These success stories are founded on the data-driven nature of the approach to learning deriving from a massive number of cases. A recent review synthesises this into a new paradigm of theory-guided data science, emphasising the need of scientific consistency in machine learning [8].

There is even a survey regarding the incorporation of knowledge into support vector machines [9]. The combination of symbolic and connectionist AI appears to be becoming more accessible. We allude to a recent assessment on graph neural networks and a research approach defined as relational inductive bias [10] in this regard. In this study machine learning with hybrid models are reviewed. There are two types of methods for flood prediction approach, short-term and long-term. Short-term flood forecasts are regarded as major research or study issue, especially within densely populated zone, in order to provide early warnings to residents and therefore limit damage [11]. Furthermore, short-term forecasts assist significantly towards water resource management sector.

In spite of recent advancements in terms of numbers, weather prediction or Numerical Weather Prediction (NWP) models, artificial intelligence (AI) and ML, short-term forecasting is still a difficult endeavour [12 -18]. Building hybrid ML approaches to increase prediction quality with regards to accuracy, generalisation, longer lead time, uncertainty, computing costs, and speed is becoming increasingly popular. There are several hybrid approaches, with more well-known models like ANFIS, which is Adaptive Network based

Fuzzy Inference model as well as adaptive neuro fuzzy inference model and WNN or Wavelet Neural Network model, along with other new algorithms.

Long-term flood forecasting is critical towards enhancing knowledge and managing water resource capability across prolonged time periods, ranging from weekly, monthly to yearly forecasts [18]. Many noteworthy ML approaches, such as ANFIS, SVR or Support Vector Regression model, SVM also known as Support Vector Machine model, ANN which is Artificial Neural Network model, bootstrap–ANN and WNN [18 - 22], have been employed for long lead-time predictions with promising results in the previous decades. Recently, the results of several ML approaches for long-term flood forecasts were placed in comparison with a number of publications [21 - 23]. However, it is still unclear whether machine learning algorithm performs the best in long-term flood prediction. A single statistical or machine learning technique is used for classification in an individual approach. Because they incorporate the complimentary advantages of more than one learning approach and overcome the weaknesses of individual techniques, hybrid and ensemble models are efficient and resilient. Stand-alone hybrid models, transformational hybrid models, strongly linked hybrid models, and completely coupled hybrid models are all possible [24].

Table 1. A Review of Flood Prediction Hybrid Machine Learning Models

Author	Title	Models	Objective	Function	Assumption	Accuracy	Conclusion
Li & Willems, 2020	A Hybrid Model for Fast and Probabilistic Urban Pluvial Flood Prediction	Logistic regression and multivariate characterisation	To create a one-of-a-kind hybrid model addressing the issue in regards to models that are data-driven.	It is ideal for real-time flood forecasting in urban catchments due to its capability in providing rapid, generally accurate flood predictions at specified inner-city locations.	Upstream instantaneous inflow depth, Upstream catchment rainfall return period, Downstream outfall rainfall return period, Current node depth, Upstream instantaneous rainfall returns period, and Downstream outfall rainfall return period	Flood alerts with a probability of 50% can have an accuracy of up to 86 %, with a computation saving of 96 %.	The created hybrid model incorporates advantage of both physics-based as well as data-driven models. The suggested hybrid model's computation performance is actually slower when compared to models that use ANN data-driven approaches. Extending the model's boundaries using intensity of rainfall episodes for calibration may result in worse accuracy and less detail than the 1-D HD model.
Kurian et al., 2020	Effective flood forecasting at higher lead times through hybrid modelling framework	M4 hydrological model with ANN	The research suggests a hybrid models of ANN combined with a hydrological model based on physical distribution.	Using observed streamflow as input for prediction accuracy while shortening forecast lead times.	Rainfall, observed streamflow, observed rainfall, forecasted rainfall, simulated streamflow, and simulated streamflow.	Model M4, a 2% variance of noise, with efficiency of 0.95, and 0.94 for a 5% variance of noise.	When paired with assumed flows from HEC-HMS or SWAT with longer lead periods, the forecast accuracy of ANN models rose considerably.

Kan et al., 2020	Hybrid machine learning hydrological model for flood forecast purpose.	ANN and K-nearest neighbour.	continually predicting discharge without diminishing accuracy	Hybridisation increases ANN performance by finding a compromise between network complexity and training accuracy by overcoming the network topology and parameters concurrent global optimization network problem with complicated topology.	Rainfall and antecedent runoff	Most flood occurrences have an NSCE greater than or equal to 0.9.	The suggested model outperforms standard neural network models in terms of continuous discharge prediction without accuracy loss due to its carefully built model structure. In real-world applications, the performance and dependability of the HML hydrological model were shown.
Xie et al., 2021	Hybrid Deep Learning Modelling for Water Level Prediction in Yangtze River	Wavelet transform and LSTM network.	This research uses hybridised deep learning technique LSTM system and discontinuous wavelet transform for daily water level prediction.	Towards better understanding of temporal properties, for the wavelet transform is used to decompose time series into detail approximate part and a novel LSTM model is used with a greedy layer wise unsupervised learning algorithm to learn	Water Level	When compared to other models, the Water Level WA-LSTM model consistently has the lowest MAE and RMSE, which are 46 %, 51 %, 39 %, 50 %, 74 %, and 72 %.	Comprehensive study demonstrates that the maximum performance is obtained by using 5 to 6 days observations lag as features and the Meyer wavelet transform with decomposition level 4., confirming the WA-LSTM model's adaptability and generalisation.

				features of water level via layer-by-layer granulation.			
Yoo et al., 2020	Data Driven Water Surface Elevation Forecasting Model with Hybrid Activation Function—A Case Study for Hangang River, South Korea	Hybrid Activation Function with LSTM	A hybrid model in the form of a blend between hyperbolic tangent function with rectified linear unit using a weighting factor towards water level and long-term forecasting period accuracy improvement.	LSTM in TensorFlow along with open-source software library on Google's deep-learning, was used to predict water level rise in Hangang River, Korea, from 2009 to 2018.	Elevation, Precipitation, Wind Speed, Outlet Discharge, Water Tidal Level and Dissolved Oxygen Concentration.	The predicting accuracy of available data was given as 0.31 m via RMSE while it is 98.9 % via NSE.	The hybrid framework or model presented in this work proved suitable for forecasting increase in water levels in the Hangang River's stated elevation despite a lower forecasting accuracy at the longer leading time interval and a less noticeable improvement in model performance for virtual flood forecasting.
Moishin et al., 2021	Designing Deep-Based Learning Flood Forecast Model With ConvLST	CNN with LSTM	To construct and assess a forecasting model that predicts the occurrence of flood occurrences in the future, this	The study employs a Flood Index (IF) as a mathematical representation, which is used by flood monitoring system to anticipate the next	Precipitation, Antecedent daily rainfall data	All predicted horizons had d values and RNSE that are more than 0.93, 0.85as well as 0.95.	The forecasting approach may not be suitable in places where the applicability of IF has not yet been demonstrated, as the predictive model built only employed two input variables.

IOP Conf. Series: Earth and Environmental Science 1091 (2022) 012040 doi:10.1088/1755-1315/1091/1/012040

	M Hybrid Algorithm		study proposes a combination of deep learning method that combines the predictive advantages of CNN and LSTM Network.	daily IF value, this forecasting model uses IF lag that is statistically significant, enhanced using antecedent and rainfall statistic in real time			
Imran & Sheikh Abdul Khader, 2020	Forecasting Water Level of Jhelum River of Kashmir Valley, Using Prediction and Early Warning System	Water level prediction using ANFIS model and Linear Multiple Regression or LMR model.	Purpose of this study is to suggest alternative framework that is easy to install, low-cost and accurate. The system's LMR method is quick to fit, easy to read, and suitable for forecasting continuous responses like water level.	The framework is made up of a command-and-control centre including base station. The base station is made up of a sensor along with a processing module that performs a localised water level forecasting and transmits projected and observed data to control centre	30 years of daily precipitation and temperature data, as well as daily water-level data from 1980 to 2019.	This hybrid approach has a daily forecast accuracy of 93.53 percent and an hourly prediction accuracy of 99.91 percent.	According to the model, the predicted as well as measured values were nearly similar., with residuals lying between 1, mild inflation variance of a coefficient, the system's accuracy in short-term forecasts is higher than its accuracy in long-term predictions.

3. Discussion and Recommendation

Current status of ML models in regard to flood prediction is still quite young and in its infancy. The purpose of this study is to present a high-level overview of hybrid machine learning system or framework used in flood prediction. The suggested hybrid models have an advantage over standard models in that some can forecast constantly without losing accuracy due to its uniquely built model structure. Though there are limitations, such as, calibration of the models requires comprehensive flood data at localities where it sometimes may be difficult to collect. However, with a growing emphasis for data that are crowdsourced and publicly available gathering as well as citizen research, a more complete flood record data will become more widely available in the future. Despite the promising outcomes of adopting machine learning approaches that have already been published, there has been substantial study and testing for future enhancement and advancement, four key trends for increasing prediction quality were documented. Hybridization, whether through the combination of either two or more ML techniques, the combination of a ML methods with more traditional approaches, with or without soft computing. Data decomposition techniques are used for the goal of increasing dataset quality greatly contributed to boosting prediction accuracy. The usage of an ensemble of techniques, which significantly enhanced the generalisation capacity of the models and reduced prediction uncertainty. The final method was to increase the quality of machine learning algorithms by using optimiser algorithms.

4. Conclusion

Certainly, the growth of these unique ML approaches is heavily reliant on the effective use of soft computing techniques in the construction of fresh learning algorithms and soft computing approaches are key contributors to constructing future hybrid ML systems. These approaches listed in this paper are only but some of the available options for hybridised machine learning system. There are still plenty more though there may have been some reviews and comparison done for hydrological and Neural Network hybrid models, others are still not that well documented or reviewed. This may be another opportunity for future study.

5. References

[1] Lal P N, Mitchell T, Aldunce P, Auld H, Mechler R, Miyan A, Romano L E, Zakaria. S 2012 *National systems for managing the risks from climate extremes and disasters Managing the Risks of Extreme Events and Disasters to Advance Climate Change Adaptation* (*Special Report of the Intergovernmental Panel on Climate Change* Vol. 9781107025) ed A Dlugolecki and T Masumoto et al (Cambridge University Press, Cambridge and New York) p 339–392.

[2] Seo D J and Breidenbach J P 2002 Real-time correction of spatially nonuniform bias in radar rainfall data using rain gauge measurements. *J. Hydrometeorology*. **3** 93

[3] Mosavi A, Ozturk P, Chau K W 2018 *Flood prediction using machine learning models: Literature review* (Water vol 10) ed Chang F. J and Hsu K et al (MDPI: Switzerland) p 1-14

[4] Silver D, Huang A, Maddison C J, Guez A, Sifre L, Van Den Driessche G, Schrittweiese J, Antonoglou L, Panneershelvam V et al. 2016 Mastering the game of Go with deep neural networks and tree search. *J. Nature.* **529** 484.

[5] Kutz J N 2017 Deep learning in fluid dynamics. *J. Fluid. Mechanics*. **814** 1.

[6] Ching T, Himmelstein DS, Beaulieu-Jones BK, Kalinin AA, Do BT, Way GP, Ferrero E, Agapow P M, Zietz M, Hoffman M M et al. 2018 Opportunities and obstacles for deep learning in biology and medicine. *J. R. Soc. Interface.* **15** 1

[7] Butler K T, Davies D W, Cartwright H, Isayev O, Walsh A 2018 Machine learning for molecular and materials science. *J. Nature* **559** 547

[8] Karpatne A, Atluri G, Faghmous J, Steinbach M, Banerjee A, Ganguly A, Shekhar S, Samatova N and Kumar V 2016 Theory-guided Data Science: A New Paradigm for Scientific Discovery from

Data. *J. IEEE. Trans. Knowl. Data. Eng.* **29** 2318
[9] Lauer F and Bloch G. 2008 Incorporating prior knowledge in support vector machines for classification: A review. *J. Neurocomputing.* **71** 1578
[10] Battaglia P W, Hamrick J B, Bapst V, Sanchez-Gonzalez A, Zambaldi V, Malinowski M, Tacchetti A, Raposo D, Santoro A, Faulkner R et al. 2018 Relational inductive biases, deep learning, and graph networks. *J. arXiv.* v **3** 1
[11] Zhang J, Hou G, Ma B, Hua W 2018 Operating characteristic information extraction of flood discharge structure based on complete ensemble empirical mode decomposition with adaptive noise and permutation entropy. *J. Vib. Control.* **24** 5291
[12] Noori R, Karbassi A, Farokhnia A, Dehghani M 2009 Predicting the Longitudinal Dispersion Coefficient Using Support Vector Machine and Adaptive Neuro-Fuzzy Inference System Techniques. *J. Environ. Eng. Sci.* **26** 1503
[13] Panda R K, Pramanik N, Bala B 2010 Simulation of river stage using artificial neural network and MIKE 11 hydrodynamic model. *J. Comput. Geosci.* **36** 735
[14] Kourgialas N N, Dokou Z, Karatzas G P 2015 Statistical analysis and ANN modeling for predicting hydrological extremes under climate change scenarios: The example of a small Mediterranean agro-watershed. *J. Environ. Manage.* **154** 86
[15] Saghafian B, Haghnegahdar A, Dehghani M 2017 Effect of ENSO on annual maximum floods and volume over threshold in the southwestern region of Iran. *J. Hydrol. Sci.* **62** 1039
[16] Kim G, Barros A P 2001 Quantitative flood forecasting using multisensor data and neural networks. *J. Hydrol.* **246** 45
[17] Badrzadeh H, Sarukkalige R, Jayawardena A 2015 Hourly runoff forecasting for flood risk management: Application of various computational intelligence models. *J. Hydrol.* **529** 1633
[18] Choubin B, KHalighi Sigaroodi S H, Malekian A 2016 Impacts of Large-Scale Climate Signals on Seasonal Rainfall in the Maharlu - Bakhtegan Watershed. *J. Range. Watershed. Management.* **69** 51
[19] Liang Z, Li Y, Hu Y, Li B, Wang J A 2018 Data-driven SVR model for long-term runoff prediction and uncertainty analysis based on the Bayesian framework. *J. Theor. Appl. Climatol.* **133** 137
[20] Elsafi S H 2014 Artificial Neural Networks (ANNs) for flood forecasting at Dongola Station in the River Nile, Sudan. *J. Alexandria. Eng.* **53** 655
[21] Teng J, Jakeman A J, Vaze J, Croke B F W, Dutta D, Kim S 2017 Flood inundation modelling: A review of methods, recent advances and uncertainty analysis. *Environ. Model. Softw.* **90** 201
[22] Choubin B, Moradi E, Golshan M, Adamowski J, Sajedi-Hosseini F, Mosavi A 2019 An ensemble prediction of flood susceptibility using multivariate discriminant analysis, classification and regression trees, and support vector machines. *J. Sci. Total. Environ.* **651** 2087
[23] Han S, Coulibaly P 2017 Bayesian flood forecasting methods: A review. *J Hydrol.* **551** 340
[24] Bahrammirzaee A, Ghatari A R, Ahmadi P, Madani K 2009 Hybrid credit ranking intelligent system using expert system and artificial neural networks. *J. Appl. Intell.* **34** 28
[25] Li X, Willems P A 2020 Hybrid Model for Fast and Probabilistic Urban Pluvial Flood Prediction. *J. Water. Resour. Res.* **56** 1
[26] Kurian C, Sudheer K P, Vema V K, Sahoo D 2020 Effective flood forecasting at higher lead times through hybrid modelling framework. *J. Hydrol.* **587** 124945.
[27] Kan G, Liang K, Yu H, Sun B, Ding L, Li J, He X, Shen C 2020 Hybrid machine learning hydrological model for flood forecast purpose. *J. Open. Geosci.* **12** 813
[28] Xie Z, Liu Q, Cao Y 2021 Hybrid deep learning modeling for water level prediction in yangtze river. *J. Intell Autom. Soft. Comput.* **28** 153
[29] Yoo H J, Kim D H, Kwon H H, Lee S O 2020 Data driven water surface elevation forecasting model with hybrid activation function-a case study for hangang river, South Korea. *J. Appl. Sci.* **10** 1424

IOP Conf. Series: Earth and Environmental Science 1091 (2022) 012040 doi:10.1088/1755-1315/1091/1/012040

[30] Moishin M, Deo R C, Prasad R, Raj N, Abdulla S 2021 Designing deep-based learning flood forecast model with ConvLSTM hybrid algorithm. *J. IEEE. Access.* **9** 50982
[31] Imran M, Sheikh Abdul Khader P 2020 Forecasting water level of Jhelum River of Kashmir Valley India, using prediction and earlywarning system. *J. Geogr. Environ. Sustain.* **13** 35

Flood River Water Level Forecasting using Ensemble Machine Learning for Early Warning Systems

Amrul Faruq*[1,2], Shamsul Faisal Mohd Hussein[2], Aminaton Marto[2,3], and Shahrum Shah Abdullah*[2]

[1] Department of Electrical Engineering, Faculty of Engineering, Universitas Muhammadiyah Malang, 65144 Malang, Indonesia
[2] Malaysia-Japan International Institute of Technology (MJIIT), Universiti Teknologi Malaysia, 54100 Kuala Lumpur, Malaysia
[3] Research Center for Soft Soil, Universiti Tun Hussein Onn Malaysia, 86400 Batu Pahat, Johor, Malaysia

*Corresponding author e-mail: faruq@umm.ac.id; shahrum@utm.my

Abstract. Flood forecasting is crucial for early warning system and disaster risk reduction. Yet the flood river water levels are difficult and challenging task that it cannot be easily captured with classical time-series approaches. This study proposed a novel intelligence system utilised various machine learning techniques as individual models, including radial basis function neural network (RBF-NN), adaptive neuro-fuzzy inference system (ANFIS), support vector machine (SVM), and long short-term memory network (LSTM) to establish intelligent committee machine learning flood forecasting (ICML-FF) framework. The combination of these individual models achieved through simple averaging method, and further optimised using weighted averaging by K-nearest neighbour (K-NN) and genetic algorithm (GA). The effectiveness of the proposed model was evaluated using real case study for Malaysia's Kelantan River. The results show that ANFIS outperforms as individual model, while ICML-FF-based model produced better accuracy and lowest error than any one of the individuals. In general, it is found that the proposed ICML-FF is capable of robust forecasting model for flood early warning systems.

1. Introduction

Flood forecasting models are essential in hazard assessment and disaster management. The research on the advancement of flood forecasting will increase since it contributes to disaster risk reduction, which is a difficult task, challenging and highly complex to model [1]. According to the Sendai frameworks 2015-2030, disaster risk reduction (DRR) is given priority numbers three and four. The framework states "investing in disaster risk reduction for resilience" and "enhancing disaster risk preparedness for effective response" among its priorities [2]. In connection with these viewpoints, hence flood modelling and forecasting is crucial for disaster risk management. In many regions of the world, flood forecasting is one of the few feasible options to manage flood disasters. Moreover, flood forecasting is essential for an early warning system (EWS), in which such EWS is an integral component of disaster risk management. A flood forecasting system provides the operating environment within which the flood forecasting model can be operated and is sometimes called the system environment [3].

The most well-known works of flood forecasting modelling include artificial neural networks (ANNs) [4], [5] support vector machines (SVM) [6], [7] and adaptive neuro-fuzzy inference system (ANFIS) [8], [9]. These models were effectively employed for both short-term and long-term flood forecasting. As a new method in ANN models, deep learning is a significant subject of interest in AI methods. Deep learning is being studied in many problems, such as image processing, speech recognition, and natural language processing. In the subject of forecasting, recent studies have reported the successful use of deep learning in various fields [10]–[12], respectively, for power load and probability density forecasting, traffic flow forecasting and rainfall forecasting.

As recently reported by Yaseen *et al.* [13] and Luo He *et al.* [12], the most successfully implemented flood forecasting model include both single and hybrid model. However, few published studies have systematically examined the concept of committee machine intelligent system technique in hydrological-engineering problems, especially for flood forecasting. Hence, there is a need to investigate the effectiveness of committee machines for flood forecasting in various flood-prone areas. More specifically, in the case of Malaysia, the committee model approach has not been applied so far in the flood forecasting problem. This combination methods among the individuals were essential to produce final result, and over there, simple averaging is the most popular one [14]. Though, the disadvantage is the important contribution of the individuals cannot be emphasized due to giving equal weights to all the individuals.

In response to these problems, this study proposes to design intelligent flood forecasting models and develop committee machine learning based methods for further improvement and advancement of flood forecasting methods. The notion is to extract the pertinent information simulated by individual models and further optimize it via weighted averaging methods. Moreover, comparison analysis of the effectiveness of the K-NN and GA as ensemble method were further investigated in this study with a case study from Kelantan River, located in Kelantan State, Peninsular Malaysia.

2. Methodology

2.1. Data Source and Case Study

The case study used is dataset from Kelantan River as the flood forecasting point (FFP). This reservoir is among the most frequent occurrence of seasonal flood disasters in Malaysia. The state of Kelantan belongs to the eastern region and is located in the northeast of peninsular Malaysia, with Kota Bharu as its capital city. About twenty-seven months of data in January 2013 – March 2015 were collected through department of irrigation and drainage (DID) supervisory control and data acquisition systems. There are 19,672 records datasets were used, employed for training and validation test. As shown in Figure 1. and Table 1., three variables indicating the river water level (WL), rainfall (RF), and river streamflow (SF) were used as independent variables, while one WL as target output used as dependent variable. The proposed intelligent machine learning network requires these input data, while the observed water level becomes the system's output/target data.

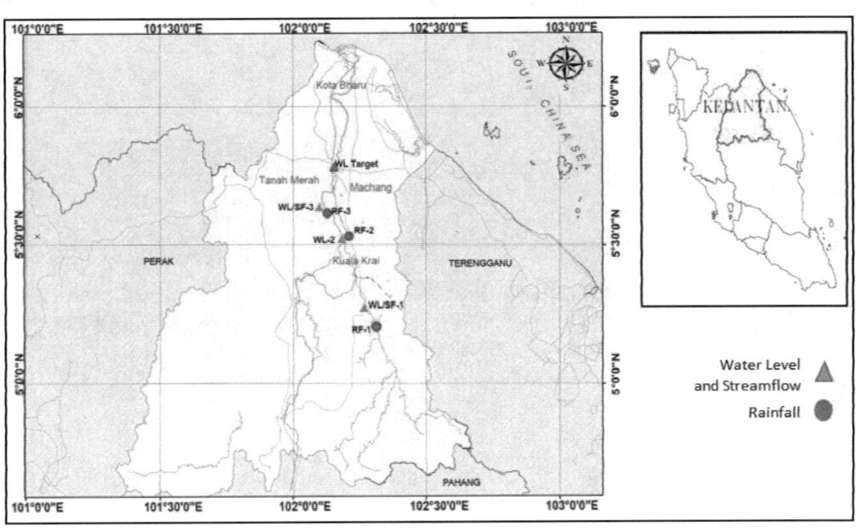

Figure 1. Kelantan river basin at Peninsular Malaysia (DID, Malaysia)

Table 1. List of hydrological observations stations of Kelantan River basin

Station ID	Station name	Observation items	Location (District)
5222452	Lebir river	Water Level-1 (WL-1)	Tualang (Kuala Krai)
5521444	Kelantan river	Water Level-2 (WL-2)	Kuala Krai
5621401	Sokor river	Water Level-3 (WL-3)	Tegawan village (Tanah Merah)
4922001	Rkt. Lebir	Rainfall-1 (RF-1)	Tualang (Kuala Krai)
5522047	JPS Kuala Krai	Rainfall-2 (RF-2)	Kuala Krai
5621051	Kenneth farm	Rainfall-3 (RF-3)	Tanah Merah
5222452	Lebir river	Streamflow-1 (SF-1)	Tualang (Kuala Krai)
5621401	Sokor river	Streamflow-2 (SF-2)	Tegawan village (Tanah Merah)
5721442	Kelantan river	Water Level (Observed/ Target)	Guillemard Bridge (Tanah Merah)

2.2. Model Development

Three (3) past inputs were considered in this study based on autocorrelation and partial autocorrelation functions, to develop intelligent-based flood forecasting models. Hence, the general flood forecasting model M, which includes input variables at upstream and downstream stations, is denoted in equation (1). Where t is variables value at t time, and i is 1, 2, 3, to 6 steps ahead of time forecasting. While Y is water level at target river, and W, R, and Q respectively are river water level, rainfall, and streamflow at upstream stations. Finally, f is the intelligent model that can be either RBFNN, ANFIS, SVM, or LSTM algorithms.

$$M(t+i) = f\begin{Bmatrix} Y(t), Y(t-1), Y(t-2), \\ W(t), W(t-1), W(t-2), \\ R(t), R(t-1), R(t-2), \\ Q(t), Q(t-1), Q(t-2) \end{Bmatrix} \tag{1}$$

2.3. Proposed ICML-FF

Individual expert intelligent systems (RBFNN, ANFIS, SVM and LSTM) will first forecast the flood water level. In this study, the ensemble method based on ICML-FF design includes ensemble averaging and optimised weighted averaging tuned by K-NN and GA to extract the optimum weights of individual models is investigated. A final forecast output is then achieved, $Y = n + t$, where n is the forecasted data, and t is time ahead forecasting horizons (1 to 6 step ahead). A schematic diagram of ICML can be illustrated in Figure 2.

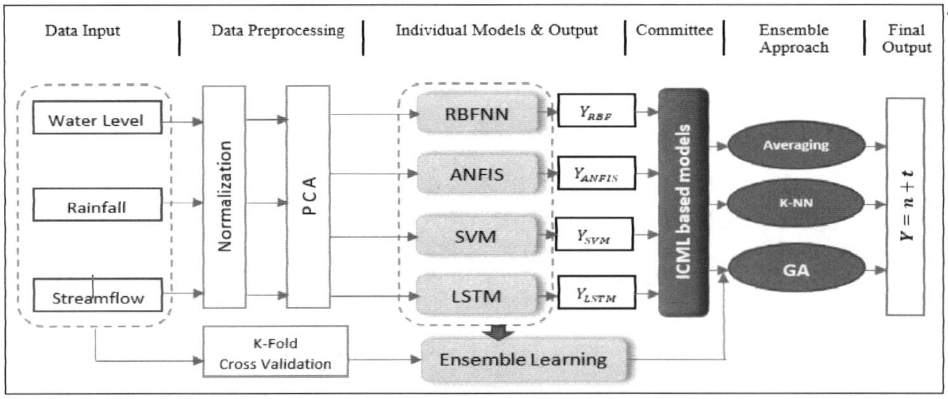

Figure 2. Proposed ensemble ICML-FF based model

2.4. Ensemble Machine Learning by K-NN and Genetic Algorithm

The Euclidian distance, D_i, between the forecasted data points, Y_j and the observed data points, X_t was calculated by equation (2). The weights for each forecasted data point and each multi-time step ahead forecasting, $\mathcal{W}_i$, were calculated by the reciprocal of the distance:

$$D_i = \sqrt{\sum_{j=1}^{K} (Y_j - X_t)^2} \quad ; \quad \mathcal{W}_i = \frac{1}{D_i} \tag{2}$$

The idea of the K-NN as weighted average method in ensemble learning is demonstrated in Figure 3 (a) [15], [16] and while Figure 3 (b) is GA optimisation algorithm used to tune the hyperparameters of the ensemble learning [17].

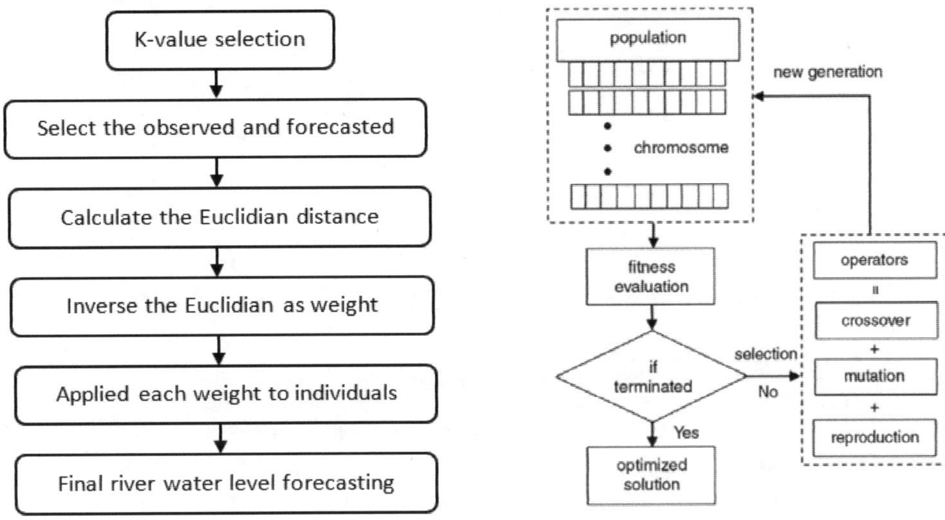

(a) (b)

Figure 3. Flow chart of ensemble machine learning by (a). K-NN and (b). GA

3. Result and Discussion

As a result, in the simple averaging method, any one of the individual intelligent models has an equal contribution in constructing ICML. The four individual base learners simply have their weight value set to 0.25. Thus, the calculated flood water level from ICML could be obtained using equation (3):

$$\begin{aligned} &EnsembleAveraging\ (E - AVG)_{(T-hr)} = \\ &\quad 0.25\left(Forecasted_{RBF(T-hr)}\right) + 0.25\left(Forecasted_{ANFIS(T-hr)}\right) + \\ &\quad 0.25\left(Forecasted_{SVM(T-hr)}\right) + 0.25\left(Forecasted_{LSTM(T-hr)}\right) \end{aligned} \tag{3}$$

Here, $T - hr$ represents each time hour ahead forecasting horizons, including one to six-hour lead-time. Applying this formula for forecasting the final output of individual models has provided the correlation coefficient value, R, and error value, mean percentage error (MPE), in each step ahead of time forecasting.

The last step of ensemble learning combines the output of individual intelligent models to get the final output using optimised weights by the genetic algorithm optimisation technique. It should be noted that the hyperparameters were tuned on the training stage using a 5-Fold cross-validation set. The average percentage error and coefficient correlation of 5-Fold cross-validation were then used as the objective function of GA. The MPE and R-value were then selected as the fitness (cost) value of GA. Thus, with a constant weight of fitness (Wcf) value of 0.5, the objective function as stated:

$$f_{GA} = 0.5\, MPE + 0.5\,(1 - R) * 100 \tag{4}$$

To extend the comparison analysis, the forecasted results emitted by individual models in this present study is examined using K-NN method as ensemble learning to produce final flood water level forecasting. However, only Kelantan dataset is selected to evaluate the compared committee machine-based model since its more complex and more features used rather than other two case studies. With the value of K-nearest neighbor set to 2, the flood forecasting performance results of committee machine-based model comprise three ensemble methods include simple averaging, genetic algorithm, and K-NN method is depicted in Table 2.

Table 2. ICML flood water level forecasting result, the best model is marked in bold.

Hour	MPE						
	RBF	**ANFIS**	**SVM**	**LSTM**	**E-AVG**	**E-GA**	**E-KNN**
1	1.7210	**0.1047**	0.4953	0.3523	0.5221	0.1969	0.1976
2	0.7575	**0.1586**	0.4637	0.9578	0.3457	0.1678	0.2223
3	2.0847	0.2605	0.6237	0.6055	0.6361	**0.2518**	0.2845
4	1.6412	0.4084	1.0490	0.9916	0.6964	**0.3886**	0.6423
5	6.8516	0.5331	1.4158	1.0052	2.2748	**0.5210**	1.1655
6	2.4323	0.6561	1.7325	1.1198	1.1942	**0.6016**	1.0065
Hour	**R**						
	RBF	**ANFIS**	**SVM**	**LSTM**	**E-AVG**	**E-GA**	**E-KNN**
1	0.9922	**0.9999**	**0.9999**	0.9996	0.9996	**0.9999**	**0.9999**
2	0.9984	**0.9999**	0.9998	0.9987	0.9997	**0.9999**	**0.9999**
3	0.9806	0.9996	0.9996	0.9992	0.9992	0.9997	**0.9998**
4	0.9740	0.9977	0.9984	0.9944	0.9981	0.9986	**0.9992**
5	0.8533	0.9961	0.9968	0.9939	0.9834	**0.9980**	0.9970
6	0.8929	0.9946	0.9950	0.9945	0.9948	0.9974	**0.9978**

As a result, committee machine-based GA demonstrates the adequacy with outperform other ensemble methods in MPE performances. Even though, R results are varied in multi-step ahead forecasting, and K-NN based method could compete with the GA. In addition, the weighted averaging ensemble method based on K-NN shows improvement over simple averaging method. This result signifies that every individual model in ICML has its own strength and domain knowledge with their proper weights related to generalisation ability [18]. This observation agrees with Azmi *et al.* [15] and Fan *et al.* [16] which has similar to that investigated in this present study.

Table 3. summarised the optimised weights attained using GA and K-NN ensemble learning-based method. In GA, the achieved weights are dominated by ANFIS model, in which ANFIS model also shows outperforms other individual in this Kelantan dataset, hence, the ANFIS's weights are contributed significantly in improving the committee machine model. This finding denotes that K-fold cross validation technique used in ensemble-GA training could benefit in generalised the committee machine model. In contrast, the weights obtained by K-NN are achieved depending on the distance of the forecasted and observed (actual) river water value. In K-NN, the nearer the distance, the greater weights assigned. Results showed that ANFIS's weights are larger than other in one to three ahead forecasting. This observation is also supported with the findings in the simulation results, in which ANFIS model is outperforms other individual in short-term ahead forecasting indicated by lowest MPE and R values.

Table 3. Optimised weights obtained by GA and ***K*-NN**

Hour	Weights by GA				Weights by *K*-NN			
	RBF	ANFIS	SVM	LSTM	RBF	ANFIS	SVM	LSTM
1	0.0653	0.8896	0.0179	0.0272	0.0277	0.2286	0.1868	0.0862
2	0.0149	0.9351	0.0169	0.0331	0.0705	0.3908	0.1875	0.0353
3	0.0100	0.9000	0.0100	0.0800	0.0129	0.1568	0.1356	0.0723
4	0.0100	0.8167	0.0113	0.1620	0.0125	0.0569	0.0717	0.0278
5	0.0100	0.7700	0.0100	0.2100	0.0076	0.0435	0.0503	0.0252
6	0.0101	0.7184	0.0104	0.2611	0.0090	0.0374	0.0404	0.0247

4. Flood Early Warning System

In connection with disaster risk management and flood warning system, according to DID [19] and Sulaiman [20] for river water level data-above sea level, it can be classified into three main categories, including "alert level", "warning level", and "danger level". The river water level at Guillemard bridge, located in the Kelantan River, was selected as a flood forecasting point in the first dataset. The alert level for the Guillemard bridge station is 12 meters, while for warning level is 14 meters, and the danger level is 16 meters – above sea levels [21].

Occurrences of heavy rainfall would flow into the reservoir and are expected to increase in river water level in just a couple of hours. It indicates when the river water level is more than the maximum observed value, it can be assumed flood will occur in that area since it could reach a dangerous level. According to Wu-Jian [22], the forecast lead time of predictions, at the very least, should be set to three hours since flood managers can issue flood warnings and notify the public to prepare for the possibility of floods. Thus, representing ensemble ICML based time series results with the flood water level threshold of the Kelantan River datasets were respectively illustrated in Figure 4. This result expressed the important things of the proposed ICML-FF. Hence, the accurate forecast models are expected to be useful for early warning systems for disaster risk reduction.

These simulation results illustrated that when the river water level significantly above the normal level, it indicates an alert level, and the flood operation room should be activated (flood warning issues). Likewise, when the river water level is near the flooding (above alert level), it's entering the warning, and hence the district flood operation room should be activated. At the end, when danger level, it shows that the river water level can cause considerable flooding, indicating that evacuation may be needed (flood commences).

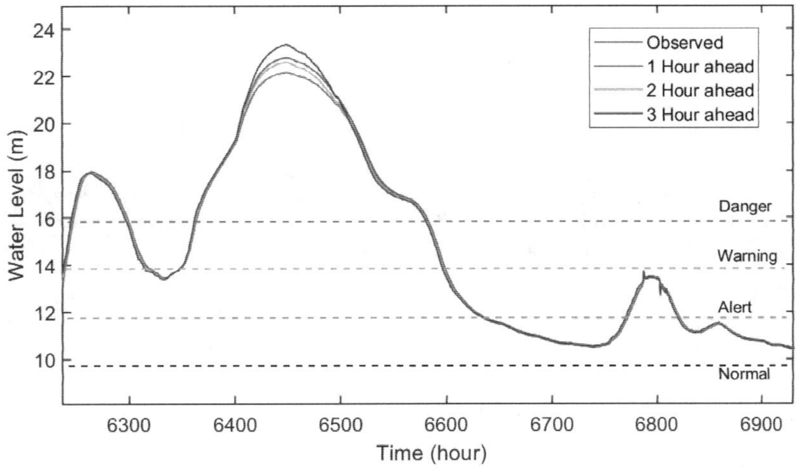

Figure 4. Water Level Threshold at Guillemard bridge in Kelantan River as a FFP using ICML Model

5. Conclusion

This study has offered a framework for exploring flood forecasting systems using an advanced ensemble machine learning approach. Returning to the purposes posed at the beginning of this study, it is now possible to state that advanced machine learning-based methods for flood forecasting problems have been developed, and their performances have been investigated. Final test results found that the ANFIS model has the best performance in the short-term forecasting horizon, for one and two hours ahead in the Kelantan River dataset. Overall, the results in multi-step ahead of time forecasting from the Kelantan River datasets show that the developed ICML-FF has a simple structure and has an easy way of constructing it. Moreover, since there are multiple directions to solve a problem, it could improve the model's performance with a high correlation coefficient value and provides smaller errors than averaging all individual experts by combining their outputs.

In addition to the flood forecasting model for early warning system and disaster risk reduction, the developed ICML-FF model consistently produces more accurate prediction results according to the designated levels criteria of flood warning analysis, including normal level, alert level, warning level, and danger level. A better prediction result is essential towards achieving better flood early warning system and disaster risk management for major problems in urban floods. Furthermore, the developed data-driven based on machine learning methods could be an alternative in terms of prediction improvement and the robustness of the models.

Acknowledgments

The authors would like to acknowledge the Malaysia-Japan International Institute of Technology (MJIIT), Universiti Teknologi Malaysia, for supporting and facilitating this research. The first author is grateful to Universitas Muhammadiyah Malang for supporting his study in undertaking this work. The sponsorship given by Japan-ASEAN Integration Fund (JAIF) is highly appreciated.

ORCID IDs

Amrul Faruq https://orcid.org/0000-0002-0702-668X
Shamsul Faisal Mohd Hussein https://orcid.org/0000-0003-0925-0906
Aminaton Marto https://orcid.org/0000-0003-3975-1269
Shahrum Shah Abdullah https://orcid.org/0000-0001-5759-1241

References

[1] S. K. Jain *et al.*, "A Brief review of flood forecasting techniques and their applications," *Int. J. River Basin Manag.*, vol. 16, no. 3, pp. 329–344, 2018.
[2] UNISDR, *Sendai Framework for Disaster Risk Reduction 2015-2030*. Japan, 2015.
[3] K. Sene, *Flood Warning, Forecasting and Emergency Response*. Springer, 2008.
[4] S. H. Elsafi, "Artificial Neural Networks (ANNs) for flood forecasting at Dongola Station in the River Nile, Sudan," *Alexandria Eng. J.*, vol. 53, no. 3, pp. 655–662, 2014.
[5] A. Faruq, S. S. Abdullah, A. Marto, M. A. A. Bakar, S. F. M. Hussein, and C. M. C. Razali, "The use of radial basis function and non-linear autoregressive exogenous neural networks to forecast multi-step ahead of time flood water level," *Int. J. Adv. Intell. Informatics*, vol. 5, no. 1, pp. 1–10, 2019.
[6] S. Zhu, J. Zhou, L. Ye, and C. Meng, "Streamflow estimation by support vector machine coupled with different methods of time series decomposition in the upper reaches of Yangtze River, China," *Environ. Earth Sci.*, vol. 75, no. 6, p. 531, 2016.
[7] W. C. Hong, "Rainfall forecasting by technological machine learning models," *Appl. Math. Comput.*, vol. 200, no. 1, pp. 41–57, 2008.
[8] A. K. Lohani, N. K. Goel, and K. K. S. Bhatia, "Improving real time flood forecasting using fuzzy inference system," *J. Hydrol.*, vol. 509, pp. 25–41, 2014.
[9] M. Ashrafi, L. H. C. Chua, C. Quek, and X. Qin, "A fully-online Neuro-Fuzzy model for flow forecasting in basins with limited data," *J. Hydrol.*, vol. 545, pp. 424–435, 2017.
[10] Z. Guo, K. Zhou, X. Zhang, and S. Yang, "A deep learning model for short-term power load and

probability density forecasting," *Energy*, vol. 160, pp. 1186–1200, 2018.

[11] L. Qu, W. Li, W. Li, D. Ma, and Y. Wang, "Daily long-term traffic flow forecasting based on a deep neural network," *Expert Syst. Appl.*, vol. 121, pp. 304–312, 2019.

[12] X. He, J. Luo, G. Zuo, and J. Xie, "Daily Runoff Forecasting Using a Hybrid Model Based on Variational Mode Decomposition and Deep Neural Networks," *Water Resour. Manag.*, vol. 33, no. 4, pp. 1571–1590, 2019.

[13] Z. M. Yaseen, S. O. Sulaiman, R. C. Deo, and K.-W. Chau, "An enhanced extreme learning machine model for river flow forecasting: state-of-the-art, practical applications in water resource engineering area and future research direction," *J. Hydrol.*, vol. 569, pp. 387–408, 2019.

[14] A. Kadkhodaie-Ilkhchi, H. Rahimpour-Bonab, and M. Rezaee, "A committee machine with intelligent systems for estimation of total organic carbon content from petrophysical data: An example from Kangan and Dalan reservoirs in South Pars Gas Field, Iran," *Comput. Geosci.*, vol. 35, no. 3, pp. 459–474, 2009.

[15] M. Azmi, S. Araginejad, and M. Kholighi, "Multi Model Data Fusion for Hydrological Forecasting using K-Nearest Neighbour Method," *Iran. J. Sci. Technol.*, vol. 34, no. B1, pp. 81–92, 2010.

[16] G. F. Fan, Y. H. Guo, J. M. Zheng, and W. C. Hong, "Application of the weighted k-nearest neighbor algorithm for short-term load forecasting," *Energies*, vol. 12, no. 5, 2019.

[17] M. M. Labani, A. Kadkhodaie-Ilkhchi, and K. Salahshoor, "Estimation of NMR log parameters from conventional well log data using a committee machine with intelligent systems: A case study from the Iranian part of the South Pars gas field, Persian Gulf Basin," *J. Pet. Sci. Eng.*, vol. 72, no. 1–2, pp. 175–185, 2010.

[18] R. Barzegar, A. A. Moghaddam, and H. Baghban, "A supervised committee machine artificial intelligent for improving DRASTIC method to assess groundwater contamination risk: a case study from Tabriz plain aquifer, Iran," *Stoch. Environ. Res. Risk Assess.*, vol. 30, no. 3, pp. 883–899, 2015.

[19] Azad, F. Fauzi, and M. Ghazali, "National Flood Forecasting and Warning System of Malaysia : Automated Forecasting for The East," *Hydrol. Water Resour. Div. Dep. Irrig. Drain. (JPS), Malaysia*, 2019.

[20] A. H. Bin Sulaiman, *Flood Management in Malaysia*. Department of Irrigation and Drainage Systems, 2009.

[21] D. of I. and D. DID Malaysia, "On-Line River Level Data (m) - above Mean Sea Level," *Department of Irrigation and Drainage, Malaysia*, 2020. http://infobanjir.water.gov.my/waterlevel_page.cfm?state=WLH.

[22] J. Wu, H. Liu, G. Wei, T. Song, C. Zhang, and H. Zhou, "Flash flood forecasting using support vector regression model in a small mountainous catchment," *Water (Switzerland)*, vol. 11, no. 1327, pp. 1–16, 2019.

Community based index of coastal erosion using ahp analysis

Wan Norshuhada Binti Wan Khairuddin[*1], **Shuib Bin Rambat**[1,2], **Aizul Nahar Bin Harun**[1,3]

[1]Malaysia-Japan International Institute of Technology (MJIIT), Jln. Sultan Yahya Petra, Kg. Datuk Keramat, 54100 Kuala Lumpur, WP Kuala Lumpur, Malaysia.
[2]Wind Engineering for (Urban, Artificial, Man-made) Environment IKhoza, MJIIT, Jln. Sultan Yahya Petra, Kg. Datuk Keramat, 54100 Kuala Lumpur, WP Kuala Lumpur, Malaysia.
[3]Innovation Management Ikhoza, MJIIT, Jln. Sultan Yahya Petra, Kg. Datuk Keramat, 54100 Kuala Lumpur, WP Kuala Lumpur, Malaysia.

Abstract. Coastal erosion can be seen on practically every beach in Malaysia, but it is especially common on the country's east coast. Depending on the severity of the erosion, coastal erosion has been addressed using a variety of methods and treatments. However, it is critical for coastal managers and responsible authorities to have an erosion index that can be used to determine the level of erosion so that suitable mitigation and treatment methods may be designed. The factors that contribute to coastal erosion must be identified and classified, and this study will use a literature review and community feedback questionnaires to identify the primary factors that contribute to coastal erosion. The AHP method will be used in this paper to assess the level of contribution of each parameter influencing coastal erosion. According to the findings of the AHP analysis, coastal structure was the most influential factor in coastal erosion, followed by human activity, waves, and wind, with weights of 0.5333, 0.2404, 0.1804, and 0.0459, respectively. This weighting of factors will be supplementary to existing guidelines such as NCES in making quick decisions, particularly in coastal areas that require immediate mitigation, and will serve as an additional guide to local agencies in planning.

1. Introduction

The world's coastlines are a complex and crucial area since they are a major industrial asset and home to a significant number of people. The coastal zone is critical for human development in terms of economic and social worth (Bagheri et al., 2021; Silva et al., 2014). Beaches in Malaysia have witnessed continual coastal erosion and land loss since 1985 as a result of the rapid growth and expansion of human activities along the coast (Ghazali, 2006). It was found that 1,324 kilometres of Malaysia's coastline were eroded by the National Coastal Erosion Survey (NCES) 2015 (DID, 2015).

The most severe coastal erosion occurs in rapidly developing areas, such as the northern Terengganu River estuary, particularly the Kuala Nerus coastline area. Over time, any human activity that alters coastal dynamics will result in a change in the shoreline's morphology (Prasad & Kumar, 2014). For example, the extension of International Airport in Terengganu, Malaysia, is a clear example of this. Since the completion of the airport's 500-meter-to-the-sea projection, coastal erosion in the northern part of the structure has been clearly visible from year to year. According to NCES, the 1.7km coastline around Kuala Nerus (particularly the area near the airport) has been eroding since 2015 due to natural and external factors. With no countermeasures in place, these eroded areas are expected to deteriorate at a faster rate, especially as the factors that cause long-term erosion increase.

Due to the aforementioned factors, determining the extent to which each parameter influences coastal erosion is an important aspect of coastal management. Recognizing the most significant factors that can affect existing coastal conditions in the event of a disaster is critical for coastal disaster risk management and long-term growth and development (Rangel-Buitrago et al., 2020; Rudiastuti et al., 2020; Sheik Mujabar & Chandrasekar, 2013). As a result, the purpose of this paper is to examine the weights for each element in order to thoroughly analyse the most influential factors. The assessment of these factors can provide clear fundamental guidance in determining the extent to which a factor's tendency to erode. Because the weights obtained in this study are based on questionnaires from local communities, the weights created are only for Malaysian beaches.

2. Methodology

2.1. Test area: Telipot beach, Kuala Nerus, Terengganu

The study area is approximately 3 kilometres long, ranging from Tok Jembal beach to Telipot beach. The northeast monsoon dominates Terengganu's coastal area, with semi-daily and annual average tidal temperatures ranging from 25 to 27 degrees Celsius (Chalabi et al., 2005). According to the National Coastal Erosion Survey (NCES) 2015, the rate of erosion in Kuala Nerus is between 2 and 11 metres per year. However, since 2016, the erosion rate in Kuala Nerus' Mengabang Telipot and Mengabang Telung has grown to 20 m/year (Figure 1). The rate of erosion in the nearby area this area has also accelerated dramatically.

Figure 1. Erosion site at the study area.

2.2. Method and materials

The main factors influencing coastal erosion are discussed in this topic, as well as the AHP method for calculating the weights of each factor influencing coastal erosion.

2.2.1. The Analytic Hierarchy Process (AHP)

The Analytical Hierarchy Process (AHP), the decision analytic method used in this study, is a theory of measurement based on pairwise comparisons that derives priority scales using expert judgement (Saaty, 2002). The essence of AHP's mathematics and computing approaches is to build a matrix describing the relative importance of a series of attributes, and each step in the AHP calculation is as follows;

- Step 1: Define the problem and select the criteria;
- Step 2: Pair-wise comparison matrices;
- Step 3: Summarize the decision and estimate the relative weight; and
- Step 4. Determine how important each choice element is in order to come up with a set of ratings for the options/strategies of decision.

The typical form of Analytic Hierarchy Process (AHP) as illustrates in figure 2.

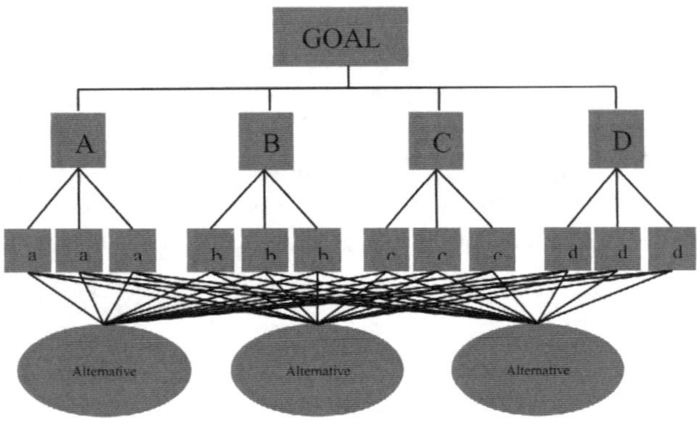

Where;

- Goal represents an Index erosion
- A, B, C, and N represent the parameter dimension of coastal erosion
- a, b, c, d and n represent the parameters or factors

Figure 2. Typical model of AHP analysis.

2.3. Parameter's identification

A literature search of previous studies was used to identify the initial parameters that influence erosion in this study. There are a number of elements that contribute to coastal erosion around the world, and this stage is critical in identifying them. As one of the most comprehensive abstract and citation databases for peer-reviewed literature, including scientific journals, books, and conference proceedings, the Scopus database has been particularly searched for credible data (Nobre & Tavares, 2017). List was then inspected in the manner depicted in figure 3. The parameters will be listed in table 1 following the findings of the comprehensive literature review.

1st STEP	**2nd STEP**	**3rd STEP**	**Final STEP**
General Topic Key word: Coastal Erosion **244,000 Documents**	Specific topic Key word: Coastal erosion factors, Causes of coastal erosion, Coastal Erosion phenomenon **139,000 Documents**	Specific topic on specific Journal Key word: Coastal erosion factors, Causes of coastal erosion, Coastal erosion phenomenon in Terengganu **905 Documents**	Specific papers Key word: Erosion causes and Coastal erosion factors, Coastal erosion in Terengganu **63 Documents**

Figure 3. Literature review key word searching mapping process.

Table 1. Summary of coastal erosion factors identified through literature review.

Dimensions	**Factors**	**Dimensions**	**Factors**
Man-made Factors	Human activities Ecosystem Destruction	Natural Factors	Wave Wind Sea Level Rise Storm Tides Water Temperature Ecosystem destruction
Socio-Economic Factors	Coastal Structure		

2.4. Parameter's selection

After identifying all relevant components, they must be further detailed to identify the key contributors to erosion. This phase ensures that the AHP analysis' input is not excessive, ensuring an accurate result. The most important factors influencing erosion were identified through a literature review and community surveys in this study.

2.4.1. Based on literature

The abstracts of the retrieved papers were examined for the keywords "erosion causes," "coastal erosion factors," and "coastal erosion in Terengganu." The present study uses 63 papers on coastal erosion that all express the same causes. In addition to scholarly resources, the articles were chosen from famous international conferences and journals with high Impact Factors (JCR). As a result, criteria will be defined based on an assessment of the principal causes of erosion. Table 2 summarizes all identified parameters as percentages. They will be classified into three categories: natural, man-made, and socio-economic. Within each of the three dimensions, only the highest percentage parameters will be considered sub-criteria. The percentage (Table 2) is calculated by multiplying the frequency of each eroding factor.

Table 2. Summary of parameters in the literature.

Dimensions	Factors	Percentage
Nature Factors	Wave	15
	Wind	10
	Sea Level Rise	2
	Storm	3
	Tides	2
	Water Temperature	1

Dimensions	Factors	Percentage
Man-Made Factors	Human activities	25
	Ecosystem Destruction	1
Socio-Economic Factors	Coastal Structure	41

2.4.2. Based on questionnaires

The perspectives and opinions of long-time residents of coastal areas are used as a second way to discover issues that contribute to local erosion. Interviewed at two beach locations, Tok Jembal and Menggabang Telipot, to ensure each person interviewed fit the study's scope, and at evening because more people were on the beach. The survey started at Tok Jembal Beach and ended at Telipot Beach (before Pengkalan Maras), with the intention of receiving 25 to 30 replies. On a scale of one to five, respondents ranked each cause of erosion. Table 3 shows the Likert scale used to measure each component's impact on coastal erosion. The questionnaire is written in Bahasa Melayu, and the aim of each question will be explained. Table 4 illustrates the questionnaire's scope.

Table 3. Likert scale.

Least relative significance	1
Have relative significance	2
Neutral	3
High relative significance	4
Highest relative significance	5

Table 4. Questionnaire of erosion considerations on coastal erosion.

Scope	Questionnaire
Description of Coastal Erosion	• Give an overview of the Malaysian shoreline. • Malaysia's current situation as a result of coastal erosion. • Identify and list the primary cause(s) of coastal erosion in Malaysia.
Coastal Erosion related to the nature factors. ▪ Scoring parameter of wind, wave, sea level rise, water temperature, and tides Coastal Erosion related to the man-made factors. ▪ Scoring parameter of coastal structure, human activities, and ecosystem destruction.	• In order to determine the score points for the AHP analysis, participants will be asked a series of questions about the level of effect for each parameter. • In addition, respondents will be asked to rate a pair-wise factor comparison.

2.4.3. Summary of parameter's selection

Based on a literature review and feedback from community questionnaires three dimensions and four parameters were chosen as AHP analysis criteria for analysing factor weighted, as below;

- Natural Factor : Wave, wind
- Man-made Factors : Human activities
- Socio-economic Factors : Coastal structure

2.4.3.1. Wave

The Terengganu coastal area is most affected by monsoon-generated waves (Ariffin, Mathew, et al., 2018; Ariffin, Sedrati, et al., 2018; Saadon et al., 2020). Annual and monthly plots from the Wave-watch III (WW3) historical model retrieved from location 6°N, 103.2°E in Terengganu waters were used to create the wave data/roses. figure 4 show the annual waves rose diagram for 1985 until 2019 in Kuala Terengganu. The majority of the prevailing wave energy looks to be originating from 55°N.

2.4.3.2. Wind

The research area is influenced by monsoon winds and wave conditions. The South China Sea shoreline is susceptible to monsoon winds during and between monsoons. Data from Kuala Terengganu were provided by Malaysian Meteorological Services Department (MET). Figure 6 depicts Kuala Terengganu's annual wind rose diagram. According to the wind rose, less than 5 m/s winds account for about 80% of the observation period, with 7.7% being quiet. These winds are widespread during the northeast and southwest monsoon seasons. The proportion of calm varies by month and monsoon, from 6.6 to 9.9% (November to March) (October).

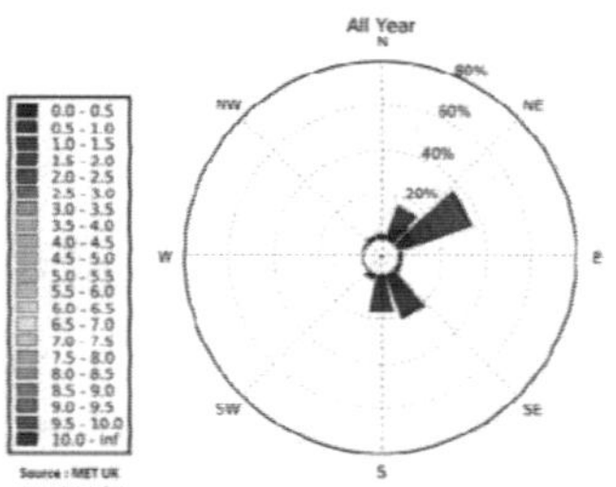

Figure 4. Annual wave rose for Kuala Terengganu (1985-2019).

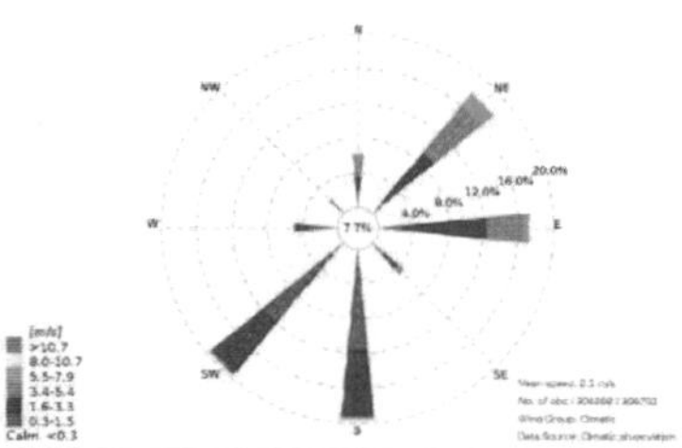

Figure 5. Annual wave rose for Kuala Terengganu (1985-2019).

2.4.3.3. Coastal structure

With all mitigation options available, the protective strategy using hard structures is often recognized as the best coastal erosion management practice, and it is the only option in many countries (Rangel-Buitrago, Williams, & Anfuso, 2018). These structures have been identified as a serious issue along many coastlines around the world because they affect longshore sediment movement or change the dynamical patterns of the nearshore current system (Hsu et al., 2007; Ndour et al., 2018; Prasetya, 2001; Rangel-Buitrago, Williams, Pranzini, et al., 2018). Changes in coastal dynamics caused by external structural disturbances were also discovered in the study area, which stretched from Sultan Mahmud Airport to Telipot Beach (Aziz et al., 2019; Chalabi et al., 2005; Yee Ling et al., 2019). Figure 6 and Table 5 show the location of the structure from the airport to the research area.

Figure 6. Location of structures along Kuala Nerus until study area.

Table 5. Structures in the Kuala Nerus beach.

Location	Location	Structures
P1	Pantai Tok Jembal	Breakwater
P2	Pantai Tok Jembal	Groyne, Revetment
P3	Pantai Menggabang Telipot	Rock cove, Revetment

2.4.3.4. Human activities

Humans and natural forces shape our beaches and coastal landforms in various areas (Sheik Mujabar & Chandrasekar, 2013). This is due to rising demand factors such as industry, trade and commerce, tourism, human population growth, and migration along the coast, which are all contributing to global pressures in the coastal zone (Zhang et al., 2000). Human activities on the coast typically involve engineering projects such as land reclamation for urban and airport expansion, navigation dredging, and the construction of ports, harbors, and jetties. As a result of disturbances inherent on the coast that act as a buffer for incoming coastal transport, such activities in coastal areas will cause sediment circulation currents to deviate from normal (Zulfakar et al., 2020). Figure 7 and table 6 below depict the activities to be considered in this study.

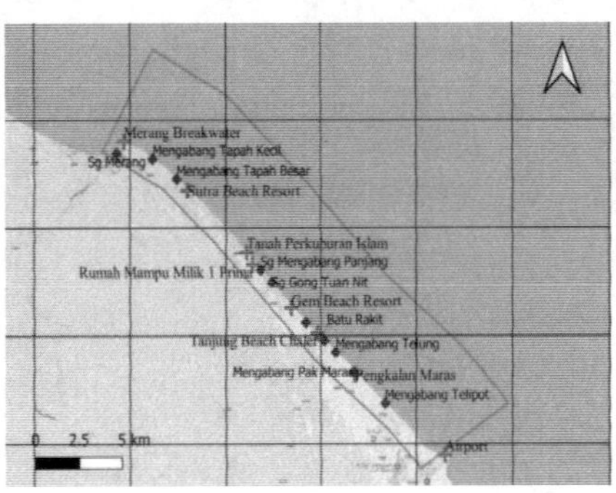

Figure 7. Extent of study area for human activities consideration.

Table 6. Existing activities summary in study area.

Location	Remarks/Activities
Batu Rakit	Airport Extension
Batu Rakit	Submerged rock 640m from shore. Erosion occurring. Scarp formation
Sg. Mengabang Panjang	Small sand mining from year 2017-2018.
Mengabang Tapah Besar	Small sand mining from year 2018-2019.
Mengabang Tapah Kecil	

3. Results and discussion

3.1. Parameters ranking

Respondents were asked to prioritize parameters based on the evaluation system using a Likert scale ranging from 1 to 5. The analysis was carried out using equation (1), and the results are shown in table 7.

$$\text{Average Value} = \frac{(v1 \times 1) + (v2 \times 2) + (v3 \times 3) + (v4 \times 4) + (v5 \times 5)}{(\text{Total number of respondents})} \quad (1)$$

Where;

v1: number of responses for least significance value
v2: number of responses for partially significance value

Table 7. The level of importance of parameter.

Parameter	Less significance	Partially Significance	Significance	More significance	Most significance	Average Value	Importance Index	Level	Ranking
	1	2	3	4	5				
	Evaluation								
Natural Factors (2)									
Wave (N1)	1	2	7	0	0	2.6	52%	M	4
Wind (N2)	0	0	3	6	0	3.3	66%	M	3
Man-made Factors (1)									
Human activities (M1)	0	1	0	6	3	4.1	82%	H	1
Socio-economic Factors (1)									
Coastal Structure (S1)	1	2	0	1	6	3.9	78%	H	2

3.2. Pair-wise comparison

The initial weights for each dimension (natural, man-made, and socioeconomic) were determined using the AHP "pair-wise" method. The results of paired comparisons between factors were obtained from the results of community questionnaires. To address the final aggregate paired comparison matrix, values with a higher number of feedback repetitions were picked (table 8). For example, if 7 respondents gave the same score of 1/2 in a paired comparison of human activities and coastal structures, a score value of 1/2 was chosen for this comparison. To help with additional computations and to evaluate the rationale and correctness of the factor comparisons, the AHP excel spreadsheet was created. Then, as shown in table 9, the fractional values in each row were converted to decimals and the averages determined.

Table 8. Normalized Matrix with a regard to the Cause Criteria.

Criteria	Wave, b_1	Wind, b_2	Coastal structures, b_3	Human activities b_4
Wave, b_1	1	5	1/5	1
Wind, b_2	1/5	1	1/7	1/7
Coastal Structures, b_3	5	7	1	2
Human Activities, b_4	1	5	1/2	1
TOTAL	**36/5**	**20**	**129/70**	**29/7**

Table 9. The normalized matrix with row average values for the cause criteria.

Criteria	Wave	Wind	Coastal Structures	Human Activities	Weightage
Wave	0.0347	0.0625	0.0795	0.0603	0.1804
Wind	0.0069	0.0125	0.0193	0.0086	0.0459
Coastal Structures	0.1736	0.0875	0.1356	0.1207	0.5333
Human Activities	0.0347	0.0875	0.0678	0.0603	0.2404
				TOTAL	**1.000**

3.3. Weightage for each aspect

The weights are supposed to be a continuous function from 0 to 1. The closer the weights are to 1, the more acceptable elements of coastal erosion become. For factors that have less impact on coastal erosion, the weights approach 0. Table 9 shows the final weighting attained for all aspects. The weight allocated to a parameter represented its importance for a certain application and has a significant impact on the index. The goal was not really to get the exact weight distribution to the decimals but to find a broad agreement on the appropriate relative weight of the different components.

Table 10. The Final Weightage for Each Criteria.

Criteria	**Weightage**
Coastal Structures	0.5333
Human Activities	0.2404
Wave	0.1804
Wind	0.0459

3.4. Consistency for AHP matrix

The next step is to double-check the overall matrix's consistency after obtaining the AHP results. This is essential because when comparing two factors side by side, respondents may lose sight of their previous responses. As a result, it's vital to double-check that all responses are genuine and consistent, as well as that the AHP method's basis is met. Inconsistencies appear frequently in AHP, particularly when respondents are requested to conduct too many pair-wise comparisons in the questionnaire. The Consistency Ratio (CR) was created to assess the matrices' consistency. Figure 8 shows the overall discrepancies of the AHP matrices constructed in excel (red box). A score of 0.1 or under (0.1) is considered acceptable, suggesting a moderate degree of matrix consistency (Franek & Kresta, 2014; Yang & Xu, 2016). By 0.1, the inconsistency is likely to be significant, and the AHP results may be inaccurate in determining the factor weightage.

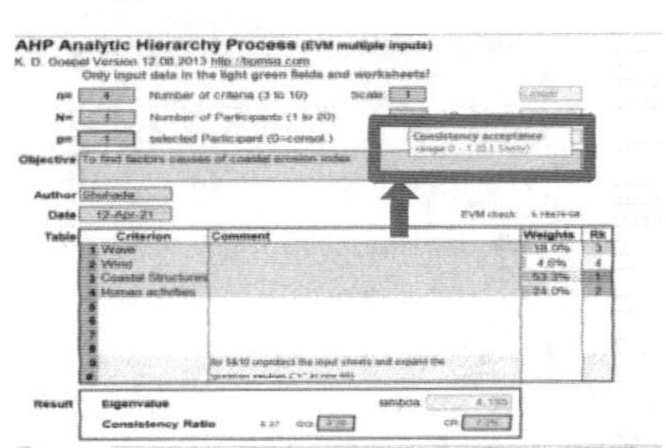

Figure 8. Overall inconsistency from AHP calculation.

The study's findings demonstrate that factors have varying influence on coastal erosion. The following are the study's findings:

i. According to community opinions gathered from questionnaires evaluated with AHP, coastal structural factors have the highest impact on coastal erosion. The existence of numerous coastal structures in the research area, such as offshore breakwaters, revetment, and groin, clearly has several consequences, including sand trapped at the structure's top and restricting sediment movement to nearby areas, causing coastal erosion along the adjacent shoreline. The most of the Kuala Terengganu/Kuala Nerus coastlines are classified as critical due to construction of erosion defense structures, (Chalabi et al., 2005; Muslim et al., 2011; Saadon et al., 2020; Zulfakar et al., 2020).

ii. Human activity in coastal areas is the second most significant contributor to the problem of coastal erosion. The Sultan Mahmud airport, which projects 500 meters into the sea, is one of the human activities that clearly has an impact on the Kuala Nerus coast. Following the completion of construction in 2008, erosion on the beach adjacent to the airport (north side) became visible, causing erosion along 290 to 360 meters in some places (Yee Ling et al., 2019).

iii. The research area faces the South China Sea. The primary wave generating mechanism for waves incident to the study area is wind conditions in the South China Sea and the Gulf of Thailand. Winds from the east and north east coincide with the north-eastern monsoon season, as can be seen. During this season, the waves and wind are at their most extreme, with wave heights reaching 3 to 4 meters. During this season, when the wave direction is perpendicular to the coast (cross-shore), the waves will carry a large amount of sediment and cause erosion.

4. Conclusion

A range of factors contribute to coastal erosion, each of which has a substantial impact. The AHP method has been found to be quite competent and simple in terms of calculation for evaluating the level of contribution of each factor to coastal erosion without the need for physical data from each factor. In this study, the AHP method was used to subjectively assess the experience and intuition of long-established coastal communities on the assessment of categories and levels of coastal erosion risk. The study findings are encouraging, implying that this method could be used in real-world situations. Through a detailed and extensive weighted analysis of several severely eroded areas in Malaysia, an index-based method of determining the erosion rate for each influential factor can be developed in the future.

Acknowledgments

This research was funded by UNIVERSITI TEKNOLOGI MALAYSIA, grant number Q.K130000.3843.20J00 and Malaysia-Japan International Institute of Technology. Appreciation also goes to my department, DID Malaysia for the data and other abilities.

References

[1] Ariffin, E. H., Mathew, M. J., Yaacob, R., Akhir, M. F., Shaari, H., Zulfakar, M. S. Z., Sedrati, M., & Awang, N. A. (2018). Beach morphodynamic classification in different monsoon seasons at Terengganu beaches, Malaysia. *Journal of Sustainability Science and Management*, *13*(Special Issue 5), 65–74.

[2] Ariffin, E. H., Sedrati, M., Akhir, M. F., Daud, N. R., Yaacob, R., & Husain, M. L. (2018). Beach morphodynamics and evolution of monsoon-dominated coasts in Kuala Terengganu, Malaysia: Perspectives for integrated management. *Ocean and Coastal Management*, *163*(June), 498–514. https://doi.org/10.1016/j.ocecoaman.2018.07.013

[3] Aziz, A., Rahman, A. B., Zainal, M. K., Aqiff, M., Wahid, A., Hamzah, M. L., Ariffin, E. H., & Awang, A. A. (2019). Multi Method Analysis for Identifying the Shoreline Erosion During Northeast Monsoon Season. *Journal of Sustainability Science and Management*, *14*(3), 43–54.

[4] Bagheri, M., Ibrahim, Z. Z., Mansor, S., Manaf, L. A., Akhir, M. F., Talaat, W. I. A. W., &

Beiranvand Pour, A. (2021). Land-use suitability assessment using delphi and analytical hierarchy process (D-ahp) hybrid model for coastal city management: Kuala terengganu, peninsular malaysia. *ISPRS International Journal of Geo-Information*, *10*(9). https://doi.org/10.3390/ijgi10090621

[5] Chalabi, A., Mohd-Lokman, H., Mohd-Suffian, I., & Karamali, M. (2005). GEomorphological features of kuala terenganu spit from ikonos imagery and aerial phographs using segmentation method. *Asian Association on Remote Sensing - 26th Asian Conference on Remote Sensing and 2nd Asian Space Conference, ACRS 2005*, *1*(January), 577–583.

[6] DID. (2015). *National Coastal Erosion Study 2015*. *1*, ES 27.

[7] Franek, J., & Kresta, A. (2014). Judgment Scales and Consistency Measure in AHP. *Procedia Economics and Finance*, *12*(March), 164–173. https://doi.org/10.1016/s2212-5671(14)00332-3

[8] Ghazali, N. H. M. (2006). Coastal erosion and reclamation in Malaysia. *Aquatic Ecosystem Health and Management*, *9*(2), 237–247. https://doi.org/10.1080/14634980600721474

[9] Hsu, T. W., Lin, T. Y., & Tseng, I. F. (2007). Human impact on coastal erosion in Taiwan. *Journal of Coastal Research*, *23*(4), 961–973. https://doi.org/10.2112/04-0353R.1

[10] Muslim, A. M., Ismail, K. I., Khalil, I., Razman, N., & Zain, K. (2011). Detection of shoreline changes at Kuala Terengganu, Malaysia from multi-temporal satellite sensor imagery. *34th International Symposium on Remote Sensing of Environment - The GEOSS Era: Towards Operational Environmental Monitoring*, *April*.

[11] Ndour, A., Laïbi, R. A., Sadio, M., Degbe, C. G. E., Diaw, A. T., Oyédé, L. M., Anthony, E. J., Dussouillez, P., Sambou, H., & Dièye, E. hadji B. (2018). Management strategies for coastal erosion problems in west Africa: Analysis, issues, and constraints drawn from the examples of Senegal and Benin. *Ocean and Coastal Management*, *156*, 92–106. https://doi.org/10.1016/j.ocecoaman.2017.09.001

[12] Nobre, G. C., & Tavares, E. (2017). Scientific literature analysis on big data and internet of things applications on circular economy: a bibliometric study. *Scientometrics*, *111*(1), 463–492. https://doi.org/10.1007/s11192-017-2281-6

[13] Prasad, D. H., & Kumar, N. D. (2014). Coastal Erosion Studies—A Review. *International Journal of Geosciences*, *05*(03), 341–345. https://doi.org/10.4236/ijg.2014.53033

[14] Prasetya, G. (2001). Chapter 4: Protection from coastal erosion. *Coastal Protection in the Aftermath of the Indian Ocean Tsunami: What Role for Forests and Trees? Proceedings of the Regional Technical Workshop, Khao Lak, Thailand, 28–31 August 2006*, 103–132. http://www.fao.org/docrep/010/ag127e/AG127E00.htm

[15] Rangel-Buitrago, N., Neal, W. J., & de Jonge, V. N. (2020). Risk assessment as tool for coastal erosion management. *Ocean and Coastal Management*, *186*(January), 105099. https://doi.org/10.1016/j.ocecoaman.2020.105099

[16] Rangel-Buitrago, N., Williams, A. T., & Anfuso, G. (2018). Hard protection structures as a principal coastal erosion management strategy along the Caribbean coast of Colombia. A chronicle of pitfalls. *Ocean and Coastal Management*, *156*, 58–75. https://doi.org/10.1016/j.ocecoaman.2017.04.006

[17] Rangel-Buitrago, N., Williams, A. T., Pranzini, E., & Anfuso, G. (2018). Preface to the special issue: Management strategies for coastal erosion processes. *Ocean and Coastal Management*, *156*(November), 1–3. https://doi.org/10.1016/j.ocecoaman.2017.11.020

[18] Rudiastuti, A. W., Rahadiati, A., Dewi, R. S., Soetrisno, D., & Maulana, E. (2020). Assessing coastal vulnerability index of tourism site: The case of Mataram Coast. *E3S Web of Conferences*, *153*. https://doi.org/10.1051/e3sconf/202015303002

[19] Saadon, M. S. I., Ab Wahida, N. S., Othman, M. R., Nor, D. A. M., Mokhtar, F. S., Nordin, N., Kowang, T. O., & Nordin, L. (2020). An evaluation of the impact of coastal erosion to the environment and economic activities at mengabang telipot, Terengganu. *Journal of Critical Reviews*, *7*(8), 1132–1136. https://doi.org/10.31838/jcr.07.08.238

[20] Saaty, T. L. (2002). Decision making with the Analytic Hierarchy Process. *Scientia Iranica*, *9*(3), 215–229. https://doi.org/10.1504/ijssci.2008.017590
[21] Sheik Mujabar, P., & Chandrasekar, N. (2013). Coastal erosion hazard and vulnerability assessment for southern coastal Tamil Nadu of India by using remote sensing and GIS. *Natural Hazards*, *69*(3), 1295–1314. https://doi.org/10.1007/s11069-011-9962-x
[22] Silva, R., Martínez, M. L., Hesp, P. A., Catalan, P., Osorio, A. F., Martell, R., Fossati, M., Da Silva, G. M., Mariño-Tapia, I., Pereira, P., Cienguegos, R., Klein, A., & Govaere, G. (2014). Present and future challenges of coastal erosion in Latin America. *Journal of Coastal Research*, 1–16. https://doi.org/10.2112/SI71-001.1
[23] Yang, H., & Xu, J. (2016). *A Kind of Analytic Hierarchy Process and Neural Network Integrated Evaluation Method of Teaching*. *Ichess*, 231–235. https://doi.org/10.2991/ichess-16.2016.50
[24] Yee Ling, S., Veerasamy, K., Hidayat Jamal, M., & Shahrizal Ab Razak, M. (2019). Preliminary Modelling Assessment of Hydrodynamic Simulation due to Terengganu Airport Runway Extension. *IOP Conference Series: Earth and Environmental Science*, *357*(1). https://doi.org/10.1088/1755-1315/357/1/012036
[25] Zhang, K., Douglas, B. C., & Leatherman, S. P. (2000). Twentieth-century storm activity along the U.S. East Coast. *Journal of Climate*, *13*(10), 1748–1761. https://doi.org/10.1175/1520-0442(2000)013<1748:TCSAAT>2.0.CO;2
[26] Zulfakar, M. S. Z., Akhir, M. F., Ariffin, E. H., Awang, N. A., Yaacob, M. A. M., Chong, W. S., & Muslim, A. M. (2020). The effect of coastal protections on the shoreline evolution at Kuala Nerus, Terengganu (Malaysia). *Journal of Sustainability Science and Management*, *15*(3), 71–85.

Lightweight deep CNN models for identifying drought stressed plant

MH. Kamarudin*[1], Zool H. Ismail[1]

[1] Malaysia-Japan International Institute of Technology, Universiti Teknologi Malaysia, Jalan Sultan Yahya Petra, Kuala Lumpur 54100, Malaysia

*Corresponding author e-mail: hider@graduate.utm.my

Abstract. Drought is one of the most severe climatological disasters that has negative impact on agricultural production around the world. Over the years, computer vision technology has been used in conjunction with machine learning applications to replace traditional destructive and time-consuming methods for real-time monitoring of drought-affected plant. Deep learning (DL) techniques have gained a stellar reputation in image classification recently, with convolutional neural network (CNN) emerging as the industry standard. However, the size of deep CNN models is frequently large due to massive number of parameters and field application is often not feasible due to limited storage and computational resources. Several lightweight CNN models have been selected based on the number of network parameters of less than 6M and were trained and tested. The EfficientNet model has achieved a classification accuracy of 88.12 and 88.97 percent for identifying severe drought, mild drought, and no drought plants on visible and near-infrared images respectively. The findings of this study can be used to assist in the development of automated early detection of drought stressed plant with model sizes suitable for real-time plant diagnosis on mobile or embedded devices.

1. Introduction

Droughts are becoming more common around the world as a result of global climate change. Drought conditions that last for an extended period can have a detrimental impact on agriculture, prompting a red alert on food security. As irregular rainfall limits water availability, most plant crops suffered from water stress, the most critical of plant abiotic stresses that can affect plant growth. An early detection system for water stressed plants is crucial for ensuring sustainable agricultural productivity. Traditionally, visual characteristics such as leaf yellowing and wilting have been manually monitored as indicators of drought conditions. However, the technique is inefficient and unsuitable for early identification of drought-stressed plants. More capable techniques, such as quantifying leaf and stem water potential at midday [1], have been introduced; however, the methods are time consuming and destructive. Recently, the development of plant sensing equipment utilising remote sensing to assess environmental and plant physiological changes in a fast and nod-destructive manner has seen a significant leap in progress [2].

Computer vision has been widely adopted as a low-cost system for identifying drought-affected plants [3]. The Internet of Things (IoTs) and robotics platforms have been used to create an automated system based on image analysis that extracts information such as morphological features (size, shape, texture), spectra (colour, temperature, humidity), and time data (growth rate, development, dynamic change of spectral and morphological modes) that show strong correlations with the plant water status [4]. Plant phenotyping using computer vision has also been implemented in drought resistance plant breeding studies [5]. Despite this, the technology is still in research and development stage and has seen little use in practise due to operational and processing difficulties. The high-throughput data of plant images necessitates fast processing mechanism to determine the useful information that reflects the plant's water stress status.

Machine learning (ML) has quickly become the standard approach in computer vision applications due to its ability to effectively process large amounts of information in a non-linear framework. Gutiérrez, Diago, Fernández-Novales and Tardaguila [6] have used machine learning models in thermal imaging analysis to assess water stress status in vineyards. Zhuang, Wang, Jiang, Li and Gong [7] have proposed a maize water stress detection model based on a supervised learning algorithm of Gradient Boosting Decision Tree (GBDT) using features extracted from plant colour images. However, conventional machine learning techniques require manual extraction of image features, which are then fed into the ML model to identify drought stressed plant using pattern recognition. This requires specific technical knowledge and is currently the limiting factor influencing the model accuracy [8]. When compared to traditional machine learning, the advanced ML technique of deep learning (DL) provides the benefits of automated feature extraction and analysis with better performance. Convolutional Neural Networks (CNNs) have been developed for quick application and can provide a much-detailed representation of the image features.

Many studies have used deep CNN models to detect drought affected plants, with such applications being solved either by finely tuned CNNs or by training from scratch. Chandel, Chakraborty, Rajwade, Dubey, Tiwari and Jat [9] tested three different pre-trained deep CNNs, AlexNet, GoogLeNet, and Inception V3 to identify drought stress in maize, okra, and soybean in which GoogLeNet performance was found to be superior compared to the other models. Zhuang, Wang, Jiang and Li [10] used CNN to identify drought stress in maize by learning features of the leaf phenotype. In their most recent study, Islam and Yamane [11] created HortNet417v1, a deep CNN model for detecting drought stress in pot-grown peach plants. Aside from exceptional performance, deep model architectures typically include large network parameters, which necessitates large storage capacity devices. Deep learning models are also extremely dependable on large amounts of data and involve significant computational resources, often supplied by the graphics processing units (GPUs). The application of such technology in the agricultural field may result in significant hardware costs. Cloud computing may be the best alternative option, but it requires a consistent internet connection, which is simply not available in some croplands.

Lightweight deep learning model structures for image classification have been developed over the years for small and mobile devices, including edge computing applications. These models have comparatively simpler and more efficient network constructions with fewer parameters, consuming less memory as well as computing power. The objective of this research is to investigate the ability of several lightweight deep CNN models, namely MobileNet, Mobile NasNet, and EfficientNet, to identify drought-affected plants. The goal is to create a real-time embedded image processing and plant diagnosis system for water stress detection using limited computer hardware.

2. Materials and devices

2.1. Plant images dataset

Plant images were obtained from the Donald Danforth Plant Science Centre's standard publicly available dataset [12]. The dataset includes images of plant shoot area of ten Setaria grass lines (*S. viridis* (accession A10), *S. italica* (accession B100), and eight randomly selected RILs derived from a cross of *S. viridis* and *S. italica*). Setaria grass is a model plant that has been used in many drought-related studies to analyse the plant phenotypes [13].

Four water treatments were performed on full-water capacity (FC): 100% FC, 66% FC, 33% FC, and 0% FC imposed 17 days after planting (DAP) and maintained for another 17 days. Images prior to 17 DAP were excluded from the sample due to low to none biomass production. Plants that received no water after 17 DAP died within 7 days thus 0% FC images were also omitted from the sample. Because early treatment has no discernible effect on the plants, the sample images were taken after 2 days of treatment, from 21 DAP until 33 DAP. One top view and four side view images were taken for each plant. We combined all four-sided images and omitted the top view images in the sample. These images were sorted by drought severity of severe drought, mild drought, and no drought.

The dataset was created using visible (VIS) and near-infrared (NIR) images, two types of images commonly used in plant water stress analysis studies. Visible image of red, green, and blue (RGB), is the most widely used imagery for measuring plant's morphological properties collected using optical device that detects light wavelengths between 400 and 700 nm. The rate of NIR absorption is the highest in the spectral range of 1400 to 1450 nm and is highly correlated with plant moisture content. We used both images to determine which image provided the best accuracy for identifying drought-stressed plants using lightweight CNN models. As NIR image is essentially a grayscale image, it should be faster to compute than an RGB image. The RGB image resolution is 2,454 x 2,056 pixels, while the NIR image resolution is 150 x 200 pixels. Number of images in each sample class are shown in Table 1.

Table 1. Number of images in each sample class

Drought severity	Number of images	
	RGB	NIR
Severe drought	2,281	2,280
Mild drought	2,286	2,285
No drought	2,382	2,300

2.2. Image pre-processing

Image cropping was performed as a pre-processing step to remove any boxes and pot pixels. This is done to eliminate the unnecessary background and focus only on the plant features, as shown in figure 1 and 2. The cropping process also reduces the size of the images, which speeds up the computation. Images for this study were taken in a laboratory setting with a white background and adequate lighting. Because the background is white, no plant segmentation is required. We trained with a single image of a plant shoot because we believe that the plant structure would provide morphological information that aids in detecting drought-stressed plants. The images were downscaled to fit the size specification of various models, and the intensity values were adjusted to fill the entire intensity range [0, 255].

(a) **(b)** **(c)**

Figure 1. RGB plant images (a) severe drought, (b) mild drought (c) no drought

(a) **(b)** **(c)**

Figure 2. NIR plant images (a) severe drought, (b) mild drought (c) no drought

2.3. Lightweight deep CNNs

The architectures of different lightweight CNN models can be defined by the number of convolutional layers, activation function of each layer, and the hidden units of each layer. Predefined models of MobileNet (versions 1 and 2), MnasNet and EfficientNet were evaluated for their accuracy in identifying drought stressed plant, as shown in Table 2. These models were chosen based on a small number of parameters (fewer than 6 million) to reduce computational complexity and memory cost.

Table 2. Lightweight CNN models evaluated in the study

Model	Parameters	Size (MB)	Depth	Time (ms) per inference step	
				CPU	GPU
MobileNet	4,253,864	16	88	22.6	3.44
MobileNetV2	3,538,984	14	88	25.9	3.83
NasNet mobile	5,326,716	23	-	27.04	6.70
EfficientNet	5,330,571	29	-	46.0	4.91

2.3.1. MobileNet

MobileNet [14] is a lightweight architecture optimized for mobile and embedded vision systems. It divides the convolution into a depth-wise separable convolution followed by a pointwise convolution to construct a lightweight deep CNN model. The parameters are much lower than those of established models such as VGG [15], but the accuracy is comparable when trained on well-known public datasets.

2.3.2. MobileNetV2

MobileNetV2 [16] is an updated version of MobileNet that improves its efficiency and effectiveness in terms of accuracy and speed. MobileNetV2 enhances the state-of-the-art performance of mobile models on a variety of tasks and benchmarks, as well as across a range of model sizes. In contrast to traditional residual models, the architecture is based on an inverted residual structure, with the input and output of the residual block being thin bottleneck layers.

2.3.3. MnasNet

Mobile NasNet or MnasNet [17] is a CNN architecture designed for mobile devices with limited computing power. Based on the Neural Architecture Search (NAS) approach, lightweight deep model is automatically designed and optimized using reinforcement learning [18]. It outperforms MobileNetV2 on ImageNet classification and COCO detection tasks, performing 1.5 times faster. The model's computational cost and parameters can be easily scaled to address a wide range of problems.

2.3.4. EfficientNet

EfficentNet's architecture [19] employs mobile inverted bottleneck convolution (MBConv), which is similar to MobileNetV2 and MnasNet but slightly larger. EfficientNet used compound scaling to create a series of EfficienNets that were both more accurate and larger in size. Only the baseline architecture was used in this study.

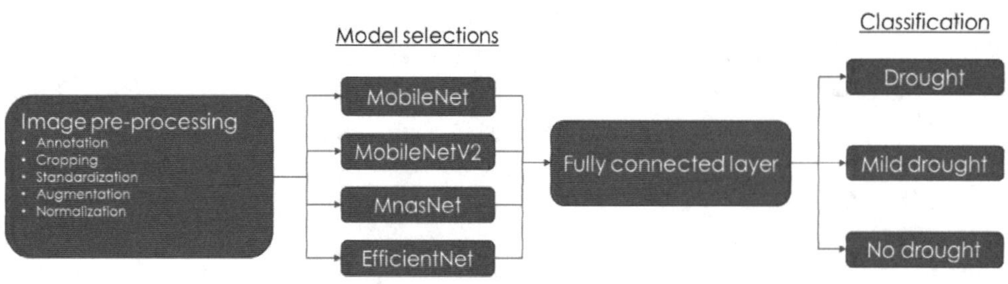

Figure 3. Schematic representation of the flow of work

2.4. Training and validation

Seventy percent of the sample images were used for training, twenty two percent for cross-validation, and ten percent for testing. To avoid overfitting, validation data was used to tune network parameters and hyperparameters. To obtain a generalized measure of classification accuracy, unseen test data was used. Flipping, adjusting aspect ratio, and intensity transformation were all part of the augmentation process as in figure 3. To evaluate training and testing accuracies, Adam optimizer was used. The learning rate was fixed to 0.001, and the models were built using the Tensorflow [13] framework in python platform. To reduce the risk of model overfitting, the callbacks function was used to perform early stopping. Fine tune layer starts at layer 134 for MobileNets (versions 1 and 2), layer 50 for NasNet, and layer 100 for EfficientNet.

3. Results and discussion

Table 3 shows that the deep CNN models were able to efficiently classify plants that had been subjected to three different levels of drought stress with respect to the accuracy and their training epochs. With 0.6875, MobileNet version 1 has the lowest accuracy. For RGB images, MobileNetV2 has an accuracy of 0.7687 and MnasNet has an accuracy of 0.8187. For RGB images, EfficientNet has the highest accuracy of 0.8812. EfficientNet also provides the best results for NIR images, with 0.8897 accuracy, followed by MnasNet model with 0.7364 accuracy, MobileNetV2 with 0.7396 accuracy, and MobileNetV1 with 0.7083 accuracy.

Table 3. Accuracy results on test dataset

Model	**Test accuracy**		**Epochs**	
	RGB	NIR	RGB	NIR
MobileNet	0.6875	0.7083	46	74
MobileNet-V2	0.7687	0.7396	67	87
NasNet mobile	0.8187	0.7364	51	74
EfficientNet	**0.8812**	**0.8897**	103	75

Overall, EfficientNet produced the best results for identifying drought-stressed plants. This could be due to EfficientNet's deeper structure being the largest in comparison to the other models. MnasNet, which had comparable parameters and size to EfficientNet, also performed well. The two MobileNet versions had lower accuracy, which could be attributed to smaller parameters and sizes. According to table 1, EfficientNet can be used in mobile devices to detect drought-stressed plants. When more data is available, the model's ability to scale up is also advantageous. In addition, when compared to other models, the results showed that EfficientNet transferred well to other datasets. Although an upscaled model can improve accuracy, the trade-off between accuracy and computational cost can be overcome by introducing an appropriate upscale coefficients multiplier.

Although the use of NIR images may simplify the process of detecting drought stressed plants through fast model convergence as referred to epochs in table 2, the results show that VIS images perform better when deep CNN models are used. This is to be expected, as colour is one of the dominant features retrieved by the CNN alongside shape and texture for drought-stressed plant identification [10]. This demonstrates that high accuracy can be obtained without the use of specialised cameras to detect drought-stressed plants. Nonetheless, both VIS and NIR systems could be used concurrently to improve detection accuracy. It should be noted that the pre-trained models are mostly trained on the publicly available ImageNet dataset, which contains colour images. This may result in a slight bias toward using colour images over grayscale images. To avoid this issue, it is recommended that the CNN model be trained from scratch using the NIR image rather than fine-tuning from the existing models.

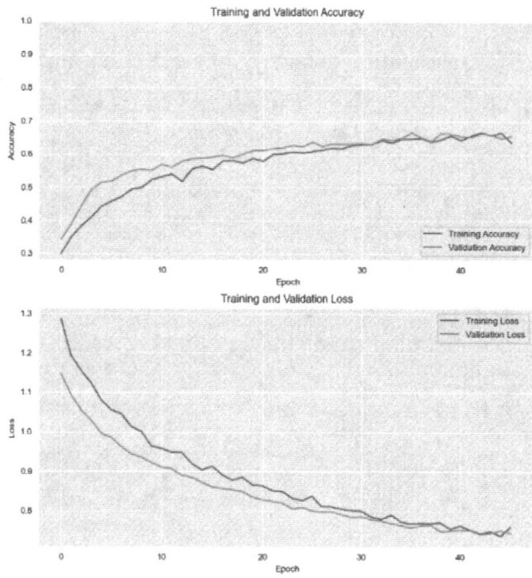

Figure 4. MobileNet training and validation accuracy and loss versus epochs for RGB image.

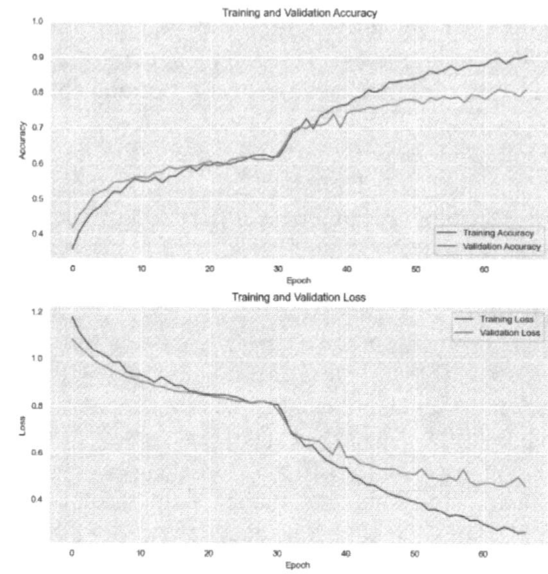

Figure 5. MobileNet-V2 training and validation accuracy and loss versus epochs for RGB image.

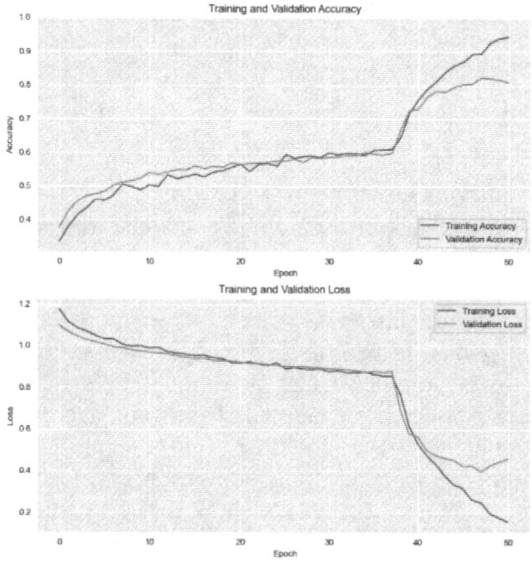

Figure 6. NasNet mobile training and validation accuracy and loss versus epochs for RGB image.

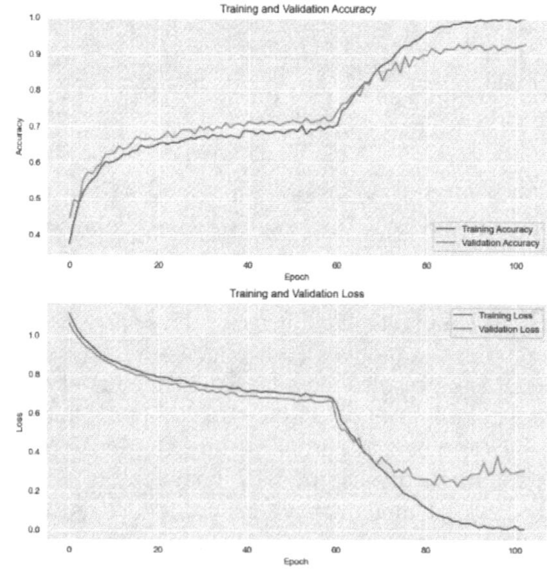

Figure 7. EfficientNet training and validation accuracy and loss versus epochs for RGB image.

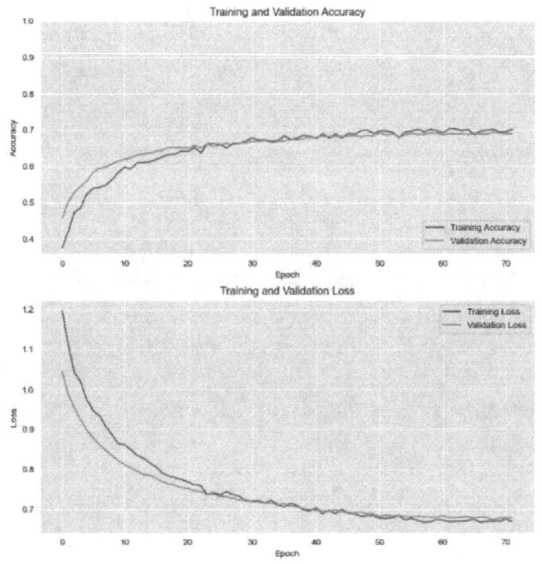

Figure 8. MobileNet training and validation accuracy and loss versus epochs for NIR image.

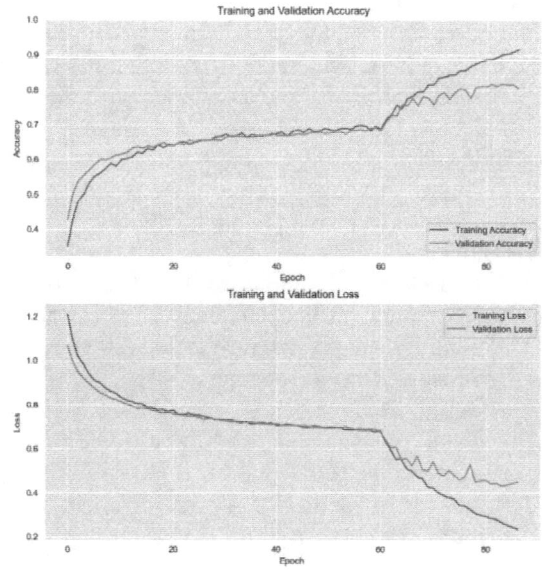

Figure 9. MobileNet-V2 training and validation accuracy and loss versus epochs for NIR image.

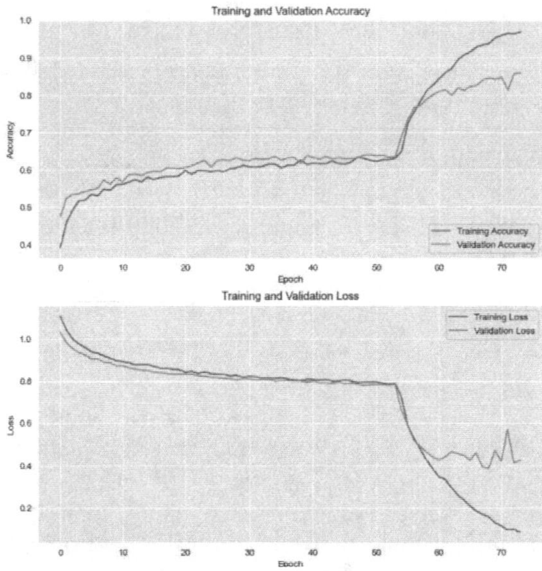

Figure 10. NasNet mobile training and validation accuracy and loss versus epochs for NIR image.

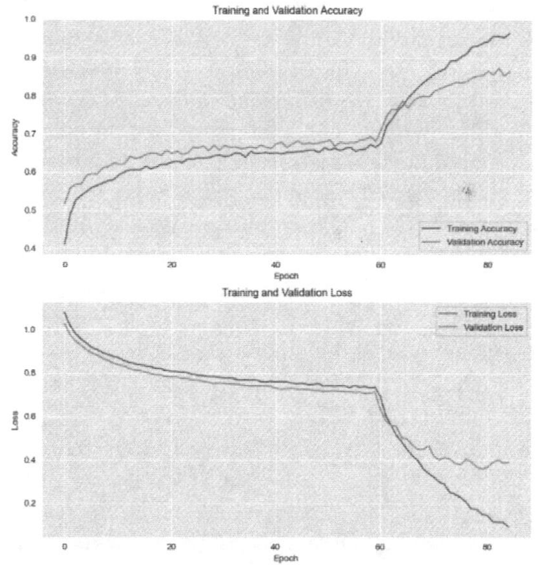

Figure 11. EfficientNet training and validation accuracy and loss versus epochs for NIR image.

4. Conclusion

This study investigated the ability of lightweight deep convolutional neural network models to identify drought-stressed plants, which can easily meet the design requirements for mobile and embedded vision applications. EfficientNet, the most successful model architecture, achieved success rates of 88.12 percent and 88.97 percent for RGB and NIR images, respectively. In the future, the trade-off between accuracy and computational cost may lead to the development of small and low-cost models, and appropriate models can be selected based on resource constraints.

Acknowledgments

This research was funded by Ministry of Higher Education and Universiti Teknologi Malaysia under research grant no. R.K130000.7843.5F348 and Q.K130000.2543.19H97. The authors gratefully thank Siti Aisyah Rohade for her assistance with the preparation of the manuscript.

ORCID iDs

MH. Kamarudin https://orcid.org/0000-0002-1187-4622

References

[1] Virnodkar S S, Pachghare V K, Patil V and Jha S K 2020 Remote sensing and machine learning for crop water stress determination in various crops: a critical review *Precision Agriculture* 1-35

[2] Ahmad U, Alvino A and Marino S 2021 A Review of Crop Water Stress Assessment Using Remote Sensing *Remote Sensing* **13** 4155

[3] Biabi H, Mehdizadeh S A and Salmi M S 2019 Design and implementation of a smart system for water management of lilium flower using image processing *Computers and Electronics in Agriculture* **160** 131-43

[4] Katsoulas N, Elvanidi A, Ferentinos K P, Kacira M, Bartzanas T and Kittas C 2016 Crop reflectance monitoring as a tool for water stress detection in greenhouses: A review *biosystems engineering* **151** 374-98

[5] Kim M, Lee C, Hong S, Kim S L, Baek J-H and Kim K-H 2021 High-Throughput Phenotyping Methods for Breeding Drought-Tolerant Crops *International Journal of Molecular Sciences* **22** 8266

[6] Gutiérrez S, Diago M P, Fernández-Novales J and Tardaguila J 2018 Vineyard water status assessment using on-the-go thermal imaging and machine learning *PloS one* **13** e0192037

[7] Zhuang S, Wang P, Jiang B, Li M and Gong Z 2017 Early detection of water stress in maize based on digital images *Computers and Electronics in Agriculture* **140** 461-8

[8] Chlingaryan A, Sukkarieh S and Whelan B 2018 Machine learning approaches for crop yield prediction and nitrogen status estimation in precision agriculture: A review *Computers and electronics in agriculture* **151** 61-9

[9] Chandel N S, Chakraborty S K, Rajwade Y A, Dubey K, Tiwari M K and Jat D 2020 Identifying crop water stress using deep learning models *Neural Computing and Applications* 1-15

[10] Zhuang S, Wang P, Jiang B and Li M 2020 Learned features of leaf phenotype to monitor maize water status in the fields *Computers and Electronics in Agriculture* **172** 105347

[11] Islam M P and Yamane T 2021 HortNet417v1—A Deep-Learning Architecture for the Automatic Detection of Pot-Cultivated Peach Plant Water Stress *Sensors* **21** 7924

[12] Fahlgren N, Feldman M, Gehan M A, Wilson M S, Shyu C, Bryant D W, Hill S T, McEntee C J, Warnasooriya S N and Kumar I 2015 A versatile phenotyping system and analytics platform reveals diverse temporal responses to water availability in Setaria *Molecular plant* **8** 1520-35

[13] Feldman M J, Paul R E, Banan D, Barrett J F, Sebastian J, Yee M-C, Jiang H, Lipka A E, Brutnell T P and Dinneny J R 2017 Time dependent genetic analysis links field and controlled environment phenotypes in the model C4 grass Setaria *PLoS genetics* **13** e1006841

[14] Howard A G, Zhu M, Chen B, Kalenichenko D, Wang W, Weyand T, Andreetto M and Adam H 2017 Mobilenets: Efficient convolutional neural networks for mobile vision applications *arXiv preprint arXiv:1704.04861*

[15] Simonyan K and Zisserman A 2014 Very deep convolutional networks for large-scale image recognition *arXiv preprint arXiv:1409.1556*

[16] Sandler M, Howard A, Zhu M, Zhmoginov A and Chen L-C 2018 Mobilenetv2: Inverted residuals and linear bottlenecks. In: *Proceedings of the IEEE conference on computer vision and pattern recognition,* pp 4510-20

[17] Tan M, Chen B, Pang R, Vasudevan V, Sandler M, Howard A and Le Q V 2019 Mnasnet: Platform-aware neural architecture search for mobile. In: *Proceedings of the IEEE/CVF Conference on Computer Vision and Pattern Recognition,* pp 2820-8

[18] Zoph B and Le Q V 2016 Neural architecture search with reinforcement learning *arXiv preprint arXiv:1611.01578*
[19] Tan M and Le Q 2019 Efficientnet: Rethinking model scaling for convolutional neural networks. In: *International Conference on Machine Learning*: PMLR) pp 6105-14

Determination of Critical Environmental, Social, Infrastructure, and Economic Indicators as Evaluation Checklist for Pre-Project Abandonment Plan Assessment in-line with Sustainable Development Goal Plan

SN Abdullah*[1], A Aznah[2], Mohd Hakim[2], MF Md Din[3], M Ponraj[4], S Mat-Taib[5]

[1]Department of Environment (DOE), 62574, Putrajaya, Malaysia
[2]Department of Chemical and Environmental Engineering, Malaysia-Japan International Institute of Technology, Universiti Teknologi Malaysia (UTM), 54100, Kuala Lumpur, Malaysia.
[3]Centre for Environmental Sustainability and Water Security (IPASA), Research Institute of Sustainable Environment (RISE), School of Civil Engineering, Faculty of Engineering, Universiti Teknologi Malaysia (UTM), 81310, Skudai, Malaysia
[4]Department of Biological Sciences, School of Mathematics and Natural Sciences, The Copperbelt University, Riverside, Jambo Drive, P. O. Box: 21692, Kitwe, Zambia
[5]Department of Environmental Engineering, School of Civil Engineering, Faculty of Engineering, Universiti Teknologi Malaysia (UTM), 81310, Skudai, Malaysia

*Corresponding author e-mail: sna@doe.gov.my

Abstract. The increasing number of project abandonment because of the Novel Coronavirus pandemic 2019 (Covid-19) calls for better management of abandonment projects. The outbreak of Covid-19 has affected the environment, economy, and society aspects globally. In a developed country like Malaysia, the requirement to submit an abandonment plan is stipulated under the Environmental Guidelines in Malaysia (EGIM) 2016. Therefore, this study aimed to develop a Sustainable Pre-Project Abandonment Plan Assessment Checklist in Malaysia that is in line with the Sustainable Development Goal (SDG) 2030 aspirations. This new Sustainable Pre-Project Abandonment Plan Assessment Checklist in Malaysia is comprehensively structured; is expected to minimise the negative impact of the abandonment process towards the environment, socio-economy, and efficient management of infrastructure; and complies with the related legislation in Malaysia. In this study, data was collected and analysed using NVivo12. It consisted of findings from documents reviewed such as legislations, journals, and books. From the NVivo12 analysis, the results showed 28 critical indicators for abandonment plan assessment and related legislation that synchronises with SDG 2030 is required to develop the Proposed Sustainable Pre-Project Abandonment Plan Assessment Checklist in Malaysia.

1. Introduction

The outbreak of Covid-19 has really impacted the economy. Various sectors were affected including agriculture, trade, tourism, textile, electronics, automobile, iron and steel mineral processing, real estate, and software companies [1]. The Project Proponent intends to abandon the Environmental Impact Assessment (EIA) project due to various reasons. In Malaysia, if the project proponent intends to implement a project abandonment, they need to send a report to the Department of Environment (DOE). In this case, the abandonment plan report submitted must be evaluated and approved by the DOE. The requirement for submission is stipulated in the Environmental Guidelines in Malaysia (EGIM), 2016 [2]. Nevertheless, the limitation of a clear framework in driving towards proper project abandonment will cause the report to be disapproved, which will affect the management of the EIA project abandonment plan. The abandonment management of imperfect EIA projects will contribute to various environmental problems. According to Sommer et al. [3], there are two major impacts on environmental problems: the impact of noise and contamination from end-point like waste. Moreover, some of the challenges, including inexperience in carrying out EIA project abandonment and unclear frameworks, will further complicate the management of the project. Under Section 34A, Environmental Quality Act 1974 [4], the Project Proponent is bound by law and will be subjected to legal action for non-compliance with that section. Furthermore, unplanned demolition collapse would lead to incidents that cause loss of life. This view is supported by Zhang [5], who mentioned that uncontrolled demolition causes deaths and injuries.

Therefore, this study is significant because it contributes to the gap due to limited information regarding the project abandonment framework. This research will also improve the Environmental Management Plan as stipulated in the EGIM [2]. This research focuses on the abandonment of the EIA project under the jurisdiction of DOE. It identifies the crucial indicators for project abandonment and clusters them through their main items of assessment that synchronise with Sustainable Development Goal (SDG) 2030 aspirations. It also outlines the related legislation for the abandonment of the project. Subsequently, this research framework is in accordance with the requirements of Section 34A (2C) of the EQA1974 (Act 127) [4] to provide an understanding of the EIA procedures, preparation, and submission of the EIA report for review and approval. In addition, compliance with the requirements will fulfil the Project Proponent's obligation stated under Section 34A (2C) of the EQA. Eventually, a framework is developed to serve as a reference in addressing project abandonment plan assessment in Malaysia.

2. Methodology

The research techniques used for the data collection process in this study involved two main approaches:

2.1. Review of documents by using NVivo12 analysis

During this phase, document review was used to triangulate the collected data. The relevant documents on existing policy or legislation, journals, and books were reviewed and further discussed. As quoted by O'Neill et al. in 2018, there are eight steps to conduct a literature review using NVivo12, divided into three modules, namely Module 1 (Preparation), Module 2 (Analysis), and Module 3 (Writing a Literature Review). Module 1 (Preparation) consists of setting up an NVivo12 Project, searching the literature, sorting literature, preparing EndNote, and importing it into NVivo12. Module 2 (Analysis) consists of classifying literature, visualisation, and topic coding of literature for themes. Finally, Module 3 (Writing a Literature Review) consists of NVivo12 Project summarisation.

3. Results and Discussion

3.1. Summary of source classifications sheet for literature captured via NVivo12

Figure 1. Summary of legislations in Malaysia related to project abandonment

3.2. Summary of cases for matrices

The cases for framework matrices were divided into four items which are topic, issues, rules, and penalty that will be imposed for non-compliance with the legislation or policy, as shown in Figure 2. For legislation, the most important point to gain was the topic or title of the Act, legislation, policy or guidelines, and penalty. In addition, the information assisted in gaining the critical indicators for the development of the Sustainable Pre-Project Abandonment Plan Assessment Checklist in Malaysia.

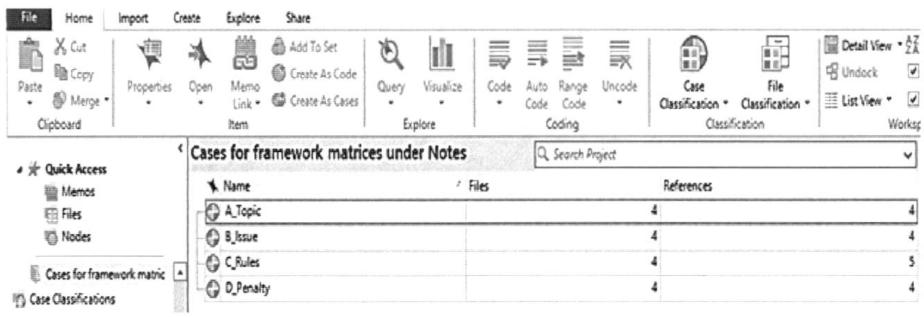

Figure 2. Summary of cases for framework matrices under notes

3.3. Literature review matrix summary

The matrix table analysis was carried out based on the list of literature review matrix summary. Figure 3 shows the detailed method for the literature matrix summary for legislation, as an example. Data gathered from this technique would provide better results because all the data were converged. This process helped the researchers scrutinise the document transcript towards finding emerging patterns. The results of the analysis assisted in identifying a theme for indicators for the development of the Proposed Sustainable Pre-Project Abandonment Plan Assessment Checklist in Malaysia

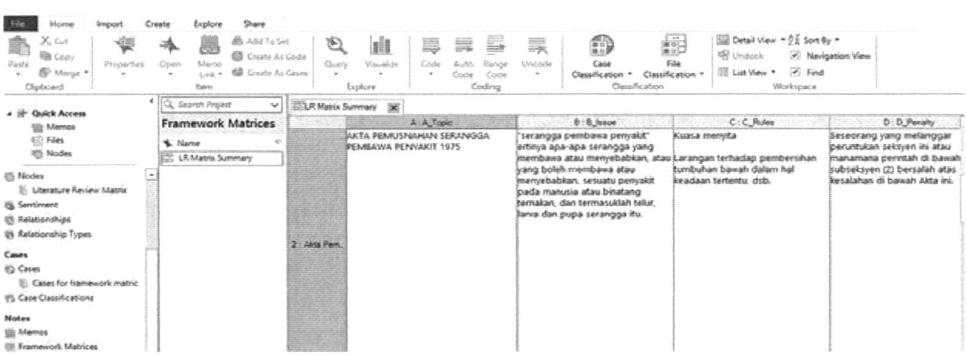

Figure 3. Example of literature matrix summary for Legislation

3.4. Content analysis based on literature matrix summary

Cross-sectional analysis was carried out based on the results obtained from the literature matrix summary. This process also helped the researchers scrutinise the transcript towards finding emerging patterns. Then, the listed free nodes before the related child nodes (sub-nodes) were clustered to the main nodes (main themes). The analysis results were presented using a cognitive mapping diagram that served to identify the initial Proposed Sustainable Pre-Project Abandonment Plan Assessment Checklist in Malaysia. As an example, the detailed method for Theme 1, which is Environmental Indicators, is shown in Figure 4. This table shows the summary of the analysis done to determine the critical indicators based on the literature review matrix summary. The indicators were clustered into four (4) main assessments: Theme 1 (Environmental Indicators), Theme 2 (Economic Indicators), Theme 3 (Social Indicators), and Theme 4 (Infrastructure Management).

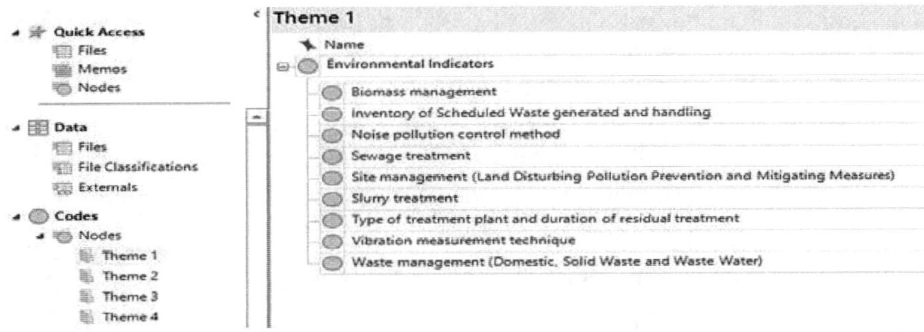

Figure 4: Summary of analysis done using NVivo12 for Theme 1

3.5. Cognitive mapping using NVivo12

The detailed content analysis resulted in the derivation of cognitive mapping, as shown in Figure 5. This was to show the summary of analysis on the illustrated figures. The cognitive mapping developments were based on each theme identified from this study.

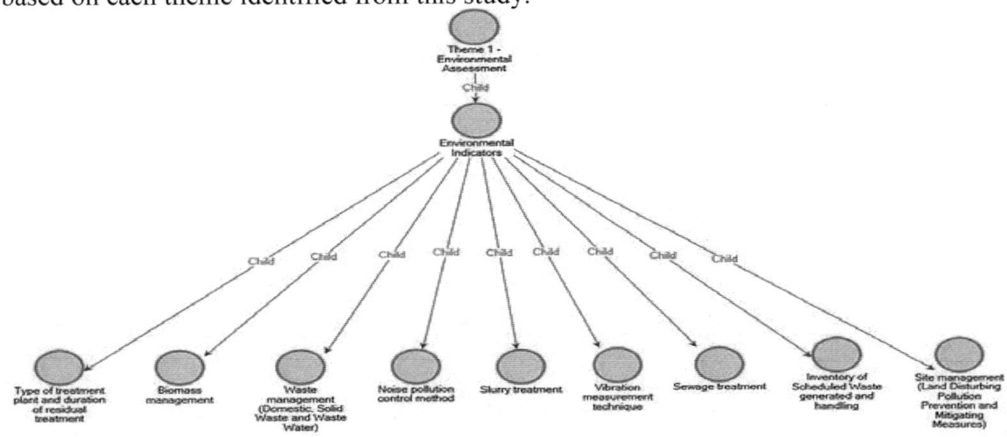

Figure 5: Cognitive mapping developed using NVivo12 for Theme 1

During the final analysis stage, the results obtained from the cross-sectional contents of all the cases were summarised. The summary of the results is presented in Tables 1, 2, 3, and 4, grouped under the related themes.

Table 1. A matrix for generating a summary of cross-sectional content analysis (Theme 1)

Theme 1 – Environmental assessment	
Environmental indicators	Inventory of scheduled waste generated and handling
	Slurry management and treatment
	Noise pollution control method
	Vibration measurement technique
	Sewage management
	Waste management (domestic/solid waste and wastewater)
	Site management (Land Disturbing Pollution Prevention and Mitigation Measures)
	Type of treatment plant and duration of residual treatment
	Biomass management

Table 2. A matrix for generating a summary of cross-sectional content analysis (Theme 2)

Theme 2 – Economic assessment

Economic Indicators	Circular economy concept (Reduce, Reuse, Recycle) Inventory of domestic waste generated Disposal of scheduled waste cost Assumption cost of residue treatment Rehabilitation cost Sustainable consumption of energy and workers

Table 3. A matrix for generating a summary of cross-sectional content analysis (Theme 3)

Theme 3 – Social assessment	
Social indicators	Safety & health of workers Site housekeeping Vermin control Heritage project special management Disease management control

Table 4. A matrix for generating a summary of cross-sectional content analysis (Theme 4)

Theme 4 – Infrastructure assessment	
Infrastructure indicators	Information on stream diversion once project abandonment occurs Method of blasting Method of demolition Dilapidation survey Buffer zone Drainage management system Type of noise barrier

3.6. Proposed Sustainable Pre-Project Abandonment Plan Assessment Checklist in Malaysia

Sustainable Development Goals 2030 has become a target for peace and prosperity for people and the planet, now and in the future. Therefore, to achieve a sustainable abandonment of the project, agencies in Malaysia play a crucial role in managing project abandonment; it requires cooperation between all relevant stakeholders. Thus, the Sustainable Pre-Project Abandonment Plan Assessment Checklist in Malaysia from this research consists of the main assessment types: environmental, economic, social, and infrastructure management that synchronise with their respective indicators and related Sustainable Development Goal 2030, as shown in Table 5. These indicators need to be considered during the report preparation and evaluation by both project proponents and related agencies before the approval of the project abandonment report.

Table 5 shows that for Environmental Assessment, there are nine types of critical indicators which are inventory of scheduled waste generated and handling, slurry management and treatment, noise pollution control method, vibration measurement technique, sewage management, waste management that comprise of domestic or solid waste and wastewater, site management, type of treatment plant and duration of residual treatment, and biomass management. The most important aspect to be considered for scheduled waste management is the inventory of scheduled waste management as stipulated in the Scheduled Waste Regulation 2005 [4]. The inventory will provide information regarding the amount of scheduled waste generated, information on the transporter, and final disposal of scheduled waste location. Thus, the track record can be used for future reference by either the project proponent or reviewer of the report. Good management of this indicator will support SDG 13, which is Climate Change.

In addition, abandonment work that generates slurry must also be managed as stipulated in the Standard Specifications for Building Works, Jabatan Kerja Raya (material generated from construction/ excavation work) [6]. This indicator must also be managed well as it can deteriorate our water quality and must be treated before disposal to any landfill approved by the local authority. A well-managed indicator will support SDG 14 and 15, which are Life Below Water and Life on Land.

If the abandonment activities involve heavy machinery produce noise pollution, an effective control method is important to control the pollution. For EIA project abandonment, noise pollution control as stipulated in the Guidelines for Environment Noise Limit and Control 2017 [7] is under the jurisdiction of the Department of Environment (DOE). This guideline provides noise acceptance criteria for quantitative assessment of noise to define disturbance or otherwise, which are useful to project proponents and decision-makers to reduce the potential impact of noise affecting public health or causing annoyance or disturbance.

Meanwhile, the vibration measurement technique is very important to measure impact assessment. The vibration limits may be set based on certain vibration parameters, such as an absolute limit based on the vibration in a particular designated direction which should not be exceeded. The governing limits depend on the repetitive nature and duration of the vibration (continuous, short term or single event). For human response and annoyance evaluation, limits may be set for different periods of the day (day and night). The management of this indicator shall fulfil the requirement as stipulated in The Planning Guidelines for Vibration Limits and Control in The Environment 2007 [8]. Proper management of noise and vibration indicators will support SDG 3, which is Good Health.

In addition, sewage management is also important to prevent a river from being polluted, as stated in Sewage Regulation 2009 [4]. The implementation of these indicators will support SDG 6, which is Clean Water and Sanitation, and ensure availability and sustainable management of water and sanitation for the workers and nearby residents and ecosystem.

Waste management is very important in order to minimise the amount of waste generated from the site and ensure proper collection, storage, and disposal of the different types of waste generated from the abandonment activities. In Malaysia, for solid waste management, the related legislation for the management of solid waste generated from construction area is Solid Waste and Public Cleansing Management (Licensing) (Undertaking of Provision) of collection services for Construction Solid Waste Regulations 2018, Solid Waste and Public Cleansing Management (Scheme for Construction Solid Waste) 2018, and Solid Waste and Public Cleansing Management (Scheme for Commercial, Industrial, Institutional Solid Waste) 2018 [9]. Meanwhile, for wastewater management, the indicator must comply with Environmental Quality Act 1974 requirement, which is Industrial Effluent Regulation 2015 or related legislation. This regulation prevents pollution in rivers and waterways within the site project and surrounding area. This is to maintain water quality at baseline conditions or better within the DOE prescribed limits.

Land disturbing refers to any project abandonment that can possibly involve activities such as clearing of trees or vegetation, excavating, raising or sloping of ground, stripping, grading grubbing, trenching, excavating, filling, logging, storing of materials, and blasting. At the site, the Project Proponent is responsible to ensure the abandonment site is well managed, including physical effects such as management of aquatic areas, private, and public lands; obstruction of stream channels and navigable rivers; and flood area management as stipulated in the Guidelines on Land Disturbing Pollution Prevention and Mitigation Measures 2017 [10].

On the other hand, the types of treatment plant and duration of residual treatment are very important due to contamination that will affect the surrounding environment. The final discharge effluent limitation from industries is under the jurisdiction of DOE Malaysia as stipulated in the Environmental Quality (Industrial Effluent) Regulations 2009, Environmental Quality Act 1974 [4]. In these regulations, industrial effluent means any waste in the form of liquid or wastewater generated from the manufacturing process, including the treatment of water for water supply or any activity occurring at any industrial premises.

Furthermore, the Project Proponent needs to identify whether the biomass generated is categorised as wood, garbage, crops, landfill gas, or alcohol fuels as stipulated in the Solid Waste and Public Cleansing Management (Scheme for Construction Solid Waste) Regulations 2018 [9]. Thus, the management of waste, project site, treatment plant, and biomass will support SDG 14 and SDG 15 to protect the life below water and life on land.

In terms of economic assessment, there are seven important indicators for project abandonment. These include the circular economy concept (reduce, reuse and recycle), inventory of domestic waste generated, disposal of scheduled waste cost, assumption cost of residue treatment, rehabilitation cost, sustainable consumption of energy and workers, and allocation of environmental conservation budgeting. In this case, for legal compliance, the implementation of circular economy and inventory of domestic waste was stipulated on the Solid Waste and Public Waste Management Act 2007[9]. Meanwhile, the circular economy concept would also support Cluster 3, National Cleanliness Act 2019 [11]. In addition, a well-managed cost related to disposal purposes of scheduled waste, residue treatment, and rehabilitation costs must be allocated as stipulated in the Guided Self-Regulation (GSR), Department of Environment (DOE) Malaysia [12]. GSR is a mainstreaming tool that can be used for project abandonment purposes that also consists of an allocation of environmental conservation and budgeting as a tool. The Project Proponent also needs to assess the impact of the project abandonment on the economy of the resident nearby the project area and ecosystem.

Sustainable consumption of energy and workers must comply with Part XI, Demolition, Factories and Machinery (Building Operations and Works of Engineering Construction (Safety) Regulations 1986) [13]. Thus, the management of the circular economy concept, inventory of domestic waste generated, disposal of scheduled waste cost, assumption cost of residue treatment, rehabilitation cost, and allocation of environmental conservation budgeting will support SDG 7 and SDG 8, providing good jobs and enhancing economic growth.

From the social assessment perspective, there are five critical indicators which are safety and health of workers, site housekeeping, vermin control, heritage project special management, and disease management control. For safety and health indicators, the related legislation is Occupational Safety and Health Act 1994 [14]. This legislation emphasises a safe and healthy approach for working at the site to avoid accidents and health problems. Furthermore, site housekeeping is also important because it will affect the cleanliness and tidiness of the site project that would affect the lives of workers and nearby residents of the project area. Solid waste management is stipulated in the Solid Waste and Public Cleansing Management Act 2007 [9]. On the other hand, vermin control is also important to avoid disease as stipulated in the Destruction of Disease-Bearing Insects Act 1975 (Act 154) and Prevention and Control of Infectious Disease Act 1988 (Act 342) [15].

In addition, heritage project management is also important to ensure that the abandonment of projects classified as heritage project is abandoned according to the National Heritage Act 2005 [16]. Meanwhile, the Covid-19 pandemic is affecting workers' movement on site projects. To control the outbreak, project proponents need to comply with the Prevention and Control of Infectious Disease Act 1988 with the implementation of Standard Operating Procedure (SOP) at the site project. Moreover, these indicators will support SDG 3, ensuring the health and promoting well-being for workers and nearby residential areas. It is because safety and health are very precious.

For infrastructure assessment, there are seven critical indicators: information on stream diversion once project abandonment occurs, method of blasting, method of demolition, dilapidation survey, buffer zone, drainage management system, and noise barrier. Each indicator is important and ensures well-managed project abandonment activities related to infrastructure. For information on stream diversion, once project abandonment occurs, the project proponent needs to ensure that the implementation of these activities comply with Chapter 5, Erosion and Sediment Control Facilities, Diversion Channel, Guideline for Erosion and Sediment Control in Malaysia, 2010 [17]. This guideline also emphasises a structured drainage layout plan to avoid floods at the project site. Blasting work is one of the critical indicators for the abandonment of project activities. Referring to the Malaysian Standard (MS) 2318:2012, Demolition Code of Practice [18], the blasting work must obtain permission or approval from Polis Diraja Malaysia (PDRM) before being carried out. In addition, the project proponent is responsible for appointing a competent person for the blasting work as stipulated in the standard. Demolition activities include dismantling, razing, destroying, or wrecking any building, structure, or part thereof by pre-planned and controlled methods. In addition, before executing the demolition work, the project proponent needs to prepare a demolition plan to do demolition work without creating risks and execute it in a safe and orderly manner with careful planning for each stage of demolition.

A dilapidation survey is an inspection carried on the structural conditions of the existing surrounding buildings before the commencement of demolition, construction, or development as stipulated in Subchapter 4.1.4, Dilapidation Survey, Malaysian Standard, (MS) 2318:2012 [18]. This survey is important to check the layout arrangement of structural elements, the state of maintenance and deterioration, and any structural implication that may affect the demolition work. The buffer zone is also a critical indicator for project abandonment, known as setbacks (EESIM, 2018) [19]. In addition, buffer zones may act as filter strips. As stipulated in Chapter 5.2, Erosion Control Facilities, Guideline for Erosion and Sediment Control in Malaysia, 2010 [17], the example of types of Best Management Practices (BMPs) used at buffer zone can be divided into several types such as vegetated swales, steep, rocky slopes, and stream banks. Besides that, to ensure good management of project abandonment, one of the important abandonment of project indicators is a drainage management system. In addition, as stipulated in Subchapter 4.3.4, Drainage Control/Runoff Management, Guideline for Erosion and Sediment Control In Malaysia, 2010 [17], the main principle of the drainage management system is to ensure that a drainage system is in sites to direct runoff water so that it does not run across disturbed and unstable areas. Thus, the drainage system's hydraulic characteristics, including overall drainage pattern, dimension and flow of any rivers and streams, springs and well including flow and well logs, subsurface conditions including aquifer type and capacity, and well maintenance work, are important elements to be considered by the project proponent.

Moreover, project abandonment requires comprehensive management, including noise management at the project site. There are a few aspects to be considered in managing the noise: identification of sensitive receptors, establishing noise acceptance criteria, baseline noise monitoring, noise modelling, noise impact assessment, and noise mitigation. Therefore, as stipulated in the Guidance on noise control, Guidelines for Environment Noise Limit and Control 2019 [7], to control noise during the abandonment of the project, several types of noise barriers should be installed at the project site such as permanent noise barriers, movable and/or temporary barriers, partial enclosures, full enclosures, silencers and mufflers, and external lagging.

Thus, well management of the infrastructure assessment indicators that consist of green innovation of machinery and equipment used, sustainable drainage system, and noise barrier for the abandonment of work will support SDG 9 (Innovation and Infrastructure). In addition, well management of these critical indicators will affect life around the project site and support SDG 15 (Life on Land).

Table 5. Proposed Sustainable Pre-Project Abandonment Plan Assessment Checklist in Malaysia

Environmental Assessment		
Environmental indicators	Legislation/Policy	Related Sustainable Development Goal, 2030 (SDG)
Inventory of scheduled waste generated and handling	Scheduled Waste Regulation 2005, Environmental Quality Act 1974 [4].	SDG 13 (Climate Action)
Slurry management and treatment	Standard Specifications for Building Works, Jabatan Kerja Raya (Material generated from construction/ excavation work) [6]	SDG 14 (Life below water) SDG 15 (Life on land)
Noise pollution control method	Guidelines for Environment Noise Limit and Control 2017 [7]	SDG 3 (Good Health)
Vibration measurement technique	The Planning Guidelines for Vibration Limits and Control In The Environment 2007 [8]	SDG 3 (Good Health)
Sewage management	Sewage Regulation 2015, Environmental Quality Act 1974 [4]	SDG 6 (Clean Water and Sanitation)
Waste management (Domestic or solid waste and Wastewater	Solid Waste and Public Cleansing Management (Scheme for Construction Solid Waste) 2018[9] Industrial Effluent Regulation 2015, Environment Quality Act 1974 [4]	SDG 14 (Life below water) SDG 15 (Life on land) SDG 14 (Life below water) SDG 15 (Life on land)
Site management (LD-P2M2)	Guidelines on Land Disturbing Pollution Prevention and Mitigation Measures, 2017[10]	SDG 14 (Life below water) SDG 15 (Life on land)
Type of treatment plant and duration of residual treatment	Industrial Effluent Regulation 2015, Environment Quality Act 1974[4]	SDG 14 (Life below water) SDG 15 (Life on land)

Biomass management	Solid Waste and Public Cleansing Management (Scheme for Construction Solid Waste) 2018[9]	SDG 14 (Life below water) SDG 15 (Life on land)
	Economic Assessment	
Economic Indicators	Legislation/Policy	Related Sustainable Development Goal, 2030 (SDG)
Circular economy concept (Reduce, reuse, recycle)	Cluster 3, National Cleanliness Policy 2019 [11], Solid Waste and Public Waste Management Act 2007 [9]	SDG 7 (Renewable Energy)
Inventory of domestic waste generated	Solid Waste and Public Waste Management Act 2007 [9]	SDG 7 (Renewable Energy)
Disposal of scheduled waste cost Assumption cost of residue treatment Rehabilitation cost	Guided Self-Regulation (GSR), Department of Environment Malaysia [12]	SDG 7 (Renewable Energy) SDG 8 (Good Jobs and Economic Growth)
Sustainable consumption of energy and workers	Part XI, Demolition, Factories and Machinery (Building Operations and Works Of Engineering Construction (Safety) Regulations 1986) [13]	SDG 8 (Good Jobs and Economic Growth)
Allocation of environmental conservation budgeting	Guided Self-Regulation (GSR), Department of Environment Malaysia [12]	SDG 7 (Renewable Energy) SDG 8 (Good Jobs and Economic Growth)
	Social assessment	
Social indicators	Legislation/Policy	Related Sustainable Development Goal, 2030 (SDG)
Safety and health of workers	Department of Occupational Safety and Health Act 1994 [14]	SDG 3 (Good Health)
Site housekeeping	Solid Waste and Public Waste Management Act 2007 [9]	SDG 3 (Good Health)
Vermin control	Destruction of Disease-Bearing Insects Act 1975 (Act 154) & Prevention and Control of Infectious Diseases Act 1988 (Act 342) [15]	SDG 3 (Good Health)
Heritage project special management	National Heritage Act 2005 [16]	SDG 3 (Good Health)
Disease management control	Prevention and Control of Infectious Diseases Act 1988 (Act 342) [15]	SDG 3 (Good Health)
	Infrastructure assessment	
Infrastructure management indicators	Legislation/Policy	Related Sustainable Development Goal, 2030 (SDG)
Information on stream diversion once project abandonment occurs	Chapter 5.0 Erosion and Sediment Control Facilities, Diversion channel, Guideline for Erosion and Sediment Control In Malaysia, 2010 [17]	
Method of blasting	Subchapter 6.6.1, Pre blast consideration, Malaysian Standard, (MS) 2318:2012 [18]	
Method of demolition	Chapter 6, Methods of demolition, Malaysian Standard (MS) 2318:2012 [18]	
Dilapidation survey	Subchapter 4.1.4, Dilapidation Survey, Malaysian Standard, (MS) 2318:2012[18]	SDG 9 (Innovation and Infrastructure) SDG 15 (Life on Land)
Buffer zone	Chapter 5.2, Erosion Control Facilities, Guideline for Erosion and Sediment Control In Malaysia, 2010 [17]	
Drainage management system	Subchapter 4.3.4, Drainage Control/Runoff Management, Guideline for Erosion and Sediment Control In Malaysia, 2010 [17]	
Type of noise barrier	Annexe E, Guidance on noise control, Guidelines for Environment Noise Limit and Control 2019 [7]	

4. Conclusion

This study has followed a rigorous research process using multiple embedded techniques to gain in-depth information on critical indicators based on legislation or policy that are also synchronised with Sustainable Development Goal (SDG) 2030. These 28 indicators that were selected will be used to develop a Sustainable Framework for Abandonment Plan Assessment. Moreover, these indicators are supported by related legislations, and the project proponents must comply with the requirements. Further research can also examine other factors like risk assessment, type of project, and best technology practices that can further refine the framework for pre-abandonment projects in Malaysia.

Acknowledgement
The researchers would like to thank the funder of this research project, Malaysia–Japan International Institute of Technology, University Teknologi Malaysia (UTM).

References

[1] Vega Falcón V, Castro Sánchez F and Romero Fernández AJ 2020 *Univ. soc.* **12** 207–216.
[2] Department of Environment Malaysia *Environmental Impact Assessment Guideline In Malaysia. 2016* (Putrajaya, Malaysia: Ministry of Natural Resources and Environment Malaysia)
[3] Sommer B, Fowler AM, Macreadie PI, Palandro DA, Aziz AC and Booth DJ 2019 *Sci. Total Environ.* **658** 973–981.
[4] Services IL 2018 *Environmental Quality Act, 1974* (Selangor: SS Graphic Printers (M) Sdn Bhd.)
[5] Zhang B, Wang H, Wang L, Mei G, Shi L, Xu N and Li J 2020 Large-scale field test on abandoned deep anhydrite mine-out for reuse as crude oil storage – A case study. Eng. Geol. **267** 105477
[6] Jabatan Kerja Raya Malaysia 2014 *Standard Specification For Building Works 2014* (Kuala Lumpur, Malaysia: Jabatan Kerja Raya).
[7] Department of Environment Malaysia 2019 *Guidelines for Environmental Noise Limits and Control* Retrieved from https://www.doe.gov.my
[8] Department of Environment Malaysia 2019 *The Planning Guidelines For Vibration Limits and Control In The Environment (*Putrajaya, Malaysia: Ministry of Natural Resources and Environment Malaysia)
[9] Department of Solid Waste Management 2018 *Solid Waste and Public Cleansing Management (Licensing) (Undertaking of Provision of Collection Services for Construction Solid Waste) Regulations. Waste Management* (Kuala Lumpur: Attorney General's Chamber)
[10] Malaysia Department of Environment 2017 *Guidelines On Land Disturbing Pollution Prevention and Mitigation Measures (LD-P2M2)* (Putrajaya: Polo Enterprise)
[11] Ministry of Housing and Local Government 2019 *National Cleanliness Policy*, Putrajaya.
[12] Department of Environment, Ministry of Environment and Water; https://emains.doe.gov.my/theme1/assets/EM_Directive_BI.pdf (retrieved 09 February 2021).
[13] Department of Safety and Health 1986 *Factories and Machinery (Building Operations and Works of Engineering Construction) (Safety) Regulations 1986.*
[14] Department of Safety and Health 1994 *Occupational Safety and Health Act 1994.*
[15] Ministry of Health Malaysia 1988 Prevention and Control of Infectious Diseases Act 1988 (Act 342) *The Commissioner of Law Revision Malaysia* (Kuala Lumpur, Malaysia).
[16] Department of National Heritage Malaysia 2006 *National Heritage Act 2005* (Kuala Lumpur, Malaysia).
[17] Department of Irrigation and Drainage Malaysia. 2010. *Guideline for Erosion and Sediment Control* (Kuala Lumpur, Malaysia: Department of Irrigation and Drainage (DID))
[18] Malaysia Standard 2012 *Malaysian Standard (MS 2318: 2012) Demolition of buildings - Code of practice* (First revision) (Kuala Lumpur, Malaysia).
[19] Department of Environment Malaysia 2017 *Environmental Essentials for Siting of Industries in Malaysia*, (October), 177.

High cadmium tolerance in *Stichoccocus*-like microalgae (*Tetratostichoccocus* sp. P1) from Malaysia

E Sahabudin[1]*, N Othman[3], I Suzuki[2],

[1]Department of Chemical and Environmental Engineering, Malaysia-Japan International Institute of Technology, Universiti Teknologi Malaysia, Jalan Sultan Yahya Petra, 54100 Kuala Lumpur, Malaysia.
[2]Graduate School of Life and Environmental Science, University of Tsukuba, 1-1-1 Tennodai, Tsukuba, Ibaraki, 305-8572, Japan.
[3]Department of Mechanical Precision Engineering, Malaysia-Japan International Institute of Technology, Universiti Teknologi Malaysia, Jalan Sultan Yahya Petra, 54100 Kuala Lumpur, Malaysia.

*Corresponding author e-mail: erisahabudin@gmail.com

Abstract. Cadmium (Cd) is a common industrial pollutant that has become a global issue due to its toxicity to living creatures, particularly aquatic organisms. Algal-based treatment offers cost-effective and environmentally friendly solutions for heavy metal removal. In this work, we studied the acid-tolerant microalgae *Tetratostichoccocus* sp. P1 isolated from a tropical peatland in Malaysia for its potential for Cd removal. The objective was to study the growth of *Tetratostichoccocus* sp. P1 strain cultivated in high-level Cd concentrations at pH 3.0. *Tetratostichoccocus* sp. P1 grew best in 20 μM and could survive in a concentration up to 100 M, according to the specific growth rate ($\mu = 0.36 \pm 0.05\ d^{-1}$) and the chlorophyll content (28.24 μg mL^{-1}). This strain was also highly resistant to Cd, evidenced by its half-maximal inhibitory concentration (IC_{50}) value, which was determined at 125 μM (14.8 mg L^{-1} Cd). This is the first study of its kind to demonstrate *Tetratostichoccocus* sp. P1's ability to absorb Cd at elevated concentrations under acidic conditions.

Keywords: *Tetratostichoccocus* sp. P1, cadmium uptake, heavy metal removal, acid-tolerant microalgae, acidic pH

1. Introduction

Industrial and human activities substantially impact the environment, resulting in the release of heavy metals into the environment. Only copper, zinc, manganese, and iron are utilized in trace amounts by the biota; others are regarded as non-essential and harmful [1]. Cadmium (Cd) is a highly toxic heavy metal that enters the environment through mining, battery disposal, and wastewater treatment. Cd is also commonly used in the galvanoplasty industry and semiconductor alloys for household products. Samadani et al. [2] estimated that wastewater treatment plants contribute 15% of the 6.4 tonnes of Cd discharged into the aquatic environment. Cd concentrations in severe ecosystems, such as acid mine drainages (AMDs), range from 300 g L^{-1} to 5000 g L^{-1}, with sulfur ions predominating [3]. In Malaysia, Cd levels in aquatic species were reported to range from 29.33 µg g^{-1} to 148.01 µg g^{-1} [4].

Pycoremediation can be defined as the use of microalgae to recover hazardous heavy metals from the contaminated medium. The use of microalgae in remediation is superior to conventional physicochemical techniques due to its advantages, such as fast recovery, low cost, high efficiency, and eco-friendliness [5]. For instance, in treating wastewater, phycoremediation produces less sludge, uses less energy, and minimizes greenhouse gas emissions [6].

In general, microalgae have high sorption capabilities, strong metal ion sorption selectivity, large surface areas, and high binding affinity to remove metals from aqueous solutions. However, not all microalgae can live in acidic and cadmium-contaminated environments (e.g., AMD, peatland, geothermal hot springs) [7]. Only acidophilic and acidic-tolerant eukaryotic algae can survive in such environments [8]. Among acidophilic microalgae that have been used for Cd and other heavy metals removal were *Spirulina platensis* (cultured in pH 1) [9], *Rhizoclonium hookeri* (pH 2) [10], *Desmodesmus* sp. and *Heterochlorella* sp. (pH 3.5) [11][10], and *Oscillatoria laete-virens* (pH 5) [12]. In terms of Cd removal percentage, acidophilic microalga *Euglena gracilis* was found to be capable of accumulating up to 50–75% of 2–25 µM CdCl at pH 3.5 [13]. *Desmodesmus* sp. MAS1 and *Heterochlorella* sp. MAS3 accumulated more than 58% of Cd in the presence of 2 mg L^{-1} in the medium of pH 3.5. It was also reported that at pH 7, *D. armatus* accumulated up to 341 µg g^{-1} Cd from 93 µM CdCl [14].

It is important to understand how microalgae respond to heavy metals in acidic conditions to help identify suitable strains for bioremediation purposes. In our preliminary work, *Stichocococcus*-like microalgae have been identified as a promising candidate for Cd removal. *Stichocococcus*-like microalgae are divided into seven genera [15], widely distributed, short-lived, and tolerant to a wide range of pH [16]. This green microalga was isolated from Malaysia's peatlands and has been registered as *Tetratostichoccocus* sp. P1 [GenBank: MT053478]. The objective of this study was to determine the algal strain's ability to live in various Cd concentrations at pH 3.0. This paper will be useful for future research on the cultivation of *Tetratostichoccocus* sp. P1 and its applications in highly acidic and Cd-contaminated conditions.

2. Materials and methods

2.1. Cd treatment and cultivation

Tetratostichoccocus sp. P1 was originally isolated from an acidic peatland in Selangor, Malaysia [17]. The acid-tolerant strain was cultivated in BG11 culture medium at pH 3.0 in a test tube under continuous illumination (70 µmol photons m^{-2} s^{-1}), aerated with 2% CO_2 (v/v) at a temperature of 30 ± 1 °C. A Cd stock solution (100 mM) was prepared by dissolving $CdCl_2$ in ultrapure water and filtered through a 0.22 µm sterile disposable syringe. Four different concentrations of Cd solutions were made: 20 µM (2.2 mg L^{-1}), 50 µM (5.6 mg L^{-1}), 100 µM (11.2 mg L^{-1}), and 200 µM (22.4 mg L^{-1}). Each Cd solution was mixed with 50 mL of BG11 in a 100-mL flask. There were triplicates of each sample.

2.2. Analysis of algal growth and chlorophyll content

Algal growth was determined by measuring the optical density (OD) at 730 nm using a UV-Vis spectrophotometer (UV-1900, Shimadzu, Japan) every 2 days. The specific algal growth rate, μ was determined at the exponential phase. For chlorophyll content analysis, cultures were pelleted by centrifugation at 15000 rpm for 3 minutes. The chlorophyll was extracted by 90% methanol. The content of chlorophyll-a and chlorophyll-b was determined at 655.2 nm and 652 nm, respectively, by using the UV-Vis spectrophotometer. Each measurement was conducted three times. The OD results were presented as averages with standard deviations, while the total chlorophyll was a sum of chlorophyll-a and chlorophyll-b. IC_{50} was examined by using OD values taken on day 9 (final OD).

3. Results and discussion

3.1. Effect of Cd concentrations on the growth and chlorophyll content

To investigate the effect of Cd concentrations on algal growth, Cd concentrations were ranged between 0 μM (control) and 200 μM. **Figure 1** shows that the growth of *Tetratostichoccocus* sp. P1 increased over time for all concentration levels except for 200 μM which was detrimental for this strain.). On day ninth, the strain reached 2.76 and 2. 61 at OD_{730} under 20 μM and 50 μM. At the same time, the control goes OD_{730} 2.31. The best growth was found at 20 μM (specific growth rate, μ: 0.36 ± 0.05 d^{-1}), followed by 50 μM (μ: 0.33 ± 0.06 d^{-1}), 0 μM (μ: 0.23 ± 0.05 d^{-1}), 100 μM (μ: 0.16 ± 0.03 d^{-1}), and 200 μM (μ: 0.02 ± 0.01 d^{-1}). These results indicate that *Tetratostichoccocus* sp. P1 was tolerant to Cd up to 100 μM and had an adverse effect at 200 μM.

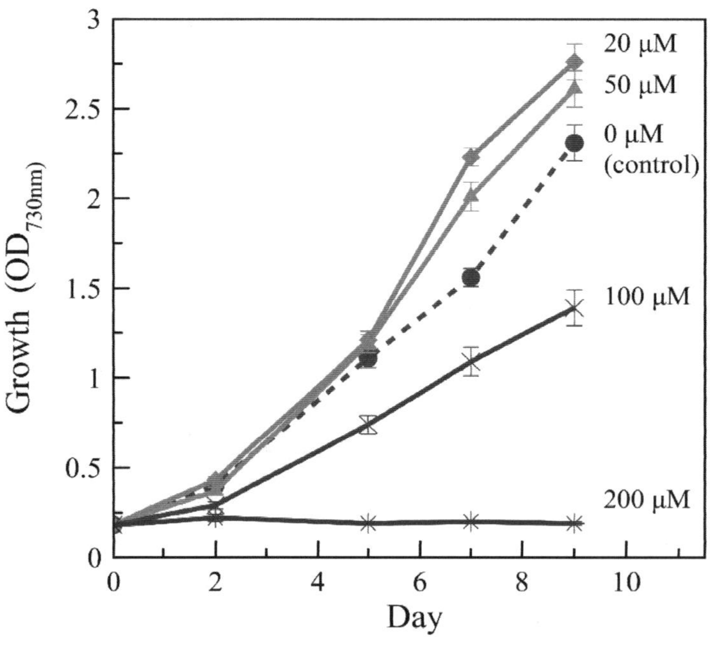

Figure 1. Response of *Tetratostichoccocus* sp. P1 in terms of optical density at 730nm to the stress imposed by different concentrations of Cd at pH 3.0.

The chlorophylls, a and b, are the pigments of photosynthesis. The primary role of chlorophyll is to absorb light energy into chemical energy for algal growth. **Figure 2** shows the total chlorophyll content for 20 μM and 50 μM Cd was comparable with the control (0 μM), implying that at this range (20–50 μM), Cd may not affect the strain's metabolism. In contrast, at 200 μM, the chlorophyll content in

Tetratostichoccocus sp. P1 was negatively affected, as evidenced by the decreasing chlorophyll content after day 2.

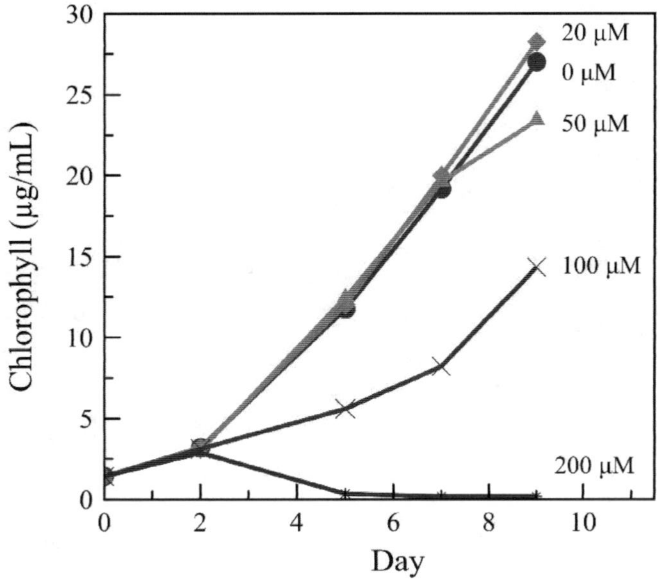

Figure 2. Total chlorophyll content in *Tetratostichoccocus* sp. P1.

3.2. IC_{50} of Tetratostichoccocus sp. P1

The tolerance of Microalagae to Cd can be assessed by measuring the IC_{50} (the concentration of Cd causing cell inhibition by 50%) or LC_{50} (lethal concentration 50%). After nine days of cultivation, results showed that the IC_{50} of *Tetratostichoccocus* sp. P1 was around 125 μM (14.8 mg L^{-1}), as shown in **Figure 3**. The algal growth was inhibited by only 39% compared to the control at 100 μM Cd. Interestingly, the growth of *Tetratostichoccocus* sp. P1 did not seem to be affected by the 20 μM and 50 μM Cd. In fact, the growth was better than the control, as mentioned earlier in Figure 2. The IC_{50} determined in this study was higher in comparison to similar past studies. For example, the IC_{50} for *Chlorella* sp. was found to be at 8 mg L^{-1} [1], *Chlorella sorokiniana* was 0.3 mg/*L* [18], IC_{50} for *Scenedesmus incrassatulus* was 7.36 mg L^{-1} [19], LC_{50} for *Ankistrodesmus* sp. was 5.43 mg L^{-1} [1], and LC_{50} for *Chlamydomonas acidophila* was 1.94 μ M [20]. Our findings have indicated that *Tetratostichoccocus* sp. P1 has a better potential for withstanding high concentration levels of Cd compared to other strains. It is noteworthy that environmental factors such as pH, temperature, light, cellular density, and culture medium are among the key factors for growth, uptake, and tolerance. In fact, experimental results can be different even for the same algal species and heavy metal types [21]. For instance, the pH of the medium is possibly the most critical factor affecting metal adsorption by microalgae. It has been reported that the tolerance of Cd increased from 12% to 51% when medium pH was decreased from 7 to 4 [2].

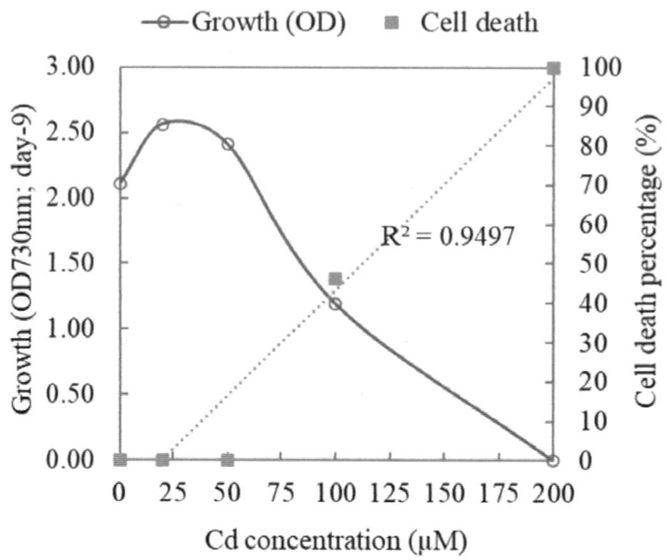

Figure 3. IC_{50} and mortality percentage for *Tetratostichoccocus* sp. P1.

Cd removal or accumulation by microalgae is generally conducted at neutral pH, and not all strains can grow under acidic and toxic conditions. According to Yu et al. [22], Cd concentrations in the contaminated-environment range between 5–50 μg L^{-1}. This study has demonstrated that *Tetratostichoccocus* sp. P1 can tolerate up to 11.2 mg L^{-1} (100 μM), which is 224 times higher than the typical concentrations found in the environment. Our findings showed that *Tetratostichoccocus* sp. P1 is a great candidate for Cd removal in aqueous conditions, given its high tolerance to Cd. Thus, it is also suggested that *Tetratostichoccocus* sp. P1 can be promising for other heavy metal removals as well.

3.3. Microalgae bioremediation mechanism

Tetratostichoccocus sp. P1 involves some particular mechanism to deal with Cd pollutants since it has high tolerance by showing fast growth and high IC_{50}. The inhibition indicated that high Cd concentration was accumulated in the cell and affected metabolism. At the same time, some Cd might be pumped out from the cell. *Tetratostichoccocus* sp. P1 may involve two possible mechanisms to deal with Cd pollutants: (i) Cd ion binding at cell surfaces and intracellular levels and (ii) metal ion transport. It has long been believed that the high metal-binding abilities of microalgae are attributed to their cell walls [23]. It has been suggested that when the cell walls or membranes block the metal cation uptake, this is the first step in heavy metal accumulation. The proteins, carbohydrates, and lipids found on the cell walls and membranes could react with metallic species [24]. Polyelectrolytes with charged groups such as carboxyl, phosphate, hydroxyl, or amine might be found in algal cell walls. The amphoteric characteristics of the algal cell walls are due to anionic and cationic sites. The functional groups are essential in the uptake of heavy metals by both living and non-living microalgal cells. Furthermore, these groups are protonated or deprotonated [5].

Metal ion transport into the cytoplasm is vital. Metal transporters in the plasma membrane help algae interact with their environment. These membrane transporters protect algal cells from being exposed to heavy metals. The study of algal metal transporter repertoires could shed light on a critical element of algal life [25]. When the concentrations of heavy metals are substantially higher than those in the intracellular, activated transport through the plasma membrane allows the movement of metal cations between the extracellular and intracellular compartments [26]. Due to the difference in concentration levels, Cd is transported and accumulates in the cells.

4. Conclusions

Tetratostichococus sp. P1, which was isolated from a tropical peatland in Malaysia, has been shown to be highly resistant to Cd under acidic conditions. Algal growth was seen to be significantly greater in the presence of Cd than in the absence of Cd, showing that this strain can survive in a Cd environment. *Tetratostichoccocus* sp. P1 grew best at 20 μM, with a specific growth rate (μ) of 0.36 ± 0.05 d^{-1} and chlorophyll content of 28.24 μg mL^{-1}, respectively. The growth of *Tetratostichoccocus* sp. P1 was inhibited at 200 μM, indicating that Cd was toxic to the strain at this concentration level. The IC_{50} for this strain was determined to be 125 μM (14.8 mg L^{-1} Cd), showing that it is very resistant to Cd when compared to other strains described in the literature. This is the first study of its kind to demonstrate the potential of *Tetratostichoccocus* sp. P1 to absorb Cd at different concentrations under acidic environments.

Acknowledgments The authors gratefully thank Prof. Hirofumi Hara (Graduate School of Agricultural and Life Sciences, The University of Tokyo Japan), Koji Iwamoto, Ali Yuzir, J.M. Kudzari. and Fatimah Azizah Riyadi (Universiti Teknologi Malaysia) for their excellent technical assistance.

Funding information AUN/SEED-Net supported this work under the Collaborative Education Program for Sustainable Environmental Engineering Network (CEP-SEEN) [Grant no: R.K130000.7343.4B403].

References

[1] Duque D, Montoya C and Botero L R 2019 Cadmium (Cd) tolerance evaluation of three strains of microalgae of the genus Ankistrodesmus, Chlorella and Scenedesmus *Rev. Fac. Ing.* 60–9

[2] Samadani M, Perreault F, Oukarroum A and Dewez D 2018 Effect of cadmium accumulation on green algae Chlamydomonas reinhardtii and acid-tolerant Chlamydomonas CPCC 121 *Chemosphere* **191** 174–82

[3] Cui M, Jang M, Cho S H, Khim J and Cannon F S 2012 A continuous pilot-scale system using coal-mine drainage sludge to treat acid mine drainage contaminated with high concentrations of Pb, Zn, and other heavy metals *J. Hazard. Mater.* **215–216** 122–8

[4] Salam M A, Paul S C, Mohamad Zain R A M, Bhowmik S, Nath M R, Siddiqua S A, Aka T Das, Iqbal M A, Kadir W R, Ahamad R B, Khaleque M A, Rak A E and Amin M F M 2020 Trace metals contamination potential and health risk assessment of commonly consumed fish of Perak River, Malaysia *PLoS One* **15** 1–18

[5] Suresh Kumar K, Dahms H U, Won E J, Lee J S and Shin K H 2015 Microalgae - A promising tool for heavy metal remediation *Ecotoxicol. Environ. Saf.* **113** 329–52

[6] Ahmad N, Mounsef J R, Tayeh J A and Lteif R 2020 Bioremediation of Ni, Al and Pb by the living cells of a resistant strain of microalga *Water Sci. Technol.* **82** 851–60

[7] Danouche M, El Ghachtouli N, El Baouchi A and El Arroussi H 2020 Heavy metals phycoremediation using tolerant green microalgae: Enzymatic and non-enzymatic antioxidant systems for the management of oxidative stress *J. Environ. Chem. Eng.* **8** 104460

[8] Hirooka, S. Hirose, Y. Kanesaki, Y. Higuchi, S. Fujiwara, T. Onuma, R. Era, A. Ohbayashi, R. Uzuka, A. Nozaki, H. Yoshikawa, H. Miyagishima S 2017 Acidophilic green algal genome provides insights into adaptation to an acidic environment *Proc. Natl. Acad. Sci.* **114** E8304–13

[9] Nithya K, Sathish A, Pradeep K and Kiran Baalaji S 2019 Algal biomass waste residues of Spirulina platensis for chromium adsorption and modeling studies *J. Environ. Chem. Eng.* **7** 103273

[10] Kayalvizhi K, Vijayaraghavan K and Velan M 2015 Biosorption of Cr(VI) using a novel

microalga Rhizoclonium hookeri: equilibrium, kinetics and thermodynamic studies *Desalin. Water Treat.* **56** 194–203

[11] Abinandan S, Subashchandrabose S R, Pannerselvan L, Venkateswarlu K and Megharaj M 2019 Potential of acid-tolerant microalgae, Desmodesmus sp. MAS1 and Heterochlorella sp. MAS3, in heavy metal removal and biodiesel production at acidic pH *Bioresour. Technol.* **278** 9–16

[12] Miranda J, Krishnakumar G and D'Silva A 2012 Removal of Pb2+ from aqueous system by live Oscillatoria laete-virens (Crouan and Crouan) Gomont isolated from industrial effluents *World J. Microbiol. Biotechnol.* **28** 3053–65

[13] Santiago-Martínez M G, Lira-Silva E, Encalada R, Pineda E, Gallardo-Pérez J C, Zepeda-Rodriguez A, Moreno-Sánchez R, Saavedra E and Jasso-Chávez R 2015 Cadmium removal by Euglena gracilis is enhanced under anaerobic growth conditions *J. Hazard. Mater.* **288** 104–12

[14] Pokora W, Baścik-Remisiewicz A, Tukaj S, Kalinowska R, Pawlik-Skowrońska B, Dziadziuszko M and Tukaj Z 2014 Adaptation strategies of two closely related Desmodesmus armatus (green alga) strains contained different amounts of cadmium: A study with light-induced synchronized cultures of algae *J. Plant Physiol.* **171** 69–77

[15] Pröschold T and Darienko T 2020 The green puzzle Stichococcus (Trebouxiophyceae, Chlorophyta): New generic and species concept among this widely distributed genus *Phytotaxa* **441** 113–42

[16] Sivakumar G, Jeong K and Lay J O 2014 Bioprocessing of Stichococcus bacillaris strain siva2011 *Biotechnol. Biofuels* **7** 1–9

[17] Sahabudin E, Lee J, Asada R, Atikah E and Nurtasbiyah M 2022 Isolation and characterization of acid - tolerant Stichococcus - like Microalga (Tetratostichococcus sp . P1) from a tropical peatland in Malaysia *J. Appl. Phycol.*

[18] Fatima N, Kumar V, Vlaskin M S, Jaiswal K K, Jyoti, Gururani P and Kumar S 2020 Toxicity of cadmium (Cd) on microalgal growth, (IC50 value) and its exertions in biofuel production *Biointerface Res. Appl. Chem.* **10** 5828–33

[19] Peña-Castro J M, Martínez-Jerónimo F, Esparza-García F and Cañizares-Villanueva R O 2004 Heavy metals removal by the microalga Scenedesmus incrassatulus in continuous cultures *Bioresour. Technol.* **94** 219–22

[20] Díaz S, de Francisco P, Olsson S, Aguilera Á, González-Toril E and Martín-González A 2020 Toxicity, physiological, and ultrastructural effects of arsenic and cadmium on the extremophilic microalga Chlamydomonas acidophila *Int. J. Environ. Res. Public Health* **17**

[21] Ouyang H L, Kong X Z, He W, Qin N, He Q S, Wang Y, Wang R and Xu F L 2012 Effects of five heavy metals at sub-lethal concentrations on the growth and photosynthesis of Chlorella vulgaris *Chinese Sci. Bull.* **57** 3363–70

[22] Yu Z, Wei H, Hao R, Chu H and Zhu Y 2018 Physiological changes in: Chlamydomonas reinhardtii after 1000 generations of selection of cadmium exposure at environmentally relevant concentrations *Environ. Sci. Process. Impacts* **20** 923–33

[23] Singh D V, Bhat R A, Upadhyay A K, Singh R and Singh D P 2021 Microalgae in aquatic environs: A sustainable approach for remediation of heavy metals and emerging contaminants *Environ. Technol. Innov.* **21** 101340

[24] Monteiro C M, Castro P M L and Malcata F X 2011 Capacity of simultaneous removal of zinc and cadmium from contaminated media, by two microalgae isolated from a polluted site *Environ. Chem. Lett.* **9** 511–7

[25] Danouche M, El Ghachtouli N and El Arroussi H 2021 Phycoremediation mechanisms of heavy metals using living green microalgae: physicochemical and molecular approaches for enhancing selectivity and removal capacity *Heliyon* **7** e07609

[26] Monteiro C M, Castro P M L and Malcata F X 2012 Metal uptake by microalgae: Underlying mechanisms and practical applications *Biotechnol. Prog.* **28** 299–311

The effect of LEDs on *Galdieria sulphuraria* cultivation at indoor and outdoor condition.

N Jusoh[1], Y S Lee[1], Y C Teo[1], F S Othman[1], S E Mohamad[1]

[1] Malaysia-Japan International Institute of Technology, Universiti Teknologi Malaysia, Jalan Sultan Yahya Petra, 54100, Kuala Lumpur, Malaysia

*Corresponding author: noorshartika@utm.my

Abstract. Conventional methods for recovering heavy metals from industrial waste stream often involve significant energy consumption and chemical use. Hence, using microalga as a biosorbent material can be a low-cost and eco-friendly solution. *Galdieria sulphuraria* is a red microalga which is known to be capable of recovering heavy metals. This algal species is resistant to extreme temperatures and acidic environment. It is also well-established that light emitting diodes (LED) is effective at promoting algal growth. Our research objectives were to study the effects of LED lightings (red, blue, green, and white) on the growth of *G. sulphuraria* and to compare the growth in indoor and outdoor cultivations. In this work, the culture medium was the modified 2x Allen's with pH 2.5. To promote the rapid growth of *G. sulphuraria,* the culture medium was supplemented with 2% CO_2. Algal growth was determined by measuring the optical density at 750nm (OD750) using UV-vis spectrophotometer. Results have shown that in indoor cultivation, the growth rate of *G. sulphuraria* was highest under red light (μ= 0.06 day^{-1}) and lowest under blue light (μ= 0.02 day^{-1}). For outdoor cultivation, algal growth rate increased by 2 to 6-fold higher compared to that in indoor cultivation. Algal cultivation in outdoors was best with white light (OD = 2.46 ± 0.09; μ = 0.13 day^{-1}). Although the use of LEDs enhanced algal growth, we found it did not make a significant difference statistically when compared to the set without LEDs. This may be attributed to the Malaysia's environment (photoperiods, temperatures, etc.). This paper will serve as a foundation for future research on cultivating *G. sulphuraria* indoors and outdoors using various LED lights.

Keywords: Algal growth; LED; artificial light, *Galdieria sulphuraria;* tropical environment

1. Introduction

This century has seen a significant increase in the production of industrial waste. Toxic components in these industrial wastes will contaminate the environment and create disposal issues [1]. Having said that, some of the wastes contain precious metals such as gold and platinum which can be used for other purposes [2]. Pyrometallurgy and hydrometallurgy are two established techniques for recovering valuable metals from various waste streams. Pyrometallurgy is metals using high temperatures, which involves roasting, smelting, and melting [3] which is necessary to achieve desired conditions [4]. Hydrometallurgy is a more cost-effective and environmentally friendly process, but it requires more time to achieve high metal recovery efficiency [5]. As a result, scientists are exploring for low-cost, more efficient, and environmentally friendly methods such as biohydrometallurgy, which employs microorganisms in the process. *G. sulphuraria* has been recognized as a promising algal species for metal recovery from wastewater due to its adaptability to a wide variety of conditions [6]. *G. sulphuraria* can grow in acidic and hot environments (up to 65 °C) [7]. In addition, according to Selvaratnam [8], the biomass yield of *G. sulphuraria* is 10% higher compared to other algal strain. It can survive in a toxic metal environment which most of the living organisms cannot grow. With these extremophilic properties, *G. sulphuraria* is an outstanding candidate for heavy metals adsorption from various waste streams [9].

Algal growth and the photosynthesis process is highly dependent on light conditions. Chlorophyll and carotenoid are the two major categories of photosynthetic pigments [10]. Generally, chlorophyll and carotenoid absorb the greatest amount of light in the red and blue wavelength regions [11]. Besides sunlight, light emitting diodes (LED) can also be a great source of light; LEDs have already been shown to improve algae growth [12]. When compared fluorescent lights, LEDs are more cost effective due to its low-wattage irradiance, less heat energy, longer shell life, and higher conversion [13-15]. In this study, the authors investigated the impact of red, blue, green, and white LEDs on the growth of *G. sulphuraria*. The growth of algae indoors and outdoors were studied and compared. This work will add to the current research into the use of LED lighting for algae cultivation in tropical regions.

2. Materials and Methods

2.1. Cultivation medium for G. sulphuraria

According to Minoda [16], the suitable cultivation medium for *G. sulphuraria* is the modified 2× Allen's medium. The following recipe was used to prepare the culture medium: $(NH_4)_2SO_4$, 2.62 g/L; KH_2PO_4, 0.54 g/L; $MgSO_4{\bullet}H_2O$, 0.50 g/L; $CaCl_2{\bullet}2H_2O$, 0.28 g/L; $FeCl_3{\bullet}6H_2O$, 4 g/L; $MnCl_2{\bullet}4H_2O$, 1.80 g/L; $ZnCl_2$, 0.105 g/L; $NaMoO_2{\bullet}2H_2O$, 0.39 g/L; $CoCl_2{\bullet}6H_2O$, 0.04 g/L; $CuCl_2$, 0.043 g/L. The modified 2x Allen's medium was adjusted to pH 2.5 using dilute H_2SO_4. *G. sulphuraria* was inoculated in a 50 mL modified 2x Allen's medium using a 100 mL conical flask.

2.2. Culture of G. sulphuraria

The experiments were carried for three weeks in indoor and outdoor conditions (**Table 1**). For indoor cultivation, LEDs were used as the light source throughout the experiment, and the control set was kept in the dark. For outdoor cultivations, the cultures were exposed to natural sunlight during the day and LEDs were turned on for only 12 hours (7pm–7am). In Malaysia, the length of the day and night are nearly equal because it is located close to the Equator. A thermo-recorder (TR-71wf, T&D Corporation, Japan) was used to measure the temperature of the outdoor cultivation, which ranged between 26–45°C.

Table 1. Experimental design for indoor and outdoor cultivations.

Experiment A	Culture set	LED color	Photoperiod (24 h)	
Indoor cultivation	A1	Blue	LED	
	A2	Green	LED	
	A3	Red	LED	
	A4	White	LED	
Experiment B	Culture set	LED color	Photoperiod (12h: 12h)	
Outdoor cultivation	B1	Blue	Sunlight	LED
	B2	Green	Sunlight	LED
	B3	Red	Sunlight	LED
	B4	White	Sunlight	LED
	Control	-	Sunlight	-

Figure 1 shows the LED strip, and setups in the indoors and outdoors. Each culture set had a triplicate. For outdoor experiments, the cultures were covered by a 95% shading net to reduce the impact of direct sunlight. The type of LEDs used in these experiments were light strips (SMD 5050, Lightstec, China). Four LED

colors were used, namely, red, blue, green, and white. The LED strips were installed beneath the culture rack. A portable pyranometer (MP-200, Apogee Instruments, Canada) was used to measure the light intensity of the LEDs, which was determined to be 50–70 mol m^{-2} s^{-1} at the surface of culture vessel. To promote the rapid growth of *G. sulphuraria*, all cultures were also supplemented with 2% CO_2.

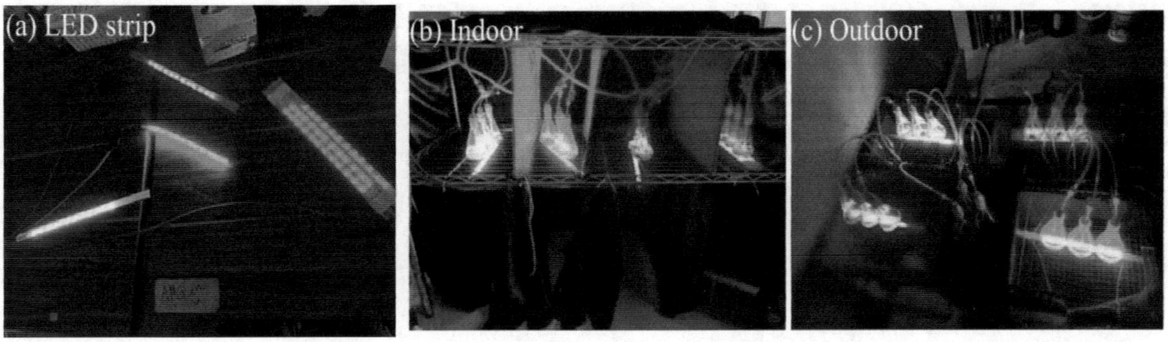

Figure 1. Experimental setups for (a) indoor cultivation and (b) outdoor cultivation, with four different LED colors.

2.3. Calculation

Algal growth was determined by reading the optical density (OD) at 750 nm using a UV-Vis spectrophotometer (UV-1900, Shimadzu, Japan) every 3–4 days. The results were presented as averages with standard deviations. The specific growth rate of *G. sulphuraria* can be calculated by using **Equation 1**. From the graph and calculation, we can obtain the maximum specific growth rate, μ_{max} of *G. sulphuraria*.

$$\mu\mathbf{max} = \frac{\boldsymbol{ln}\frac{\boldsymbol{OD_2}}{\boldsymbol{OD_1}}}{\boldsymbol{t_2 - t_1}} \quad (1)$$

The T-test [two-tailed sample assuming unequal variances; statistical software, Excel (Microsoft)] was used to evaluate significant variations in the OD of *G. sulphuraria* between two growth conditions. If the *p*-values were less than 0.05, the results were considered significant.

3. Result and Discussion

3.1. Effect of LED lights on algae growth

3.1.1. Indoor cultivation

The initial OD taken on the first day of cultivation was 0.255. The OD values increased steadily from day 1 until day 20. Results showed that the growth rate of *G. sulphuraria* was highest under red light (OD = 1.315 ± 0.33; μ= 0.06 day^{-1}), followed by white (1.148 ± 0.11; μ= 0.04 day^{-1}), green (0.965 ± 0.18; μ= 0.03 day^{-1}), and blue (0.898 ± 0.24; μ= 0.02 day^{-1}), as shown in **Figure 2**.(a) This is in line with a previous study on *Chlorella kessleri* by Koc et al. [17], which found that the number of cells and biomass were best when cultured under red LED, and it was recommended that red light could be used to produce the desired cell concentrations. Additionally, Yan et al. [18] proposed that red wavelength light has the optimal effect on the biomass of *Chlorella* sp.

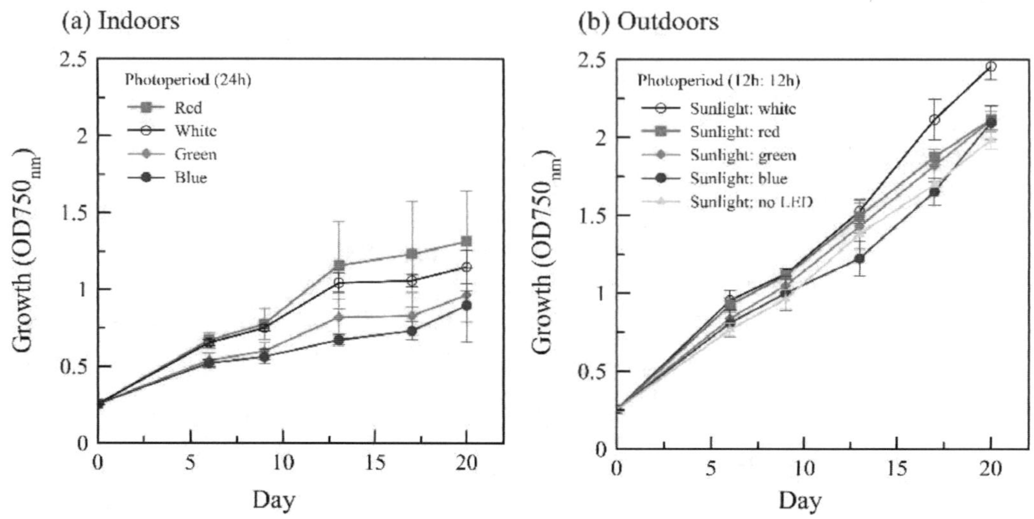

Figure 2. The growth of *G. sulphuraria* in (a) indoor and (b) outdoor cultivations.

The performance of red light can be linked to the dominant pigment produced in *G. sulphuraria* [19]. *G. sulphuraria* has potential to produce phycocyanin which is a bluish pigment found in the cytoplasm of microalgae. It has been reported that in the exponential phase, *G. sulphuraria* contains 2–4 mg phycocyanin per gram of total dry weight [20]. Since the pigment of *G. sulphuraria* appears in blue green [21] the respective color absorbed by the algal pigment is red. This can be understood by a complementary color scheme [12] as shown in **Table 2.** This also explains why red light is more favored by *G. sulphuraria* compared to other lights. Despite the use of blue light being recommended to promote algal growth [22], our results showed the opposite; algal growth was lowest with blue light. In addition to phycocyanin (blue pigment). *G. sulphuraria* also produces phycoerythrin (red pigment), but in small amounts. This explains why blue and green lights were not absorbed efficiently by *G. sulphuraria.*

Table 2. Complementary color scheme.

Color observes	Wavelength/color absorbed by pigments
Yellow green	400–435 nm/ violet
Yellow	435–480 nm/ blue
Orange	480–490 nm/ green blue
Red	490–500 nm/ blue green
Purple	500–560 nm/ green
Violet	560–580 nm/ yellow green
Blue	580–595 nm/ yellow
Green blue	595–605 nm/ orange
Blue green	605–700 nm/ red

3.1.2. Outdoor cultivation

Figure 2b shows that algal growth rate in outdoor cultivation was higher by 2 to 6-fold compared to indoor cultivation. This indicates that the outdoor settings were more favorable for *G. sulphuraria.* Algal growth in outdoors was found to be the highest with white (OD = 2.458 ± 0.09; μ= 0.13 day^{-1}), followed by red (2.120 ± 0.08; μ= 0.09 day^{-1}), green (2.112 ± 0.06; μ= 0.10 day^{-1}), and blue (2.097 ± 0.11; μ= 0.12 day^{-1}) LEDs. The alternation of sunlight with white light was more favorable to algal development than the other colors. White light is a combination of wavelengths of different colors, ranges between 400 nm and 700 nm. In other words, the algae were exposed to multiple wavelengths at the same time. It was also expected that the growth in a control set (without LED) to be low since it was fully dependent on the sunlight. Nonetheless, the OD values of the control set were comparable to those sets supplemented with LEDs. This implies that the use of shading helped reduce the excessive intensity of sunlight which potentially inhibit algal growth.

Conclusion

We have successfully demonstrated that *G. sulphuraria* can grow under different light conditions both indoors and outdoors. The best algal cultivation indoors was achieved with red light (OD 1.315 ± 0.33; μ = 0.06 day^{-1}), while that in the outdoors was achieved with white light (OD 2.46 ± 0.09; μ = 0.13 day^{-1}). The growth of *G. sulphuraria* in outdoor cultivation was increased by 2 to 6-fold compared to that in indoor cultivation. The outdoor cultivation was also benefited from the natural sunlight, tropical temperatures (26–45°C), and daylight photoperiod (12 hours). It can also be suggested that being dependent on natural sunlight was sufficient for algal cultivation in Malaysia's environment because the use of artificial lights did not make a significant difference statistically. Future outdoor cultivation studies may explore different photoperiods as well in addition to LED lights effects on the growth of *G. sulphuraria.*

Acknowledgement

The authors gratefully acknowledge the International Industry Incentive Grant from Universiti Teknologi Malaysia (Q.K130000.3643.03M39 Enhancement of Algal Growth by LED Light Supplement). The author is sincerely thankful for the great contribution from Assoc. Prof. Dr. Koji Iwamoto, Assoc. Prof. Dr. Norhayati Abdullah and Dr. Jauharah Md Khudzari.

References

[1] Jadhav, U. and H. Hocheng, *A review of recovery of metals from industrial waste.* Journal of Achievements in Materials and Manufacturing Engineering, 2012. **54**(2): p. 159-167.
[2] Canda, L., T. Heput, and E. Ardelean. *Methods for recovering precious metals from industrial waste.* in *IOP Conference Series: Materials Science and Engineering.* 2016. IOP Publishing.
[3] Mishra, S. and R. Dattab, *Encyclopedia of Materials.* Science and Technology, 2nd ed., Ed. KHJ Buschow, RW Cahn, MC Flemings, B. Ilschner, EJ Kramer, S. Mahajan, and P. Veyssière, Elsevier, 2001: p. 2761-68.
[4] Carelse, C., et al., *Au and Ag distribution in alloys produced from the smelting of printed circuit boards-an assessment using SEM-EDS, EPMA, and LA-ICP-MS analysis.* Journal of the Southern African Institute of Mining and Metallurgy, 2020. **120**(3): p. 203-210.
[5] Al-Harahsheh, M. and S. Kingman, *Microwave-assisted leaching—a review.* Hydrometallurgy, 2004. **73**(3-4): p. 189-203.
[6] Graziani, G., et al., *Microalgae as human food: chemical and nutritional characteristics of the thermo-acidophilic microalga Galdieria sulphuraria.* Food & function, 2013. **4**(1): p. 144-152.
[7] Yoshimura, E., et al., *Mechanism of aluminium tolerance in Cyanidium caldarium.* Hydrobiologia, 2000. **433**(1): p. 57-60.
[8] Selvaratnam, T., et al., *Feasibility of algal systems for sustainable wastewater treatment.* Renewable Energy, 2015. **82**: p. 71-76.

[9] Čížková, M., M. Vítová, and V. Zachleder, *The red microalga Galdieria as a promising organism for applications in biotechnology.* Microalgae-From physiology to application, 2019.
[10] Ra, C.H., et al., *Effects of light-emitting diode (LED) with a mixture of wavelengths on the growth and lipid content of microalgae.* Bioprocess and biosystems engineering, 2018. **41**(4): p. 457-465.
[11] Richmond, A. and Q. Hu, *Handbook of microalgal culture: applied phycology and biotechnology.* 2013: John Wiley & Sons.
[12] Selim, Y. and A. Mohamed, *Role of dyestuff in improving dye-sensitized solar cell performance.* Renewable Energy and Sustainable Development, 2017. **3**(1): p. 79-82.
[13] You, T. and S.M. Barnett, *Effect of light quality on production of extracellular polysaccharides and growth rate of Porphyridium cruentum.* Biochemical Engineering Journal, 2004. **19**(3): p. 251-258.
[14] Wang, C.-Y., C.-C. Fu, and Y.-C. Liu, *Effects of using light-emitting diodes on the cultivation of Spirulina platensis.* Biochemical Engineering Journal, 2007. **37**(1): p. 21-25.
[15] Chen, W., M. Sommerfeld, and Q. Hu, *Microwave-assisted Nile red method for in vivo quantification of neutral lipids in microalgae.* Bioresource technology, 2011. **102**(1): p. 135-141.
[16] Minoda, A., et al., *Recovery of rare earth elements from the sulfothermophilic red alga Galdieria sulphuraria using aqueous acid.* Applied microbiology and biotechnology, 2015. **99**(3): p. 1513-1519.
[17] Koc, C., G.A. Anderson, and A. Kommareddy, *Use of red and blue light-emitting diodes (LED) and fluorescent lamps to grow microalgae in a photobioreactor.* 2013.
[18] Yan, C., et al., *Effects of various LED light wavelengths and light intensity supply strategies on synthetic high-strength wastewater purification by Chlorella vulgaris.* Biodegradation, 2013. **24**(5): p. 721-732.
[19] Gross, W. and C. Schnarrenberger, *Heterotrophic growth of two strains of the acido-thermophilic red alga Galdieria sulphuraria.* Plant and Cell Physiology, 1995. **36**(4): p. 633-638.
[20] Sloth, J.K., M.G. Wiebe, and N.T. Eriksen, *Accumulation of phycocyanin in heterotrophic and mixotrophic cultures of the acidophilic red alga Galdieria sulphuraria.* Enzyme and Microbial Technology, 2006. **38**(1-2): p. 168-175.
[21] Kuddus, M., et al., *Recent developments in production and biotechnological applications of C-phycocyanin.* BioMed research international, 2013. **2013**.
[22] Teo, C.L., et al., *Enhancing growth and lipid production of marine microalgae for biodiesel production via the use of different LED wavelengths.* Bioresource technology, 2014. **162**: p. 38-44.

IOP Conf. Series: Earth and Environmental Science 1091 (2022) 012047 doi:10.1088/1755-1315/1091/1/012047

Potential Application and Future Prospects of Algal-Bacterial Aerobic Granular Sludge for Wastewater Treatment

L D A Purba[1,*], K Shimizu[2], J Hermana[3]

[1] Malaysia-Japan International Institute of Technology (MJIIT), Universiti Teknologi Malaysia (UTM), Jalan Sultan Yahya Petra, Kuala Lumpur, 54100, Malaysia
[2] Graduate School of Life and Environmental Sciences, University of Tsukuba, Ibaraki, 305-8572, Japan
[3] Department of Environmental Engineering, Faculty of Civil, Planning and Geoengineering, Sepuluh Nopember Institute of Technology, Surabaya, 60111, Indonesia

*Corresponding author e-mail: amalia.l@graduate.utm.my

Abstract. Algal and bacterial consortia have a high potential to enhance wastewater treatment efficiencies. Algae have been embedded into aerobic granulation process allowing interactions between algal and bacterial cells within the sludge whereby granular flocs were successfully formed. The unique structure of algal-bacterial aerobic granules leads to easier separation between sludge and treated wastewater. The bibliographic data shows increasing interest in the field of algal-bacterial aerobic granular sludge due to the high treatment efficiency in treating different types of wastewaters, including those containing heavy metals and antibiotics. Moreover, the algal-bacterial aerobic granules may be harvested by the end of treatment process for extraction of biolipid for possible resource recovery in form of biofuel. The applications and resource recovery of this system is thoroughly elucidated in this article.

1. Introduction

The integration of microalgae or algae and bacteria in wastewater treatment system was originated to improve the treatment efficiency, whereby algae could co-work with heterotrophic bacteria in removing excess nutrients and organic matters [1]. The interaction between algae and bacteria is divided into three categories, namely commensalism, parasitism and mutualism interactions [2–4]. However, in wastewater treatment system, mutualism interaction is desired to achieve best treatment performance.

The mutualism interaction of algae and bacteria is based on exchange of substrates, cell-to-cell communication and horizontal gene transfer [4]. In the process of substrate exchange, microalgae produce O_2 through photosynthesis that can be used by bacterial cells, whereas bacteria produce CO_2 through respiration that help in sustaining algae growth [5]. The interaction between O_2 and CO_2 alone may reduce the capital cost for oxygenation and also reduce the greenhouse gas emission of wastewater treatment plants [6]. Meanwhile, in cell-to-cell communication, it was recently found that the chemical mediators in quorum sensing activity, i.e. N-acyl homoserine lactones, autoinducers 2 and autoinducing peptides are not limited for intra-bacterial communications, instead, these chemical mediators may be involved in communications between algae and bacteria [7]. Despite the recent finding on interkingdom signaling between algae and bacteria, there are still huge missing information to understand the underlying mechanisms of the signaling communications between these organisms. Therefore, intensive study is needed to expand the knowledge of this matter, which is useful to provide economic advantages of this system.

Algal-bacterial aerobic granular sludge is the granular form of algae-bacteria consortia. Due to the superior characteristics of aerobic granular sludge as compared to the conventional activated sludge, algae-bacteria consortia in granular forms has numerous advantages, including lower energy demands, high nutrient removal efficiency [8], simple microalgae harvesting process and high potential for resource recovery [9]. Nevertheless, algal-bacterial interaction in granular form is a considerably new topic of research that requires extensive experimental proof and comprehensive discussion. This article elucidates the bibliographic data on algal-bacterial aerobic granular sludge, along with the application in wastewater treatment system and its prospect for resource recovery purposes.

2. Bibliographic data of algal-bacterial aerobic granular sludge

The bibliographic data of algal-bacterial aerobic granular sludge was obtained from Scopus database as it serves as one of the largest databases for scientific articles. The search string resulted in 27 research articles on algal-bacterial aerobic granular sludge within 2015 to 2021 (data was obtained on 1st December 2021). The number of published research articles were insignificant for the first few years and increased sharply from 2018 onwards, as depicted in Figure 1. From the available data, it is expected that the number of publications on algal-bacterial aerobic granular sludge will continue to increase following the research trends. It is worth noting that majority of published research articles were originated from China, Japan and Taiwan, as shown in Figure 2. This was supported by data on funding sponsor which highlighted top three funders as Japan Society for the Promotion of Science, Natural Science Foundation of China and University of Tsukuba. These indicates that the respective countries have dedicated tremendous efforts in advancing the research on algal-bacterial aerobic granules for wastewater treatment.

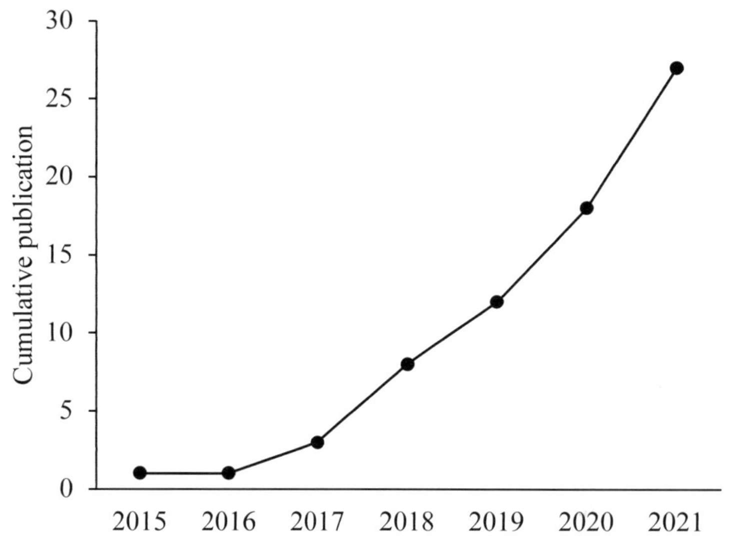

Figure 1. The number of cumulative publications on algal-bacterial aerobic granular sludge.

In terms of the top affiliated institutes, University of Tsukuba owns 15 published research articles, which was distantly followed by National Taiwan University (9) and Chinese Academy of Science (4). However, most affiliation were mainly consisted of Asian institutes that might indicate the leading of Asian countries in this research field. The countries outside Asia region include Canada, Netherlands, Poland and Spain that have ventured into the field of algal-bacterial aerobic granular sludge. Figure 2 shows the co-authorship relationship between the highest producing countries, whereby dominance of Asian countries was highlighted as red color. A strong collaboration between Japan, China and Taiwan was indicated by high link strength. Japan has collaborated with Taiwan in 9 research articles while Japan and China have collaborated in 7 articles. Besides, China and Taiwan were related with a total of 3 research articles. Apart from the top 3 countries, Netherlands and Canada have also made collaboration with the other countries. These highlighted countries were appeared to be the leaders in the field of aerobic granular sludge [10]. Nevertheless, some countries, including Spain, Poland and India have not performed collaboration with other countries in algal-bacterial aerobic granular sludge. Research articles in this field were successfully published in numerous journals, in which among highest producing journals were Bioresource Technology (8), Bioresource Technology Reports (4), Journal of Hazardous Materials (4) and Journal of Water Process Engineering (4). These bibliographic data were hoped to intrigue the researchers for publication and collaboration opportunities.

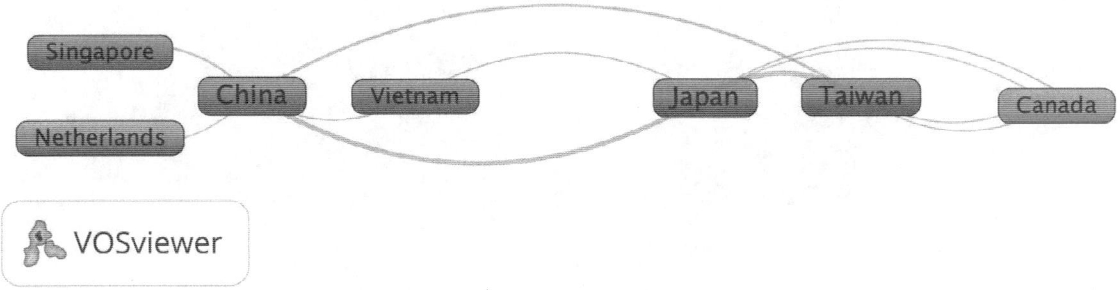

Figure 2. A bibliometric map on countries co-authorship on research articles of algal-bacterial aerobic granular sludge (URL: https://bit.ly/3rxpHwq).

3. Applications of algal-bacterial aerobic granular sludge

3.1 Development mechanisms of algal-bacterial aerobic granular sludge
Self-aggregated algal-bacterial aerobic granular sludge is commonly generated in photobioreactor modified with sequencing batch operation mode, called photo-sequencing batch reactor (PSBR) [9]. The matured granular size may be ranging from 0.2 to 5 mm depending on the operational conditions. Because of its considerably large size, the granular sludge can be harvested easily without the aid of chemical flocculant, thus, reducing the cost of downstream processes up to 30% [4]. A proposed mechanism justified that bacteria may act as the initiator during the granulation process, whereby bacteria aids in protecting algal cells from possible harmful species [11]. Moreover, bacteria readily settle algal cells in suspension form. The extracellular polymeric substance (EPS) produced by bacterial cells was also found to help in granular maturation process, thus enhancing the settleability of the granules [12]. The other way around, breakdown of algal cells also contribute to granulation process as it supplies more nutrients for bacterial growth [13]. Figure 3 depicts the development of algal-bacterial aerobic granular sludge as reported by Tiron et al. (2017) [14].

The granulation process of algal-bacterial consortia is determined by numerous factors, including seed strains, inorganic/organic carbon concentration, reactor configuration, light intensity and external aeration [3]. During development of algal-bacterial aerobic granular sludge, there are two major approaches, firstly with the addition of specific algae species and secondly by using indigenous algae in wastewater. A range of algae strain was studied, including *Leptolyngbya*, *Chlorella vulgaris* and *Scenedesmus obliquus* [17, 18]. Cai *et al.* (2019) successfully synthesized algal-bacterial aerobic granular sludge with excellent settling velocity of 12.2 m/h by inoculating *Leptolyngbya* into the seed sludge. Another study by Liu *et al.* (2017) reported that inoculated species remained dominant in matured granules. However, a competition for occupying the granular surface is expected between photoheterotrophic bacteria and algae species. Nevertheless, it is still difficult to deduce the conclusion on the effect of seed strains towards the performance of algal-bacterial aerobic granular sludge and its further downstream application.

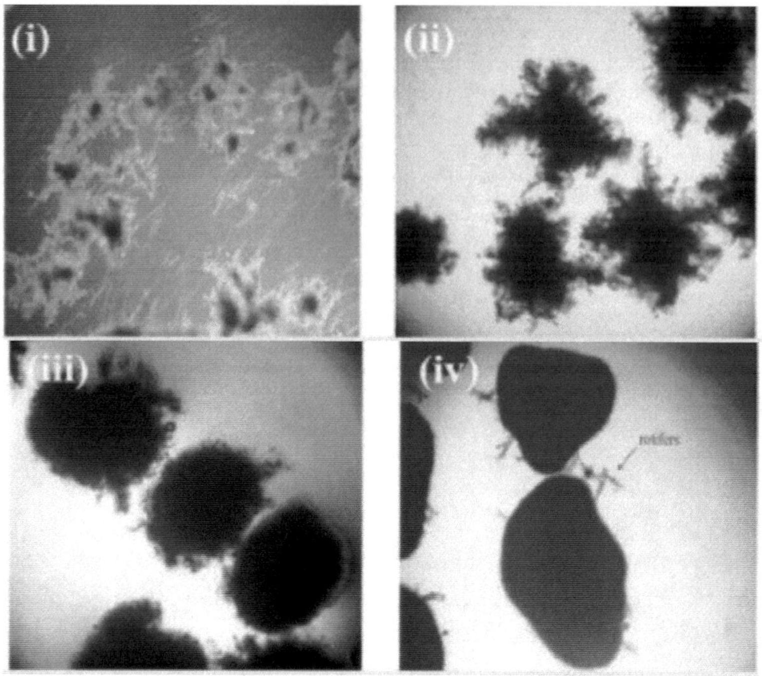

Figure 3. Development of microalgae-bacteria aerobic granular sludge; (i) microalgae bioflocs, (ii) and (iii) activated microalgae aggregates, (iv) mature microalgae-bacteria aerobic granular sludge (source: Tiron et al. (2017) [14]).

Development of algal-bacterial aerobic granular sludge has been reported using various types of wastewater, including synthetic, municipal and industrial wastewaters [19, 20]. It was discovered that high ratio of organic carbon (i.e., glucose) to inorganic carbon sources (i.e., acetate) may enhance the aggregation process of algal-bacterial aerobic granules [18]. This is because of the growth inhibition of toxic filamentous cyanobacteria in the system. The effect of various carbon sources on granular formation has also been extensively studied [19]. The presence of multivalent ions in wastewater, such as Ca^{2+} may help granular formation via neutralization of cell surface charge. However, further study is required to clarify the detailed effects on multivalent cations on formation of algal-bacterial aerobic granular sludge.

Moreover, illumination is one of the important factors in developing algal-bacterial granular sludge as it is needed for photosynthesis process in microalgae cells. Light may be considered as a substrate as well as inhibitors for microalgae growth, whereby too intense illumination may lead to photoinhibition affecting the overall nutrient removal efficiencies in wastewater treatment system [20]. A wide range of light intensities have been applied for algal-bacterial aerobic granular sludge, including high intensity of more than 20,000 lx, intermediate strength ranging between 2,000 to 20,000 lx and low strength of less than 2,000 lx [17, 19]. Variety of light sources also have been used, i.e. natural light, LED light and indoor light [21,22]. Moreover, the light/dark cycle was also tested from 12h/12h to 24h/00h. However, recent study discovered that a minimum of 8,000 lx light intensity is favorable in the development of algal-bacterial granular sludge [23]. Huang *et al.* (2015) investigated the impact of natural sunlight on development of aerobic granular sludge and it was concluded that under natural sunlight, the algal-bacterial aerobic granular sludge has a looser structure [24]. Although the natural sunlight is the most preferable choice due to its cost-effective operations, the fact that natural sunlight is not stable and changes depending on weathers and seasons may hinder its application in development of algal-bacterial aerobic granular sludge.

Algae are known to supply oxygen to bacterial cells through photosynthesis process. However, studies revealed that the oxygen from algal cells to be insufficient for bacterial use [23]. Therefore, addition of external aeration may be needed although, ideally, external aeration should be kept at minimum level to fully utilize the use of algal-bacterial aerobic granular sludge in wastewater treatment. Ahmad *et al.* (2017) applied an air flow rate of 0.5 cm/s in the continuous flow reactors and the results highlighted that algal-bacterial aerobic granular sludge can sustain its structures for long term operation [17]. It was concluded that the algal-bacterial aerobic granular sludge has a compact structure and can withstand the shear force from external aeration.

Numerous other factors affecting the development of algal-bacterial aerobic granular sludge include temperature and pH values. Room temperature and neutral pH are typically monitored in previous studies. However, the effect of high or low pH needs to be investigated to explore the feasibility of algal-bacterial aerobic granular sludge in treating industrial effluents with more complex characteristics. Moreover, a study on impacts of temperature variations towards wastewater treatment efficiency may be advantageous for application of this system in a variety of climate conditions. Many aspects of algal-bacterial aerobic granular sludge in wastewater treatment remain undiscovered and demand further studies.

3.2 Wastewater treatment performance using algal-bacterial aerobic granular sludge

Although the symbiotic relationship between algae and bacteria have been discovered from 1950s, the application of algal-bacterial consortia in a granular form have only started recently. The applications of algal-bacterial aerobic granular sludge have been investigated in various kinds of wastewater, such as synthetic wastewater by using acetate, glucose, valerate and butyrate as carbon source, municipal and industrial wastewaters, i.e., paper mill effluent and liquid digestate. The treatment performance of algal-bacterial aerobic granular sludge is summarized in Table 1.

Table 1. Performance of algal-bacterial aerobic granular sludge system in treating different types of wastewaters.

Substrate	Initial conc.	Removal efficiency	Ref.
Synthetic wastewater	COD: 300 mg/L NH_4^+-N: 35 mg/L Phosphorus: 11 mg/L	COD: 93% TN: 66% Phosphorus: 35%	[16]
Synthetic wastewater	COD: 300 mg/L NH_4^+-N: 35 mg/L Phosphorus: 10 mg/L	COD: 98% TN: 50% Phosphorus: 35%	[25]
Synthetic wastewater	COD: 300 mg/L NH_4^+-N: 100 mg/L Phosphorus: 10 mg/L	COD: 95% NH_4^+-N: 99% Phosphorus: 50%	[17]
Synthetic wastewater	COD: 400 mg/L NH_4^+-N: 50 mg/L Phosphorus: 10 mg/L	COD: 94% TN: 71% TP: 54%	[21]
Municipal wastewater	COD: 321 mg/L NH_4^+-N: 49.5 mg/L TN: 55 mg/L TP: 5.6 mg/L	COD: 88-89% NH_4^+-N: 99% TN: 45-52% TP: 20-33%	[26]
Paper mill effluent	TOC: 356 mg/L TN: 30 mg/L TP: 7.5 mg/L	TOC: 57% TN: 76% TP: 73%	[18]
Domestic wastewater	COD: 189 mg/L NH_4^+-N: 24 mg/L	COD: 72% NH_4^+-N: 72%	[27]

At present, majority studies on algal-bacterial aerobic granular sludge are still limited on the treatment of synthetic wastewater. Liu *et al.* (2017) used glucose as carbon source in synthetic wastewater with COD concentration of 300 mg/L, 35 mg/L ammonia and 11 mg/L phosphate [16]. A similar concentration of synthetic wastewater was also found in different studies [17,25]. This range of organic concentration was chosen to specifically mimic the concentration of municipal and domestic wastewaters. Mature aerobic granular sludge and two species of microalgae (*Chlorella* and *Scenedesmus*) were used as the seed sludge resulting in 98%, 50% and 35% ammonia, total nitrogen (TN) and total phosphorus (TP) removal efficiencies, respectively [25]. These were significantly better than removal efficiencies by using conventional aerobic granular sludge which may be due to the ability of microalgae to directly assimilate nitrogen, especially ammonia from the wastewater. Furthermore, higher total nitrogen removal was supported through denitrification process by photosynthetic bacteria.

Treatment of synthetic wastewater containing 400 mg/L COD was investigated in a shaking PSBR [21]. The removal of dissolved organic carbon was always more than 94% with almost complete removal of ammonia (>99%). Phosphorus removal was found to be better during illumination period, resulting in effluent containing 4.5 mg/L phosphorus. This phenomenon explained that microalgae were responsible for phosphorus uptake rather than the abiotic precipitation of the phosphorus. Moreover, the algal-bacterial aerobic granular sludge possessed good granular integrity in shaking PSBR. Performance of algal-bacterial aerobic granular sludge was also tested on low strength wastewater as low as 50 mg/L COD concentration. In order to evaluate the performance and stability of algal-bacterial aerobic granular sludge in low strength wastewater, a decreasing COD concentration was applied [28], whereby COD concentration was decreased from 400 mg/L to 200 mg/L, 100 mg/L and 50 mg/L while the nitrogen concentration was remained constant at 50 mg/L resulting in decreased COD/N ratio (8:1, 4:1, 2:1, 1:1). Throughout the experimental period, the DOC removal efficiency was stable at 95-96%. However, low COD/N ratio, specifically 4:1, 2:1 and 1:1 negatively impact the nitrification and denitrification process. This result explained that ammonia uptake by microalgae is highly affected by COD/N ratio, whereby low COD/N ratio might inhibit ammonia uptake by microalgae as well as the microbial activity of ammonia oxidizing bacteria (AOB). Low COD/N ratio also inhibited the growth and activity of phosphate accumulating organism (PAO) to some extent, whereby a slight decrease of phosphorus removal efficiency was observed.

A small number of studies have reported the application of algal-bacterial aerobic granular sludge in treatment of actual wastewater. A previous study by Petrini *et al.* (2020) demonstrated treatment of municipal wastewater using two types of seed sludge whereby one of them was inoculated with *Chlorella vulgaris* [26]. The COD removal efficiency was approximately 89% while almost complete ammoniacal nitrogen removal was observed in both reactors. Purba *et al.* (2021) investigated the treatment of low-strength actual domestic wastewater using algal-bacterial aerobic granular sludge containing *Scenedesmus obliquus* which recorded 72% and 75% of COD and ammoniacal nitrogen removal efficiencies, respectively [27,29]. It was also found that biomass washout occurred in the system treating low-strength wastewater. Moreover, the varying concentration of domestic wastewater stream posed a challenge to maintain stable performance of algal-bacterial aerobic granular sludge. The treatment of industrial effluent, specifically paper mill effluent was also studied using algal-bacterial aerobic granular sludge [18]. The system was able to remove 58% organic carbon, 27% inorganic carbon, 77% total nitrogen and 73% total phosphorus from the influent wastewater, respectively.

Recent studies have expanded the application of algal-bacterial aerobic granular sludge for removal of hexavalent chromium [Cr(VI)]. Yang *et al.* (2021) demonstrated up to 99% reduction of Cr(VI) and 89% total Cr removal within 6 hours at pH 2 and 6, respectively [30]. Moreover, salinity at 5 g/L was reported to inhibit the reduction of Cr(VI) and removal of total Cr [31]. Further study by Van Nguyen *et al.* (2021) highlighted that 1% salinity may impacted the removal of total inorganic nitrogen (TIN) and total phosphorus, especially when Cr(VI) was added at 2 mg/L [32]. A 1% salinity was also reported to decrease Cr(VI) removal, although the difference was insignificant. In addition, supply of external electron donor (i.e., glucose) has successfully increased the biosorption of Cr(VI) up to 93.8% biosorption efficiency [31]. These performances were able to be maintained under addition of levofloxacin and metabolic inhibitor (i.e., NaN_3) in the influent. On the other hand, addition of tetracycline antibiotic in the influent of algal-bacterial aerobic granular sludge system resulted in deterioration of chlorophyll content, NH_3-N removal efficiency and relative abundance of algae and cyanobacteria [33]. Nevertheless, reports on algal-bacterial aerobic granular sludge for heavy metals removal are still limited to removal of Cr(VI) while removal of other heavy metals have not been studied.

Besides, the performance of this system under dosing of antibiotics have not been fully understood, which calls for further research.

Despite the extensive investigation on treatment of wastewater using algal-bacterial aerobic granular sludge, the process is still facing several challenges, for instance its long-term stability and performance in bioreactors. The stability of granular sludge may be disrupted due to internal hydrolysis of EPS, low porosity that limiting the substrate and oxygen diffusion, and growth of filamentous bacteria. Many fields of algal-bacterial aerobic granular sludge remained unexplored starting from molecular and microbiological aspects, economic analysis, and the full-scale implementation of the system. Moreover, further studies are required to clarify the separation of bacteria from algae cells and whether the bacterial contamination will affect the post-treatment processes for resource recovery.

4. End-use of algal-bacterial aerobic granular sludge

Microalgae are well-known to be able to accumulate high amount of lipid in its cell which can be used as alternative green renewable fuel source. In comparison to the sole microalgae system, the cultivation of algal-bacterial aerobic granular sludge for lipid production may be more advantageous, whereby bacteria may induce microalgae growth to some extent [13]. Moreover, by using algal-bacterial aerobic granular sludge system, the harvesting process will be more efficient and cost-effective [4]. The summary of lipid production from algal-bacterial aerobic granular sludge is shown in Table 2.

Table 2. Lipid production from algal-bacterial aerobic granular sludge treating synthetic wastewater.

Substrate	System used	Lipid production	Ref.
Synthetic domestic wastewater COD: 300 mg/L	PSBR	63.9 mg/g SS	[16]
Synthetic municipal wastewater COD: 300 mg/L	PSBR	47.9 mg/g SS	[25]
Synthetic wastewater COD: 600 mg/L	PSBR	59.6 mg/g SS	[23]
Synthetic wastewater COD: 600 mg/L	PSBR	57.7 mg/g SS	[34]

Previous studies reported that lipid production was proportionally related to the granulation process, whereby higher granulation rate leads to higher lipid yield [16]. Moreover, a study by Liu *et al.* (2018) showed that higher fatty acid methyl ester (FAME) yield can be achieved in algal-bacterial aerobic granular sludge seeded with *Chlorella* and *Scenedesmus* rather than conventional aerobic granular sludge [25]. It was also found that there is a highly positive relationship between light intensity and lipid production in algal-bacterial granular consortia [23]. High salinity has also been proven to induce accumulation of mono-unsaturated FAMEs while negatively impact the accumulation of saturated FAMEs and poly-unsaturated FAMEs [34].

In order to understand the effect of light intensity towards lipid accumulation in algal-bacterial aerobic granular sludge, different light intensity was applied starting from 45 $\mu mol/m^2s^1$ to 90, 135, 180 and 225 $\mu mol/m^2s^1$ [23]. The results implied that there is highly positive relationship between light intensity to lipid production in granular consortia, whereby highest lipid content of 59.6 mg/g SS was detected in reactor with highest light intensity of 225 μmol photon/m^2s^1. Moreover, the main composition of FAME from extracted lipids were analyzed and it was found that stronger illumination promotes the accumulation of mono-unsaturated FAME (C16:1 and C18:1) and poly-unsaturated FAME (C18:2 and C18:3). However, according to the results, it was predicted that biodiesels from this system may not be stable during long-term storage as poly-unsaturated fatty acids are easily oxidized.

Moreover, to discover the effect of different salt concentration to lipid production in algal-bacterial aerobic granular sludge, different concentration of NaCl (0 g/L, 10 g/L, 20 g/L and 30 g/L) was used in the synthetic wastewater [34]. It has been reported that lipid accumulation can protect microalgae cells from salt stress. Therefore, highest lipid content of 57.7 mg/g SS was found in highest salinity of 3% (30 g/L NaCl). FAME composition showed high dominance of oleic acid (C18:1) followed by hexadecenoic acid (C16:1), hexadecanoic acid (C16:0) and pentadecanoic acid (C15:1). High salinity has been proven to enhance the production of mono-unsaturated FAMEs (C16:1 and C18:1) while reducing the production of saturated FAMEs and poly-unsaturated FAMEs.

Upon literature reviews, it was found that algal-bacterial aerobic granular sludge has high potential to be applied in wastewater treatment system and for future resource recovery as compared to conventional activated sludge and conventional aerobic granular sludge systems. However, the major issue hindering the application of algal-bacterial aerobic granular sludge for lipid production or biofuel production for commercialization is the proof whether the process is economically sustainable. Therefore, a comprehensive economic assessment is required to further expand the knowledge in this field. Furthermore, the use of life cycle assessment (LCA) is important to accurately determine the potential impact on environmental sustainability and renewability of biofuels produced from algal-bacterial aerobic granular sludge.

5. Conclusion

This article highlighted the application of algal-bacterial aerobic granular sludge for wastewater treatment and lipid production. This field has gained increasing attention indicated by the rapid growing number of publications. However, despite the extensive investigation, many aspects have remained unexplored in algal-bacterial aerobic granular sludge. For instance, the long-term stability and performance of this system have not been reported, followed by its application in larger scale of wastewater treatment system. In addition, the possible resource recovery from algal-bacterial aerobic granular sludge requires experimental proof for commercialization and economic analysis to confirm the sustainability of this process.

ORCID IDs

L D A Purba https://orcid.org/0000-0001-8198-1055
K Shimizu https://orcid.org/0000-0002-8274-0410
J Hermana https://orcid.org/0000-0002-0142-3297

Acknowledgments

The authors wish to thank Collaborative Education Program for Sustainable Environmental Engineering Network (CEP-SEEN) program from AUN/SEED-Net by JICA for fully funding this research under grant No. 4B403.

References

[1] Oswald W J, Gotaas H B, Golueke C G, Kellen W R, Gloyna E F and Hermann E R 1957 Algae in waste treatment [with discussion] *Sewage Ind. Waste.* **29** 437–57
[2] Fuentes J L, Garbayo I, Cuaresma M, Montero Z, González-Del-Valle M and Vílchez C 2016 Impact of microalgae-bacteria interactions on the production of algal biomass and associated compounds *Mar. Drugs* **14**
[3] Lee Y and Lei Z 2019 Microalgal-bacterial aggregates for wastewater treatment : A mini-review *Bioresour. Technol. Reports* **8** 100199
[4] Zhang B, Li W, Guo Y, Zhang Z, Shi W, Cui F, Lens P N L and Tay J H 2020 Microalgal-bacterial consortia: From interspecies interactions to biotechnological applications *Renew. Sustain. Energy Rev.* **118** 109563
[5] Liu J, Wu Y, Wu C, Muylaert K, Vyverman W, Yu H Q, Muñoz R and Rittmann B 2017 Advanced nutrient removal from surface water by a consortium of attached microalgae and bacteria: A review *Bioresour. Technol.* **241** 1127–37
[6] Acién F G, Gómez-Serrano C, Morales-Amaral M del M, Fernández-Sevilla J M and Molina-Grima E 2016 Wastewater treatment using microalgae: how realistic a contribution might it be to significant urban wastewater treatment? *Appl. Microbiol. Biotechnol.* **100** 9013–22
[7] Zhou J, Lyu Y, Richlen M L, Anderson D M and Cai Z 2016 Quorum sensing is a language of chemical signals and plays an ecological role in algal-bacterial interactions *CRC. Crit. Rev. Plant Sci.* **35** 81–105
[8] Purba L D A, Tijani H, Yuzir A, Mohamad S E, Iwamoto K, Zamyadi A and Abdullah N 2020

Various Applications of Aerobic Granular Sludge : A Review *Environ. Technol. Innov.* **20** 1–44
[9] Quijano G, Arcila J S and Buitrón G 2017 Microalgal-bacterial aggregates: Applications and perspectives for wastewater treatment *Biotechnol. Adv.* **35** 772–81
[10] Purba L D A, Md Khudzari J, Iwamoto K, Mohamad S E, Yuzir A, Abdullah N, Shimizu K and Hermana J 2022 Discovering future research trends of aerobic granular sludge using bibliometric approach *J. Environ. Manage.* **303** 114150
[11] Montemezzani V, Duggan I C, Hogg I D and Craggs R J 2015 A review of potential methods for zooplankton control in wastewater treatment High Rate Algal Ponds and algal production raceways *Algal Res.* **11** 211–26
[12] Ramanan R, Kim B H, Cho D H, Oh H M and Kim H S 2016 Algae-bacteria interactions: Evolution, ecology and emerging applications *Biotechnol. Adv.* **34** 14–29
[13] Wang H, Hill R T, Zheng T, Hu X and Wang B 2016 Effects of bacterial communities on biofuel-producing microalgae: stimulation, inhibition and harvesting *Crit. Rev. Biotechnol.* **36** 341–52
[14] Tiron O, Bumbac C, Manea E, Stefanescu M and Lazar M N 2017 Overcoming microalgae harvesting barrier by activated algae granules *Sci. Rep.* **7** 1–11
[15] Cai W, Zhao Z, Li D, Lei Z, Zhang Z and Lee D 2019 Algae granulation for nutrients uptake and algae harvesting during wastewater treatment *Chemosphere* **214** 55–9
[16] Liu L, Fan H, Liu Y, Liu C and Huang X 2017 Development of algae-bacteria granular consortia in photo-sequencing batch reactor *Bioresour. Technol.* **232** 64–71
[17] Ahmad J S M, Cai W, Zhao Z, Zhang Z, Shimizu K, Lei Z and Lee D J 2017 Stability of algal-bacterial granules in continuous-flow reactors to treat varying strength domestic wastewater *Bioresour. Technol.* **244** 225–33
[18] Van Den Hende S, Rodrigues A, Hamaekers H, Sonnenholzner S, Vervaeren H and Boon N 2017 Microalgal bacterial flocs treating paper mill effluent: A sunlight-based approach for removing carbon, nitrogen, phosphorus, and calcium *N. Biotechnol.* **39** 1–10
[19] Cai W, Huang W, Lei Z, Zhang Z, Lee D and Adachi Y 2019 Granulation of activated sludge using butyrate and valerate as additional carbon source and granular phosphorus removal capacity during wastewater treatment *Bioresour. Technol.* **282** 269–74
[20] Zhang R, Kong Z, Chen S, Ran Z, Ye M, Xu J, Zhou C, Liao K, Cao J and Yan X 2017 The comparative study for physiological and biochemical mechanisms of Thalassiosira pseudonana and Chaetoceros calcitrans in response to different light intensities *Algal Res.* **27** 89–98
[21] Zhao Z, Liu S, Yang X, Lei Z, Shimizu K, Zhang Z, Lee D J and Adachi Y 2019 Stability and performance of algal-bacterial granular sludge in shaking photo-sequencing batch reactors with special focus on phosphorus accumulation *Bioresour. Technol.* **280** 497–501
[22] He Q, Chen L, Zhang S, Chen R, Wang H, Zhang W and Song J 2018 Natural sunlight induced rapid formation of water-born algal-bacterial granules in an aerobic bacterial granular photo-sequencing batch reactor *J. Hazard. Mater.* **359** 222–30
[23] Meng F, Xi L, Liu D, Huang W, Lei Z, Zhang Z and Huang W 2019 Effects of light intensity on oxygen distribution, lipid production and biological community of algal-bacterial granules in photo-sequencing batch reactors *Bioresour. Technol.* **272** 473–81
[24] Huang W, Li B, Zhang C, Zhang Z, Lei Z, Lu B and Zhou B 2015 Effect of algae growth on aerobic granulation and nutrients removal from synthetic wastewater by using sequencing batch reactors *Bioresour. Technol.* **179** 187–92
[25] Liu L, Zeng Z, Bee M, Gibson V, Wei L, Huang X and Liu C 2018 Characteristics and performance of aerobic algae-bacteria granular consortia in a photo-sequencing batch reactor *J. Hazard. Mater.* **349** 135–42
[26] Petrini S, Foladori P, Beghini F, Armanini F, Segata N and Andreottola G 2020 How inoculation affects the development and the performances of microalgal-bacterial consortia treating real municipal wastewater *J. Environ. Manage.* **263** 110427
[27] Purba L D A, Yuzir A, Zamyadi A, Ab Halim M H, Zeca E D and Abdullah N 2021 Cultivation of aerobic granular sludge by modification of seeding condition *Desalin. Water Treat.* **223** 280–9

[28] Zhao Z, Yang X, Cai W, Lei Z, Shimizu K, Zhang Z, Utsumi M and Lee D J 2018 Response of algal-bacterial granular system to low carbon wastewater: Focus on granular stability, nutrients removal and accumulation *Bioresour. Technol.* **268** 221–9

[29] Purba L D A 2019 *Treatment of sewage by aerobic granulation process enhanced with microalgae* (Universiti Teknologi Malaysia)

[30] Yang X, Zhao Z, Zhang G, Hirayama S, Nguyen B Van, Lei Z, Shimizu K and Zhang Z 2021 Insight into Cr(VI) biosorption onto algal-bacterial granular sludge: Cr(VI) bioreduction and its intracellular accumulation in addition to the effects of environmental factors *J. Hazard. Mater.* **414** 125479

[31] Yang X, Zhao Z, Nguyen B Van, Hirayama S, Tian C, Lei Z, Shimizu K and Zhang Z 2021 Cr(VI) bioremediation by active algal-bacterial aerobic granular sludge: Importance of microbial viability, contribution of microalgae and fractionation of loaded Cr *J. Hazard. Mater.* **418** 126342

[32] Nguyen B Van, Yang X, Hirayama S, Wang J, Zhao Z, Lei Z, Shimizu K, Zhang Z and Le S X 2021 Effect of Salinity on Cr(VI) Bioremediation by Algal-Bacterial Aerobic Granular Sludge Treating Synthetic Wastewater *Processes* **9** 1400

[33] Wang S, Ji B, Zhang M, Gu J, Ma Y and Liu Y 2021 Tetracycline-induced decoupling of symbiosis in microalgal-bacterial granular sludge *Environ. Res.* **197** 111095

[34] Meng F, Liu D, Huang W, Lei Z and Zhang Z 2019 Effect of salinity on granulation , performance and lipid accumulation of algal-bacterial granular sludge *Bioresour. Technol. Reports* **7** 100228

IOP Conf. Series: Earth and Environmental Science 1091 (2022) 012048 doi:10.1088/1755-1315/1091/1/012048

Analysis of Rock Mass Surface Colour using CIELAB and Munsell Chart

M. Razali[1,2], M.A.M Ismail[*3], I. N. Yusoff[1], H. Tobe[4], T. Miyoshi[4], K. Date[5] and Y. Yokota[5]

[1]School of Civil Engineering, Engineering Campus, Universiti Sains Malaysia, Penang, Malaysia

[2] School of Civil Engineering, College of Engineering, Universiti Teknologi MARA, Cawangan Pulau Pinang, Kampus Permatang Pauh, Penang, Malaysia

[*3]School of Civil Engineering, Engineering Campus, Universiti Sains Malaysia, Penang, Malaysia

[4]Rock Mechanics and Hydro-geology Group, Kajima Technical Research Institute, Kajima Corporation, Japan

[5]Kajima Technical Research Institute Singapore, Kajima Corporation, Singapore

*Corresponding author e-mail: ceashraf@usm.my

Abstract. Rock mass surface colour is crucial for weathering assessment. CIELAB colour space and Munsell chart are methods to express the surface colour of rock masses. This study analysed the L*a*b* value from both approaches and examined the effectiveness and reliability of rock slope weathering assessment method. The Munsell Rock chart is a conventional method primarily used in practice. However, due to the constraint of individual vision and variation in the optical environment, errors and bias in rock surface colour can easily occur using the conventional method. The colourimeter was used to determine the CIELAB colour values of the samples, while the rock colour chart determined the colour coordinates and Munsell values. The Munsell hue and chroma were assigned using linear interpolation by Munsell conversion software CMC 2.1 to convert Munsell colour systems to L*a*b* value. A total of six granitic rock slopes were investigated. For quantitatively method, the Munsell optical system was less effective in assessing the rock mass surface colour. The positive correlation of a* and b* with R^2=0.9407 demonstrated the efficiency of CIELAB system. The a*b* result of the rock slope indicated that most of the slopes were considered medium to strong with a moderately weathered surface. This result is strongly supported by a geomechanical test of Schmidt-UCS and index properties (Id_2) with a range of 50-100 MPa and 2-4 MPa respectively. The finding verified mineralogical assessment which showed the correlation of rock mass surface colour to mineral composition. The study recommended any geotechnical and technical purpose in enhancing the method of weathering assessment by using the CIELAB colour space to rapidly and efficiently measure the rock surface colour.

1. Introduction

The rock mass surface colour is one of the parameters used to describe the weathering grades and mineral composition. It is one of the most crucial components in weathering classification. Weathering layers gradually cover the surface of exposed rock masses which slowly change when exposed to sunlight and water over time. Physical disintegration, chemical degradation, and biological factors contribute to rock weathering. Almost all minerals are affected by chemical decomposition except most quartz which is largely unaffected. Oxidation, reduction, hydration, hydrolysis, carbonation, and solution are the processes involved in the disintegration of rock slopes [1]–[3]. The rock slope surface in the study area is predominantly light grey to light brown in hue. The typical tropical weathering results into weathering products and dissolution, which reduce the amount of most minerals' constituents, but alum and iron remain [4]. The oxidized iron in the weathering products creates red or yellowish-red oxides and hydroxides. These factors cause the soils to change colour to red [4]–[6]. Rock mass surface colour is one of the crucial things in weathering assessment. This study used CIELAB and Munsell's chart of a surface colour method to differentiate weathered and unweathered rock surfaces. The natural rock surfaces are classified into various parameters, including colour and colour homogeneity. Climate change has altered the surface colours of outcrop rock masses, which may differ from the colours of the original rock masses. Naked eyes can differentiate between numerous colours of visible light, allowing them to visually identify the hues of various items, such as rocks [6]–[8]. For example, the magnetic minerals have blackish colours, weathered copper minerals are greenish, and salt minerals are pure white.

The Munsell Rock chart is commonly used to determine the rock mass's surface colour. However, there are certain limitations to obtaining a good outcome due to different individuals' interpretations. This study used CIELAB surface colour testing to digitalize colour variations on rock surfaces to address the issue [7]–[9]. The effectiveness of weathering and colour changes in rocks mass using the digital method was compared to the conventional method of Munsell colour chart [10]–[13]. The correlation with point load test and weathering grade was used to verify the weathering grade of rock mass slope surface. The point of the study is to evaluate the efficiency of the CIELAB system's colour by comparing the L*a*b* value with the Munsell system conversion for rock mass weathering surface colour in the study area.

2. Rock mass surface colour

Surface colour variation is one of the signs of deterioration. The primary surface colour of the rock mass can be described as brown, green, red, pink, or khaki. A secondary colour is described as an adjective, such as reddish-brown, greyish green, or yellowish khaki. The colour should be increased further where it is significant by utilizing the following value descriptions: dark reddish-brown, light yellowish khaki, very light, light, medium, dark, extremely dark and so on.

2.1. CIELAB colour space

In CIELAB colour space, the value of L* indicates the lightness. Meanwhile, the value of a* and b* indicates the colour space. L* stands for lightness, an achromatic scale that ranges from 0 (extremely dark) to 100 (extremely light). The a* axis represents a green-red continuum, green is negative, and red represents positive. Meanwhile, the b* axis displays a blue-yellow continuum, with blue indicating a negative value and yellow indicating a positive value. The precise geological structure of the rock surfaces may be included in the high-resolution data obtained using these approaches [7]. A colourimeter is a device that performs a type of perceptual sample analysis, which means that its readings are comparable to human perception [14]–[16]. The luminance or brightness of an image is determined by L*. The values are [0, 100], with 0 indicating black and 100 indicating white. The colour difference is determined using Equation 1.

$$\text{Colour difference, } \Delta E^*_{ab} = \sqrt{(L^*_2 - L^*_1)^2 + (a^*_2 - a^*_1)^2 + (b^*_2 - b^*_1)^2} \quad (1)$$

Delta Eab is a total colour difference, L is lightness, red/green, and b is yellow/blue.

In Figure 1, for example, both apples have L* a* b* colour differences. Apple 2 appears lighter, less red, and more yellow than Apple 1. When the values of L*=+4.03, a*=-3.05, and b*=+1.04 are entered

IOP Conf. Series: Earth and Environmental Science 1091 (2022) 012048 doi:10.1088/1755-1315/1091/1/012048

into the colour difference equation as shown in Equation 1, the total colour difference between the two apples is 5.16. The L*a*b* colour space ensures that products are manufactured to exact colour specifications by allowing precise colour communication between a company and its supply chain. The colour difference provides a simple quality check Eab calculated from the L*a*b* values of control and sample. The mechanism used to measure the apple is similar to the method used in this study for weathering assessment of rock slope with a colourimeter. The colour of the rock sample represents the degree of weathering. The higher a* and b* values, the more weathered the rock is, and this provides a good weathering assessment [8]. a* and b* values of heavily weathered samples are higher than those of unweathered samples, as shown in Figure 1 below.

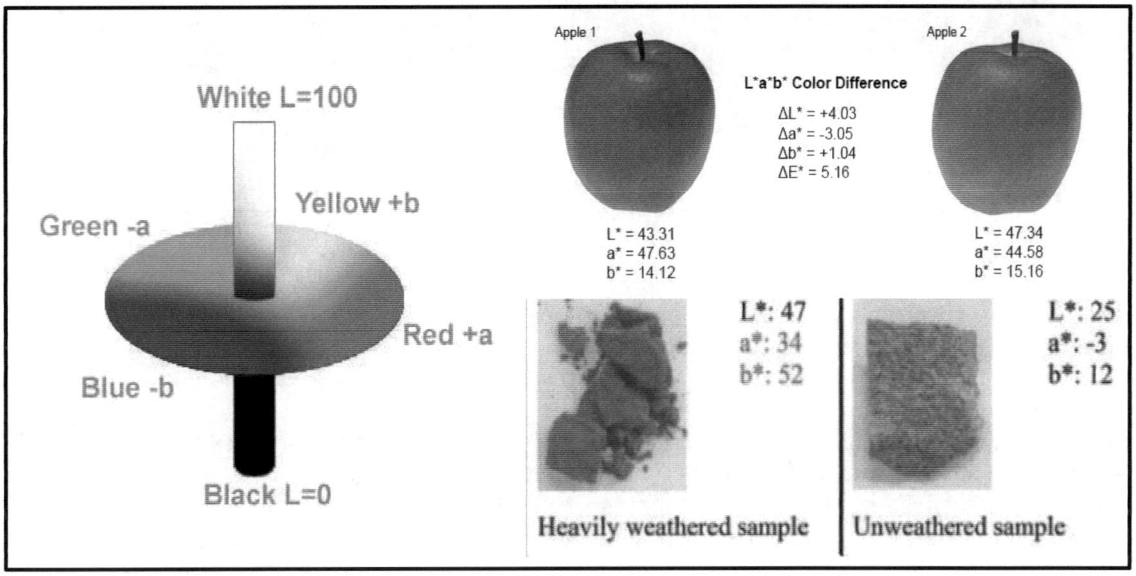

Figure 1. L*, a*, and b* Colour Difference Concept From The Spectrophotometer

2.2. Munsell colour chart

The Munsell colour system [12], [13]was designed to be perceptually uniform. Perceptual uniformity refers to a change in a colour component anywhere in the system that causes a human observer to perceive the same perceptual difference. Hue, value, and chroma are the three elements that make up the colour in the Munsell system, as shown in Figure 2. The first component hue is represented by a number between 0 and 10 and a one- or two-letter abbreviation of the hue's name. Aside from the colour hues in Figure 3, there is a special abbreviation N for neutral, reserved for grey shades. On a scale of 0 (black) to 10 (white), value determines a colour's lightness (white). In theory, the third component chroma determines saturation and has no upper bound, but it is limited by the colours available in practice. A neutral colour has a chroma value of 0. The highest value in the chart is 8, and a value of 20 indicates a highly saturated colour. Hue, Value, and Chroma are the three components of the Munsell rock colour chart that describes the colour of the rock. Hue refers to the primary colours of red, yellow, and blue and hues that are intermediate between them, such as yellow and red. On a scale of 0-10, the value represents the degree of rock lightness or darkness, while 0-8 chroma represents the rock's intensity colour, saturation, and vibrance. The Munsell colour chart can be used for in-situ tests and laboratory work. However, the Munsell colour chart has difficulties in determining the precise colour of the rock due to the limitation of human vision. Figure 3 illustrates the Munsell colour concept and Munsell Rock Colour Chart which are used to determine the colour properties of soil and rock.

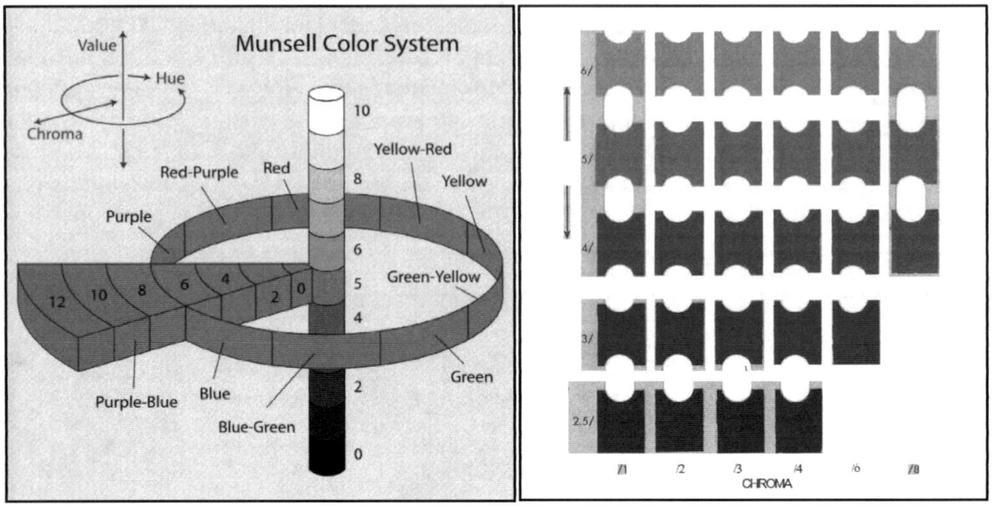

Figure 2. The Munsell Soil-Colour Concept and Munsell Rock Colour Chart

2.3. Comparison of CIELAB and Munsell colour system

Munsell proposed developing a set of colour chips based on his system. Each chip's hue, value, and chroma values defines its colour. Munsell also emphasizes that the colour can be determined using a spectrophotometer from the spectral curve obtained under laboratory conditions. On the other hand, the Munsell cylindrical system is not well suited to colour quantification [17], [18]. In 1976, the CIEL*a*b* colour space was developed as a hardware-neutral reference model. It was based on the CIE 1931 XYZ colour space for linearizing colour differences. The CIE L*a*b* system in cartesian coordinates represents a universal colour space; its advantage is that it allows for distinguishing colour characteristics and differences between soil and rock types [15]. In contrast to the Munsell system, the L*a*b*colour space distinguishes between the luminance L* and the chromatic colour component (tone, saturation). The value of L* varies between 0 and 100, from darkest black to lightest white, and the chromatic component is defined by two cartesian coordinates a* and b*. The former refers to the colour range between green (–a*) and red (+a*), while the latter refers to the colour range between blue (–b*) and yellow (+b*)[12], [19]–[22].

3. Methods

3.1. Insitu testing for CIELAB weathering surface colour

The studied rock slope is near Jalan Mengkuang Ulu in Kuad Quarry (P129), Penanti, Pulau Pinang (5 24' 55.45" N 100 28'8.00"). This study chose the weathered granitic rock slope based on its predominant rock formation in the Malaysian Peninsula. The Kuad Quarry at Penanti is composed of acid intrusive granitic rock (Figure 3). This study used a colourimeter to assess the rock slope's surface weathering colour [10], [23]–[26]. The FRU Colourimeter WR18 with a 40 mm aperture was used in this study because it can measure largely textured, uneven, and rough surfaces (Figure 4). This study chose six slopes from the study area, namely slopes 2, 3, 4, 5, 6, and 7, as shown in Figure 5. The colourimeter test was used to obtain L*a*b* value while the Munsell rock colour chart was used to obtain Hue value and chroma. The values of hue and chroma were converted to L*a*b* value to verify the efficiency of the methods.

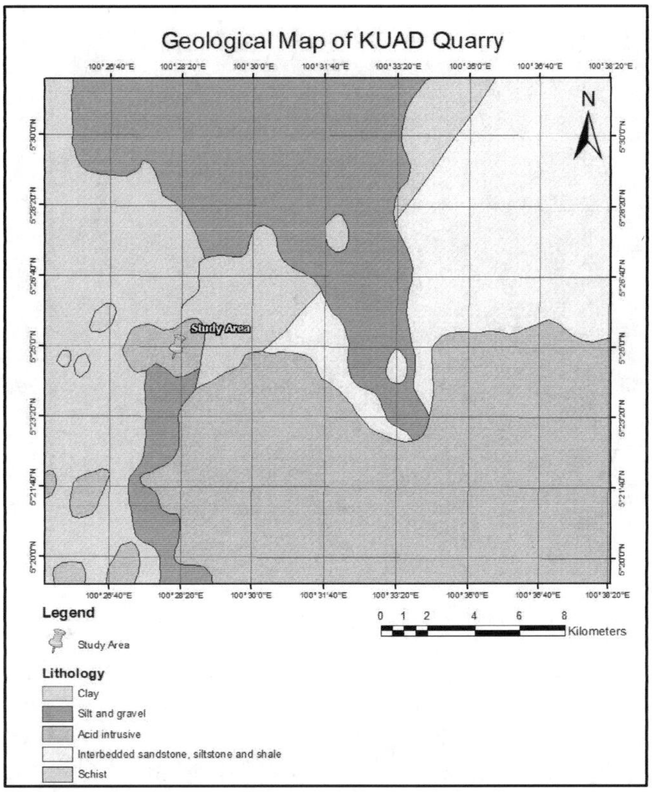

Figure 3. Geological Map of Kuad Quarry

Figure 4. CIELAB Site Testing Using Colourimeter WR18

Figure 5. Site Viewing For All Rock Slope Surfaces

3.2. Conversion from Munsell system to CIELAB colour system

Munsell colour codes were converted using the Munsell Conversion Program (Munsell Conversion CMC21). The Munsell colourimetric scale of rock was also converted to the CIELAB colour as referred to previous study results of Munsell conversion table [12]. Munsell Conversion makes use of the constants determined for light source C, whereas current algorithms make use of light source D65 values. Therefore, the Munsell conversion table is recommended to convert XYZ data to CIEL*a*b* coordinates. It is necessary to undergo a stage of chromatic adaptation: the transition from light to darkness.

The ΔL*, Δa*, and Δb* were calculated to show the differences between both methods. The differences were calculated from the equations 2, 3 and 4.

$$\Delta L^* = L^* \text{ (Lab Value)} - L^* \text{ (Conversion of Munsell Colour)} \quad (2)$$

$$\Delta a^* = a^*\text{(Lab Value)} - a^* \text{ (Conversion of Munsell Colour)} \quad (3)$$

$$\Delta b^* = b^* \text{ (Lab Value)} - b^* \text{ (Conversion of Munsell Colour)} \quad (4)$$

4. Results and discussion

4.1. Rock mass colour using CIELAB and Munsell chart

Six different slopes were examined to determine the value of CIE colour (L*a*b*) and Munsell colour codes (Hue, Value, and Chroma values). The slope samples are represented in Fig 6 as points of colour were tested. In Table 1, the examples of the surface colour of slope id No 2 are described by the average value of measured L*, the a* and b* by colourimeter readings, while the Hue, Value, and Chroma values are from the rock Munsell colour chart conversion.

Figure 6. Marked Point Samples At Slope Surface

Weathering colours of red and green at rock surface are separated by the a* positive and negative value, which indicates that the colour is in the middle. Positive and negative values of a* are used to determine which colour is formed, with the negative value being green and the positive being red. The surface colour is more likely to be red because the rock slopes measured on site have a* positive values. b* is a measure of the colour between blue and yellow. In this case, the negative value of the b* colour is blue, and the positive value of the b* colour is yellow. The yellow colour appears on rock slopes surface because b* values are all positive. a* and b* values represent the weathering colour of rock which was high since a* and b* values showed the rock surface became more weathered while small value was unweathered area. The CIELAB colour space of L*, a*, and b* values positively correlated with weathering grades. The results (Figure 7) reflect the theory where rock is more weathered with the increasing value of a* and b*. Similar results supported the hypothesis in predicting the tunnel face weathering grades [8].

The finding showed 130 measurement points on the six rock slopes. Table 1 summarises the data for rock slope No. 2 on an average basis. According to the results, the a* value ranged from 2 to 15 with an average of 5.46, while the b* value ranged from 7 to 25 with an average of 11.64. With R^2=0.94079, which was nearest to one, the total correlation of the a* and b* values showed significant results (for all slopes). The positive a* value represents the redness of slope surface in the CIELAB colour space and Munsell conversion value. The results showed how the colour of the rock slope changed from unweathered to highly weathered. The finding indicated that heavily weathered rock samples had higher a* and b* values than unweathered rock samples. The results were compared to Munsell colour a* and b* value using CMC21 as described in Vodyaniiski's table [12]. A significant correlation with Munsell's description was found as tabulated in Table 1. The selected samples from 130 points of measurement in the study area demonstrated a reliable rock mass colour analysis, in which the increase of a* and b* values increased the level of weathering from slightly weathered (strong rock) to highly weathered (weak rock).

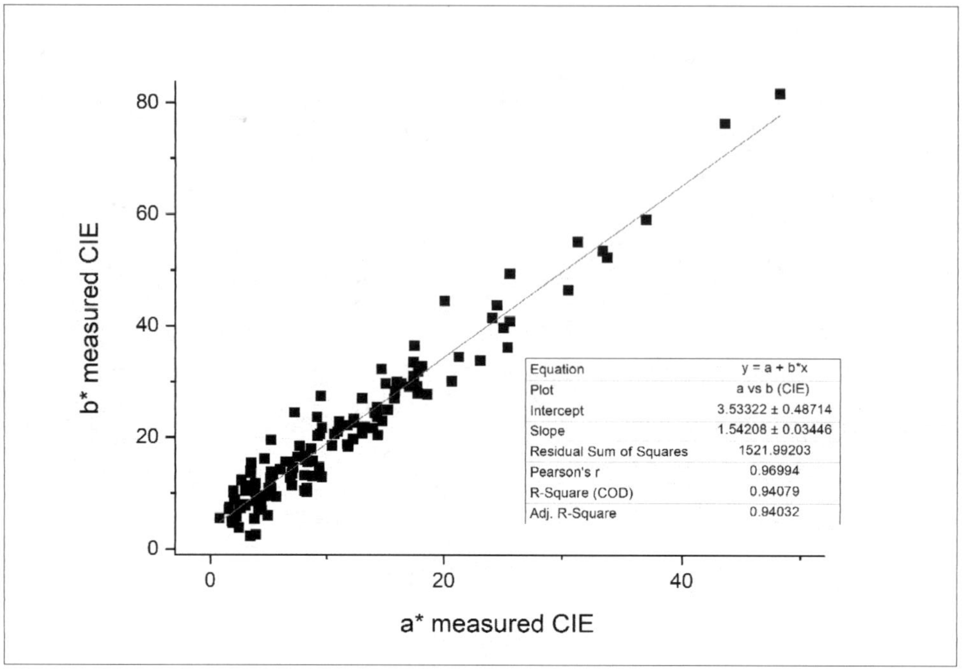

Figure 7. Correlation of a* to b* of the Study Area

4.2. Comparison of CIELAB surface colour and Munsell colour

It is impossible to compare the CIELAB colour space directly to the Munsell colour space because of significantly different value parameters. Thus, the values obtained from the Munsell colour chart, namely hue, value, and chroma, must be converted to the CIELAB colour system to compare the CIELAB and Munsell colour charts. The Munsell colour system is converted to the CIELAB colour system using the Munsell Conversion Program CMC2.1 [12], [22]. The comparison was done by calculating the differences between the L*, a*, and b*parameters. The differences between the L*a*b* value and the Munsell colour conversion to the CIELAB Colour System are tabulated in Table 2. The more significant difference between the L*a*b* value and the Munsell colour conversion indicates that the value is less precise than the actual L*a*b* value obtained from the colourimeter. Meanwhile, the more negligible difference between the L*a*b* value and the Munsell colour conversion indicates that the value is closer to the actual L*, a*, and b* values obtained from the colourimeter.

Table 1. Sample results L*a*b* value for granitic rock from slope No. 2, for CIELAB and Munsell conversion method

ID	CIE (Measured)			Munsell Conversion			Rock Colour	Colour Differences		
	L*	a*	b*	L*	a*	b*	Munsell Chart	ΔL*	Δa*	Δb*
2a	68.77	14.75	22.92	71.60	-0.04	0.05	N7	-2.83	14.79	22.87
2b	60.98	14.35	20.37	71.60	-0.04	0.05	N7	-10.62	14.39	20.32
2c	50.89	5.28	10.90	71.60	-1.92	15.25	5Y 7/2	-20.71	7.20	-4.35
2d	60.92	15.79	27.00	41.22	26.54	13.24	5R 4/6	19.70	-10.75	13.76
2e	52.07	3.62	11.26	71.60	-1.92	15.25	5Y 7/2	-19.53	5.54	-3.99
2f	64.12	15.22	24.95	41.22	23.99	21.69	10R 4/6	22.90	-8.77	3.26
2g	51.98	2.68	7.28	61.70	2.44	13.08	10YR 6/2	-9.72	0.24	-5.80
2h	56.54	3.83	11.37	41.22	23.99	21.69	10R 4/6	15.32	-20.16	-10.32
2i	50.98	2.20	8.85	51.57	-2.92	-3.03	5B 5/1	-0.59	5.12	11.88
2j	57.21	3.10	10.44	71.60	-1.92	15.25	5Y 7/2	-14.39	5.02	-4.81
2k	51.80	1.65	7.48	51.57	15.12	14.17	10R 5/4	0.23	-13.47	-6.69

ID	CIE (Measured)			Munsell Conversion			Rock Colour	Colour Differences		
	L*	a*	b*	L*	a*	b*	Munsell Chart	ΔL*	Δa*	Δb*
2l	56.94	5.21	12.80	71.60	-1.92	15.25	5Y 7/2	-14.66	7.13	-2.45
2m	56.18	5.25	10.17	41.22	-0.03	0.03	N4	14.96	5.28	10.14
2n	55.13	2.56	7.75	41.22	23.99	21.69	10R 4/6	13.91	-21.43	-13.94
2o	57.05	1.99	10.17	41.22	-0.5	7.29	5Y 4/1	15.83	2.49	2.88
2p	56.55	1.99	10.40	71.60	-3.34	43.47	5Y 7/6	-15.05	5.33	-33.07
2q	52.06	4.80	9.56	61.70	-5.84	2.18	5G 6/1	-9.64	10.64	7.38
2r	47.36	2.81	7.79	61.70	-5.84	2.18	5G 6/1	-14.34	8.65	5.61
2s	48.75	4.66	9.46	41.22	-5.12	1.81	5G 4/1	7.53	9.78	7.65
2t	47.98	4.46	7.00	61.70	-5.84	2.18	5G 6/1	-13.72	10.30	4.82
2u	51.41	7.01	11.37	30.77	-0.02	0.03	N3	20.64	7.03	11.34
2v	48.11	4.02	8.57	41.22	-0.03	0.03	N4	6.89	4.05	8.54
2w	46.13	4.19	7.40	41.22	-5.12	1.81	5G 4/1	4.91	9.31	5.59
2x	50.51	2.19	6.16	51.57	-2.92	-3.03	5B 5/1	-1.06	5.11	9.19
2y	53.18	3.00	7.91	51.57	-2.92	-3.03	5B 5/1	1.61	5.92	10.94
2z	47.99	5.43	13.33	41.22	23.99	21.69	10R 4/6	6.77	-18.56	-8.36

The finding in Table 1 shows that the CIELAB value correlates well with a* and b* compared to the Munsell Conversion value. The high a* and b* values may indicate that the slope has been subjected to highly weathering. The value of a* and b* for Munsell conversion showed that the data collected were almost similar to others and the trend did not progressive well for each test point. According to the R^2 values for CIELAB and Munsell Conversion (Figure 7 and Figure 8), the CIELAB had a strong correlation than the Munsell Conversion value for the values a* and b*, which was nearest to one. The correlation of Munsell conversion showed a weak positive correlation with $R^2 = 0.296$. The findings showed a weak correlation because the data contained multiple similar values.

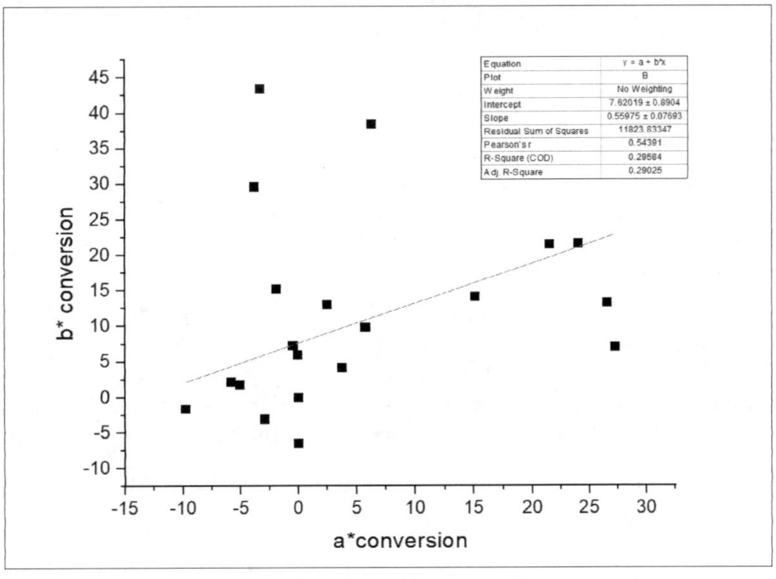

Figure 8. Correlation of a* to b* Conversion From Munsell Chart of the Study Area

4.3. Verification of Rock Mass Colour With Index Properties and Weathering Grade

In this study, the rock slope surface colour should be compared to the results of a geomechanical test to validate the findings. This study included geomechanical testing such as uniaxial compressive strength (UCS) from an N-type Schmidt rebound hammer and a point load test (index properties). Referring to Franklin (1985), the methodology applied for the index properties was ASTM D5731-08, 2008; and the

IOP Conf. Series: Earth and Environmental Science 1091 (2022) 012048 doi:10.1088/1755-1315/1091/1/012048

samples tested were cylindrical cores with a width (D) to height (H) ratio of 0.3 DHD. The correlation UCS = C x Is50 was used to convert the point load strength to UCS [27], [28]. Referring to Brown (1981), the weathering grade was correlated to CIELAB colour [29]. According to Table 2, the high value of a* b* for granitic slope indicates that the rock was moderately weathered with a low Schmidt-UCS and an average weathering grade III or R4, which was a medium-strong to strong rock mass.

Table 2. Verification of CIELAB rock surface colour with index properties (Id2) and weathering Grade grades

Slope Name	L*	a*	b*	Weathering Description	Average Schmidt UCS (MPa)	Id_2 (MPa)	Weathering Grade
2	53.91	5.46	11.64	Moderately weathered, medium-strong rock mass, medium fractured,	91.7	2-4	III (R4)
3	54.52	7.13	12.41	Moderately weathered, medium-strong to strong rock mass,	70.6	2-4	III (R4)
4	60.93	9.62	18.43	Moderately weathered, medium-strong to strong rock mass, medium fractured	71.9	2-4	III (R4)
5	57.28	14.15	23.28	Moderately weathered, medium-strong rock mass, medium fractured	56.8	2-4	III (R4)
6	73.52	14.22	26.26	Moderately weathered, strong rock mass, medium fractured.	68.5	2-4	III (R4)
7	76.37	16.64	32.36	Moderately weathered, medium-strong rock mass, medium fractured	66.2	2-4	III (R4)

4.4. Verification of rock mass colour with mineralogy characteristic

Feldspars (30%-35%), plagioclase (25%-35%), and quartz (20%-50%) are the predominant mineral families in granitic rock, while biotite (5%-10%) is usually prominent than matrix/ground mass (Table 3). Due to the natural heterogeneity in terms of size and grain colour, the colour parameters of the granites must be analysed. The granite from the study area has a light colour because of the small quantities of mafic minerals, with L* values between 40 to 70, which have a higher biotite quantity. The amount of biotite determines the lightness value; thus, only black mineral major in the rocks is analysed. The lightness is not only influenced by the proportion of biotite, as the hue of the light minerals is also crucial. For example, the quartz minerals in the study area have a grayish hue and for this reason, the L* parameter is lower. Another aspect revealed by the results is that the higher value of the b* parameter in the study area reflects their yellowish colour, which is more evident in verifying the significance of L*a*b* results. These results must be considered only indicative due to natural variations in outcrops and quarry.

Table 3. Mineralogical Assessment of granitic from Kuad Quarry

Samples	Weathering Grade	Thin Section Images	Mineral Composition
S1	II		Quartz (50%), K.Feldspar (35%) Plagioclase (30%), Biotite (5%), Matrix/ Ground Mass (10%)
S2	II		Quartz (20%), K. Feldspar (35%), Plagioclase (25%), Biotite (10%), Matrix/ground mass(10%)
S3	I		Quartz (20%), K. Feldspar (30%), Plagioclase (30%), Biotite (10%), Matrix/ground mass(10%)
S4	I		Quartz (25%), K. Feldspar (30%), Plagioclase (28%), Biotite (7%), Matrix/ground mass (10%)
S5	II	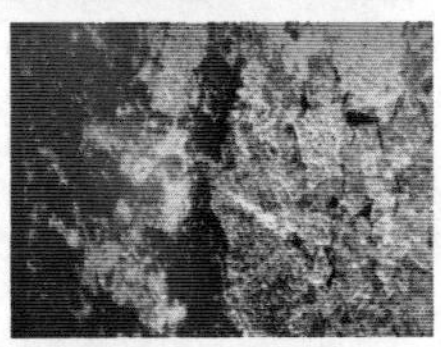	Quartz (22%), K. Feldspar (30%), Plagioclase (28%), Biotite (10%), Matrix/ground mass (10%)
S6	I		Quartz (20%), K. Feldspar (30%), Plagioclase (30%), Biotite (10%), Matrix/ground mass (10%)
S7	I		Quartz (22%), K. Feldspar (35%), Plagioclase (28%), Biotite (5%), Matrix/ground mass (10%)

5. Conclusion

In conclusion, the Munsell visual system is less effective in quantitatively assessing the rock mass surface colour compared to CIELAB colour space. The CIELAB colour space shows the efficiency of L*a*b* results in weathering assessment with a strong correlation. The Munsell Rock chart method is challenging to operate on site, but the surface colour can be precisely characterized using a colourimeter, as in the case of CIELAB colour space. Munsell Conversion Software CMC2.1 is reliable and easy to use for L*a*b* value conversion results. Both techniques are reliable for sorting and giving rock mass surface colour at the study area for granitic slope and are practically used for weathering assessment. The geomechanical test results of the rock samples from the study area verified the results of L*a*b* for both techniques of rock mass surface colour. The findings also showed a more significant contribution of digital value by CIELAB colour space on the characterization of rock mass surface colour as compared to Munsell system. Mineralogical studies also verified the findings. The darkness of surface colour represents the major value of feldspar and plagioclase while the grayish colour represents the amount of quartz, which is also most important. This investigation was carried out based on samples obtained in the quarry; for this reason, the results could be helpful in the implementation of a weathering assessment in another cut slope. Nevertheless, the difference in rocks texture must be considered under evaluation before applying this methodology.

Acknowledgments
This study is funded by an AUNSEED-Net/JICA Project under the Alumni Support Program for Research on the topic 'Development of New Evaluation Methods for Geomechanical Characterization'. The first author expresses gratitude to USM GRA Assist for the first-year financial support of Ph.D. study. The authors would also like to express their gratitude to the anonymous reviewers for the constructive suggestions in improving this article.

ORCHID iDs

M. Razali https://orcid.org/0000-0002-6994-2149
M.A.M Ismail https://orcid.org/0000-0003-0541-0693

References

[1] V. Ghiasi, H. Omar, and B. K. Huat, "A study of the weathering of the Seremban granite," *Electronic Journal of Geotechnical Engineering*, vol. 14 D, 2009.

[2] M. H. Ghobadi and A. A. Momeni, "Assessment of granitic rocks degradability susceptive to acid solutions in urban area", DOI: 10.1007/s12665-010-0895-6.

[3] L. Borrelli, S. Coniglio, S. Critelli, A. La Barbera, and G. Gullà, “Weathering grade in granitoid rocks: The San Giovanni in Fiore area (Calabria, Italy),” *Journal of Maps*, vol. 12, no. 2, 2016, doi: 10.1080/17445647.2015.1010742.

[4] T. V. Tran, D. Alkema, and R. Hack, "Weathering and deterioration of geotechnical properties in time of groundmasses in a tropical climate," *Engineering Geology*, vol. 260, p. 105221, Oct. 2019, DOI: 10.1016/J.ENGGEO.2019.105221.

[5] S. V. A. N. K. Abad, E. T. Mohamad, I. Komoo, and R. Kalatehjari, "Assessment of weathering effects on rock mass structure," *Jurnal Teknologi*, vol. 72, no. 1, pp. 71–75, 2015, DOI: 10.11113/jt.v72.2875.

[6] S. v. Alavi Nezhad Khalil Abad, E. T. Mohamad, and I. Komoo, "Dominant weathering profiles of granite in southern Peninsular Malaysia," *Engineering Geology*, vol. 183, pp. 208–215, 2014, DOI: 10.1016/j.enggeo.2014.10.019.

[7] M. Razali, M. Ashraf, M. Ismail, T. Miyoshi, K. Date, and Y. Yokota, "Geo Engineering International Conference 2020 INFLUENCE OF CIELAB COLOUR SPACE," 2020.

[8] H. Tobe, Y. Miyajima, S. Shirasagi, T. Yamamoto, and J. Kawabata, "A rapid image analyzing method for determining crack distribution and interval on tunnel faces," *ISRM International Symposium - 10th Asian Rock Mechanics Symposium, ARMS 2018*, 2018.

[9] H. Tobe, Y. Miyajima, S. Shirasagi, and T. Yamamoto, "A new method for predicting of rock fall from the tunnel face," *Rock Mechanics for Natural Resources and Infrastructure Development-Proceedings of the 14th International Congress on Rock Mechanics and Rock Engineering, ISRM 2019*, pp. 1761–1768, 2020.

[10] D. B. Kemp, "Colourimetric characterisation of flatbed scanners for rock/sediment imaging," *Computers and Geosciences*, vol. 67, pp. 69–74, 2014, DOI: 10.1016/j.cageo.2014.03.002.

[11] D. J. Croft and K. Pye, "Colour theory and evaluation of an instrumental method of measurement using geological samples for forensic applications," *Geological Society Special Publication*, vol. 232, pp. 49–62, 2004, DOI: 10.1144/GSL.SP.2004.232.01.07.

[12] Yu. N. Vodyanitskii and N. P. Kirillova, "Conversion of Munsell colour coordinates to Cie-L*a*b* system: Tables and calculation examples," *Moscow University Soil Science Bulletin*, vol. 71, no. 4–5, pp. 139–146, Oct. 2016, DOI: 10.3103/S0147687416040074.

[13] M. Potočnik, B. Klemenc, F. Solina, and U. Herlec, "Computer aided method for colour calibration and analysis of digital rock photographs," *Geologija*, vol. 58, no. 2, pp. 247–260, 2015, DOI: 10.5474/geologija.2015.020.

[14] U. O. Cinko and B. Becerir, "Dependence of colour difference formulae on regular changes of colour coordinates in CIELAB colour space," *Industria Textila*, vol. 70, no. 3, pp. 248–254, 2019, DOI: 10.35530/IT.070.03.1525.

[15] Y. Yeerken, "Applicability of the CIELAB and CIEDE2000 Formulae for Detection of Colour Changes in Colour-Changeable Chewing Gum for Evaluating Masticatory Function," *Journal of Clinical and Diagnostic Research*, pp. 119–123, 2017, DOI: 10.7860/jcdr/2017/23950.9754.

[16] L. Lovisolo, "Improvement of Objective Image Quality Evaluation Applying Colour Differences in the CIELAB Colour Space," *International Journal of Image Processing*, no. 5, pp. 236–243, 2011.

[17] M. K. Gokay, "Analyses of Rock Surface Colour Changes Due To Weathering," *Selcuk University Journal of Engineering, Science and Technology*, vol. 6, no. 1, pp. 1–13, 2018, DOI: 10.15317/scitech.2018.111.

[18] A. K. Roy Choudhury, *Colour-difference assessment*, vol. 2. Woodhead Publishing Limited, 2015. DOI: 10.1533/9781782423881.55.

[19] R. A. Viscarra Rossel, B. Minasny, P. Roudier, and A. B. McBratney, "Colour space models for soil science," *Geoderma*, vol. 133, no. 3–4, pp. 320–337, Aug. 2006, DOI: 10.1016/J.GEODERMA.2005.07.017.

[20] Z. H. Yue, Q. Huang, L. Xiao, J. Li, and C. M. Huang, "Quantitative Conversion of Soil Colour from CIELAB to Munsell System," *Guang Pu Xue Yu Guang Pu Fen Xi/Spectroscopy and Spectral Analysis*, vol. 39, no. 9, 2019, DOI: 10.3964/j.issn.1000-0593(2019)09-2842-05.

[21] L. M. O. Sousa and B. M. M. Gonçalves, "Colour assessment of granitic rocks and implications for their ornamental utilization," *Colour Research and Application*, vol. 37, no. 5, pp. 375–383, 2012, DOI: 10.1002/col.20681.

[22] N. P. Kirillova, Y. N. Vodyanitskii, and T. M. Sileva, "Conversion of soil colour parameters from the Munsell system to the CIE-L*a*b* system," *Eurasian Soil Science*, vol. 48, no. 5, pp. 468–475, 2015, DOI: 10.1134/S1064229315050026.

[23] I. Yoshitake, T. Maeda, and M. Hieda, "Image analysis for the detection and quantification of concrete bugholes in a tunnel lining," 2018, DOI: 10.1016/j.cscm.2018.01.002.

[24] Y. Liang, "Rock fracture skeleton tracing by image processing and quantitative analysis by geometry features," *Journal of Geophysics and Engineering*, 2016, DOI: 10.1088/1742-2132/13/3/273.

[25] N. Ohta and A. R. Robertson, "Colourimetry: Fundamentals and Applications," 2005.

[26] "ISO - ISO/CIE 11664-4:2019 - Colourimetry — Part 4: CIE 1976 L*a*b* colour space." https://www.iso.org/standard/74166.html (accessed Jul. 29, 2021).

[27] J. A. Franklin, "Suggested method for determining point load strength," *International Journal of Rock Mechanics and Mining Sciences and*, vol. 22, no. 2. pp. 51–60, 1985. DOI: 10.1016/0148-9062(85)92327-7.

[28] Astm D5731-08, "Standard Test Method for Determination of the Point Load Strength Index of Rock and Application to Rock Strength Classifications," *ASTM International, West Conshohocken, PA,* vol. 22, no. 2, pp. 51–60, 2008.

[29] E. T. Brown, *Rock characterization testing and monitoring. ISRM suggested methods.* 1981. DOI: 10.1016/0148-9062(81)90524-6.

Rapid method of aerobic granular sludge bioreactor start-up for domestic wastewater treatment

Karn Tanavarotai*[1], Aznah Nor Anuar*[1], Alijah Mohd Aris[2], Zhongfang Lei[3] and Mohd Hakim Ab Halim*[1]

[1] Malaysia-Japan International Institute of Technology, Universiti Teknologi Malaysia, Kuala Lumpur, Malaysia.
[2] Indah Water Research Center, Bunus Sewage Treatment Plant, Kuala Lumpur, Malaysia.
[3] Graduate School of Life and Environmental Sciences, University of Tsukuba, Tsukuba, Japan.

*Corresponding author e-mail: karngunn@hotmail.com; aznah@utm.my; hakim_90@rocketmail.com

Abstract. This study presents a rapid method on how to speed up aerobic granular sludge (AGS) cultivation and ensure excellent and stable removal performance during bioreactor operation for domestic wastewater treatment. This new strategy consists of start-up the bioreactor using only anaerobic granular sludge (AnGS) as a seed and feeding with crude sewage extracted from a full-scale Extended Aeration Plant. This experiment used a 2.5 L lab-scale sequencing batch reactor (SBR). The bioreactor operated at low dissolved oxygen (DO) concentration controlled at the value of 2.0 mg/L and below. After 60 days of operation, it clearly showed that almost 90% of AnGS seeds turned from black color to brown. The physical characterization analysis showed that the average sizes were unchanged, and the granules remained compact. Also, the SBR operation monitored with brown granules showed stable removal performance. Average removal efficiencies during steady-state cycles at room temperature of COD, ammoniacal nitrogen, and phosphate reached 84%, 92%, and 100%, respectively.

1. Introduction

Aerobic granular sludge (AGS) technologies are one of the biological wastewater treatment technologies [1, 2] that have lower requirements of land used and operation cost since the whole treatment process can be done within a single sequence batch reactor [3-7]. AGS technology has developed in both treating performance and variety of domestic [8, 9] and industrial [3, 10-14] wastewater in the recent decade. However, one of the main bottlenecks of this technology is the long granulation period during start-up. The cause of AGS long start-up time due to AGS usually cultivated from activated sludge which required several months to develop into granules [15]. Several methods had proposed to overcome the long granulation time during the start-up of AGS technology, such as adding additive [16-18], storing AGS in different forms [2, 19-23], and transforming AnGS to AGS [24, 25].

Transforming AnGS to AGS method was conducted in this research since AnGS is easy to find and purchase compared to AGS. Furthermore, AnGS did not require a complicated condition to store, and transforming AnGS into AGS consumes less time than granulating the AGS from activated sludge. The cultivating of AnGS in a reactor feeding with wastewater for a while was how to obtain the AGS. The AnGS transforming methods that have been proposed by converting AnGS to AGS cultivating in a continuous up-flow reactor can be done within 45 days [25], while cultivating in an SBR can be done in 60 days [24] by both methods used the synthetic wastewater to develop.

Based on the research of Chen *et al.* [26], the AGS that develop with actual wastewater can adapt faster and better to the change of the wastewater during the operation compared to AGS cultivated from synthetic wastewater. Resulting in more sustainable on treating the actual domestic wastewater where there was always a fluctuation in the composition.

The success of transforming AnGS into AGS to lower the start-up time of AGS technology had indicated in previous research. However, during cultivation, the influent used was synthetic wastewater. The use of synthetic wastewater led to high costs and a low possibility to adopt it into full scale. This research investigated the possibility of granulating the AGS in the SBR by using only the AnGS as the seed sludge with the actual domestic wastewater influents to shorten the long start-up period of AGS technology. As a result, the research can enhance the development of AGS with less time consuming, lower cost, the possibility to be used in full scale, and the obtained AGS be able to sustain and adapt to the fluctuation of wastewater.

2. Materials and Methods

2.1. Reactor setup and operation

An automated 2.5 L lab-scale SBR operated in the laboratory room at Indah water research center, Kuala Lumpur, Malaysia; the reactor had a diameter of 6 cm and 100 cm in height. The working volume of the reactor was 2350 mL with an exchange volume of 850 mL. The influent port and the bubble diffuser were at the bottom of the reactor. The effluent port was at the middle of the reactor, and the whole reactor was operated automatically with the time relay switches. The reactor operated at lab room temperature. The aeration was controlled with an airflow meter and made sure not to be too harsh to lower the break of granules. The DO in the reactor was maintained to be around 1.5-2.0 mg/L. The schematic diagram of the SBR used in this research was as shown in Figure 1.

The reactor runs 3 h per cycle, including influent feeding for 1 h, followed by aerating for 1 h 50 min, then 5 min of settling, and 5 min of discharge. The reactor runs continuously for 60 days.

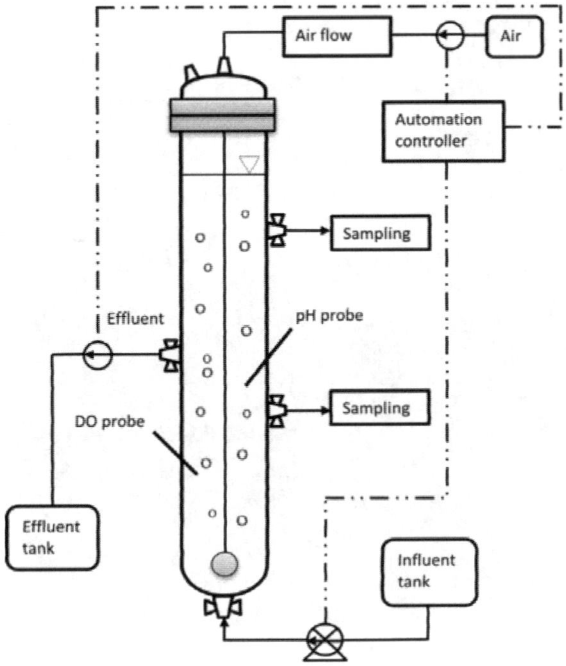

Figure 1. Schematic diagram of 2.5L lab-scale SBR.

2.2. Influent and seeding sludge
The AnGS was purchased from Thailand and used as seed sludge with an average diameter of 2.0-3.0 mm. The granules have a round shape and black color.

The influent used in this research was an actual domestic wastewater pass-through screening and grit removal from a full-scale extended aeration type plant. The influent tank was empty and refilled manually every two days.

2.3. Analytical
The granules characteristic was done both before and after transforming to AGS. The analysis of MLSS, MLVSS, SVI_5, and settling velocity using the standard method [27]. A Leica EZ4W stereomicroscope with a 5-megapixel camera was used to determine the granules' size. The granule's pictures took using Huawei P30 Pro in super macro mode with a 20-megapixel camera and regular mode with a 40-megapixel camera.

The DO in the reactor was monitored during the transformation process using YSI Pro ODO optical dissolved oxygen instrument. The influent and effluent samples for performance analysis after 60 days of operation. COD, BOD_5, ammoniacal nitrogen ($NH_3 - N$), nitrate (NO_3^-), phosphate ($PO_4^{3-} - P$), and total suspended solids (TSS) were analyzed using the standard method [27]. The COD and $PO_4^{3-} - P$ were digested by HACH COD 45600 reactor, then measured with HACH DR 3900 spectrophotometer. Moreover, $NH_3 - N$ and NO_3^- were also measured using HACH DR 3900 spectrophotometer. BOD_5 was measured using HACH BODTRAK II at 20°C.

3. Result
3.1. General observation
The SBR was started up by seeding with the AnGS purchased from Thailand. After 30 days of cultivation, the granule's color becomes brown except the bottom part where the granules were clogged and not exposed to enough aeration, as shown in Figure 2. According to the visual inspection, the granules remain round and compact. The DO in the reactor was maintained to be around 1.5-2.0 mg/L.

Figure 2. Black granules at the bottom of the reactor.

3.2. AnGS and AGS characteristics

The randomly sampling AnGS had an overall diameter of 2.0-3.0 mm with a dense, round, and compact size. The granules had black color with the MLSS and MLVSS of 3600 mg/L and 2500 mg/L. While the SVI_5 and settling velocity of the AnGS were 59 mL/g and 98.77 m/h. Figure 3 shows the picture of AnGS color and size, and Table 1 indicates the characteristics of AnGS.

After 60 days of operation, the granules were sampled randomly for characteristic analysis. The result showed that granules had an average diameter around 2.0-3.0 mm, still round, compact, and dense as before transformed into AGS. The granule's color became brown. MLSS and MLVSS were 3450 mg/L and 2760 mg/L. The AGS SVI_5 and settling velocity are 62 mL/g and 62.18 m/h, respectively. Figure 3 shows the color and size of the AGS, and Table 1 shows the characteristics of AnGS and AGS.

Figure 3. Color of (a) AnGS and (b) AGS. Size of (c) AnGS and (d) AGS.

Table 1. Characteristics of AnGS compared with AGS.

Parameter	AnGS	AGS
pH	7.2	6.85
MLSS (mg/L)	3600	3450
MLVSS (mg/L)	2500	2760
SVI_5 (ml/g)	59	62
Size (mm)	2-3	2-3
Settling velocity (m/h)	98.77	62.18

3.3. Removal performance

After 60 days of operation, the removal performance of the transformed granules analyzes by sampling the influent and effluent from the reactor. The result indicated that the COD removal rate was over 80%, the BOD_5 removal rate was around 80%, the ammoniacal nitrogen removal rate was above 90%, and the phosphate removal rate could reach 100%. Table 2 shows the characteristics of influent and effluent after 60 days.

Table 2. The characteristics of influent and effluent after 60 days of operation.

Parameter	Influent	Effluent	Removal Percentage
pH	6.85	6.28	-
COD (mg/l)	317	52	83.56%
BOD_5 (mg/l)	123	24	80.49%
$NH_3 - N$ (mg/l)	30.1	2.3	92.36%
NO_3^- (mg/l)	6.7	9.2	No removal
$PO_4^{3-} - P$ (mg/l)	4.3	0	100%
TSS (mg/L)	210	80	61.90%

4. Discussion

4.1. Granules transforming observation

During the transformation process, after 30 days, most of the granules started to turn from black to brown color. In contrast, granules in the bottom of the reactors remain black. The granules at the bottom of the reactor were not transforming because they did not expose to enough aeration. The issue was solved by every 2-3 days manually stirring was performed during aeration to stir up the granules. With the solution, black granules were disappeared from the reactor. Figure 5 indicated the picture of black granules mixed with brown granules after stirring the bottom of the reactor.

According to the visual inspection, the granules remain round and compact. It indicated that maintaining the DO at around 1.5-2.0 mg/L where the aeration rate was not too harsh and hydrodynamic shear force is at optimum rate results in granules can maintain their morphology while transforming to AGS and lowering both the damage and the breaking off of the granules as expected. In contrast, granules still lost some of the volumes while operating due to the broken granules that could not be neglected but could reduce.

Figure 4. A mixture of black and brown granules after stirring.

4.2. AnGS and AGS comparison

According to the observation and analysis result, the AnGS successfully transform into AGS. The change in color from black to brown indicated that anaerobic microorganisms in the reactor started to phase out then aerobic microorganisms took place instead. The round shape and the same average size of granules showed that maintaining the DO to be around 1.5-2.0 mg/L to lower the damage on the granules was a success. The settling velocity decreased from 98.77 m/h to 62.18 m/h. The decrease in settling velocity occurred as supposed referred to Linlin *et al.* [24], decreasing in the settling velocity also indicated that the water content in the granules increased. The SVI_5 value remained the same, indicating no change in granules size.

4.3. Transformed AGS performance

According to the performance, analysis results indicated that AnGS was successfully transformed into AGS using actual domestic wastewater. In addition, the overall removal rate above 80% indicated that the transformed granules could treat wastewater. However, the evaluated removing performance seems not high compared to granules cultivated with synthetic wastewater, caused by fluctuation of composition within the influent water the granules must expose. In contrast, the removal performance of over 80% shows that the granules can sustain and adapt with the fluctuating composition of the actual domestic wastewater.

5. Conclusion

The research indicates that transforming AnGS into AGS with actual domestic wastewater is one of the possible ways to cultivate AGS that reduce the long cultivation time of AGS technology. The research also shows that the transformation of AnGS into AGS on a full scale was possible since the influent used was actual domestic wastewater. Using actual domestic wastewater also lowered the cost compared to using synthetic wastewater. The obtained AGS could maintain its shape and the removal performance is over 80%. The AGS obtained should be able to adapt and sustain to the change in influent composition. Maintaining the DO within the reactor to be around 1.5-2.0 mg/L lower the damage and break of the granules, led to no change in the average size after the AnGS successfully transformed into AGS.

Acknowledgments

This work is funded by the AUN/SEED-Net Grant no: R.K130000.7343.4B403 under Collaborative Education Program for Sustainable Environmental Engineering Network (CEP-SEEN)

References

[1] Show K Y, Lee D J and Tay J H 2012 Aerobic granulation: advances and challenges *Appl Biochem Biotechnol* **167** 1622-40

[2] Hu J, Zhang Q, Chen Y Y and Lee D J 2016 Drying and recovery of aerobic granules *Bioresour Technol* **218** 397-401

[3] Abdullah N, Ujang Z and Yahya A 2011 Aerobic granular sludge formation for high strength agro-based wastewater treatment *Bioresour Technol* **102** 6778-81

[4] Nancharaiah Y and Reddy G K K 2018 Aerobic granular sludge technology: mechanisms of granulation and biotechnological applications *Bioresource technology* **247** 1128-43

[5] Li J, Ding L B, Cai A, Huang G X and Horn H 2014 Aerobic sludge granulation in a full-scale sequencing batch reactor *Biomed Res Int* **2014** 268789

[6] Pronk M, de Kreuk M K, de Bruin B, Kamminga P, Kleerebezem R and van Loosdrecht M C M 2015 Full scale performance of the aerobic granular sludge process for sewage treatment *Water research* **84** 207-17

[7] Swiatczak P and Cydzik-Kwiatkowska A 2018 Performance and microbial characteristics of biomass in a full-scale aerobic granular sludge wastewater treatment plant *Environ Sci Pollut Res Int* **25** 1655-69

[8] Sguanci S, Lubello C, Caffaz S and Lotti T 2019 Long-term stability of aerobic granular sludge for the treatment of very low-strength real domestic wastewater *Journal of Cleaner Production* **222** 882-90

[9] Hamza R A, Iorhemen O T, Zaghloul M S and Tay J H 2018 Rapid formation and characterization of aerobic granules in pilot-scale sequential batch reactor for high-strength organic wastewater treatment *Journal of Water Process Engineering* **22** 27-33

[10] Rosman N H, Nor Anuar A, Othman I, Harun H, Sulong Abdul Razak M Z, Elias S H, Mat Hassan M A, Chelliapan S and Ujang Z 2013 Cultivation of aerobic granular sludge for rubber wastewater treatment *Bioresour Technol* **129** 620-3
[11] Harun H and Anuar A N 2014 Development and Utilization of Aerobic Granules for Soy Sauce Wastewater Treatment: Optimization by Response Surface Methodology *Jurnal Teknologi* **69**
[12] Othman I, Anuar A N, Ujang Z, Rosman N H, Harun H and Chelliapan S 2013 Livestock wastewater treatment using aerobic granular sludge *Bioresour Technol* **133** 630-4
[13] Jiang Y, Shang Y, Gong T, Hu Z, Yang K and Shao S 2020 High concentration of Mn(2+) has multiple influences on aerobic granular sludge for aniline wastewater treatment *Chemosphere* **240** 124945
[14] Paulo A M S, Amorim C L, Costa J, Mesquita D P, Ferreira E C and Castro P M L 2021 Long-term stability of a non-adapted aerobic granular sludge process treating fish canning wastewater associated to EPS producers in the core microbiome *Sci Total Environ* **756** 144007
[15] Lin H, Ma R, Hu Y, Lin J, Sun S, Jiang J, Li T, Liao Q and Luo J 2020 Reviewing bottlenecks in aerobic granular sludge technology: Slow granulation and low granular stability *Environmental Pollution* **263**
[16] Basri H, Anuar A, Yuzir A and Ab Halim M H 2020 Diatomite carrier for rapid formation of Aerobic Granular Sludge. In: *IOP Conference Series: Earth and Environmental Science*: IOP Publishing) p 012028
[17] Liu S, Zhan H, Xie Y, Shi W and Wang S 2018 Rapid cultivation of aerobic granular sludge by xanthan gum in SBR reactors *Water Sci Technol* **2017** 360-9
[18] Pijuan M, Werner U and Yuan Z 2011 Reducing the startup time of aerobic granular sludge reactors through seeding floccular sludge with crushed aerobic granules *Water Res* **45** 5075-83
[19] Gao D, Yuan X and Liang H 2012 Reactivation performance of aerobic granules under different storage strategies *Water Res* **46** 3315-22
[20] Yuan X, Gao D and Liang H 2012 Reactivation characteristics of stored aerobic granular sludge using different operational strategies *Appl Microbiol Biotechnol* **94** 1365-74
[21] Zhao Z, Wang S, Shi W and Li J 2013 Recovery of stored aerobic granular sludge and its contaminants removal efficiency under different operation conditions *BioMed research international* **2013**
[22] Wan C, Lee D J, Yang X, Wang Y and Lin L 2014 Saline storage of aerobic granules and subsequent reactivation *Bioresour Technol* **172** 418-22
[23] Duarte K L S, Castellanos R M, Costa R C, Mahler C F and Bassin J P 2021 Start-up of an aerobic granular sludge system from stored granules: Evaluating the impact of storage period on biomass activity and stability and the effect of temperature on nitrification and phosphorus removal rates *J Environ Manage* **296** 113200
[24] Linlin H, Jianlong W, Xianghua W and Yi Q 2005 The formation and characteristics of aerobic granules in sequencing batch reactor (SBR) by seeding anaerobic granules *Process Biochemistry* **40** 5-11
[25] Sun H, Yu P, Li Q, Ren H, Liu B, Ye L and Zhang X X 2017 Transformation of anaerobic granules into aerobic granules and the succession of bacterial community *Appl Microbiol Biotechnol* **101** 7703-13
[26] Chen Y, Zhu J-Y, Qin Y, Zhang Z-M and Yuan S-C 2017 Reactivation of hypersaline aerobic granular sludge after low-temperature storage *Tecnología y ciencias del agua* **8** 61-70
[27] Rodger Baird; Andrew D Eaton; Eugene W Rice; Laura Bridgewater; American Public Health Association A W W A, ; Water Environment Federation, 2017 Standard Methods for the Examination of Water and Wastewater. ed A W W A Rodger Baird; Andrew D Eaton; Eugene W Rice; Laura Bridgewater; American Public Health Association, ; Water Environment Federation,: American Public Health Association, American Water Works Association, Water Environment Federation) p 1504

IOP Conf. Series: Earth and Environmental Science 1091 (2022) 012050 doi:10.1088/1755-1315/1091/1/012050

Development of Microalgae-Bacteria Aerobic Granular Sludge for Landfill Leachate Treatment

C Khen[1], D Ilmasari [1], A Yuzir*[1]

[1] Malaysia-Japan International Institute of Technology, Universiti Teknologi Malaysia, Jalan Sultan Yahya Petra, 54100, Kuala Lumpur, Malaysia

*Corresponding author e-mail: muhdaliyuzir@utm.my

Abstract. Over the past few decades, aerobic granular sludge has become favorable biotechnology in wastewater treatment. Combining the AGS process with microalgae technology is desired to achieve higher treatment efficiency, especially for high-strength wastewater. This study aims to investigate the effectiveness of microalgae-bacteria AGS in treating landfill leachate. The photo-sequencing batch reactor, inoculated with *Acutodesmus obliquus*, was fed with old synthetic leachate, and run for 30days. The result showed that chemical oxygen demand (COD) removal up to 96%. The initial ammonia concentration was 130mg/L and achieved an efficiency of up to 80.77%. At the same time, the average removal of the phosphorus was 19.5mg/L, equal to 48.68%. The result of microalgae-bacteria AGS can produce the fast-settling properties of the granules throughout the experiment.

1. Introduction

Malaysia hugely relies on landfilling as a primary disposal method in dealing with municipal solid waste (MSW), which has become one of the most significant environmental issues. There were 300 operated and non-operated solid waste disposal sites in Malaysia, and only 13 were classified as sanitary landfills [1]. Approximately 95% of the collected waste was disposed in open-dumpling and landfill sites, making most sites exceed their full operating capacity [2]. Later, the degradation of the waste in landfills over time releases gaseous products and liquid waste called leachate. Leachate is considered a very toxic and highly contaminated solution product, and its composition varies widely depending on landfill's age, climate condition, type of waste [3].

The constituent of leachate consists of ammonia, dissolved solids, heavy metals, biodegradable, and refractory organic matter. Most importantly, landfills will continue to produce leachate for 30 to 50 years even after the landfills site is closed [4]. It is significant to collect leachate and treat to prevent the impact on the environment and human health, usually by removing the hazardous substances before discharging them to the water system. It is pretty challenging to treat leachate due to its variety of characteristic composition and quantity [5]. Several different technologies, such as chemical, biological, physiochemical, and physical methods, are available to treat landfill leachate to achieve the standards established by legislation. However, these conventional methods generally require various processes which are expensive and complicated [6].

Despite the limitation, the biological treatment process is still used worldwide to treat leachate. Though it is only effective for young leachate as it contains an easily biodegradable organic matter, old leachate is predominantly constituted by refractory organic matter and high ammonia nitrogen concentration which is not easily degraded [6]. In the meantime, microalgae are receiving great attention due to their ability to produce valuable metabolites and perform wastewater treatment at once. The worthiness of microalgae is due to their capability to uptake metal, time and energy-saving, eco-friendliness, and enable easy recovery [7]. Nevertheless, harvesting is one of the most challenging aspects of microalgae biotechnology, accounting for 20 to 30 percent of total production expenses [8]. Recent studies discovered that microalgae could integrate into the aerobic granular sludge system to induce its formation and stable performance. Therefore, cultivating microalgae with the AGS system was favored. Aerobic granulation for wastewater treatment is a recent innovation inside biofilm reactors constitutes a providential technology that does not require material support for biofilm growth which is considered a particular type of biofilm compromised of self-immobilized cells [9]. It is known to have a strong microbial structure, compact and dense, with excellent settleability and high biomass retention, which is suitable for treating toxic substrates and a wide range of wastewater types [10]. However, the ability of AGS to treat leachate wastewater that might contain toxic substrates is still unknown.

Considering the information of leachate treatment by microalgae-bacteria AGS is lacking. The purpose of this study is to cultivate microalgae-bacteria aerobic granular sludge for landfill leachate treatment. *Acutodesmus obliquus* was used as a supplementary microalga in this investigation. The inclusion of microalgae species was intended to improve treatment efficiency while also promoting the granular process.

2. Materials and methods

2.1. Reactor setup and operational condition

The PSBR, which consists of a plexiglass cylinder with a diameter of 60mm and a height of 650 mm, was utilized to develop microalgae-bacteria aerobic granular sludge. The given diameter and height provide the H/D ratio of 10.83, and the total volume of the reactor is 1.8 L. The reactor had a working volume of 1L and a volume exchange ratio of 50 percent (VER). This PSBR system was run in an eight-hour cyclic time, consisting of 10 minutes feeding, 90 minutes of non-aeration, 365 minutes of aeration, settling time of 10 minutes settling time, and 5 minutes for wastewater to be discharged [11]. The reactor was illuminated by white LED light at 130 $\mu mol/m^2s$ light intensity under constant dark/night cycles of 12h/12h and operated at room temperature ($25 \pm 2°C$). The diagram of the SBR is shown in figure 1. A 500 mL seed sludge, including activated sludge and microalgae, was used to inoculate the reactor. 500 mL of wastewater was supplied and released at the beginning and end of each cycle.

2.2. Seed sludge

In the study, seed sludge consisted of activated sludge from Indah Water Konsortium (IWK) Sewage Treatment Plant in Bunus, Kuala Lumpur, and target microalgae species *Acutodesmus obliquus* from algae biomass ikozha, Malaysia-Japan International Institute of Technology. The collected activated sludge was filtered to eliminate massive residue to avoid clogging the reactor tube, followed by pre-conditioning, which involved acclimating the activated sludge for 24 hours at room temperature. *Acutodesmus obliquus* was selected regarding its remarkable growth rate in the presence of leachate and positive growth rate in all nutrient-rich settling [12]. The microalgae were cultured in AF6-medium for two weeks to reach the exponential phase of microalgae growth. [13]. Subsequently, it was inoculated with activated sludge with approximately 10^8 cells. The seed sludge biomass initial concentration was 9575mg/L, and the sludge volume index was (SVI_{30}) was 69.8mL/g.

2.3. Wastewater preparation

This synthetic wastewater was prepared to follow the characteristic of old leachate. This leachate was formulated with tannic acid to stimulate the refractory organic matter contained in old leachate. The ingredients to prepare 1 liter of synthetic old leachate were listed as follows (mg/L): tannic acid 200, NaCl 2000, $CaCl_2$ 700, $NaHCO_3$ 2000, NaOH 297, K_2HPO_4 32.5, NH_4Cl 120-150, and trace metal 0.02mL ($MnSO_4.7H_2O$ 500mg, $CoSO_4.7H_2O$ 150mg, $NiSO_4.6H_2O$ 500 mg, $CuSO_4.5H_2O$ 40mg, H_3BO_4

50mg, $ZnSO_4.7H_2O$ 50mg, $(NH_4)_6Mo_7O_{24}.4H_2O$ 50mg, $Al_2(SO_4)_3.16H_2O$ 30mg, and 96%H_2SO_4 1ml) [14].

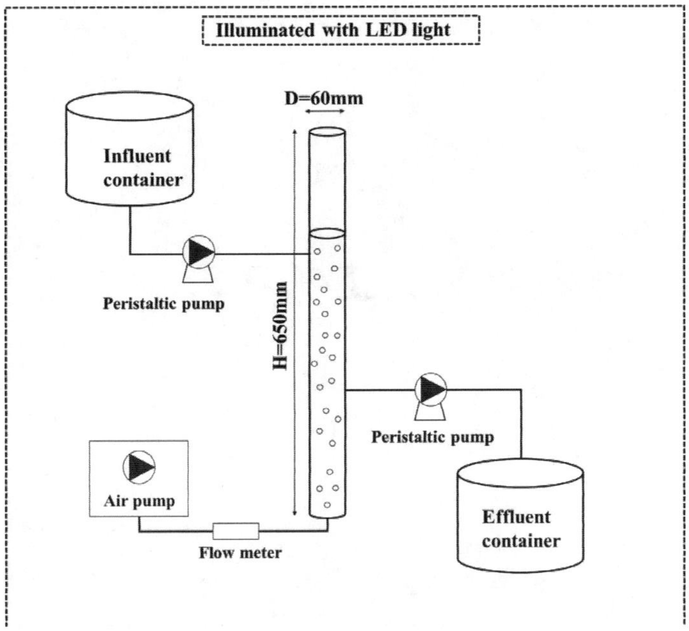

Figure 1. The simplified diagram of PSBR.

2.4. Analytical methods
The wastewater and sludge analyses were carried out using parameters including chemical oxygen demand (COD), total phosphorus (TP), and ammoniacal nitrogen (NH3-N). The sample was taken twice a week and analysed immediately for a better result, and the effluent was not filtered before being analysed. The system's development was evaluated on mixed liquor suspended solids (MLSS) and sludge volume index (SVI_{30}) following standard methods [15]. The granular sludge settling velocity and aspect ratios were used to undertake physical characterization of the microalgae-bacteria aerobic granular sludge. The average time it took granular sludge to reach the bottom of a cylindrical column filled with tap water was used to calculate velocity. At the same time, the aspect ratios were determined as the ratio of the granular sludge's shortest and longest dimensions to reflect the granular sludge's roundness.

3. Results and Discussion

3.1. Development of microalgae-bacteria aerobic granular sludge

3.1.1. Morphological observation. The initial seed sludge was microalgae culture, and the activated sludge had an uneven form and flossy structure with extremely tiny bioflocs. After seven days, the sludge began to form minute granules ranging from 1.3 to 2.3mm. Due to the sheer presence of microalgae, the colour of the sludge also changed to light brown. On day 14, the granule size was slightly more noticeable, with diameters ranging from 1.45 to 2.61mm. It is obvious during the trial period that the granule size began to dominate with a more compact shape on day 30th. In the reactor, granular sludge with an average size of 1.54 to 2.41mm and a dark brown colour was observed. Figure 2 depicts the maturation of the granular sludge.

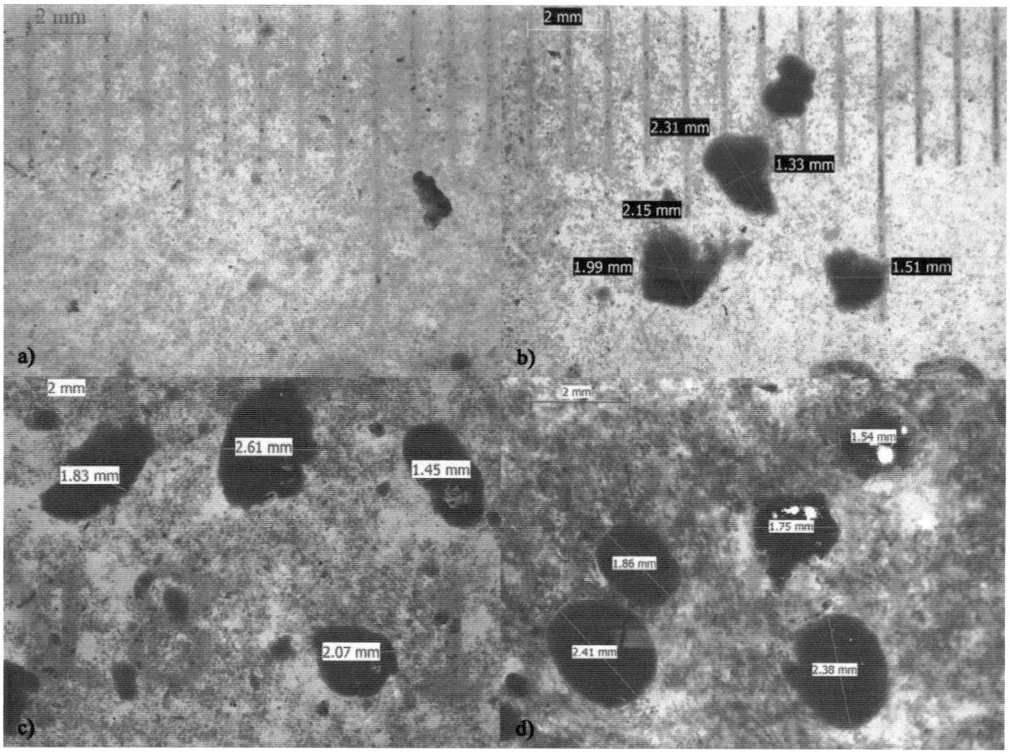

Figure 2. The development of microalgae-bacteria AGS inside the PSBR system **a)** Seed sludge; **b)** At 7th day of operation; **c)** At the 14th day of operation; and **d)** At 30th day of the experiment.

3.1.2. Biomass growth and settleability. The seed sludge contained activated sludge and microalgae at an initial MLSS of 9575mg/L. During the experimental period, weekly evaluations of biomass growth and settleability were performed, as shown in figure 2. During the first week, the MLSS dropped considerably to 4200mg/L. It seemed that less compacted sludge with poor settling abilities was discharged with the wastewater. This process happened as a result of the desire to keep sludge with strong settling capabilities in order to adapt and develop a granular sludge [13]. Hence, the MLSS concentration started to recover from day 7 to day 21 with an average concentration of 5525mg/L. Despite that, the final MLSS remained only 3350mg/L by day 30. It appeared that some granules might be too big and started to disintegrate and wash out through the effluent. Moreover, the amount of sludge in the reactor was not monitored during the experimental study. Therefore, it is essential to pay attention to the sludge contained in the reactor to avoid the major drop for the whole development.

The settleability of the granular sludge is represented by the sludge volume index for 30mins (SVI_{30}). The initial value of SVI_{30} was 69.8mL/g, which suggested poor settling properties of the seed sludge. The SVI_{30} rapidly dropped to 29.07 mL/g over the first 14 days of reactor operation as the seed sludge changed into granular sludge. The result showed the importance of microalgae and filamentous bacteria in the granulation process [16]. Even though the SVI_{30} was minor increased on day 21 to 37.21 mL/g due to the disintegration of the granule and biomass washout, the SVI_{30} was able to attain 35.82 mL/g by day 30 of the experiment.

3.1.3. Physical characteristic. The physical characteristic of aerobic granular sludge was addressed for the mature granules' settling velocity and aspect ratio. The average aspect ratio of microalgae-bacteria aerobic granular sludge was 0.8, demonstrating the capacity to swiftly produce a round-shaped granular sludge utilizing synthetic leachate over the 30-day experimental period. Additionally, the settling velocity of the mature granule found after 30 days achieved 40.65m/h, which was higher than a previous study from Arcila and Buitrón [17] with the formation of microalgae-bacteria aggregate for treating municipal wastewater of 18m/h. At the same time, Cai et al [18] achieved a settling velocity of 12.8m/h of the granules treating synthetic domestic wastewater. Thereby, this result could indicate the fast-settling velocity of the mature granules. Consequently, the result of both SVI_{30} and settling velocity could show excellent settling properties throughout the experiment. Regarding this result, it is worth mentioning that although microalgae-bacteria AGS can produce the fast-settling properties of the granules, it is also essential to maintain the concentration of MLSS along with time.

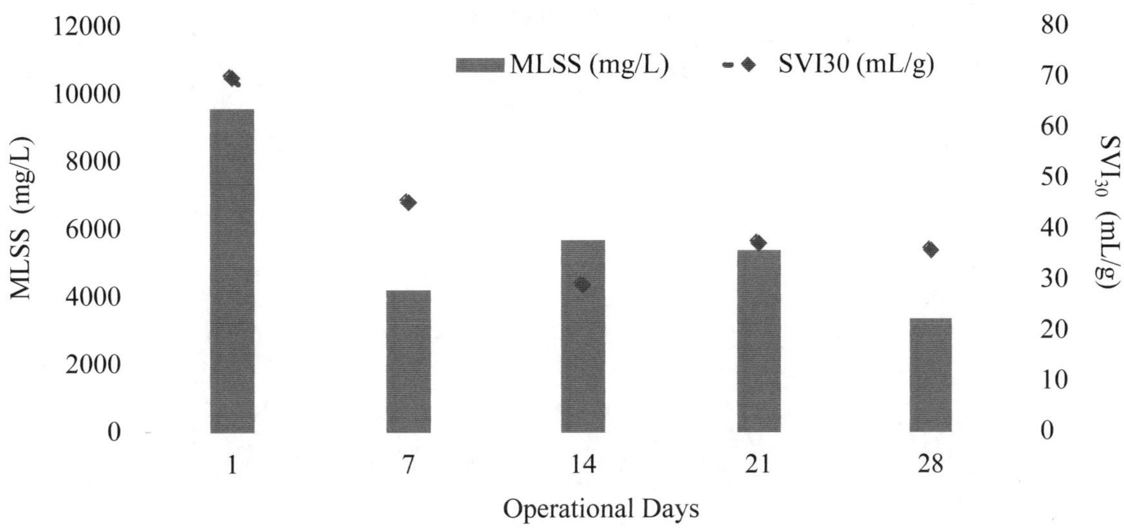

Figure 3. Dynamic of the biomass concentration (MLSS) and settle ability (SVI_{30}) throughout the 30 days of experiment.

3.2. Performance of microalgae-bacteria granular sludge

3.2.1. COD removal. The initial concentration of COD in the influent was 3690mg/L. As shown in figure 4 (a), stable COD removal was observed throughout the 30days of the experimental period. Despite a biomass washout at the beginning of the experiment, COD removal remained greater than 90%. The excellent potential for organic removal was performed mainly by the bacterial consortium in the activated sludge, which converts organic carbon as an energy source for their growth and proliferation. As a result, the heterotopic bacteria proliferate faster than microalgae, implying that the removal of organic matter was mostly attributable to bacteria in the co-culture system [19]. By the end of the experiment, the average COD removal rate was 96 percent, resulting in a COD concentration of 160.61 mg/L in the effluent. This finding was shown to be comparable to prior publications dealing with lower COD concentrations [20-22]. According to Huang *et al.* [20], COD removal from synthetic wastewater containing 600mg/L was around 96 percent, with a final COD content of less than 30mg/L.

In comparison, Zhang *et al.* [21] achieved 96% COD removal with the initial concentration of 1200mg/L COD. A prior investigation utilizing synthetic wastewater yielded a similar result, with an average COD removal effectiveness ranging from 72-89% [13, 19, 23]. According to the findings of this investigation, granular sludge has a high potential for organic removal.

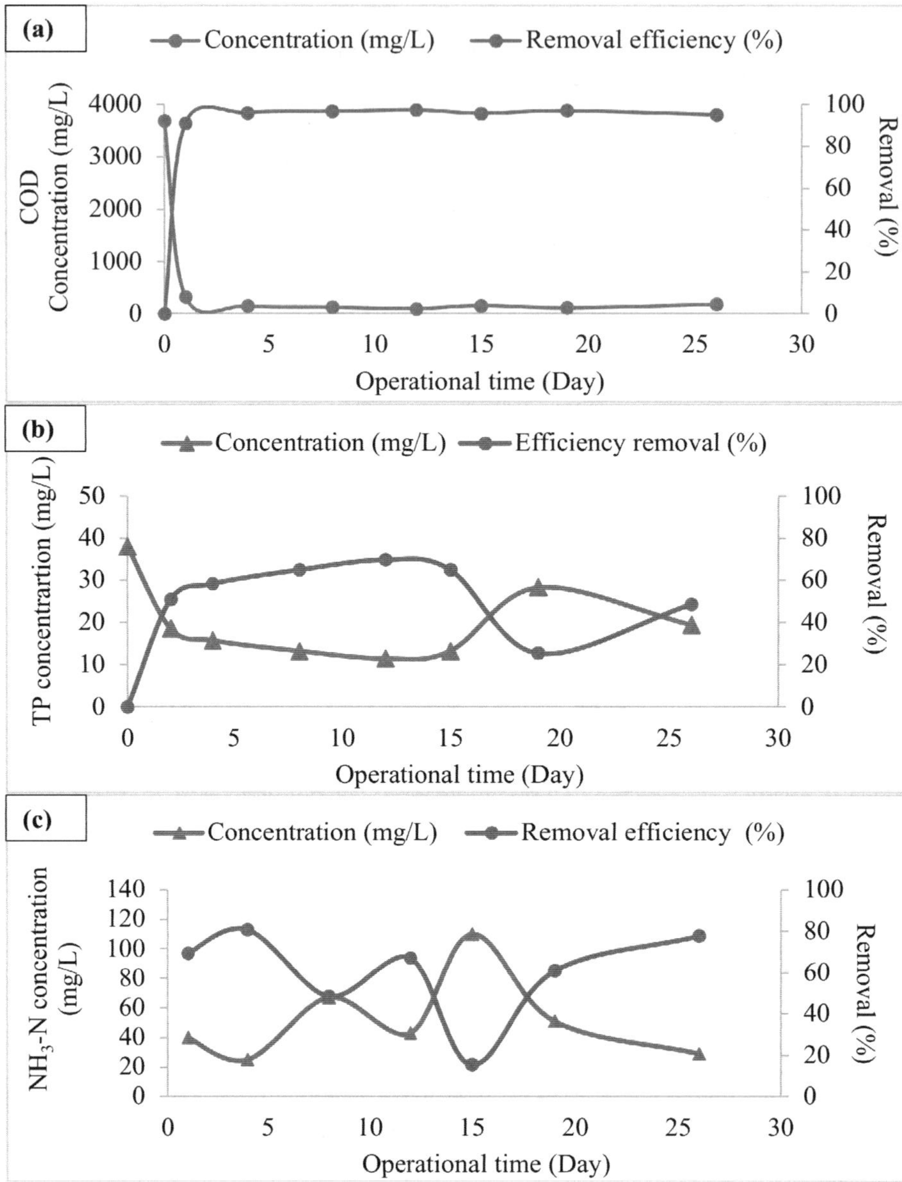

Figure 4. (a) COD concentration and percentage removal in PSBR system; **(b)** TP concentration of the effluent in PSBR; **(c)** Ammonia concentration of the effluent in PSBR; throughout 30days experiment.

3.2.2. Performance of nutrient removals. The biological nutrient removal from synthetic leachate wastewater was determined in terms of TP and ammonia. The concentration of influent P was 38mg/L; thus, TP removal can be analyzed by PO_4-P removal. The result shows that the average effluent PO_4-P concentration during the first 15 days was 13.3mg/L, equal to 65% removal efficiency. Subsequently, from day 15 to 20, PO_4-P in the effluent was slightly increased to 28.3mg/L, equivalent to 25.53% in removal efficiency. However, by day 30, this concentration remains only 19.5mg/L, equal to 48.68%. The Average PO_4-P concentration was 16.8mg/L resulting in 55.79% PO_4-P removal from synthetic leachate wastewater. These findings were found to be lower than previous studies using synthetic wastewater [19, 21]. Zhang *et al.* [21] and Zhu *et al.* [19] achieved PO_4-P removal to 89% and 100%, respectively. According to Purba *et al.* [13], poor PO_4-P removal efficiency may be attributable to the absence of sludge retention time (SRT) control in the system; therefore, no intentional biomass discharge was conducted. Furthermore, the organic loading rate (OLR) in the system is not favorable for nutrient removal in this study.

The ammonia concentration in the effluent was unstable throughout the experiment, as illustrated in figure 4 (c). The initial concentration of ammonia in synthetic leachate was 130mg/L. In the first 5 days of operation, the removal efficiency of ammonia in the treated effluent was 80%. However, the result varied and achieved the lowest removal efficiency after 15 days at a concentration of 110mg/L, corresponding to 15% removal efficiency. The ammonia fluctuations in the effluent greatly increase due to biomass washout in the system [13]. However, at the end of the experiment, the removal efficiency was able to reach 78%, which was equal to 29mg/L. Through 30days experimental period, the average concentration of ammonia in the effluent was 52.14mg/L with a removal efficiency of 60%. As indicated in table 1, the outcome of ammonia removal in this study was lower than in earlier publications. Meanwhile, Zhu *et al.* [19] demonstrated an ammonia removal efficiency of 75%, corresponding to an ammonia concentration of 65mg/L. According to Liu *et al* [24], the ability of microalgae cells to directly assimilate ammoniacal nitrogen in wastewater has the potential for microalgae-bacteria aerobic granular sludge to attain optimum efficiency.

Table 1. Summary of wastewater performance using microalgae-bacteria aerobic granular sludge.

Wastewater type	Percentage removal of the effluent			
	COD	TP	Ammonia	References
Synthetic leachate	96% (160.61mg/L)	55.79% (16.8mg/L)	60% (52.14mg/L)	This study
Synthetic domestic wastewater	95.2% (<30mg/L)	44%	_	[20]
Domestic wastewater	72%	5% (7mg/L)	72% (6mg/L)	[13]
Synthetic domestic wastewater	_	17-35%	40-92%	[25]
Municipal wastewater	88-89%	20-33%	99%	[23]
Synthetic wastewater	96%	89%	99%	[21]
Synthetic municipal wastewater	_	34%	98%	[24]
Domestic wastewater	96%	44%	99%	[22]
Synthetic wastewater	82%	100%	75%	[19]

4. Conclusions

Microalgae-bacteria aerobic granular sludge was successfully developed using old synthetic leachate within 30days of the experiment. The produced granules had a satisfactory settling property, with a settling velocity of 40.65m/h and an SVI_{30} value of 35.82mL/g. Additionally, the wastewater treatment performance of microalgae-bacteria aerobic granular sludge remained steady during the whole experiment, with a COD removal efficiency of 96%, PO_4-P removal efficiency of 55.79%, and an ammoniacal nitrogen removal efficiency of 60%. A further study should focus on higher ammonia nitrogen and heavy metal removal from the system. Moreover, the chemical characteristic of the granule should also be analysed.

Acknowledgments
This work is funded by the AUN/SEED-Net Grant no: R.K130000.7343.4B403 under Collaborative Education Program for Sustainable Environmental Engineering Network (CEP-SEEN).

References

[1] JPSPN MoHaLg 2015 Department Project (Technical).Available from: http://goo.gl/FzF0ng. (Accessed on: November 28, 2021)

[2] NSP 2005 National Strategic Plan for Solid Waste Management. Ministry of Urban Wellbeing, Housing and Local Government, Malaysia.Available from: http://goo.gl/OsVlGG. (Accessed on: December 15, 2021)

[3] Luo H, Zeng Y, Cheng Y, He D, and Pan Xs 2020 Recent advances in municipal landfill leachate: A review focusing on its characteristics, treatment, and toxicity assessment *Science of The Total Environment* **703** 135–468

[4] Gao J, Oloibiri V, Chys M, Audenaert W, Decostere B, He Y, et al.s 2015 The present status of landfill leachate treatment and its development trend from a technological point of view *Reviews in Environmental Science and Bio/Technology* **14** 93–122

[5] Peng Ys 2017 Perspectives on technology for landfill leachate treatment *Arabian Journal of Chemistry* **10** S2567–S74

[6] Costa AM, Alfaia RGdSM, and Campos JCs 2019 Landfill leachate treatment in Brazil–An overview *Journal of environmental management* **232** 110–6

[7] Mennaa FZ, Arbib Z, and Perales JAs 2015 Urban wastewater treatment by seven species of microalgae and an algal bloom: biomass production, N and P removal kinetics and harvestability *Water research* **83** 42–51

[8] Rodolfi L, Chini Zittelli G, Bassi N, Padovani G, Biondi N, Bonini G, et al.s 2009 Microalgae for oil: Strain selection, induction of lipid synthesis and outdoor mass cultivation in a low-cost photobioreactor *Biotechnology and bioengineering* **102** 100–12

[9] Bassin JP. Aerobic granular sludge technology. Advanced Biological Processes for Wastewater Treatment: Springer; 2018. p. 75–142.

[10] Sarma S, and Tay Js 2018 Aerobic granulation for future wastewater treatment technology: challenges ahead *Environmental Science: Water Research & Technology* **4** 9–15

[11] Ren Y, Ferraz F, Kang AJ, and Yuan Qs 2017 Treatment of old landfill leachate with high ammonium content using aerobic granular sludge *Journal of biological engineering* **11** 1–8

[12] Sforza E, Al Emara MK, Sharif A, and Bertucco As 2015 Exploitation of urban landfill leachate as nutrient source for microalgal biomass production *Chemical Engineering Transactions* **43** 373–8

[13] Purba LD, Abdullah N, Yuzir A, Zamyadi A, Shimizu K, and Hermana Js 2021 Rapid Development of Microalgae-Bacteria Granular Sludge Using Low-Strength Domestic Wastewater *Journal of Water and Environment Technology* **19** 96–107

[14] Ren Y, Ferraz F, Kang AJ, and Yuan Qs 2017 Treatment of old landfill leachate with high ammonium content using aerobic granular sludge *Journal of Biological Engineering* **11** 42

[15] APHA. Standard methods for the examination of water and wastewater2012.

[16] He Q, Chen L, Zhang S, Chen R, Wang H, Zhang W, et al.s 2018 Natural sunlight induced rapid formation of water-born algal-bacterial granules in an aerobic bacterial granular photo-sequencing batch reactor *Journal of hazardous materials* **359** 222–30

[17] Arcila JS, and Buitrón Gs 2017 Influence of solar irradiance levels on the formation of microalgae-bacteria aggregates for municipal wastewater treatment *Algal Research* **27** 190-7

[18] Cai W, Zhao Z, Li D, Lei Z, Zhang Z, and Lee D-Js 2019 Algae granulation for nutrients uptake and algae harvesting during wastewater treatment *Chemosphere* **214** 55–9

[19] Zhu S, Qin L, Feng P, Shang C, Wang Z, and Yuan Zs 2019 Treatment of low C/N ratio wastewater and biomass production using co-culture of Chlorella vulgaris and activated sludge in a batch photobioreactor *Bioresource Technology* **274** 313–20

[20] Huang W, Li B, Zhang C, Zhang Z, Lei Z, Lu B, et al.s 2015 Effect of algae growth on aerobic granulation and nutrients removal from synthetic wastewater by using sequencing batch reactors *Bioresource Technology* **179** 187–92

[21] Zhang B, Lens PN, Shi W, Zhang R, Zhang Z, Guo Y, et al.s 2018 Enhancement of aerobic granulation and nutrient removal by an algal–bacterial consortium in a lab-scale photobioreactor *Chemical Engineering Journal* **334** 2373–82

[22] Ahmad JSM, Cai W, Zhao Z, Zhang Z, Shimizu K, Lei Z, et al.s 2017 Stability of algal-bacterial granules in continuous-flow reactors to treat varying strength domestic wastewater *Bioresource technology* **244** 225–33

[23] Petrini S, Foladori P, Beghini F, Armanini F, Segata N, and Andreottola Gs 2020 How inoculation affects the development and the performances of microalgal-bacterial consortia treating real municipal wastewater *Journal of environmental management* **263** 110–427

[24] Liu L, Zeng Z, Bee M, Gibson V, Wei L, Huang X, et al.s 2018 Characteristics and performance of aerobic algae-bacteria granular consortia in a photo-sequencing batch reactor *Journal of Hazardous Materials* **349** 135-42

[25] Liu L, Fan H, Liu Y, Liu C, and Huang Xs 2017 Development of algae-bacteria granular consortia in photo-sequencing batch reactor *Bioresource technology* **232** 64–71

IOP Conf. Series: Earth and Environmental Science 1091 (2022) 012051 doi:10.1088/1755-1315/1091/1/012051

The applications of aerobic granular sludge for leachate treatment: A review

D Ilmasari[1], A Yuzir*[1],

[1] Malaysia-Japan International Institute of Technology (MJIIT), Universiti Teknologi Malaysia, Jalan Sultan Yahya Petra, 54100, Kuala Lumpur, Malaysia

*Corresponding author e-mail: muhdaliyuzir@utm.my

Abstract. The vast majority of the world's daily municipal solid waste (MSW) is dumped in landfills without being treated first. Leachate generated from these landfills is defined as dark-coloured, highly contaminated wastewater that has become a problem for the environment, posing a threat to soil, surface water, and groundwater quality and having a detrimental influence on human health. Thus, leachate treatment is strongly recommended prior to final discharge. The application of aerobic granular sludge (AGS) technology for wastewater treatment has increased in recent years, especially for industrial wastewater such as leachate. Due to its significant advantages over conventional activated sludge technology, such as denser structure, improved settleability, faster effluent separation, higher biomass content, and improved shock loading resistance, AGS is a viable option for leachate treatment. This article provides detailed discussions of the leachate characteristics along with the available treatment systems, focusing on the AGS system. The efficacy of AGS technology in the treatment of landfill leachate was elucidated by highlighting its benefits, governing factors affecting its performance, and limitations. Based on the latest literature, this paper attempts to identify the research gaps and obstacles in using AGS technology for landfill leachate treatment.

1. Introduction

Municipal solid waste (MSW) management is a severe environmental, social, and economic concern worldwide [1]. Most MSW is dumped in landfills, which are well-acknowledged for their simplicity and economic benefits [2]. However, the generation of leachate is one of several issues linked with landfills. Landfill leachate is a dark-coloured complex effluent generated from landfills that contains high levels of organic materials, ammonia, dissolved solids, heavy metals, and xenobiotic organic compounds [3]. It has been reported that leachate can cause a threat to soil, surface water, and groundwater quality and have a detrimental influence on human health [4]. Thus, leachate treatment is strongly recommended prior to final discharge.

Numerous methods are available for treating landfill leachate. One such approach is the biological treatment by activated sludge which has been extensively reported in the literature, achieving COD and ammonia removal up to 90% and 99%, respectively [5–8]. This approach, however, has been found to be inefficient as a stand-alone process, whereby nitrification is aided by activated sludge, while denitrification is reported to be less effective in reducing the nitrogen contents in leachate [9]. Disadvantages include excessive sludge production, high demand for operation cost, and microbial population inhibition due to leachate's high ammonia levels continue to be a severe challenge for its implementation [10].

Aerobic granular sludge (AGS) technology, which has been the subject of intensive research over the last decade, has been proven to be a novel and promising development in the field of wastewater treatment processes. Aerobic granulation is often accomplished in sequencing batch reactors (SBRs), with a cycle arrangement set to provide a rigorous selection for rapid settling sludge and frequent recurrence of feast and famine regimes resulting in the formation of dense and stable granules [11]. Due to the unique structure of AGS, including a distinct aerobic and anaerobic/anoxic layer within the granules, this system outperforms the conventional activated sludge, which has been confirmed by both lab-scale and full-scale studies [12]. Moreover, their excellent settling ability compared to sludge flocs is essential for successful sludge-effluent separation. AGS also has a higher and more stable metabolic rate, is resistant to shocks and toxins, and has long biomass residence periods [13]. In terms of treatment efficiency, AGS has shown a stable performance for BOD as well as nitrogen removal [14]. Furthermore, previous studies have reported high COD removal rate (>90%) from various types of wastewater, including leachate by AGS [15–17]. Therefore, AGS offers significant potential for landfill leachate treatment which its application has been studied in recent years. However, until the present, no comprehensive review of AGS utilization in leachate treatment has been reported.

This review presents detailed discussions of the leachate characteristics along with the available treatment systems, focusing on the AGS system. The efficacy of AGS technology in the treatment of landfill leachate was elucidated by highlighting its benefits, governing factors affecting its performance, and limitations. Based on the latest literature, this paper attempts to identify the research gaps and obstacles in using AGS technology for landfill leachate treatment, which shall substantially contribute to future studies of landfill leachate management.

2. Leachate production and characteristics

MSW is a general term for all waste generated by households [18]. The amount of MSW generated yearly on a worldwide scale is expected to hit 2.2 billion tonnes by 2025 [19]. Besides, the world's population is estimated to increase to 10 billion people in 2057 with an annual growth rate of 1.05%. This demographic increase adds significantly to the production of MSW, posing a serious environmental concern [20]. In terms of MSW management, landfilling, incineration, and composting are three of the most common disposal techniques [21]. Across many countries, particularly developing countries, sanitary landfills and open dump landfills are the most favored methods due to its economic benefits [22]. However, one of the most significant issues associated with landfills, in addition to numerous other drawbacks, is the production of toxic leachate [23].

Leachate is liquid that has seeped through waste piles that may have decomposed aerobically or anaerobically. The ensuing leachate is heavily polluted, possibly infiltrating into the ground and contaminating soil and groundwater. [24]. Some of the pollutants found in the leachate from landfills can be divided into four categories: dissolved organic compounds, inorganic macro-compounds, xenobiotic organic compounds, and heavy metals [25]. Leachate characteristics may vary from each site, whereby factors affecting their characteristics include landfill age, waste type and composition, seasonal variations, precipitation, and rainfall [26,27]. Several physicochemical parameters that are used to assess the quality of leachate such as biochemical oxygen demand (BOD), chemical oxygen demand (COD), total organic carbon (TOC) total suspended solid (TSS), total dissolved solid (TDS), ammonia (NH_3), total nitrogen (TN), total phosphorus (TP), chloride, sulfur compounds, alkalinity, and heavy metals [20–22]. Complex pollutants such as PFCs, PAHs, alkaline earth metals, and phthalic esters complicate the leachate composition even further [28].

Because leachate characteristics change over time, landfill age becomes crucial in leachate composition [29]. For example, BOD and COD concentrations continue to decrease with landfill age, possibly caused by organic waste decomposition in leachate [30]. Therefore, leachate is classified as young, intermediate, and mature based on landfill age [10]. As presented in Table 1, young acidogenic leachate is commonly characterized by a high level of COD (>10,000 mg/L) and has more easily biodegradable organic matter, indicated by BOD/COD ratio of >0.3. Meanwhile, mature leachate shows more stable leachate with a low concentration of biodegradable organic matter (BOD/COD ratio of <0.1). It is also observed that ammonia concentration does not decrease significantly over time and due to its toxicity, it may disturb the biological unit in leachate treatment, which indicate ammonia in leachate is a major environmental issue [31]. Because the quality and quantity of leachate varies, it is critical to implement dependable and effective treatment systems capable of dealing with such complex effluent before it can be safely discharged to the environment.

Table 1. Landfill leachate composition [3,32,33]

Parameters	Young (<5 years)	Intermediate (5-10 years)	Mature (>10 years)
COD (mg/L)	More than 10,000	4000-10,000	Less than 4000
NH_3 (mg/L)	Less than 400	-	More than 400
BOD/COD	More than 0.3	0.1-0.3	Less than 0.1
Total phosphorus (mg/L)	5-100	-	5-10
Heavy metals (mg/L)	More than 2.0	Less than 2.0	Less than 2.0
pH	Less than 6.5	6.5-7.5	More than 7.5
Biodegradability	Low to medium	Low	Low
Organic compound	80% VFA	VFA 5-30%, Humic acid, Fulvic acid	Humic acid and Fulvic acid

*VFA: Volatile Fatty Acids

3. Overview of landfill leachate treatment

Pollution of groundwater and surface water is a significant environmental issue arise from the discharge of landfill leachate [34,35]. The treatment of leachate is substantially more challenging than municipal wastewater because it contains humic compounds released during the decomposition of the organic matter of the waste [36]. Moreover, two main elements to consider while treating leachate are the volumetric flow rate and its composition [3]. Prior to releasing leachate into water bodies, it is critical to remove COD, BOD, and ammonium levels [25].

The treatment of leachate can be classified into two groups, including conventional and advanced treatment. Conventional treatment can further be divided into three sub-groups. First is the leachate transfer [37] as well as recycle or reinjection to landfill cell [38]. Second is the biological degradation processes which include aerobic and anaerobic process [39]. The third sub-group is the physicochemical processes such as adsorption, air stripping, coagulation/flocculation, advanced oxidation, chemical precipitation, etc [40]. Most conventional treatment approaches require multistage processes. Depending on the leachate characteristics, the treatment technique applied might be biological and/or chemical [3]. Meanwhile, leachate treatment utilizing membrane technology is considered as advanced treatment [41].

Biological degradation processes were reported to successfully remove 96% of organic matters [42] and 90% of inorganic content [43], respectively. Despite their ease of operation and low cost, their efficacy is limited to leachate with a high degree of biodegradability, which corresponds to young leachate [9]. Regarding physicochemical processes, chemical precipitation is one of the most applied processes that is easy to use and has a high reaction rate, usually faster than biological processes. This method, on the other hand, is often not effective at removing organic matters and total solids [44]. Occasionally, combining biological and physicochemical processes appears to be a viable option for reaping the benefits of both technologies.

4. Performance of AGS

The advantages of AGS have been highlighted as having a dense structure (Figure 1) with a high settling velocity, rich biomass retention, superior organics and ammonia removal, and resistance to shock loadings and hazardous substances [45]. AGS relies on microbial activity to remove organics, nitrogen, phosphorus, and other pollutants all at once in a single reactor (mostly in SBR), rather than different processes taking place in multiple treatment stages [46–48]. AGS does seem to be a more promising and energy-saving treatment alternative in the wastewater treatment industry, as well as for leachate treatment, given the occurrence of simultaneous nitrification denitrification and phosphorus removal (SNDPR) due to the presence of several redox conditions inside the granules (Figure 2) [17,49,50].

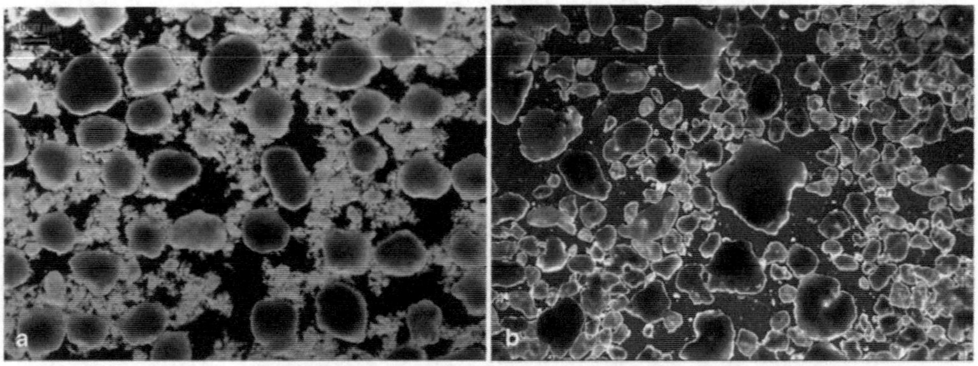

Figure 1. Stereoscopic view of AGS [51]

4.1. Performance of AGS for leachate treatment

For the treatment of leachate, AGS has demonstrated its effectiveness at removing organic matters and nutrients, e.g., NH_3 (Table 2). The system's ability to survive toxic loading allows it to be used for leachate containing a variety of contaminants. In general, the overall results of AGS treatment for landfill leachate are satisfactory. According to Ren et al. [52] AGS can remove 64% of COD, 95-100% of NH_3, and 84% of TP from leachate whereby the compact and distinctive shape of AGS may have aided in the almost complete NH_3 removal.

Wei at al. [53] conducted one of the earliest research on the treatment of leachate using AGS. The effect of varying NH_3 levels in influent on COD and nitrogen removal was investigated. By applying a modified kinetic model, COD removal efficiency was 76-91%, indicating that increasing NH_3 levels in the influent decreased the COD removal efficiency. Meanwhile, depending on the NH_3 levels in the influent, the nitrogen removal efficiency attained ranged from 40 to 92%.

Another study regarding the development of AGS for leachate treatment was conducted by Ren et al. [54]. By diluting leachate with municipal wastewater at a volumetric ratio ranging from 10 to 100%, leachate was gradually introduced to SBR. In comparison to the activated sludge system, the AGS system had a better settleability. Due to the refractory organic content in leachate, COD removal efficiency was low in both activated sludge and AGS. It was further revealed that AGS is more resilient than suspended-activated sludge in treating leachate with high NH_3 and free ammonia (FA) levels.

According to Wei et al. [55], the appearance of AGS seemed to increase the microbial diversity in the system. Some key and dominating populations in the activated sludge seed died out slowly, whereas a few microbial species emerged as the aerobic granules formed and matured. The newly emerged species were assumed to be reliable bacteria responsible for the high pollutant removal efficiency.

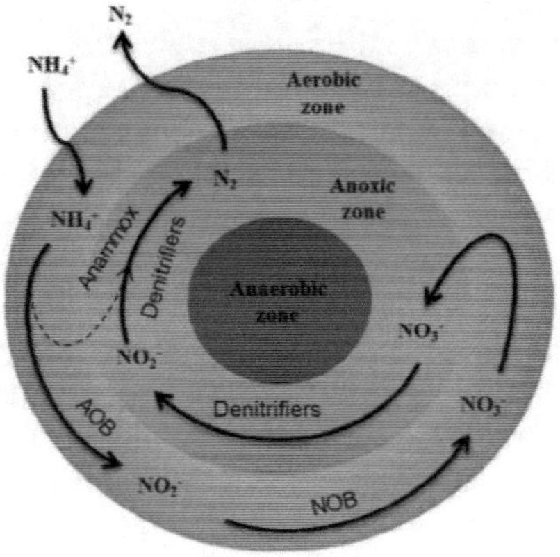

Figure 2. Nitrogen removal pathways inside AGS

Table 2. AGS for leachate treatment

Reactors	Leachate type (Influent feeding)	Removal efficiency	References
SBR	• Young leachate & synthetic wastewater (gradually added to SBR by dilution)	• COD:87-89% • NH_3:99% • TN:98% • TP:37-64%	[56]
SBR	• Old leachate & municipal wastewater • COD:1600-2850 mg/L • NH_3:1200 mg/L	• COD:31% • NH_3:99,7% • TN: 26%	[54]
SBR	• Synthetic young leachate • COD:498 mg/L • NH_3:120-500 mg/L	• COD:67-87% • NH_3:99% • TN:44% • TP:49%	[49]
SBR	• Synthetic old leachate & sewage (gradually added to SBR by dilution) • COD:1960-2850 mg/L • NH_3:704-1088 mg/L	• COD:64% • NH_3:95-100% • TN:42% • TP:84%	[52]
SBR	• Synthetic old leachate • sCOD:448-654 mg/L • NH_3:135-465 mg/L • TP:6 mg/L	• COD:55-73% • NH_3:61-95% • TN:39% • TP:54%	[50]
SBR	• Leachate diluted with tap water • COD:3996 mg/L • NH_3:1339 mg/L • TN:63%	• COD:84% • NH_3:89% • TN:63%	[55]
SBR	• Intermediate leachate • COD:4502-5992 mg/L • NH_3:257 mg/L (with pre-treatment) • NH_3:783-1079 mg/L (without pre-treatment)	• COD:82-84% • NH_3:92% (with pre-treatment) • NH_3:44-61% (without pre-treatment)	[53]
SBR	• Leachate diluted with tap water (1:1) and additional inorganic carbon added • COD:937 mg/L • BOD:589 mg/L • NH_3:303 mg/L	• COD:18-27% • TN:19-36% • NH_3:100%	[57]

4.2. Factors governing AGS treatment process for leachate

Several parameters, such as leachate toxicity, concentrations of organic matters and nutrients (i.e., NH_3) and leachate stability have been identified as the significant factors influencing AGS performance in leachate treatment process. The following sub-sections elaborate the factors governing AGS treatment efficiency.

4.2.1. Leachate age

Leachate aging causes the declining concentration of BOD; thus leachate becomes more stable due to the presence of refractory organic matter (e.g., fulvic and humid acid) as well as the decrease value of BOD/COD ratio [3]. Moreover, as landfills age, the non-biodegradable organic content of old leachate rises. As a result, biological treatment techniques may not treat old leachate as successfully as when it is applied for young leachate [30].

When AGS was used to treat old leachate, it was found that COD removal efficiency was as low as 31%. The findings indicated that the majority of COD in the old leachate was not biodegradable.

AGS, on the other hand, still outperformed activated sludge in terms of COD removal [54]. Similar findings were also obtained by Ren et al. [52] suggesting that low COD removal (64%) was associated with low BOD/COD ratios of old leachate varying from 0.15 to 0.2 (low biodegradability). On the other hand, when AGS was used to treat young and intermediate leachate, satisfactory COD removal was attained with 84% and 89% removal recorded (COD influent 4502-5992 mg/L) [53,56]. Based on these findings, AGS has the potential for old leachate treatment, yet there has been so no literature that study the improvement scheme to maintain the high performance of AGS in treating old leachate.

4.2.2. Ammonia nitrogen concentration
Under alkaline conditions, the existence of free ammonia (FA) is unavoidable due to the high concentration of ammonium in wastewater [58]. Meanwhile, various important bacteria essential for wastewater treatment are believed to be inhibited by the presence of FA [59]. Extracellular polymeric substances (EPS) which regarded responsible for the granules' chemical and physical properties, have been shown to be broken down by FA as well [60,61].

According to di Bella et al. [17], when ammonium concentrations in leachate rise to 2000 mg/L, organic and nitrogen removal tend to fall. Furthermore, a study by Wei et al. [55] revealed that ammonium removal from leachate by AGS was hindered when the high nitrogen loading rate (NLR) (1000-1500 mg/L) was applied, but ammonium removal was excellent at the low NLR (±500 mg/L). At the high NLR stage, the diversity and activity of microorganisms, particularly those of the NOB species, were suppressed. On the other hand, the microbial population stabilised at low NLR, ensuring efficient ammonium removal.

4.2.3. Temperature
An increase in temperature has been shown to increase wastewater treatment efficacy by aerobic granules [62,63]. Song et al. [64] reported that 30°C was the optimum operating temperature for better settleability, microbial activity, and granule compaction. The authors also imply that at various temperatures, different types of bacteria may play crucial roles. According to de Kreuk et al. [65], the growth of filamentous bacteria at 8-15°C causes granules to have an irregular structure, resulting in inadequate denitrification and other biodegradation processes. Furthermore, temperature influences the rivalry between phosphate accumulating organisms (PAO) and glycogen accumulating organisms (GAO) in granules [66].

Mieczkowski et al. [57] tested the effect of temperature on the removal of nitrogen and organic compound from landfill leachate. It was revealed that gradual adjustment of the granules to lower operational temperatures (29-20°C) allowed the granules to completely oxidise ammonium from leachate at ambient temperature (20°C), considerably improving denitrification efficiency. Aerobic and anaerobic ammonium-oxidizing bacteria both participated in ammonium oxidation. It was also revealed that the highest TN removal was achieved when the system was operated under 25°C condition. However, from all the temperature tested, COD removal was under 50%, indicating there is no significant effect of temperature on organic removal.

4.2.4. Leachate co-treatment
In the literature, the co-treatment of leachate and municipal wastewater using AGS has been discussed. This approach is being proposed as an alternative to overcome the shortcomings of biological leachate treatment [3]. Moreover, since it can be implemented at existing wastewater-treatment plants (WWTPs), this has been regarded as a practical and relatively low-cost option [3,67].

AGS treating young leachate that used the volumetric ratios of 5, 10, and 20% removed 89%, 99%, 98%, and 64% of COD, NH_3, TN, and TP, respectively. The COD and TN levels in the effluents increased as the leachate concentration increased from 5% to 20%. However, removal efficiencies were closely similar in all ratios tested, indicating that the increase in leachate had no significant effect on the treatment processes [56].

5. Limitations and future prospects

To this date, there is no treatment system that has been widely accepted and applied for leachate treatment. However, it is projected that AGS will likely overtake the conventional activated sludge system in the wastewater treatment industry. However, the implementation of AGS for industrial wastewater treatment, particularly for landfill leachate, still faces certain challenges.

- At present, most of AGS research has been done in lab-scale SBRs with defined substrates and well-controlled working conditions, utilizing synthetic wastewater. There are some studies that use actual leachate, but they are still deemed limited. Furthermore, the viability of adopting another type of reactor, such as continuous flow reactors, which are commonly used in WWTPs for leachate co-treatment, must be evaluated.
- Earlier research using AGS for leachate mainly focused on COD, nitrogen, and phosphate removal. The use of AGS to remove other pollutants such as heavy metals and phenolic compounds has yet to be investigated. Exploring the scopes of AGS application for leachate treatment is thus one of the next research directions.
- The operation parameters must be explored for varied leachate types. There are no standard operating criteria or specifications for AGS leachate treatment. Thus, future study should continue to focus on standardising treatment process.

6. Conclusion

AGS has good settling ability, low sludge generation, and high organic, nutrients, and toxic compound removal efficiency compared to conventional activated sludge. Furthermore, AGS has been widely utilized in the wastewater treatment industry and it is applicable for a wide range of wastewater types, including for industrial wastewater such as landfill leachate. However, the studies of AGS for landfill leachate treatment is still deemed limited. Future works should be conducted and focus on: (1) utilization of actual landfill leachate and various types of reactors, (2) exploring the application scope, (3) standardizing the operating parameters.

Acknowledgments

This work is funded by the AUN/SEED-Net Grant no. 4B403 under Collaborative Education Program for Sustainable Environmental Engineering Network (CEP-SEEN).

References

[1] Chien Bong C P, Ho W S, Hashim H, Lim J S, Ho C S, Peng Tan W S and Lee C T 2017 Review on the renewable energy and solid waste management policies towards biogas development in Malaysia *Renew. Sustain. Energy Rev.* **70** 988–98

[2] Abuabdou S M A, Ahmad W, Aun N C and Bashir M J K 2020 A review of anaerobic membrane bioreactors (AnMBR) for the treatment of highly contaminated landfill leachate and biogas production: Effectiveness, limitations and future perspectives *J. Clean. Prod.* **255** 120215

[3] Renou S, Givaudan J G, Poulain S, Dirassouyan F and Moulin P 2008 Landfill leachate treatment: Review and opportunity *J. Hazard. Mater.* **150** 468–93

[4] Azmi N B, Bashir M J K, Sethupathi S and Ng C A 2016 Anaerobic stabilized landfill leachate treatment using chemically activated sugarcane bagasse activated carbon: kinetic and equilibrium study *Desalin. Water Treat.* **57** 3916–27

[5] Boonnorat J, Kanyatrakul A, Prakhongsak A, Ketbubpha K, Phattarapattamawong S, Treesubsuntorn C and Panichnumsin P 2021 Biotoxicity of landfill leachate effluent treated by two-stage acclimatized sludge AS system and antioxidant enzyme activity in Cyprinus carpio *Chemosphere* **263** 128332

[6] Pirsaheb M, Hossini H, Secula M S, Parvaneh M and Ashraf G M 2017 Application of high rate integrated anaerobic-aerobic/biogranular activated carbon sequencing batch reactor (IAnA-BioGACSBR) for treating strong municipal landfill leachate *Sci. Rep.* **7** 1–11

[7] Saleem M, Lavagnolo M C, Campanaro S and Squartini A 2018 Dynamic membrane bioreactor (DMBR) for the treatment of landfill leachate; bioreactor's performance and metagenomic insights into microbial community evolution *Environ. Pollut.* **243** 326–35
[8] Campos F, de Freitas Bueno R and Piveli R P 2020 Influence of landfill leachate on organic matter removal in activated sludge and activated sludge with carrier processes *Brazilian Arch. Biol. Technol.* **63** 1–11
[9] Lebron Y A R, Moreira V R, Brasil Y L, Silva A F R, Santos L V de S, Lange L C and Amaral M C S 2021 A survey on experiences in leachate treatment: Common practices, differences worldwide and future perspectives *J. Environ. Manage.* **288**
[10] Luo H, Zeng Y, Cheng Y, He D and Pan X 2020 Recent advances in municipal landfill leachate: A review focusing on its characteristics, treatment, and toxicity assessment *Sci. Total Environ.* **703** 135468
[11] Nancharaiah Y V. and Kiran Kumar Reddy G 2018 Aerobic granular sludge technology: Mechanisms of granulation and biotechnological applications *Bioresour. Technol.* **247** 1128–43
[12] Sarma S J and Tay J H 2018 Aerobic granulation for future wastewater treatment technology: Challenges ahead *Environ. Sci. Water Res. Technol.* **4** 9–15
[13] Ni B J and Yu H Q 2010 Mathematical modeling of aerobic granular sludge: A review *Biotechnol. Adv.* **28** 895–909
[14] Fall C, Barrón-Hernández L M, Gonzaga-Galeana V E and Olguín M T 2022 Ordinary heterotrophic organisms with aerobic storage capacity provide stable aerobic granular sludge for C and N removal *J. Environ. Manage.* **308** 114662
[15] Long B, Xuan X, Yang C, Zhang L, Cheng Y and Wang J 2019 Stability of aerobic granular sludge in a pilot scale sequencing batch reactor enhanced by granular particle size control *Chemosphere* **225** 460–9
[16] Cai W, Huang W, Lei Z, Zhang Z, Lee D J and Adachi Y 2019 Granulation of activated sludge using butyrate and valerate as additional carbon source and granular phosphorus removal capacity during wastewater treatment *Bioresour. Technol.* **282** 269–74
[17] Bella G Di and Torregrossa M 2014 Aerobic granular sludge for leachate treatment *Chem. Eng. Trans.* **38** 493–8
[18] Moh Y C and Abd Manaf L 2017 Solid waste management transformation and future challenges of source separation and recycling practice in Malaysia *Resour. Conserv. Recycl.* **116** 1–14
[19] Hoornweg D and Bhada-Tata P 2012 What a waste: a global review of solid waste management *Urban Dev. Ser. Pap. no. 15* 116
[20] Shah A V., Srivastava V K, Mohanty S S and Varjani S 2021 Municipal solid waste as a sustainable resource for energy production: State-of-the-art review *J. Environ. Chem. Eng.* **9** 105717
[21] Abdel-Shafy H I and Mansour M S M 2018 Solid waste issue: Sources, composition, disposal, recycling, and valorization *Egypt. J. Pet.* **27** 1275–90
[22] Oakley S M and Jimenez R 2012 Sustainable sanitary landfills for neglected small cities in developing countries: The semi-mechanized trench method from Villanueva, Honduras *Waste Manag.* **32** 2535–51
[23] Zhang T, Shi J, Qian X and Ai Y 2019 Temperature and Gas Pressure Monitoring and Leachate Pumping Tests in a Newly Filled MSW Layer of a Landfill *Int. J. Environ. Res.* **13** 1-19
[24] Bashir M J K, Aziz H A, Yusoff M S and Aziz S Q 2012 Color and chemical oxygen demand removal from mature semi-aerobic landfill leachate using anion-exchange resin: An equilibrium and kinetic study *Environ. Eng. Sci.* **29** 297–305
[25] Chua K H, Jati E, Sahid M and Leong Y P 2011 Sustainable Municipal Solid Waste Management and GHG Abatement in Malaysia *Green Energy Manag.* 1–8
[26] Moody C M and Townsend T G 2017 A comparison of landfill leachates based on waste composition *Waste Manag.* **63** 267–74
[27] Han Z, Ma H, Shi G, He L, Wei L and Shi Q 2016 A review of groundwater contamination near municipal solid waste landfill sites in China *Sci. Total Environ.* **569–570** 1255–64

[28] Reshadi M A M, Bazargan A and McKay G 2020 A review of the application of adsorbents for landfill leachate treatment: Focus on magnetic adsorption *Sci. Total Environ.* **731** 138863
[29] Kulikowska D and Klimiuk E 2008 The effect of landfill age on municipal leachate composition *Bioresour. Technol.* **99** 5981–5
[30] Torretta V, Ferronato N, Katsoyiannis I A, Tolkou A K and Airoldi M 2017 Novel and conventional technologies for landfill leachates treatment: A review *Sustain.* **9** 1–39
[31] Fernandes A, Pacheco M J, Ciríaco L and Lopes A 2015 Review on the electrochemical processes for the treatment of sanitary landfill leachates: Present and future *Appl. Catal. B Environ.* **176–177** 183–200
[32] Aziz H A and Ramli S F 2018 Recent development in sanitary landfilling and landfill leachate treatment in Malaysia *Int. J. Environ. Eng.* **934** 201-229
[33] Kamaruddin M A, Yusoff M S, Rui L M, Isa A M, Zawawi M H and Alrozi R 2017 An overview of municipal solid waste management and landfill leachate treatment: Malaysia and Asian perspectives *Environ. Sci. Pollut. Res.* **24** 26988–7020
[34] Salem Z, Hamouri K, Djemaa R and Allia K 2008 Evaluation of landfill leachate pollution and treatment *Desalination* **220** 108–14
[35] Postacchini L, Ciarapica F E and Bevilacqua M 2018 Environmental assessment of a landfill leachate treatment plant: Impacts and research for more sustainable chemical alternatives *J. Clean. Prod.* **183** 1021–33
[36] Ghaedi M, Ansari A and Sahraei R 2013 ZnS: Cu nanoparticles loaded on activated carbon as novel adsorbent for kinetic, thermodynamic and isotherm studies of Reactive Orange 12 and Direct yellow 12 adsorption *Spectrochim. Acta - Part A Mol. Biomol. Spectrosc.* **114** 687–94
[37] Ferraz F M, Bruni A T, Povinelli J and Vieira E M 2016 Leachate/domestic wastewater aerobic co-treatment: A pilot-scale study using multivariate analysis *J. Environ. Manage.* **166** 414–9
[38] Yuan Q, Jia H and Poveda M 2016 Study on the effect of landfill leachate on nutrient removal from municipal wastewater *J. Environ. Sci. (China)* **43** 153–8
[39] Kurniawan T A, Lo W, Chan G and Sillanpää M E T 2010 Biological processes for treatment of landfill leachate *J. Environ. Monit.* **12** 2032–47
[40] Kurniawan T A, Lo W H and Chan G Y S 2006 Physico-chemical treatments for removal of recalcitrant contaminants from landfill leachate *J. Hazard. Mater.* **129** 80–100
[41] Xu Y, Chen C, Li X, Lin J, Liao Y and Jin Z 2017 Recovery of humic substances from leachate nanofiltration concentrate by a two-stage process of tight ultrafiltration membrane *J. Clean. Prod.* **161** 84–94
[42] Hashemi H, Jasemizad T, Derakhshan Z and Ebrahimi A 2017 Determination of Sequencing Batch Reactor (SBR) Performance in Treatment of Composting Plant Leachate *Heal. Scope* **In Press**
[43] Sun H, Peng Y, Wang S and Ma J 2015 Achieving nitritation at low temperatures using free ammonia inhibition on Nitrobacter and real-time control in an SBR treating landfill leachate *J. Environ. Sci. (China)* **30** 157–63
[44] Gomes A I, Santos S G S, Silva T F C V, Boaventura R A R and Vilar V J P 2019 Treatment train for mature landfill leachates: Optimization studies *Sci. Total Environ.* **673** 470–9
[45] Nancharaiah Y V. and Reddy G K K 2018 Aerobic granular sludge technology: Mechanisms of granulation and biotechnological applications *Bioresour. Technol.* **247** 1128–43
[46] Bengtsson S, de Blois M, Wilén B M and Gustavsson D 2019 A comparison of aerobic granular sludge with conventional and compact biological treatment technologies *Environ. Technol. (United Kingdom)* **40** 2769–78
[47] Bengtsson S, de Blois M, Wilén B M and Gustavsson D 2018 Treatment of municipal wastewater with aerobic granular sludge *Crit. Rev. Environ. Sci. Technol.* **48** 119–66
[48] Purba L D A, Ibiyeye H T, Yuzir A, Mohamad S E, Iwamoto K, Zamyadi A and Abdullah N 2020 Various applications of aerobic granular sludge: A review *Environ. Technol. Innov.* **20** 101045
[49] Ren Y, Ferraz F, Lashkarizadeh M and Yuan Q 2017 Comparing young landfill leachate treatment efficiency and process stability using aerobic granular sludge and suspended growth activated sludge *J. Water Process Eng.* **17** 161–7

[50] Ren Y, Ferraz F, Kang A J and Yuan Q 2017 Treatment of old landfill leachate with high ammonium content using aerobic granular sludge *J. Biol. Eng.* **11** 1–8

[51] Pronk M, Abbas B, Al-zuhairy S H K, Kraan R, Kleerebezem R and van Loosdrecht M C M 2015 Effect and behaviour of different substrates in relation to the formation of aerobic granular sludge *Appl. Microbiol. Biotechnol.* **99** 5257–68

[52] Ren Y, Ferraz F M and Yuan Q 2017 Landfill Leachate Treatment Using Aerobic Granular Sludge *J. Environ. Eng.* **143** 04017060

[53] Wei Y, Ji M, Li R and Qin F 2012 Organic and nitrogen removal from landfill leachate in aerobic granular sludge sequencing batch reactors *Waste Manag.* **32** 448–55

[54] Ren Y, Ferraz F M and Yuan Q 2018 Biological leachate treatment using anaerobic/aerobic process: suspended growth-activated sludge versus aerobic granular sludge *Int. J. Environ. Sci. Technol.* **15** 2295–302

[55] Wei Y, Ye Y, Ji M, Peng S, Qin F, Guo W and Ngo H H 2021 Microbial analysis for the ammonium removal from landfill leachate in an aerobic granular sludge sequencing batch reactor *Bioresour. Technol.* **324** 124639

[56] Bueno R de F, Faria J K, Uliana D P and Liduino V S 2020 Simultaneous removal of organic matter and nitrogen compounds from landfill leachate by aerobic granular sludge *Environ. Technol. (United Kingdom)* **0** 1–30

[57] Mieczkowski D, Cydzik-Kwiatkowska A, Rusanowska P and Świątczak P 2016 Temperature-induced changes in treatment efficiency and microbial structure of aerobic granules treating landfill leachate *World J. Microbiol. Biotechnol.* **32**

[58] Yadu A, Sahariah B P and Anandkumar J 2018 Influence of COD/ammonia ratio on simultaneous removal of NH4+-N and COD in surface water using moving bed batch reactor *J. Water Process Eng.* **22** 66–72

[59] Liu Y, Ngo H H, Guo W, Peng L, Wang D and Ni B 2019 The roles of free ammonia (FA) in biological wastewater treatment processes: A review *Environ. Int.* **123** 10–9

[60] Zhang C, Qin Y, Xu Q, Liu X, Liu Y, Ni B J, Yang Q, Wang D, Li X and Wang Q 2018 Free Ammonia-Based Pretreatment Promotes Short-Chain Fatty Acid Production from Waste Activated Sludge *ACS Sustain. Chem. Eng.* **6** 9120–9

[61] Felz S, Al-Zuhairy S, Aarstad O A, van Loosdrecht M C M and Lin Y M 2016 Extraction of structural extracellular polymeric substances from aerobic granular sludge *J. Vis. Exp.* **2016** 1–8

[62] Bassin J P, Pronk M, Muyzer G, Kleerebezem R, Dezotti M and van Loosdrecht M C M 2011 Effect of elevated salt concentrations on the aerobic granular sludge process: Linking microbial activity with microbial community structure *Appl. Environ. Microbiol.* **77** 7942–53

[63] Ab Halim M H, Nor Anuar A, Azmi S I, Jamal N S A, Wahab N A, Ujang Z, Shraim A and Bob M M 2015 Aerobic sludge granulation at high temperatures for domestic wastewater treatment *Bioresour. Technol.* **185** 445–9

[64] Song Z, Ren N, Zhang K and Tong L 2009 Influence of temperature on the characteristics of aerobic granulation in sequencing batch airlift reactors *J. Environ. Sci.* **21** 273–8

[65] De Kreuk M K, Pronk M and Van Loosdrecht M C M 2005 Formation of aerobic granules and conversion processes in an aerobic granular sludge reactor at moderate and low temperatures *Water Res.* **39** 4476–84

[66] Winkler M K H, Bassin J P, Kleerebezem R, de Bruin L M M, van den Brand T P H and Van Loosdrecht M C M 2011 Selective sludge removal in a segregated aerobic granular biomass system as a strategy to control PAO-GAO competition at high temperatures *Water Res.* **45** 3291–9

[67] Brennan R B, Clifford E, Devroedt C, Morrison L and Healy M G 2017 Treatment of landfill leachate in municipal wastewater treatment plants and impacts on effluent ammonium concentrations *J. Environ. Manage.* **188** 64–72

IOP Conf. Series: Earth and Environmental Science 1091 (2022) 012052 doi:10.1088/1755-1315/1091/1/012052

Development and characteristics of microalgae-bacteria aerobic granules treating low strength municipal wastewater

S A Zahra[1], L D A Purba[1], N Abdullah*[1]

[1] Department of Chemical and Environmental Engineering, Malaysia-Japan International Institute of Technology (MJIIT), Universiti Teknologi Malaysia, Jalan Sultan Yahya Petra, 54100, Kuala Lumpur, Malaysia

*Corresponding author e-mail: norhayati@utm.my

Abstract. This study aims to assess the feasibility of cultivating microalgae-bacteria aerobic granular sludge using low-strength municipal wastewater (<200 mg COD/L) as a substrate. A laboratory-scale photo-sequencing batch reactor was inoculated with activated sludge and fed with municipal wastewater. The development and characteristics of microalgae-bacteria aerobic granular sludge were observed for 90 days. The average COD removal efficiency of this system was 71%. The granulation process was observed to occur on day 28, indicated by a high biomass accumulation and decreasing sludge volume index (SVI). After 90 days of cultivation, MLSS value was found to be 3.7 g/L and SVI_{30} at 18.9 mL/g. In addition, the diameter of the granules was significantly increased. Mature microalgae-bacteria aerobic granules were developed with an average and a maximum diameter of 0.65 mm and 6 mm, respectively. Moreover, cocci-shaped bacteria and microalgae cells were observed to reside on the granular surface during observation using field emission scanning electron microscopy (FESEM). The granules exhibited an excellent settling velocity at 80 m/h. The findings suggest the possibility of using low-strength wastewater to develop microalgae-bacteria aerobic granular sludge.

1. Introduction

Aerobic granular sludge is a dense and self-immobilized aggregate consisting of microbial consortiums that concurrently remove carbon, nitrogen, phosphorus, and other contaminants only in a single reactor [1]. It has been shown that aerobic granular sludge is a more efficient and novel method of treating wastewater than conventional activated sludge [2,3]. Aerobic granular sludge system demonstrated several advantages such as high rates of organic matter and nutrient removals, lower land footprint, less investment cost, lower energy, and better effluent quality [4,5]. At present, aerobic granular sludge has been successfully developed using various wastewater sources such as domestic, municipal, and industrial wastewaters, including palm oil mills effluent (POME), paper mills effluent, dairy, textile, and leachate [6]. Currently, several full-scale aerobic granular sludge technologies have been developed and are mainly used to treat municipal wastewaters with medium to high COD concentration (>500 mg/L) [7]. However, a full-scale aerobic granular sludge system for the treatment of low-strength wastewater remains understudied [8].

Low COD concentrations due to rain and sewer leakage are prevalent in countries with combined sewer systems, such as China [9]. Similarly, low influent COD concentrations of 300–400 mg/L were frequently reported in Malaysia [8]. However, the implementation of aerobic granular sludge for low-strength municipal wastewater remains a challenge due to the long-term granular stability and the lengthy start-up phases. While the latter problem may be solved by adding biodegradable organic substrates or/and inoculating with mature granules, complete knowledge of the fundamental factors that affect granule stability is still inadequate, posing the primary disadvantage of long-term aerobic granular sludge application for the treatment of actual low-strength wastewater.

Recent research has shown that microalgae can be integrated into aerobic granular sludge systems, namely the microalgae-bacteria aerobic granular sludge system [10]. This newly established microalgae-bacteria aerobic granular sludge system was reported to enhance system stability, induce the rapid formation of aerobic granular sludge, and require less carbon source [11,12]. Up to the present, several studies have developed microalgae-bacteria aerobic granular sludge systems using low-strength and synthetic wastewater [12,13]. However, limited studies have been conducted to investigate the feasibility of developing microalgae-bacteria aerobic granular sludge using actual low-strength wastewater as a substrate. Synthetic wastewater may not accurately represent how the system will perform when using actual wastewater, as actual wastewater is more likely to contain complex organic compounds. Moreover, actual wastewater has varied concentrations that can affect system stability. Therefore, the purpose of this study is to develop and characterize microalgae-bacteria aerobic granular sludge using actual low-strength municipal wastewater as an influent.

2. Materials and methods

2.1. Reactor design and operational conditions

A single cylindrical column type photo-sequencing batch reactor (PSBR) column with a total working volume of 1.5 L and a height/diameter ratio of 16.67 (6 cm of internal diameter and 100 cm of height) was employed to develop microalgae-bacteria aerobic granular sludge. A microsparger at the bottom of the reactor was used to introduce air at a rate of 2.5 L/min. The reactor was operated at room temperature, and illumination was provided at 113 μmol/m^{2}s with 12hr:12hr (light:dark) photoperiod. The batch cycle lasted 6 hours, comprising 5 min feeding period, 232-247 min reaction period (90 min of non-aeration and 227-242 min of aeration), 15-20 min settling period, and 3 min period of effluent discharge. The column was seeded with 750 ml of activated sludge and was fed with municipal wastewater collected from a local wastewater treatment plant (WWTP). The characteristics of municipal wastewater employed in this study are listed in Table 1.

Table 1. Characteristics of municipal wastewater

Elements	Concentrations (mg/L)
COD	184 ± 73
BOD	67.9 ± 6.9
Total Nitrogen (TN)	16.6 ± 2.3
Total Phosphorus (TP)	2.4 ± 1.5
pH	6.9 ± 0.1

2.2. Analytical methods

Parameters indicating the development of microalgae-bacteria aerobic granular sludge were evaluated, including sludge volume index (SVI_{30}), mixed liquor suspended solids (MLSS), and chemical oxygen demand (COD) removal efficiency, respectively. All analyses were carried out in compliance with the Standard Methods for the Examination of Water and Wastewater [14]. The shape and diameters of microalgae-bacteria aerobic granules were determined using a stereomicroscope (LEICA EZ4W, Germany). Additionally, aerobic granular sludge's microscopic structure and chemical elemental content were observed using a field emission scanning electron microscope-energy dispersive spectroscopy (FESEM-EDS; JSM-7800F, Japan). Prior to FESEM-EDS analysis, the granules were fixated with 2.5

percent glutaraldehyde mixed with phosphate buffer saline (PBS) for 2 hours and then dried with ethanol gradients (10%, 30%, 50%, 70%, and 100%).

3. Results and discussions

3.1. Development of microalgae-bacteria aerobic granular sludge

Figure 1 depicts the changes in MLSS concentration and SVI_{30} values throughout the study. When the PSBR was first inoculated with activated sludge and fed with municipal wastewater, the initial MLSS was 2.1 g/L with an SVI_{30} of 166.67 mL/g, indicating poor settleability. MLSS values were first observed to increase but then sharply decreased due to biomass washout after day-14. Moreover, the decrease in biomass concentration may be attributed to a scarcity of carbon sources. To combat this issue, settling time was increased from 10 minutes to 15 minutes. MLSS concentration was observed to gradually increase after day-35, and 3.7 g/L of MLSS was obtained at the end of the experiment. The MLSS concentration in this study is comparable to previous reports in treating low-strength municipal and domestic wastewater [15,16]. Following a similar pattern, the SVI_{30} value fluctuated until day 35; then, a decreasing trend was observed as MLSS concentration was stably increasing. A final SVI_{30} value of 18.9 mL/g was recorded, indicating an excellent settleability. An SVI_{30} value of 30-80 mL/g is generally reported for the settleability of mature aerobic granules [5].

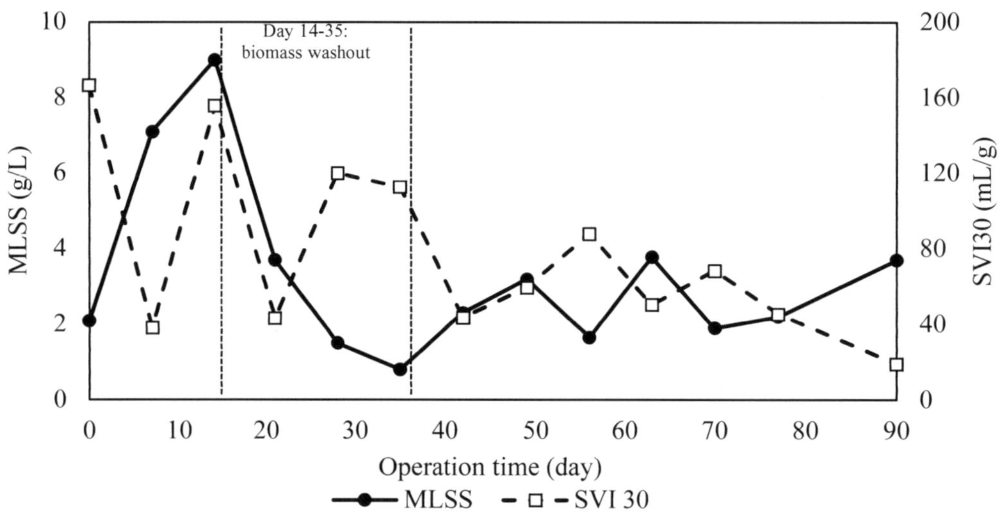

Figure 1. MLSS and SVI30 value throughout the microalgae-bacteria aerobic granular sludge system.

As this study utilized actual low-strength wastewater, the influent COD concentration (184 ± 73 mg/L) fluctuated during the entire experimental period, causing the fluctuation in effluent COD concentration. The municipal wastewater taken from a local WWTP used an open tank system; therefore, unforeseen weather circumstances such as heavy rainfall can directly affect the organic content of the wastewater. However, the effluent COD concentration was consistently less than 50 mg/L, albeit greater in some cases, as shown in figure 2. The average COD removal and effluent COD content after 90 days of the experimental period were 71% and 48.5 mg/L, respectively. This result is in accordance with Malaysia Sewage and Industrial Effluent Discharge Standard A (≤50 mg/L) [17].

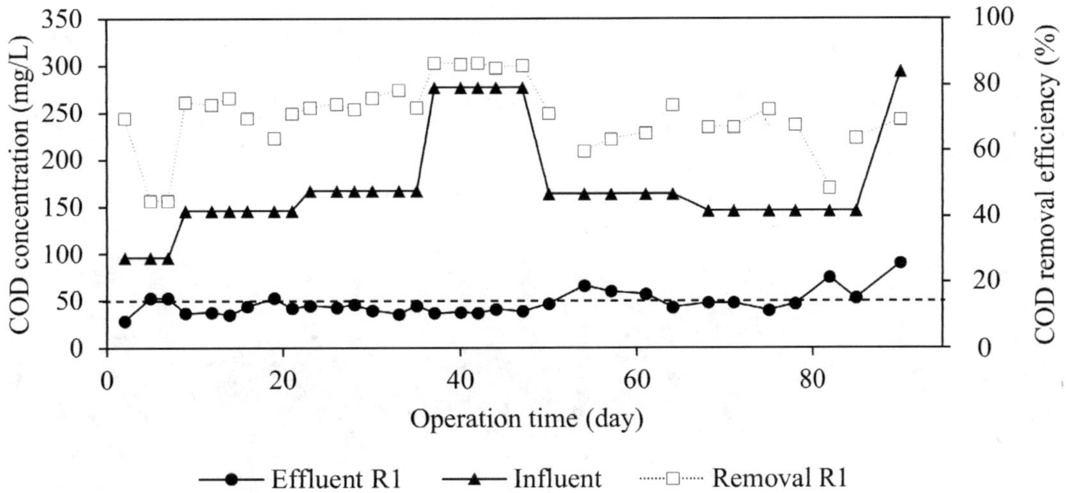

Figure 2. COD removal in the microalgae-bacteria aerobic granular sludge system.

In this study, the COD removal efficiency is reported to be lower than prior findings in treating actual low-strength wastewater at ≈80-85% removal efficiency [5,15,18,19]. However, it should be emphasized that the COD concentration of the municipal wastewater employed in this study (184 mg/L) is much less than the previous study (218-600 mg/L). This shows that microalgae-bacteria aerobic granular sludge may be suitable for treating very low-strength municipal wastewater.

3.2. Characteristics of microalgae-bacteria aerobic granular sludge

3.2.1. Morphological and physical characteristics

The mean size of the granules was increased significantly after day 28 and was stable at around 0.6 mm, as shown in figure 3. This indicates that granulation occurs roughly on day 28, and this observation is supported by the gradual increase in MLSS value and the decrease in SVI_{30} value after day 30. Similarly, a rather small granule size of 0.25-0.63 mm was reported by Sguanci et al. [18]. When the diameter of aerobic granules reaches above 1 mm during low-strength wastewater treatment, they become loose, porous, and unstable; therefore, they are more likely to undergo disintegration [20,21].

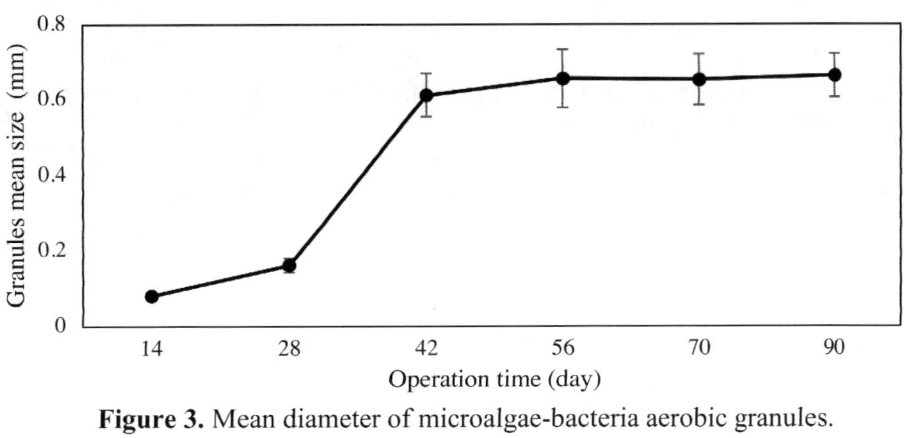

Figure 3. Mean diameter of microalgae-bacteria aerobic granules.

At the end of the experiment, microalgae-bacteria aerobic granules were harvested from the PSBR column. Figure 4 shows the notable difference between the size of the granules on day-0 and at the end of the experimental period. The obtained mature microalgae-bacteria aerobic granular sludge had a maximum diameter of approximately 6 mm. Furthermore, the microalgae-bacteria aerobic granular sludge has a remarkable settling velocity of 80 m/h, which is about ten times faster than the settling velocity of conventional activated sludge used in this study (7.9 m/h). This corresponds to the low SVI_{30} value at the end of the experiment, indicating an excellent settleability as well as full granulation.

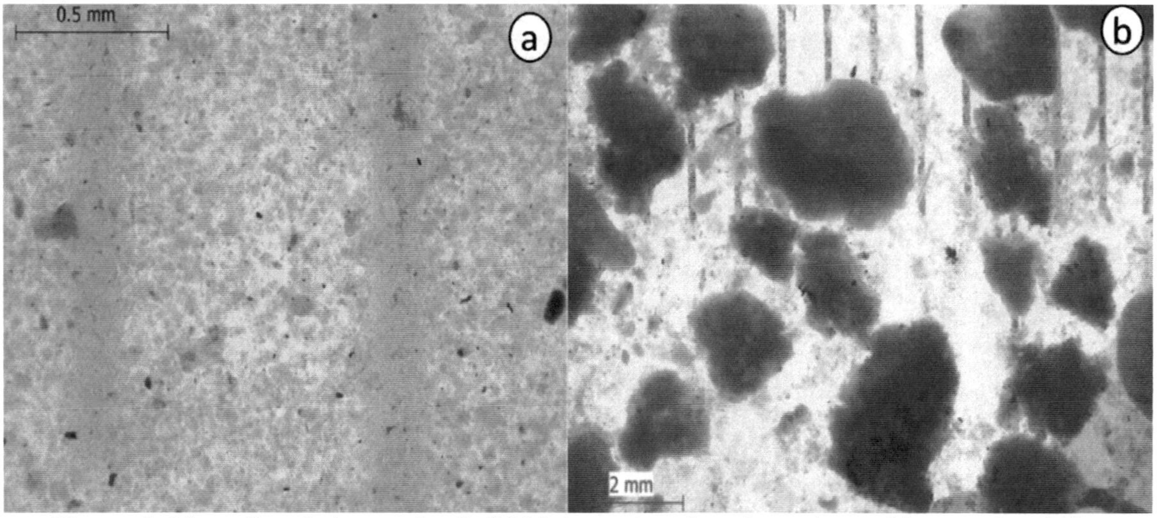

Figure 4. (a) seeding sludge on day-0 (b) Mature microalgae-bacteria aerobic granular sludge harvested after 90 days.

3.2.2. SEM images and elemental composition

FESEM analysis revealed many cocci-shaped bacteria residing on the exterior layer of microalgae-bacteria aerobic granules as shown in figure 5 (a). Cocci-shaped bacteria have been shown to function as a support in the aerobic granulation's microbial attachment process [22]. In addition, microalgae cells and a glue-like material termed extracellular polymeric substances (EPS) were also found to be embedded on the surface of microalgae-bacteria aerobic granules, as indicated in figure 5 (b). EPS is a complex substance consisting of a mixture of different macromolecules such as carbohydrates, proteins, lipids, nucleic acids and glycoproteins. EPS has been recognized as the primary component leading to the production of mature granules because it serves as a 'glue' to trigger biofilm formation among microorganisms [23]. Moreover, numerous different-sized micropores were seen on the granular surface. Micropores are required for substrate transportation into the inner layer of the granules as well as for excretion of the metabolic product out from the granules [24].

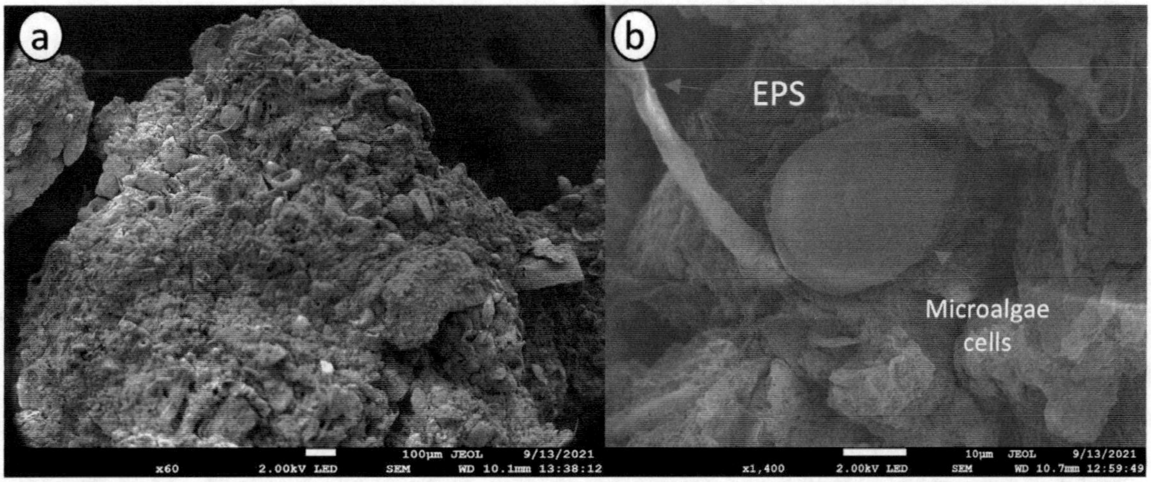

Figure 5. (a) Overview of microalgae-bacteria aerobic granular sludge (b) EPS and microalgae cells.

EDS analysis detected 9 types of elements (figure 6); carbon (C), oxygen (O), iron (Fe), sodium (Na), magnesium (Mg), aluminum (Al), silicon (Si), potassium (K), and calcium (Ca). The results showed a dominating proportion of carbon and oxygen in the microalgae-bacteria aerobic granular sludge of 44% and 26%, respectively. Calcium, potassium, and magnesium (Ca^{2+}, K^{+}, and Mg^{2+}) are all strongly linked to the composition of the EPS matrix. [25]. Ca^{2+} is required for the gelation of the alginate-like exopolysaccharides (ALE) and ALE provides mechanical stability and elasticity to the microalgae bacteria aerobic granular sludge [26]. Potassium content in microalgae-bacteria aerobic granular sludge was 0.6%, and none was detected in the activated sludge. However, calcium and magnesium content in microalgae-aerobic granular sludge was significantly lower than in activated sludge. These findings contradict the excellent settleability showed by microalgae-bacteria aerobic granular sludge and the significant increase in granule diameter. Therefore, complementary spectroscopic analysis such as x-ray photoelectron spectroscopy (XPS) is recommended to confirm and elucidate the elemental composition of microalgae-aerobic granular sludge.

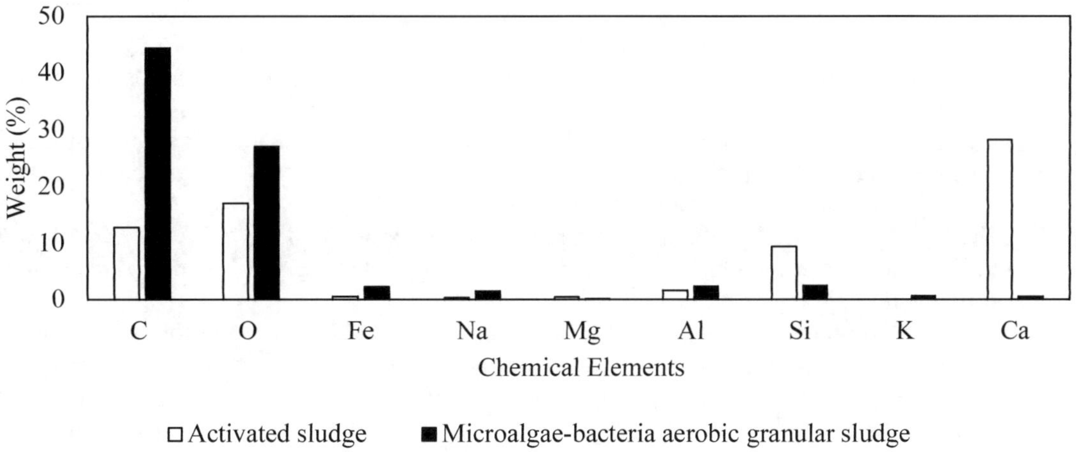

Figure 6. Elemental composition of microalgae-bacteria aerobic granular sludge.

4. Conclusion

Microalgae-bacteria aerobic granular sludge was successfully developed using low-strength municipal wastewater. Granulation started after 28 days of the experimental period, indicated by a stable increase of MLSS concentration up to 3.7 g/L, excellent settleability indicated by 80 m/h settling velocity and low SVI of, as well as a significant increase in granule diameter. The average effluent COD concentration and COD removal efficiency is >50 mg/L and 71% respectively, indicated the capability of microalgae-bacteria aerobic granular sludge to be used for the treatment of low-strength wastewater. Finally, complementary spectroscopic analysis is recommended to characterize microalgae-aerobic granules' element composition in detail.

Acknowledgments

The authors would like to thank Collaborative Education Program for Sustainable Environmental Engineering Network (CEP-SEEN) program under ASEAN University Network/Southeast Asia Engineering Education Development Network (AUN/SEED-Net) by Japan International Cooperation Agency (JICA) for funding this research under grant No. 4B403.

ORCID iDs

S A Zahra https://orcid.org/0000-0002-7166-3544
L D A Purba https://orcid.org/0000-0001-8198-1055
N Abdullah https://orcid.org/0000-0002-9734-5560

References

[1] Nancharaiah Y v. and Kiran Kumar Reddy G 2018 Aerobic granular sludge technology: Mechanisms of granulation and biotechnological applications *Bioresource Technology* **247** 1128–43

[2] Boltz J P, Johnson B R, Takács I, Daigger G T, Morgenroth E, Brockmann D, Kovács R, Calhoun J M, Choubert J M and Derlon N 2017 Biofilm carrier migration model describes reactor performance *Water Science and Technology* vol 75 (IWA Publishing) pp 2818–28

[3] Sarma S J and Tay J H 2018 Aerobic granulation for future wastewater treatment technology: Challenges ahead *Environmental Science: Water Research and Technology* **4** 9–15

[4] Pronk M, Abbas B, Al-zuhairy S H K, Kraan R, Kleerebezem R and van Loosdrecht M C M 2015 Effect and behaviour of different substrates in relation to the formation of aerobic granular sludge *Applied Microbiology and Biotechnology* **99** 5257–68

[5] Derlon N, Wagner J, da Costa R H R and Morgenroth E 2016 Formation of aerobic granules for the treatment of real and low-strength municipal wastewater using a sequencing batch reactor operated at constant volume *Water Research* **105** 341–50

[6] Purba L D A, Ibiyeye H T, Yuzir A, Mohamad S E, Iwamoto K, Zamyadi A and Abdullah N 2020 Various applications of aerobic granular sludge: A review *Environmental Technology and Innovation* **20**

[7] Sepúlveda-Mardones M, Campos J L, Magrí A and Vidal G 2019 Moving forward in the use of aerobic granular sludge for municipal wastewater treatment: an overview *Reviews in Environmental Science and Biotechnology* **18** 741–69

[8] We A C E, Aris A and Mohd Zain N A 2020 A review of the treatment of low-medium strength domestic wastewater using aerobic granulation technology *Environmental Science: Water Research and Technology* **6** 464–90

[9] Cao Y, van Loosdrecht M C M and Daigger G T 2020 The bottlenecks and causes, and potential solutions for municipal sewage treatment in China *Water Practice and Technology* **15** 160–9

[10] Guo D, Zhang X, Shi Y, Cui B, Fan J, Ji B and Yuan J 2020 Microalgal-bacterial granular sludge process outperformed aerobic granular sludge process in municipal wastewater treatment with less carbon dioxide emissions *Environmental Science and Pollution Research*

[11] Zhang Y, Dong X, Liu S, Lei Z, Shimizu K, Zhang Z, Adachi Y and Lee D J 2020 Rapid establishment and stable performance of a new algal-bacterial granule system from conventional bacterial aerobic granular sludge and preliminary analysis of mechanisms involved *Journal of Water Process Engineering* **34** 101073

[12] Yang X, Zhao Z, Zhang G, Hirayama S, Nguyen B van, Lei Z, Shimizu K and Zhang Z 2021 Insight into Cr(VI) biosorption onto algal-bacterial granular sludge: Cr(VI) bioreduction and its intracellular accumulation in addition to the effects of environmental factors *Journal of Hazardous Materials* **414**

[13] Dong X, Zhao Z, Yang X, Lei Z, Shimizu K, Zhang Z and Lee D J 2021 Response and recovery of mature algal-bacterial aerobic granular sludge to sudden salinity disturbance in influent wastewater: Granule characteristics and nutrients removal/accumulation *Bioresource Technology* **321** 124492

[14] Anon 1999 *Standard Methods for the Examination of Water and Wastewater* APHA/AWWA/WEF 2012 Standard Methods for the Examination of Water and Wastewater *Standard Methods* **541**

[15] Yu C, Wang K, Tian C and Yuan Q 2021 Aerobic granular sludge treating low-strength municipal wastewater: Efficient carbon, nitrogen and phosphorus removal with hydrolysis-acidification pretreatment *Science of the Total Environment* **792**

[16] Campo R, Sguanci S, Caffaz S, Mazzoli L, Ramazzotti M, Lubello C and Lotti T 2020 Efficient carbon, nitrogen and phosphorus removal from low C/N real domestic wastewater with aerobic granular sludge *Bioresource Technology* **305**

[17] Ministry of Environment and Water 2009 Environmental Quality Sewage Regulations *Environmental Quality Act 1974* P.U.A 432-2009

[18] Sguanci S, Lubello C, Caffaz S and Lotti T 2019 Long-term stability of aerobic granular sludge for the treatment of very low-strength real domestic wastewater *Journal of Cleaner Production* **222** 882–90

[19] Petrini S, Foladori P, Beghini F, Armanini F, Segata N and Andreottola G 2020 How inoculation affects the development and the performances of microalgal-bacterial consortia treating real municipal wastewater *Journal of Environmental Management* **263**

[20] Zhang H, Dong F, Jiang T, Wei Y, Wang T and Yang F 2011 Aerobic granulation with low strength wastewater at low aeration rate in A/O/A SBR reactor *Enzyme and Microbial Technology* **49** 215–22

[21] Peyong Y N, Zhou Y, Abdullah A Z and Vadivelu V 2012 The effect of organic loading rates and nitrogenous compounds on the aerobic granules developed using low strength wastewater *Biochemical Engineering Journal* **67** 52–9

[22] Zhang H, Dong F, Jiang T, Wei Y, Wang T and Yang F 2011 Aerobic granulation with low strength wastewater at low aeration rate in A/O/A SBR reactor *Enzyme and Microbial Technology* **49** 215–22

[23] Tay J H, Liu Q S and Liu Y 2001 The effects of shear force on the formation, structure and metabolism of aerobic granules *Applied Microbiology and Biotechnology* **57** 227–33

[24] Xu G, Xu X, Yang F and Liu S 2011 Selective inhibition of nitrite oxidation by chlorate dosing in aerobic granules *Journal of Hazardous Materials* **185** 249–54

[25] Dahalan F A, Mohamed Najib M Z, Salim M R and Ujang Z 2015 Characteristics of developed granules containing phototrophic aerobic bacteria for minimizing carbon dioxide emission *International Biodeterioration and Biodegradation* **102** 15–23

[26] Lin Y M, Sharma P K and van Loosdrecht M C M 2013 The chemical and mechanical differences between alginate-like exopolysaccharides isolated from aerobic flocculent sludge and aerobic granular sludge *Water Research* **47** 57–65

One-way flow system for improvement of the acid sulfate soil reclamation process in the Belanti II tidal swamp irrigation network, Central Kalimantan, Indonesia.

A Setiawan[1] , B S Wignyosukarto[2], A P Rahardjo[2]
[1]Master Student, Civil and Environmental Engineering Department, Universitas Gadjah Mada, Indonesia
[2]Civil and Environmental Engineering Department, Universitas Gadjah Mada, Indonesia,
*) Corresponding author email: budiws@ugm.ac.id

Abstract. Belanti II tidal irrigation network was developed in 1980 on acid sulphate soil with an area of approximately 3976 ha. Belanti II has one primary channel and several secondary channels, without any water structures that regulate the direction of flow. Two-way flow occurs in the channel, supplying fresh water from the river at high tide and flowing back into the river at low tide. This two-way flow system inhibits the leaching process of acid sulphate soil. Up to now, acid sulfate soils are still found at a depth of 50-75 cm with a pH of 2.5 ~ 3.5. A one-way flow system is expected to improve water circulation, assisted by channel rehabilitation, collector channels and water control structures. The spillway structure at the secondary channel regulates the freshwater supply at high tide and the flap gate at the end regulates the flow of water to the collector channel at low tide. Hydraulic analysis was carried out using HEC-RAS modelling. The results showed the ability of the system to supply fresh water, flushing/leaching acid sulfate water and drainage the toxic water. The effectiveness of the system is evaluated regarding the direction and the volume of freshwater flowing into the secondary channel.

1. Introduction

Tidal swamp is a potential land for the development of agricultural land whose water management has been developed in Indonesia since 1970, especially in Kalimantan and Sumatra [1]. In Kalimantan the development of tidal swamp land is carried out to support the transmigration program through the Tidal Rice Field Development Project (P4S) [2]. One of them is the Belanti II tidal irrigation area with an area of about 3976 ha which was started to be developed in 1980. In the first stage of development, the water system was a two-way open channel system, where river water flows into the rice fields at high tide and flows into rivers. at low tide. The network scheme is not yet equipped with a structure to regulate water circulation and maintain the leaching process.

Improvement of the network system was carried out in the second stage of development by constructing several water structures to control water circulation, to prevent saltwater intrusion, to maintain surface water levels and groundwater levels in secondary channels. After a long period of operation, some of the water control structures, such as stoplogs and sliding gates, were damaged and did not function. The next problem is efforts to flush out sediment deposits and toxic materials at the ends of the primary and secondary channels. Fresh water supply from rivers and rainfall are not able to support the sulfuric acid leaching process because the flow capacity is reduced. The two-way flow system is considered ineffective for dissolving and diluting acidity and toxicity.

The entire irrigation area in Belanti II is tidal irrigation which is spread over 3 villages. Agricultural productivity in this area is 5.5 ~ 6.5 tons/ha with a cropping index of IP 200. This number will be increased to meet the Pulang Pisau government target of 7 tons/ha with a cropping index of IP 300. The causes of low agricultural productivity are cultivation capital, market access, inadequate irrigation facilities, inadequate availability of agricultural tools and technology, and relatively infertile soil.

Acid sulphate soils (ASS) are commonly found in tropical and subtropical coastal areas located in swampy areas because of optimal environmental conditions for the formation of seawater sulphide minerals [3]. These soils are formed through the oxidation of sulphide minerals in the soil and often have low soil pH values <4 [3]. In Belanti II, until now, acid sulphate soils are still found at a depth of 50 ~ 75 cm with a pH of 2.5 ~ 3.5. This indicates that the land reclamation process has not yet been fully completed. Therefore, to improve the quality of agricultural land, a reclamation process is still needed, by supplying some fresh water as washing or leaching acid water in the land or in the channel. Problems due to oxidation of acid sulphate soils are the toxicity of aluminium, iron, and hydrogen sulphide [4]. One of the recommended alternatives in the management of acid sulphate soil is water management. Sulfuric acid is treated by reducing its toxicity by diluting it with freshwater [5]. The recommended water volume for dilution is 5 cm or 500 m3/ha/day. This water leaching process should be carried out with a one-way flow system to increase water circulation, including increasing channel capacity, adding collector channels and water control buildings. The spillway in the secondary channel regulates the supply of fresh water at high tide and the sliding gate/flap gate at the end of the secondary channel regulates the flow of water to the collector channel at low tide.

This study intends to evaluate the application of a water structure in the secondary channel to regulate one-way flow from the primary channel to the collector channel. This one-way flow is expected to flush the accumulated sulfuric acid in the channel. The hydraulic simulation uses HEC-RAS software to simulate the tidal flow phenomenon, knowing the direction of flow and the cumulative volume of fresh water entering the channel system for dilution of toxic materials.

2. Material methods

2.1. Existing condition

The Belanti II Scheme is a tidal swamp area located in Pulang Pisau Regency, Central Kalimantan Province. Built in the 1980s as part of the Tidal Rice Field Development Project (P4S) and Block D Peatland Development Project (PLG). Currently the Belanti II Scheme is one of the irrigation areas included in the national strategic program set by the Ministry of Works Public and Public Housing to support food security programs in Central Kalimantan Province [3]. The Belanti II scheme is situated between the Kahayan River in the west and the Kapuas River in the east (Figure 1). The source of tidal water comes from the Kahayan River which enters through primary and handil channels as well as collector channels in border areas. The Belanti II irrigation network was initially equipped with a tidal pond with an area of 12 ha, a primary channel with a length of 8.70 km and a width of 45 m, 100 secondary channels (41 channels on the right with a length between 0.40 ~ 2.70 km and 59 channels on the left with a length between 2.0 ~ 2.80 km), the collector channel is 14.22 km long and 6 m wide, with a width of 5.6 m. The distance between secondary channels is 200 m serving ± 3976 Ha of irrigation area. The water flow system is a two-way open channel system, where river water flows into the system at high tide and flows back into the river at low tide. The reclamation process has been ongoing since it was used until now. Currently, several existing networks and channels have decreased their function in water management due to reduced capacity.

The various objectives of water management in this area are removal of excess water (rainwater), provision of irrigation water, prevention of soil subsidence, dilution and flushing of sulfuric acid in soil and channels. The ability of drainage and irrigation as well as plant suitability is determined simply based on the hydro topographic classification (A to D). The hydro-topographical classification are identified as : A: tidal irrigated areas, in both wet and dry season; B: tidal irrigated areas, only in the wet season; C: areas just above (< 0.50 m) the average high water level in wet season; D: high areas, water levels independent from tidal influence [6]. Considering the hydro-topographic conditions associated with the tides of the Kahayan River, it can be concluded that 75% of the study area is class A and 25% class B [7]

The supply of fresh water for rinsing and leaching is planned in sufficient quantities. It is hoped that the leachate can flow back into the river quickly at low tide [1]. In addition, flushing can also avoid sedimentation. The collector channel is made to help drain water at the end of the secondary channel. However, with two intake channels from the river (primary channel and collector channel), the tidal influence of the two intakes will meet at a point in the channel and the velocity at that point is relatively zero, causing sedimentation (dead zone). To overcome this, it is recommended to adjust the flow in one direction to increase circulation in the network to encourage flushing by using a weir and flap gate to regulate the flow direction [1].

2.2. Proposed improvement

Water management in swamp irrigation networks must meet at least four objectives, first to drain excess water during the rainy season, second to provide irrigation water, third to provide sufficient water for leaching and dilution of acid water, and fourth to maintain the potential of acid sulphate soil under reduced conditions to avoid oxidation [8].

The results of a survey in 2021 on the Belanti II scheme show that after decades of reclamation, pyrite layers at a depth of 0.50~1.0 m below the ground surface are still found in several places. The acidity of the channel water (pH) ranged from 2.55~4.40, indicating that pyrite oxidation was continuing. Improvement of water management is proposed to prevent excessive oxidation and allow freshwater supply at high tide and rainwater to meet the needs of irrigation water and leaching of acid water in the channel due to pyrite oxidation. The proposed water management improvement can be done by increasing the capacity of primary, secondary and collector channels through normalization. Increasing the channel capacity will increase the amount of water to meet the needs of the dilution process. As suggested by Konsten et al. [5,9], the need for fresh water is more than 1000 cm for 200 days, or about 5 cm/day (Table 1). The amount of fresh water needed for dilution/washing of the acid solution in the Belanti scheme with an irrigation area of 3976 ha is about 500 m3/day/ha of water. This means that the collector channel must provide 1988 million m3/day of fresh water.

Table 1. Effects of leaching on aluminium concentration in soil solution at depths 20~25 cm of undisturbed severe acid sulphate soils by various authors [5].

Authors	Days of leaching	Water used cm	Al^{3+} in soil solution at depth **20~25** cm	
			Before leaching mmol(+)l^{-1}	After leaching mmol(+)l^{-1}
Sen 1988a	30	400 ~ 450	33	6 ~18
Tuong et al. 1993	15	500	10	4.5
	30	1300	10	4.5
SIWRR 1990	63	450	9.3	2.5
Konsten et al. 1991	200	>1000	>20	App 2

2.3. Hydraulic Simulation

Flow simulation imposed by tidal fluctuations should be carried out using unsteady flow analysis. The flow simulation in the channel can be solved with the help of the HEC-RAS program which is able to solve the continuity and momentum equations numerically using the finite difference method with an implicit scheme [1,6,10]. The simulation was carried out using tidal data measurements during 2 days in the Kahayan River as the water level boundary condition (Figure 1). The hydraulic simulation uses a scheme as shown in Figure 2 to evaluate the performance of the channel to support the leaching process using fresh water from the Kahayan River. This simulation does not evaluate the influence of the extreme rainfall on the channel network.

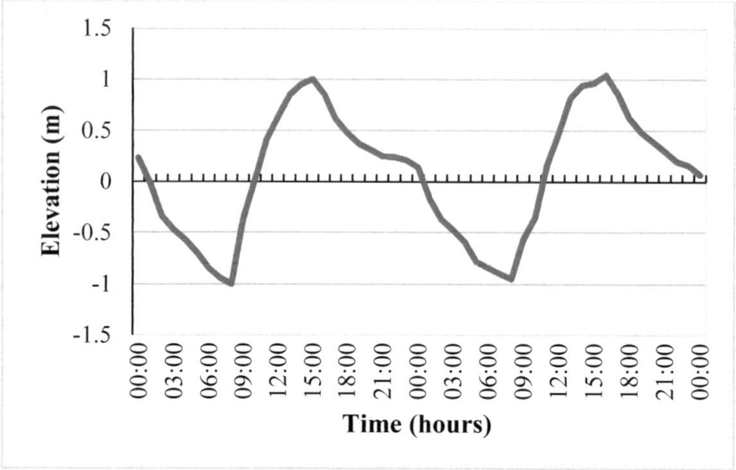

Figure 1. Measured tide on November 16-18,2020

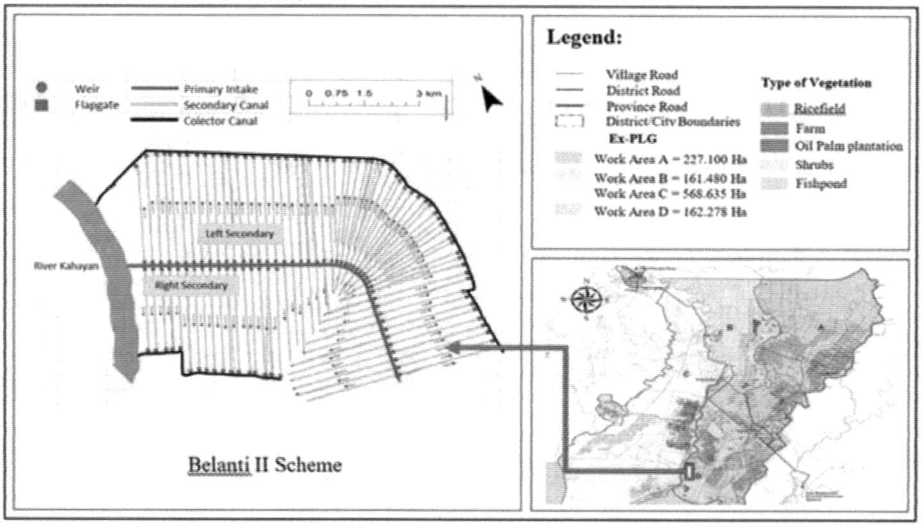

Figure 2. Schematic of Belanti II irrigation network

Two simulation scenarios were used to evaluate the hydraulic performance of the scheme. The first simulation using the existing conditions. The second scenario simulates normalization conditions with the addition of a water regulating structure. In Figure 3 the flow from the primary channel to the left secondary and right secondary is regulated by a weir, so that fresh water can flow at a certain elevation when the water level begins to rise. While in Figure 4 the flap gate is installed at the end of the left secondary channel and the right secondary channel, to prevent reverse flow from collector channel during high tide but allowing flow out into the collector channel during low tide. It is hoped that freshwater flow through the weir will dilute the acidic water in secondary channel and drain it, and encourage a faster flow out of the system through the flap gate.

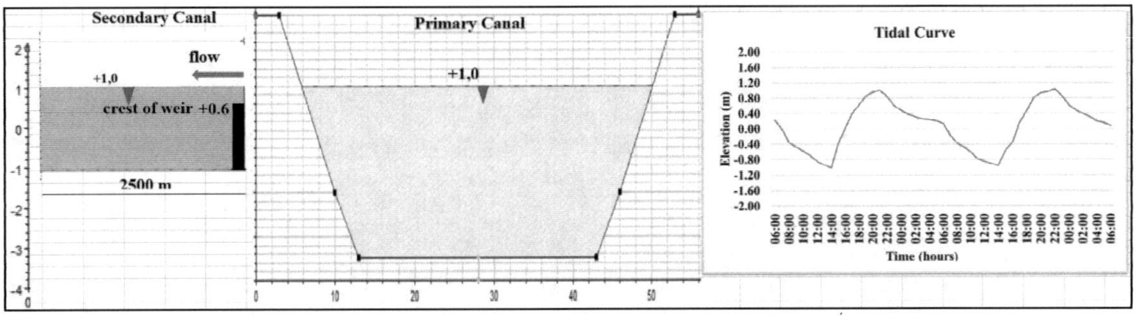

Figure 3. Weir crest elevation to tidal elevation in the mouth of primary channel

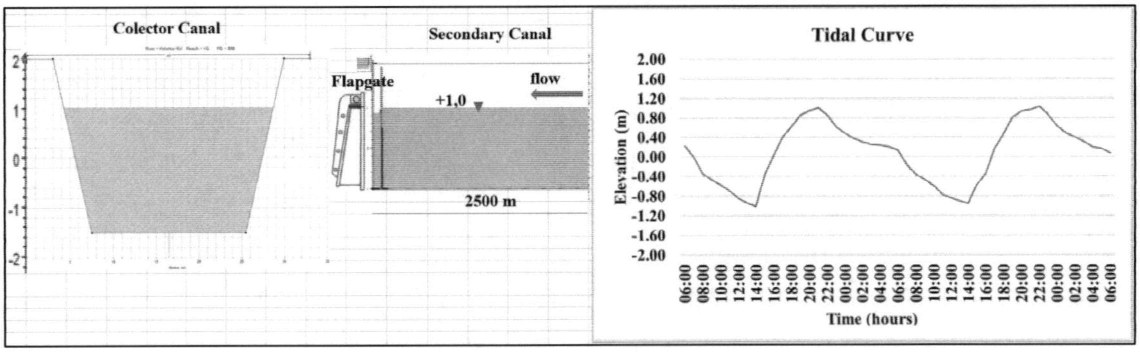

Figure 4. Position of the flap gate on the secondary channel section

2.4. Evaluation of the results of the research parameters

In general, three evaluations were carried out in this study, to evaluate the ability to supply leaching and flushing water. First evaluation calculating the nett cumulative volume of water entering to the network through the primary channel during one full tidal cycle. The nett commulative volume must be equal to or exceed the amount of water supply needed per day. Second evaluation observe the capability of leaching that focuses on two things, namely the direction of flow and flow rate. The positive direction of flow is the flow in the secondary channel from primary channel to collector channel. One-way flow occurs when more volume of water enters through the secondary channel and exits through the collector channel without any reverse flow (negative direction). While the evaluation of the flow rate in terms of the discharge value in the secondary channel. The greater the discharge value, the better the network circulation.

3. Result and discussion

3.1. Scenario 1 Existing Condition

The results of the hydraulic simulation on the existing channel indicate attenuation of tidal fluctuations along the channel. The attenuation in the middle channel is 30 cm, while the attenuation in the upstream channel is 40 cm, see Figure 5. This attenuation will affect the ability of tidal water to enter the secondary channel. The result of simulation that depicted in Figure 6, shows the volume fresh water entering and leaving the scheme in one tidal cycle. The net cumulative water entering to the scheme through primary channel is about 1,219,554 m^3 while the current demand for leaching water in the irrigation area is about 1,988,000 m^3. The existing system (scenario 1) cannot meet the supply criteria due to a deficit of -38.65% of the total demand.

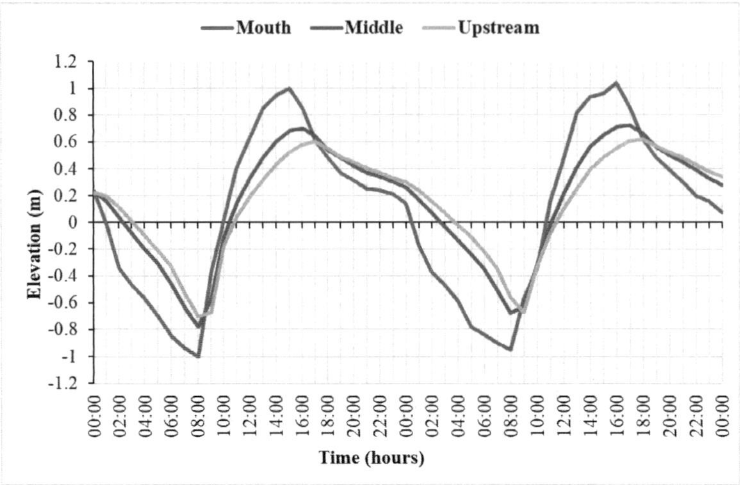

Figure 5. Water level elevation in scenario 1

This water deficit is caused by channel capacity or water storage in the network due to sedimentation or siltation on the channel. Normalize the capacity of the channel could increase the volume of entering freshwater flow.

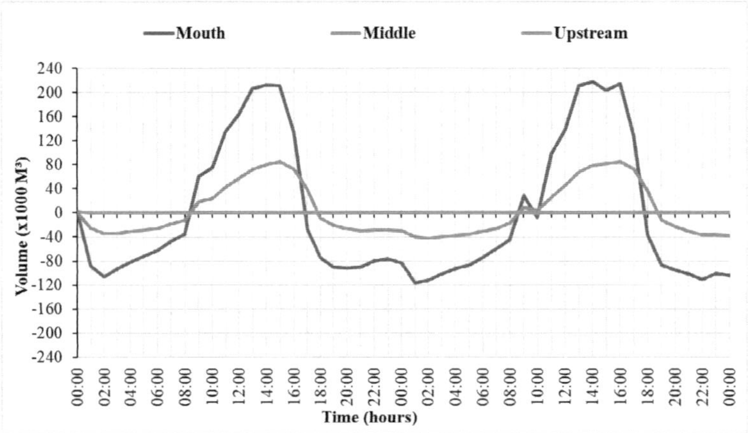

Figure 6. Volume of water entering and leaving the scheme in scenario 1

3.2. Scenario 2 Normalization with Weir and Flap gate

In this scenario, the primary channel is widened from 45 m to 50 m width, the secondary channels are widened from 6~8 m to 10 m width and collector channel widened from 8 m to 27 m width. The weir and flap gate designs in scenario 2 determine the supply and flushing capability of the system. Considering the need for leaching water, the crest of weir elevation was installed at elevation of +0.60 m, 0.40 cm below the peak tide elevation +1.00 m, while the flap gate was installed at the end of the secondary channel with the dimension of 2.5 m x 2.5 m and bottom elevation of -0.80 m. The flap gate will only open if the water level in the collector channel is lower than water level in the secondary channel. Figure 7 show the discharges flowing through combination of weir and flap gate in the left secondary channel 5 and 50 and the right secondary channel 15 during 2 tidal cycles. At high tide, the flow over the weir of the left secondary 1 and 50 also the right secondary 15 have a maximum discharge of 2.14 m3/s, 2.35 m3/s and 2.16 m3/s respectively while at low tide, its secondary flap gate can release a maximum discharge of 0.65 m3/s, 0.66 m3/s and 0.65 m3/s respectively.

Weir and flap gate dimensions can be adjusted to meet the leaching and flushing requirements. Alternatively, lowering the weir crest elevation will help to increase the water supply at high tide. While the flap gate will intensify the drainage flow to the collector channel.

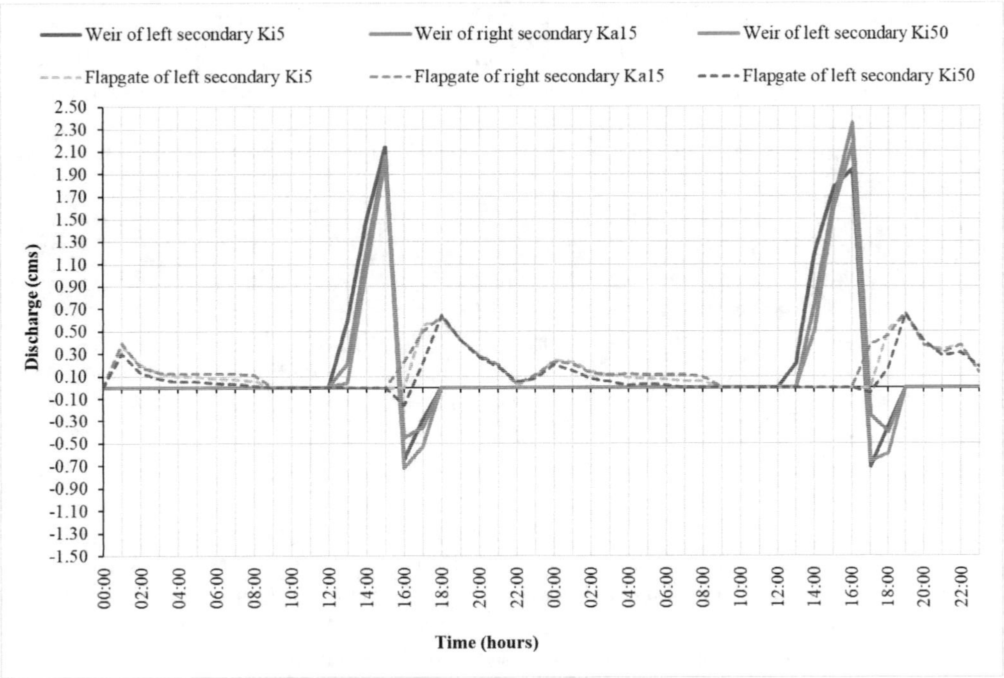

Figure 7. Weir and flap gate discharges

4. Conclusion

In tidal irrigation areas with shallow acid sulphate soil such as the Belanti II irrigation area, preventing further oxidation of potential acid sulphate soil and leaching of the sulphate solution causing by soil reclamation is recommended to meet the need of increasing agriculture production. Water management that regulates the flow direction and allowing large volume of freshwater entering to the system is already evaluated in this paper. A couple of structure i.e., weir and flap gate installed in the secondary channel is one of the alternatives to regulate one way flow system that is needed to prevent the pollution of supply water by toxic water resulting from oxidation of acid sulphate soils. Adjusting crest elevation and width of weir could increase the supply of freshwater, as well as widening the primary, secondary and collector channels. Channel widening is preferred rather than deepening with the aim of preventing further oxidation of acid sulfate soils.

Acknowledgement

This research was conducted in one of the swamp schemes under BWS Kalimantan II, and utilized some data from consultants PT Yodya Karya, PT Wahana Krida Konsulindo, PT Rancangtama Bangunmandiri Indracipta. Appreciation and thanks to the leadership and staff for the favors and assistance during the research and preparation of this paper.

References

[1] Wignyosukarto B S 2013 Leaching and flushing of acidity in the reclamation of acid sulphate soil, Kalimantan, Indonesia *J. Irrigation and Drainage 62* 75–81.
[2] Subagjo H and Widjaja-Adhi I P G 1998 Opportunities and use of swamp land for agricultural development in Indonesia: the case of South Sumatra and Central Kalimantan. Main Paper

IOP Conf. Series: Earth and Environmental Science 1091 (2022) 012053 doi:10.1088/1755-1315/1091/1/012053

Discussion and Communication Meeting on Soil and Agroclimate Research Results (Bogor)

[3] Konsten C J M, Suping S, Aribawa I B and Widjaja-Adhi I P G 1990 Chemical processes in Acid sulphate soils in Pulau Petak, South and Central Kalimantan, Indonesia, In: *AARD & LAWOO Papers workshop on acid sulphate soils in the humid tropics*; 20-22 November 1990; Bogor. Indonesia. p. 109-135

[4] Dent D 1986 Acid sulphate soils: a baseline for research and development, *The Netherland: ILRI-Wageningen Publication 39*

[5] Tuong T P 1993 An overview of water management of acid sulphate soils Ho Chi Minh City Symp. on acid sulphate soils,*The Netherland: ILRI-Wageningen Publication 53* pp 265-279.

[6] Suryadi F X 2020 Soil and Water Management Strategies for Tidal Lowlands in Indonesia. *CRC Press. London*, doi.org/10.1201/9780429333231

[7] Balai Wilayah Sungai (BWS) Kalimantan II 2021 Survey, investigation and design of rehabilitation and network improvement swamp irrigation Block D Working Area (Package 4), Central Kalimantan. Final Report Yodya Karya Consultant Wahana Krida Konsulindo Consultant and Rancangtama Bangunmandiri Indracipta Consultant, Palangkaraya, Indonesia.

[8] Houterman J and Ritzema, H 2009 Land and Water Management in the Ex-Mega Rice Project Area in Central Kalimantan, Technical Report 4, Jakarta: Euroconsult Mott MacDonald and Deltares.

[9] Konsten C J M, Suping S, Aribawa I B and Widjaja-Adhi I P G 1991 Column experiments Indonesia. In: Simulation model of physical and chemical processes in acid sulphate soils to evaluate water management strategies, 27-58. AARD & LAWOO, Bogor.

[10] Roelse K 1988 Land reclamation of lowlands in Indonesia Thesis (The Netherland: TU Delft)

IOP Conf. Series: Earth and Environmental Science 1091 (2022) 012054 doi:10.1088/1755-1315/1091/1/012054

Management of Waste Cooking Oil and its Potential for Value-added Materials: A Mini Review

N. A. Hamdi[1*], S. Sha'arani[1], N. F. Azman[1], S. B. Mohd. Rafi[1], E. Norsin[1], N. Othman[1]

[1]Malaysian-Japan International Institute of Technology, Universiti Teknologi Malaysia, Jalan Sultan Yahya Petra, 54100, Kuala Lumpur, Malaysia

*Corresponding author e-mail: aliya.hamdi@yahoo.com

Abstract. Improper management of waste cooking oil in food industries like restaurants, hotels, and the use of retort technology in food processing is causing a devastating impact towards the environment. This is observed through odor problems and oil film formation on the surface of water bodies that increases the biological oxygen demand. Additionally, different types of cooking oil represent different physicochemical properties. This allows for the potential use of waste cooking oil as a carbon source for fungal growth thus contributing to the production of value-added products at a low cost and eco-friendly process. Biodegradation may take place by the enzymatic reaction which involves the hydrolysis of triglycerides and metabolized through the β-oxidation pathway to acetyl-CoA. Many fungal and bacterial strains capable of degrading vegetable cooking oils have been reported with most from the genus *Aspergillus, Pseudomonas, Enterobacter,* and others. From the industrial perspective, with the correct technology and proper recycling technique, waste cooking oil can be readily utilized as chemical building blocks in lubricant, candle, and fuel productions.

1. Introduction

Cooking oil is a significant component in our daily lives and the growth of the population has led to an increasing demand of cooking oil in the last few years. Aside from home cooking, cooking oil is used in various food and beverage industries such as the retort industries, hotels, catering services as well as fast food chain restaurants. Additionally, it is also widely used in industrial applications such as paints, soaps, biofuels, and lubricants [1]. According to the Population Reference Bureau [2], there are approximately 7.8 billion people around the world, with the numbers expected to increase exponentially, leading to an increasing demand for food. While the population dietary guidelines are somewhat different between countries, in terms of reducing risk of cardiovascular disease, the dietary guidelines generally reduced the intakes of saturated fats to approximately 10% of dietary energy while average fat intake is suggested to be around 30-35% dietary energy [3]. The total fat intake, on the other hand, is recommended to be at least 30% of energy [4].

Cooking oil originates from various vegetable oils, mainly palm oil, palm kernel oil, rapeseed oil, canola oil, cottonseed oil, soybean oil, peanut oil, sunflower oil, castor oil, coconut oil, olive oil and some rare types of cooking oils such as tiger nut oil, rice bran oil as well as prima oil [5]. Despite having a bad reputation, cooking oils do contain some amount micronutrients, for instance vitamin E, sterols, phospholipids, carotenes, squalene and glycolipids, all of which are found in fresh palm oil (see Table 1 for details) [6]. Additionally, cooking oil can be broken down by microorganisms like fungi, actinomycetes and bacteria since they primarily consist of fatty acids; or triacylglycerols (TAGs) [7]. However, the high usage of cooking oil due to the increasing global demands may lead to improper waste cooking oil (WCO) management by some industries, making it a major environmental threat. When disposed of improperly, WCO creates oil film on water surface, causing the aquatic lives to suffocate due to lack of oxygen. This could also cause issues if WCO is disposed in the piping system, where solidified WCO can cause blockages and congest treatment filters [5], leading to various issues, such odor. Unfortunately in Malaysia, disposal of WCO is not standardized, with some communities not even equipped for communal WCO collection system. Therefore, bioremediation through microbial degradation is seen as a potential alternative to reduce the impact of improper WCO disposal to the environment.

Table 1. Element in Palm Oil

Component	Description
Carotenes	Precursors for vitamin A, helps in promoting growth and tissue maintenance as well as prevents night blindness
Phytosterols	These lipophilic sterols are easily absorbed in the gastrointestinal tract and converted into enzymatic reactions to cholesterol, a major precursor of steroid hormones
Squalene	Possesses negative feedback inhibition activity on an enzyme (HMG-CoA reductase) involved in the production of cholesterol in the liver.
Vitamin E	Act as antioxidants that could reduce plasma cholesterol and inhibit or delay oxidative deterioration of cellular membranes.

2. Physical and Chemical Properties of Cooking Oil

Since oils are an important part of our diet, providing not only energy but also a transporter for lipid-soluble vitamins, it is important to include oils in a balanced meal. The physical and chemical properties of oils help to determine its quality and present state [8]. The physical properties of cooking oil are evaluated based on their viscosity, density, stability, peroxide value (PV), iodine value (IV) and saponification value (SV) [9]. Other physical characteristics that are also analyzed includes flow point, corrosion as well as cloud point [10]. The chemical properties, especially the type of fatty acids of different position and their comparative proportions, on the other hand, is the key factor that determines their physical characteristics, and it is crucial in the determination of oil stability and overall performances [10]. Not only that, the presence of saturated or unsaturated fatty acids also affects the cloud and flow points of the cooking oils [11]. Cooking oils are composed of triglycerides, molecules that consist of hydrogen (H), oxygen (O) and carbon (C), following their structures that are made up of three fatty acids and one glycerol [11].

The high usage of cooking oil coupled with the lack of proper WCO management is causing various issues. As mentioned earlier, WCO causes damage to the aquatic environment and the dumping of WCO on land or via drain pipes can cause pest infestations and odor problems. Coupled with other environmental issues such as oil spillage, various alternatives are being carried out to curb the problem. One of which is bioremediation, the microbial degradation of lipid materials that will be discussed in further detail throughout this paper.

3. Cooking Oil as Carbon Source for Fungal Growth

Cooking oil, in its fresh state, will be degraded by the actinomycetes or fungi, some of which are known to produce lipases for lipid digestion [7]. However, WCO usually releases harmful compounds making its usage limited. Regardless, various studies have proven that fungi could take up carbon sources for their potential growth. These studies promote the significance of biovalorization of the oily materials as a substrate for fungal growth, especially since the demands and usage of fatty acids are increasing [12]. In fact, one study recommends the conversion of waste olive oil to biomass by *Aspergillus* and

Penicillium strains in order to produce added-value compounds from their substrates. Meanwhile, another study had provide evidence that biosurfactants like *Mucor circinelloides* could reduce surface tension of culture medium and form a clear zone in an oil spreading test [13]. The best condition was at the presence of 8% used cooking oil, 2g/L yeast extract as nitrogen source and 0.01 mM $FeSO_4$ within 5 days of incubation and 87.6% of crude oil in 12 days of incubation [13].

In an earlier research, Cihangir and Sarikaya reported that *Aspergillus sp.* needed olive oil as carbon source and peptone as its nitrogen source to produce lipase and enhance its growth, while aeration is one of the factors to improve better growth and lipase production [14]. Not only that, a study using a single-cell protein (SCP), the *Yarrowia lypolitica,* found that the highest reported *Yarrowia lypolitica* biomass concentration (57.37g/L) was achieved when WCO was used as the main carbon source [15]. Thevenieau and Nicaud (2013) infers that yeasts may consume several types of carbon sources for their production of cell mass and lipids [16]. The sources could be glucose, xylose, glycerol, cellulose hydrolysates, starch, as well as industrial and municipal organic wastes. The accumulation of lipids takes place under limited circumstances caused by a nutrient other than carbon in all cases [16].

4. Cooking oil degradation and its metabolites by fungi

Engaging the biochemical abilities of microorganisms is one of the popular strategies in biological treatment of contaminated environment, including wastewater and water treatment. Several filamentous fungi and yeast species were reported by numerous studies on their capability of cooking oil degradation. They form a distinctive species known to deteriorate a vast variety of compound and materials. Fungi have contributed significantly to human welfare since ancient times [17]. Biodegradation is a degradation process which involves the actions of microorganisms converting or altering the materials or mineralization into products such as carbon dioxide, water and biomass. It is a biotechnological process that is environmentally acceptable and utilizes valuable genetic material. Meanwhile, bioaugmentation improves the degradation process by metabolic or enzymatic reaction of the microorganism towards the contaminant found in the environment. Furthermore, the chemical process may take place in transforming the organic materials (complex) into simpler molecules [18]. In fungi, cooking oil is hydrolyzed to yield free fatty acids, glycerol, diacylglycerol (DAG) and monoacylglycerol which is then taken up and reassemble within the microorganism cells. In fact, the secretion of by-product of the degradation process could be potentially improved by genetic engineering [19], making this a valuable alternative to WCO management.

Biodegradation can take place in either one of two ways: 1] primary biodegradation and 2] complete degradation. Complete degradation involves the mineralization of organic pollutant as sole source of energy and carbon by certain microorganism such as bacteria, fungi and yeast. Meanwhile, primary degradation refers to the small alteration of the physical characteristic of the substance by microorganism thus leaving most of the compound unchanged. Therefore, the intermediary metabolite or the end product resulting from primary degradation may be more toxic than the actual compound [20]. On the other hand, ultimate degradation is attained with complete utilization of contaminant by microorganism to produce mineral salts, methane, water, and carbon oxide as well as their new cellular component [21].

Aspergillus sp., Aspergillus niger, Rhizophus oryzae, and *Yarrowia lipolytica* are reported to hydrolyze the triglycerides into glycerol and fatty acid by secreting the extracellular lipases [22]. The utilization of the cooking oil as a substrate for the enzymatic hydrolysis of triglycerides releasing the glycerol and fatty acids such as palmitic acid, stearate acid, oleic acid, and linoleic acid in culture medium by extracellular lipase [23]. Based on Figure 1, free fatty acids will undergo a lipolysis in the cytoplasm where it will metabolize in β-oxidation pathway to acetyl-CoA by transporting across the cell membrane. In addition, ATP-citrate lyase (ACL) is the enzyme involved in fatty acid metabolism by fungi in cytosol to yield acetyl-CoA. It will further convert the free fatty acid (FFA) in three consecutive stages by acetyl-CoA carboxylase, palmitoyl-ACP thioesterase and fatty acid synthases [24]. Acyl-CoA is produced by the activation of FFA by fatty acyl-CoA synthetase (FAAI). In lipid bodies, TAGs (85%) and steryl esters (8%) accumulated in oleaginous yeast. In endoplasmic reticulum, the acyl-CoA isc converted to DAG, then the DAG1 and DAG2 genes are responsible for transforming the DAG to TAGS through the Kennedy pathway. The expression of ARE1 gene, which is located in lipid bodies, stimulate the sterol esterification and acyl-CoA, forming the steryl esters [24]. Meanwhile, *Candida bambicola, Candida tropicalis, Candida utilis, Pichia jadinii* have been reported to promote lipid solubilization by synthesizing the biosurfactant compound. Some filamentous fungi such as *Aspergillus* sp., *Candida tropicalis, Cryptococus curvatus, Penicillium expansum, Wickerhamomyces anomalus, Yarrowia lipolytica* have also been reported to allow the adhesion of lipids by modifying the cell surface [24-29].

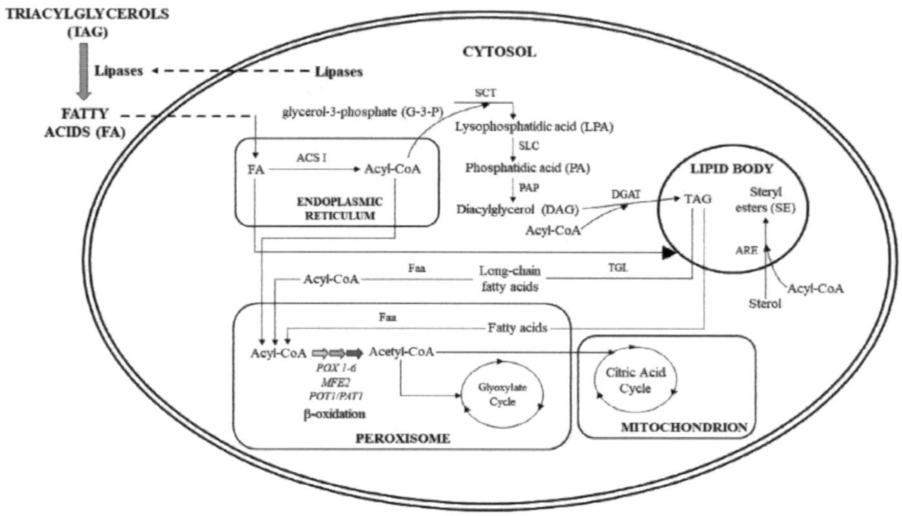

Figure 1. Metabolic pathways on lipid degradation and assimilation by *Yarrowia lipotica*. Abbreviations for listed enzymes: ACS (fatty acyl-CoA synthetase), ARE (acyl- CoA:cholesterol acyltransferase), DGAT (acyl-CoA:diacylglycerol acyltransferase), Faa (fatty acyl- CoA synthetase), PAP (phosphatidate phosphatase), SCT (glycerol-3-phosphate acyltransferase), SLC (acyl glycerol-3-phosphate acyltransferase) and TGL (triacylglycerol synthetase); and for genes which encode the following enzymatic activities: POX (acyl-CoA oxidase); MFE (multifunctional enzyme); POT (thiolase) [24].

5. Industrial usage of added value compound by cooking oil

WCO if managed well, can be a useful compound that promotes the sustainable management and use of food waste. Traditionally, WCO is mainly recycled to produce biodiesel, but this process also produces a considerable amount of oleochemical that is also beneficial for other usage, such as lubricants and cosmetics. Not only that, the utilization of WCO as microorganism feedstock will help to mitigate the environmental and social impacts regarding the edible oil and the oleochemical industries [30]. Generally, the chemical composition of cooking oil is composed of TAGs and fatty acid. Several studies have been carried out to identify specific yeast and fungi that can use WCO as carbon sources to produce value-added green product such as polymers, plasticizer, lubricants, biosurfactants, biomaterials, microbial lipids, and more [31-33]. Polyhydroxyalkanoates (PHAs) is one of the main metabolites produces from microorganism degradation of cooking oil which assemble in the form of intracellular granules as an alternative to the more environmentally friendly synthetic polymers [34]. These biopolymers have an extensive range of application in various field like food industry, agriculture, wastewater treatment, tissue engineering and pharmacology. Cruz and the coresearchers addressed the use of waste cooking oil for the accumulation of PHAs by the yeast *Wickerhamomyces anomalus* [35].

They found that the addition of molasses as a co-carbon substrate in the WCO has been proved to increase the PHAs production. Furthermore, the amount of nitrogen content as well as presence of food residues in waste cooking oil can cause a good environment for the microorganism growth thus, increase the production of PHAs consequently [30].

Another value added is biosurfactant which is classified as neutral lipids, phospholipids, lipopeptides and glycolipids. These compounds attract the recent global market due to their environmental advantages than synthetic surfactant [37]. This surface-active compound can decrease the surface and interfacial tension between air, water, and oil by the action of its amphiphilic molecules (hydrophilic and hydrophobic) [38]. Based on their chemical structure, the biosurfactant able to enhance the solubility between fluid of difference polarities, biodegradation, and bioavailability of substrates. In the literature, most studies done on *Candida* species to produce biosurfactant from WCO [39]. It is reported that, waste cooking oil with high nitrogen concentration, carbon nitrogen ratio (C/N) are the major factor that affecting the production of biosurfactant. The high availability of nitrogen content in medium may promote the growth of the microorganism hence limit the production of biosurfactant [40]. Several authors recognize that some organic acids resulting from the bioconversion of the WCO were also reported as added value compounds with great market potential [41-43]. *Aspergillus* sp., *Penicillium expansum, Yarrowia lipolytica* able to produce oxalic acid, citric acid and erythritol which successfully use in food and beverage industry. In addition, the WCO is the most favorable than the fresh cooking oil for production of these metabolites [44, 45]. The WCO concentration as well as the C/N ratio were the critical factor on the production of the natural sweetener erythritol. Besides, magnesium and nitrogen were needed in order to achieve maximum production of citric acid [46].

6. Conclusion

Cooking oil had become one of the most important food ingredients that have huge industrial demands. However, there are serious challenge on managing the WCO disposal that may lead to detrimental effects on marine life and ecosystem. Cooking oil degradation by fungi provides a reasonable solution for the ecological issue as well as the value-added compounds which can give a new great opportunity for the global market. The biological approaches of cooking oil valorization have given a great promising approach on biotechnology especially on its production economics.

References

[1] Kumar, A., Sharma, A., and Upadhyaya, K.C. 2016 Vegetable oil: Nutritional and industrial perspectives *Current Genomics* **17** 230-240

[2] Population Reference Bureau 2020 ISBN 978-0-917136-14-6 9

[3] Boateng, L., Ansong, R., Owusu, W.B. and Steiner-Asedu, M. 2016 Coconut oil and palm oil's role in nutrition, health, and national development: A review *Ghana Medical Journal* **50** 189-196

[4] Hooper, L., Abdelhamid, A., Moore, H.J., Douthwaite, W., Skeaff, C.M. and Summerbell, C.D. 2012 Effect of reducing total fat intake on body weight: systematic review and meta-analysis of randomised controlled trials and cohort studies *British Medical Journal* 345 e7666

[5] Ibrahim, S., Abd. Shukor, M.Y., Yazid, N.A. and Ahmad, S.A. 2018 Microbial degradation of vegetable oils: A review *Malaysian Journal of Biochemistry and Molecular Biology* **3** 45-55

[6] Wattanapenpaiboon, N. and Wahlqvist, M.L. 2003 Phytonutrient deficiency: The place of palm fruit *Asia Pacific Journal of Clinical Nutrition* **12** 363-368

[7] Popoola, B.M. and Onilude, A.A. 2017 Microorganisms associated with vegetable oil polluted soil *Advances in Microbiology* **7** 377-386

[8] Kumar, R., Chandra, S., Kumar, K. and Kumar, V. 2018 Analysis of the physical and chemical characteristics of edible vegetable blended oil *International Journal of Chemical Studies* **6** 10-15
[9] Zahir, E, Saeed, R., Abdul Hameed, M. and Yousuf, A. 2017 Study of physicochemical properties of edible oil and evaluation of frying oil quality by Fourier Transform-Infrared (FT-IR) Spectroscopy *Arabian Journal of Chemistry* **10** 3870-3876
[10] Devi, A. and Khatkar, B.S. 2016 Physicochemical, rheological, and functional properties of fats and oils in relation to cookie quality: A review *Journal of Food Science and Technology* **53** 3633-3641
[11] Tulcan, O.E.P., De Andrade, D.O., De Andrade, E.T. and Pereira, R.G. 2008 Analysis of physical characteristics of vegetable oils *CIGR-International Conference of Agricultural Engineering* xxx
[12] Papanikolaou, S., Dimou, A., Fakas, S., Diamantopoulou, P., Philippoussis, A., Galiotou-Panayotou, M. and Aggelis, G. 2011 Biotechnological conversion of waste cooking olive oil into lipid-rich biomass using *Aspergillus* and *Penicillium* strains *Journal of Applied Microbiology* **110** 1138-1150
[13] Hasanizadeh, P., Moghimi, H. and Hamedi, J. 2017 Biosurfactant production by *Mucor circinelloides* on waste frying oil and possible uses in crude oil remediation *Water Science and Technology* **76** 1706-1714
[14] Cihangir, N. and Sarikaya, E. 2004 Investigation of lipase production by a new isolate of *Aspergillus sp. World Journal of Microbiology and Biotechnology* **20** 193-197
[15] Spalvins, K., Geiba, Z., Kusnere, Z., and Blumberga, D. 2020 Waste cooking oil as substrate for single cell protein production by yeast *Yarrowia lypolitica Environmental and Climate Technologies* **24** 457-469
[16] Thevenieau, F. and Nicaud, J. 2013 Microorganisms as sources of oils *Oilseeds and fats, Crops and Lipids* **20** 1-8
[17] Al-Hawash, A. B., Zhang, J., Li, S., Liu, J., Ghalib, H. B., Zhang, X., and Ma, F. 2018 Biodegradation of n-hexadecane by Aspergillus sp. RFC-1 and its mechanism *Ecotoxicology and environmental safety* **164** 398-408
[18] Sharma, I. 2020 Bioremediation Techniques for Polluted Environment: Concept, Advantages, Limitations, and Prospects *in Trace Metals in the Environment-New Approaches and Recent Advances*. IntechOpen. xxxx
[19] Lopes, M., Miranda, S. M., Alves, J. M., Pereira, A. S., and Belo, I. 2019 Waste cooking oils as feedstock for lipase and lipid-rich biomass production *European Journal of Lipid Science and Technology* **121** 1800188-1800196
[20] Das, N., and Chandran, P. 2011 Microbial degradation of petroleum hydrocarbon contaminants: an overview. *Biotechnology Research International* xxxxx
[21] Zahri, K. N. M., Zulkharnain, A., Sabri, S., Gomez-Fuentes, C., and Ahmad, S. A. 2021 Research trends of biodegradation of cooking oil in Antarctica from 2001 to 2021: A bibliometric analysis based on the scopus database. *International Journal of Environmental Research and Public Health* **18** 2050.
[22] Weyda, I., Sinha, M., Sørensen, A., Lübeck, P. S., and Ahring, B. K. 2018 Increased production of free fatty acids and triglycerides in *Aspergillus carbonarius* by metabolic engineering of fatty acid biosynthesis and degradation pathways *JSM Microbiology* **6**
[23] Domínguez, A., Deive, F. J., Angeles Sanromán, M., and Longo, M. A. 2010 Biodegradation and utilization of waste cooking oil by *Yarrowia lipolytica* CECT 1240 *European Journal of Lipid Science and Technology* **112** 1200-1208.
[24] Rao, U., Sridhar, R., and Sehgal, P. K. 2010 Biosynthesis and biocompatibility of poly (3-hydroxybutyrate-co-4-hydroxybutyrate) produced by *Cupriavidus necator* from spent palm oil *Biochemical Engineering Journal* **49** 13-20.

[25] Lopes, M., Miranda, S. M., and Belo, I. 2020 Microbial valorization of waste cooking oils for valuable compounds production–a review. *Critical Reviews in Environmental Science and Technology*, **50** 2583-2616.
[26] Arous, F., Atitallah, 1. B., Nasri, M., and Mechichi, T. 2017 A sustainable use of low-cost raw substrates for biodiesel production by the oleaginous yeast *Wickerhamomyces anomalus*. 3 *Biotech* **7** 268-277
[27] Nanou, K., Roukas, T., Papadakis, E., and Kotzekidou, P. 2017 Carotene production from waste cooking oil by *Blakeslea trispora* in a bubble column reactor: The role of oxidative stress *Engineering in Life Sciences* **17** 775–780
[28] El Bialy, H., Gomaa, O. M., and Azab, K. S. 2011 Conversion of oil waste to valuable fatty acid using oleaginous yeast *World Journal of Microbiology and Biotechnology* **27** 2791-2798
[29] Csutak, O., Corbu, V., and Vassu, T. 2017 Studies on the correlation between biosurfactant and lipid synthesis in *Candida tropicalis* CMGB114 using hydrocarbons and vegetable oil wastes *Revisfa de Chimie (B charest)* **68** 255-259.
[30] Patel, A., and Matsakas, L. 2018 A comparative study on de novo and ex novo lipid fermentation by oleaginous yeast using glucose and sonicated waste cooking oil *Ultrasonics Sonochemisfry*, **52** 364-374
[31] Papanikolaou, S., Dimou, A., Fakas, S., Diamantopoulou, P., Philippoussis, A., Galiotou-Panayotou, M., and Aggelis, G. 2011 Biotechnological conversion of waste cooking olive oil into lipid-rich biomass using *Aspergillus and Penicillium strains Journal of Applied Microbiology* **110** 1138-1150
[32] Panadare, D. C. (2015). Applications of waste cooking oil other than biodiesel: a review *Iranian Journal of Chemical Engineering (IJChE)* **12** 55-76.
[33] Athenaki, M., Garden, C., Diamantopoulou, P., Tchakouteu, S. S., Sarris, D., Philippoussis, A., and Papanikolaou, S. 2017 Lipids from yeasts and fungi: Physiology, production, and analytical considerations *Journal of Applied Microbiology* **124** 336-367
[34] Beligon, V., Christophe, G., Fontanille, P., and Larroche, C. 2016 Microbial lipids as potential source to food supplements *Current Opinion in Food Science* **7** 35-42
[35] Katre, G., Joshi, C., Khot, M., Zinjarde, S., and Ravikumar, A. 2012 Evaluation of single cell oil (SCO) from a tropical marine yeast *Yarrowia Lipolytica* NCIM 3589 as a potential feedstock for biodiesel *AMB Express* **2** 36-49
[36] Cruz, M. V., Sarraguc¸a, M. C., Freitas, F., Lopes, J. A., and Reis, M. A. M. 2015 Online monitoring of P(3HB) produced from used cooking oil with near-infrared spectroscopy *Journal of Biotechnology* **194** 1–9.
[37] Ojha, N., and Das, N. 2018 A Statistical approach to optimize the production of polyhydroxyalkanoates from Wickerhamomyces anomalus VIT-NNO1 using Response Surface Methodology *International Journal of Biological Macromolecules*, **107** 2157-2170
[38] Kumar, P., and Kim, B. S. 2019 Paracoccus sp. strain LLI as a single cell factory for the conversion of waste cooking oil to polyhydroxyalkanoates and carotenoids *Applied Food Biotechnology* **6** 53-60
[39] Kanna, R. 2018 Enhanced and cost-effective biosurfactant production for marine remediation contaminated with oil spill *International Journal of Civil Engineering and Technology* **9** 373–381
[40] Almeida, D. G., Silva, R. C. F. S., Luna, J. M., Rufino, R. D., Santos, V. A., and Sarubbo, L. A. 2017. Response surface methodology for optimizing the production of biosurfactant by *Candida tropicalis* on industrial waste substrates *Frontiers in Microbiology*, **8** 1–13

[41] Csutak, O., Corbu, V., Stoica, I., and Vassu, T. 2018 Fatty acids effect on lipase and biosurfactant induction in Rhodotorula Glutinis CMGB-RG5 *"Agriculture for Life, Life for Agriculture Conference Proceedings* **1** 515–522.
[42] Batista, R. M., Rufino, R. D., Luna, J. M., Souza, J. E. G., and Sarubbo, L. A. 2010 Effect of medium components on the production of a biosurfactant from *Candida tropicalis* applied to the removal of hydrophobic contaminants in soil *Water Environment Research* **82** 418–425
[43] Bellou, S., Makri, A., Triantaphyllidou, I.-E., Papanikolaou, S., and Aggelis, G. 2014 Morphological and metabolic shifts of *Yarrowia lipolytica* induced by alteration of the dissolved oxygen concentration in the growth environment *Microbiology* **160** 807–817
[44] Beopoulos, A., Haddouche, R., Kabran, P., Dulermo, T., Chardot, T., and Nicaud, J.-M. 2012 Identification and characterization of DGA2, an acyltransferase of the DGAT1acyl-CoA: Diacylglycerol acyltransferase family in the oleaginous yeast *Yarrowia lipolytica.* New insights into the storage lipid metabolism of oleaginous yeasts *Applied Microbiology and Biotechnology*, **93** 1523–1537
[45] Carsanba, E., Papanikolaou, S., and Erten, H. 2018 Production of oils and fats by oleaginous microorganisms with an emphasis given to the potential of the nonconventional yeast *Yarrowia lipolytica Critical Reviews in Biotechnology*, **38** 1230–1243
[46] Xiaoyan, L., Xinjun, Y., Jinshun, L., Jiaxing, X., Jun, X., Zhen, W., Yuanfang, D. 2017 A cost-effective process for the coproduction of erythritol and lipase with *Yarrowia lipolytica* M53 from waste cooking oil *Food and Bioproducts Processing*, **103** 86–94.
[47] Liu, X., Lv, J., Xu, J., Zhang, T., Deng, Y., and He, J. 2015 Citric acid production in *Yarrowia lipolytica* SWJ-1b yeast when grown on waste cooking oil *Applied Biochemistry and Biotechnology* **175** 2347–2356.
[48] Abghari, A., and Chen, S. 2017 Engineering *Yarrowia lipolytica* for enhanced production of lipid and citric acid *Fermentation*, **3**, 34–56

Acknowledgement

The authors thank Assoc. Prof. Dr. Hirofumi Hara (University of Tokyo) for his expert advice and endless support that led to this paper. We also thank the Ministry of Higher Education Malaysia (MOHE) and Universiti Teknologi Malaysia (UTM) under the Fundamental Research Grant Scheme (FRGS) (R.K.1300000.7843.4F991).

Backswamp Development's Effects on Water Management System of Tidal Swamp Irrigation Palingkau SP1 SP2 SP3, Kalimantan, Indonesia

Athanasius Donny Firstyadi[1], Budi Santosa Wignyosukarto*[2], Istiarto Istiarto[2], Yakubson Yakubson[3]

[1] Master Student, Civil and Environmental Engineering Department,
Universitas Gadjah Mada, Indonesia
[2]Department of Civil and Environmental Engineering, Faculty of Engineering,
Universitas Gadjah Mada, Yogyakarta, 55281, Indonesia
[3]Ministry of Public Works and Housing, Palangka Raya, Indonesia
*Corresponding author e-mail: budiws@ugm.ac.id

Abstract. The Palingkau SP1 SP2 SP3 swamp irrigation is an expansion of handil's agricultural area, which is influenced by tides from the Kapuas River on the left and the Kapuas Murung river on the right. The new area that exploits backswamp and the buffer zone upstream of Handil, gives an additional burden of dispersion of toxic materials resulting from the oxidation of acid sulfate soils and the loss of fresh water supply from backswamp. The actual leaching process is indicated by the water pH value in the collector and primary channels ranging from 2.58 ~ 3.6. Meanwhile, the water quality of Handil, Kapuas and Kapuas Murung rivers is better, ranging from 3.7 ~ 5.34. The confluence of tidal from two opposite sides has resulted in the disruption of pollutant discharge. Instead of discharged into rivers, its flow to Handil's agricultural area and reduce agricultural productivity. The HEC RAS software simulation model is used to evaluate the optimal water management system. One-way flow system is introduced hydraulic structures for regulating the flow of freshwater water at high tide from the left handil into the new irrigation area and regulating the drainage at low tide through the right handil.

Keywords: *Acid Sulphate Soil, Backswamp, Water Management, Leaching*

1. Introduction

The Swamp Irrigation Area of Palingkau SP1 SP2 SP3 is located in Kapuas Regency, Central Kalimantan Province, Indonesia. This Swamp Irrigation Area is one part of the block A area of the former Peatland Development Project (ex-PLG) which was developed by the government starting in 1995 [1]. The Swamp Territorial Irrigation Area SP1 SP2 SP3 is located between the Kapuas River on the left and the Kapuas Murung River on the right. At first, local communities and early transmigrants cultivated agricultural areas on the banks of the Kapuas Murung River which were affected by the tides and then made a two-way channel as a water source as well as a drainage channel called handil [2,3]. Considering that the agricultural productivity in this handil area is quite good, then the government carried out development by opening a backswamp area behind the handil, which was originally a forest into a new irrigation area. With the development of the area behind the handil, it is hoped that the tides of the Kapuas River and the Kapuas Murung river can enter the network system behind the handil and support the reclamation process. Currently, DIR Palingkau SP1 SP2 SP3 has a irrigation area of approximately 9,000 Ha consisting of a right and left handil agricultural area of 5,000 Ha and a development area behind handil of 4,000 Ha [4].

The development of the back swamp area has changed the use of the area which was originally a forest and buffer area into an agricultural area with open channels. The channel system in the network consists of collector channels, primary channels, secondary channels, and tertiary channels. The development of the back swamp area will go through a land reclamation process. The lowering of the groundwater level as part of the reclamation process triggers the occurrence of pyrite oxidation in acid sulfate soils which produces toxins and pollutants. At the beginning of the rainy season, toxins and pollutants dissolve and enter the canal which causes the pH of the water in the channel to become acidic and harmful to plants [6,7]. Handil is currently a two-way channel, freshwater from the river flows in at high tide and flows out at low tide. Handil and channels in the current network system are not functioning properly due to sedimentation and are covered with vegetation. The tidal effect of the two rivers adds to the non-smoothness of the leaching process. Pollutants from within the network system that cannot be discharged into rivers will enter productive agricultural areas and reduce agricultural productivity. The loss of the back swamp also means the loss of freshwater sources from the buffer zone behind the handil. For this reason, sufficient freshwater sources are needed to support the leaching process in the network system.

Water regulation and management is key in the treatment of acid sulfate soils [8]. This study wanted to find out the alternative for improving water management in Palingkau SP1 SP2 SP3, to support the leaching process. Water management transforms a two-way flow system into a one-way flow. Fresh water from the Kapuas River on the left and the Kapuas Murung River on the right, flows into the new irrigation area via tidal energy using tidal energy. To ensure a one-way flow, a weir is installed on the left handil, and a flap gate is installed on the right handil. Another alternative is to increase the carrying capacity by normalizing the channel. The location of the Palingkau SP1 SP2 SP3 Swamp Irrigation Area is shown in Figure 1 below.

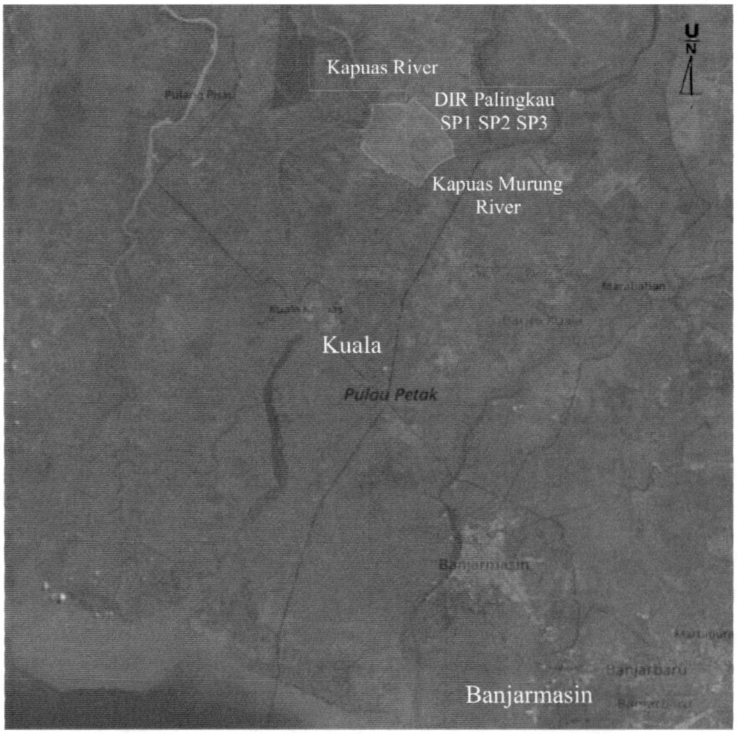

Figure 1. Location of Palingkau SP1 SP2 SP3 Scheme

IOP Conf. Series: Earth and Environmental Science 1091 (2022) 012055 doi:10.1088/1755-1315/1091/1/012055

2. Material methods

2.1 Existing Condition

The Palingkau SP1 SP2 SP3 channel network consists of collector channels, primary channels, and secondary channels. Meanwhile, the handil agricultural areas on the right and left of the network system get fresh water from the Kapuas Murung and Kapuas rivers. Handil's condition, both on the right and on the left, is sedimented and covered in vegetation. Some handils are not connected to the collector canal because they are covered by sediment and also dense vegetation, some other handil are in better condition and are connected to the collector and primary canals. The two main handil connecting the right and left sections are also used as a means of transportation for the local community. The length of the handil on the left varies between 2.2~2.9 km with a channel width of 7~14 m and the pH value of the water varies between 3.85~5.08. The handil on the right is longer than the handil on the left, varying between 3.9~4.1 km with a channel width of 4~12 m. The pH value of the water on the right handil ranged from 3.7~5.34. The pH value of the water at the point close to the river is closer to 5, the farther from the river and close to the network system, the lower the pH value.

The collector channel Palingkau SP1 SP2 SP3 surrounds the network system and acts as a barrier between the primary channels and handil. Some sections of the collector canal are sedimented and covered with vegetation, resulting in deadlocks. The transverse collector channel has a channel length of between 4.2~4.5 km, while the collector channel on the right and left has a length of 9.6~9.8 km with a channel width of 3~11 m. The pH value of the water in the collector channel ranges from 2.58~3.84. The primary canal has a length that varies between 4.2~4.5 km with a channel width of 6~11 m and the pH value of the water varies from 2.8~3.6. The condition of the primary canal in general also experiences sedimentation and is covered with vegetation, however, there are 2 primary canals in good condition which are connected with handils on the right and left. Secondary canals also have similar conditions, namely experiencing sedimentation and being covered with vegetation. The sedimentation that occurs results in obstruction of drainage and flooding of the land. The channel network on the Palingkau SP1 SP2 SP3 can be seen in Figure 2. The existing condition of the handil channel, collector channel, primary channel and secondary channel can be seen in Figures 3 and 4.

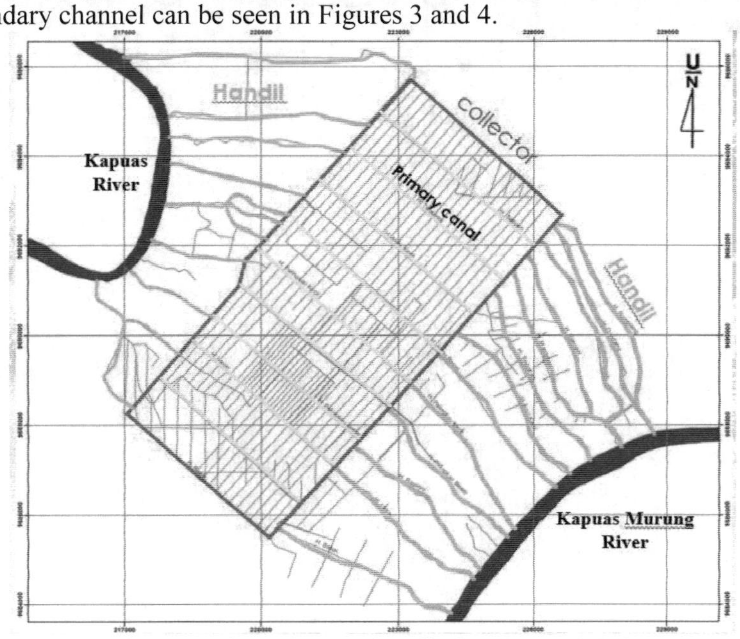

Figure 2. Channel network Palingkau SP1 SP2 SP3

IOP Conf. Series: Earth and Environmental Science 1091 (2022) 012055 doi:10.1088/1755-1315/1091/1/012055

Most of the agricultural land is in the right handil area, while the left handil is dominated by shrubs and mixed gardens. In the new irrigation area, most of the agriculture area is in the SP1 as rainfed rice fields and a little area in the SP2, while the SP3 area is dominated by mixed gardens, oil palm, and watermelon plants. The local community's cropping pattern is a one-planting per year with local varieties of rice and productivity of around 1.145 tons/Ha. This is related to the overflow capability of tidal water that does not reach the relatively higher land. Agriculture in the handil area is still affected by the tides, but due to toxic materials and pollutants from the network system, agricultural productivity is also low. The land use map on the Palingkau SP1 SP2 SP3 can be seen in Figure 5. In the new irrigation area, most of the pyrite can be found at a shallow depth of 0~50 cm. Likewise, peat is mostly found at a depth of 0~50 cm. Based on data on water quality (pH) and depth of pyrite in the new irrigation area, it can be said that the reclamation process has not been completed and affects the quality of water and land in the handil area

(a) (b)

Figure 3. Collector channel and handil on the right area (a) and handil on the left area (b)

(a) (b) (c)

Figure 4. Collector channel on upper area (a) primary channel (b) and secondary channel (c)

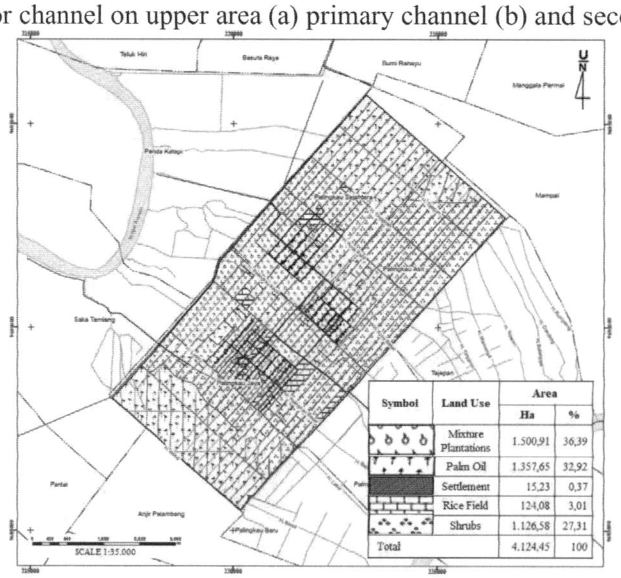

Figure 5. Land use of Palingkau SP1, SP2, dan SP3 scheme [4]

2.2. Water Management

One of the recommended methods for reclamation of acid sulphate soil is with good water management [8]. Managing the groundwater level, flow in channels, and the leaching process are efforts that can be made to overcome the problem of acid sulphate soils [9]. Maintaining groundwater levels or wetting the land can inhibit pyrite oxidation, reduce acidity, and also reduce the risk of dispersion toxicities [10]. The leaching process can be done by making a good and controlled drainage system [11]. Freshwater sources for the leaching process come from tidal rivers or rainwater which are supported by a good drainage system [12]. The efficiency of reclamation measures is determined based on water management balances including maximum pollution coefficients estimation [13]. The effect of leaching can be greatly enhanced by land preparation and sun-drying [14]. Toxicities were removed faster, and water was used more efficiently.

The source of freshwater for Palingkau SP1 SP2 SP3 comes from the tidal waters of the Kapuas and Kapuas Murung rivers which have a pH value of around 5. However, with handil's length of more than 4 km, it is doubtful that freshwater from the river can flow to the end of the channel. The source of freshwater for areas far from rivers comes from rainwater. However, the availability of rainwater is also quite limited, especially during the dry season. According to previous research, the water requirement for the leaching process is around 1,000 cm in 200 days or the equivalent of 50 cm per day [9]. To support the leaching process in a new irrigation area of 4,124.25 Ha with a water requirement of 500 m3/Ha/day, 2.06 million m3/day is needed. To meet these water needs, it is necessary to manage water flows that can ensure the flow of freshwater as much as possible through the handil to new irrigation areas and prevent reverse flow that carries toxic solutions.

This study examines the leaching process by adjusting the system from a two-way flow system to a one-way flow system. One-way flow is regulated by placing a weir on the left handil and a flap gate on the right handil. Freshwater enters through the handil on the left side overflowing through the weir at high tide, flowing into the primary channel in the new irrigation area. During low tide, the water drains from the primary channel to the right handil through the flap gate arrangement. In addition, channel capacity was also increased by normalizing handil and primary channel in new irrigation areas. It is hoped that the water flowing to the new irrigation areas can meet the water requirements for the leaching process, both in terms of quality and quantity.

2.3. Hydraulic simulation

Unsteady flow model (HEC-RAS) is used to measure the effectiveness of flow in the channel network. The geometry of the model is made based on topographic measurement data from previous studies. Tidal data on the Kapuas and Kapuas Murung rivers were used as boundary conditions for unsteady flow analysis.

The hydraulic simulation was carried out in three scenarios, namely the existing condition (scenario 1), normalization of the handil and channel (scenario 2), and the addition of a flow control structure (weir and flap gate) in the normalized channel 9scenario 3). In the existing condition, several handils on the right side are covered with vegetation and sedimentation, so that water cannot flow into the collector canal. In channel normalization conditions, the channel width is adjusted to the largest channel cross-section in one channel section length, while channel excavation is only limited to removing sedimentation according to existing cross-section data. Channel deepening was not carried out to avoid excessive pyrite oxidation. An overview of the channel normalization carried out is depicted in Figure 6.

In the third scenario, the addition of a flow control structure is carried out on the normalized channel and handil. The flow control structure that is carried out in the simulation is a weir on the left handil and a flap gate on the right handil. All of the flow control structure is located upstream of the handil. The location of the flow control structure can be seen in Figure 7. The Weir crest is about 90 cm above the bottom of the left handil (adjusted to the fluctuation of tide) and the weir width is adjusted to the handil's width. Flap gate structures at the right handil are 1 m in width and 1,5 m in height. The number of the

IOP Conf. Series: Earth and Environmental Science 1091 (2022) 012055 doi:10.1088/1755-1315/1091/1/012055

flap gates is varying 4~5 gates per structure, adjusted to the right handil's width. Flap gate structures at the right handil in the simulation are used as unsteady flow data with elevation controlled gates in the boundary condition. When the difference of the water level on the upstream and downstream flap gate is bigger than 0.1 m then the gates will open, and when the difference is 0 m the gates will be closed. The effectiveness of the canal network is evaluated by knowing tidal fluctuations, flow direction, and the volume of flow entering the new irrigation area.

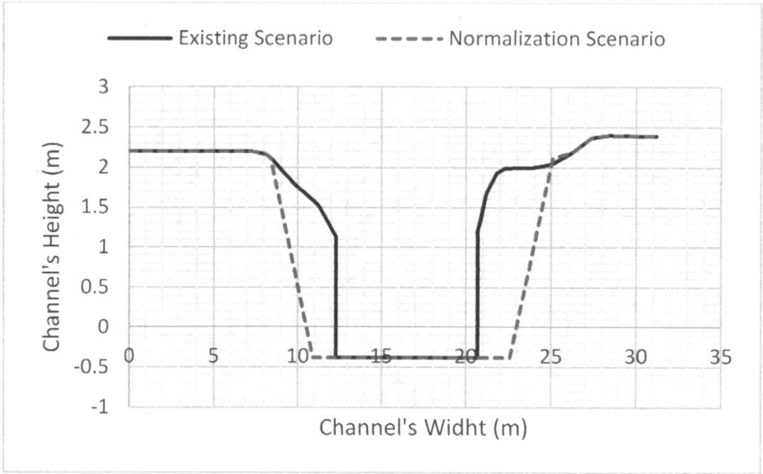

Figure 6. Cross section at the right handil

Figure 7. Location of the flow structures

3. Results and discussion

The hydraulic simulation results of the three model scenarios with the HEC RAS program can be seen in Figures 8, 9, and 10 below.

The tidal fluctuation of the left handil in the existing condition is between +0.09 m ~ +0.84 m (Figure 8). In the normalized channel, with widening and deepening, tidal fluctuations are around -0.4 m ~ +0.9

m. The influence of the tidal river that enters through the handil can be more pronounced compared to the existing condition where the handil is heavily sedimented and covered with vegetation. With a smaller damping effect, the water flowing into the handil and the channel can be faster and larger. In the scenario of normalization and flow control structures, a water structure is made in the form of a weir with a crest elevation of +0.4 m or 0.9 m above the bottom of the channel (-0.5 m). The flap gate is 1 m wide and 1.5 m in height. The number of gates varies from 4 to 5 according to the width of the channel.

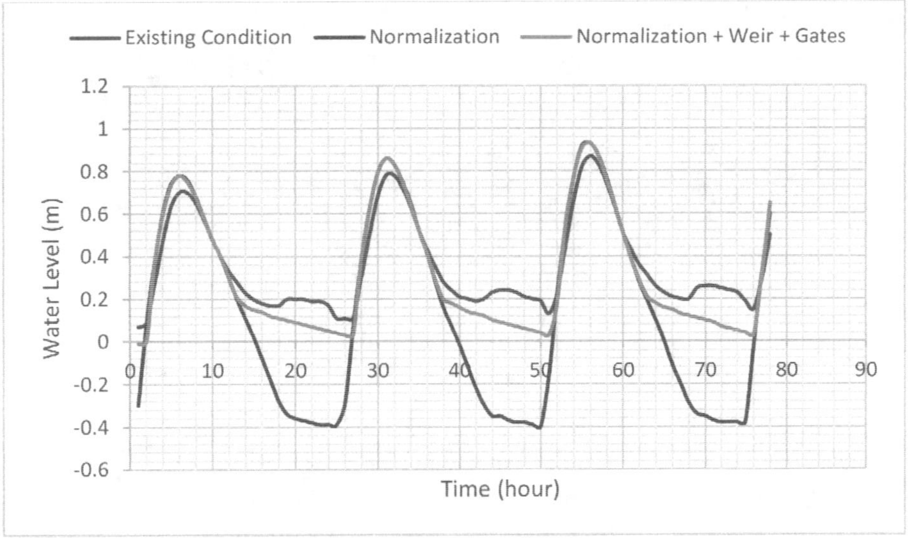

Figure 8. Water level on the left handil with 3 scenarios

The tidal fluctuation in primary cahnnel for the third scenario, can be controlled in certain elevation, flutuated above +0.10 m, to prevent oxidation of potential acid sulphate soil (figure 9). The effect of the crest of weir elevation setting can be evaluated by considering the amount of negative flow in the primary channel (negative flow indicates the direction of reverse flow from the primary channel to the left handil). With crest elevation of +0.40 m, there is no reverse flow in primary channel (Figure 10). The crest elevation should be adjusted according to the tidal elevation so that there is no reverse flow from the primary channel to the left handil.

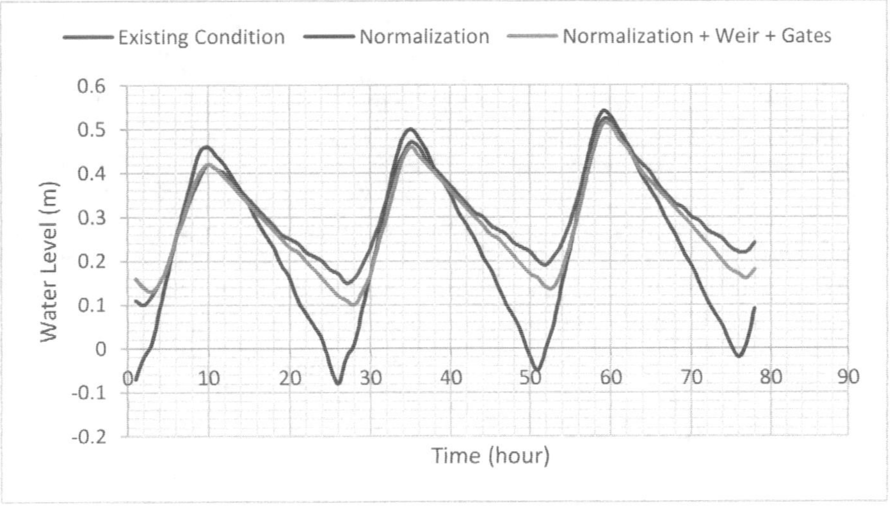

Figure 9. Water level on the primary channel

Comparing the tidal fluctuation on the right handil (Figure 11) and the direction of flow in the primary channel(Figure 10), for the third scenario, the tide of the right handil give no influence to the flow of the primary channel, due to the protection of the flap gate. This shows the third scenario guaranteeing a one-way flow.

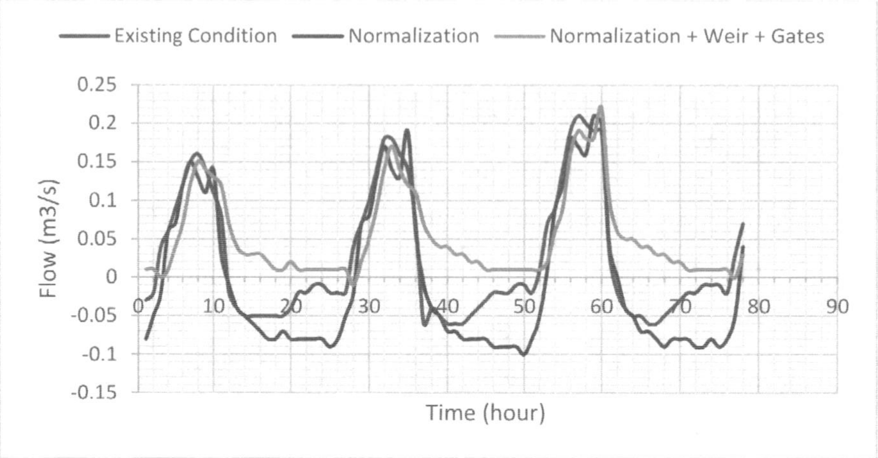

Figure 10. Flow on the Primary Channel

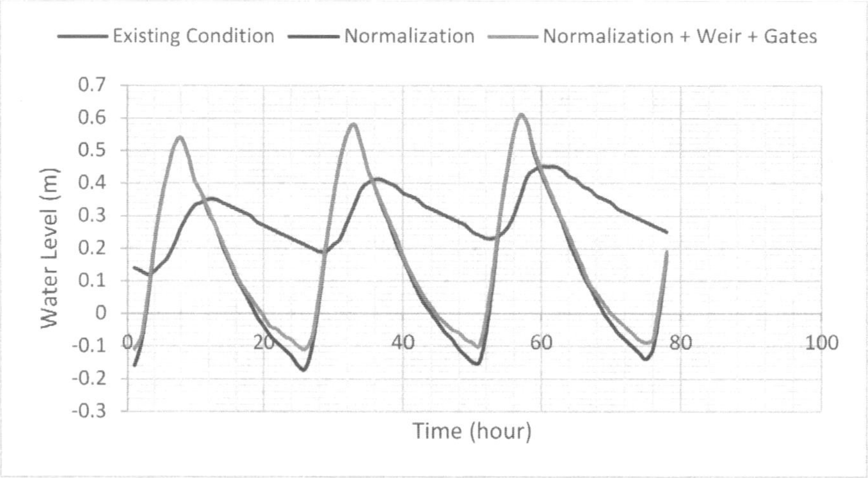

Figure 11. Water level on the right handil

4. Conclusions

The development of the backswamp area behind the handil which was originally a forest area and buffer zone has a significant impact on the environment and agriculture in the existing handil area. The reclamation process due to land cultivation accompanied by lowering of the groundwater table has caused the oxidation of pyrite in acid sulphate soils. Many years after the opening of the Rawa Palingkau Irrigation Area SP1 SP2 SP3, the reclamation process in the network system is still ongoing. Dispersion of toxic water into the handil area causes water quality degradation and decreased agricultural productivity. The water source from the area which was originally a forest buffer zone has disappeared and has become a source of toxic water. For this reason, it is necessary to regulate the flow of water that can bring freshwater into the primary canal in the new irrigation area and push toxic water out of the new irrigation area through the handil to the main river. The one-way water system regulates freshwater from the Kapuas River at high tide through the left handil, pushing toxins and pollutants from the new irrigation system out through the right handil to the Kapuas Murung river. The one-way flow system is realized with a reliable normalization scenario and the installation of weirs and flap gates. The elevation and width of the weir crest on the left handil ensures that there is no reverse flow at low tide and the quantity of freshwater entering the primary channel. The setting of the flap gate opening is also very important to ensure the one-way flow system can run and accelerate the disposal of toxic water through the right handil to the main river.

Acknowledgement
This research was conducted in one of the swamp schemes under BWS Kalimantan II, and utilized some of the data from consultant PT Virama Karya. Appreciation and gratitude to the leaders and staff on favors and assistance for conducting research and preparing this paper

References

[1] Wignyosukarto B S 2000 Review of Lamunti, Dadahup, and Palingkau water management Peatland Development Project, Central Kalimantan Indonesia, *Forum Teknik Volume 24 No 3*
[2] Suryadi F X 2020 Soil and Water Management Strategies for Tidal Lowlands in Indonesia. *CRC Press. London*, doi.org/10.1201/9780429333231
[3] Houterman, J., dan Ritzema, H., 2009, Land and Water Management in the Ex-Mega Rice Project Area in Central Kalimantan, *Technical Report 4, Jakarta: Euroconsult Mott MacDonald and Deltares.*
[4] Virama Karya. 2020. Survey investigation and design (SID) of rehabilitation and improvement of swamp irrigation network in block A (Package 1). *Kalimantan River Basin II Ministry of Public Works and Public Housing Palangka Raya Kalimantan Indonesia.*
[5] Subagjo, H. dan I P.G. Widjaja-Adhi. 1998. Opportunities and use of swamp land for agricultural development in Indonesia: the case of South Sumatra and Central Kalimantan. Main Paper Discussion and Communication Meeting on Soil and Agroclimate Research Results (Bogor)
[6] Roelse K 1988 Land reclamation of lowlands in Indonesia Thesis (*The Netherland: TU Delft*)
[7] Wignyosukarto B S 2013 Leaching and Flushing of Acidity in the Reclamation of Acid Sulphate Soil, Kalimantan, Indonesia *J. Irrigation and Drainage 62* 75–81.
[8] Dent D 1986 Acid sulphate soils: a baseline for research and development (The Netherland: *ILRI-Wageningen) Publication 39*
[9] Tuong T P 1993 An overview of water management of acid sulphate soils Ho Chi Minh City Symp. on acid sulphate soils *(The Netherland: ILRI-Wageningen) Publication 53* pp 265-279.
[10] Toivonen J and Österholm P 2011 Characterization of acid sulphate soils and assessing their impact on a humic boreal lake. *Journal of Geochemical Exploration Volume 110*, Issue 2, August 2011, Pages 107-117
[11] Sutela T and Vehanen T 2017 The effects of acidity and aluminium leached from acid-sulphate soils on riverine fish assemblages. *Boreal Environment Research. 22*: 385–391. ISSN 1797-2469
[12] Kselik R A L, Smilde KW, Ritzemal H P, Subagyono K, Saragih S, Damanik M and Suwardjo H 1993 Integrated research on water mangement, soil fertility and croping systems on acid sulphate soils in South Kalimantan, Indonesia, The Ho Chi Minh City Symposium on Acid Sulphate Soils, *Publication 53, hal. 177-194, The Netherland: ILRI-Wageningen*
[13] Glazunova I V and Karpenko N P 2021 Analysis of the reclamation measures efficiency within watershed and their impact on the water *bodies IOP Conf. Ser.: Earth Environ. Sci. 817 012035 doi:10.1088/1755-1315/817/1/012035*
[14] Tuong T P 1993 Effect of land preparation on leaching of an acid sulphate soil at Cu Chi, Vietnam. Selected Papers on the Ho Chi Minh City Symposium on Acid Sulphate Soils. *Publication 53. International Institute for Land Reclamation and Improvement (ILRI)*: Wageningen, the Netherlands.

Energy efficiency and nutrient removal performance: comparison between several types of activated sludge process

Wan Mohd Bukhary W Muzaffar[1]*, A. Aznah[1] and Hakim Halim[1]

[1] Malaysia-Japan International Institute of Technology (MJIIT), Universiti Teknologi Malaysia, Kuala Lumpur.

*Corresponding e-mail: wanzbukhary@gmail.com; wanmohdbukhary@graduate.utm.my

Abstract. The operation of a wastewater treatment plant entails a huge amount of electricity. The majority of energy inputs are consumed by aeration systems to support biological processes for treated wastewater. Urban wastewater treatment plants are energy-intensive facilities that consume significant amounts of energy. For conventional activated sludge systems, 25% to 60% of the operating costs are associated with energy use. Malaysia's wastewater treatment plants have fallen short in terms of technological advancement in the sewerage industry. The goal of this research is to analyse and make a comparison of the capabilities of the sequencing batch reactor (SBR) and other activated sludge (AS) system treatment plants in Selangor, Malaysia. High energy electricity consumption was an important issue that affected the operational cost and development of wastewater treatment plants (WWTP). The result and discussion for the analysis had been presented into comparison result for each objective of this research studied which was to determine lowest energy efficiency, to assess highest nutrient removal efficiency and compliance rate for each WWTP process plant. This paper presents best practices that can be implemented and adapted by operators in their pursuit of energy and reduce cost or expenses in the sewage treatment plant.

Keywords: Energy efficiency, wastewater treatment plant; nutrient removal; compliance rate; Malaysia.

1. Introduction

Urban wastewater treatment plants (WWTPs) are energy-intensive facilities that consume significant amounts of energy [1]. For conventional WWTPs, 25% to 60% of the operating costs are associated with energy use. Electricity was the main energy source to operate the wastewater plant and to ensure all equipment ran without fail. It has been reported that electric power consumption was 15% - 30% of the total running costs for the larger WWTP compared to 30% - 40% for the smaller WWTP. The Malaysian government makes significant investments to ensure that our drinking water is indeed clean and safe. The wastewater treatment plant (WWTP) was one of the investments because it produces safer treated water from wastewater. Because of the high cost of operation and maintenance, the investment in WWTP was quite high. Sewage treatment plants (STP) in municipalities are energy-intensive facilities that use a lot of power consumption [2]. Energy efficiency has been becoming an issue of global concern. Energy efficiency was introduced and implemented in this study by using an energy performance index (EPI) method. More advanced treatment technologies are required in sewage treatment plants (STP) to improve the removal efficiency of organics and nutrients to meet increasingly stringent discharge regulations.

The energy consumed, largely in the form of electricity, was mainly used for two main functions, which were to provide mechanical equipment (usually powered by pumps and mixers) and to provide oxygenated air (also called aeration). Reports indicate that aeration was a major contributor to energy consumption, estimated at 50% – 75% of total treatment plant energy expenditure. Malaysia is currently lagging in terms of innovations and technologies for mechanizing sewage treatment plants. There was a great deal of old and conventional activated sludge that needed to be properly maintained because of the high operation costs due to industrial tariffs, high electricity usage, and low efficiency of the equipment in the plant. Most importantly was the Indah Water Konsortium (IWK), the nation's largest wastewater treatment operator, which operates all wastewater treatment plants throughout Malaysia except Sabah and Sarawak, spending huge expenses annually on maintenance and operational costs. More than 50% of the annual budget was spent purposely on electricity consumption.

The main objective of wastewater treatment plants (WWTP) is to purify treated water. Because energy cost is not a major concern and a big issue for government officials, contractors, and developers, most wastewater treatment plants in operation today are constructed and built with no regard for energy efficiency criteria in mind. Wastewater treatment plants are energy-intensive, especially in terms of electrical energy demands, which contribute to more than 60% of overall electricity usage [3]. A wastewater treatment plant is a sophisticated system for eliminating contaminants from homes and municipal waste. Biological and chemical techniques were employed in a wastewater treatment facility to remove pollutants before the treated water was discharged into a receiver or bodies of water.

The study looked into and compared the efficiency and treatment capacity of sequencing batch reactors (SBR) and other activated sludge (AS) treatment plants in Selangor, Malaysia. The goal of this research was to compare and find the best energy efficiency using the energy performance index (EPI) for each type of process, the most efficient nutrient removal and compliance rate to meet the requirements of Standard A and Para (1) the Department of Environment (DOE) Malaysia. This study will identify which is the best wastewater treatment process that is more efficient, has high nutrient removal, and high compliance rate for WWTP design consideration. The objectives of the study are:

1) To determine the highest compliance rate of between sequencing batch reactor (SBR) plant and several activated sludge (AS) process of intermittent decanted extended aeration (IDEA) plant and extended aeration (EA) plant.
2) To quantify overall nutrients removal percentage and efficiency of organic load for biochemical oxygen demand (BOD), chemical oxygen demand (COD), ammonia (NH3N), nitrates (NO3N), oil and grease (O&G), and total suspended solids (TSS) with different type of activated sludge (AS) system.
3) To identify energy efficiency and energy performance index (EPI) between sequencing batch reactor (SBR) plant and other plant types of extended aeration (EA) and intermittent decanted extended aeration (IDEA).

2. Wastewater Treatment Technologies

Wastewater treatment plants (WWTP) are an energy-intensive industry that consume high electricity. The wastewater process includes three main stages; physical treatment is used to remove the suspended solids and other impurities. Around 25% of energy is consumed to drive the influent pumps and aerated grit chamber [3]. Much of the energy used in this stage is for the aeration process. Only a small amount of energy is used for pumping water. About 60-70% of energy is used in biological treatment [4]. The process of aeration, which is the most energy-consuming part of the daily operation of wastewater treatment, uses up to 70% of the energy in the facilities. Sludge conditioning and dewatering processes are also significant energy users in conventional wastewater treatment processes. About 10% of wastewater sector energy is used. Preliminary and primary treatment stages are less energy-intensive than conventional secondary treatment, while tertiary treatment can be energy-intensive as secondary treatment depending on the type and quantity of pollutants being removed and the desired and/or regulated effluent quality. The energy efficiency of the wastewater treatment plant is influenced to a great extent by the wastewater treatment technology, plant type, treatment process design, plant categories, and the sewage inflow and effluent quality. Studies of the main process and equipment that contribute to energy consumption are critical to implementing target strategies for energy reduction plan action. Comparison between SBR and other types of plant & energy consumption is shown in table 1.

Table 1. Technology comparison between SBR and other plant types

Parameter	EA	IDEA	MBR	SBR
BOD, mg/l	<30	<30	<3 - <5	<5
COD, mg/l	<250	<250	<100	<100
TSS, mg/l	<100	<100	<5	<10
Area, Acre	10.9	9.0	5.0	6.3
Power Cost, RM/m3	1.71	1.60	3.0	1.14
Maintenance cost, RM/m3	0.22	0.20	1.1	0.27

*Note: EA – Extended Aeration
IDEA – Intermittent Decanted Extended Aeration
MBR – Membrane Batch Reactor
SBR– Sequential Batch Reactor

The comparison technology shows that the SBR and MBR can remove higher pollution loading than the average treatment plant systems in Malaysia, such as EA and IDEA. The SBR could achieve efficiencies between 90% - 96% for BOD and COD removal, 95% for SS removal, and up to 92% for nutrient removal. More advantages for SBR are less electricity consumption and a small area for constructing the wastewater treatment plant compared to others. The demand for wastewater treatment plants has gradually increased because of urbanization, population growth, and improved living standards in many countries. Wastewater treatment plants (WWTPs) have been designed and operated to reduce the pollution of wastewater and to minimize the adverse impacts on environmental quality and human health [5]. Wastewater treatment plants (WWTP) have substantial environmental impacts during their life cycle due to energy consumption, chemical usage, and gas emissions, as well as sludge generation which requires additional treatment for disposal.

Many technologies have been developed for wastewater treatment plants (such as membrane bioreactor, sequencing batch reactor, and others). The evaluation of technology is important for obtaining better economic efficiency, as well as for reducing the environmental lifetime impacts [7]. Properly treated wastewater can be reused for various purposes to provide ecological benefits, reduce the demand for potable water and augment water supplies [8]. Several environmental and health impacts resulting from insufficient wastewater treatment have been identified in the scientific literature and actions need to be taken to reduce these impacts. These impacts can include negative effects on aquatic life and wildlife populations, oxygen depletion, beach closures and restrictions on recreational water use, restrictions on fish and shellfish harvesting.

3. Methodology

Not much research was studied and journals were found comparing the energy saving of electricity consumption of the SBR system with other types of wastewater treatment plants in Malaysia. The primary source of energy or power required in wastewater treatment plants (WWTP) was electricity, which accounted for approximately around 25% to 50% of the operating cost of conventional activated sludge (AS) plants. The available data of wastewater characterization from the treatment plant and data about electricity consumption have been analysed into a table set using Microsoft Excel. The first step involved a detailed analysis of the nutrient removal which showed the efficiency of removing biological and chemical microbes from incoming influent to effluent discharge with a suitable treatment process for the wastewater. The result was, later on, it had was analysed and identified as a baseline or parameter set for the upper limit and lowered the limit to maintain suitable microbe of sewage in the wastewater treatment process.

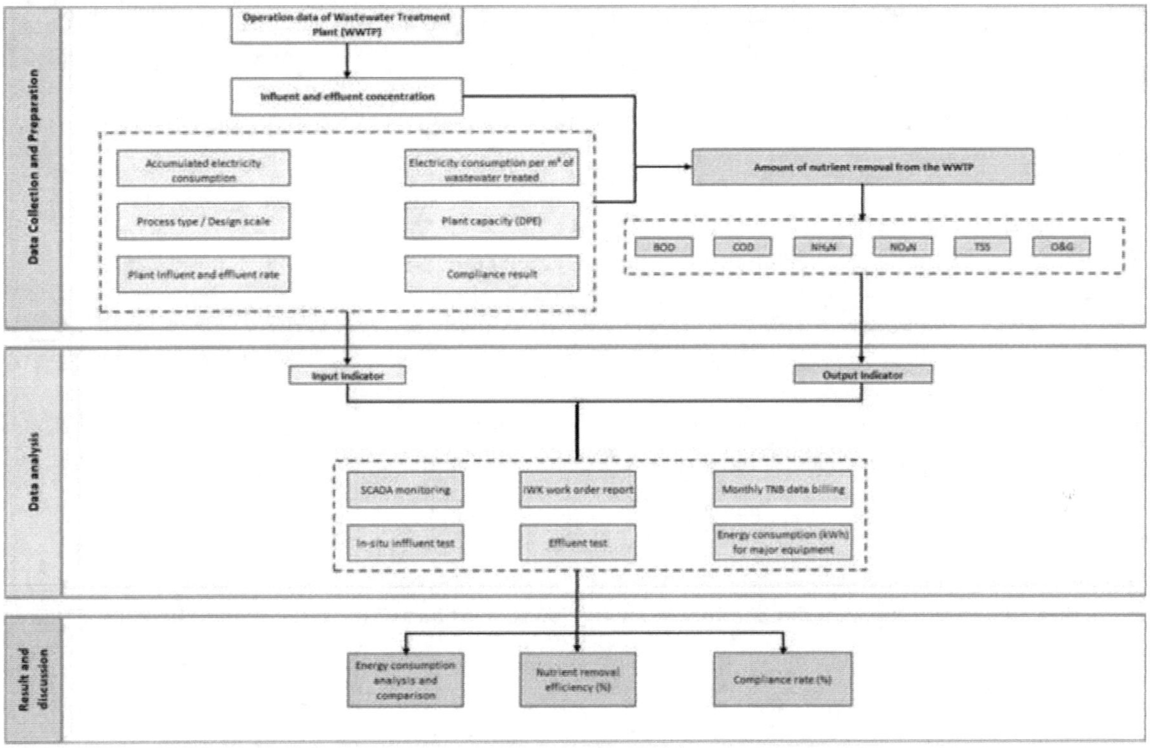

Figure 1. The technical framework of this study

In Figure 1, the evaluation framework is presented. Three steps make up the framework, in which performance data for the three types of wastewater treatment systems (WWTP) for SBR, EA, and IDEA processes were categorized into six inputs. The first step was data collection and preparation, in which the operational data of three types of wastewater treatment system (WWTP) for SBR, EA, and IDEA processes were categorized into six inputs consisting of accumulated electricity consumption data, process plant type, plant influent and effluent rate data, electricity consumption per m^3, plant capacity and sludge production data. Meanwhile, six output indicators of biological removal in the wastewater process consist of BOD, COD, NH_3N, NO_3N, TSS, and O&G. Then, the indicators of all wastewater treatment plants (WWTP) were calculated and analyzed with Microsoft excel for comparison results of electricity consumption, nutrient removal efficiency, and sludge production. For the data collection in this research study, there were six inputs that needed analysis. That consisted of SCADA monitoring data logger. The IWK supervisor ordered the report, monthly TNB data billing for electricity usage, in-situ test result for influent sewage, effluent result compliance, and energy consumption (kWh) for major equipment at each plant. Finally, the result and discussion for the analysis are presented in comparison with each objective of this research study, which was to determine energy consumption, to assess nutrient removal efficiency and compliance rate for each WWTP process typed. The duration of this research study would take 2 years and start from January 2019 until December 2020.

3.1. Study area

The sewage treatment plant (STP) in this case study was located on Peninsular Malaysia's west coast. Sequencing batch reactor (SBR), intermittent decanted extended aeration (IDEA), and extended aeration (EA) were chosen for the three-type mechanized wastewater treatment plant (WWTP). All the STP were categorized as Standard 'A' and Para (I) under the Environmental Quality (Sewage) Regulation 2009 (EQSR 2009). Effluent from the plant was measured by a flowmeter when discharged into the nearest river. Wastewater treatment typically involves one of several types of processes, known as preliminary treatment, primary treatment, secondary treatment, and tertiary treatment. The wastewater treatment plant that was selected was located in Selangor, Malaysia, and under Indah Water Konsortium Sdn Bhd operation to undergo this research study. Table 2 shows that the capacity loading rate for SBR is 50%, EA is 69%, and IDEA is 53% of the total ultimate plant loading. Detailed description of other plants that were selected for this comparison study as per Table 2 below;

Table 2. Description and Capacity of Sewage Treatment Plant for this case study

Plant Type	Sequential Batch Reactor (SBR)	Extended Aeration (EA)	Intermittent Decanted Extended Aeration (IDEA)
Design Population Equivalent (DPE)	150,000 PE	108,000 PE	134,000 PE
Current Population Equivalent (CPE)	75,000 PE	75,000 PE	70,500 PE
Design treatment capacity (m^3 /day)	16,875.00 m^3 /day	16,875.00 m^3 /day	15,873.75 m^3 /day

* Note: PE = Population Equivalent

4. Result and Discussion

There was a clear relationship between the plant operators in designing wastewater inflow, energy, and nutrient removal efficiency. A sewage treatment facility with a low load rate, defined as the volume of sewage applied to a particular treatment process, would have poor energy performance, and such low-capacity utilization would result in increased energy expenditure. It indicates that equipment and processes are most efficient when working under design flow circumstances, for example, when the wastewater treatment facility is at full capacity. When the load rate reached its optimum value, equipment and process operations were more efficient because process operations were more stable than when the loading rates were reduced. Indeed, the operation of the treatment process shows minimal changes in wastewater volume and pollutant concentrations, and conditions are more favourable for microorganisms and sludge growth.

4.1. Plant Performance and Nutrient Removal Efficiency

To argue objectively about the performance of each process technology, relevant data on plant operations were collected and analysed from January 2019 to December 2020. Table 3 shows the comparison of maximum nutrient removal efficiency (BOD, COD, NH_3N, NO_3N, TSS, and O&G) result that has been collected as the following;

i. With an average efficiency of 97.5%, SBR and IDEA wastewater treatment plants have the highest removal efficiency for BOD and COD.
ii. The SBR wastewater treatment plant has the highest NH_3N removal efficiency with an average efficiency of 94.5 percent, followed by IDEA (86.5%) and EA (82.5%).
iii. SBR treatment plant has the highest removal efficiency for NO_3N with an average efficiency of 98%, followed by EA with 89%, and IDEA with 85.5%.
iv. EA treatment plant has the highest removal efficiency for O&G with an average efficiency of 95.5%, followed by IDEA with 91.5% and SBR with 88%.

Based on the results, the SBR treatment plant had the highest score for nutrient removal of BOD, COD, NH_3N, NO_3N, and TSS compared to other activated sludge systems. EA and IDEA have the highest scores of O&G nutrient removal due to the extra retention time and clarifier in both processes, so both activated sludge systems managed to remove more oil and grease. It can be concluded that the SBR treatment system achieves the highest efficiency in treating the wastewater treatment system in terms of water quality; additionally, the SBR method has the added advantage of simplicity and lower cost when compared to other activated systems, according to the qualitative assessment. Nitrogen removal from wastewater can assist to lessen these impacts. Biological nutrient removal (BNR) is a sludge-based method that involves careful environmental management to promote nitrification, denitrification, and nitrate removal, which results in nutrient removal. The sequencing batch reactor (SBR) system was the best process of nutrient removal in removing the BOD, COD, NH_3N, NO_3N, and TSS with average efficiency from 2019 to 2020, is 97.5%, 91.5%, 94.5%, 98 %, and 96.5% Meanwhile extended aeration process (EA) was the best in removing BOD and O&G with the efficiency is 97% and 95.5%. Intermittent decanted extended aeration (IDEA) was the best in removing BOD, COD, O&G, and TSS with efficiency of 97.5%, 91.5%, 91.5%, and 92%. The result that can be concluded from Table 3 shows that the best treatment plant for nutrient removal efficiency is SBR followed by IDEA and EA system processes.

Table 3. Nutrient removal efficiency from 2019-2020

			Nutrient Removal Efficiency (%)					
Wastewater Treatment Plant type	**DOE Para**	**Years**	**BOD**	**COD**	**NH_3-N**	**NO_3-N**	**O&G**	**TSS**
Sequential Batch Reactor (SBR)	1	2019	98	91	94	98	88	96
		2020	97	92	95	98	88	97
		Avg (%)	97.5	91.5	94.5	98	88	96.5
Extended Aeration (EA)	1	2019	98	91	88	88	96	84
		2020	96	89	77	90	95	86
		Avg (%)	97	90	82.5	89	95.5	85
Intermittent Decanted Extended Aeration (IDEA)	1	2019	97	92	85	85	91	93
		2020	98	91	88	86	92	91
		Avg (%)	97.5	91.5	86.5	85.5	91.5	92

4.2. Compliance Rate Efficiency

As can be seen in Table 4, the main objective of determining the wastewater treatment plant (WWTP) with the highest compliance rate has been achieved. For SBR, from a total sample of 49 in 2019, that plant managed to comply 48 times, with an equivalent to a 98% compliance rate. Meanwhile, in 2020, there were a total of 48 samples being taken and SBR only complied 47 times, which was equivalent to 98% of the compliance rate. This result shows that SBR systems are very efficient in removing nutrients and are the most efficient process system. The total sample for compliance was taken every week (12 Months x 4 weeks = 48 samples) since it was the requirement from the DOE to ensure that the wastewater treatment plant meets the regulations and standards to discharge treated wastewater into the river. For IDEA, from a total sample of 49 in 2019, that plant managed to comply 47 times, with an equivalent to a 96% compliance rate. Meanwhile, in 2020, there is a total of 48 samples sampling have been taken and IDEA only complied 43 times. That was equivalent to 90% of the compliance rate due to the major equipment being broken down and under repair. Lastly for EA, from a total sample of 49 in 2019, that plant managed to comply 45 times, with an equivalent to a 92% compliance rate. Meanwhile, in 2020, there were a total of 50 samples taken and EA only complied 45 times that were equivalent to 90% of the compliance rate due to some major equipment being broken down.

Table 4. Compliance result from 2019-2020

			Compliance Result			
Wastewater Treatment Plant type	**DOE Para**	**Years**	**Total Sample**	**Comply**	**Non-Comply**	**%**
Sequential Batch Reactor (SBR)	1	2019	49	48	1	98
		2020	48	47	1	98
Extended Aeration (EA)	1	2019	49	45	4	92
		2020	50	45	5	90
Intermittent Decanted Extended Aeration (IDEA)	1	2019	49	47	2	96
		2020	48	43	5	90

The best wastewater treatment plant was a sequencing batch reactor (SBR), followed by intermittent decanted extended aeration (IDEA) and finally extended aeration (EA). This was because a technology used in SBR was more efficient in removing high-impact wastes with an average incoming of BOD = 92 mg/l and COD = 198 mg/l. The final discharge effluent for SBR managed to reduce and eliminate the nutrients in BOD = 2 mg/l and COD = 17 mg/l. The main advantages of SBR system type parameters were their flexibility, which allows the working parameters to be adjusted according to various conditions or situations needed. On the other hand, such a plant was very complex and required great knowledge of biological aspects as well as an advanced control system to control the treatment process in efficient work.

4.3. Energy Efficiency and Energy Performance Index (EPI)

The cubic meter of wastewater treated per year was used to calculate the energy performance index (EPI) in this study. EPI was a measure of energy intensity used to gauge the effectiveness of the energy management efforts and can be measured as kWh/m3/year. The disaggregated electrical consumption can be measured or estimated, combining the rated power in kilowatt (kW) and the working hours in a year to provide an estimation of kWh used in each stage per unit of time to calculate the total energy consumption in Equation 1. As such, energy units have been measured in kWh/m3 of treated wastewater per year. All the increasing factors of electricity consumption were due to an increase in water demands which was associated with population growth scenarios in an urban area.

$$\text{Energy Consumption} = \text{Operating Hours Daily (h)} \times \text{Power Rating (kW)} = \text{kWh} \quad (1)$$

$$\text{Energy Performance Index (EPI)} = \frac{\text{Total Energy Consumption (kWh)}}{\text{Total Incoming Influent (Flowrate, m3)}}\ \text{kWh/m3} \quad (2)$$

The energy performance indicator (EPI) calculated in this study is the cubic meter of sewage treated per year (Equation 2). As such, the energy units will be measured in kWh/m3 sewage treated per year. Table 5 show that energy consumption of electricity usage for treatment plants at SBR, EA, and IDEA gradually increased by 6%, 10%, and 23% from the year 2019 to 2020. All the increasing factors of electricity consumption are due to an increase in water demands which is associated with population growth scenarios in an urban area. Total incoming influent flow shows each wastewater treatment received flow increasing which is around 8% for SBR, 13% for EA, and 1% for IDEA from the previous year in 2019 compare to the year 2020.

Table 5. Summary of Energy Performance Index from 2019-2020

Wastewater Treatment Plant type	Years	Energy Consumption (kWh/year)	Flow Rate, Flowmeter (m^3)	Energy Performance Index, EPI (kWh/m^3)
Sequential Batch Reactor (SBR)	2019	1,810,204.00	6,469,959.44	0.28
	2020	1,916,385.00	7,060,808.00	0.27
Extended Aeration (EA)	2019	3,030,778.00	5,184,084.00	0.58
	2020	3,382,053.00	5,933,426.32	0.57
Intermittent Decanted Extended Aeration (IDEA)	2019	1,221,859.00	1,323,925.00	0.92
	2020	1,577,482.00	1,340,859.70	0.85

The benchmark for the energy performance indicator (EPI) will be around 0.2 kWh/m^3 – 0.8 kWh/m^3 for large wastewater treatment plants because the organic pollution load correlates best with energy consumption and also ensures a better comparison between different plants. Factors that influence the energy consumption-total influent flow rate-related energy performance index are known to include catchment area of wastewater treatment plant, treatment plant system type, type of customer service, and operational efficiency. Table 5 shows that the treatment plant SBR system was the highest for energy efficiency with energy performance index. EPI value was 0. 27 kWh/m^3 in the year 2020 with 96% energy efficiency compared with 2019. The second highest in the EA system with 0.57 kWh/m^3 with 98% energy efficiency from 2019 and the lowest energy efficiency for EPI result was the IDEA system with 0. 85 kWh/m^3 with 92% energy efficiency compared with 2019. EPI for 2020, all treatment plants have shown increases in terms of energy efficiency when each plant set the benchmarking to optimum electricity usage.

Other than that, it also shows that the EA plant has high energy consumption with a total average of 3,206,415.50 kWh/year with around a 10% increase from 2019. Followed by SBR and IDEA with each plant's energy consumption around 6% and a 23 % increase from the previous year in 2019. The annual flow was significant because it had the potential to influence the performance of the plant through economies of scale. Furthermore, influent characteristics and pollutant concentrations are regarded as important factors for energy consumption since they influence the treatment process and aeration needs. Table 5 also shows the Energy Performance Index (EPI) scores. The SBR system showed the lowest score of 0.28 kWh/m^3 in 2019 and it was reduced to 0.27 kWh/m^3 in 2020, which shows in terms of energy efficiency, consistency is important. In other words, in terms of energy efficiency, both wastewater treatment plants were comparable. In contrast, the IDEA plant presented the highest score of 0.92 kWh/m³ in 2019 and reduced to 0.85 kWh/m³ in 2020. It shows the inconsistency in the energy efficiency values.

5. Conclusion

This study was successful in comparing the energy efficiency, nutrient removal efficiency, and compliance rate of several activated sludges (AS) systems from three different wastewater treatment plant types (SBR, EA, and IDEA) in Selangor, Malaysia. The conclusion of the study was summarized as follows;

1) The sequencing batch reactor (SBR) plant performed the highest compliance rate based on the regulatory requirements of Environmental Quality (Sewage) Regulation 2009 (EQSR, 2009) compared to others plant types.
2) The sequencing batch reactor (SBR) plant achieved the highest percentage in overall nutrient removal percentage and efficiency of organic load for biochemical oxygen demand (BOD), chemical oxygen demand (COD), ammonia (NH3N), oil and grease (O&G), and total suspended solids (TSS) with a different type of activated sludge (AS) system.
3) Based on the energy performed index (EPI) calculated, the results also show that sequencing batch reactor (SBR) plants are the most energy-efficient system.

Acknowledgments

The authors have gratefully acknowledged the support from Indah Water Konsortium Sdn Bhd and Malaysia-Japan International Institute of Technology (MJIIT), Universiti Teknologi Malaysia, Kuala Lumpur for the cooperation, support, and contribution to the data collection of this study.

ORCID iD

W M Bukhary W Muzaffar https://orcid.org/0000-0003-2860-5370

References

[1] Danièle Mousel, Laurence Palmowski and Johannes Pinnekamp. (2017). Energy demand for elimination of organic micropollutants in municipal wastewater treatment plants. *Science of The Total Environment*, **575** 1139-1149.

[2] Guo Z, Sun Y, Pan SY, Chiang PC. (2019). Integration of Green Energy and Advanced Energy Efficient Technologies for Municipal Wastewater Treatment Plants. *Int J Environ Res Public Health*, **16**(7) 1282.

[3] Indah Water Konsortium Sdn Bhd. (2021). The evolution of sewage treatment in Malaysia. *Indah Water Website*.102.

[4] K. Smith and S. M. Liu. (2017). Energy for conventional water supply and wastewater treatment in urban China. *Global Challenges*, **1** 1–6.

[5] T. Xie and C. W. Wang. (2012). Energy consumption in wastewater treatment plants in China.*World Congress on Water, Climate, and Energy*, **34** 1–6.

[6] Wang, H.T, Yang, Y, Keller, A.A, Li, X, Feng, S, Dong, Y.N and Li. (2016) Comparative analysis of energy intensity and carbon emissions in wastewater treatment in USA, Germany, China and South Africa. *Appl. Energy*, **184** 873–881.

[7] Indah Water Konsortium Sdn Bhd. (2019). Sustainability report 2019/2020. *Indah Water Website*, 63.

[8] Al-Shididi, S. Henze, Mogens and Ujang, Z. (2003). Feasibility study of sequencing batch reactor system for upgrading wastewater treatment in Malaysia. *Water science and technology: a journal of the International Association on Water Pollution Research*, **48** 327-35.

[9] Ll. Corominas, J. Foley, J.S. Guest, A. Hospido, H.F. Larsen, S. Morera and A. Shaw. (2013). Life cycle assessment applied to wastewater treatment State of the art. *Water Research*, **47** 5480-5492.

[10] Weiwei Mo and Qiong Zhang. (2013). Energy–nutrients–water nexus: Integrated resource recovery in municipal wastewater treatment plants. *Journal of Environmental Management*, **127** 255-267.

[11] L. Luo, M. Dzakpasu, B. Yang, W. Zhang, Y. Yang, X.C. Wang. (2019). A novel index of total oxygen demand for the comprehensive evaluation of energy consumption for urban wastewater treatment. *Appl. Energy*, **236** 253-261.

[12] P. Foladori, M. Vaccari, F. Vitali. (2015). Energy audit in small wastewater treatment plants: methodology, energy consumption indicators, and lessons learned. *Water Sci. Technol.* **72** 1007-1015.

[13] Asano, T., Burton, F., & Leverenz, H. (2007). Water reuse: issues, technologies, and applications. *McGraw-Hill Education.*

[14] Robert L. Irvine, Lloyd H. Ketchum Jr. & Takashi Asano. (1989). Sequencing batch reactors for biological wastewater treatment. *Critical Reviews in Environmental Control*, 255-294.

[15] Uzma Showkat, Ishtiyaq Ahmed Najar. (2018). Study on the efficiency of sequential batch reactor (SBR)-based sewage treatment plant. *Applied Water Science*, **9**.

Sustainable indicators for designing the wastewater treatment plant

Azila Sembuil*, Aznah Nor and Md. Hakim A-Halim

Department of Chemical and Environmental Engineering, Malaysia-Japan International Institute of Technology (MJIIT), Universiti Teknologi Malaysia, Kuala Lumpur

*Corresponding author e-mail: azilasems95@gmail.com

Abstract. As rapid population and urbanisation in our country have put pressure on our water resources, it is vital to consider sustainable solutions in water and wastewater system, especially to our wastewater treatment plant design. Several studies have shown the importance of having an indicator in both plannings, designing and operating a wastewater treatment plant to help develop a sustainable method in wastewater treatment plants. Therefore, this paper presents a study of crucial sustainable indicators for designing Malaysia's sewage wastewater treatment plant. This paper focused on three (3) aspects: environment, social and economic, followed by significant data to emphasise sustainable indicators. This study was conducted using a quantitative methodology (survey questionnaire online form) and distributed to wastewater engineers or any groups related to the wastewater treatment plant in Malaysia. The survey conducted showed that most crucial sustainable indicators in sewage WWTPs design were (i) nutrient removal, (ii) worker's health and safety, (iii) public health and safety, (iv) operational and maintenance costs, (v) energy consumption and (vi) greenhouse gases emissions.

1. Introduction

Wastewater is defined as used water from various sources such as residential areas and industrial sectors. Wastewater treatment is a process of removing contamination in the water that harmful to the environment and humans [1]. This wastewater treatment aims to produce clean water before releasing it into the environment and used by living organisms, especially human beings. However, wastewater management has become a significant problem due to drastically increased population growth and urbanisation. The high demand on natural resources such as water has put pressure on the environment and threatened the sources of water supplies. According to [2], many rivers in Malaysia suffer from sewage pollution and cause eutrophication. Besides that, human activities such as discharging the untreated wastewater directly into the river or water body also contribute to pollution. Therefore, access to sanitation and drinking water will be quite tricky for the engineers to improve the sanitation system. More than 95% of wastewater in many countries were unproperly discharging the untreated wastewater into the environment [3]. According to [3] there were still many people throughout the world without access to better sanitation. Besides that, focusing on the sanitation target causing the wastewater treatment rise as a critical issue.

Throughout the year, many existing sewages wastewater treatment plants have evolved from basic sanitation facilities to advanced treatment facilities to address sewage pollution. For example, septic tank, imhoff tank, oxidation ponds, aerated lagoons, activated sludge system, package system and mechanical plant [4]. Despite the evolution of the sewage wastewater treatment plant, it remains a major pollutant of inland waterways in Malaysia [4]. Therefore, the development of sewage treatment needs to be more proactive and maintain the sustainability of the treatment system in aspects of the environment, economic and social to provide a better future towards the environment and economy and community.

One of the targets in promoting this effort is the introduction of Sustainable Development Goals (SDGs), in which the untreated wastewater should halved from 2015 [3]. Sustainable Development Goals defined as the '*development that meets the present generation's needs without compromising the ability of future generations to meet their own needs'*. The meaning of this definition is to show that all human beings in the present or future should have equal rights. One of sustainable development purposes is to promote water protection, conservation, and water efficiency. In SDGs, there are 17 goals, including clean water and sanitation (Goal 6). This goal aims for clean water accessibility and the ability for all people to access drinking water and protect water resources such as rivers. Relation sustainability to wastewater treatment is to urges all people to protect the aquatic organisms and increase public awareness of the impact of pollution on the environment and human safety and health. According to [5] many countries face excessive pollution in water reservoirs and causing severe environmental problems, including disturbing the ecological balance. Thus causing environmental damage and affects the cost operation of the wastewater treatment plant.

Besides that, sustainable indicators in wastewater treatment plants are also important to estimate the sustainability in wastewater treatment, such as environmental impact and energy saving. A sustainable indicator is a statistic used to measure environment protection, social equity and economic growth. Besides that, an indicator can help policy and decision-making and provide quantitative and qualitative information [6]. The selection of good indicator sustainability can alert to a problem that may occur in the system and provide practical actions on how to fix it before it becomes a severe problem.

This study aims to determine crucial indicators for sustainability in wastewater treatment system design. The findings in this study will also determine sustainable indicators that should be considered in wastewater treatment system design and propose relevant qualitative and quantitative data to translate the sustainable indicators.

2. Methodology

This chapter will include the procedure for this study. The methodology for this study is by using quantitative method. The procedures in this work were data collection from a scientific database such as ScienceDirect, Google Scholar, Scopus, Springer and EZproxy from UTM. Besides that, identifying crucial indicators was done through a survey questionnaire. The data from survey questionnaire were analysed using SPSS software.

2.1. Data collection

Data collection for this study includes all secondary sources such as journals, published research articles and review papers conducted in Malaysia and other countries. Relevant information and data were selected to aid in constructing the survey questionnaire before distributing it to targeted participants.

2.2. Questionnaire

The purposes of this questionnaire are to identify the crucial indicator for designing the sewage wastewater treatment plant and to propose the qualitative and quantitative data for emphasised the identified sustainable indicator. The questionnaire is addressed to the wastewater engineer or any groups experienced in the wastewater or water industry. The number of respondents in this questionnaire is 31 respondents.

There were three (3) sections in the questionnaire, namely Demographic (Section A), crucial sustainable indicator in sewage treatment plant design (Section B) and lastly Section C, which is the significant data for emphasised the identified sustainable indicator. Section A consists of the respondents' background section, while section B consists of a crucial sustainable indicator for designing the wastewater treatment plant. Section C consists of the qualitative and quantitative data to emphasise the identified sustainable indicators. Besides that, the survey questionnaire was implemented via Google Form online application.

Four steps have been taken to conduct the questionnaire in this study. The first step is to conduct a library study from the scientific database. The purpose is to identify the sustainable indicators in the wastewater treatment plant and propose qualitative and quantitative data to emphasise the identified indicators. The information from the literature review was used to form Section B and Section C of the questionnaire. The second step is to validate the questionnaire that has been made. For the validation, three Indah Water Konsortium consultants were selected to validate and comment on the questionnaire before distributing it to the target participants. The third step is to improve the questionnaire based on comments from the consultants. The last step is a pilot test to check the reliability and improve the quality and efficiency of the questionnaire.

2.3. Data analysis
Data from the questionnaires are analysed and interpreted using SPSS software (Statistical Package for the Social Sciences). The data analysis is divided into two (2) parts.
(a) Part one: Descriptive analysis for a demographic section such as frequency distribution and percentage.
(b) Part two: Section B and C data were analysed using one-sample t-test analysis and interpreted into a statistical significant method where the p-value was obtained. Data with a significant level of $p > 0.05$ is specified as not significant, while p-value less or equal to 0.05 were specified as significant. In other words, the smaller the significant level (p-value), the more significant the results.

3. Results and discussion

3.1. Survey participant profile
Two questions were designated to get demographic data from participants, including experience working in the wastewater or water industry and experience on the project design of the wastewater treatment plant.

Table 1. Respondent's demographics analysis results

Question	Distribution	Frequency	Percentage (%)
Experience in working in wastewater or any industry related to wastewater treatment plant	0-5 years	19	61.3
	6-10 years	8	25.8
	11-15 years	2	6.5
	More than 15 years	2	6.5
Experience in wastewater treatment plant design project	1-5 times	14	45.2
	6-10 times	2	6.5
	11-15 times	1	3.2
	More than 15 times	3	9.7
	None	11	35.5

Based on the data obtained from the respondents, most of the respondents have experience working in wastewater for five (5) years and below, with a percentage of 61.3%. In comparison, 25.8% of the respondents have experience for 6 to 10 years, followed by 6.5% for both working experiences of 11 to 15 years and 15 years above, respectively. However, 45.2% of the respondents have experience on working in designing the wastewater treatment plant, followed by 6.5% (6-10 times), 3.2% (11-15 times), 9.7% (more than 15 times) and 35.5% (none experience) as shown in Table 1. Based on these results, although most of the respondents have working experience for five years and below, they have experience in designing the wastewater treatment plant at least one (1) to five (5) times.

3.2. Crucial indicator for designing the sewage wastewater treatment plant
In order to determine the crucial indicators for designing the WWTP, the indicators are divided into three aspects (environmental, social and economic). One sample t-test was used to analyse each

indicator. One sample t-test was used to compare a single sample's mean to a predetermined value to determine if the sample mean is significantly greater or less than the value. In this study, the t-test value used is 3, indicating a neutral response, either important or not important.

Table 2. The significant indicators

Aspects	Indicator	Mean	*p*
Environment	Nutrient removal (BOD, COD, TSS, Nitrate, etc)	4.23	0.00
	Energy consumption	3.61	0.01
	GHG emissions	3.74	0.00
Social	Worker's health and safety	4.16	0.00
	Public health and safety	4.03	0.00
Economic	Operational and maintenance costs	3.81	0.00

Table 2 shows three crucial indicators in the environmental aspects: nutrient removal indicator, energy consumption and GHG emissions. The majority of wastewater treatment systems used in Malaysia for nutrient removal is activated sludge. However, the operating costs are high and contribute to the sludge disposal problem at the end. Therefore, one of the approaches to address this issue is by using an environmentally friendly method such as phytoremediation. Phytoremediation is a method of removing, stabilising, transferring, or reducing contaminants in soil or water media using green plants, making it an attractive biological treatment method [7]. Besides that, it is found that phytoremediation agents can remove organic and inorganic contaminants, and it requires low cost, low energy consumption and minimal environmental disturbance [8]. According to [9] reported that the phytoremediation agents such as Eichhornia crassipes (Water hyacinth) could remove 83%-92% of COD and TSS. Hence, nutrient removal technologies or methods used in wastewater treatment plants should be chosen wisely to achieve sustainability in the wastewater treatment plant. This is especially for supporting SDG 6, which is clean water and sanitation, and SDG 14 to sustainably manage and protect marine and coastal ecosystems from pollution and comply with Environment Quality (Sewage) Regulation 2009.

The second indicator in the environmental aspect is the energy consumption indicator. The energy used in the WWTP continuously increased as clean water demand increased and eventually affected the total cost operational expenditure (OPEX) [10]. Besides that, [11] stated that most WWTP found low energy efficiency performance levels due to high water demand and higher discharge parameters in the treated effluent. In addition, the energy used from fossil fuels contributes to environmental issues such as the greenhouse effect and acid rain. Thus, reducing the dependency on fossil fuels must be considered to achieve SDGs and support Green Technology Masterplan Malaysia (2016-2020). The goal of GTMP is to promote sustainable development and strengthen the green technology and economy focusing on energy, manufacturing, waste, building, water and transportation. Besides, to support the GTMP in terms of energy, WWTP must reduce their dependency on fossil fuels by utilising renewable energy like biogas. For example, biogas produced from IWK's sewage wastewater treatment with an anaerobic digester system was reported to be capability to generate 20 MW hour per day [12]. Hence, it synchronised with SDG 7 (Affordable and Clean energy) and SDG 13 (Climate action) and supported the National Green Technology.

The operation of WWTP causes direct emissions of GHG such as carbon dioxide (CO_2), nitrous oxide (N_2O) and methane (CH_4), where carbon dioxide is emitted from the energy consumption during plant operation [13]. Well-managed WWTP operation protects the environment and supports SDG 7 (Renewable Energy) and SDG 8 (Climate Action) and compliance with the National Policy of Climate Change. Besides, [3] stated that energy use influences GHG emissions in the wastewater treatment plant. The more WWTP depends on fossil fuels as energy source, the more they contribute to GHG emission. Therefore, the need to use renewable energy and well-managed wastewater treatment plant operation in terms of energy use is needed. In addition, the Malaysian government is considering climate change in the infrastructure system engineering design for water to ensure sustainable management and encourage the private sector to include net-zero GHG in their sustainable operation agenda [14].

In the social aspect, public health and safety indicators and worker health and safety indicator are selected as crucial indicators in WWTP. Various activities in WWTP such as toxic air contaminants, detection of chemicals and control of pollutants present in wastewater impact human health [15]. In fact, producing clean water, proper wastewater treatment, and disposal are essential for community health. The treated wastewater in urban areas should meet with Environment Quality Act 1974 and its regulations Environmental Quality (Sewage) Regulation 2009 before being discharged to surface water. According to [16] stated that an analysis made by the International Occupational Safety and Health Information Centre (CIS) found many accident hazards in the sewage wastewater treatment industry, such as water-born disease, hearing loss as well as slip and fall. The awareness of safety and health in the WWTP is important to reduce occupational accidents in the workplace. Besides that, in Malaysia, there are laws enforced by the Department of Occupational Safety and Health, such as the Occupational Safety and Health Act (OSHA) 1994. Thus, provide safety and healthy work environment for the wastewater operators or other workers by implementing safety guidelines, provide personal protective equipment and training are needed. Besides that, compliance to OSHA Act 1994 and Environmental Quality Act 1974 reduce the accidence level, avoid disease, and support SDG 3 (Good health and well-being) by ensuring healthy lives for all workers and the public. SDG 14 (Life below water) and SDG 11 (Sustainable Cities & Communities) also are supported by preventing any disease and pollution that may harm the public and environment.

Lastly, operational and maintenance cost indicators are selected as crucial indicators for the economic aspect. The investment spends on the sewage wastewater treatment plant for its operation and maintenance is high. The electricity cost of the WWTP contributes the highest expenses, followed by contractor costs and office rental and premises. According to Malaysia's Indah Water Konsortium (IWK), the electricity cost in 2010 was around RM 152.50 million and has been increased since 2007 [10]. This shows that the operating cost is influenced by electricity usage in the WWTP. Therefore, while planning a new WWTP in the future, it is important to consider using equipment with energy-saving, implementing periodical maintenance of equipment, and using alternative energy sources.

3.3. Identifying the significant data for emphasised the identified crucial indicators

Table 3. Significant data for identified sustainable indicators

Environment	Qualitative and quantitative data	Respondents Acceptance/Recommendation	Mean	SD	*p*
Energy consumption	Energy usage in WWTP (Quantitative)	Accepted	3.84	1.15	0.00
	Usage of fuels and electricity from renewable sources (Qualitative)	Rejected	3.35	1.25	0.12
	Legal standard to meet	New monitoring data is proposed			
Nutrient removal	Contaminant removal efficiency to mitigate environment risks (Qualitative)	Accepted	4.32	0.74	0.00

Table 3. Significant data for identified sustainable indicators (continued)

Environment	Qualitative and quantitative data	Respondents Acceptance/Recommendation	Mean	SD	p
Nutrient removal	Quality level of discharge water compliance with DOE (Qualitative)	Accepted	4.48	0.76	0.00
	Profiling of incoming waste	New monitoring data is proposed			
	Compliance to local regulatory	New monitoring data is proposed			
GHG emissions	Carbon dioxide emission intensities (Quantitative)	Accepted	3.74	1.06	0.00
	Index of greenhouse gas emissions (Quantitative)	Accepted	3.74	1.06	0.00

Based on Table 3, the proposed quantitative and qualitative data for energy consumption is energy usage in WWTP per volume of treated wastewater and usage of fuels and electricity from renewable energy. The data analysis showed that the qualitative data is significant where the p-value is less than 0.05. However, for quantitative data, the p-value is greater than 0.05, which indicates not statistically significant. This means the proposed qualitative data for energy consumption indicator is not chosen as important by most respondents. For the energy usage in WWTP, other than renewable energy as an energy source, the size of plant, design, and technologies used should also be considered [17]. This is because some of the technology used in the WWTP required high energy consumption, such as aerators and blowers in the aeration tank. The need to improve energy efficiency in WWTP is essential to reduce the cost of energy consumption and minimise the operation cost. Even though increasing energy efficiency in the plant requires a high cost, it has been proven that the cost of energy consumption is reduced [18]. Therefore, to emphasise the energy consumption in WWTP, renewable energy, types of technology used and energy efficiency need to be considered. However, another suggestion by respondents to emphasise energy consumption is a legal standard to meet.

Next is nutrient removal (BOD, COD, TSS and Nitrate) indicator. Contaminant removal efficiency to mitigate environmental risks and quality level of discharge water compliance with the Department of Environment (DOE) is qualitative data to emphasise the nutrient removal indicator. The p-value for these qualitative data was statistically significant, with a p-value of less than 0.05. The efficiency of the technology used in the wastewater treatment plant to remove the contaminant in the wastewater before being discharged into the water bodies are important to prevent pollution and avoid spreading disease to public and aquatic organisms. The wastewater discharge should also comply with Environmental Quality (Sewage) Regulation 2009 under Environment Quality Act 1974 regulated by DOE to ensure sustainable growth among the community and protect the environment. The neglect of the Environment Quality Act 1974 caused damage to the country's environmental, social, and economic sectors. Malaysia faced various issues due to the discharge of undesirable effluents and waste products into the environment, leading to natural resources degradation, pollution, and human health risks for many years. Therefore, considering the efficiency of nutrient removal in WWTP and compliance to DOE is crucial to achieving sustainability in WWTP. However, the respondents also suggested that profiling incoming waste and compliance with local regulations should be considered to emphasise the nutrient removal indicators for sustainable WWTP.

The proposed quantitative data for the GHG emissions indicator is carbon dioxide emission intensities and the index of GHG emissions. Table 4 shows that both quantitative data for GHG emissions are statistically significant. One of the concerns of the Malaysian government to achieve sustainable development is climate change due to GHG emissions in many sectors, including wastewater treatment. Excessive greenhouse gases emission into the atmosphere has led to climate warming. The CO_2 emission is produced by the emission of carbon dioxide directly and indirectly. The combustion of CH_4 and the organic matter oxidation occur in biological reactors emitted CO_2 directly while indirectly in energy consumption in the WWTP [13]. High carbon dioxide intensities and GHG index will increase GHG emissions. Therefore, reducing CO_2 and other greenhouse gas emissions in the WWTP will reduce the GHG emission into the atmosphere.

Table 4. Statistical survey results on significant data for Social and Economic aspects indicator

Social	Qualitative and quantitative data	Respondent Acceptance/Recommendation	Mean	SD	*p*
Worker's health and safety	Awareness of safety in the workplace (Qualitative)	Accepted	4.65	0.66	0.00
	Implementation of SOP and guidelines (Quantitative)	Accepted	4.52	0.72	0.00
	Work safety training provided (Qualitative)	Accepted	4.55	0.72	0.00
	Awareness of health in the workplace (Qualitative)	Accepted	4.58	0.72	0.00
Public health and safety	Potential impacts of the plant design on public (Quantitative)	Accepted	4.16	0.86	0.00
	Awareness of health issues (Qualitative)	Accepted	4.29	0.82	0.00
	Awareness on the safety (Qualitative)	Accepted	4.29	0.82	0.00
Economic	Qualitative and quantitative data	Respondent Acceptance/Recommendation	Mean	SD	*p*
Operational and maintenance costs	Operating costs per volume unit of wastewater treated (Quantitative)	Accepted	4.29	0.97	0.00
	Maintenance cost of facility per year (Quantitative)	Accepted	4.23	0.88	0.00

Table 4 shows the statistical survey for proposed qualitative and quantitative data for the social and economic sectors. The proposed qualitative and quantitative data for worker health and safety indicators are awareness of safety in the workplace, awareness of health in the workplace, implementation of SOP and guidelines and work training provided. All these qualitative and quantitative data emphasise the work health and safety indicator for sustainable WWTP. The awareness of safety and health in the workplace should be taken as a serious matter in WWTP as the workers have been involved in various activities in wastewater treatment plants and exposure to chemical and toxic air contaminants. According to [15], safety, such as providing PPE to workers and good work training, should be implemented to avoid any hazardous situations and accidents. Besides, [19] also added that conducting awareness programs about safety and possible risk in the workplace will decrease accidents. Besides that, implementing SOP and guidelines in the workplace aids in preventing any accidents or hazards in the workplace. Besides, the SOP and guidelines in the workplace should be updated if have changes in operation in the workplace. Therefore, these qualitative and quantitative data emphasise the importance of worker health and safety for sustainable WWTP.

Meanwhile, the proposed qualitative and quantitative data for emphasised public health and safety indicator is the potential impact of WWTP plant on public, awareness of safety and awareness of health issue. A wastewater treatment plant is one of the sources of toxic compounds and gases contaminant, which can lead to environmental pollution to its surrounding. Therefore, when designing the wastewater treatment plant, it is crucial to consider its effects, especially local health and safety. According to [20], one of the potential impacts of WWTP on the public environment is microbial air pollution and can spread for 800 meters from its source. Therefore, it is important to identify potential impacts to prevent or control any issue. Besides that, public concern on clean water quality for their safety is needed to be evaluated. According to [21] a better understanding of the public concern of drinking water contributes to water management, such as the use of treatment processes and customer services. Besides that, the awareness of health issues also needs to be evaluated, especially for maintaining public life quality. For example, well facilities management such as proper treatment and disposal of wastewater is crucial to protect the environment and public health. Besides that, risk assessment related to the WWTP needs to be implemented to obtain the risk that may happen during the operation of WWTP, which may affect human health [22].

For the economic aspect, the proposed quantitative data for emphasised operational and maintenance cost is the operational cost per volume unit of treated wastewater and facility maintenance cost per year. Operations are any activities conducted in the WWTP, while maintenance involves activities or actions to ensure the equipment used in the WWTP is efficient to reach their objectives [23]. Usually, the operation cost is affected by the equipment used and the characteristic of the wastewater. Therefore, to achieve sustainability in the wastewater treatment plant, it is very important to choose the high performance of equipment used during the operation, such as can treat wastewater at a lower cost [24]. Lower operation costs will reduce the expenditure facility, thus producing more profit and can help to improve or maintain the facility. Besides, if the maintenance cost increases due to the lack of preventive maintenance, this will decrease the profit and asset cost of the facility.

4. Conclusion

This study significantly gives a new perspective in the decision-making, planning and operation of wastewater management services, especially in developing sustainable wastewater treatment plants. In conclusion, the survey questionnaire has shown that these five indicators for designing the sewage wastewater treatment plant are crucial and should be considered. The identification of crucial indicators is to embrace the impact of the development and operation towards achieving environmental, social and economic sustainability, which can significantly be a green baseline for designing a future wastewater treatment plant.

Acknowledgements
In preparing this paper, I contacted many people, researchers, and academicians. They have contributed towards my understanding and thoughts. In particular, I wish to express my sincere appreciation to my main supervisor for encouragement, guidance, critics and friendship. My sincere appreciation also extends to all my colleagues and others who have assisted at various occasions. Their views and tips are useful indeed.

ORCID iD
S. Azila https://orcid.org/0000-0001-7758-8074

References

[1] Nuralhuda AJ 2020 The design for wastewater treatment plant (WWTP) with GPS X modelling *Cogent Engineering* **7(1)**

[2] Ariffin M and Sulaiman SNM 2015 Regulating sewage pollution of Malaysian rivers and its challenges *Procedia Environmental Sciences* **30** 168-173

[3] Cossio C, Norrman J, McConville J, Mercado A and Rauch S 2020 Indicators for sustainability assessment of small-scale wastewater treatment plants in low and lower middle income countries *Environmental and Sustainability Indicators* ***6*** 100028

[4] Karunanithi A 2018 Institutional and technological reforms in urban wastewater management story of Malaysia. (New Delhi: Centre for Policy Research)

[5] Dereszewska A and Cytawa S 2016 Sustainability considerations in the operation of wastewater treatment plant 'Swarzewo' In *E3S Web of Conferences* 10, EDP Sciences

[6] Garrett RD and Latawiec AE 2015 What are sustainability indicators for? in *Sustainability Indicators In Practice, Warsaw, Poland,* De Gruyter Open Poland, 12- 22

[7] Al-Thani RF and Yasseen BT 2020 Phytoremediation of polluted soils and waters by native Qatari plants: Future perspectives *Environment Pollution* **259** 113694

[8] Idris M, Abdullah SRS, Titah, HS, Latif MT, Abasa AR, Husin AK, Hanima RF and Ayub R 2016 Screening and identification of plants at a petroleum contaminated site in Malaysia for phytoremediation *Journal of Environmental Science and Management* **19** 27–36.

[9] Safauldeen SH, Hasan HA and Abdullah SRS 2019 Phytoremediation efficiency of water hyacinth for batik textile effluent treatment *Journal of Ecology Engineering* **20** 177–187

[10] Ramli NA and Abdul Hamid, MF 2017 Analysis of energy efficiency and energy consumption costs: a case study for regional wastewater treatment plant in Malaysia *Journal of Water Reuse and Desalination* **7(1)** 103-110

[11] Oliveira P, Fernandes B, Analide C and Novais P 2021 Forecasting energy consumption of wastewater treatment plants with a transfer learning approach for sustainable cities *Electronics* **10(10)** 1149

[12] Indah Water Konsortium Sdn Bhd 2021 Renewable energy *Indah Water website*

[13] Campos JL, Valenzuela-Heredia D, Pedrouso A, Val del Rio A, Belmonte M and Mosquera-Corral A 2016 Greenhouse gases emissions from wastewater treatment plants: minimisation, treatment, and prevention *Journal of Chemistry* **9** 1-12

[14] The Star 2021 *PM: Malaysia plans to reduce greenhouse gas emission by 45%.* The Star

[15] Popovic T and Kraslawski A 2018 Quantitative indicators of social sustainability and determination of their interdependencies. Example analysis for a wastewater treatment plant *Periodica Polytechnica Chemical Engineering* **62(2)** 224-235

[16] Malakahmad A, Downe AG and Fadzil SDM 2012 Application of occupational health and safety management system at sewage treatment plant In *2012 IEEE Business, Engineering & Industrial Applications Colloquium (BEIAC)* 347- 350.

[17] Capodaglio AG and Olsson G 2019 Energy issue in sustainable urban wastewater management: use, demand reduction and recovery in the urban water cycle *Sustainability* **12** 266

[18] Azis NF, Ramli NA and Abd Hamid MF 2019 Energy efficiency of wastewate treatment plant through aeration system *Desalination and Water Treatment* **156** 38- 45

[19] Upadhyaya JK and Moore G 2012 Sustainability indicators for wastewater reuse systems and their application to two small systems in rural Victoria, Australia *Journal of Civil Engineering* **39**

[20] Adamus-Bialek W, Wawszczak M and Swiercz A 2015 Impact of sewage treatment plant on local environment *Proceeding of ECOpole* **9(2)**

[21] Wang L, Zhang L, Lv J, Zhang Y and Ye B 2018 Public awareness of drinking water safety and contamination accidents: a case study in Hainan Province, China *Water* **10(4)**

[22] Tuser I and Oulehlova A 2021 Risk assessment and sustainability of wastewater treatment plant operation *Sustainability* **13** 5120

[23] Noufal M 2014 The 'o' and the 'm' in operation and maintenance of water and wastewater treatment plants Retrieved from https://www.wateronline.com/doc/the-o-and-the-m-in-operation-and- maintenance-of-water-and-wastewater-treatment- plants-0001

[24] Altin S, Altin A and Dogru S 2020 Investigation of operating costs at an urban wastewater treatment plant *Turkish bulletin of hygiene and experimental biology* **77** 49- 56

IOP Conf. Series: Earth and Environmental Science 1091 (2022) 012058 doi:10.1088/1755-1315/1091/1/012058

Physical Characteristics and Removal Performance of Aerobic Granular Sludge in Biological Treatment of Domestic Wastewater

Mohd Hakim Ab Halim[1*], Aznah Anuar[1], Nur Syahida Abdul Jamal[2], Siti Izaidah Azmi[2], Wan Nur Atiqah Wan Hassan[1,2,3], Zaini Ujang[4] and Mustafa M Bob[5]

[1] Department of Chemical and Environmental Engineering, Malaysia-Japan International Institute of Technology, Universiti Teknologi Malaysia, Kuala Lumpur, Malaysia
[2] Department of Water and Environmental Engineering, School of Civil Engineering, Faculty of Engineering, Universiti Teknologi Malaysia, Johor Bahru, Malaysia
[3] Department of Biosciences, Faculty of Science, Universiti Teknologi Malaysia, Johor Bahru, Malaysia
[4] Centre for Environmental Sustainability and Water Security, Research Institute for Sustainable Environment, Universiti Teknologi Malaysia, Johor Bahru, Malaysia
[5] Department of Civil Engineering, College of Engineering, Taibah Universiti, Al Madinah Al Monawarah, Saudi Arabia

*Corresponding author e-mail: hakim_90@rocketmail.com

Abstract. This study aims to demonstrate the formation of aerobic granular sludge (AGS) for domestic wastewater treatment applications in hot climate and low humidity conditions (e.g. Saudi Arabia) as well as to characterize and examine the bioreactor performance in removing organic matter and nutrients. Based on the study, aerobic granular sludge (seeded with domestic activated sludge from Al-Madina Wastewater Treatment Plant and fed with synthetic wastewater with an organic loading rate (OLR) of 1.6 kg COD $m^{-3}d^{-1}$) was successfully cultivated with average size ranging from 1 to 2 mm in a laboratory scale sequencing batch reactor (SBR) system operated at 30 days with a complete cycle time of 3 h and at the working temperature of 30 ± 1°C. Results also showed that the aerobic granular sludge formed had excellent settling ability (settling velocity, 38.98 mh^{-1} and MLSS, 7.8 gL^{-1}) and great removal performance of COD and nutrient (i.e. more than 90%).

1. Introduction

Biological processes are a cost-effective and environmentally sound alternative to the chemical treatment of wastewaters. The biological process based upon suspended biomass (i.e., Activated sludge processes) is effective for organic carbon and nutrient removal in municipal and industrial wastewater plants (Borkar et al., 2013). Recently, interesting alternatives for the conventional activated sludge systems were developed relying on compact and self-immobilized granulated biomass. Examples are for instance Aerobic Granular Sludge (AGS), Sequenced Batch Reactor (SBR), Upflow Anaerobic Sludge Bed (UASB), as well as Expanded Granular Sludge Bed technology (EGSB) (Beun et al., 1999; Lettinga and Hulshoff Pol, 1991; Morgenroth et al., 1997; Rinzema et al., 1993). Aerobic granulation in SBRs has been reported by many researchers and is widely used in treating high-strength wastewaters containing organics, N and P, and toxic substances as the SBR operation conditions (cyclic feeding and starvation, high shear stress and short settling time) promote development of granules (Morgenroth et al., 1997; Beun et al., 1999; Peng et al., 1999; Etterer and Wilderer, 2001; Tay et al., 2001a; Liu and Tay, 2002, Adav et al., 2008b; Adav et al., 2009d).

Aerobic granular sludge process has been proposed as a promising approach to biological wastewater treatment (de Kreuk et al., 2007). Compared with conventional bioflocs, aerobic granular sludge has dense microbial structure, good settling ability, high biomass retention and the ability to withstand a high organic loading rate (Morgenroth et al., 1997; Y. Liu and J.-H. Tay, 2004, Adav et al., 2008b). Some studies of aerobic granular sludge had been operated on high temperature, e.g., 27-35°C (Liu et al., 2007; Song et al., 2009; Ebrahimi et al., 2010; Muda et al, 2010; Abdullah et al., 2011; Nor-Anuar, 2012; Bassin et al., 2012; Rosman et al., 2013; Othman et al., 2013, Cui F. et al., 2014). A stable and compact AGS structure with good settling ability is reported to be obtained at temperature 30-35°C both by using synthetic (Song et al., 2009; Muda et al., 2010; Nor-Anuar et al. 2012, Bassin et. al, 2012; Cui et al., 2014) and real wastewater (Liu et al., 2007, Nor-Anuar et al., 2012; Othman et al., 2013; Rosman et al., 2013). Besides that, excellent removal performance of COD, Nitrogen and Phosphorus also reported by most researchers with removal efficiency more than 80%.

In the hot climate countries such as in Asia or Arab countries, the temperature of sewage usually varies between 30 to 35°C. Thus, it is necessary for investigation of high temperature effects on aerobic granular sludge be done before the technology can be efficiently scaled-up and put into practice in the countries with high climate. Therefore, the aim of this study was to develop an AGS system for domestic wastewater treatment applications in hot climate and low humidity conditions (e.g. Arab Saudi). In this study, a 4-liter laboratory-scale SBR was designed and used to study the characterization of AGS seeded with domestic sludge from Al-Madina Wastewater Treatment Plant and fed with synthetic wastewater at temperature 30°C. Besides that, the study also was aim to investigate the performance of AGS to remove organic matter (COD) and nutrients (Nitrogen and Phosphorus) simultaneously at 30°C in the AgSBio system.

2. Materials and Methods

2.1. Experimental Procedure and Reactor Operation

Experiments were carried out in a cylindrical column bioreactor (internal diameter of 10 cm and total height of 68.5 cm) with a working volume of 4 L. During the start-up period, 2000 mL of sludge from a municipal sewage treatment plant will be added as the inoculums in bioreactor. The bioreactor will be operated in SBR mode with a cycle of 3 h (180 min and 3 cycles a day) at temperature of 30 ± 1 °C, and DO values be kept below 2 mgL^{-1} without control of pH value. The 3 h cycle of bioreactor operation consists of: 5 min of feeding, 140 min of aerobic reaction, 15-30 min of settling, and 5 min of effluent discharge. A set of two peristaltic pumps were to be used to feed and discharge the wastewater in the bioreactor system. During the filling phase, the influent will be introduced in the reactor through a port located at the bottom of the column. Air will be introduced via a fine bubble aerator at the bottom of the reactor during reaction phase at a flow rate of 2 $Lmin^{-1}$. The wastewater effluent will be withdrawn through an outlet port located at medium height of the bioreactor column, which results in a volumetric exchange ratio (VER) of 50%. The sludge retention time will be set by the discharge of suspended solids with the effluent.

2.2. Synthetic Wastewater Characteristics and Seed Sludge

The bioreactor was fed with synthetic wastewater with composition of the influent media (A) NaAc 65.1 mM, $MgSO_4.7H_2O$ 3.7 mM, KCl 4.8 mM and (B) NH_4Cl 35.2 mM, K_2HPO_4 2.2 mM, KH_2PO_4 4.4 mM (Nor Anuar et. Al., 2007) and 10 mlL^{-1} trace element solution (Vishniac and Santer, 1957). 150 mL of medium A and B would be dosed together with 1500 mL distilled water per cycle. Sodium acetate will be used as the only organic substrate for its biodegradability. The seed sludge would be obtained from Al-Madina Wastewater Treatment Plant, Arab Saudi in order to provide ideal inoculums with ability to acclimatize with low and high temperature conditions. The accepted activated sludge sample will be stored in a cooling room until needed (10 °C). One and half liter of this sample will be seeded in the reactor for the startup stage. It will be aerated for 24 hours and then it will be fed continuously with synthetic wastewater.

2.3. Analytical Methods
The pH and DO were continuously monitored throughout the experiment using an Orion 4-Star Benchtop pH/DO Meter. Parameters such as chemical oxygen demand (COD), BOD_5 ,total nitrogen (TN), total phosphorus (TP), mixed liquor suspended solids (MLSS), mixed liquor volatile suspended solids (MLVSS) were carried out according to Standard Methods for the Examination of Water and Wastewater (APHA, 2005). The aerobic granules developed in the SBR were analyzed for their physical characteristics such as size, settling velocity and sludge volume index (SVI). The morphological and structural observations of the granules were conducted periodically by using a stereo microscope equipped with digital image processing and analyzer (PAX-ITv6, ARC PAX-CAM). The microstructure compositions within the granule were observed with a scanning electron microscope (FESEM-Zeiss Supra 35 VPFE-SEM). For the pre-treatment procedure for SEM image, the granules were left dried at room temperature prior to gold sputter coating (Biorad Polaron Division SEM Coating System).

3. Results and Discussion
3.1. Morphology of Aerobic Granular Sludge
Observations of aerobic granules were done periodically under stereomicroscope as shown in Figure 1. At beginning, flocculent sludge with fluffy, loose and irregular shape is predominant in the reactor. Some tiny granules start to appear in the reactor after 7 days of operation. After 2 weeks, dark brown granules which were small in size (average diameter = 1.0–1.5 mm) were observed in the reactor while the sludge flocks gradually disappeared. The granules at this stage were unstable and easily broken up. After 30 days cultivation, large amounts of granules showed near-round shape structure and presented predominantly in the reactor. Cui et al. (2014) reported the shape of the mature granule was close to spherical, which was evidently different from the activated sludge flocs. Attachment of microbes to each other would form a smooth spherical shape of aerobic granules with the EPS distributed through the granules (Zhu et al., 2012). The mature granular structure had a more compact and dense structure and excellent settling ability with maximum diameter 2.87 mm and settling velocity of 38.98 mh^{-1}.

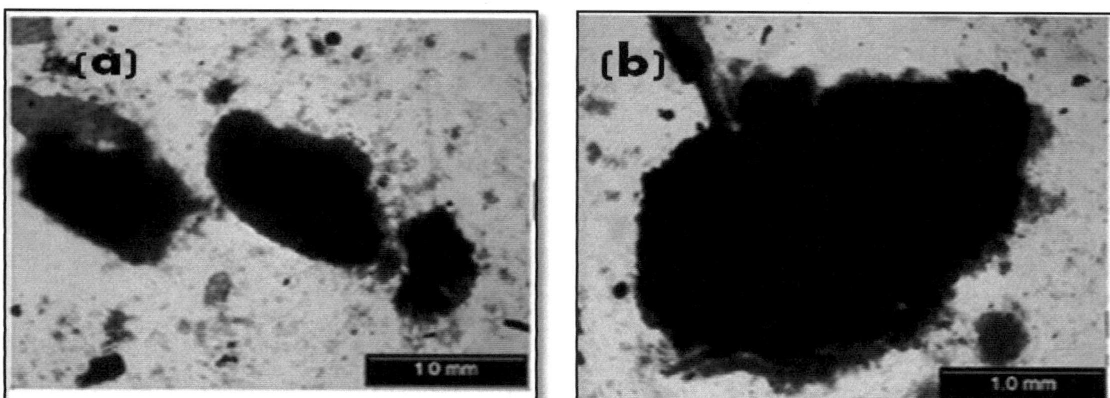

Figure 1. Microscopic observation of granular sludge a) Day 15 (Average ϕ = 1.41 mm, υAGS15 = 21.63 mh^{-1}); b) Day 30 (Average ϕ = 2.87 mm, υAGS15 = 38.98 mh^{-1}).

3.2. Biomass Profile and Settling Characteristics

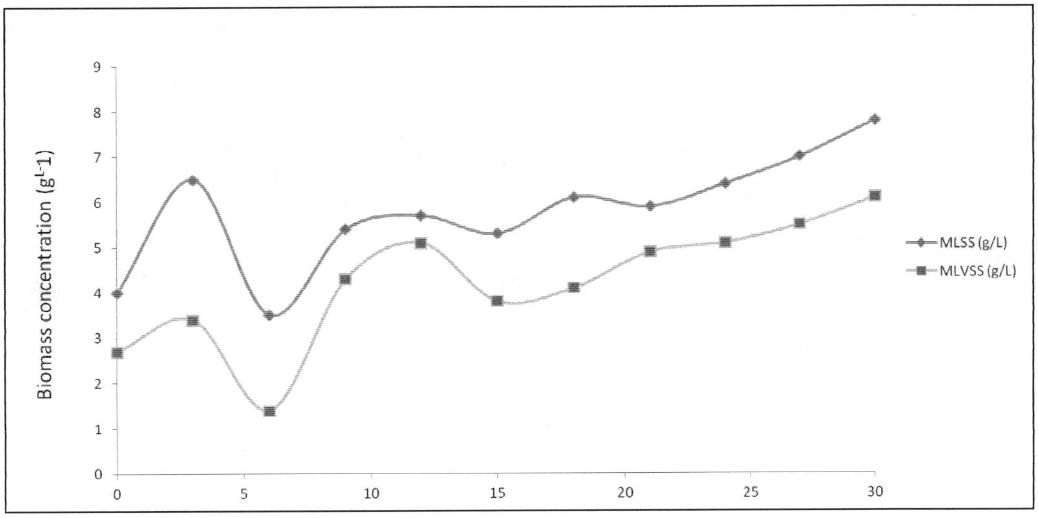

Figure 2. Profiles of biomass concentration for 30 days experiment (◆) MLSS; (■) MLVSS.

Figure 2 illustrates the profile of biomass concentration throughout 30 days of the experimental period. The initial values of MLSS and MLVSS concentration of seed sludge were 4.0 gL^{-1} and 2.7 gL^{-1} respectively. In the first few days, most of the sludge was washed-out from the reactor causing a rapid decrease in the biomass concentration and increased the effluent solids concentration. From day 3 to 9, value of MLSS reduced drastically from 6.5 gL^{-1} to 3.5 gL^{-1} presumably because of the sludge washout and its poor settling properties. The MLSS then increased to 5.7 gL^{-1} on day 12 indicating the high growth of biomass in the reactor. On day 15, value of MLSS reduced again due to washout of the sludge presumably due to formation of granular sludge where small granules (1.0-1.5 mm) were observed in the reactor. In this study, the MLVSS profile showed a similar trend to MLSS. The value of MLSS and MLVSS increased steadily after 15 days of operation and achieved steady state at about 7.8 gL^{-1} and 6.1 gL^{-1}, respectively during days 30, where mature granules with maximum diameter 2.87 mm were observed in the bioreactor. Abdullah et al. (2011) reported that the formation of stable biomass concentration indicates good accumulation of biomass in the reactor. The settling velocity of granules formed increased from 21.63 mh^{-1} to 38.98 mh^{-1}, confirming the fast settling ability of granular sludge compared to conventional bioflocs.

3.3. Removal Efficiencies

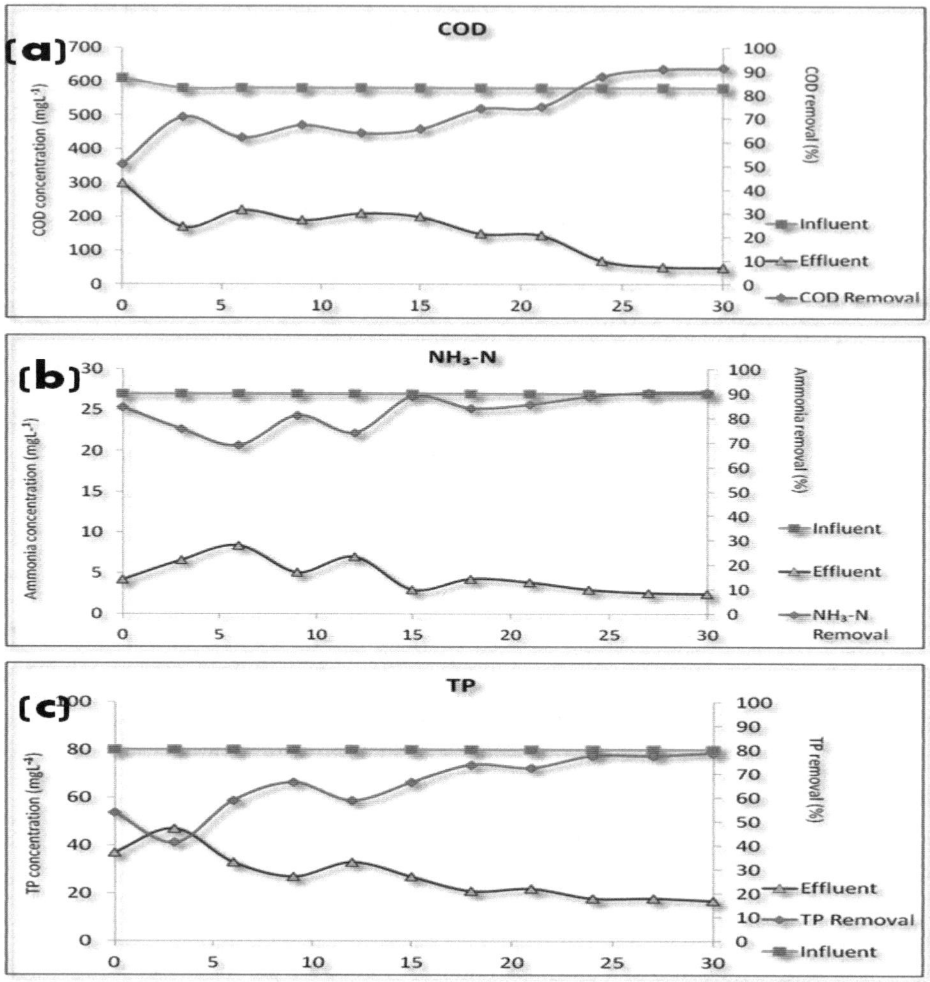

Figure 3. Profile of removal performances in the SBR system within 30 days for (a) COD, (b) Ammonia and (c) TP. (■) Influent Concentration; (▲) Effluent Concentration; (◆) Percentage Removal.

The performance of the reactor with respect to removal efficiencies of COD, NH_3-N, and TP were illustrated in Figure 3. At the beginning, the COD, NH_3-N and TP removal efficiencies fluctuated but became stable after formation of aerobic granules in the second week of the experiment. The COD concentration in the effluent was reduced from 300 mgL^{-1} to 50 mgL^{-1} at the end of experiment, indicating high efficiency of COD removal with percentage up to 91% which is comparable to previous result obtained by Song et al. (2005) in treating domestic wastewater. Rosman et al. (2013) reported that high percentage of COD removal indicates the high biological activity occurred during microbial aerobic degradation process in the bioreactor. The percentage removal of TP was low and fluctuated at the beginning of experiment (53.8% - 66.3%) and then increased as granules were formed and reached removal efficiency up to 78%. The effluent concentration of ammonia was lower than 10 mgL^{-1} and reduced to 2.5 mgL^{-1} with removal efficiency reaching 90%. In most cases, low effluent concentration of COD, Ammonia and TP in wastewater indicates high quality of effluent.

4. Conclusions

The stable aerobic granular sludge fed with synthetic wastewater at 30°C was successfully developed in the SBR system with domestic sludge after 30 days of operation. Mature granules with size ranging from 2.30 to 2.87 mm were observed appearing in the reactor at the end of experiment. Following granulation, the aerobic granules formed a good accumulation of biomass in the reactor with MLSS of 7.8 gL^{-1} and settling velocity of 38.98 mh^{-1}. When tested for COD, ammonia and TP removal, the granules showed 91.38%, 90.74%, 78.75% maximum removal at an OLR of 1.6 kg COD $m^{-3}d^{-1}$ and cycle time of 3 hour operation. The results indicate the feasibility of one single SBR system for the treatment of domestic wastewater.

Acknowledgements

This research project was funded by the National Plan for Science, Technology and Innovation (MAARIFAH) - King Abdel Aziz City for Science and Technology - the Kingdom of Saudi Arabia, award number (10WAT104705). Partial funding for the project was provided by the Ministry of Higher Education Malaysia and Universiti Teknologi Malaysia (UTM) under the professional development research university grant scheme (Grant No.: Q.J130000.21A2.05E22). The authors are very grateful to the authorities at the Madinah City Wastewater Treatment Plant for facilitating samples collection.

References

[1] Abdullah N, Ujang Z and Yahya A 2001 Aerobic Granular Sludge Formation for High Strength Agro-Based Wastewater Treatment *Bioresour. Technol.* **102** 6778-81

[2] Adav S S, Lee D J, Show K Y and Tay J H 2008 Aerobic Granular Sludge: Recent Advances. *Biotechnol. Adv.* **26** 411-23

[3] Beun J J, Hendriks A, vanLoosdrecht M C M, Morgenroth E, Wilderer P A, Heinen J J 1999 Aerobic Granulation in a Sequencing Batch Reactor *Water Res.* **33**(10) 2283-90

[4] Beun J J, van Loosdrecht M C M and Heijnen J J 2002 Aerobic Granulation in a Sequencing Batch Airlift Reactor *Water Res.* **36** (3) 702 -12.

[5] Bassin J P, Kleerebezem R, Dezotti M, van Loosdrecht M C M 2012 Simultaneous Nitrogen and Phosphate Removal in Aerobic Granular Sludge Reactors Operated at Different Temperatures *Water Res.* **46** 3805-16

[6] Cui F, Park S, Kim M 2014 Characteristics of Aerobic Granulation at Mesophilic Temperatures in Wastewater Treatment *Bioresour. Technol.* **151** 78–84

[7] de Kreuk M K, Heijnen J J and van Loosdrecht M C M 2005a Simultaneous COD Nitrogen and Phosphate Removal by Aerobic Granular Sludge *Biotechnol and Bioeng* **9** 761-9.

[8] de Kreuk M K, Pronk M and van Loosdrecht M C M 2005b Formation of Aerobic Granules and Conversion Processes in an Aerobic Granular Sludge Reactor at Moderate and Low Temperatures *Water Res.* **39** 4476-84.

[9] Ebrahimi S, Gabus S, R-Brandt E, Hosseini M, Rossi P, Maillard J and Holliger C 2010 Performance and Microbial Community Composition Dynamics of Aerobic Granular Sludge from Sequencing Batch Bubble Column Reactors operated at 20 °C, 30 °C, and 35 °C *Appl. Microbiol. Biotechnol.* **87** 1555–68

[10] Liu Y and Tay J H 2002 The Essential Role of Hydrodynamic Shear Force in the Formation of Biofilm and Granular Sludge - Review *Water Res.* **36** (7) 1653-65

[11] Liu QS, Tay J H, Liu Y 2003 Substrate Concentration-Independent Aerobic Granulation in Sequential Aerobic Sludge Blanket Reactor *Environ. Technol.* **24** 1235–42.

[12] Liu Y Q, Liu Y, Tay J H 2004 The Effects of Extracellular Polymeric Substances on The Formation and Stability of Biogranules *Appl. Microbiol. Biotechnol.* **65** 143–8.

[13] Liu Y Q, Yan-Pui M B, Tay J H 2007 COD Removal and Nitrification of Low- Strength Domestic Wastewater in Aerobic Granular Sludge Sequencing Batch Reactors *Enzyme Microb. Technol.* **42** 2007 23–8

[14] Muda K, Aris A, Salim M R, Ibrahim Z, Yahya A, van Loosdrecht M C M, Ahmad A, Nawahwi M Z 2010 Development of Granular Sludge for Textile Wastewater Treatment *Water Res.* **44** 4341-50

[15] Nor-Anuar A 2008 Development of Aerobic Granular Sludge Technology for Domestic Wastewater Treatment in Hot Climates. Doctor of Philosophy, Universiti Teknologi Malaysia, Skudai

[16] Song Z, Ren N, Zhang K and Tong L 2009 Influence of Temperature on the Characteristics of Aerobic Granulation in Sequencing Batch Airlift Reactors *J Environ Sci.* **21** 273–8

[17] Tay J H, Liu Q S and Liu Y 2001 Microscopic Observation of Aerobic Granulation in Sequential Aerobic Sludge Blanket Reactor *Appl. Microbiol.* **91** 168-75

[18] Tay J H, Pan S, He Y, Tay S T L 2004 Effect of Organic Loading Rate on Aerobic Granulation. II: Characteristics of Aerobic Granules J. *Environ. Eng.* **130** (10) 1102–9

[19] Vishniac W and Santer M 1975 The Thiobacili *Bacteriol. Rev.* **21** 195-213

[20] Yang S F, Tay J H, Liu Y 2004 Respirometric Activities of Heterotrophic and Nitrifying Populations in Aerobic Granules Developed at Different Substrate N/COD Ratios. *Curr. Microbiol*. **49** (1) 42–96

[21] Yang S F, Tay J H, Liu Y 2005 Effect of Substrate Nitrogen/Chemical Oxygen Demand Ratio on The Formation of Aerobic Granules *J. Environ. Eng.* **131** (1) 86–92

[22] Zitomer D H, Duran M, Albert R and Guven E 2007 Thermophilic Aerobic Granular Biomass for Enhanced Settleability *Water Res.* **41** 819-25

[23] Zhu L, Lv M, Dai X, Yu Y, Qi H, Xu X 2012 Role and Significance of Extracellular Polymeric Substances on the Property of Aerobic Granule *Bioresour Technol.* **107** 46–54

To Optimize Sequencing Batch Reactor (SBR) and Monitor Aerobic Granular Sludge (AGS) Development for Domestic Wastewater Treatment

Norazah Mohd Kassim[1*], Aznah Nor Anuar [1*], Hakim Halim[1*], Noor Atiqah Sharani[2], Alijah Md Aris[2], Khor Bee Chin[2]

[1] Department of Chemical and Environmental Engineering, Malaysia-Japan International Institute of Technology, Universiti Teknologi Malaysia, Kuala Lumpur, Malaysia
[2] Indah Water Research Centre, Kuala Lumpur, Malaysia

*Corresponding Author: norazahmohdkassim@gmail.com, aznah@utm.my, mohdhakim@utm.my

Abstract. The aim of this study is to determine design and process parameters of the aerobic granular sludge system to treat actual domestic wastewater. The aerobic granular sludge was developed in a 2.5 L lab-scale bioreactor which uses the operational system based on the sequencing batch reactor (SBR). The bioreactor was fed with influent from local sewerage treatment plant and seeding with fresh inoculums (seed sludge) from activated sludge tank. The operation of the reactor was based on the SBR system with a complete cycle time of 4 hours and specifically designed to be operated for 24 hours continuously. Two sets of experiments were carried out in a single-walled cylindrical glass column bioreactor. These two sets of experiments are referred to as SBR_A and SBR_B with different time setting. SBR systems have four stages which is feeding, aeration, settling and decant. It was carried out in sequence: (1) 'feeding with aeration' for SBR_A (feed and aerate pumps operated simultaneously); 'feeding without aeration' for SBR_B (feed and aerate pumps operated separately), (2) 'aeration only', (3) 'settling', and (4) 'discharge'. Removal performance of the SBR such as BOD, COD, AMN, $NO3^-$, TSS, VSS and TP along with the other several parameters of the SBR such as DO, SOUR, ORP and settling profile are monitored daily. Starting from day 1 until day 26 the average DO values were around 1 - 2 mg/l but after 1.5 hours the DO values increased drastically. It shows that the microbe did not consumed air during the process. Day 13 onwards SOUR profiles show that the microbe consumed oxygen slowly. These results suggest that low amount of readily oxidized organic available in the bioreactor. The SVI reading shows good settling performance which is less than 100 mL/g. The settling profile shows the sludge has been further compact day by day. From the finding, it is suggested that it would be better for feed and aerate pumps to be operated separately.

1. Introduction

The dramatic increase of human population and industrialization has resulted in excess scale of wastewater to be handled before it becomes reusable and enters into ecosystems, which means new development of sewage technique are needed [1]. Aerobic granulation is described as a self-aggregation process that transform loose biomass into dense and compact granules under control conditions [2]. AGS is well known for its physical characteristics including regular in shape and compact structure [3]. As a dense microbial aggregate, AGS has great settling ability which can enhance the effectiveness of sludge separation from treated effluent. It also has high biomass levels that would ensure faster and efficient removal of pollutants [4]. The high surface area and porosity of AGS also provide excellent resistance towards toxic substances. Besides, other notable features of AGS are the ability to tolerate high organic loads and carry out the simultaneous removal of organic substances and nutrients [5]. AGS technology demonstrates a 20-25% reduction in operation costs, 23-40% less electricity requirement and a 50-75% reduction in space requirements compared to the conventional activated sludge treatment system [6].

In an existing activated sludge system, a mixed culture of suspended biomass is growing and removing organic carbon and nutrients from the influent. In such a process the biomass (the activated sludge), which are usually present as flocs, are mixed with the wastewater in a large aerated basin. Supply of fresh wastewater to the bioreactor and discharged of treated wastewater from the bioreactor occur continuously. The discharged wastewater is led to a settling tank. In the settler, separation of activated sludge from the treated wastewater is carried out by means of gravitational forces. The treated wastewater can then be discharged into surface waters, or is used for further treatment. Existing activated sludge plants produce surplus sludge. Part of the settled activated sludge is recycled to the bioreactor. The remainder of the sludge is usually treated anaerobically and later disposed in landfills, or is used as fertilizer in agriculture.

Performance of activated sludge system deteriorates due to sludge separation problem caused by sludge bulking. Bulking activated sludge which settles slowly and compacts poorly is still one of the most common operational problems plaguing the wastewater treatment plant. Moreover However, existing activated sludge plants take up a substantial footprint. In order to treat large amounts of wastewater completely, large aeration tanks are needed. The settling tanks cover a large area, because the settling velocity of the activated sludge flocs is very low, normally <1 mh^{-1}.

AGS cultivation has been preferentially carried out in sequencing batch reactors (SBRs). In this system, all phases (filling, reaction, settling, decanting, idling) take place in the same tank. Therefore, a secondary clarifier is not needed as in activated sludge system. AGS has a small footprint and a low energy demand compared to conventional municipal wastewater treatment technologies [7]. Since the available ground area to build the treatment plant is usually limited, especially in Malaysia, there is a need for a more compact and high-performance reactor. This need for more compact reactors and short hydraulic retention time (HRT) directed the study towards the development of systems with high biomass concentrations. Furthermore, it demonstrated unique features including granules without carrier material, excellent settling properties, great biomass enrichment, simple single-tank concept with simultaneously biochemical oxygen demand (BOD), chemical oxygen demand (COD), nitrate ($NO3^-$), ammoniacal nitrogen (AMN) and total phosphate (TP) removal and low costs compared with conventional activated sludge processes.

The objectives of this study are (i) to optimize sequencing batch reactor (SBR) and (ii) to monitor aerobic granular sludge (AGS) development for domestic wastewater treatment whereby this study will utilize fully hundred percent actual domestic wastewater to comply with the existing wastewater treatment operation. This could be a benchmark for local sewerage services to start-up their full-scale aerobic granulated sludge system.

2. Materials and Methods

2.1. Designed and Setting Up Bioreactor

Cyclic Aerobic Granular Sludge Bioreactor (CAgABio); a 2500 mL laboratory-scale reactor have been used to develop the granules as shown in Figure 1. The basis of this reactor set-up was developed according to the reactor configuration built in the Kluyver Laboratory for Biotechnology, Delft University of Technology, the Netherlands [8]-[10]. The schematic diagram of the reactor set-up is shown in Figure 2. The operation of the reactor was based on the normal SBR system with a complete cycle of 4 hours and specially designed to be operated for 24 hours continuously. Two sets of experiments were carried out in a single-walled cylindrical glass column bioreactor with a working volume of 2.5 L. The two sets of experiments are referred to as SBR_A and SBR_B. The process flow diagram of the CAgABio system is shown in Figure 3.

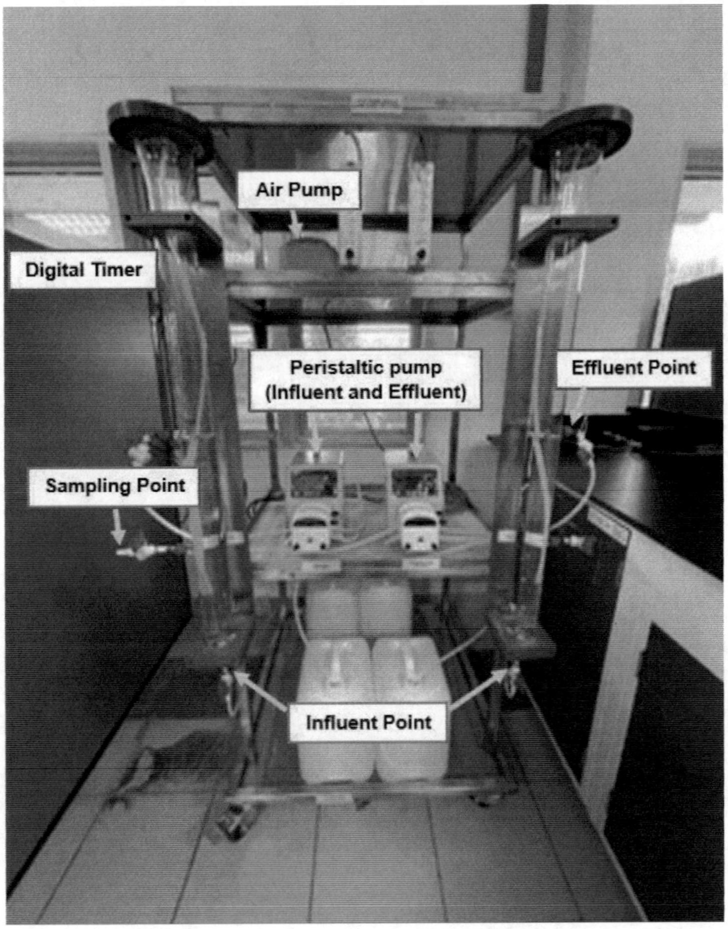

Figure 1. 2500 mL laboratory-scale SBR.

2.2. Seeding Sludge

A fresh activated sludge taken from a Sequencing Batch Reactor Treatment Plant (Bunus STP – KLR 403, Kuala Lumpur) was used as seed sludge in the reactor. The seed sludge characteristics used in this experiment is given in Table 1. The characterisation study is essential to provide physical and morphological information for the initialisation of the granulation process where at the end of the study, the final characteristics of granulated seed sludge could be figured out. The physical characteristics were determined in terms of sludge volume index (SVI), size and settling velocity (v).

2.3. Influent Feeding

The influent for the experiments with actual domestic wastewater were collected from wastewater treatment plant (BUNUS STP – KLR 403 Kuala Lumpur). For every cycle, 1000 mL of actual domestic wastewater was dosed to the reactor. The composition of the influent dosed to the laboratory-scale is given in Table 2.

2.4. Analytical Procedures

All analytical measurements performed in this study were conducted according to Standard Methods for the examination of water and wastewater [11].

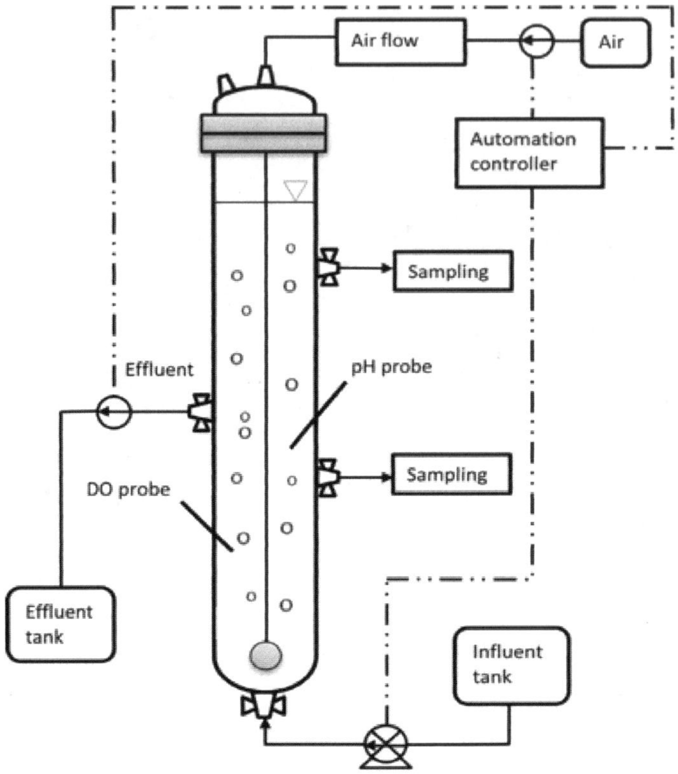

Figure 2. Schematic diagram of reactor.

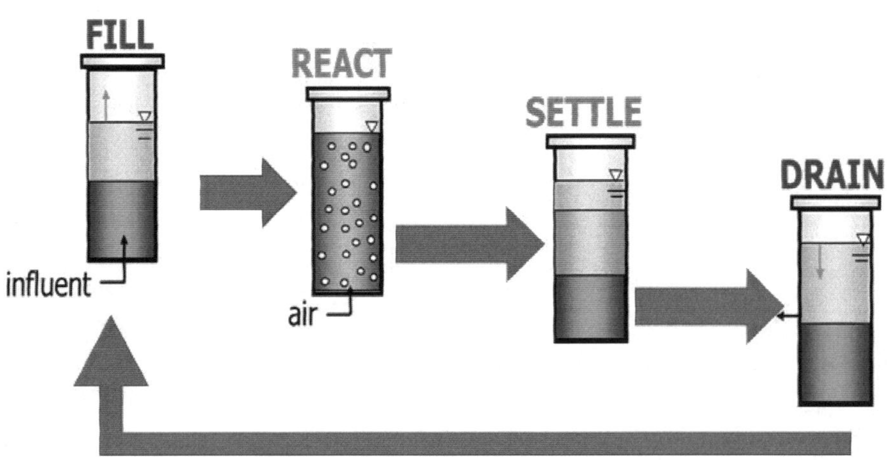

Figure 3. Process flow diagram.

Table 1. Physical characteristic of activated sludge (seed sludge).
(BUNUS STP – KLR 403 Kuala Lumpur).

No	Set Value	Descriptions
1.	pH	7.0
2.	DO (mg/L)	2.0
3.	SV_{30} (mL/g)	150
4.	Size (mm)	<0.2
5.	Settling velocity (m/h)	0.6

Table 2. Composition of influent from BUNUS STP – KLR 403 Kuala Lumpur.

No	Set Value	Descriptions
1.	pH	6.7
2.	COD (mg/L)	304
3.	BOD_5 (mg/L)	123
4.	Ammoniacal Nitrogen (AMN), NH_3-N (mg/L)	20
5.	Nitrate, $NO3^-$ (mg/L)	0.5
6.	Total Phosphate (TP), PO_4^{3-} (mg/L)	5

3. Results and Discussion

3.1. Bioreactor time setting

SBR systems have four processes and it was carried out in sequence feeding, aeration, settling and decant: (1) 'feeding with aeration' (feed and aerate pumps operated simultaneously), (2) 'aeration only and without feeding', (3) 'settling', and (4) 'discharge'. Starting from day 1 until day 26 the average Dissolved Oxygen (DO) values were around 1 - 2 mg/l but after 1.5 hours the DO values increased drastically. It shows that the microbe did not consumed air during the process. Day 13 onwards SOUR profiles show that the microbe consumed oxygen slowly. These results suggest that low amount of readily oxidized organic available in the bioreactor. Oxygen Reduction Potential (ORP) reading shows that no denitrification happened along the cycle. The lowest value of ORP monitored was 111 mV whereby the normal (ORP) value for denitrification should be ranges between -50 mV to +50 mV. Oxidation-reduction potential or ORP has been used for many years in facilities that process wastewater generated by metal finishing plants, but only recently has it become prominent in municipal wastewater treatment plants [12]. To meet the ORP corresponding values, setting time for aeration has changed by separated operating of the aeration pump during feeding (feeding without aerate).

3.2. Formation and settling velocity of Granules

Day 32 first batch of AGS was harvested using 0.3mm, 0.6mm and 1.0mm sieve. The diameter of the granules was obtained under microscope observation with an average diameter of 1.64mm (irregular shapes of granules were developed). The developed granules at 32 days are shown in Figure 4. Morphology characteristic are shown in Figure 5 (micro colonies Cocci-shaped bacteria acted as a supporting consortia during the formation of granular), Figure 6 (extracellular polymeric substances EPS acts as a sticky glue between the microbial cells and strengthen the structure of the granules) and

Figure 7 (cavities act as a passage for the transportation of substrate, oxygen, and nutrients into the inner cores of the granules and ensure the stability within the granules composition). Different methods to measure the granule density include the pycnometer method, the Percoll density gradient method, the dextran blue method, and the settling velocity method [13]. Settling velocity method was used in this study. The granule's settling velocity is approximately thirty times faster than the actual activated sludge (actual activated sludge 0.6 m/h and AGS 18.04 m/h). The granules minimum settling velocity is 30 m/h and can reach up to 70 m/h depending on the compactness and density of the granules [14]. The SVI_{30} of granules (average 60 mL/g is lower than actual activated sludge with SVI_{30} value of 150 mL/g). The SVI of AGS is lower than 100 mL/g while activated sludge flocs, 150 mL/g and above [15]. Figure 8 and Figure 9 shows SVI_{30} for 1st and 2nd aeration setting time.

3.3. Removal Efficiency of Granules

During the 32 days development parameter such as BOD, COD, Ammoniacal Nitrogen (AMN), $NO3^-$, TSS, TP, MLSS and MLVSS were monitored daily and weekly. The performance of the reactor from 1st to 2nd aeration setting time improved satisfactory. Average percentage BOD removal was about 33% to 76.7%. COD removal 42% to 89.7%. These results indicate the removal performance of BOD and COD improved at different operation setting time, due to the transformation of the flocculent sludge into granular sludge took place in the bioreactor system. These results are comparable to the previous study by Ab Halim et al. (2018) [16] who also conducted experiment by using actual domestic wastewater for the biological activity of granules. AMN removal 95.5% to 74.2% slightly decreased (it may due to the influent which has high AMN content) but still complied with Malaysia Environmental Quality Act (design effluent values discharge to river/stream for Standard A < 5 mg/L). The removal of NO3 was decreased 22.6% to -95.9%. However, the value of 6.5 mg/L of $NO3^-$ still complied with Malaysia Environmental Quality Act (design value required was 10 mg/L). TSS and TP the removal performance percentage were 40% to 91.9% and -74% to 96.3%, respectively (refer Figure 10, Figure 11 and Table 3).

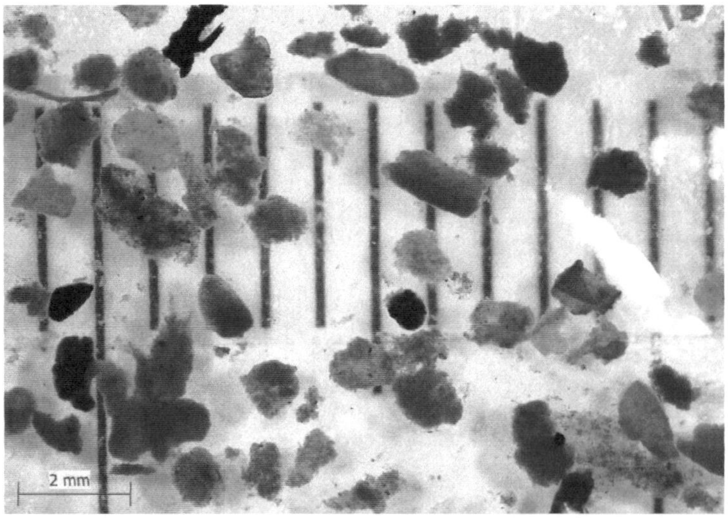

Figure 4. Microscope images of granules at 32 days.

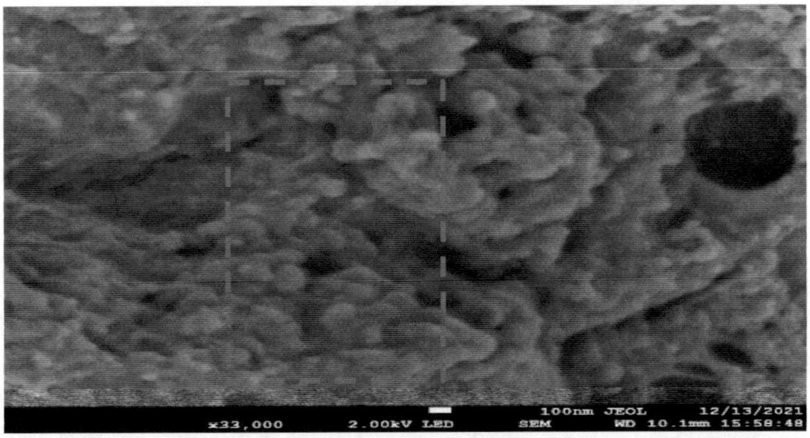

Figure 5. Photo of microbes captured within the granules (Cocci-shaped) using scanning electron microscope.

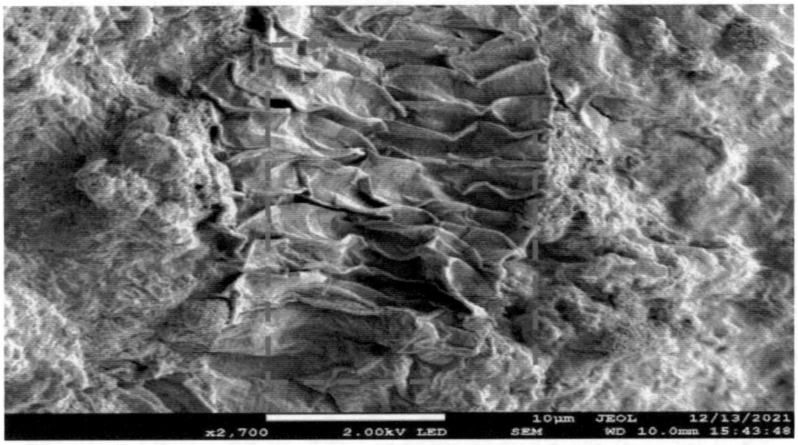

Figure 6. Photo of EPS captured within the granules using scanning electron microscope.

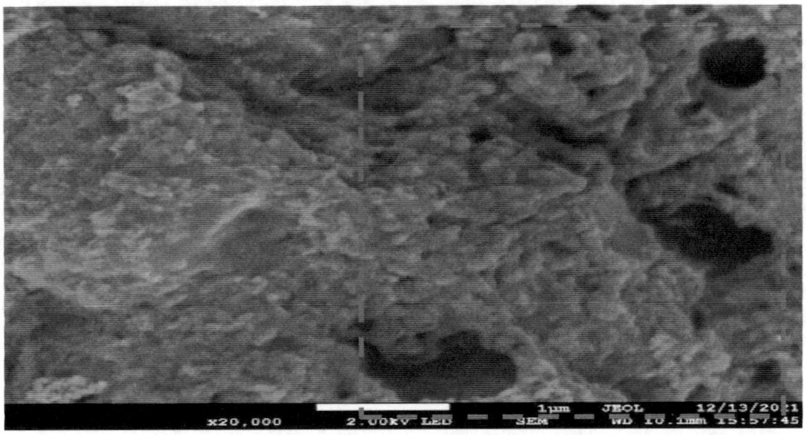

Figure 7. Photo of cavities captured within the granules using scanning electron microscope.

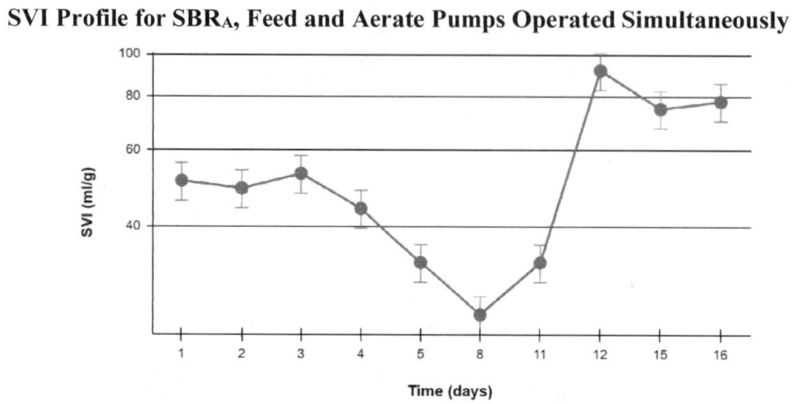

Figure 8. SVI_{30} value until day 16 with 1st aeration setting time.

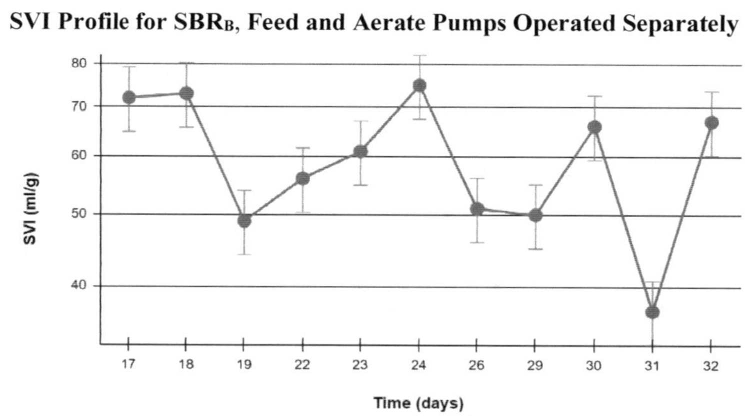

Figure 9. SVI_{30} value until day 32 with 2nd aeration setting time.

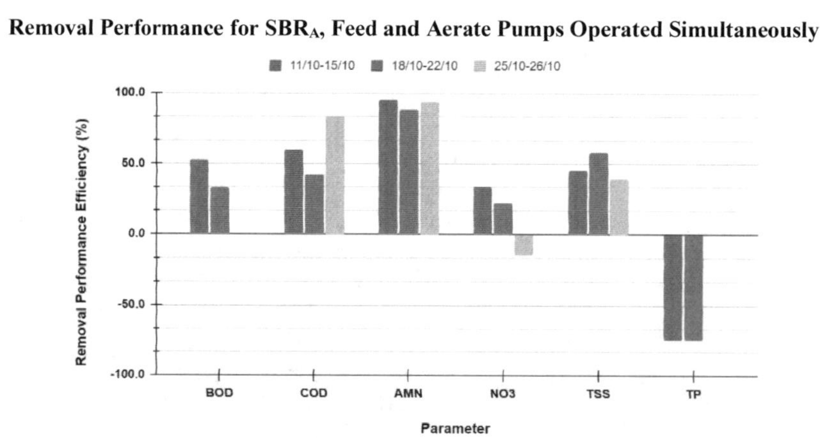

Figure 10. SBR removal performance for 1st aeration setting time.

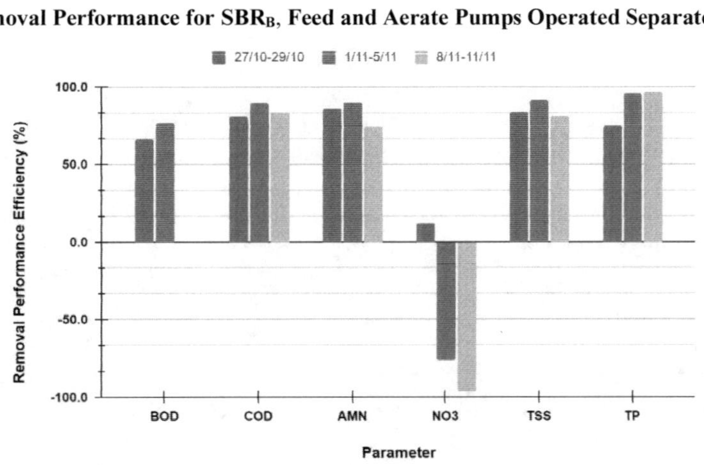

Figure 11. SBR removal performance for 2nd aeration setting time.

Table 3. Removal Performance for SBR_A and SBR_B

No	Parameters	SBR_A	SBR_B
1.	BOD	33%	76.7%
2.	COD	42%	89.7%
3.	AMN	95.5%	74.2%
4.	NO3	22.6%	-95.9%
5.	TSS	40%	91.9%
6.	TP	-74%	96.3%

4. Conclusion

The developed aerobic granular sludge (AGS) in SBR_A and SBR_B were harvested using 0.3mm, 0.6mm and 1.0mm sieve after 32 days and settling velocity for both reactors were analysed. From the finding, it is suggested that it would be better for feed and aerate pumps to be operated separately.

Acknowledgements

This research was collaborated between MJIIT Universiti Teknologi Malaysia and Indah Water Research Centre. Indah Water Konsortium (IWK) is Malaysia's national sewerage company which has been entrusted with the task of developing and maintaining sewerage system. The authors are very grateful to IWK for facilitating samples collection and lab facilities.

References

[1] Liu, W., Wu, Y., Zhang, S., Gao, Y., Jiang, Y., Horn, H., & Li, J. (2020). Successful granulation and microbial differentiation of activated sludge in anaerobic/anoxic/aerobic (A2O) reactor with two-zone sedimentation tank treating municipal sewage. Water research, 178, 115825.

[2] Yin, Y., Sun, J., Liu, F., & Wang, L. (2019). Effect of nitrogen deficiency on the stability of aerobic granular sludge. *Bioresource technology*, *275*, 307-313.

[3] Awang, N. A., Shaaban, M. G., Lee, C. W., & Bong, C. W. (2017). Characterization of aerobic granular sludge developed under variable and low organic loading rate. *Sains Malays*, *46*(12), 2497-2506.

[4] Jalali, S., Shayegan, J., & Rezasoltani, S. (2015). Rapid start-up and improvement of granulation in SBR. *Journal of Environmental Health Science and Engineering*, *13*(1), 1-11.

[5] Khan, M. Z., Mondal, P. K., & Sabir, S. (2013). Aerobic granulation for wastewater bioremediation: a review. *The Canadian Journal of Chemical Engineering*, *91*(6), 1045-1058.

[6] Basri, H. F., Anuar, A. N., Yuzir, A., & Ab Halim, M. H. (2020, June). Diatomite carrier for rapid formation of Aerobic Granular Sludge. In *IOP Conference Series: Earth and Environmental Science* (Vol. 479, No. 1, p. 012028).

[7] Bengtsson, S., de Blois, M., Wilén, B. M., & Gustavsson, D. (2019). A comparison of aerobic granular sludge with conventional and compact biological treatment technologies. *Environmental technology*, *40*(21), 2769-2778.

[8] Beun, J. J., Hendriks, A., Van Loosdrecht, M. C. M., Morgenroth, E., Wilderer, P. A., & Heijnen, J. J. (1999). Aerobic granulation in a sequencing batch reactor. *Water research*, *33*(10), 2283-2290.

[9] De Kreuk, M. K. (2006). Aerobic granular sludge: scaling up a new technology.

[10] Nor Anuar, A., Ujang, Z., van Loosdrecht, M. C. M., & De Kreuk, M. (2006). Aerobic granular sludge technology for wastewater treatment—an overview. In *Paper on Asiawater Conference*.

[11] APHA/AWWA/WEF (2012). *Standard Methods for the Examination of Water and Wastewater Standard Methods 541*

[12] Environmental, Y. S. I. (2008). ORP Management in wastewater as an indicator of process efficiency. *YSI, Yellow Springs, OH http://www. ysi. com/media/pdfs/A567-ORP-Management-in-Wastewater-as-an-Indicator-of-Process-Efficiency. pdf (accessed on 15.08. 13).*

[13] Van den Berg, L., Pronk, M., van Loosdrecht, M. C., & de Kreuk, M. K. (2021). Density measurements of aerobic granular sludge. *Environmental Technology*, (just-accepted), 1-27.

[14] Rosman, N. H. B. (2017). *Biogranular Sludge for Rubber Processing Wastewater In A Sequencing Batch Reactor* (Doctoral Dissertation, Universiti Teknologi Malaysia).

[15] Rosman, N. H. B. (2017). *Biogranular Sludge for Rubber Processing Wastewater In A Sequencing Batch Reactor* (Doctoral Dissertation, Universiti Teknologi Malaysia).

[16] Ab Halim, M.H. (2018). *Development of Aerobic Granules in Sequencing Batch Reactor System for Treating High Temperature Domestic Wastewater*. Doctor of Philosophy, Universiti Teknologi Malaysia, Skudai.

Current state and the application of energy conversion of sewage sludge treatment in Malaysia

L M Quan[1*], I Warmadewanthi[2] and Y Helmut[3]

[1] Malaysia Japan International Institute of Technology, Universiti Teknologi Malaysia, Jalan Sultan Yahya Petra, 54100 Kuala Lumpur, Malaysia.
[2] Department of Environmental Engineering, Faculty of civil environment and Geo-Engineering Institute Teknologi Sepuluh Nopember, Surabaya 60111, Indonesia.
[3] University of Tsukuba, 1-1-1 Tennodai, Tsukuba 305-8572, Ibaraki, Japan.
*Corresponding author: quanmle1505@gmail.com

Abstract. Following the rapid growth of Malaysia's population, the sewage sludge disposal capacity makes a big movement up to 7 million m^3 annually. It proves that the conventional method, landfilling, causes negative issues related to the secondary pollution as well as high electricity consumption of sewage sludge treatment unit at 544,900 GWh in 2016-2019. The energy consumption also plays important role in the high fossil fuel-dependent energy production in Malaysia. The integration of energy conversion with sludge treatment is the potential method solving problems and global warming prevention. Thanks to the relative high heating value of sludge, the integration of energy conversion with sewage sludge treatment was suitable to reach two purpose including total waste destruction and energy generation through heat recovery, that results in the operating cost reduction and decrease of carbon footprint. The paper provided the overview of sewage sludge disposal and treatment status in Malaysia, it came with the potential of sludge treatment such as anaerobic digestion, combustion, gasification into useful energy was presented.

1. Introduction

The population of Malaysia increased from 31.19 million people to 32.66 million people within 5 years (2016-2021) that equals to the population growth rate 1.4-0.2%/year [1] . The more population growth, the more waste disposal capacity such as solid waste and wastewater disposed from the daily activities. Sewage sludge is a by-product discharged from the wastewater treatment plant (WWTP) that contained water, non- toxic organic matters, nutrients, heavy metal, toxic organic matters, particularly, the high amount of pathogen and microorganism [2]. Landfill is the conventional method for municipal solid waste in general and sewage sludge in particular. There are 171 active landfills with 12 sanitary landfills in operation, which handles 80% of the municipal solid waste capacity in Malaysia [3]. However, the drawback of the method is more obviously and more harmfully when the growth of disposal capacity is too rapid. It causes the secondary pollution related to the incompletely decomposition of sewage sludge, the leaching of organic and heavy metal into groundwater and soil, inefficient land use which should be utilized for more necessary purpose such as agriculture and living. As a result, various countries have been implemented strictly regulation to manage the sewage sludge disposal issue so that a replacement method becomes urgent demand.

One of the barriers of sewage sludge treatment was the enormous energy consumption of process, which is mainly utilized for the moisture removal. The sewage sludge treatment process of IWK utilized around 540-550 million kWh electricity in the period 2016-2019, which contributed mostly to the mechanical treatment system (figure 2). Following the growth of electricity generation capacity was the rise of CO_2 emission capacity from 380,542 metric tons to 381,256 metric tons [4]. In 2018, electricity generation facilities of Malaysia produced 160 gigawatthours (GWh) which depended on fossil fuel such as coal (21% of the total fuel), natural gas (36% of the total fuel) or petroleum and other liquids (37% of the total fuel). The utilization of renewable energy just accounted for 6% of fuel resources, but it is intended to be increased to 25% by 2025 to reduce CO_2 emission and diversify fuel resources. The calorific value of sewage sludge, 11-22 kJ/kg, was likely most of the popular biofuels utilized for the energy generation purpose such as electricity and heating value production [5]. The combination between the sewage sludge treatment and energy recovery from the treatment process becomes the potential and efficient solution. The integration brings not only the sustainable waste treatment but also plays a role in the operating cost reduction and CO_2 emission control. Moreover, it complies with the sustainable development goals (SDGs) that ensures sustainable consumption and production patterns as well as takes urgent action to combat climate change and its impacts.

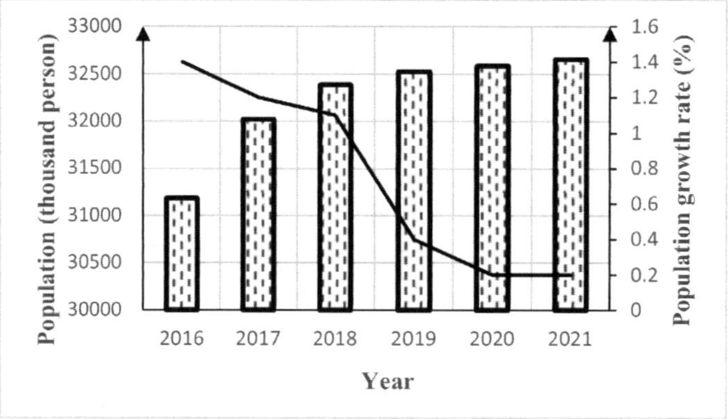

Figure 1. The population and population growth rate of Malaysia from 2016-2021.

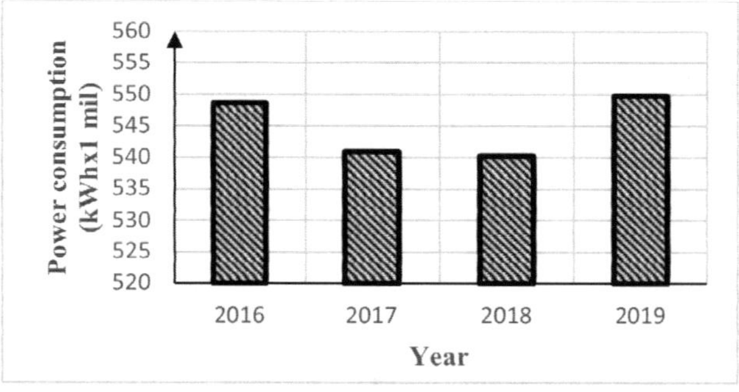

Figure 2. Power consumption of sewage sludge treatment plant of IWK from 2016-2019.

The main method for sewage sludge treatment is classified into biological treatment and thermal treatment. In details, the anaerobic digestion (AD), combustion and gasification were considered as the potential solution, the coincidence of these method is the conversion the energy in sewage sludge into the energy in the gas product, a main product of process. It is convenient to use gas stream for electricity and heating utility production without further complicated treatment. The paper aimed to present the status of using waste to energy (WtE) integrated anaerobic digestion, combustion, gasification for sewage sludge treatment Malaysia. It was said that the WtE for sewage sludge have not been popular in Malaysia so the feasible solution under multi-view including process configurations of integration process, operating condition, challenge of energy conversion method was necessary to be put under consideration in paper before utilizing for a specified case in Malaysia. In addition, the modification improves the energy efficiency of the whole process accompanying with economic view was provided.

2. Current state of sewage sludge treatment

The sewage sludge properties including proximate analysis and ultimate analysis were important factors to select the suitable application for the integration between WTE and sewage sludge treatment. The majority of the energy conversion from waste into energy in industrial scale were combustion and anaerobic digestion process to produce electricity.

2.1 Sewage sludge properties

The proximate analysis and ultimate analysis play important role in contribution of the gas product composition resulting in the variant of low heating value (LHV). Zakaria et al [6] and Ibrahim et al [7] characterized the treated sludge, which was discharged from sludge treatment in Malaysia (Sample No. 1). In addition, Abbas et al analysed the non-treated sludge from mechanical wastewater treatment plant in Malaysia (Sample No.2). It is evident that the high moisture content of wet sludge negatively affects the thermal treatment efficiency due to heat loss for moisture evaporation. To decrease the moisture content of sludge, the disc dryer and oven drying was utilized as pre-treatment method for sample No.1 and No.2, respectively. The result of analysis was described as the Table 1. The figure provided by Syed-Hassan et al [5] showed that the portion of VM of sludge in Malaysia were lower than other type of biomass (average 79.56%). However, the FC content of sludge was similar with the figure of wood pellet (15.4 ~ 18.4%), which was also a popular biomass. However, the ash content of sludge was higher than most of biomass, which took account in the decrease of LHV of sludge as well as high amount of ash disposal. Regarding of composition, the carbon, hydrogen content of sludge was lower than figures of biomass, which was around 48.43% and 6.1%, respectively. On the other hand, the high amount of nitrogen and sulphur resulted in the high contaminant such as NO_x, and SO_2 as well as organic gaseous compound in the flue gas. One of the most important values of thermal treatment for energy production is heating value. This value of sludge in Malaysia was a half as much as its in coal. It was a bit lower than other biomass, but the gap was åound 1-2 MJ/kg attributed by the high ash content.

Table 1. Proximate analysis of sewage sludge disposed from WWTP in Malaysia.

Proximate analysis						
Sample No.	Moisture (M)	Volatile matter (VM)	Fixed carbon (FC)	Ash (A)		
1	5.034	52.483	17.613	24.870		
2	12.00	48.90	19.00	32.00		
Ultimate analysis						
Sample No.	Carbon (C)	Hydrogen (H)	Oxygen (O)	Nitrogen (N)	Sulfur (S)	HHV ((MJ/kg)
1	32.43	3.21	59.01	4.62	0.73	16.04
2	33.00	5.00	6.00	54.8	1.20	15.70

2.2 Status of sludge disposal and treatment in Malaysia

IWK, the national sewerage company, there have been 65.1% of local authority's area in Malaysia put under the management of this company until 2021. In the period 2016-2020, the number of sewage sludge treatment plant being under operation have been 7003. The sewage sludge put under in charge of IWK's management was reported at 23.92 million connected population equivalents (cPE) in 2016, then the figure rose by 2.91 million cPE in 2020. Following design flow rate at 225 l/capita/day (lpcd), the sewage sludge disposal capacity equalled to 5.4 metric tons annual to 6.0 metric tons annual within 4 years [4]. According to the environment quality report of Malaysia, the stabilized sludge disposed in 2019 (257.57 metric tons/year) [8] was twice as much as the figure in 2018 (125.95 metric ton/year)[9]. The majority of sewage treatment unit was the mechanized plants that contributed 50% of total waste disposal capacity while remaining waste was treated by septic tanks and oxidation ponds [10]. The AD produced biogas with CH4 content integrated with the internal combustion engine for electricity production was successfully by the IWK. 6 sludge treatment plants managed by IWK are able to generate 20 MWh/day [4]. Take a look in the using of WTE for municipal solid waste, which sewage sludge have the similar properties, there were 1 integrated WtE with incineration plant in Langkawi (100 ton/day) and 5 mini-incinerators (10-60 ton/day) in the operation [3]. Additionally, the WtE plant converting 600 ton/day have been under commercial operation early in Negri Sembilan, that showed the effort of Malaysia in sustainable waste management.

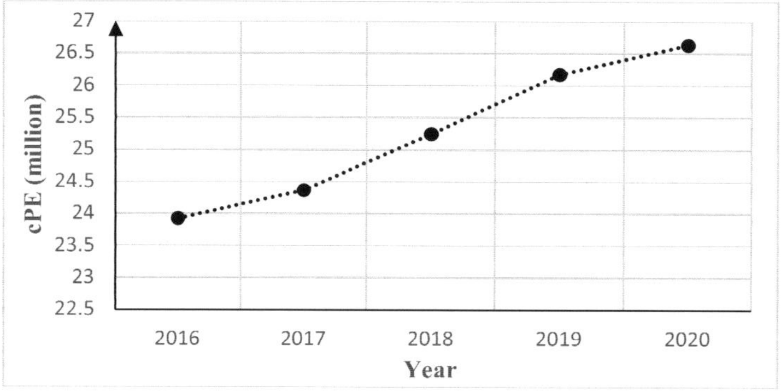

Figure 3. Sewage sludge disposal capacity to IWK's STP from 2016-2019.

3. Status and potential solution for sewage sludge treatment

3.1 Anaerobic digestion

The utilization of bacteria decomposing the sewage sludge through complex stages comprises of hydrolysis, acidogenesis, acetogenesis and methanogenesis in inert tank equipped the heating source. The CH_4 composition contributed to 60 mol% of total biogas volume and 40% of remaining gas composition such as CO_2, trace of CO, H_2S that produces the product gas containing the total calorific value 28.03-38.92 MJ/Nm^3. It was reported that the amount of biogas yield was 0.17 Nm^3/kg and 0.09 Nm^3/kg produced by different type of sludge such as raw primary sludge and digested sludge, respectively [11]. The operating temperature for the sludge biodegradation using mesophilic digestion was at 30-37°C. Although the mild condition does not require significant heating utility, it lasts long the processing time to 20 days [12]. The digestion at higher temperature 52-55°C known as thermophilic digestion motivates the degradable rate of sludge. As a result, the retention time is shortened to 10-15 days but there is the energy penalty due to heating process and electricity consumption for low viscosity stirring process [13]. The limitation existing in the anaerobic digestion such as long retention time and low CH_4 content biogas requires the consideration. The long processing time is the remarkable problems of anaerobic digestion because it spent the space for storage and energy [14]. Thermal pre- treatment method at 70-90°C within 3 hours for sewage sludge improved the methane content to 65.7 vol% [15]. The growth of CH_4 content was improved to 97.55% by the installation of water scrubber, that made the calorific value of biogas to be 28.03-51.31 MJ/Nm^3. The fat content in municipal sludge exceeded than 65% produced the lower CH_4 yield than fat content below 60% [10]. Mixing other type of raw material with sewage sludge also improved the ratio C/N and showed the better performance than the mono-digestion. The co-digestion between sewage sludge and maize straw/ cow manure increased the CH_4 composition by 20 vol%; 10 vol% than the application of mono-digestion [16]. Mixing 85% of activated sewage sludge and 15% mixed fruit waste and cheese whey improved the CH_4 composition to 31%. The high contribution of food waste (FW) in Malaysia was the good mixing material with sewage sludge (SS) to enhance the CH_4 composition to 65% at the SS:FW: poultry litter ratio 2:1:1 [17].

The figure 4a and figure 4b illustrated the integration between sewage sludge anaerobic digestion, the gas turbine and combined gas turbine/steam cycle to generate the electricity and heating utility. The rich CH_4 biogas was supplied for the turbine to produce the electricity while the heat of flue gas was recovered to form the heating utilities. The raw primary sludge conversion resulted in the highest electricity generation capacity at 1.3 (MJ/kg dried), which was reduced to 0.8 (MJ/kg dried) in case digested sludge [11]. The biogas produced from anaerobic digestion of sewage sludge supplied to electricity generator saved 27% of electricity consumption of WWTP (2000 m^3/d) instead of exporting energy from the main grid network [18]. The 15,000 tonnes sewage sludge anaerobic digestion plant produced the 7.3 m^3 biogas/PE at 20 MJ/m^3 generated the amount of electricity 2.6 MW. It was able to cover 60% of WWTP's electricity consumption [19]. The plant capacity plays important role in the economics of biogas conversion into electricity generation. The integration just showed the economic benefit when the plant capacity reached 20000 PE [11]. The energy output and organic content of sludge were 2 main factors affecting the environmental impacts and economic aspect. There was not any configuration showed the positive Net Present Value (NPV) for the low organic content excepting the value of thermophilic anaerobic digestion (THSAD) reaching 1161 million CNY. In the case of high organic content as raw sewage sludge, THSAD and anaerobic digestion with thermal hydrolysis pre-treatment (THPAD) performed the high profitable scenario with 28063 million CNY and 18614 million CNY, respectively for 100 t sludge/day treatment capacity plant [20].

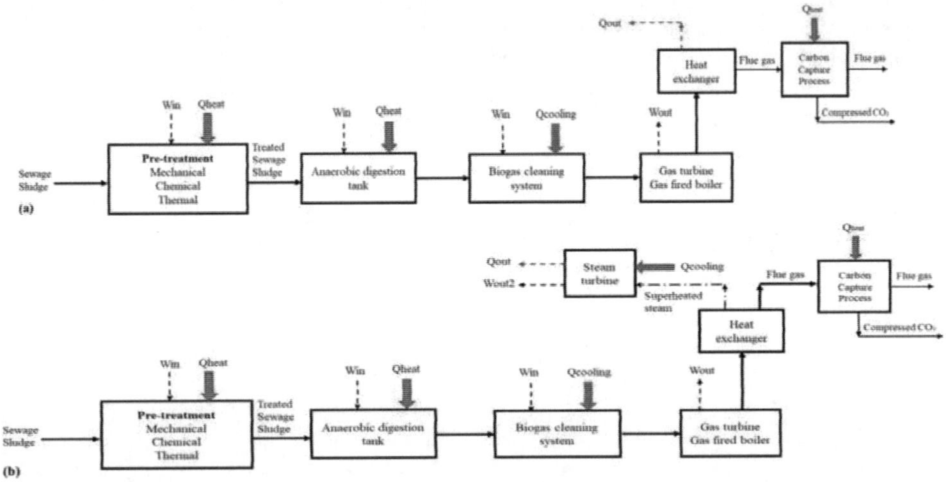

Figure 4. General configuration of waste to energy application to the anaerobic digestion for sewage sludge treatment (a) combustion engine only (b) the combined combustion engine and steam cycle W_{in}: Electricity consumption; W_{out}: Electricity generation; Q_{heat}: Heating utility consumption.

3.2 *Combustion*

In combustion processing, sewage sludge was decomposed in the excessive oxygen condition at high temperature releasing high temperature flue gas at 850°C-1300°C and inorganic solids such as CO_2, H_2O, NO_x and ash or slag [21]. Combustion process was popular in clinical and municipal solid waste because heat generation, volume of waste reduction as well as saving land for agricultural purpose. The optimized excessive air flowrate was 50% with feeding rate 0.72 kg/h reaching the highest combustion efficiency 99.6% accompanying low content of harmful gas such as SO_2, NO_x at 100 ppm, despite of relative high CO content 1300 ppm [22]. Moving bed combustor [23] and grate boiler [24] were conventional combustion technology to treat sewage sludge. As the configuration of combustor made the gas phase and solid phase to be better, fluidized bed combustor showed a high combustion efficiency 98- 99% and become the most popular equipment [22, 25]. The rich oxidation condition produces fully oxidation product gas SO_x and NO_x. Consequently, flue gas emission was specified as the remarkable disadvantages of combustion technology. The application of circulating fluidized bed reactor decreased the SO_2, NO, NO_2 to be lower than 1040 ppm, 901 ppm, 36 ppm, respectively. In addition, utilization of oxygen enriched fuel was also attractive method because it motivated the oxidation reaction leading to CO concentration to be half as much as it of the air combustion [26]. The growth of oxygen content from 28 vol% - 30 vol% showed the reduction of CO concentration from 1000-1500 ppm to 9-15 ppm, that converted to rich CO_2 flue gas with high combustion efficiency 97.1% [26]. The carbon capture process was designed at the end of process to remove CO_2 composition from the flue gas for storage, which not only reduces CO_2 emission but also produces valuable product. Combustion temperature plays important role in the completely combustion of sewage sludge, it was said that the combustion rate and the concentration of CO and organic compounds in the flue gas were lower if the temperature in combustion zone was increased from 780°C to 900°C [27]. Coal was the important fuel for electricity production; however, it caused the growth of CO_2 composition leading to global warming. The co-combustion of sewage sludge and coal indicated the remarkable performance with the combustion efficiency above 97% despite of high mixing ratio of sewage sludge to 30 wt% [28]. There are 3 mains barriers need to be removed in sewage sludge combustion. The first one is melting and agglomeration of ash at high temperature leading incompletely combustion. The second one is high requirement in the flue gas cleaning system because composition of flue gas consists of greenhouse gas CO_2; gaseous forming acid rain NO_x, SO_2 as well as toxic gas affecting human health such as dioxin and furan [5]. The final issue is high moisture of sludge resulting in heat loss for water evaporation so drying step was required to decrease moisture content to be lower than 50 wt%.

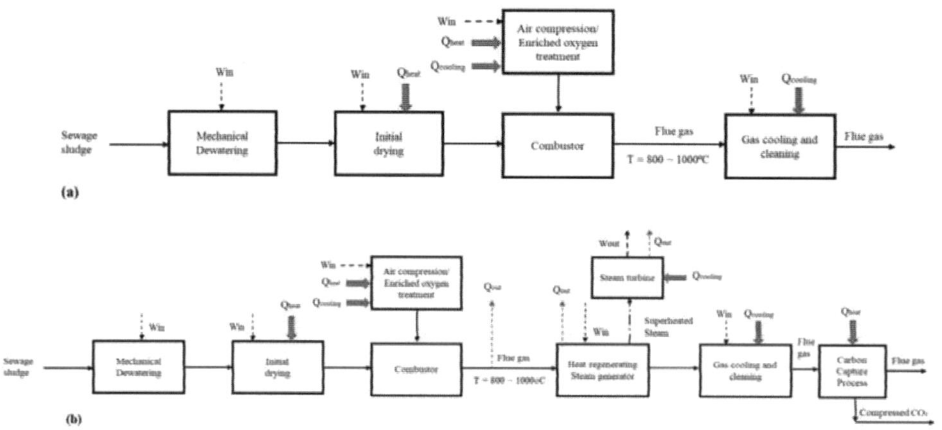

Figure 5. General configuration of waste to energy application to the combustion for sewage sludge treatment (a) original process (b) the combined combustion engine and Rankine cycle.

The figure 5a described the conventional combustion technology attached with the electricity, heating and cooling utility consumption for sewage sludge treatment. High temperature flue gas was potential heating source which was utilized to pre-heat the combustion air as well as to produce steam in boiler before generating electricity by steam turbine as figure 5b. The produced energy and heat from combined heat power unit (CHP) was utilized for the operation of unit and commercial purpose. Jiang et al [29] proposed model to simulate whole process converting sewage sludge into electricity generation in China resulting in 1300-1500 kW with generation efficiency at 14.7-17.4%. The use of dual steam turbine that operated at two stage steam pressure (13.2 MPa/2.3 MPa) improved the energy efficiency 38.61%-38.54% and exergy efficiency 37.72%-32.27% [30]. To enhance the combustion efficiency for CO_2 capture purpose before the flue gas was released out the atmosphere, the CLOF-CCS process (chemical looping oxy-firing – carbon capture process) was introduced that showed the overall efficiency, annual power yield, CO_2 removal yield was 35.7%, 1443.7 x 10^9 kWh, 1.191 x 10^9 t [31]. Horttanainen et al [32] evaluated the economic benefit of integrated sewage sludge treatment with combined heat power (WtCHP) less favourable than waste to heat (WtH) due to high investment cost of power cycle, which made the payback time longer. It was suitable that there was large electricity demand, so electricity cost is more profitable than value of steam. Additionally, WtCHP was able to produce enough power to cover the power consumption of WWTP. Mixing sewage sludge with the natural gas was considered as the solution to decrease the CO_2 emission in the study of Coskun et al [33]. It was reported that each municipal dewatered sewage sludge generated 121 Wh of electricity and released 888 g CO_2/kWhe. The sewage sludge drying consumed the large amount of energy to remove the high moisture content of sludge. It made the integrated CHP with combustion process for sewage sludge treatment to be a loss around 315.08 CNY/d [34]. To overcome the economic issue, the utilization of flue gas as drying media or co-combustion with the municipal solid waste was perfomed to increase the profit of process to 199.23 CYN/d [35].

3.3 Gasification

On the other hand, gasification processing converts the wasted solid into combustible gases CO, H_2, CH_4, small amount of C_nH_m as well as liquid tar and ash in the lack of oxygen condition [14]. Basically, the gasification processing is similar with the combustion processing excepting the amount of oxygen supplied to reactor, so reduction reactions are enhanced rather than oxidation reactions. It is well-known as heat self-consumption processing, the combustion reaction provided heat for endothermic reaction to produce combustible gas. Seggiani et al [36] and Werle et al [37] used moving bed reactor to gasify the sewage sludge while the fluidized bed reactor was used by Andrés et al, Qin et al and Schweitzer et al [38-40]. The updraft gasification using air as the gasifying agent produced the syngas with the LHV 3.5-6 MJ/Nm^3_{dry} for sewage sludge or mixture sewage sludge and biomass treatment [36, 37]. The equivalence ratio (ER) has to be controlled carefully in air gasification, which should be maintained in the range of 0.2 – 0.5. Low oxygen consumption tended to low conversion of tar and char into gas whist excessive oxygen produces high concentration of CO_2 in the syngas, which reduces the calorific value of produced gas. The higher LHV syngas reached 12.8-11.4 MJ/Nm^3_{dry} was reported when the superheated steam was used as the reactant agent due to the content of H_2 rising to above 50 mol% [41]. Additionally, increasing the temperature of steam to 1000°C produced the highest content of H_2 up to 55%vol [42] Other type of gasifying agent was also considered for sewage sludge gasification such as enriched oxygen flue. As report of [43], the LHV of syngas increased from 3.19 to 7.00 MJ/kg and efficiency of process was improved through increasing the oxygen content in reactant. The combination of air and superheated steam in gasification resulted in LHV 4.12 to 6.2 MJ/Nm^3 and H_2/CO in the range of 1.46-3.25 [44]. Tar, which were mixture of condensable gas in syngas, causes gas yield reduction, clogging and fouling in equipment and piping in downstream. It contributed 2.4-15.6 g/Nm^3 in syngas, which shall be removed by the catalyst addition, the temperature of reactor, reactor design and the cyclone/scrubber installation. Similar with the combustion process, moisture of sewage sludge should be dried first to decrease the water content to below 30% before it was fed into the reactor. Regarding of the environment protection, the reactions in the lack of oxygen condition do not form the NO_x, SO_2 and toxic gas as it does in combustion process. Figure 6a showed the gasification for sewage sludge treatment without any energy recovery, it consumed energy mainly for the gasifying agent generation. One of main distinction of heat recovery for gasification process is combustible gas was directly used in the combustion equipment as well as it was suitable to be fabricated with solid oxide fuel cell (SOFC) to produce electricity [45]. Syngas was directly burned in the gas turbine to produce electricity while the heat from syngas and flue gas were recovered for heat exchange purpose as figure 6b. To increase the energy recovery efficiency, the steam cycle was designed to increase the electricity capacity as figure 6c. The generated electricity and heating utility was used for drying and gasifying agent treatment and carbon capture process was integrated to reduce the released CO_2. Yang et al [21] proposed model that there were 0.071 kWh electricity energy to be generated on 1 m^3 of treated wastewater, the electricity capacity was able to cover 25.4-28.4% of WWTP's total energy consumption. Elsner et al [45]combined gasification system and micro capacity CHP unit producing syngas with the calorific value at 3.4 MJ/Nm^3, which could generated 35 kW electricity. Total efficiency of process reached the highest point 63.48% at 700°C attached with the H_2 production. As the syngas contain high amount of tar causing the clogging the equipment, the configuration including air gasification/external fuel gas turbine got the highest exergy efficiency 37.1%. The superheated steam (500-700°C; 240 bara) gasification was used to treat dewatered sewage sludge that integrated with the gas-fired boiler and steam cycle resulting in the energy efficiency 13.5 – 18.8% [46]. To diminish the CO_2 emission due to fossil fuel, sewage sludge was considered as mixed material with coal to be feedstock in the steam gasification. Mixture between sewage sludge (10-30 wt%) and lignite coal (0-50 wt%) was used in gasification integrated with the combined gas turbine, steam turbine and fuel cell increasing power generation efficiency to 63.48%, additionally, the H_2 gas was produced to enhance the profit of process [47]. The economic analysis described the integrated of power generation with the sewage sludge gasification as a profitable solution for sludge treatment. The positive cashflow of integrated power system for the 2.1-million-gallon raw sewage sludge per day gasification reached $3.5 million throughout 20-year life cycle, which was showed better performance than conventional method using thermal drying and landfills [48]. Electricity and heat were product for commercial purpose, that rose the economic value of the integration process for sewage sludge treatment. It shortened the payback time of project from 9 years (self-consumed for STP) to 6 years with internal return rate (IRR) at 7.52% [45]. H_2 production is the advantage of gasification despite of the growth of investment cost and longer payback time of project (12 years). The H_2 cost was high enough to made net present value (NPV). reaching 1801700 € at the end of lifetime, additionally, the profit from the H_2 production overtake the electricity generation when the H_2 cost exceeded than 4.74 €/kg H_2.

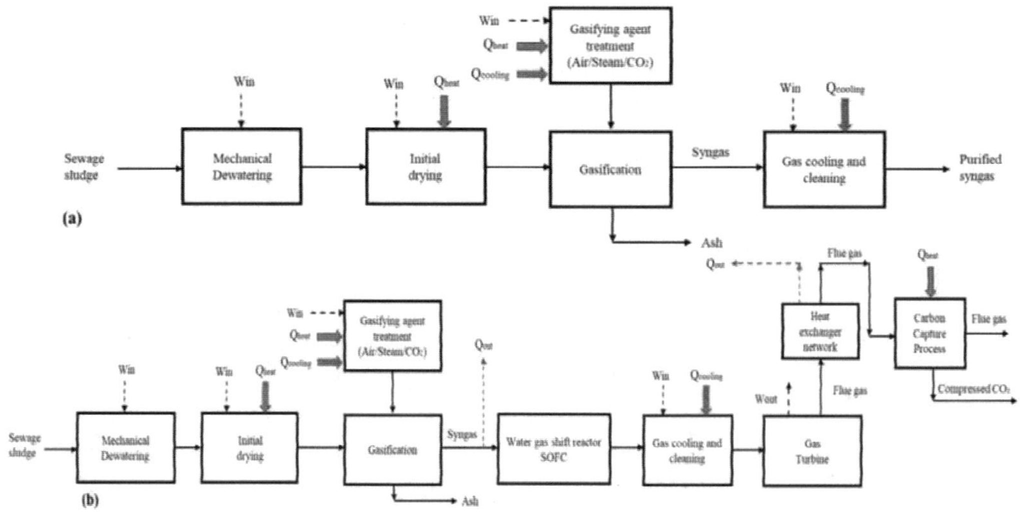

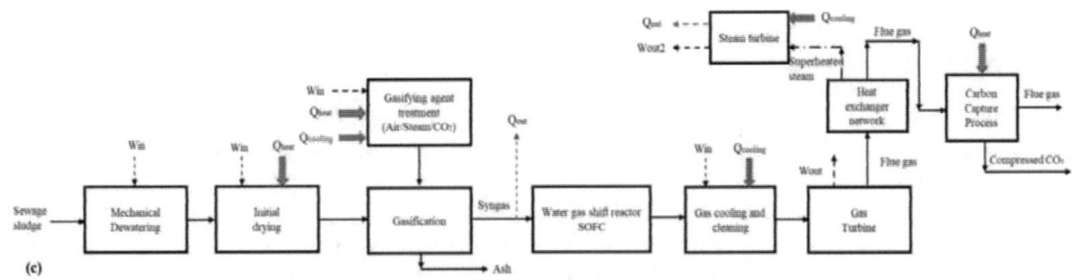

Figure 6. General configuration of waste to energy application to the gasification for sewage sludge treatment (a) conventional process (b) combustion engine only (b) the combined combustion engine and Rankine cycle.

4. Conclusion

The rapid growth of waste, particularly, sewage sludge was the main objective of the paper. The energy conversion from sewage sludge into the gas including anaerobic digestion, combustion and gasification was considered as the potential and effective solution. At the present, the integrated WTE with the anaerobic digestion have been successful applied by IWK, that produced around 20 MWh/day. To diversify the solution and to solve the long processing time issue of anaerobic digestion, it is suitable to apply thermal treatment including combustion and gasification, that relied on proximate, ultimate analysis and heating value of sludge. Various configurations have been introduced to improve the energy efficiency of the treatment process. The integration of WtE with anaerobic digestion was able to cover up to 60% of energy consumption of WWTP but it was just attractive for the high organic content sludge (> 60 wt%). Although combustion technology was rather conventional method, the combination with operating condition such as steam condition and modern process (chemical closed loop combustion, carbon capture) performed high energy efficiency to 37-38% and CO_2 removal yield at 1.191×10^9 t. However, the integration of gasification with CHP and SOFC overtook the efficiency of both mentioned method at 63.48%. Take observation on the economy aspect, the integrated gasification process performed more effective solution with the positive cashflow $3.5 million at the end of project's lifetime than combustion. Finally, the co-combustion process for sewage sludge and coal, biomass was more favourable the mono-combustion. There was the remarkable milestone in the application of energy conversion in Malaysia. However, many work shall be proceeded to get closer the sustainable development and the utilization of proposed WTE configuration with energy conversion method was powerful tools, that needs to be put under consideration seriously.

Acknowledgments

The work was sponsored by Grant no: R. K130000.7343.4B403, Collaborative Education Program for Sustainable at MJIIT-UTM. The authors gratefully acknowledge support from the Universiti Teknologi Malaysia and Post-Doctoral fellow (Teaching & Learning) Scheme under MJIIT-UTM.

Reference

[1] Malaysia DoS 2021.
[2] Hu M ,Ye Z ,Zhang H ,Chen B ,Pan Z and Wang J 2021 *Environmental Pollutants and Bioavailability* **33** 145-63.
[3] Bashir M ,Ng C ,Sethupathi S and Lim J 2019 *IOP Conference Series: Earth and Environmental Science* **268** 012044.
[4] Bhd IWKS 2020.
[5] Syed-Hassan SSA ,Wang Y ,Hu S ,Su S and Xiang J 2017 *Renewable and Sustainable Energy Reviews* **80** 888-913.

[6] Akademia Baru P ,Zakaria M ,Hassan S and Faizairi M 2015 24-9.
[7] Abbas Atya A ,Ibrahim A ,Nor F and Aris M 2011 *Proceedings of the IEEE*.
[8] Malaysia DoE 2019.
[9] Malaysia DoE 2018.
[10] Hanum F ,Yuan LC ,Kamahara H ,Aziz HA ,Atsuta Y ,Yamada T and Daimon H 2019 *Frontiers in Energy Research* **7**.
[11] Facchini F ,Mummolo G and Vitti M 2021 *Energies* **14**.
[12] Makisha N and Semenova D 2018 *MATEC Web of Conferences* **144** 04016.
[13] Nordell E ,Moestedt J ,Österman J ,Shakeri Yekta S ,Björn A ,Sun L and Schnürer A 2020 *Waste management* **119** 11-21.
[14] Oladejo J ,Shi K ,Luo X ,Yang G and Wu T 2018 *Energies* **12** 60.
[15] Ruffino B ,Campo G ,Genon G ,Lorenzi E ,Novarino D ,Scibilia G and Zanetti M 2015 *Bioresource Technology* **175** 298-308.
[16] Wei L ,Qin K ,Ding J ,Xue M ,Yang C ,Jiang J and Zhao Q 2019 *Scientific Reports* **9** 2374.
[17] Lohani SP ,Shakya S ,Gurung P ,Dhungana B ,Paudel D and Mainali B 2021 *Energy Sources, Part A: Recovery, Utilization, and Environmental Effects* 1-16.
[18] Ayoub M ,Rashed I and El-Morsy A 2016 *Civil Engineering Journal* **2** 637-45.
[19] Đurđević D ,Blecich P and Jurić Ž 2019 *Energies* **12** 1927.
[20] Li H ,Jin C ,Zhang Z ,O'Hara I and Mundree S 2017 *Energy* **126** 649-57.
[21] Yang Q ,Dussan K ,Monaghan RFD and Zhan X 2016 *Water Science and Technology* **74** 672-80.
[22] Vamvuka D ,Alexandrakis S and Galetakis M 2019 *Frontiers in Energy Research* **7**.
[23] Batistella L ,Silva V ,Suzin RC ,Virmond E ,Althoff CA ,Moreira RFPM and José HJ 2015 *Waste Management* **46** 430-9.
[24] Xiao Z ,Yuan X ,Jiang L ,Chen X ,Li H ,Zeng G ,Leng L ,Wang H and Huang H 2015 *Energy* **91** 441-50.
[25] Cammarota A ,Cammarota F ,Chirone R ,Ruoppolo G ,Solimene R and Urciuolo M 2019 *Combustion Science and Technology* **191** 1661-76.
[26] Kumar R and Singh RI 2017 *Waste Management* **70** 114-26.
[27] Moško J ,Pohořelý M ,Zach B ,Svoboda K ,Durda T ,Jeremiáš M ,Šyc M ,Václavková Š ,Skoblia S ,Beňo Z and Brynda J 2018 *Energy & Fuels* **32** 2355-65.
[28] Sever Akdağ A ,Atak O ,Atimtay AT and Sanin FD 2018 *Energy* **158** 417-26.
[29] Jiang D ,Huo X ,Dong C ,Zhang J and Yang Y. 1-5 p.
[30] Ye B ,Shi B ,Shi M ,Zhang L and Zhang R 2020 *Waste Management & Research* 0734242X20953494.
[31] Yan L ,Wang Z ,Cao Y and He B 2020 *Renewable Energy* **147** 1188-98.
[32] Horttanainen M ,Kaikko J ,Bergman R ,Pasila-Lehtinen M and Nerg J 2010 *Applied Thermal Engineering* **30** 110-8.
[33] Coskun C ,Oktay Z ,Koksal T and Birecikli B 2020 *Energy* **211** 118615.
[34] Chen L ,Liao Y and Ma X 2021 *Waste Management* **121** 11-22.
[35] Chen G ,Wang X ,Li J ,Yan B ,Wang Y ,Wu X ,Velichkova R ,Cheng Z and Ma W 2019 *Science of The Total Environment* **647** 1433-43.
[36] Seggiani M ,Vitolo S ,Puccini M and Bellini A 2012 *Fuel* **93** 486-91.
[37] Werle S 2015 *Energies* **8**.
[38] de Andrés Almeida JM ,Roche E ,Narros A and Rodríguez E 2016 *Fuel* **180** 116-26.
[39] Qin ZY ,Xie LP ,Wang YF ,Du JS and Pi JS 2017 *Ranliao Huaxue Xuebao/Journal of Fuel Chemistry and Technology* **45** 761-8.
[40] Schweitzer D ,Gredinger A ,Schmid M ,Waizmann G ,Beirow M ,Spörl R and Scheffknecht G 2017 *Biomass and Bioenergy* **111**.
[41] Zaccariello L and Mastellone ML 2020 *Chemical Engineering Transactions* **80** 175-80.
[42] Lee U ,Dong J and Chung JN 2016 *International Journal of Energy Research* **40** 1474-88.

[43] Calvo LF ,García AI and Otero M 2013 *The Scientific World Journal* **2013** 479403.
[44] Gil-Lalaguna N ,Sánchez JL ,Murillo MB ,Rodríguez E and Gea G 2014 *Chemical Engineering Journal* **248** 373-82.
[45] Elsner W ,Wysocki M ,Niegodajew P and Borecki R 2017 *Applied Energy* **202** 213-27.
[46] Ruya PM ,Purwadi R and Lim SS 2020 *Energy Conversion and Management* **206** 112458.
[47] Hantoko D ,Yan M ,Kanchanatip E ,Adnan MA ,Antoni ,Mubeen I and Hamid FS 2019 *International Journal of Hydrogen Energy* **44** 24459-70.
[48] Lumley NPG ,Ramey DF ,Prieto AL ,Braun RJ ,Cath TY and Porter JM 2014 *Bioresource Technology* **161** 385-94.

IOP Conf. Series: Earth and Environmental Science 1091 (2022) 012061 doi:10.1088/1755-1315/1091/1/012061

Anammox bacterium 'Candidatus *Kuenenia stuttgartiensis*': a review

Thilagavathi Arumugham*[1], Shaza Eva Mohamad[2]

[1] Malaysia-Japan International Institute of Technology (MJIIT), Universiti Teknologi Malaysia, 54100 Kuala Lumpur, Malaysia.
[2] Department of Chemical and Environmental Engineering, Malaysia-Japan International Institute of Technology (MJIIT), Universiti Teknologi Malaysia, 54100 Kuala Lumpur, Malaysia.

*Corresponding author e-mail: thilagavathi@graduate.utm.my

Abstract. Anaerobic ammonium oxidation (Anammox) process is a type of biological nitrogen removal technology which is known to directly convert ammonium and nitrite to nitrogen gas. The freshwater 'Candidatus *Kuenenia stuttgartiensis*' anammox under the phylum of Planctomycetes is used to study the parameters that affect the anammox development and the metabolic pathways alongside the associated enzymes. These observations were made using state-of-the art techniques for detecting anammox bacteria based on their small-subunit ribosomal RNA genes, functional genes and unique reaction pathways. This review systematically summarizes up-to-date studies on the parameters affecting the growth of the anammox bacteria and metabolic networks driving anammox bacterial anabolism and mixotrophy beyond genome-based predictions. The *K. stuttgartiensis* survives in summer and winter conditions besides in the aerobic zones (dissolved oxygen >2 mg/L), which consequently contribute to better nitrogen removal in the wastewater treatment. Furthermore, the *K. stuttgartiensis* utilizes the Wood–Ljungdahl pathway to directly assimilate extracellular formation instead of oxidising it completely to CO_2 prior reassimilation.

1. Introduction

Anaerobic ammonium oxidation (anammox) is being implemented in wastewater treatment system for nitrogen removal as excessive nitrogen compound in the wastewater may induce eutrophication, which consequently promotes the growth of green algae and cyanobacteria [1-2]. Anammox process was introduced to oxidize ammonium into dinitrogen gas by consuming nitrite as electron acceptor under anaerobic conditions [3]. Anammox is recognized as the energy-saving technology compared to conventional nitrification–denitrification due to short-cut pathway of nitrogen removal in wastewater treatment system [4]. The key advantages of anammox over conventional process are reduction of energy consumption for oxidation of soluble organics and ammonia via aeration, terminates the carbon source addition for nitrite reduction heterotrophically and deduces the volume of sludge produced from the treatment process [5].

Anammox process is mediated by the chemolithoautotrophic bacteria affiliated with a monophyletic group in the phylum Planctomycetales [6]. To date, 15 species of anammox have been successfully deciphered in wastewater treatment plants, freshwater and marine environment which categorized into six Candidatus genera [7]. Since, a classical pure culture of anammox species is challenged to be obtained, though all have the taxonomical status of Candidatus, which are classified into six different genera: (1) *Kuenenia*, represented by *Kuenenia stuttgartiensis* [8]; (2) *Brocadia* (with four species: *B. anammoxidans, B. fulgida, B. sinica*, and *B. caroliniensis*) [9-12]; (3) *Anammoxoglobus*, represented by *A. propionicus* [10]; (4) *Jettenia* with three species reported (*J. asiatica, J. moscovienalis* and *J. caeni*) [13-15]; (5) *Scalindua* including five species (*S. brodae, S. sorokinii, S. wagneri, S. profunda* and *S. rubra*) [16-19] and (6) *Anammoxomicrobium*, represented by *A. moscowii* [15].

The first genome sequence of the anammox model organism that was used to determine the detailed analyses of substrate conversions in bioreactors was Candidatus *Kuenenia stuttgartiensis* (hereafter, *K. stuttgartiensis*), which eventually lead to the discover of hydrazine [3,8, 20-21]. The species *K. stuttgartiensis* was recognized as being capable to survive and develop during winter and summer despite being grown in the aerobic zones (i.e., dissolved oxygen >2 mg/L) [22]. Furthermore, the *K. stuttgartiensis* (27.6%) that was developed in the micro-granule structure had succeeded to achieve the nitrogen removal efficiencies up to 71.8 ± 9.9% [23]. This shows that the *K. stuttgartiensis* may survive at different temperature while simultaneously resulting in excellent nitrogen removal efficiency within the smaller structure of granules. Additionally, *K. stuttgartiensis* possesses a complex network of regulatory systems to confer cells with the ability against Zn (II) toxicity. These systems include functions related to substrate degradation and Zn (II) efflux, chelation, DNA repair, protein degradation, protein synthesis, and signal transduction, respectively [24].

Anammox is one of the key biogeochemical processes in marine and estuarine environments, hence studies on natural habitats have forced us to reassess the global nitrogen cycle [22,25]. Extensive amazing and fascinating results have been reported regarding *K. stuttgartiensis* ubiquity in wastewater treatment, owing to fast evolving molecular approaches that primarily focus on small-subunit ribosomal RNA (16S rRNA) and functional genes in a culture-independent manner [26]. The distinct characteristics of *K. stuttgartiensis* have piqued the interest of many researchers around the world, who are working to improve the performance of the_anammox in the nitrogen removal process. Therefore, beyond the genome-based predictions, this review aims to summarise the parameters affecting *K. stuttgartiensis* development and the metabolic networks driving anammox bacterial anabolism and mixotrophy.

2. Parameters effect on Candidatus *Kuenenia* development

2.1. Nitrogen loading rate

According to Ma *et al.* [27], successful anammox sludge enrichment may be achieved through the sequencing batch reactor (SBR) by feeding the synthetic wastewater containing ammonium and nitrite concentration 280 mg N/L and 360 mg N/L, respectively. After 270 days of cultivation, the specific anammox activity (SAA) of the reactor was stabilized at approximately 0.31 ± 0.01 g N/g VSS d by the *K. stuttgartiensis* as the dominant anammox bacteria (81 ± 9%) among the total bacteria based on fluorescence in situ hybridization (FISH). The higher the concentration of ammonia and nitrite in the SBR medium, simultaneously longer stabilization period required by the *K. stuttgartiensis*. Moreover, the enhancement of the nitrogen loading rate by decreasing the HRT and increasing the ammonium and nitrite concentration enables the *K. stuttgartiensis* along with *Candidatus* "Brocadia" anammox community to be developed [28-29].

The enrichment of the *K. stuttgartiensis* bacteria in the anammox sludge may be achieved by varying the nitrogen loading rate along with the parameters of the reactor. This shows the adaption of the targeted anammox with the nitrogen loading rate consumes time in conjunction with the reactor configuration, parameters and the selection of the seed sludge. This subsequently causes industrial applications of anammox to be limited and challenging due to a long start-up period owing to extend doubling times of anammox bacteria (2.1–25 days) [27, 30-32].

Whilst, Izati *et al.,*[33] study utilizing upflow anaerobic sludge bed (UASB) reactor seeded with municipal returned activated sludge (RAS) able to detect the *K. stuttgartiensis* after 270 days though. The concentration of ammonium and nitrite were 52.02 and 62.42 mg N/L, respectively in the synthetic wastewater. Furthermore, the target of the study was to observe anammox micro-granular biomass enrichment performance. Thus, the nitrogen loading rate for the reactor was low compared to [27] for the feeding concentration. The anammox granules size distribution was between 253.2 to 465.7 μm for 75% of the total sample volume. The polymerase chain reaction (PCR) analyses have proven the *K. stuttgartiensis* is the second major genera in mature anammox sample whilst *Candidatus* "Brocadia" as predominant in the inoculum. This study proves the *K. stuttgartiensis* may be cultivated in the micro-granular by introducing low level concentration of ammonia and nitrite. Additionally, the niche of the seed sludge plays a crucial role for the development of the *K. stuttgartiensis* bacteria.

The enrichment of anammox sludge in different configuration of reactors consumes approximately 270 days to develop the matured anammox granules. The *K. stuttgartiensis* able to be the dominant bacteria in sample sludge with the nitrogen loading rate of ammonium to nitrite ratio of 0.7 to 0.8. Anammox bacteria requires ammonium as the electron donor and nitrite as an electron acceptor [34]. Therefore, sufficient amount of ammonium and nitrite should be supplied for the development of the *K. stuttgartiensis* in conjunction to carry out the anammox process in the anaerobic condition [35].

2.2. Seed sludge

In order to achieve quick start-up of anammox, the appropriate selection of seed sludge is crucial, which consequently may lead to the wide practical application [36]. Anammox biomass, notably anammox granule, is the most frequently used inoculum or seed sludge for anammox reactors, as it has several advantages over flocculent anammox sludge [7]. However, the utilization of solely anammox bacteria as an inoculum hinders the implementation of the anammox technology to treat nitrogen-rich wastewater systems [36]. Therefore, a mixture of more than one type of sludge is considered as the best inoculum to nurture anammox bacteria and enhance the anammox process [37].

The anammox granule that commonly used as the inoculum for the reactor start-up usually consist of *K. stuttgartiensis* as the dominant anammox bacterium [38-39]. Moreover, the *K. stuttgartiensis* can be dominant anammox species in the enrichment cultures using inoculum from different types of wastewater treatment plant activated sludge such as being sourced from landfill leachate treatment plant, municipal sewage treatment plant and monosodium glutamate (MSG) wastewater treatment plant, respectively [40]. Consequently, this demonstrates the potential of *K. stuttgartiensis* to be determined in the seed sludge sourced from the activated sludge and anammox granule. Additionally, the start-up period of the reactor is primarily influenced by the type of seed sludge selected [37]. However, the seed sludge consisting of anammox bacterium namely *K. stuttgartiensis* was found to be able to initiate the anammox process in approximately 10 to 20 days [36-37, 41]. Furthermore, different configuration of reactors utilising various options of biomass as the inoculum also may affect the start-up period. Therefore, the seed sludge selection for the anammox process is mostly focused on the maturity of the granules and the anammox bacterium consortium [41].

Many studies are being conducted to determine the appropriate seed sludge for the rapid start-up of the anammox process. However, the type of anammox bacterium in different seed sludge consequently affects the nitrogen removal efficiency in conjunction with the parameters for the reactor operation like temperature, pH and dissolved oxygen. This is because seed sludge sourced from different areas like marine, brackish and freshwater may consist of different anammox bacterium consortium according to its preferable condition to survive [42]. Table 1. summarizes the type of seed sludge used for anammox start-up resulting in the discovery of *K. stuttgartiensis* as the developed anammox bacterium.

Table 1. The type of seed sludge used for anammox start-up.

Reactor Type	Seed sludge	Start-up period (days)	N-removal efficiency (%) achieved at the end of start-up	Microbial analysis	References
Air-lift reactor	From a pilot-scale PN-anammox reactor.	94	31.28	qPCR: 8.47 x 10^{10} (floc) & 1.68 x 10^{10} (granule) (copies/gVSS)	[43]
Expanded granular sludge bed reactor	Preserved anammox consortia	20	Ammonium: 94 and nitrite: 99	Metatranscriptomic analysis: Candidatus *Kuenenia stuttgartiensis* (dominant)	[41]
Upflow anaerobic sludge blanket reactor	Municipal returned activated sludge	55	85	qPCR: Candidatus *Brocadia* (0.46 %) and Candidatus *Kuenenia* (0.24 %) relative abundance	[33]
Sequencing batch reactor	Mixed activated sludge (aerobic sludge, anaerobic sludge, simultaneous partial nitrification, anammox and denitrification (SNAD) sludge and anammox sludge)	14		qPCR of AnAOB 53 sequences: Candidatus *Brocadia*, Candidatus *Jettenia* and Candidatus *Kuenenia*	[36]
Anaerobic sequencing batch reactor	Anammox sludge and activated sludge	10	80	qPCR: Candidatus *Brocadia* and Candidatus *Kuenenia*	[37]
Anaerobic membrane bioreactor	Conventional activated sludge	125	20	FISH: Candidatus *Brocadia anammoxidans* and Candidatus *Kuenenia stuttgartiensis*	[44]
Rotating flat-sheet membrane bioreactor	Anammox activated sludge	16	Ammonium: 90 and nitrite: 90	FISH: Anammox bacteria 90% of total bacteria	[45]

2.3. Temperature

The most crucial parameter for the biological wastewater treatment plant especially for the microbial growth is the temperature. According to Oshiki *et al.* [46], the temperature range between 30 to 40 °C is been recognized as the optimum temperature for the growth of anammox bacteria. However, in many countries the average temperature is quite lower than this range. Moreover, it is a big challenge and may not be cost effective for the wastewater treatment plant to maintain an optimum temperature for anammox process. Thus, many solutions are being studied to overcome this problem, therefore, the best possible may comply for wider application of anammox technology [47].

The growth temperature preferred by the *K. stuttgartiensis* bacterium is between 25 to 37 °C, whilst the optimum temperature is 37 °C [48-49]. Additionally, in a study Arumugham *et al.,*[50] succeeded in determining the *K. stuttgartiensis* bacterium at the temperature range of 29.6 to 32.0°C at the domestic wastewater treatment plant. This indicates that the *K. stuttgartiensis* bacteria is able to survive below the optimum temperature within the open wastewater treatment system. The temperature condition reflexes upon the nitrogen removal process of the anammox because relatively high ammonia removal rate was achieved under the mesophilic temperature conditions of 30 to 40 °C [32,51]. Therefore, many studies have studied upon different temperatures in order to enhance nitrogen removal [52].

According to Ma *et al.* [52], the temperature drops from 30 to 16 °C has resulted in a drop in the nitrite and ammonium removal efficiencies from 94.35% and 92.81% to 92.31% and 78.45%, respectively. The drop in temperature subsequently resulted in the production of nitrate in the anammox UASB reactor as the ammonium removal decreased at 16 °C. This is because at the low temperature in conjunction with the high dissolved oxygen, the potential for nitrite-oxidizing bacteria (NOB) and activity is higher compared to ammonia-oxidizing bacteria (AOB)[53-54]. However, the *K. stuttgartiensis* genome survived at low temperature and succeeded in achieving stable anammox operation after 28 days [26]. Furthermore, the NOB was inhibited at low temperature by intermittent high strength feeding (IHSF) strategy. Therefore, the anammox community was predominant in the treatment compared to NOB and AOB was not significantly affected by the temperature change [26]. This shows the *K. stuttgartiensis* able to survive at low temperature and enable to carry out the anammox process without inhibition from NOB by ISHF the strategy.

3. Anammox mechanism

The application of a combination molecular techniques on *K. stuttgartiensis* to explore key enzymes and processes involved in anammox catabolism revealed hydrazine and nitric oxide as volatile intermediates in the anammox bacterium [55-57]. In the central of the anammox cell is surrounded by a unique ladderane lipid membrane, in where the nitrite is initially reduced to nitric oxide before combining with the ammonium and yield hydrazine. Lastly the hydrazine oxidizes into dinitrogen gas [58]. Then, a pH gradient is created over the anammoxosome membrane when hydrazine is oxidised, which fuels the generation of adenosine triphosphate (ATP) [59].

The anammoxosome, which is recognized as the specialized intracellular organelle is responsible for the reactions contributes for the energy conservation [60-61]. Additionally, the anammoxosome of *K. stuttgartiensis* contains membrane-bound respiratory complexes for electron transport chain, including complex I, ATP synthase and an NAD+: ferredoxin oxidoreductase (RNF)[62]. These nitrite reductase, hydrazine synthase (HZS) and hydrazine dehydrogenase (HDH) are soluble enzymes that is not membrane-bound in the anammoxosome and responsible for anammox catabolism [60,63].

According to de Almeida *et al.* [64], the most easily detected soluble enzymes on the anammoxosome of *K. stuttgartiensis* were HDH and HZS with high expression of proteins. Meanwhile, the candidate nitrite reductase in few studies, namely nitrite reductase NirS was hardly detectable, as expected from its low transcription levels [8,65]. The HDH is also recognized as an octaheme protein connected to hydroxylamine oxidoreductase (HAO) from the aerobic ammonium-oxidizing bacteria

[66-68]. There were nine more HAO-like octaheme proteins besides HDH, which were encoded in the *K. stuttgartiensis* genome and was determined in the proteome [64].

The HAO is established as the alternative catalyst for the hydroxylamine oxidation. However, in anammox species namely the *K. stuttgartiensis* generates nitric oxide using nitrite reductases as the catalyst encodes cytochrome cd_1-containing (cd1-NIR) [8,65,69]. In order to convert ammonium into nitrogen gas using nitrite as the terminal electron acceptor in the anaerobic condition, the anammox bacteria utilizes the oxidative power of nitric oxide to carry out the process [57]. Therefore, the anammox bacteria community oxidizes the hydroxylamine into nitric oxide through encoded octaheme HAO as the catalyst for the process [70]. However, each anammox species has diverse nitric oxide generation pathways. Thus, it was suggested that nitrite reduction to nitric oxide might be a feature that evolved after the core anammox catabolism was already in place.

The *K. stuttgartiensis* have revealed the anammox bacteria conserve energy and grow by using nitric oxide as its terminal electron acceptor in the anaerobic condition for ammonium oxidation [71]. Besides that, nitrate production was inhibited and the sole end product is nitrogen gas. This was determined using the comparative transcriptomics and proteomics techniques. The *K. stuttgartiensis* suppresses the transcription of proteins involved in nitric oxide production and nitrite oxidation during the development of nitric oxide-dependent ammonium oxidation [57]. It is proven that anammox bacteria is not responsible for producing any nitrous oxide, although enormous quantities of nitric oxide are supplied. This concurrently shows the anammox bacteria consumes nitric oxide in natural and manmade ecosystems to produce harmless nitrogen gas instead of the greenhouse gas nitrous oxide. As a result, anammox maintains the regulation of nitric oxide and nitrous oxide emissions, which are two key chemicals in atmospheric chemistry.

The ATP synthesis in anammox bacteria is commonly catalyzed by the ATPase-1. In *K. stuttgartiensis* has four membrane-bound ATPases encoded in its genome: the H^+-dependent F_1F_O type (ATPase-1, kustc3787–3796) and two types (ATPase-2, kuste4592–4600; ATPase-3; kustc0572–0579) that lack the δ subunit but have one or two extra membrane subunits (AtpQ, kuste4594 and kustc0574; AtpR, kuste4595) instead [8,64,72-73]. The ATPase-2 and ATPase-3 are thought to be Na^+-pumping ATP hydrolases (N-ATPase) other than that a prokaryotic V-type ATPase (ATPase-4, kuste3864–3871) is also encoded [74]. According to the previous research, ATPase-1 was recommended to be the primary ATP synthase in *K. stuttgartiensis* and the migratory profiles of the ATPases analysis have confirmed those findings [75]. Most of the N- and V-ATPases subunits which were not vary as ATPase-1, able to recover as high-molecular-mass complexes and subcomplexes, but their expression was much lower than ATPase-1. It should be highlighted that the existence of N-ATPases was only hypothesised based on genetic analysis because their protein expression had never been determined before [74]. This shows *K. stuttgartiensis* capable of producing energy through ATPase-1 besides recovering high-molecular-mass by utilizing N- and V-ATPases.

The experimental research and genomic evidence have determined that the anammox bacteria are more adaptable than originally envisaged. This study revealed substitution electron donors to ammonium, such as formate, acetate, and propionate for energy conservation with nitrite or nitrate as electron acceptors [10,76]. Besides that, the *K. stuttgartiensis* also utilizes ferrous iron as electron donor meanwhile iron and manganese oxides as electron acceptors [8,10,18,60]. This shows the *K. stuttgartiensis* is capable of substituting the electron donor for ammonium and electron acceptors for nitrite or nitrate with many other alternatives.

Intriguingly, it's been discovered that these organic substrates are entirely oxidised to carbon dioxide rather than being directly incorporated into cell biomass, implying that anammox bacteria maintain their autotrophic existence [72]. Therefore, a study was conducted upon the central carbon metabolism of a planktonic *K. stuttgartiensis* through time-series ^{13}C isotope tracing, metabolomics and isotopically nonstationary metabolic flux analysis (INST-MFA) [55]. According to previous studies [8,72], the Wood-Ljungdahl route and pyruvate:ferredoxin oxidoreductase (PFOR) would fix carbon dioxide (CO_2) on a regular basis, resulting in rapid labelling of acetyl-CoA and pyruvate, followed by phosphoenolpyruvate and other downstream metabolites. In the TCA cycle and gluconeogenesis, acetyl-CoA and pyruvate are predicted to enter to generate biomass precursors.

The planktonic *K. stuttgartiensis* on the other hand, lacks the citrate synthase necessary to initiate the oxidative TCA cycle. As a result, it's hypothesized that primary precursor metabolites like succinyl-CoA and alpha-ketoglutarate are synthesised in a reductive manner, resulting in high ^{13}C-labelling of oxaloacetate, succinate, and alpha-ketoglutarate. [72]. Alternative biomass precursor is predicted to be generated in *K. stuttgartiensis* via gluconeogenesis and the pentose phosphate pathway [8]. There are several acyltransferase candidates that exist as substitutes for citrate synthase in *K. stuttgartiensis.* For instance, genes annotated as (R)-citramalate synthase (KSMBR1_RS19040) responsible for isoleucine biosynthesis and redundant copies of 2-isopropylmalate synthase (KSMBR1_RS18315 and KSMBR1_RS10820) [71,77].

The isotopically nonstationary metabolic flux analysis (INST-MFA) was performed by fitting measured, time-resolved metabolite mass isotopomer distributions from ^{13}C-formate tracer experiments to an isotopomer network model in order to resolve the metabolic pathways used for the biosynthesis of sugar phosphates in *K. stuttgartiensis* [78]. The ^{13}C-formate tracer analysis demonstrated that a significant fraction of fructose 6-phosphate was present as M+ 1 mass isotopomers, which was unpredicted as gluconeogenesis produces mostly M+ 2 mass isotopomers [71]. As a result, different routes for the generation of fructose 6-phosphate may exist. According to the genome annotation of *K. stuttgartiensis*, the hexulose 6-phosphate synthase and 6-phospho-3-hexuloisomerase (KSMBR1 RS05220 and KSMBR1 RS18790, respectively) are recognised as primary enzymes of the ribulose monophosphate (RuMP) pathway, a formaldehyde assimilation mechanism in many methylotrophic bacteria [79]. Furthermore, these mechanisms utilize a hexulose 6-phosphate intermediate to fix formaldehyde to fructose 6-phosphate. Therefore, these processes and the undiscovered formaldehyde dehydrogenase, were demonstrated in this study could explain the abundance of M+ 1 pentose and hexose phosphate isotopomers found during ^{13}C-formate labelling [71].

In the reverse Wood–Ljungdahl pathway, the acetyl-CoA will be oxidized in order to label the formate as a pathway intermediate. However, in the study utilizing the *K. stuttgartiensis* genome, the formate remained unlabelled and the result was that the reverse Wood–Ljungdahl pathway was not involved in acetyl-CoA oxidation to CO_2 [71]. In contrast, the ^{13}C-labelling of TCA cycle and gluconeogenic metabolites has taken place to a much lesser extent compared to results from ^{13}C-formate tracing. The multiple cycles of the oxidative TCA cycle using *Si*-citrate synthase instead of *Re*-citrate synthase may possibly explain the observed unique patterns for TCA cycle metabolites. The finding of the *Si*-citrate synthase gene candidate in the *K. stuttgartiensis* genome, indicates a low abundance bacterium in the bioreactors side population may have oxidised the acetate. However, the low relative isotope abundance of *K. stuttgartiensis* proteins revealed by ^{13}C-acetate labelling, combined with the *Si*-citrate synthase activity proved by ^{13}C- and ^{2}H-acetate tracer experiments, shows that *K. stuttgartiensis* was unable to use acetate as an electron donor or carbon source in situ. Therefore, the study suggests that *K. stuttgartiensis* utilizes the Wood–Ljungdahl pathway to directly assimilate extracellular formate instead of oxidising it completely to CO_2 and then reassimilation.

4. Perspectives and further research

Anammox process is recognised as the key step in the biogeochemical nitrogen cycle and has been established as a novel treatment technology for removing nitrogen from wastewater in a sustainable way. There are various anammox species in the environment, however, the *K. stuttgartiensis* was selected to be studied in detail. This is due to the feasibility of *K. stuttgartiensis* to survive in summer and winter conditions besides in the aerobic zones (dissolved oxygen >2 mg/L). Additionally, the microbial detection methods for anammox bacteria and metabolic networks drive anammox bacterial anabolism and mixotrophy beyond genome-based predictions. Therefore, the genomics investigation on *K. stuttgartiensis* is being explored globally, with the goal of revealing new anammox-specific genes, enzymes and reactions involved in anammox catabolism.

Moreover, the ATP synthase of *K. stuttgartiensis* was explained thoroughly. The substitute of electron donors for anammox instead of ammonium was discovered, for instance, like formate, acetate, and propionate for energy conservation, along with nitrite or nitrate as electron acceptors. The organic substrates in the anammox are entirely oxidised to carbon dioxide rather than being directly incorporated into cell biomass, implying that anammox bacteria maintain their autotrophic existence. The study has revealed that *K. stuttgartiensis* utilizes the Wood–Ljungdahl pathway to directly assimilate extracellular formate instead of oxidising it completely to CO_2 and then reassimilation. In conclusion, the *K. stuttgartiensis* anammox bacteria is able to contribute greatly to wastewater treatment for nitrogen removal by developing the bacteria at suitable parameters. Thus, *K. stuttgartiensis* able to carry out catabolism and anabolism in the wastewater efficiently.

The anammox bacteria, *K. stuttgartiensis* is unique as compared to other species in the genera due to its capability to survive at different environmental condition besides various substitution for electron donor. As a result, by promoting *K. stuttgartiensis*, the adaptation speed of the anammox to surrounding environment in actual wastewater may be improved towards achieving stable nitrogen removal performance in the system. . This review reveals the important connection between the reactor parameters and the development of *K. stuttgartiensis* as well as the changes in the mechanism process.

Acknowledgments
The authors would like to thank Malaysia-Japan International Institute of Technology (MJIIT) Universiti Teknologi Malaysia (UTM) and the Ministry of Higher Education for funding this study under Fundamental Research Grant Scheme (FRGS/1/2021/STG04/ UTM/ 01/1)(No;5F003).

ORCID iDs:
Thilagavathi Arumugham: https://orcid.org/ 0000-0002-0447-097X
Shaza Eva Mohamad: https://orcid.org/ 0000-0003-3199-9884

References

[1] Terada, A., Zhou, S., and Hosomi, M. (2011). Presence and detection of anaerobic ammonium-oxidizing (anammox) bacteria and appraisal of anammox process for high-strength nitrogenous wastewater treatment: A review. *Clean Technol. Environ. Policy*, **13**(6), 759–781.

[2] Winkler, M. K., and Straka, L. (2019). New directions in biological nitrogen removal and recovery from wastewater. *Curr. Opin. Biotechnol.*, **57**, 50–55.

[3] Strous, M., Heijnen, J., Kuenen, J., and Jetten, M. (1998). The sequencing batch reactor as a powerful tool for the study. *Appl. Microbiol. Biotechnol.*, **50**, 589–596.

[4] Arumugham, T., Husaini, N. A., Yuzir, A., Mohamad, S. E., Iwamoto, K., and Abdullah, N. (2020). Anammox reactor treating low strength domestic wastewater: A review. *IOP Conf. Ser. Earth Environ. Sci.,* **479**(1).

[5] Cho, S., Kambey, C., and Nguyen, V. K. (2020). Performance of anammox processes for wastewater treatment: A critical review on effects of operational conditions and environmental stresses. *Water (Switz.)*, **12**(1).

[6] Jetten, M. S. M., Wagner, M., Fuerst, J., Van Loosdrecht, M., Kuenen, G., and Strous, M. (2001). Microbiology and application of the anaerobic ammonium oxidation ('anammox') process. *Curr. Opin. Biotechnol.*, **12**(3), 283–288.

[7] Ibrahim, M., Yusof, N., Mohd Yusoff, M. Z., and Hassan, M. A. (2016). Enrichment of anaerobic ammonium oxidation (anammox) bacteria for short start-up of the anammox process: a review. *Desalin. Water Treat.*, **57**(30), 13958–78.

[8] [8] Strous, M., Pelletier, E., Mangenot, S., Rattei, T., Lehner, A., Taylor, M. W.,and Le Paslier, D. (2006). Deciphering the evolution and metabolism of an anammox bacterium from a community genome. *Nature*, **440**(7085), 790–794.

[9] Strous, M., Fuerst, J. A., Kramer, E. H. M., Logemann, S., Muyzer, G., Van De Pas- Schoonen, K.T., Webb, R., and Jetten, M. S. M. (1999). Missing lithotroph identified as new planctomycete. *Nature*, **400**(6743), 446–449.

[10] Kartal, B., Van Niftrik, L., Rattray, J., Van De Vossenberg, J. L. C. M., Schmid, M. C., Sinninghe Damsté, J., and Strous, M. (2008). Candidatus "Brocadia fulgida": An autofluorescent anaerobic ammonium oxidizing bacterium. *FEMS Microbiol. Ecol.*, **63**(1), 46–55.

[11] Oshiki, M., Shimokawa, M., Fujii, N., Satoh, H., and Okabe, S. (2011). Physiological characteristics of the anaerobic ammonium-oxidizing bacterium "Candidatus Brocadia sinica." *Microbiology*, **157**(6), 1706–13.

[12] Park, H., Brotto, A. C., van Loosdrecht, M. C. M., and Chandran, K. (2017). Discovery and metagenomic analysis of an anammox bacterial enrichment related to Candidatus "Brocadia caroliniensis" in a full-scale glycerol-fed nitritation-denitritation separate centrate treatment process. *Water Res.*, **111**, 265–273.

[13] Quan, Z. X., Rhee, S. K., Zuo, J. E., Yang, Y., Bae, J. W., Park, J. R., and Park, Y. H. (2008). Diversity of ammonium-oxidizing bacteria in a granular sludge anaerobic ammonium-oxidizing (anammox) reactor. *Environ. Microbiol.*, **10**(11), 3130–39.

[14] Ali, M., Oshiki, M., Awata, T., Isobe, K., Kimura, Z., Yoshikawa, H., and Okabe, S. (2015). Physiological characterization of anaerobic ammonium oxidizing bacterium "Candidatus Jettenia caeni." *Environ. Microbiol.*, **17**(6), 2172–89.

[15] Nikolaev, Y. A., Kozlov, M. N., Kevbrina, M. V., Dorofeev, A. G., Pimenov, N. V., Kallistova, A. Y., and Bumazhkin, B. K. (2015). Candidatus "Jettenia moscovienalis" sp. nov., a new species of bacteria carrying out anaerobic ammonium oxidation. *Microbiology*, **84**(2), 256–262.

[16] Schmid, M., Walsh, K., Webb, R., Rijpstra, W. I. C., Van De Pas-Schoonen, K., Verbruggen, M. J., and Strous, M. (2003). Candidatus "Scalindua brodae", sp. nov., Candidatus "Scalindua wagneri", sp. nov., Two New Species of Anaerobic Ammonium Oxidizing Bacteria. *Syst. Appl. Microbiol.*, **26**(4), 529–538.

[17] Woebken, D., Lam, P., Kuypers, M. M. M., Naqvi, S. W. A., Kartal, B., Strous, M., and Amann, R. (2008). A microdiversity study of anammox bacteria reveals a novel Candidatus Scalindua phylotype in marine oxygen minimum zones. *Environ. Microbiol.*, **10**(11), 3106–19.

[18] Van De Vossenberg, J., Rattray, J. E., Geerts, W., Kartal, B., Van Niftrik, L., Van Donselaar, E. G., and Jetten, M. S. M. (2008). Enrichment and characterization of marine anammox bacteria associated with global nitrogen gas production. *Environ. Microbiol.*, **10**(11), 3120–29.

[19] Speth, D. R., Lagkouvardos, I., Wang, Y., Qian, P. Y., Dutilh, B. E., and Jetten, M. S. M. (2017). Draft Genome of Scalindua rubra, Obtained from the Interface Above the Discovery Deep Brine in the Red Sea, Sheds Light on Potential Salt Adaptation Strategies in Anammox Bacteria. *Microb. Ecol.*, **74**(1), 1–5.

[20] Van De Graaf, A. A., De Bruijn, P., Robertson, L. A., Jetten, M. S. M., and Kuenen, J. G. (1997). Metabolic pathway of anaerobic ammonium oxidation on the basis of 15N studies in a fluidized bed reactor. *Microbiology*, **143**(7), 2415–21.

[21] Schalk, J., Oustad, H., Kuenen, J. G., and Y, M. S. M. J. (1998). The anaerobic oxidation of hydrazine: a novel reaction in microbial nitrogen metabolism, *FEMS Microbiol. Lett.*, **158**, 61–67

[22] Wang, S., Peng, Y., Ma, B., Wang, S., and Zhu, G. (2015). Anaerobic ammonium oxidation in traditional municipal wastewater treatment plants with low-strength ammonium loading: Widespread but overlooked. *Water Res.*, **84**, 66–75.

[23] Chen, R., Ji, J., Chen, Y., Takemura, Y., Liu, Y., Kubota, K., and Li, Y.-Y. (2019). Successful operation performance and syntrophic micro-granule in partial nitritation and anammox reactor treating low-strength ammonia wastewater. *Water Res.*, **155**, 288–299.

[24] Ma, X., Yan, Y., Wang, W., Guo, J., and Wang, Y. (2020). Metatranscriptomic analysis of adaptive response of anammox bacteria Candidatus 'Kuenenia stuttgartiensis' to Zn(II) exposure. *Chemosphere*, **246**, 125682.

[25] Francis, C. A., Beman, J. M., and Kuypers, M. M. M. (2007). New processes and players in the nitrogen cycle: The microbial ecology of anaerobic and archaeal ammonia oxidation. *ISME J.*, **1**(1), 19–27.

[26] Zhang, B., Zhao, J., Zuo, J., Shi, X., Gong, J., and Ren, H. (2019). Realizing stable operation of anaerobic ammonia oxidation at low temperatures treating low strength synthetic wastewater. *J Environ Sci (China)*, **75**, 193–200.

[27] Ma, X., Wang, Y., Zhou, S., Yan, Y., Lin, X., and Wu, M. (2017). Endogenous metabolism of anaerobic ammonium oxidizing bacteria in response to short-term anaerobic and anoxic starvation stress. *Chem. Eng. J.*, **313**, 1233-41.

[28] Wang, T., Zhang, H., Gao, D., Yang, F., Yang, S., Jiang, T., and Zhang, G. (2011). Enrichment of Anammox bacteria in seed sludges from different wastewater treating processes and start-up of Anammox process. *Desalination*, **271**(1–3), 193–198.

[29] Schmid, M. C., Maas, B., Dapena, A. D., van de Pas-Schoonen, K., van de Vossenberg, Jack Kartal, B., and van Niftrik, L. (2005). Biomarkers for in situ detection of anaerobic ammonium-oxidizing (Anammox) bacteria. *Appl. Environ. Microbiol.*, **42**(4), 231–253.

[30] Lotti, T., Kleerebezem, R., Abelleira-Pereira, J. M., Abbas, B., and van Loosdrecht, M. C. M. (2015). Faster through training: The anammox case. *Water Res.*, **81**, 261–268.

[31] Lotti, T., Kleerebezem, R., Lubello, C., and van Loosdrecht, M. C. M. (2014). Physiological and kinetic characterization of a suspended cell anammox culture. *Water Res.*, **60**, 1–14.

[32] van der Star, W. R. L., Abma, W. R., Blommers, D., Mulder, J. W., Tokutomi, T., Strous, M., and van Loosdrecht, M. C. M. (2007). Startup of reactors for anoxic ammonium oxidation: Experiences from the first full-scale anammox reactor in Rotterdam. *Water Res.*, **41**(18), 4149–63.

[33] Izadi, P., Izadi, P., and Eldyasti, A. (2020). Influence of microbial population dynamics on accelerated development of anammox micro-granular consortium. *J. Water Process. Eng.*, **36**(June), 101384.

[34] Chen, F., Li, X., Yuan, Y., and Huang, Y. (2019***b***)). An efficient way to enhance the total nitrogen removal efficiency of the Anammox process by S0-based short-cut autotrophic denitrification. *J. Environ. Sci. (China)*, **81**(Xiang Li), 214–224.

[35] Karthik, O., and Joseph, K. (2007). Anaerobic Ammonium Oxidation (Anammox) Process for Nitrogen Removal - a Review. *Proc. of Biological methods of waste treatment and management in South India Symp.*, Chennai, TN, India, February 15.

[36] Ye, L., Li, D., Zhang, J., and Zeng, H. (2018). Fast start-up of anammox process with mixed activated sludge and settling option. *Environ. Technol. (U.K.)*, **39**(23), 3088–95.

[37] Wang, Q., Wang, Y., Lin, J., Tang, R., Wang, W., Zhan, X., and Hu, Z. H. (2018). Selection of seeding strategy for fast start-up of Anammox process with low concentration of Anammox sludge inoculum. *Bioresour. Technol.*, **268**(June), 638–647.

[38] Cheng, Y. F., Zhang, Z. Z., Li, G. F., Zhang, Q., Zheng, X. P., Cai, S., and Jin, R. C. (2020). Anammox Granules Acclimatized to Mainstream Conditions Can Achieve a Volumetric Nitrogen Removal Rate Comparable to Sidestream Systems. *Environ. Sci. Technol.* (**54**).

[39] An, P., Xu, X., Yang, F., and Li, Z. (2013). Comparison of the characteristics of anammox granules of different sizes. *Biotechnol. Bioprocess Eng.*, **18**(3), 446–454.

[40] Shen, L. D., Hu, A. H., Jin, R. C., Cheng, D. Q., Zheng, P., Xu, X. Y., and Hu, B. L. (2012). Enrichment of anammox bacteria from three sludge sources for the startup of monosodium glutamate industrial wastewater treatment system. *J. Hazard. Mater.*, **199–200**, 193–199.

[41] Zhou, Y. yuan, Shao, W. li, Liu, Y. di, Li, X., Shan, X. yu, Jin, X. bai, and Li, W. (2020). Genome-based analysis to understanding rapid resuscitation of cryopreserved anammox consortia via sequential supernatant addition. *Sci. Total Environ.*, **744**, 140785.

[42] Date, Y., Isaka, K., Ikuta, H., Sumino, T., Kaneko, N., Yoshie, S., and Inamori, Y. (2009). Microbial diversity of anammox bacteria enriched from different types of seed sludge in an anaerobic continuous-feeding cultivation reactor. *J. Biosci. Bioeng.*, **107**(3), 281–286.

[43] Zhang, G., Zhang, L., Han, X., Zhang, S., and Peng, Y. (2021). Start-up of PN-anammox system under low inoculation quantity and its restoration after low-loading rate shock. *Front. Environ. Sci. Eng.*, **15**(2).

[44] Gutwiński, P., Cema, G., Ziembińska-Buczyńska, A., Surmacz-Górska, J., and Osadnik, M. (2016). Startup of the Anammox Process in a Membrane Bioreactor (AnMBR) from Conventional Activated Sludge. *Water Environ. Res.*, **88**(12), 2268–74.

[45] Jiang, T., Zhang, H., Qiang, H., Yang, F., Xu, X., and Du, H. (2013). Start-up of the anammox process and membrane fouling analysis in a novel rotating membrane bioreactor. *Desalination*, **311**, 46–53.

[46] Oshiki, M., Satoh, H., and Okabe, S. (2016). Ecology and physiology of anaerobic ammonium oxidizing bacteria. *Environ. Microbiol.*, **18**(9), 2784–96.

[47] Daverey, A., Chei, P. C., Dutta, K., and Lin, J. G. (2015). Statistical analysis to evaluate the effects of temperature and pH on anammox activity. *Int. Biodeterior. Biodegradation*, **102**, 89–93.

[48] Egli, K., Fanger, U., Alvarez, P. J. J., Siegrist, H., Van der Meer, J. R., and Zehnder, A. J. B. (2001). Enrichment and characterization of an anammox bacterium from a rotating biological contactor treating ammonium-rich leachate. *Arch. Microbiol.*, **175**(3), 198–207.

[49] Tomaszewski, M., Cema, G., and Ziembińska-Buczyńska, A. (2017). Influence of temperature and pH on the anammox process: A review and meta-analysis. *Chemosphere*, **182**, 203–214.

[50] Arumugham, T., Husaini, N. 'Amirah, Abdullah, N., Koji, I., and Mohamad, S. E. (2021). Determination of anaerobic ammonium oxidation (Anammox) bacteria in the domestic wastewater. *Desalination Water Treat.*, **214**, 355–366.

[51] Tang, C. J., Zheng, P., Wang, C. H., Mahmood, Q., Zhang, J. Q., Chen, X. G., and Chen, J. W. (2011). Performance of high-loaded ANAMMOX UASB reactors containing granular sludge. *Water Res.*, **45**(1), 135–144.

[52] Ma, B., Peng, Y., Zhang, S., Wang, J., Gan, Y., Chang, J., and Zhu, G. (2013). Performance of anammox UASB reactor treating low strength wastewater under moderate and low temperatures. *Bioresour. Technol.*, **129**, 606–611.

[53] Hellinga, C., Schellen, A. A. J. C., Mulder, J. W., Van Loosdrecht, M. C. M., and Heijnen, J. J. (1998). The SHARON process: An innovative method for nitrogen removal from ammonium-rich waste water. *Water Sci. Technol.*, **37**(9), 135–142.

[54] Bae, W., Baek, S., Chung, J., and Lee, Y. (2001). Optimal operational factors for nitrite accumulation in batch reactors. *Biodegradation*, **12**(5), 359–366.

[55] Lawson, C. E., de Graaf, R. M., Nuijten, G., Jacobson, T. B., Pabst, M., Stevenson, and D. M., Lucker, S. (2021). Autotrophic and mixotrophic metabolic network fluxes suggest versatile lifestyle for the anammox bacterium Candidatus Kuenenia stuttgartiensis. *The ISME J.*, **15**, 673–687.

[56] Oshiki, M., Ali, M., Shinyako-Hata, K., Satoh, H., and Okabe, S. (2016(*a*)). Hydroxylamine-dependent anaerobic ammonium oxidation (anammox) by "Candidatus Brocadia sinica." *Environ. Microbiol.*, **18**(9), 3133–31.

[57] Hu, Z., Wessels, H. J. C. T., van Alen, T., Jetten, M. S. M., and Kartal, B. (2019). Nitric oxide-dependent anaerobic ammonium oxidation. *Nat. Commun.*, **10**(1).

[58] Fuerst, J. A., Webb, R. I., Van Niftrik, L., Jetten, M. S. M., and Strous, M. (2006). Anammoxosomes in anaerobic ammonium oxidizing Planctomycetes. *Complex Intracellular Structures in Prokaryotes.* (Shively JM, Ed.). Berlin and Heidelberg: Springer.

[59] Van Der Star, W. R. L., Dijkema, C., De Waard, P., Picioreanu, C., Strous, M., and Van Loosdrecht, M. C. M. (2010). An intracellular pH gradient in the anammox bacterium Kuenenia stuttgartiensis as evaluated by 31P NMR. *Appl. Microbiol. Biotechnol.*, **86**(1), 311–317.

[60] van Niftrik, L., and Jetten, M. S. M. (2012). Anaerobic Ammonium-Oxidizing Bacteria: Unique Microorganisms with Exceptional Properties. *Microbiol. Mol. Biol. Rev.*, **76**(3), 585–596.

[61] Neumann, S., Wessels, H. J. C. T., Rijpstra, W. I. C., Sinninghe Damsté, J. S., Kartal, B., Jetten, M. S. M., and van Niftrik, L. (2014). Isolation and characterization of a prokaryotic cell organelle from the anammox bacterium Kuenenia stuttgartiensis. *Mol. Microbiol.*, **94**(4), 794–802.

[62] Carvajal-Arroyo, J. M., Puyol, D., Li, G., Sierra-Álvarez, R., and Field, J. A. (2014). The role of pH on the resistance of resting- and active anammox bacteria to NO2- inhibition. *Biotechnol. Bioeng.*, **111**(10), 1949–56.

[63] van Niftrik, L., Geerts, W. J. C., van Donselaar, E. G., Humbel, B. M., Yakushevska, A., Verkleij, A. J., and Strous, M. (2008). Combined structural and chemical analysis of the anammoxosome: A membrane-bounded intracytoplasmic compartment in anammox bacteria. *J. Struct. Biol.*, **161**(3), 401–410.

[64] de Almeida, N. M., Wessels, H. J. C. T., de Graaf, R. M., Ferousi, C., Jetten, M. S. M., Keltjens, J. T., and Kartal, B. (2016). Membrane-bound electron transport systems of an anammox bacterium: A complexome analysis. *Biochimica et Biophysica Acta - Bioenergetics*, **1857**(10), 1694–1704.

[65] Kartal, B., Maalcke, W. J., de Almeida, N. M., Cirpus, I., Gloerich, J., Geerts, W., Janssen-Megens, E.M., and Francoijs, K. J. (2011). Molecular mechanism of anaerobic ammonium oxidation. *Nature*, **479(**7371), 127–130.

[66] Maalcke, W. J., Reimann, J., De Vries, S., Butt, J. N., Dietl, A., Kip, N., and Kartal, B. (2016). Characterization of anammox hydrazine dehydrogenase, a Key -producing enzyme in the global nitrogen cycle. *J. Biol. Chem.*, **291**(33), 17077–92.

[67] Hooper, A. B., Vannelli, T., Bergmann, D. J., and Arciero, D. M. (1997). Enzymology of the oxidation of ammonia to nitrate by bacteria *Antonie van Leeuwenhoek*, **71**, 59–67.

[68] Cedervall, P., Hooper, A. B., and Wilmot, C. M. (2013). Structural studies of hydroxylamine oxidoreductase reveal a unique heme cofactor and a previously unidentified interaction partner. *Biochemistry*, **52**(36), 6211–18.

[69] Simon, J., and Klotz, M. G. (2013). Diversity and evolution of bioenergetic systems involved in microbial nitrogen compound transformations. *Biochimica et Biophysica Acta - Bioenergetics*, **1827**(2), 114–135.

[70] Kartal, B., and Keltjens, J. T. (2016). Anammox Biochemistry: a Tale of Heme c Proteins. *Trends Biochem. Sci.*, **41**(12), 998–1011.

[71] Lawson, C., Nuijten, G., de Graaf, R., Jacobson, T., Pabst, M., Stevenson, D., and Lücker, S. (2020). Autotrophic and mixotrophic metabolism of an anammox bacterium revealed by in vivo 13 C and 2 H metabolic network mapping. *The ISME Journal*. **15**, 673-687.

[72] Kartal, B., De Almeida, N. M., Maalcke, W. J., Op den Camp, H. J. M., Jetten, M. S. M., and Keltjens, J. T. (2013). How to make a living from anaerobic ammonium oxidation. *FEMS Microbiol. Rev.*, **37**(3), 428–461.

[73] Van Niftrik, L., Van Helden, M., Kirchen, S., Van Donselaar, E. G., Harhangi, H. R., Webb, R. I., and Strous, M. (2010). Intracellular localization of membrane-bound ATPases in the compartmentalized anammox bacterium "Candidatus Kuenenia stuttgartiensis." *Mol. Microbiol.*, **77**(3), 701–715.

[74] Dibrova, D. V., Galperin, M. Y., and Mulkidjanian, A. Y. (2010). Characterization of the N-ATPase, a distinct, laterally transferred Na+-translocating form of the bacterial F-type membrane ATPase. *Bioinformatics*, **26**(12), 1473–76.

[75] Söding, J., Biegert, A., and Lupas, A. N. (2005). The HHpred interactive server for protein homology detection and structure prediction. *Nucleic Acids Res.*, **33**(SUPPL. 2), 244–248.

[76] Narita, Y., Zhang, L., Kimura, Z. ichiro, Ali, M., Fujii, T., and Okabe, S. (2017). Enrichment and physiological characterization of an anaerobic ammonium-oxidizing bacterium 'Candidatus Brocadia sapporoensis.' *Syst. Appl. Microbiol.*, **40**(7), 448–457.

[77] Howell, D. M., Xu, H., and White, R. H. (1999). (R)-citramalate synthase in methanogenic archaea. *J. Bacteriol.*, **181**(1), 331–333.

[78] Young, J. D., Walther, J. L., Antoniewicz, M. R., Yoo, H., and Stephanopoulos, G. (2008). An Elementary Metabolite Unit (EMU) Based Method of Isotopically Nonstationary Flux Analysis. *Biotechnol. Bioeng.*, **99**(3), 686–699.

[79] Kato, N., Yurimoto, H., and Thauer, R. K. (2006). The physiological role of the ribulose monophosphate pathway in bacteria and archaea. *Biosci. Biotechnol. Biochem.*, **70**(1), 10–21.

Isolation and characterization of dye-degrading bacteria from batik wastewater

Rakhmania[1], Z Ibrahim [2], M Abba[3]

[1]Department of Chemical and Environmental Engineering, Malaysia-Japan International Institute of Technology, Universiti Teknologi Malaysia 54100, Kuala Lumpur Malaysia
[2]Department of Bioscience, Faculty of Science, Universiti Teknologi Malaysia, 81310 Skudai, Johor Bahru Malaysia
[3]Microbiology Department, Bauchi State University, 751105 Gadau, Nigeria

*Corresponding author e-mail: rakhmania99@gmail.com

Abstract. Azo dye is typically dye used in batik industries that is toxic to aquatic organisms and humans. Chemical methods can effectively degrade the dye effluent however they produce more toxic compounds. Biological methods offer an alternative that is more environmentally friendly and produce readily degradable compounds. Dye degrading bacteria can degrade the dyes in the effluent under anaerobic for decolorization followed by aerobic condition to produce readily degradable compounds. The batik wastewater sample was collected in Kelantan, Malaysia and was analyzed for COD, pH, and color. The Horikoshi-II media pH 10 was used to propagate growth of bacteria in the dye effluent. The dye degrading bacteria then was isolated from the dye effluent using serial dilution spread plate and streak plate method; One bacterium was finally selected based on faster growth. The specific growth rate and generation time of isolated bacteria were 0.069 h^{-1} and 14 h respectively. The isolated bacteria have characteristic producing red pigment and gram-positive bacilli. Overall reduction of color and COD by isolated bacteria were 23% and 41% respectively. The isolated bacteria were identified as Bacillus sp. strain ARNZ2707 via 16S rRNA sequence analysis.

1. Introduction

Batik industries is one of the local industries that have been expanded in Malaysia. Most batik industries in Malaysia are operated in the region of Terengganu and Kelantan Batik industries contribute to high development of the economy in Malaysia because they have high demand from local and abroad [1]. Batik industry is usually operated in small scale by workshops and factories throughout the country which usually developed near to the river to easily dispose of the effluent [2]. The textile dye effluent resulting from batik making process is usually discarded directly into the waterways, prompting both pollution and toxicological concerns [3]. The characteristic of batik wastewater highly contains organic compounds including suspended solid, chemical oxygen demand, biochemical oxygen demand, grease, wax, dye, surfactant etc [4]. Thus, an effective and low-cost method is required for the removal of colour and organic substance from batik wastewater.

The presence of the dye in wastewater raised a big issue and caused several environmental problems. There are several methods available for batik wastewater treatment such as physical, chemical and biological methods [5]. However, physical and chemical methods have limitations on high cost, require post-treatment handling, production of sludge and high energy consumption [6]. On the other hand, biological treatment offers a greater alternative in batik effluent treatment containing dyes. Biological treatment has advantages including environmentally friendly, easy to operate, produce less toxic sludge, and cost-effective treatment [7].

Azo dye is typically used as dying process in batik industry and it has varying effects to the water environment which include depletion of dissolved oxygen, inhibition the activity of aquatic photosynthesis and toxic to plant, animal and as well as humans [8]. Basically, azo dyes are xenobiotic compounds that have azo bonds (-N=N-) representing that they are electron deficient. Azo dye contains sulphide (SO_3^{2-}) or other electron donor groups which result in deficiency of electrons; this reaction causes the degradation of azo dye by microorganisms [9-11]. Biological treatment may involve the use of microorganisms in either aerobic or anaerobic conditions to degrade organic dyes by using fixed or suspended growth systems. Microorganisms are able to metabolize the biodegradable organic to be converted into water, carbon dioxide (CO_2), and energy for their growth and reproduction. Bacteria have enzymes called azo-reductase that can degrade azo dye in coloured wastewater [12]. In general, the degradation of azo dye in bacteria is divided into two processes. Firstly, in anaerobic condition the azo bond cleavage by enzyme azo-reductase. This condition is also known as dye reduction. This condition is divided into two stages that transfer two electrons for each stage which also acts as final electron acceptor as shown on reaction (1) and (2) below [13]:

Anaerobic condition:

$$(R_1\text{-N=N-}R_2 + 2e^- + 2H^+ \rightarrow R_1\text{-HN-NH-}R_2 \text{ (Hydrazo Intermediate)} \quad (1)$$

$$(R_1\text{-HN-NH-}R_2) + 2e^- + 2H^+ \rightarrow R_1\text{-}NH_2 + R_2\text{-}NH_2 \text{ (Reductive cleavage of azo bond)} \quad (2)$$

The end product (aromatic amines) from reaction in anaerobic condition is considered more toxic than the parent compound. Second process acts as an agent to degrade the aromatic amines into less toxic products. In aerobic conditions, the enzymes monooxygenase and di-oxygenase enhanced the incorporation of oxygen in aromatic compounds during ring opening. The product from this process is less toxic to the environment [14]. Due to the ability of bacteria to degrade the dye-containing wastewater into less toxic products, this study aims to isolate the dye-degrading bacteria from batik wastewater in its original condition. Followed by characterization of bacteria using the 16S rRNA method.

2. Materials and Method

2.1 Sample Collection and Analysis

The batik wastewater sample was collected from the batik industry in Kelantan, Malaysia. The sample from batik wastewater was analyzed in terms of pH, COD and color. The analysis of pH was using pH probe OHAUS 3100. The analysis of COD and color were conducted by using DR5000 HACH spectrophotometer.

Table 1. Raw analysis of batik wastewater

Parameters (Unit)	Value
pH	11
COD (mg/L)	1390
Color (units)	950

The color of batik wastewater was green. Table 1 shows the characteristics of batik wastewater in terms of COD, pH and color. Batik wastewater has a high number of COD and color intensity. The pH also indicates that the batik wastewater has an alkaline condition.

2.2 Preparation of Enrichment Selective Medium
Horikoshi-II modified medium was used to provide nutrient for bacterial growth contains 10.00 g of glucose, 5.00 g of yeast extract, 5.00 g of peptone, 1.00 g of K_2HPO_4 and 0.20 g of $H_{14}MgO_{11}S$ was prepared in 1 L solution. The pH was adjusted until 10 using 10% (w/v) of NaOH [15].

2.3 Enrichment of Dye-degrading Microbe
For enrichment of dye degrading bacteria in batik wastewater, 22 mL of dye effluent (90% v/v) is prepared by adding 2.5 ml of Horikoshi-II medium (10% v/v) in a 25 ml tube. The 22 mL dye effluents were used as control. The tubes were incubated at 37°C with static condition for 7 days with 24-hour observation of decolorization followed by shaking conditions at 35°C and 150 rpm. Furthermore, the turbidity was observed 24 hours for 7 days. The tubes were also analyzed by measuring the COD, pH and color [16].

2.4 Isolation of Dye-degrading Bacteria
Dye-degrading bacteria were isolated by using serial dilution on spread plate technique. The serial dilution was done by adding 0.1 mL into 9.9 mL of distilled water until the concentrations 10^{-6}. For the spread plate method 0.1 mL of diluted sample was spread on the Horiksohi-II agar plate. The plate then was incubated at 37°C for 60 hours. The plate then was streaked until obtaining the single colony and pure culture [17].

2.5 Bacterial Growth Profile
The bacterial growth profile was obtained by using a 250 mL flask; 120 mL of Horikoshi-II medium was used in the flask. The isolated pure culture of bacteria from the Horikoshi-II agar plate was inoculated into the flask. The flask further was incubated in a shaker 35°C and 150 rpm. For the reading of OD, 1 ml of the broth was used to estimate the absorbance in UV-Vis DR6000 HACH Spectrophotometer for every 2 hours until the estimated growth profile was achieved [18].

2.6 Dry Cell Weight
The dry cell weight method was obtained by growing the bacteria in broth and incubated in 37°C. The filter membrane pore size 0.45 µm was used and weighed. For this method 2 ml of broth was centrifuged to generate the cell pellet. The cell pellet was then rinsed out by distilled water into the filter membrane. The filter membrane was incubated in the oven at 70°C for 2 hours until constant dry cell weight was achieved [19].

2.7 Colony Morphology Characterization
The isolated bacteria were characterized based on its morphology by observing the colony characteristics on solid medium, colony pigmentation, elevation, form and margin. Gram staining was used to differentiate the Gram-negative and Gram-positive bacteria. In this method, Gram-positives

retain the crystal violet and hence, after decolorizing remain purple. After decolorizing with alcohol, Gram negative bacteria become colorless and appear pink when counterstaining with the alcohol [20].

The bacterial culture was spread gently into the slide by using an inoculating loop to create emulsion. The smear was heated over the flame. Crystal Violet was stained on the heat-fixed smear and was incubated for 1-2 minutes. The stain was washed with distilled water and was flooded with iodine followed by incubation within 1-2 minutes. The stain was decolorized by using ethanol followed by rinsed with distilled water. Safranin was used as counterstain followed by incubation within 1 minute. The slide was washed with distilled water. The slide was then blotted and dried. The smear was observed under light-microscopy for further investigation [21].

2.8 16S rRNA Sequence Analysis

Genomic DNA was extracted using Genomic DNA Purification Kit (Promega) and was conducted according to the manufacturer's protocol followed by polymerase chain reaction (PCR). The PCR reaction tube (50 μL) was prepared which consisted of mixture 44 μL Master Mix, 1 μL forward primer 27F, 1 μL reverse primer 1492R and 4 μL DNA template.

After genomic DNA extraction, the 16S rRNA gene was amplified using Mastercycler® PCR thermal cycler (Eppendorf, Hamburg, Germany) with temperature cycle as follows: Initial denaturation at 94°C for 4 minutes. Then 30 cycles were set of denaturation at 94°C for 1 minute, annealing at 48°C for 30 seconds and extension at 68°C for 2 minutes, followed by final extension at 68°C for 5 minutes. After PCR cycles were completed then the PCR product was further running on gel electrophoresis to estimate the presence and size of DNA.

Agarose gel was prepared by dissolving 0.8 g agarose in 80 mL of 1X TAE buffer. The mixture was then heated for 2 minutes in the microwave. After the mixture was completely dissolved, the mixture was then poured in the gel casting tray and a comb was placed into the agarose mixture. The agarose mixture took 1 hour to be solidified. The TAE buffer was then flooded into the solidified gel until it covered the solidified gel. After that 5 μL of 1 kb DNA ladder was loaded into the first well of the gel. The 4 μL of loading dye and 1 μL of DNA sample were mixed on parafilm and were then loaded into the other of the gel. The gel electrophoresis was run using 80 V for 4 minutes. The gel was stained with EtBr for 15 minutes and then destained with distilled water for 5 minutes. The gel was viewed under UV light to detect the presence of a band which represented the presence of desired DNA [22].

The PCR products were sent to First Base Laboratories, Malaysia for purification and sequencing. The obtained sequences were queried against the GenBank and Ribosomal Database Project databases using Blast algorithms. Phylogenetic relationships were analyzed by the evolutionary distance matrix calculated using the neighbor-joining method. Evolutionary analyzes were developed in MEGA7 software [23].

3. Result and Discussion

3.1 Isolation of Dye-degrading Bacteria

The Horikoshi-II media was used as enrichment media to propagate the growth of dye degrading bacteria from the alkaline batik wastewater sample. The Horikoshi-II modified media adjusted pH 10 was reported to selectively isolate bacteria from the alkaline condition of the sample. The Horikoshi-II media component was suitable to be modified until pH 10 as compared with Nutrient Broth media was not effective to be adjusted until alkaline condition [24]. The turbidity of the broth media can be used as the signal to indicate the bacteria start to growth [25].

Figure 1. Enrichment media for batik wastewater sample

Figure 1 shows the change of the color of batik wastewater from green to cyan after observation in the static condition (facultative anaerobic) for 7 days. The shaker condition was done for 3 days to increase the turbidity of the broth. The serial dilution spread plate method and streak plate method was conducted to obtain the pure colony. There are two bacteria isolated from the serial dilution spread plate method which white pigment bacteria and red pigment producing bacteria. The white pigment has unstable growth therefore the red pigmented bacteria was selected based on their faster growth. The selected bacteria were then labeled as A1.

The reading in terms of COD, pH and color also was obtained for both control and A1 after the change of the color from green to cyan. This method was conducted to determine the ability of the isolated bacteria to reduce the color and remove COD.

Table 2. Batik wastewater sample analysis after change of the color

	COD (mg/L) and removal efficiency (%)	**pH**	**Color (ADMI) and removal efficiency (%)**
Control	700 (49%)	10.08	862 (9%)
A1	862 (41%)	9.69	730 (23%)

Table 2 represents the analysis of COD, pH and color after the batik wastewater sample changed in terms of color. For control results which contain batik wastewater only, the removal of COD reaches 49% compared to A1 which contains batik wastewater and 10% of Horikoshi-II media reach 41%. For COD removal they have slightly significant differences. In terms of color reduction, the A1 has greater ability which reaches 23% rather than control only reaching 9% as well as pH reduction which decreases from 11 to 9.69.

Even the bacteria A1 show small reduction of color and COD removal but it shows great ability to degrade batik wastewater. This is due to the bacteria A1 being in un-optimized condition. Previous study reported that the efficiency of the bacterial culture immobilized into light expanded clay aggregate (LECA) could increase the performance of batik wastewater treatment resulting in 98.2% dye removal [26].

3.2 Growth Profile Bacteria

Generally, the growth profile of the bacteria is divided into four phases which include lag phase, log phase, stationary phase, and death phase. The bacteria start to adapt with the growth condition called lag phase. In this phase the individual bacteria are maturing and unable to divide. Exponential phase or log phase is the condition when the bacteria can divide, and biomass increase exponentially with time. When the growth rate and death rate become equal, this is also called the stationary phase. This stage occurs due to depletion of availability nutrients, competition within the bacteria and production of inhibitory products. Furthermore, in the death phase the bacteria are dead due to accumulation of waste products and insufficient nutrients [27].

The growth profile of bacteria A1 has a typical standard growth profile based on Figure 2. The bacteria A1 has a short lag phase. Starting from hour 2 the bacteria A1 undergo the exponential phase until hour 4. The log phase of A1 was 1.106 at 600 nm and lasted about 2 hours.

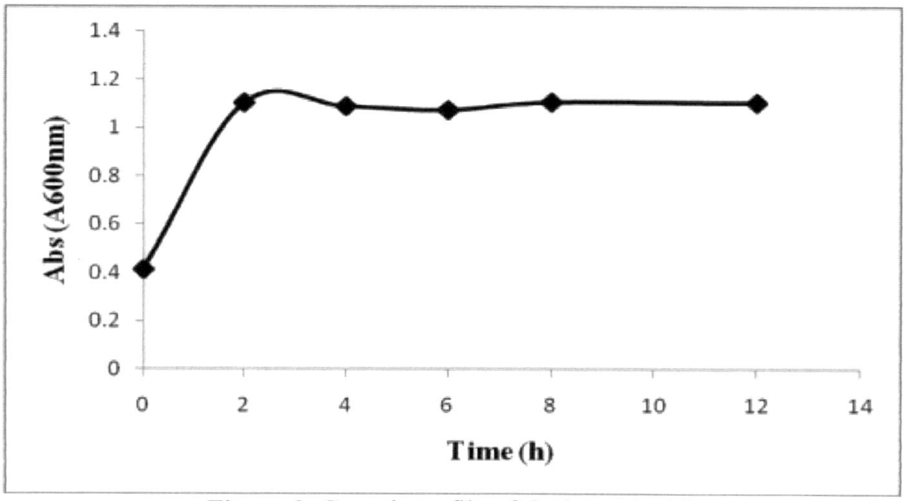

Figure 2. Growth profile of the bacteria A1

3.3 Dry Cell Weight

The dry cell weight was calculated and plotted to determine the specific growth rate and generation of time. The dry cell weight is the weight measured when the water content removed.

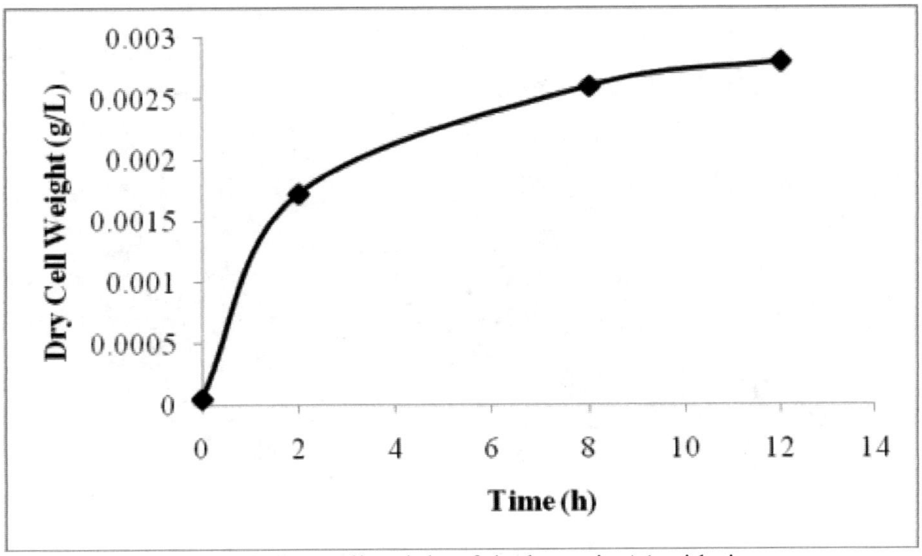

Figure 3. Dry cell weight of the bacteria A1 with time

Based on Figure 3 the maximum of dry cell weight bacteria A1 was 0.0028 g/L. The dry cell weight has a different pattern with the growth profile specifically in the stage after the hour 4. In the growth profile after hour 4 the bacteria enter the stationary phase however in the dry cell weight the number of biomasses increases. Basically, the relationship between growth profile and dry cell weight is very good in dilute suspensions bacterial culture however for higher concentrated suspensions this relationship no longer holds [28]. The specific growth rate and generation time was generated from the dry cell weight. The bacteria A1 has specific growth rate 0.069 h^{-1} and generation time 14.49 h.

3.4 Colony Morphology Characterization

The colony morphology was used to characterize the bacterial culture.

Table 3. Colony morphology of bacteria A1

Pigmentation	**Form**	**Elevation**	**Margin**
Red	Circular	Convex	Entire

From the Table 3 the colony of bacteria A1 has red pigmentation, circular form, convex elevation, and entire margin. Gram staining was used to differentiate between gram negative and gram-positive bacteria. Gram positive has properties thick peptidoglycan which can absorb the purple color of Crystal Violet however the gram negative has absence of peptidoglycan therefore it become pink color when counterstain with Safranin [29].

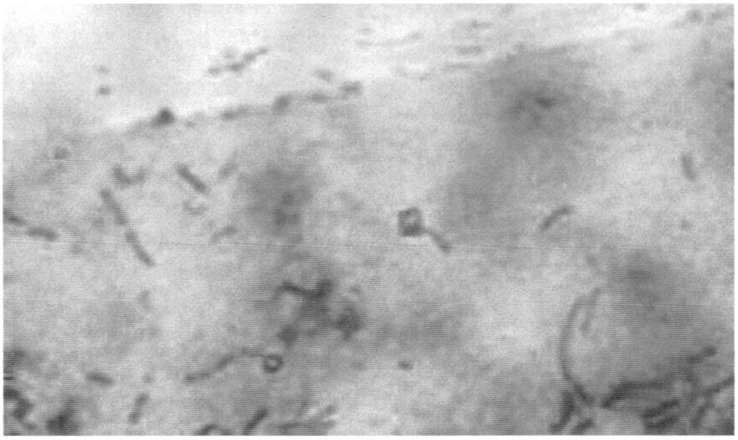

Figure 4. Gram staining observation of the bacteria under light microscope (40X magnification)

Based on Figure 4 the bacteria A1 was dark purple color which indicate the bacteria A1 are gram positive bacteria. The cellular morphology was rods as observed under the light microscope with 40X magnification.

3.5 16S rRNA Sequence Analysis

Prior to conducting the 16S rRNA the bacteria A1 was characterized as having the ability to remove COD and reduce color of batik wastewater samples. The extracted genomic DNA of bacteria A1 was used as template of PCR. The 16S rRNA gene was amplified using the universal primer with forward sequence (27F) 5'-AGAGTTTGATCMTGGCTCAG-3' and reverse sequence (1492R) 5'-TACGGYTACCTTGTTACGACTT-3'.

The 1 kb of ladder was used to estimate the molecular weight of the PCR product. Based on gel electrophoresis of PCR product, a band near 1500 suggesting successful amplification of the 16S rRNA gene [30].

3.6 BLAST Analysis

The purified PCR products were sent for DNA sequencing. The obtained result sequence for ARNZ 2707 strain was then analyzed using the BLAST software on the NCBI database (www.blast.ncbi.nlm.nih.gov). ARNZ 2707 was found to be closely related to *Bacillus cohnii* strain NBRC 15565 with 96% homology and *Bacillus cohnii* strain DSM 6307 with 96% homology. The strain was then designated as *Bacillus* sp. ARNZ 2707.

3.7 Phylogenetic Analysis

Bacteria A1 was then labeled as strain ARNZ 2707. Phylogenetic analysis based on the 16S rRNA gene sequences was constructed to prove the similarity between clusters of Bacillus sp. with strain ARNZ 2707. The phylogenetic tree was constructed based on the neighbor-joining method. Figure 5 represents the phylogenetic tree of strain ARNZ 2707 within same genus.

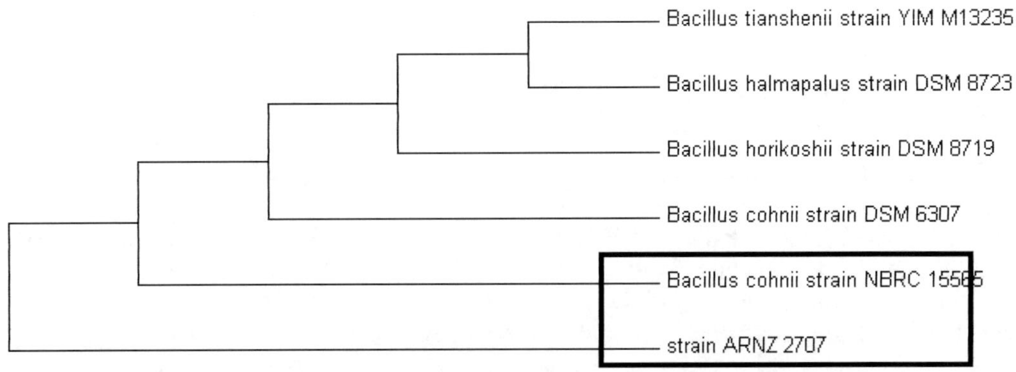

Figure 5. Phylogenetic tree analysis for ARNZ 2707 strain

The previous study on *Bacillus cohnii* RAPT1 shows a gram positive alkaliphile and successfully degrades reactive red (RR120) dye with high efficiency in immobilized packed bed reactors [31]. As a conclusion, ARNZ 2707 was isolated and determined to be a Dye-degrading bacteria. The strain ARNZ 2707 was identified to be *Bacillus* sp. via 16S rRNA.

4. Conclusions

One type of bacterium labelled as A1 was isolated from batik wastewater samples. The A1 was selected using Horikoshi-II medium with alkaline condition. The A1 has the ability to reduce 23% of color and 41% of COD. After characterization using the 16S rRNA method, the A1 is identified as *Bacillus* sp. ARNZ2707. The performance of the isolated bacteria can be increased in future work such as optimization at optimum condition, immobilization, mixed culture, etc.

Acknowledgments

The authors are grateful for the support from all staff of Malaysia-Japan International Institute of Technology and School of Biology, Faculty of Science, Universiti Teknologi Malaysia and especially for Prof. Dr. Muhammad Ali Yuzir for the fully support.

References

[1] Lim SL, Chu WL and Phang SM 2010. Use of Chlorella vulgaris for bioremediation of textile wastewater *Bioresource Technology* **101** 7314-7322.
[2] Subki N and and Rohasliney H 2011. A preliminary study on batik effluent in Kelantan state: A water quality perspective *International conference on management, Social Science, Biology and Pharmaceutical* 274-276.
[3] Koppel N, Rekdal VM and Balskus EP 2017. Chemical transformation of xenobiotics by the human gut microbiota *Science* **365**.
[4] Yaacob MR, Ismail M, Zakaria MN, et al. Environmental awareness of batik entrepreneurs in Kelantan, Malaysia: An early insight. *International Journal Academic Resource Business Social Science* 2015; 4: 273-286.
[5] Khalik W, Ngee LH and An SO 2015. Decolorization and mineralization of batik wastewater through solar photocatalytic process *Sains Malaysia* **44** 607-612.
[6] Ngulube T, Gumbo JR and Masindi V 2017. An update on synthetic dyes adsorption onto clay-based minerals: a state-of-art review *Journal Environmental Management* **191** 35-57.
[7] Guo G, Li X, Tian F and Liu T 2020. Azo dye decolorization by a halotolerant consortium under microaerophilic conditions *Chemosphere* **244**.
[8] Vasconcelos GMD, Mulinary J and Souza SMAG 2021. Biodegradation of azo dye-containing wastewater by activated sludge: a critical review *World Journal of Microbiology and Biotechnology* **37**.

[9] Barragan BE, Costa C and Marquez MC 2007. Biodegradation of azo dyes by bacteria inoculated on solid media. *Dyes pigment* **75** 73-81.
[10] Hsueh CC, Chen BY and Yen CY 2009. Understanding effects of chemical structure on azo dye decolorization characteristics by Aeromonas hydrophila. *Journal Hazardous Material* **167** 995-1001.
[11] Kuberan T, Anburaj J and Sundaravadivelan C 2011. Biodegradation of azo dye by Listeria sp. *International Journal Environmental Science* **1** 1760-1770.
[12] Leelakriangsak M and Sukallaya B 2012.Characterization of decolorizing activity of azo dye by Bacillus subtilis azoreductase AzoR1 *Journal Science Technology* **34(5)** 509-516.
[13] Mercier C, Chalansonnet V and Orenga S 2013. Characteristics of major Escherichia coli reductases involved in aerobic nitro and azo reduction *Journal Applied Microbiology* **115** 1012-1022.
[14] Vitor V and Corso CR 2008. Decolorization of textile dye by *Candida albicans* isolated from industrial effluents *Journal Industrial Microbial Biotechnology* **35** 1353-1357.
[15] Sarethy PI, Saxena Y and Kapoor A 2012. Amylase produced by Bacillus sp. SI-136 isolated from sodic-alkaline soil for efficient starch desizing *Journal Biochemistry Technology* **2** 604-609.
[16] Guadie A, Tizazu S, Melese M and Guo 2017. Biodecolorization of textile azo dye using Bacillus sp. strain CH12 isolated from alkaline lake *Biotechnology Reports* **15** 92-100.
[17] Rathod MG and Pathak AP 2016. Optimized production, characterization and application of alkaline proteases from taxonomically assessed microbial isolate from Lonar soda lake, India *Biocatalyst Agricultural Biotechnology* **7** 164-173.
[18] Tung PS, Ransangan J and Hatai K 2019. Identification, Growth Profile and Probiotic Properties of Autochthonous Intestinal Bacteria of Sagor catfish (Hexanematichthys sagor) *Biocontrol Science* **24** 1-11.
[19] Ismail TM, Ramzy K and Abdel-Wahab MN 2018. Performance of hybrid compression ignition engine using hydroxy (HHO) from dry cell *Energy Conversion and Management* **155** 287-300.
[20] Azhar M, Uniyal V and Chauhan N 2014.Isolation and biochemical characterization of Halophiles from Sahastradhara region, Dehradun, India *International Journal Current Microbiology Applied Science* **3** 753-760.
[21] Silva TP, Noyma NP and Duque TLA 2014. Visualizing aquatic bacteria by light and transmission electron microscopy *Antonie van Leeuwenhoek* **105** 1–14.
[22] Kaufmann ME and Pitt TL 2018. *Pulsed-Field Gel Electrophoresis of Bacterial DNA* CRC Press p.10.
[23] Ghalyanchilangeroudi, A., Hosseini, H., Jabbarifakhr, M., Mehrabadi, M.H.F., Najafi, H., Ghafouri, S.A., Mousavi, F.S., Ziafati, Z. and Modiri, A. 2018. Emergence of a virulent genotype VIIi of Newcastle disease virus in Iran. *Avian Pathology*, 47(5), 509-519.
[24] Shiroosi M, Amoozegar MA and Khajeh K 2016. Purification and characterization of an alkaline chloride-tolerant laccase from a halotolerant bacterium, Bacillus sp. strain WT *Journal of Molecular Catalysis B: Enzymatic* **134** 89-97.
[25] Thakur B, Amarnath CA and Mangoli SH 2015. Polyaniline nanoparticle based colorimetric sensor for monitoring bacterial growth *Sensors and Actuators B: Chemical* **207** 262-268.
[26] Yanto DHY, Guntoro MA and Nurhayat OD 2021. Biodegradation and biodetoxification of batik dye wastewater by laccase from Trametes hirsuta EDN 082 immobilised on light expanded clay aggregate *3 Biotech* **11** 247.
[27] Nanung AF, Ditasari A and Yuny E 2016. Isolation of Indigenous denitrifying bacteria from odorous region of a layer farm and its potential as nitrate reducing agent. In: *Proceeding of 17th Asian-Australasian Association of Animal Production Societies Animal Science Congress*, Fukouka, Japan, 22-25 August 2016, pp.685-690.
[28] Scott S. Measurement of Microbial Cells by Optical Density 2011. *Journal of Validation Technology* **17** 46-49.

[29] Yang J, Gao G and Zhang X 2019. One-step synthesis of carbon dots with bacterial contact-enhanced fluorescence emission: Fast Gram-type identification and selective Gram-positive bacterial inactivation *Carbon* **146** 827-839.
[30] Srinivasan R, Karaoz U and Volegova M 2015.Use of 16S rRNA Gene for Identification of a Broad Range of Clinically Relevant Bacterial Pathogens *PLOS One* 1-22.
[31] Padmanaban VC, Geed SR and Achary A 2016. Kinetic studies on degradation of Reactive Red 120 dye in immobilized in packed bed reactor by Bacillus cohnii RAPT1 *Bioresource Technology* **213** 39-43.

Production of Ligninolytic Enzymes from Thermophilic Bacterial Strains Isolated from Palm Oil Wastes

F A Riyadi*, **N F Azman, N Yusof, and F N M Akhir**

Department of Chemical and Environmental Engineering, Malaysia-Japan International Institute of Technology, Universiti Teknologi Malaysia, Jalan Sultan Yahya Petra, 54100 Kuala Lumpur, Malaysia

*Corresponding author e-mail: arfatimah@graduate.utm.my

Abstract. The conversion of lignocellulosic biomass into bioethanol or biochemical products requires a crucial pre-treatment process to break down the recalcitrant lignin structure. Biological depolymerization of lignin using microbial enzyme appeared to be a promising pre-treatment alternative as it offers environmentally friendly treatment with lower energy requirements. In this study, three (3) thermophilic bacterial strains (S2, S11Y, S23) with lignin-degrading potential were previously isolated from palm oil wastes and identified as *Stenotrophomonas* sp., *Bacillus subtilis*, and *Aeribacillus* sp., respectively. These isolates demonstrated the capability to grow in a medium containing AL as the sole carbon source. Most isolates also demonstrated enzymatic activities toward lignin peroxidase (LiP), manganese peroxidase (MnP), and laccase (Lac) enzymes within a short period of 72 h. The highest LiP was secreted by strain S2 (70.97 ± 0.059 U/L), followed by S11Y (63.570 ± 0.021 U/L) and S23 (60.880 ± 0.042 U/L). MnP was generated with low activities of approximately 1.313 ± 0.037 U/L by S11Y, 0.364 ± 0.004 U/L by S2, and 0.215 ± 0.021 U/L by S23. For the Lac enzyme, low activities were detected only in S2 and S11Y, with 2.075 ± 0.612 U/L and 1.463 ± 0.116 U/L, respectively. This preliminary study appeared to show that the isolated thermophilic bacteria have the potential to be applied for lignin degradation application.

1. Introduction

Malaysia's palm oil sector has contributed to the most significant biomass generation in the country, with nearly 143 million tons annual production and predicted to increase further with the expansion of palm oil plantations [1]. Utilizing the palm oil biomass, which currently has little use in the industries, for higher value-added downstream applications, such as bioethanol and bio-based chemicals, can significantly contribute to gross national income. However, the initial pre-treatment step of breaking down the lignin component of the biomass structures before transforming it into industrial products is challenging. The major components of lignocellulosic biomass are cellulose, hemicellulose, and lignin. [2]. Among them, lignin has the most complex structure, which causes it to be challenging to degrade.

Microbial enzymes to degrade lignin appear to be a potential pre-treatment strategy since it provides an environmentally friendly treatment with lesser energy requirement. Some microorganisms are predicted to have the ability to develop a complex system of oxidative enzymes that could subsequently degrade the lignin structures [3]. The enzymatic decomposition of lignin is primarily mediated by two types of ligninolytic enzymes: phenol oxidase (laccase, Lac) and peroxidases such as lignin peroxidase,

LiP, and manganese peroxidase, MnP [4]. Some white-rot and brown-rot fungi have been proven to decompose lignin, and various researches on the synthesis of fungal lignin-degrading enzymes have been reported [5,6]. However, fungi have more complex genetic and protein expression characteristics [7] than bacteria. Bacteria are well-known for their rapid growth, metabolic flexibility, and their lignin metabolism can be investigated further. However, most enzymatic lignin degradation from bacteria, especially thermophilic bacteria, is underexplored. Thermophilic bacteria are expected to secrete enzymes with better stability and resistivity to denaturation than the mesophiles [8], which require further investigation. This study evaluates three isolated thermophilic bacterial strains (S2, S11Y, S23), previously identified as *Stenotrophomonas* sp., *Bacillus subtilis,* and *Aeribacillus* sp., respectively, for their lignin-degrading potential. The strains were further examined to evaluate their ligninolytic enzyme activities.

2. Materials and Methods

2.1. Cultivation of bacteria strains and Determination of cell growth

In this study, three thermophilic bacteria strains (S2, S11Y, S23) were previously isolated from Palm Oil Empty Fruit Bunch and identified as *Stenotrophomonas* sp. strain S2, *Bacillus subtilis* strain S11Y, and *Aeribacillus* sp. strain S23, respectively. These bacterial strains were utilized to evaluate their lignin-degrading potentials. Bacterial isolates were grown on agar plates comprising of 2.5 g/L of Alkali lignin (AL) in minimal salt medium (AL-MSM) for bacterial growth [9], where AL was used as the only carbon source. Further screening was performed with Azure-B dye [6] by supplementing the agar plate with 0.01 %w/v Azure B. The appearance of a clear zone around the colony was used to indicate peroxidase enzyme activity [10].

2.2. Inoculum Preparation

A single colony of all bacteria strains was initially grown on 20 mL Luria-Bertani (LB) broth at 50 °C, shaking at 160 rpm for 24 hours to obtain a final OD600 of 1.0 [9]. The cells were harvested by sequentially centrifuged at 8,000 ×g for 10 mins. The cell pellets were washed twice with a 10 mL MSM. The bacteria cells were then resuspended in 5 mL containing the same washing media. 0.25 mL of the bacteria cells were collected and added to 10 mL of lignin fermentation medium containing AL-MSM medium to set the initial $OD_{600} \sim 0.1$. The reaction was performed by incubating at 160 rpm and 50 °C for 7 days. Sample (1 mL) was taken daily to evaluate the cell growth by evaluating the OD_{600} of the degraded samples. Samples collected were centrifuged to separate the supernatants from the pellets at 8,000 ×g for 5 mins. The supernatants were collected and used as crude enzymes to measure the ligninolytic enzyme activity.

2.3. Enzymatic Assays for Ligninolytic Enzyme Production

Lignin peroxidase (LiP) enzyme was measured based on the oxidation of dye Azure B [11] at 651 nm ($\epsilon = 48800\ M^{-1}cm^{-1}$). The reaction mixture (total volume, 1.0 mL) contained 200 μl of culture filtrate in 125 mM Na-tartrate buffer (pH 3.0) and 0.160 mM azure B. The addition of 2 mM H_2O_2 initiated the reaction. One unit of enzyme activity is expressed as a decrease of 0.1 units per minute of the culture filtrate. Manganese peroxidase (MnP) activity was determined based on Phenol red assay [10]. A total reaction volume of 1.0 mL contained 50 mM Na-tartrate buffer (pH 4.5), 0.2 mM $MnSO_4$, 0.1 mM H_2O_2, 0.0025% phenol red and culture filtrate. The reaction was monitored at 431 nm ($\epsilon = 22000\ M^{-1}cm^{-1}$). Laccase (Lac) activity was assayed through the oxidation of 2,2'-azinobis-(3-ethylbenzethiazoline-6-sulphonate) (ABTS) [12]. The enzymatic mixture consisted of 0.03% ABTS, 0.1 M Na-acetate buffer (pH 5.0), and culture filtrate, making a total volume of 1 mL. Oxidation of ABTS was measured by an increase in absorbance at 420 nm ($\epsilon = 36000\ M^{-1}cm^{-1}$).

IOP Conf. Series: Earth and Environmental Science 1091 (2022) 012063 doi:10.1088/1755-1315/1091/1/012063

3. Results and Discussion

3.1. Bacterial growth and screening for ligninolytic activity

In this study, three thermophilic bacterial strains (S11Y, S23, and S2) were evaluated for their ability to degrade lignin structure. Ligninolytic bacteria were reported as the bacteria that can utilize lignin as their primary carbon source and do not require the addition of other carbon sources as co-substrates. [13]. In this study, all bacterial strains belonging to different genera (*Stenotrophomonas* sp., *Bacillus subtilis*, and *Aeribacillus* sp.) were grown on solid and liquid medium containing MSM with AL as the sole carbon source. AL is a semi-degraded (average $M_w \sim 10{,}000$) chemically modified derivative of high-molecular-mass lignin, used as a lignin model [14]. Based on Figure 1, all bacterial strains demonstrated growth on AL-MSM agar plate medium when grown for 7 days, indicating their capability in utilizing AL for growth. Strain S11Y demonstrated prominent growth, followed by strain S2 and S23.

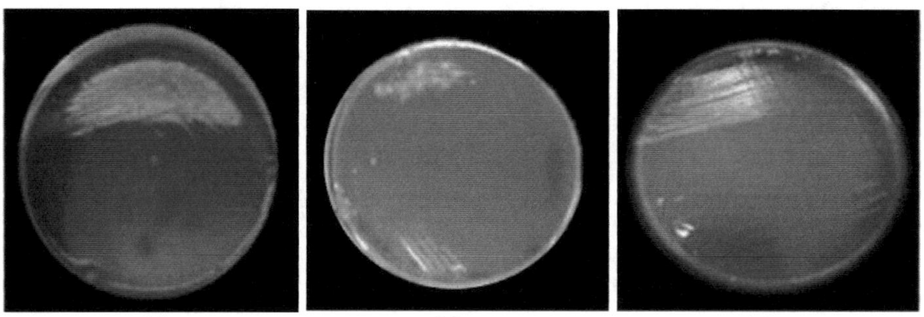

(a) (b) (c)

Figure 1. Growth of the isolated bacterial strains in solid agar plate for (a) *Stenotrophomonas sp.* strain S2, (b) *Bacillus subtilis* strain S11Y, and (c) *Aeribacillus sp.* strain S23, using AL as sole carbon source.

Growth on the liquid culture in OD_{600} was also conducted to confirm the bacterial capability in utilizing lignin. Figure 2 demonstrates the growth patterns for all bacterial strains in the AL-MSM. The rapid growth of S11Y and S23 was observed in the first 72 hours. Maximum growth was seen after 144 and 168 hours for both strains, respectively (Figure 2). On the contrary, strain S2 showed slower growth, but maximum growth was seen at 120 hours incubation period. The results indicated that the bacteria metabolized AL during the exponential growth phase to provide carbon and energy sources. However, in the previous reports, most bacteria strains, such as *Nonomuraea gerenzanensis* [14] and *Bacillus ligniniphilus* L1[15], and *Bacillus* sp. [16] utilize glucose or mannose as carbon sources at the initial growth stage and subsequently use lignin as co-substrate. Besides, compared to the alkaliphilic bacterium *Bacillus ligniniphilus* L1 [15], *Bacillus subtilis* strain S11Y in this study showed better growth when AL was used as the sole carbon source. Thus, in contrast to the previous reports, AL can be used as the sole carbon source by all strains in this study. Besides, since these bacterial strains can metabolize AL throughout the entire life cycle for their growth, better efficiency and lignin degradation rate might be attained than the bacteria strains that use lignin as co-substrate.

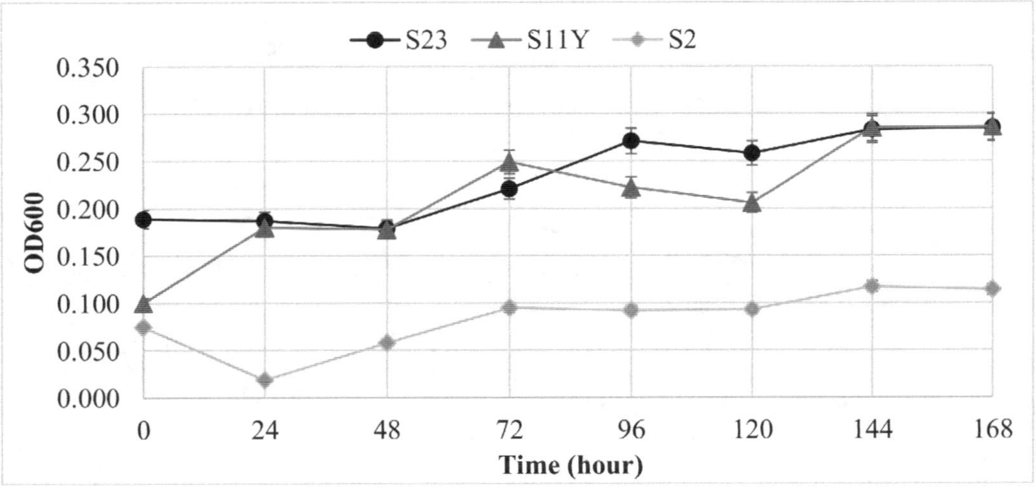

Figure 2. Growth of the isolated bacterial strains (*Stenotrophomonas* sp. strain S2, *(2.) Bacillus subtilis* strain S11Y and *(3.) Aeribacillus* sp. strain S23) using AL as sole carbon media.

The Azure B plate decolorization is commonly used as a qualitative method to screen for bacterial strains with the lignin-degrading ability [17]. Thus, this analysis was performed on the isolated bacterial strains, as presented in Figure 3.

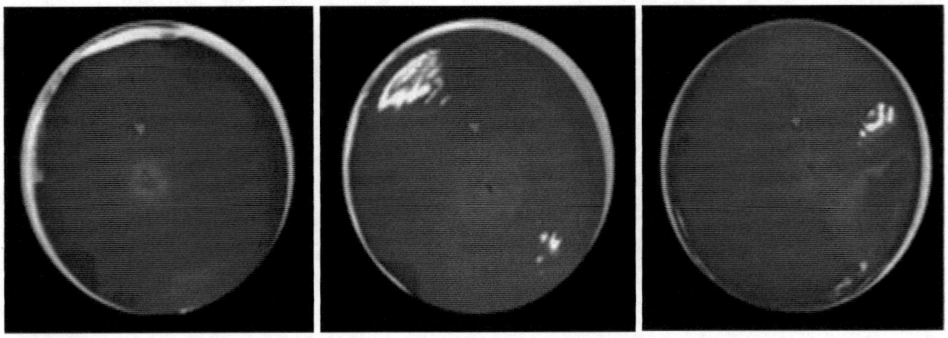

Figure 3. Azure B agar plate decolorization of (a) *Stenotrophomonas* sp. strain S2; (b) *Bacillus subtilis* strain S11Y; (c) and *Aeribacillus* sp. strain S23.

In this study, all bacterial strains showed decolorization on agar plates containing MSM incorporated with Azure B as the lignin polymeric dye (Figure 3). Based on the results, strain S2 and S11Y demonstrate a clear zone of decolorization. A similar observation was also reported for *Enterobacter hormaechei* and *Bacillus licheniformis* [18] and *Nonomuraea gerenzanensis* [14], demonstrating the presence of ligninolytic enzymes in each isolate.

3.2. Production of ligninolytic enzymes by bacterial strains

Several ligninolytic enzyme assays were performed for each isolate to ensure the lignin-degrading ability of each strain. The studied enzymes include LiP, MnP, and Lac, which were previously reported as ligninolytic enzymes. Among all the enzyme assays performed, LiP and MnP were secreted by all bacterial strains (Table 1), with LiP being the predominant enzyme being detected. Highest LiP were secreted by strain S2 (70.968 ± 0.175 U/L), followed by S11Y (63.570 ± 0.021 U/L), and S23 (60.880 ± 0.042 U/L). Based on the observation, strain S23 showed an absence of Lac activity and generated low enzyme activities compared to the rest of the strains. In most bacterial strains, maximum enzyme production was generated at 72 hours incubation period, except for the production of MnP by S11Y, where the highest enzyme production was secreted at 168 hours of incubation.

Table 1. Highest LiP, MnP, and Lac enzyme activities produced by bacterial strains studied.

Isolates	LiP Enzyme Activity (U/L)	Period (h)	Lac Enzyme Activity (U/L)	Period (h)	MnP Enzyme Activity (U/L)	Period (h)
S23	60.880 ± 0.042	72	ND*	-	0.215 ± 0.021	72
S11Y	63.570 ± 0.021	72	1.463 ± 0.116	72	1.313 ± 0.037	168
S2	70.968 ± 0.059	72	2.075 ± 0.612	72	0.364 ± 0.004	72

*ND: Not detected

Table 2. Comparison of bacterial LiP, MnP, and Lac enzyme activities with previous studies.

Isolates	Enzyme activities (U/L)			References
	LiP	MnP	Lac	
Bacillus subtilis WPI	-	-	2.28	[19]
Bacillus sp. strain CS-1	-	-	0.60	[20]
Brevibacillus agri AN-3	3.10	-	1.90	[21]
Stenotrophomonas maltophilia	9.22	7.84	3.77	[22]
S23 (*Aeribacillus* sp.)	60.88	0.22	-	This study
S11Y (*Bacillus subtilis.*)	73.57	1.31	1.46	
S2 (*Stenotrophomonas* sp.)	70.97	0.36	2.08	

Table 2. shows the comparison of the studied bacterial LiP, MnP, and Lac enzyme activities with the previous reports on ligninolytic enzyme productions by bacteria of similar species with this study. Most of these strains only secreted Lac, and no apparent LiP activity was observed when strains were grown on lignin substrate [19] [20]. In contrast, our results demonstrated the highest enzyme activity for LiP by each strain when grown in AL, and only low Lac activities were observed in S11Y and S2. Besides, most of the previous related studies for *Bacillus* sp. did not detect the presence of MnP activities [19] [20] [21], but it was detected in our study. Thus, it can be seen that the enzyme activities studied are comparable with the previous studies, indicating that the studied bacterial strains present promising results towards the production of lignin-degrading enzymes. Besides, limited reports on the ligninolytic enzymes and their function in lignin degradation by bacteria of genus *Stenotrophomonas* sp. are available [22]. No previous study on lignin degradation by *Aeribacillus* sp. was also reported. Therefore, there is a new potential for future research on attaining novel sources of lignin-degrading enzymes.

4. Conclusion

In conclusion, the thermophilic bacteria reported in this study demonstrated potential for lignin degradation since they can generate the main ligninolytic enzymes (LiP, MnP, Lac). Future work on analyzing the draft genome sequencing of the isolated bacteria could be done to validate the presence of the genes corresponding to ligninolytic enzymes and study the actual mechanism of lignin degradation by bacteria. Exploring bacterial enzymes for lignin degradation will yield fundamental insights into necessary enzymes and could be applied by industrial enzymes for the lignin degradation process.

Acknowledgment
The authors gratefully thank Dr. Hirofumi Hara (University of Tokyo) and Dr. Nor'azizi bin Othman (Universiti Teknologi Malaysia) for their excellent technical assistance. The authors would also like to express appreciation for the support of the sponsors Vot No.: R.K130000.7343.4B400.

References

[1] Agensi Inovasi Malaysia 2013 *National Biomass Strategy 2020: New wealth creation for Malaysia's biomass industry* vol 2
[2] Gillet S, Aguedo M, Petitjean L, Morais A R C, Da Costa Lopes A M, Łukasik R M and Anastas P T 2017 Lignin transformations for high value applications: Towards targeted modifications using green chemistry *Green Chem.* **19** 4200–33
[3] Martínez Á T, Speranza M, Ruiz-Dueñas F J, Ferreira P, Camarero S, Guillén F, Martínez M J, Gutiérrez A and Del Río J C 2005 Biodegradation of lignocellulosics: Microbial, chemical, and enzymatic aspects of the fungal attack of lignin *Int. Microbiol.* **8** 195–204
[4] Malherbe S and Cloete T E 2002 Lignocellulose biodegradation: Fundamentals and applications *Rev. Environ. Sci. Bio/Technology* **1** 105–14
[5] Salvachúa D, Prieto A, Martínez Á T and Martínez M J 2013 Characterization of a novel dye-decolorizing peroxidase (DyP)-type enzyme from *Irpex lacteus* and its application in enzymatic hydrolysis of wheat straw *Appl. Environ. Microbiol.* **79** 4316–24
[6] Sharma A and Aggarwal N K 2017 Isolation and screening of lignolytic fungi from various ecological niches **5** 25–34
[7] Ahmad M, Roberts J N, Hardiman E M, Singh R, Eltis L D and Bugg T D H 2011 Identification of DypB from *Rhodococcus jostii* RHA1 as a lignin peroxidase *Biochemistry* **50** 5096–107
[8] Peng X, Misawa N and Harayama S 2003 Isolation and characterization of thermophilic bacilli degrading cinnamic, 4-coumaric, ferulic acids *Appl. Environ. Microbiol.* **69** 1417–27
[9] Azman N F, Megat Mohd Noor M J, MD Akhir F N, Ang M Y, Hashim H, Othman N, Zakaria Z and Hara H 2019 Depolymerization of lignocellulose of oil palm empty fruit bunch by thermophilic microorganisms from tropical climate *Bioresour. Technol.* **279** 174–80
[10] Archibald F S 1992 A new assay for lignin-type peroxidases employing the dye Azure B *Appl. Environ. Microbiol.* **58** 3110–6
[11] Arora D S and Gill P K 2001 Comparison of two assay procedures for lignin peroxidase *Enzyme Microb. Technol.* **28** 602–5
[12] Lai C M T, Chua H B, Danquah M K and Saptoro A 2017 Isolation of thermophilic lignin degrading bacteria from oil-palm empty fruit bunch (EFB) compost *IOP Conference Series: Materials Science and Engineering* **206** 206
[13] Olajuyigbe F M, Fatokun C O and Oyelere O M 2018 Biodelignification of some agro-residues by *Stenotrophomonas* sp. CFB-09 and enhanced production of ligninolytic enzymes *Biocatal. Agric. Biotechnol.* **15** 120–30
[14] Casciello C, Tonin F, Berini F, Fasoli E, Marinelli F, Pollegioni L and Rosini E 2017 A valuable peroxidase activity from the novel species *Nonomuraea gerenzanensis* growing on alkali lignin *Biotechnol. Rep.* **13** 49–57
[15] Zhu D, Zhang P, Xie C, Zhang W, Sun J, Qian W J and Yang B 2017 Biodegradation of alkaline lignin by *Bacillus ligniniphilus* L1 *Biotechnol. Biofuels.* **10** 44
[16] Raj A, Reddy M M K, Chandra R, Purohit H J and Kapley A 2007 Biodegradation of kraft-lignin by *Bacillus* sp. isolated from sludge of pulp and paper mill *Biodegradation* **18** 783–92
[17] Falade A O, Eyisi O A L, Mabinya L V, Nwodo U U and Okoh A I 2017 Peroxidase production and ligninolytic potentials of fresh water bacteria *Raoultella ornithinolytica* and *Ensifer adhaerens Biotechnol. Rep.* **16** 12-17.
[18] Zhou H, Guo W, Xu B, Teng Z, Tao D, Lou Y and Gao Y 2017 Screening and identification of lignin-degrading bacteria in termite gut and the construction of LiP-expressing recombinant *Lactococcus lactis Microb. Pathog.* **112** 63–9

[19] Sheikhi F, Roayaei Ardakani M, Enayatizamir N and Rodriguez-Couto S 2012 The Determination of Assay for Laccase of *Bacillus subtilis* WPI with Two Classes of Chemical Compounds as Substrates *Indian J. Microbiol.* **52** 701–7
[20] Brown M E and Chang M C Y 2014 Exploring bacterial lignin degradation *Curr. Opin. Chem. Biol.* **19** 1–7
[21] Shah T A, Ali S, Afzal A and Tabassum R 2018 Simultaneous Pretreatment and Biohydrogen Production from Wheat Straw by Newly Isolated Ligninolytic *Bacillus* Sp. Strains with Two-Stage Batch Fermentation System *BioEnergy Res.* 1–15
[22] Nevita T, Sharma · G D and Pandey · Piyush 2018 Composting of rice-residues using lignocellulolytic plant-probiotic *Stenotrophomonas maltophilia*, and its evaluation for growth enhancement of Oryza sativa L. **1** 185–96

Preparation Method of Titanium Dioxide Nanoparticles and Its Application: An Update

N Seman[1], Z I Tarmizi [1*], R R Ali[1], S H M Taib[1], M S N Salleh[1,2], J C Zhe[1], S N A Mohamad Sukri[1]

[1] Department of Chemical Engineering and Environmental. Malaysia Japan International Institute of Technology, Universiti Teknologi Malaysia, 54100 Kuala Lumpur, Malaysia

[2] School of Chemical Engineering, College of Engineering, Universiti Teknologi MARA Cawangan Terengganu, Bukit Besi Campus, 23200 Dungun, Terengganu, Malaysia.

*Corresponding author e-mail: zatil.izzah@utm.my

Abstract. Titanium dioxide (TiO_2) is widely used because of its good biocompatibility and stability. Various methods were used to prepare TiO_2 by using chemical, biological, and physical methods. In this paper, the currents methods to prepare TiO_2 were evaluated either with or without using plant extract. The average particle size produced for different methods of technique and its application was also analyzed. Chemical methods use toxic chemical while physical methods such as gamma produce high energy and using biological method with plant extract is more environmentally friendly. This review gives an up-to-date summary of TiO_2 produced using various synthesis methods with the focus on their particle sizes properties. Recent publications on this topic were extracted from related journals obtained from Science Direct online database from the year 2010 to 2021. Preparation methods, average particles sizes, and the applications of the synthesized TiO_2 were evaluated and further discussed in this review. It was found that TiO_2 of smaller particle sizes were obtained when synthesized without using plant extract. Furthermore, many researchers tend to combine TiO_2 with other materials or composites to be mixed with polymers for various purposes of application especially wastewater treatment application.

1. Introduction

Nanotechnology has become one of the most popular topics in the recent decade as it evolves into becoming a key technology in the 21st century. The development of nanotechnology has shaken the world as they witnessed its capabilities in numerous innovative applications in various fields. Generally, nanotechnology can be defined as the creation of nanoparticles with a dimension ranging from 1 to 100 nm by manipulating the atomic and molecular matter through new techniques or devices [1]. Nanoparticles are known to possess unique physical and chemical characteristics and the application of nanoparticles has been mainly concentrated in the fields of chemical, energy, health care, and cosmetics [2].

Among all kinds of nanoparticles, there are metal oxide nanoparticles such as AgO, ZnO, MgO, CuO, and TiO_2[3][4]. Each type of metal oxides has its own unique properties like electrical, magnetic, optical, thermal, effective against drug-resistant bacteria, viral, and fungal strain as well as catalytic activity[5][6][7].For example of silver nanoparticles, it is used due to its antimicrobial properties and can be applied in wide applications [4]. Compared to other metal oxide nanoparticles, TiO_2 is the most astounding nanomaterial because of its chemical stability, photocatalytic properties, high stability, biocompatibility, and non-toxicity [7][8]. Other than that, TiO_2 is an essential semiconductor photocatalyst that covers a broad range of applications in water treatment, dye-sensitized solar cells, cosmetics, pigments and for seed germination [9] [10]. In the application of pigment making, TiO_2 is used to provide opacity and whiteness to products such as paints, plastics, coatings, sunscreen, inks, medicines for most toothpaste and food [11].

TiO_2 can be synthesized through three approaches which are physical, chemical, and biological methods. Physical methods include thermal decomposition, laser irradiation, and electrolysis while chemical methods use chemicals such as sodium borohydride or sodium citrate as reducing agents. Physical methods typically use highly intensive costly equipment that involves vacuum and chemical methods are considered as the most commonly used technique to synthesize nanoparticles [12]. Usually, the shape, size, and other properties of TiO_2 can be controlled depending on the methods of synthesis [13]. Preparation through chemical and physical methods can lead to environmental problems upon discharge due to the usage of dangerous chemicals [13][7]. On the other hand, biological method or green synthesis is considered to be less poisonous and more friendly to the environment as the usage of toxic chemical can be removed and reduced because plant extract has a stabilizing and reducing agent properties during the reaction[14][15]. In addition, TiO_2 has been successfully synthesized using green synthesis by utilizing plant extract from different plants such as *Jatropha curcas*, *Azadirachta indica*, and *Psidium guajava* [14][7][16] whereby the extract is obtained from the leaves, roots, twigs and, the seeds.

In this review paper, comparative analysis of the preparation methods of TiO_2 nanoparticles using plant extract and without plant extract will be discussed. The scope of the referred journal is using a Science Direct online database and research papers published from the year 2010 until 2021 only. Throughout the study, the particle sizes for TiO_2 were compared based on the preparation method and presence or absence of plant extract during synthesis. Lastly, the potential application of TiO_2 is also discussed and explored in this paper.

2. Preparation Method of Titanium Dioxide

Based on the findings from the referred journal, there are various techniques that can be adopted for the preparation of TiO_2 nanoparticles such as sol-gel method, instantaneous synthesis method, solvothermal method, microwave-assisted synthesis, simple mixing, and precipitation method. Figure 1 shows the illustration of these commonly used synthesis methods of preparing TiO_2. Meanwhile, Table 1 lists recently published research papers conducted on the preparation of titanium dioxide with different particle sizes and its application.

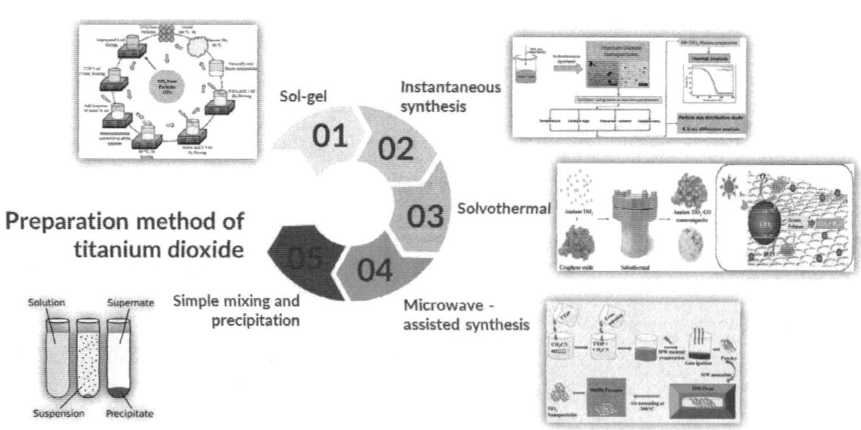

Figure 1. Preparation method of titanium dioxide

Table 1. Previous research conducted to prepare TiO_2 using different techniques which produced different average particle size and its suitable application.

Nanocomposite	Plant Extract	Preparation Technique	Average particle size	Application	References
TiO_2/Carbon (C)/polyaniline (PANI)	-	TiO_2/C-Prepare from sol-gel method TiO_2/C/PANI - in situ chemical oxidative polymerization	TiO_2 - 40 nm TiO_2/C - 31 nm TiO_2/C/PANI - 20 nm	Provide optimum photocatalyst for degradation of Methylene Blue and Crystal violet	[17]
Polypropylene (PP)/TiO_2	-	TiO_2-instantaneous synthesis method	TiO_2 - 10-20 nm	Can be used as pigment in paints, cosmetics, and food additives product	[18]
Chitosan/polyethylene oxide-Ag/TiO_2	Ag- fresh leaves of *Chenopodium murale*	Ag- aqueous extract TiO_2 - sol-gel technique	Ag - 20 nm TiO_2 - 15 nm	Has a potential to use nanocomposites for food packaging application	[19]
poly(sodium styrene sulfonate)/ TiO_2	-	TiO_2 - solvothermal method previous hydrolysis of titanium (IV) isopropoxide (TTIP). Grafting method based on UV radiation	Nanosphere TiO_2 – 7 to 10 nm Nano rods TiO_2- 15 to 22 nm	Used for bio-polymer grafting and in nanomedicine applications	[20]
COP-Ps(Covalent porphyrin polymers)/TiO_2	-	in situ hydrothermal method	TiO_2- $<$ 200 nm	Improved the photocatalytic activity to reduce CO_2	[21]
poly(3-hexylthiophene) (P_3HT)-TiO_2	-	TiO_2-microwave assisted synthesis P_3HT-TiO_2 - in-situ polymerization	TiO_2-7nm	Application in photovoltaic devices	[22]
Ag-TiO_2/porous polymer composite	-	Ag-TiO_2/porous polymer composite - in-situ solvothermal process. (Porous polymer - co-polymerization of divinylbenzene (DVB) and bismaleimide)	TiO_2- 10.3 nm TS -18.2 nm PPTS - 13.2 nm	It has a potential as a sustainable solution for pollution abatement	[23]

IOP Conf. Series: Earth and Environmental Science 1091 (2022) 012064 doi:10.1088/1755-1315/1091/1/012064

Table 1. Continued

Nanocomposite	Plant Extract	Preparation Technique	Average particle size	Application	References
TiO_2	-	Sol gel reaction from variable amounts of waste derived polymeric biosurfactant.	TiO_2 Anatase + Brookite - 18 nm Anatase - 10-10.5 nm	Photocatalytic. Also, can be used for the remediation of aqueous effluents contaminated by organics	[24]
Polypyrrole(PPy)/Ag-TiO_2	-	Ag/TiO_2 - Sol-gel method PPy/Ag-TiO_2 – In situ oxidative polymerization.	PPy/Ag- TiO_2 - 1 nm	Used for sensor application to detect ammonia gas at room temperature	[25]
TiO_2	-	Modified sol gel methods using hyperbranched polyethyleneimine polymer template (HPEI)	TiO_2 - 200 nm	Applied in environmental application in fields such as solar cells, water treatment and air purification.	[13]
Cellulose/ TiO_2	-	Coagulation in sodium hydroxide-thiourea-urea aqueous solution medium by precipitation method.	FESEM Images TiO_2 - 2 nm cellulose/ TiO_2- 1 nm TEM Images TiO_2 - 0.2 nm cellulose/ TiO_2- 50 nm	Application in many variable fields - especially in pharmacology and biotechnology	[26]
TiO_2	Extract of *Azadirachta indica* leaf	TiO_2-Precipitation	Spherical shaped TiO_2- 15 to 50 nm (TEM) SEM image – 25 to 87 nm	Used to fight against vast range of pathogen due to its antibacterial properties	[7]
Poly (methyl methacrylate) (PMMA)/2-aminopyridine/ 2-nitroaniline doped with TiO_2	-	TiO2- precipitation	TiO_2- 5 nm	Synthesized nanocomposites are low-cost material for water treatment, industrial effluents- Congo Red removal and preservatives for food	[27]
TiO_2	Extract of *Aazadirachta indica* twigs, *Ficus benghalensis* and *Syzygium aromaticum*	TiO2 - precipitation	TiO_2-20 nm (tetragonal shapes)	Application as antibacterial, antibiofilm, antifogging and mosquitocidal activity	[28]

IOP Conf. Series: Earth and Environmental Science 1091 (2022) 012064 doi:10.1088/1755-1315/1091/1/012064

Table 1. Continued

Nanocomposite	Plant Extract	Preparation Technique	Average Particles Size	Application	References
borneol-based polymer/ TiO_2 composite film	-	Borneol-based polymer - prepared by free radical polymerization. Composite film - borneol- based polymer dispersed with TiO_2 particles and polytrifluoro chloroethylene (PCTE) binder in tetrahydrofuran to form dispersion	TiO_2 - 5 to 10 nm	Suitable for wastewater treatment	[29]
TiO_2	-	Sol-gel procedure with certain ratio of brookite phase	BJH Method TiO_2- 8.36 nm pei-TiO_2 - 8.85 nm dt-TiO_2 - 8.93 nm	Photodegradation of methylene blue in water and as basic heterogeneous catalyst carbon-carbon forming reactions (act as a catalyst).	[30]
TiO_2	*Jatropha curcas L. latex* (Hedge castor oil)	Precipitation method	TiO_2- 10-20 nm spherical in shape	Used as pigment in paints, removal of toxins from air and water, biomedical sciences such as bone tissue engineering and pharmaceutical industries.	[14]
TiO_2	*Psidium guajava* (Aqueous leaves extract)	Precipitation method	TiO_2 spherical and cluster - 32.58 nm average.	Used as antibacterial coatings and wastewater disinfection	[16]
TiO2	*Glycyrrhiza glabra (licorice) (roots extract)*	Simple mixing and precipitation method	TiO2- 69 nm	Used as multifaceted therapeutic agent demanding in vivo studies	[31]

First of all, sol-gel method is one of the routes for the synthesis of nanoparticles and nanocomposites. Under mild conditions, the sol-gel method allows the mixing of two or more dissimilar phases to form hybrids [32].In sol-gel methods, the production of homogenous nano-sized materials of complex composition can be easily produced with a cost-effective production [33]. From the name of the method, the sol represents the chemical transformation of the liquid into a gel phase and subsequently transition into solid oxide material [34]. The three-dimensional network structure of sol-gel is prepared by freeze-drying and calcination [35].The synthesis of TiO_2 NPs from sol gel method is a wet chemical technique that mostly used in ceramic engineering and material sciences [15] Heshmapour and his co-researcher use sol-gel method for the synthesis of TiO_2 and then mixing with ceramic and PANI. From their experiment, spherical shaped and relatively agglomerated large particles of TiO_2 were produced. After the addition of ceramics and PANI, the particles gradually become smaller and show the disappearance of agglomeration. The reason for the disappearance of agglomeration is due to the porous surface of ceramics that increase the number of catalytic active sites. For this reason, the photocatalytic activity of TiO_2 can be amplified when it is modified with ceramic and polyaniline (PANI) which leads to the least agglomeration and clustered TiO_2. [17].

In addition to ceramic and PANI, modification of TiO_2 other nanocomposites such as waste-derived polymeric biosurfactant and hyperbranched polyethyleneimine polymer template (HPEI) have also been conducted using sol-gel method [24][36]. The sol-gel method can produce highly pure crystalline nanoparticles powder at low synthesis temperature without requiring a washing step since it is a waste-free synthesis and [37][34]

Another alternative method to synthesize TiO_2 is through the instantaneous synthesis method. As the name of the process suggests, the synthesis of TiO_2 is only in one-step process. [18]. The process reaction time of a few seconds, which is 3-5 seconds, was performed at 25 °C with no external added energy. Prakash and his co-researcher conducted their research using TiO_2 compounded with polypropylene and the time taken for production of TiO_2 only took around 5 seconds at 25 °C [18]. This technique is proven faster than the sol-gel method which takes at least 4 hours of stirring and further drying at room temperature to prepare TiO_2 [38].

The third method used for TiO_2 synthesis is by using the solvothermal method. The solvothermal method has the ability to direct precipitate arbitrary phases from the liquid phase [39] and it also appraises as a very appealing method to produce materials with different morphology [40]. Besides, this method is called the solvothermal method because it uses organic solvents in the synthesis procedure. Based on Faline-Daudre and teams, the fabrication of TiO_2 were prepared by a solvothermal method that uses the hydrolysis of titanium (IV) isopropoxide in the presence of benzyl alcohol and acetic acid at 180°C which resulted in a different shape and size of TiO_2 [20]. Alternatively, there is another reaction which is called hydrothermal. The hydrothermal method is analogous to that solvothermal method whereby it uses water to act as solvents instead of organic solvents [40]. According to Zhang and teams, nanoparticles were prepared by heating titanium sulfate as metal sources with ammonium hydrogen fluoride as a shape control agent. From this technique, the relatively pure TiO_2 is obtained which exhibits the improvement of photocatalytic activity [21].

Microwave-assisted synthesis is included as one of the methods for the synthesis of TiO_2. In accordance with this process, the reactants are directly heated by microwave. The chemical reaction is faster because electromagnetic radiation occurs at a relatively low value [41]. Moreover, due to dipole rotation, microwave irradiation offers instant localized superheating in which the transfer of energy occurs in 10^{-9} seconds per cycle of electromagnetic energy. Thus, the molecules' movement becomes more energetic which leads to a quicker reaction in less time [42]. According to Almeida and teams, the precipitation obtained from the previous step will be inserted into a microwave reactor for 4 minutes at 80 °C [22]. Thus, on a lab-scale, microwave-assisted heating is suitable and easy to be used as the outcomes will increase the reaction rate, less reaction time, increase in the product conversion, and less formation of undesired products. For this reason, this technique has gained importance due to its efficiency compared to the traditional heating route [22].

The simple mixing and precipitation method is also one of the methods that can be chosen to produce TiO_2. These methods are very common and convenient. The conversion of solution into solid by converting the substances into an insoluble form or making the solution supersaturated is called precipitation [43]. Furthermore, the precipitation method does not require any complicated process and complex preparation conditions in addition to its usefulness and attractiveness that can easily control the particles' size and shapes [44].As stated by Achudhan and the co-workers, plant extract is used for the synthesis of TiO_2 nanoparticles. $TiCl_4$ is added into the plant extract under continuous stirring until the color changes from transparent to pale white. The pale white precipitation indicates the formation of TiO_2 [28]. Other than that, simple mixing and precipitation were also being conducted without plant extract which according to Singh and the teams whereas titanium IV isopropoxide was stirred with deionized water for 35 minutes at 40 °C . During the stirring process, the mixture underwent hydrolysis, and precipitation of hydrous titanium oxide was formed. By centrifugation and washing with deionized water and methanol, white precipitation of TiO_2 was obtained [27]. Hence, using this method, the precipitation is able to calcined at a lower temperature without using any harmful surfactant [43].

3. Average Particle size

Based on the journal findings, two methods were identified for the preparation of titanium dioxide nanoparticles which are using plant extract and without using plant extract. First of all, using plants requires the preparation of the extract either from the twigs, leaves, fruit, or the roots of the plant. According to Achudhan and team, the synthesis of titanium dioxide nanoparticles was green synthesized from multiple plant extracts of *Azadirachta indica* twigs, *Ficus benghalensis* and *Syzygium aromaticum*. The resulting nanoparticle morphology obtained from the synthesis with plant extract are of crystalline and tetragonal shapes. Meanwhile, through HR-TEM images, the particle sizes show the average sizes of 20 nm [28]. Plant leaves can also be used to produce green extract for the preparation of TiO_2 nanoparticles. Based on Subhapriya and partner, the synthesis for TiO_2 nanoparticles is by using aqueous leaf extract of *Trigonella foenum-graecum*. The preparation of TiO_2 nanoparticles resulted with an average particle size of 20 nm based on HR-TEM images that also displayed the spherical shapes of polydisperse nanoparticles [45]. Besides, using leaves as the extract is also mentioned by Santhokumar and co-workers whereas aqueous leaf extract of *Psidium guajava* is used for the synthesis of TiO_2 nanoparticles. Through the FESEM images, the nanoparticles are considered smooth and spherical in shape with an average particle size of 32.58 nm [16]. After that, the extraction also used the plant latex as one of the materials for the synthesis. As stated by Hudlikar and teams, 0.3% of aqueous extract *Jatropha curcas L.* latex is used for the green synthesis of TiO_2 nanoparticles. In TEM images, it shows that the average particle size of TiO_2 obtained is in the range of 25 to 100 nm. Another source of plant extracts can be obtained from the plant roots [46]. Based on Bavanilatha and co-workers, the roots extracts from *Glycyrrhiza glabra* are used for the preparation of TiO_2 nanoparticles. Based on the FESEM images, it shows a polydisperse grain size of nanoparticles with an average size of 69 nm [31].

Furthermore, according to findings, another method for the synthesis of TiO_2 nanoparticles is without using any plant extract. There are several techniques that can be used to prepare the TiO_2 nanoparticles. First, Heshmatpour and co-workers produced the TiO_2 by performing a sol-gel method. In addition to that, the TiO_2 nanoparticles also will be composited with ceramics also through the sol-gel method and polyaniline (PANI) by in situ chemical oxidative polymerization to attain n-TiO_2/C/PANI nanocomposite. Through the TEM studies, the nanoparticles of pure TiO_2 and n-TiO_2/C/PANI nanocomposite have demonstrated the average size of 40 nm and 15 nm respectively [17]. Another technique used to acquire TiO_2 nanoparticles is via the instantaneous synthesis method. As stated by Prakash and co-workers based on the TEM images, TiO_2 in small clusters demonstrated particle sizes within the range of 10 nm to 20 nm while the individually rounded shape particle sizes are range within 5 nm to 12 nm [18]. The third method is the solvothermal method as the TiO_2 nanoparticles are fabricated along with hydrolysis of titanium (IV) isopropoxide in the presence of benzyl alcohol and acetic acid at 180 °C into different shapes and sizes. This method has resulted in two shapes of nanosphere and nanorods for TiO_2 nanoparticles whereas the size is 7 nm to 10 nm and 15 nm to 22 nm respectively based on TEM images [20]. The microwave-assisted synthesis method also is included as one of the techniques to synthesize TiO_2 nanoparticles without using plant extract. Almeida and colleagues use the microwave-assisted synthesis method. Before using the microwave, precipitation is obtained when titanium tetrachloride is mixed with distilled water under vigorous stirring along with the addition of ammonia hydroxide. After that, the precipitation is then inserted into the microwave reactor in order to acquire the TiO_2 nanoparticles. Based on HRTEM images, it indicates that the particle size for the TiO_2 nanoparticles is 20 nm [22]. Another method used to procure TiO_2 nanoparticles without plant extract is by the simple mixing and precipitation method. As mentioned by Singh and teams TiO_2 was obtained by precipitation. After that, TiO_2 is used to act as doping for poly (methyl methacrylate). Thus, based on the result of SEM images, synthesized TiO_2 nanoparticles are somehow spherical and consist of clusters and agglomerates with a size of 5 nm. Meanwhile, through TEM images, it is revealed that the shapes of TiO_2 nanoparticles are spherical, cubic, and were well dispersed. Also, the particles sizes obtained are in the range of 20 to100 nm and the average particles sizes of 65 nm [27].

The comparison of TiO_2 nanoparticles sizes using two methods which are using plant extract and without plant extract, shows that the average particle sizes of nanoparticles obtained from the synthesis of plant extract are relatively larger than without plant extract which is 15 nm - 69 nm and 5 nm - 40 nm respectively. However, the size is still acceptable because all the nanoparticles are still less than 100 nm. There are two mechanisms to drive the growth of colloids: the dissolution of smaller and soluble particles to ensuing precipitation and growth of larger particles. The second mechanism is via the agglomeration of smaller particles to form larger particles [31]. According to Santoshkumar and his teams, the TiO_2 nanoparticles with an average size of 32.58 nm still have their adsorption capability even though the size is not relatively small. It is proven when there are extracellular organic moieties on the surface of the metallic nanoparticles adsorbed when carbon, oxygen, magnesium, and chlorine are present [16]. After that, the particle sizes were found smaller by not using any plant extract than using plant extract. Based on Heshmatpour and co-workers, particle size with high surface energy tends to agglomerate more. This usually happens on pure TiO_2 nanoparticles that exhibit cluster growth. Thus, by adding ceramic and polyaniline the agglomeration diminishes, and the particle size becomes smaller. For this reason, the porous surface area of ceramic increases the number of catalytic sites, and with the formation of polyaniline, the smallest clusters arise from repulsion forces between the nanoparticles. Hence, smaller particle sizes can be produced through the modification of ceramic nanocomposite of TiO_2 by polymer without using any plant extract [17].

4. The Application of Titanium Dioxide

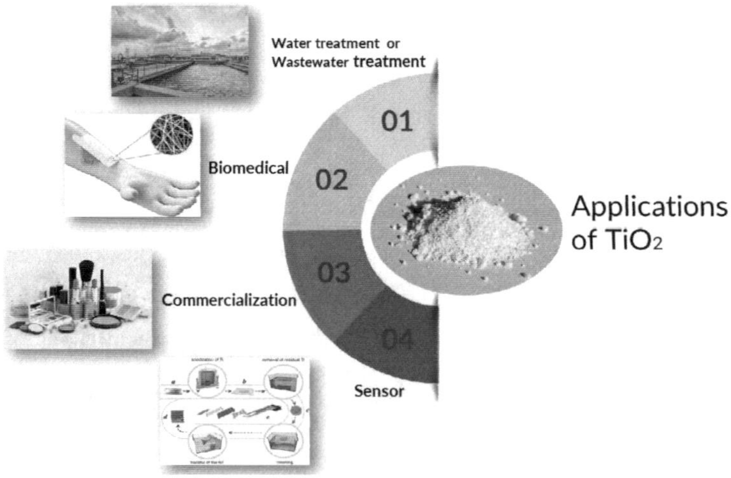

Figure 2. The application of titanium dioxide nanoparticles

The application of titanium dioxide is based on the findings journal in the tabulated data is shown in Figure 2. There are four major applications of TiO_2 which are applied in water treatment or wastewater treatment, biomedical, commercialization and sensor application. Based on the findings journal, water treatment or wastewater treatment has the largest application compared to other applications. Commonly TiO_2 was combined with other polymers such as PANI, or borneol-based polymer for pollution abatement and easy recovery as well as water purification. As an instance for water and wastewater treatment, TiO_2 is used for the removal of organic contaminants such as methylene blue, Congo red, and crystal violet. Then, the other application of TiO_2 is in commercialization. TiO_2 is used as pigment for paints and cosmetics. TiO_2 is also used for food additives or preservatives and along with for food packaging applications. In addition, it is also being used as the biofilm for photovoltaic devices such as solar cells for the generating of electricity. Another application is in biomedical and as well in pharmaceuticals. For this application, it is used as a therapeutic agent for the treatment of diseases such as infectious disease and cancer. Apart from that, TiO_2 has antibacterial properties that make it possible to use for grafting in nanomedicine for demanding in vivo studies. Then, TiO_2 is also used for employment in sensor applications. In sensor application, it is often used for the removal of toxins. For example, TiO_2 is used for the removal of ammonia gas at room temperature.

5. Conclusion

In conclusion, the preparation of TiO_2 by using different methods has resulted in different particle sizes and shapes. Besides, the time taken to obtain the synthesized product also varies depending on the methods and equipment used. Of all the methods, sol-gel method took the longest time to synthesize TiO_2 nanoparticles. Apart from that, the particle size is also affected by whether it is produced in the presence of plant extract or without plant extract, and upon the evaluation from the journal, it is discovered that the synthesis of TiO_2 without plant extract has resulted in smaller particle size as compared by using plant extract. The reason for this is because using plant extract tends to make the particles agglomerate and clustered together. In addition, through the application of TiO_2 nanoparticles, it is revealed that TiO_2 is mainly used for water treatment followed by biomedical, sensor and, commercialization such as pigments in paints and as well as cosmetics.

Acknowledgement

The authors would like to express thanks to Universiti Teknologi Malaysia for UTM Fundamental Research Grant (UTMFR) (Q.K130000.3843.22H15), SHE Empire Sdn. Bhd. for contract DTD research grant (R.K130000.7643.4C616) and PPMU-MJIIT Service Laboratories/IURL Laboratories for financial support during the research project.

References

[1] Malik A, Alshehri M.A, Alamery S.F and Khan J.M 2021 *Int. J. Biol. Macromol.* **188** 576-585
[2] Irshad M.A, Nawaz R, Rehman M.Z, Adress M, Rizwan M, Ali S, Ahmad S, Tasleem 2021 *Ecotoxicol. Environ. Saf.* **212** 111978.
[3] Karuppiah M and Alshehri M.A 2013 *Mater. Lett* **97** 141-143
[4] Taib S.H.M, Shameli K, Ali R.R, Izadiyan Z and Tarmizi Z.I 2021 *J.Res.Nanosci. Nanotechnol.* **3** 76-81
[5] Salleh M.S.N, Sangetha C, Ali R.R, Shameli K, Tarmizi Z.I and Zhe J.C 2021 *J.Res.Nanosci. Nanotechnol.* **2** 42-50
[6] Yadav R, Purwar R 2021 *Polym.Test.* **93** 106916
[7] Thakur B.K, Kumar A, Kumar D 2019 *South African J.Bot* **124** 223-227
[8] Abdel Fadeel D.A, Hanafy M.S, Kelany N.A and Elywa M.A 2021 *Heliyon* **7** e07370
[9] Rostami-Vartooni A, Nasrollahzadeh M, Salavati-Niasari M, Atarod M 2016 *J. Alloys. Compd* **689** 15-20
[10] Sazonov R, Kholodnaya G, Ponomarev D, Zhuravlev M, Pyatkov I, Konusov F, Lapteva O and Gadirov R 2021 *Phys. B Condens.Matter* **619** 413208
[11] Mathew S.S, Sunny N.E and Shanmugam V 2021 *Inorg. Chem. Commun* **126** 108485

[12] Aslam M, Abdullah A.Z and Rafatullah M 2021 *J. Ind. Eng. Chem.* **98** 1-16
[13] Mathumba P, Kuvarega A.T, Dlamini L.N and Malinga S.P 2017 *Mater Lett* **195** 172-177
[14] Goutam S.P, Saxena G, Singh V, Yadav A.K, Bharagava R.N and Thapa K.B 2017 *Chem. Eng. J.* **336** 386-396
[15] John.C.P, Ali R.R, Shameli K, Tarmizi Z.I, Salleh M.S.N, Zhe J.C 2021 *J.Res.Nanosci. Nanotechnol.* **4** 35-48
[16] Santhoshkumar T, Rahuman A.A, Jayaseelan C, Rajakumar G, Marimuthu S, Kirthi A.V, Velayutham K, Thomas J, Venkatesan J and Kim S.K 2014 *Asian Pac. J. Trop. Med* **7** 968-976
[17] Heshmatpour F and Zarrin S 2017 *J. Photochem. Photobiol. A Chem.* **346** 431-443
[18] Prakash M and Ghosh A.K 2021 *Solid State Sci.* **120** 106707
[19] Abutalib M.M and Rajeh A 2021 *Polym. Test.* **93** 107013
[20] Falentin-Daudre C, Baumann J.S, Migonney V and Spadavecchia J 2017 *Front Lab. Med.* **1** 217-223
[21] Zhang Y, Zhang G.L, Wang Y.T, Ma Z, Yang T.Y, Zhang T and Zhang Y.H 2021 *J.Colloid Interface Sci.* **596** 342-351
[22] Almeida A.I.A, Ferreira L.D.L, Almeida G.C, Calado H.D.R and Viana M.M 2020 *Synth. Met.* **269** 116544
[23] Sabir A. Sherazi T.A and Xu Q 2021 *Surfaces and Interfaces* **26** 101318
[24] Boffa V. Perrone D.G, Magnacca G and Montoneri E 2014 *Ceram. Int.* **40** 12161-12169
[25] Ramesan M.T, Santhi V, Bahuleyan B.K and Maghrabi M.A 2018 *Mater. Chem. Phys.* **211** 343-354
[26] Arularasu M.V, Harb M and Sundaram R 2020 *Carbohydr. Polym.* **249** 116868
[27] Singh P, Mirza U, Bhat S.A, Kareem A and Nishat N 2021 *Inorg. Chem. Commun.* **132** 108817
[28] Achudhan D, Vijayakumar S, Malaikozhundan B, Divya M, Jothirajan M, Subbian K, Gonzalez-Sanchez Z.I, Mahboob S, Al-Ghanim K.A and Vaseeharan B. 2020 *J. Environ. Chem. Eng.* **8** 104521
[29] Chen X, Nie Q, Shao Y, Wang Z and Cai Z 2021 *Environ. Technol.Innov.* **21** 101304
[30] Ortiz-Bustos J, Fajardo M, Hierro I and Perez Y 2019 *Mol. Catal.* **475** 110501
[31] Bavanilatha M, Toshitha L, Nivedhitha S and Sahithya S 2019 *Biocatal. Agric. Biotechnol.* **19** 101131
[32] Hernandez-Gonzalez A.C, Tellez-Jurado L and Rodriguez-Lorenzob L.M 2020 *Carbohydr. Polym.* **250** 116877
[33] Mandic V, Kurajica S and Ocko T 2020 *Ceram. Int.* **46** 29388-29401
[34] Suarez-Vega A, Agustin-Saenz C, O'Dell L.A, Brusciotti F, Somers A and Forsyth M 2021 *Appl. Surf. Sci.* **561** 149881
[35] Chang Y, He P, Wei Z, Chen Y, Wang H, Wu C, Zhou Z, Huang H, Kowalska E, Dong S 2020 *Mater. Lett.* **268** 127592
[36] Nasrollahzadeh M and Sajadi S.M 2015 *Ceram. Int.* **41** 14435-14439
[37] You Y.F, Xu C.H, Xu S.S, Cao S, Wang J.P, Huang Y.B and Shi S.Q 2014 *Ceram. Int.* **40** 8659-8666
[38] Hreniak A, Gryzlo K, Boharewicz B, Sikora A, Chmielowiec J and Iwan 2015 *A. Opt. Mater. (Amst)* **46** 45-51
[39] Astaraki H, Masoudpanah S.M and Alamolhoda S 2021 *J. Mater. Res. Technol.* **14** 229-241
[40] Kurian J and Mathew M.J 2018 *J. Magn. Magn Mater.* **451** 121-130
[41] Liu H, Zhao Y, Zhou C, Mu B and Chem L 2021 *Chem. Phys. Lett.* **780** 138906
[42] Jaiswal K.S and Rathod V.K 2021 *J. Indian Chem Soc* **98** 100020
[43] Sharma R.K and Ghose R 2016 *J.Alloys Compd.* **686** 64-73
[44] Zhang M, Sheng G, Fu J, An T, Wang X and Hu X 2005 *Mater. Lett.* **59** 3641-3644
[45] Subhapriya S and Gomanthipriya P 2018 *Microb Pathog.* **116** 215-220
[46] Hudlikar M, Joglekar S, Dhaygude M and Kodam K. 2012 *Mater. Lett.* **75** 196-199

IOP Conf. Series: Earth and Environmental Science 1091 (2022) 012065 doi:10.1088/1755-1315/1091/1/012065

Modified Surface Polymer Prepared by using Radiation-Induced Grafting Co-polymerization Method and Its Application: A Mini Review

N Seman[1], Z I Tarmizi [1*], R R Ali[1], M S N Salleh[1,2]

[1] Department of Chemical Engineering and Environmental. Malaysia Japan International Institute of Technology, Universiti Teknologi Malaysia, 54100 Kuala Lumpur, MALAYSIA

2 School of Chemical Engineering, College of Engineering, Universiti Teknologi MARA Cawangan Terengganu, Bukit Besi Campus, 23200 Dungun, Terengganu, MALAYSIA

*Corresponding author e-mail: zatil.izzah@utm.my

Abstract. The use of radiation-induced grafting (RIG) polymerization techniques is an appealing way to create and develop polymerization of polymer. The method of copolymerization preparation was evaluated in this paper based on the types of radiation-induced grafting polymerization and their prospective applications. Based on the method of preparation, the optimum grafting yield of the monomer in the polymer backbone is described. The approach used to summarise this review was to go through polymer-related papers from the science direct online database from 2010 to 2021. Then, from those selected journals, the technique of preparation, the grafting yield, and the application were reviewed. Furthermore, the majority of researchers from reviewed journals employed gamma irradiation to prepare the modified polymer, followed by electron beam irradiation, plasma irradiation and UV irradiation. Gamma irradiation is popular because it has a better penetration rate and generates a purer result. As a result of this review paper, the radiation-induced graft polymer from the conducted study is ideal for use in biomedical applications, as a material for wastewater treatment fibre membranes, as an absorbent, and as a gas remover.

1. Introduction

Polymer is a key substance in today's culture that has been widely utilized in a variety of industries, including as clothes, packaging materials, home appliances, and wastewater treatment. Furthermore, by merging modern technology with polymer, a wide range of polymeric substrates can be created, resulting in a new material for artificial skin. The qualities required for skin tissue engineering, including as non-immunogenicity, biocompatibility, biodegradability, and hydrophilicity, can be created by modifying the polymeric properties. [1]. Thus, by applying RIG co-polymerization, the polymer modification can be changed.

The use of advanced modification techniques to modify polymeric materials has increased dramatically in recent decades. Based on their design structure, the changed materials can be applied in a wide variety of applications. RIG polymerization is a method of designing and developing additional functionality for polymer materials while maintaining the shape and physical qualities of the trunk polymer [2][3]. Apart from that, while utilizing RIG polymerization to transfer functional groups to polymers, there is no unnecessary use of chemicals or catalysts. As a result, this process can be regarded both cost-effective and environmentally friendly, as it does not involve the use of a catalyst or a hazardous chemical that could harm human skin or the environment [4]. Gamma rays, electron beams, plasma treatment, and ultraviolet treatment are the four basic methods for initiating graft polymerization.

Grafting yield is used as a criterion to represent the degree of polymerization to determine the appropriateness of the radiation-induced graft polymerization reaction. The grafting yield determines the degree of graft polymerization response. Because it demonstrates the presence of a significant number of monomer functions in the trunk polymer, a higher grafting yield suggested the superiority of the functional materials. Apart from being associated with its application, the grafted polymer with a high grafting yield is the most desired while producing the polymer [2].

2. Surface modification of polymer using radiation-induced grafting polymerization method

According to the Science Direct database, there are four common ways to change the polymer using RIG techniques. There is gamma irradiation, electron beam irradiation, plasma irradiation, and UV irradiation (Figure 1).

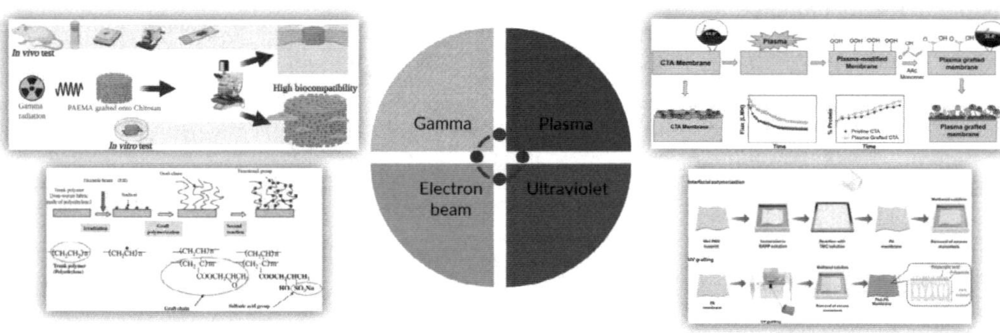

Figure 1. The types of commonly used radiation-induced grafting techniques.

First, utilizing cobalt-60 as a source of radiation, gamma radiation-induced grafting is more preferable and successful for altering and functionalizing polymers. As a result, gamma radiation-induced grafting can create instantaneous and uniform active radical sites, resulting in a high and clean grafting yield free of catalytic contamination and other harmful initiators. On a contrary, in the biomedical field, gamma is the preferred option [5]. However, because gamma radiation uses a radioactive isotope element, it is extremely damaging to one's health. Furthermore, waste management is challenging due to radioactive emissions, which may have a negative impact on human health and the environment [6].

The second most popular approach for polymer modification is electron beam radiation. On a polymeric surface, electron beam radiation can be utilized to create active sites or free radicals that can react with monomers to generate a graft copolymer, which is also known as a branched copolymer made of the main chain [7]. The gamma and electron beam methods have proven to generate a more stable product; nevertheless, gamma radiation is deemed notably superior to electron beam radiation since the approach achieved great grafting efficiency based on the copolymerization of natural rubber and polystyrene [8]. Then, another method to prepare the polymer is using plasma radiation techniques. Recently, Plasma irradiation has recently gained popularity as a surface modification technology that introduces desired functional groups of material structure that is ecologically friendly, does not use solvents, and takes less time [6]. According to Yang and colleagues, the plasma-induced grafting technique was used to modify carbon nanotubes (CNTs). The procedures were simple to use in revealing the active surface of CNTs and generating defects on the framework of prime CNTs without causing damage [9]. Furthermore, plasma irradiation can induce particular functional groups, such as carboxyl groups, to form on the membrane's surface, increasing the degree of hydrophilicity [10][11]. The membrane surface was coated with various energies and exposure times to activate gas plasma, and then the activated membrane was exposed to air to generate a peroxide group during plasma grafting polymerization. On the membrane surface, monomers such as hydroxyl, amine, or carboxyl groups were produced [11][10].

Finally, ultraviolet (UV) is one of the ways for polymer modification. UV radiation from sunlight was formerly thought to be a prevalent stressor on human skin. UVA (320-400 nm), UVB (290-320 nm), and UVC (320-400 nm) are the three forms of an ultraviolet radiation based on their spectrum [12]. The most damaging UV is UVC, which has the ability to harm human life due to its short wavelength and strong energy. UVB, on the other hand, is an excellent choice for breaking the polymer and causing photodegradation. UV irradiation grafting technologies have the advantages of milder reaction conditions, ease of operation, and the ability to be used on a larger industrial scale [13]. In addition, compared to alternative methods, UV treatment offers simplicity, speed, and inexpensive fabrication costs [14]. Table 1 summarises the research that was conducted based on the journal that was obtained.

3. Grafting yield

The fraction of monomer attached to the trunk polymer is referred to as grafting yield. The value of the percentage degree of grafting (DG) must be assessed in order to determine grafting yields. The types of solvents utilized, irradiation time or reaction time, temperature, absorbed dose, monomer concentration, and irradiation distance are all factors that influence the value of the DG during the polymerization process.

The different types of solvents have different effects on grafting yields, as shown in Table 1. Poly [(R)-3-hydroxybutyric acid] (PHB) was grafted with the n-hydroxyethyl acrylamide (HEAA) monomer produced using ^{60}Co gamma radiation at a dosage rate of 1.5 kGy/h at a dose of 10 kGy, according to Ochoa-Segundo and colleagues. Several solvents, including ethanol (EtOH), hexane (Hex), ethyl acetate (EtOAc), chloroform ($CHCl_3$), acetone (Ace), sodium acetate (SA), silica nanoparticles (SN), and acetic acid, were used in this experiment (Ac). Based on the type of solvent used, it was discovered that ethyl acetate produced the highest grafting yield of 83.06% when compared to other solvents. The propensity of ethyl acetate to create radicals during polymerization is the reason behind this. It is because ethyl acetate has the ability to produce radicals during the polymerization that affects the result of the gravimetric grafting degree [1].

The reaction time and irradiation temperature, on the other hand, also play a crucial effect in grafting yields. According to Aoki and others, electron beam radiation-induced grafting is used to create an antimicrobial fabric from NVP grafted onto polyolefin nonwoven fabric and iodine adsorption. First, the polyolefin nonwoven fabric is irradiated with a total dose of less than 160 kGy in a nitrogen atmosphere. It is then dipped in a 30wt % aqueous solution of (NVP) monomer and immediately transferred to a glass vessel for heating. The grafted fabric is then washed and dried, yielding PVP fabric as a result. The grafting process was seen in this publication to be continuous and did not manipulate the types of solvents utilised; instead, it measured the grafting degree based on reaction time and heating polymerization temperature. As a result of the findings, it can be concluded that a high grafting yield may be obtained when the reaction time is 30 minutes and the heating temperature is 337 °C, which resulting in a grafting yield of 98 percent [15]. As a corollary, the grafting yield is considered to be influenced by the reaction temperature. Selambakkannu and his colleagues suggest that this is because high temperatures accelerate the kinetics response, resulting in a larger degree of crosslinking. Furthermore, the temperature condition has an impact on crosslinker diffusion inside the solution. As a result, the crosslinker will have easy access to the active sites produced by irradiation on the polymer surface [38].

IOP Conf. Series: Earth and Environmental Science 1091 (2022) 012065 doi:10.1088/1755-1315/1091/1/012065

Table 1. Previous research conducted for the modification of polymer by using radiation-induced grafting techniques.

Polymer	Methods of Preparation	Optimum grafting yield	Application	References
N-Vinyl pyrrolidone irradiated onto polyolefin nonwoven fabric and subsequent adsorption of iodine	Primary process -RIG using electron beam under nitrogen atmosphere Secondary process - Adsorption of Iodine	The degree of grafting (DG) is 98%, 90% and 70% depend on the temperature which is 317 K, 327 K, and 337 K respectively	antibacterial fabric	[15]
Styrene and chloromethyl styrene are irradiated onto poly(ethylene-co-tetrafluoroethylene)	Irradiated with gamma rays	50 % monomer produce 68% DG 25% monomer produce 31%DG	The production of polymer electrolyte fuel cells to be used as a power source for portable electronic devices and automobiles.	[16]
Bio-adsorbent modified natural rubber (NR) was prepared by RIG polymerization of glycidyl methacrylate (GMA) onto NR seed particle in a latex	gamma irradiation	Grafting percentage - above 70% at 90% glycidyl methacrylate emulsion/10% NR seed particles in latex form at 5 kGy absorbed dose	Remove Th^{4+} ion from the aqueous solution	[17]
Poly [(R)-3-hydroxybutyric acid] (PHB) grafted with the n-hydroxyethyl acrylamide (HEAA) monomer	gamma radiation	Poly [(R)-3-hydroxybutyric acid] (PHB)-g-n-hydroxyethyl acrylamide (HEAA) - 83.06%	Used for potential skin tissue engineering applications	[1]
Methyl methacrylate (MMA), acrylic acid (AA) and styrene (St) were used as the monomers.	electron beam	N/A	Used for textile dyeing	[18]
Grafting of N, N-dimethyl aminoethyl methacrylate from the surface of polyethylene/polypropylene nonwoven fabric	gamma irradiation	40% N, N-dimethyl aminoethyl methacrylate (DMAEMA) with DMF as a solvent - 12 % DG 50% of DMAEMA with water/acetone - 5% GD	To act as antibacterial properties	[19]
Grafting the cationic and anionic monomers onto radiation crosslinked chitosan through mutual radiation grafting technique.	Gamma irradiation	[2-(methacryloxy)ethyl] trimethylammonium chloride-g- CR-chitosan (Grafting extent~ 12.44%)	Used for the removal of cationic and anionic dyes from an aqueous medium with fast kinetics.	[20]

IOP Conf. Series: Earth and Environmental Science 1091 (2022) 012065 doi:10.1088/1755-1315/1091/1/012065

Table 1. Continued

Polymer	Methods of Preparation	Optimum grafting yield	Application	References
Polyolefin-based sheath/core type composite fibers such as polyethylene/polypropylene, polyethylene/polyethylenerephtarate, PE.PE(HDPE) grafted with glycidyl methacrylate	Gamma radiation	N/A	Used as filters for the purification of contaminants in the environment Fe oxide - better performance for arsenic removal	[21]
Nylon-6 and vinyl benzyl chloride	Electron beam	Optimum DG - 130 wt.% at 20% vol% of (mixtureof3-and4-isomers,97%contains700–1100ppm Nitromethane as an inhibitor,50–100ppm tert-butylcatecholasan inhibitor)	Adsorbent applications	[22]
Alicyclic polyimides (APIs) derived from 1,2,4,5-pyromellitic dianhydride (PMDA) bis (4-aminocyclo-hexyl) methane (DCHM) onto styrene	Gamma irradiation	N/A	Used for the polymer electrolyte membrane for fuel cells	[23]
Grafting the of 2,2,2-trifluoroethyl methacrylate (TFEM) onto polypropylene (PP)	Gamma irradiation	based on types of solvent used: Tetrahydrofuran – 124% Dimethyl sulfoxide – 108%	It has a hydrophobic monomer thus it is a good water repellence that can be used for face masks, protective clothing for medical/outdoor application.	[4]
Glycidyl methacrylate grafted onto vinyl modified sepiolite	Gamma radiation	Effect on the absorbed dose at 1% vinyl modified sepiolite and 5% glycidyl methacrylate – 350%	Used for metal ion adsorbent for the recovery of metals from wastewater	[24]
Glycidyl methacrylate (GMA) onto ultra-high molecular weight polyethylene (UHMWPE) fiber and further modifying with triethylenetetramine (TETA) and glycidyl trimethyl-ammonium chloride (GTA)	Gamma-ray pre-irradiation	447.5% DG at absorbed dose 10kGy, glycidyl methacrylate at 20% and temperature at 50 °C	Has the potential application for Cr (VI) removal from wastewater	[25]

Table 1. Continued

Polymer	Methods of preparation	Optimum grafting yield (%)	Application	References
Polysaccharides grafted onto poly(3-hydroxybutyrate)	Gamma-ray radiation	Ethyl acetate solvent – 32.11%	Application as biomaterials	[26]
N-vinylpyrrolidone onto segmented polyurethane samples based on isophorone diisocyanate	Gamma-ray radiation	N-vinylpyrrolidone – 53% DG at 1.2kGy	Modification of the polyurethane surface is used for biomedical application	[27]
N-vinyl-2-pyrrolidone is grafted onto polyethylene-coated polypropylene fiber	Electron beam	149% DG at 200kGy dose –	Adsorbent for volatile iodine adsorption from fission products in nuclear power plant reactors	[28]
PVDF (polyvinylidene fluoride) membrane modified with glycidyl methacrylate (GMA) and sodium sulfite	Gamma radiation	16% DG at concentration of glycidyl methacrylate 0.5wt%	Ammonia removal for membrane bioreactor	[29]
Acrylamide (AM) onto carboxymethyl cellulose (CMC)in the presence of a cross-linking agent, N, N-methylene bisacrylamide (MBA)	Gamma radiation	N/A	To control the release of agrochemicals by reducing irrigation water consumption and improve the physical properties of soil	[30]
Polystyrene (PS) grafted sepiolite nanohybrid (MS-g-PS) in the presence of dichloromethane (DCM) as a solvent	Gamma radiation	170% DG at 58% mmol/g of sepiolite nanohybrid with adsorbed dose at 70kGy	Used as a low cost and efficient ion exchange resins	[31]
Poly(N-iso-polyacrylamide) (PNIPAAm) onto an ethylene-tetrafluoroethylene copolymer (ETFE)	Electron beam	16% DG at 30wt% of NIPAAm solution -	Used to generate tissue and organ through lamination and transplantation	[32]
Grafting of zwitterionic 2-methacryloyloxyethyl phosphorylcholine (MPC) polymer brushes (5–8 nm) on the silicone (polydimethylsiloxane, PDMS)	Initiator free atmospheric plasma-induced polymerization	N/A	Used as silicone-based contact lenses, wound dressing, and artificial organs	[33]

Table 1. Continued

Polymer	Methods of preparation	Optimum grafting yield (%)	Application	References
Iodine grafted into poly (*N*-vinyl pyrrolidone) coatings (pPVP) with controllable thicknesses	Plasma induced grafting polymerization	N/A	Used as coating for the medical implants' devices and wound dressing	[34]
Glycidyl methacrylate (GMA) was grafted onto polypropylene (PP) fibers	Plasma induced grafting polymerization	N/A	Used as the selective adsorption of Cr (VI)	[35]
Acrylic acid (AA) monomer was grafted on the Low-Density Polyethylene (LDPE) film	UV radiation	AA onto LDPE films in different solvent solutions (water) – 550% under 4 min treatment AA onto LDPE films in various concentrations of water/acetone solutions – approximately 520%	Application in food packaging and food cling wrap by maintaining food quality with reduced additive	[36]
Polyethylene glycols (PEG) grafted onto the surface of Poly (ether ether ketone) (PEEK) membranes	UV radiation	Optimum grafting degree 12% DG at Irradiation time at 30 min and 60 min	Used as the antifouling for the hollow fiber membrane	[13]
Cellulose triacetate (CTA) membrane grafted with acrylic acid	Plasma grafting polymerization using Argon and carbon dioxide gases	13% DG at time 5s (Acrylic acid + CO_2) and 12% DG at15s (Acrylic acid + Ar)	Used as an anti-protein fouling properties in forward osmosis for protein recovery application.	[10]
(Poly (ethylene glycol) methyl ether methacrylate) (poly (PEGMA)) grafted onto polysulfone (PSf) substrates	UV/ozone induced graft polymerization	N/A	For CO_2 separation and remarkable for other membrane application.	[14]
Hydroxyethyl methacrylate (HEMA) was grafted onto the surface of PAN fiber fabric (PAN-g-HEMA)	UV radiation	80% DG at monomer concentration 40wt%. 75% DG at concentration of Benzophenone at 2wt%	Used as fire retardant	[37]

Other criteria that affect grafting yield are depend on the irradiation dose and monomer concentrations. N-vinyl-2-pyrrolidone (NVP) was grafted into polyethylene-coated polypropylene skin-core (PE/PP) by Ye and his team. NVP and ethanol were introduced to PE/PP in a PE bag before the irradiation procedure, with the mass ratio of the NVP/ethanol solution ranging from 50% to 100%. After the leftover monomer was removed, the PE bag was heat-sealed. The NVP-containing PE/PP was then grafted using an electron beam of 1.5 MeV energy with varied absorbed doses ranging from 25 to 200 kGy at ambient temperature and an irradiation time of 5 path – 40 path with an absorbed time of roughly 40s to 320s. It is determined that the highest grafting yield was obtained when the absorbed doses were at 100 kGy with 100% monomer concentrations. However, the grafting yield increased sharply with 100% NVP and reached 146% at the absorbed dose of 200 kGy. However, high energy consumption seems to make it an unsuitable condition for polymerization. Thus, it is determined that the materials with grafting yield of 100 to 200 % were considered as the best adsorbents and the adsorbents prepared from 70% NVP at the adsorbed doses of 100 kGy are appraise as the finest condition [28].

Other than that, according to Taimur and teams, polystyrene (PS) was grafted with sepiolite nanohybrid (MS) using gamma rays' radiation grafting from the ^{60}Co irradiator in the presence of dichloromethane (DCM) as the solvent [31]. To get the highest grafting yield and kinetics, researchers are investigating the effects of varied irradiation dosages and monomer concentrations [39] [40]. Increases in grafting yield were observed as the absorbed dosages were increased. The increase in grafting yield is attributed to a rise in the number of radicals generated by radiation on monomers [41] [42]. Nonetheless, when grafting yield reaches particular doses, it can stop increasing because it has attained homo-polymerization, which causes the system's viscosity to increase, making it difficult to approach the monomer radicals to the grafting chain on MS [31] [43]. In terms of monomer concentration, the higher the monomer concentration, the higher the grafting yield. This is because the monomers are capable of reacting with the active sites of the irradiation materials, resulting in a corresponding increase in grafting yield [44]. In addition, according to Madrid and co-workers, the increasing of grafting yield along with the increasing of the monomer concentration is due to the higher diffusivity of monomers into the polymer matrix which amplifies the grafting yield [42].The irradiation distance is the last parameter that affects the value of the grafting yield. According to Ren and colleagues' research, UV induced photo graft polymerization was used to graft hydroxyethyl methacrylate (HEMA) onto the surface of a flame retardant polyacrylonitrile (PAN) fibre fabric. The grafting yield was affected by the irradiation distance. As the radiation distance reduces, the intensity of the radiation light on an object increases. Distance plays such an important role in establishing radical spots on the fabric. However, if the radiation is too close, the energy emitted by the radiation becomes too strong, causing the fabric to yellow and become brittle. [37].

4. Application of polymer prepared from RIG method.

According to the data in Table 1, biomedical and pharmaceutical applications are some of the applications. In this case, the polymer is used to make an antibacterial fabric that can be used as a wound dressing. It's also utilized for new skin tissue engineering, which involves laminating and transplanting new tissue and artificial organs. Aside from that, polymer modification to increase hydrophobicity and waterproof characteristics can be used to make face masks and contact lenses. Then, the other application that can be produced by using radiation-induced grafting technique is for electrical devices and automobiles where the trunk polymer was grafted to form polymer electrolyte fuel cells that act as power sources for the electrical devices and automobiles. An adsorbent and gas remover is another application. Hazardous gases such as TH^{4+} and ammonia may be easily removed, and it has the potential to be used as an adsorbent in water treatment for the removal of Cr (VI) and dyes from the water body. Apart from that, radiation-induced grafting polymer has commercial potential, as it may be utilized for textile dyeing. In food packaging application, it allows the reduction of food additives while retaining food quality. The modified polymer can also be used as an antifouling agent for hollow fiber membranes and as a fire retardant to inhibit fire spread.

5. Conclusion

Finally, the gamma, electron beam, plasma, and ultraviolet methods for preparing polymer via radiation-induced grafting are identified. For the synthesis of polymer utilising radiation induced grafting, most researchers prefer gamma. Furthermore, electron beam radiation is gaining popularity because it produces and contains no radioactive elements such as gamma. Plasma and UV are also used by some researchers. Furthermore, the most significant aspect in determining the optimum degree of grafting during polymerization is grafting yield. The types of solvent, irradiation time, irradiation temperature, irradiation dosages, monomer concentration, and irradiation distance are all parameters that affect grafting yield. These are the most critical elements to consider in order to maximize the grafting yield. Furthermore, it has been discovered that the polymer created by radiation-induced grafting can be used in a variety of applications, including biomedical, electrical applications, adsorbent for gas removal, commercialization as a waterproof face mask, antifouling for hollow fibre membranes, and fire retardant.

Acknowledgement

The authors would like to express thanks to Malaysia-Japan International Institute of Technology, Universiti Teknologi Malaysia (MJIIT UTM) for UTM Fundamental grant (Q.K130000.3843.22H15), SHE Empire Sdn. Bhd. for contract DTD grant (Q.K130000.7643.4C616) and PPMU-MJIIT Service Laboratories/IURL Laboratories for financial support during the research project.

References

[1] Ochoa E.L, Gonzalez M, Cabrere A, Shanchez R, Huerta B.M, Melgarejo Y, Leyva G, Rivera E.M, Cortes H, Velasquillo C, Vargas S, Rodriguez R 2020, *Mater. Sci. Eng. C.* **116.** 111176

[2] Ueki Y, Seko N, Maekawa Y 2021 *Appl. Mater. Today.* **25** 101158

[3] Nasef M.M, and Guven O. 2012. *Prog. Polym. Sci.* **37** 1597-1656

[4] Thinkohkaew K, Piroonpan T, Jiraborvornpongsa N, and Potiyaraj P 2021 *Surface and Interfaces* **24** 101125

[5] Nisar S, Pandit A.H, Nadeem M, Rizvi M.M.A, and Rattan S 2021 *Int.J.Bio. Macromol.* **182** 37-50.

[6] Drobny J.G 2013 *Ioniz. Radiat. Polym.* William Andrew 237-243

[7] Lim S.J, and Shin I.H 2019 *Nucl. Eng. Technol.* **52** 373-380

[8] Khamplod T, Loykulnant S, Kongkaew C, Sureeyatanapas P, and Prapainainar P 2015 *Polymer (Guildf)* **79** 135-145

[9] Yang S, Han C, Wang X and Nagatsu M 2014 *J.Hazard Mater* **274** 46-52

[10] Khongnakorn W, Bootluck W, and Jutaporn P 2020 *J.Water Process Eng.* **38** 101633

[11] Vatanpour V, Zoqi N 2017 *Appl. Surf. Sci* **396** 1378-1489.

[12] Chen X, Li L, Xu S, Bu W, Chen K, Li M, and Gu H 2017 *J.Photochem Photobiol B.Biol* **178** 152-164

[13] Chen Y, Huang T, Jiang C, Zhong T, Su Z, Xu Q, Jiang M, Liu P 2021 *Mater. Today Commun.* **27** 102326.

[14] Lee J.Y, Park C.Y, Moon S.Y, Choi J.H, Chang B.J, and Kim J.H 2019 *J.Memb.Sci* **589** 117214

[15] Aoki S, Fujiwara K, Sugo T and Suzuki K 2012 *Radiat, Phys.Chem* **84** 242-245

[16] Sawada S, Hasegawa S, Zhao Y, Maekawa Y 2017 *J.Memb.Sci* **532** 105-114

[17] Jaffary N.A.M, Mohamed N.H, Siong K.K, Fadzil S.M, Hamdan S.N.S, Azly U.A.F.U, Aziz N.M, Hashim A and Um W 2021 *Radiat. Phys. Chem.* **187** 109546

[18] Enomoto I, Katsumura Y, Kudo H and Soeda S 2011 *Radiat. Phys. Chem.* **80** 169-174

[19] Madrod J.F, Barsbay M, Abad L and Guven O 2016 *Radiat, Phys. Chem.* **124** 140-154

[20] Kumar S, Tiwari A, Chaudhari C.V and Bhardwaj Y.K 2021 *Radiat. Phys. Chem.* **182** 109377.

[21] Fujiwara K, Masubuchi T, Miyata K, Shiozawa M, Takato T and Harakawa H 2010

Radiat. Phys. Chem. **79** 238-240
[22] Ting T.M, Nasef M.M and Hashim K 2015 **109** 54-62
[23] Park J, Enomoto K, Yamashita T, Takagi Y, Todaka K and Maekawa Y 2013 *J. Memb.Sci.* **438** 1-7
[24] Tahir M, Raza A, Nasir A and Yasin T 2021 Radiat. Phys. Chem. 179. 109259.
[25] Gao Q, Hua J, Li R, Zing Z, Pang L, Zhang M, Xu L, Wu G 2017 *Radiat. Phys. Chem* **130** 92-102
[26] Toressa M.G, Munoz S.V, Rosales S.G.S, Carreon M.P, Munoz R.A.E, Gonzalez R.O, Gonzalez M.R.E, Talavera R.R.2015 *Carbohydr. Polym.* **133** 482-492
[27] Walo M, Przybytniak G, Kavakli P.A and Guven O 2013 *Radiat. Phys. Chem.* **84** 85- 90
[28] Ye F, Huang C, Jiang X, He W, Gao X, Ma L, Ao J, Xu L, Wang Z, Li Q, Li J, Ma H 2020 *Ecotoxicol Environ. Saf.* **203**. 111021.
[29] Shin I.H, Hong S, Lim S.J, Son Y.S, Kim T.H 2017 *J. Ind. Eng.Chem.* **46** 103-110
[30] Hemvichian K, Chanthawong A, and Suwanmala P 2014 Radiate.Phys. Chem **103** 167-171
[31] Taimur S, Hassan M.I, Yasin T and Ali S.W. 2017 *Radiat. Phys. Chem.* **148** 19-24
[32] Shimura R. Suematsu Y, Horiuchi H, Takeoka S, Oshima A and Washio M 2020 *Radiat. Phys.Chem.* **171** 108741
[33] Chen J.S, Ting Y.S, Tsou H.M and Liu T.Y 2017 *Surf.Coatings Technol.* **344** 621-625
[34] Gillam T.A, Goh C.K, Ninan N, Billimoria K, Shirazi H.S, Saboohi S, Al-Bataineh S, Whittle J and Blencowe A 2020 *Appl. Surf. Sci.* **537** 147866
[35] Luo Z, Guo M, Jiang H, Geng W, Wei W and Lian Z 2020 *React. Funct. Polym.* **150** 104552
[36] Shin J, Liu X, Chikthimmah N and Lee Y.S 2016 *Appl. Surf. Sci.* **386** 276-284
[37] Ren Y, Gu Y, Zeng Q and Zhang Y 2017 *Eur. Polym. J,* **94.** 1-10
[38] Selambakkannu S, Othman N.A.F, Ting T.M, Mohamed N.H, Hashim A and Karim Z.A 2020 *J.Environ.Chem. Eng.* **8** 103737
[39] Othman N.A.F, Selambakkannu S and Abdullah T.A.T 2018 *Mater. Today Proc.* **29** 207-211.
[40] Ueki Y, Dafader N C, Hoshina H, Seko N and Tamada M 2012 *Radiat. Phys. Chem* **81** 889–898
[41] Goel N.K, Kumar V, Pahan S, Bhardwaj K and Sabharwal S 2011 *J. Hazard. Mater* **193** 17–26
[42] Madrid J F, Nuesca G M and Abad L.V 2013 *Radiat. Phys. Chem* **85** 182–188
[43] Salimi P, Aroujalian A, and Iranshahi D 2021 *Sep. Purif. Technol* **258** 117939
[44] Kavakli P.A, Kavakli C and Güven O 2010 *Radiat. Phys. Chem* **79** 233–237

AUTHOR INDEX

Abba, M 585
Abdullah, N 490
Abdullah, Shahrum Shah 390
Abdullah, Sn 418
Adi, A D 237, 247
A-Halim, Md. Hakim 534
Akhir, F N M 596
Alam, M N 100
Alfaro-Alejo, R 144
Alghiffary, R. 174
Ali, Mohamed Sukri Mat 380
Ali, R R 603, 613
Allan, Christian Paul 125
Andiny, A N 247
Anggoro, R. 134
Anh, L T 227
Anuar, Aznah Nor 464, 551
Anuar, Aznah 544
Aquino, D H 355
Arif, M Ihsan 338
Aris, Alijah Md 464, 551
Arumugham, Thilagavathi 572
Asanuma, J. 304
Azharudin, M A A 16
Azman, N F 507, 596
Azmi, Siti Izaidah 544
Aznah, A. 418, 524
Balinas, Vicente 110
Belizario-Quispe, G 144
Bob, Mustafa M 544
Budianta, W 119
Bunmi, M R 164
Calatayud-Mendoza, A P 144
Camena, Jerilee 110
Chan, Kimsan 294
Chan, Ratboren 284, 294
Chheang, P 267
Chin, Khor Bee 551
Chui-Betancur, H N 144
Chum, Kimleang 284
Ciptadi, R 216
Colegio, Karlo Daniel Q 191
Cruz, Eric 125
Daag, Arturo S. 69
Date, K. 451
Daud, N N Nik 1, 16
David, Marjorie 125
De Jesus, Richard 79
Decena, Eunice R. 69
Defiesta, Gay 110
Din, Mf Md 418
Diyana, F A Noor 210
Do, Quang Khanh 29
Domingo, Heinrich Joyce S. 69
Dungca, J. 200
Evangelista, M K 355
Faris, F 237, 247
Faruq, Amrul 390
Fathani, T F 257
Fathi, M. S. 39
Fatmah 89
Fauzan 274
Firstyadi, Athanasius Donny 515
Galasso, Carmine 79
Galupino, J 200
Garciano, Lessandro 79
Grio, Mary Earl 79, 110
Habieb, A. B. 134
Hakim, Mohd 304, 418
Halim, Hakim 524, 551
Halim, Mohd Hakim Ab 464, 544
Hamdi, N. A. 348, 507
Hardiyatmo, H C 363
Harun, Aizul Nahar Bin 398
Hassan, Wan Nur Atiqah Wan 544
Helmut, Y 561
Hen, Chhordaneath 284
Hermana, J 441
Ho, Nghia H 154
Hoang, Trong Quang 29
Huaquisto-Ramos, E 144
Hussein, Shamsul Faisal Mohd 390
Huynh, Hoai C 154
Ibabao, Rhodella A. 110
Ibabao, Rhodella 79
Ibrahim, Z 585
Ich, Ilan 284, 294
Ilmasari, D. 471, 480
Imamura, F 316
Iranata, D. 134, 174
Irawan, D. 174
Ismail, M. A. M 451
Ismail, Zool H. 409
Ismanti, S 274
Istiarto, Istiarto 515
Istiarto 59

Iwamoto, K 267, 316
Jamal, Nur Syahida Abdul 544
Jayadi, R 59, 328
Jayasinghe, S 100
Joshi, A 100
Jusoh, N 435
Kamarudin, Mh. 409
Kamulyan, B 216
Kassim, Norazah Mohd 551
Khairuddin, Wan Norshuhada Binti Wan 398
Khe, Sotheanea 294
Khen, C 471
Khoeun, Chanseyma 294
Khudzari, J Md 267, 316
Kiet, N 227
Kong, Leangkim 304
Kristijanto, H. 174
Laila, N 100
Lee, Y S 435
Legono, Djoko 338
Lei, Zhongfang 464
Luknanto, Djoko 338
Maas, A 328
Maligon, Ardeth Joy E. 69
Marpaung, Maruddin F 338
Marto, Aminaton 390
Mat-Taib, S 418
Mendoza, Rodolfo 79
Min, F T Y 267
Miyoshi, T. 451
Mohamad, S E 435, 572
Muhanifah, H 237
Muzaffar, Wan Mohd Bukhary W 524
Nguyen, Bay T 154
Nguyen, T C T 316
Nguyen, Thang 29
Nor, Aznah 304, 534
Norsin, E. 507
Nurlaela, S. 134
Oeurng, Chantha 284, 294
Ohta, H 119
Ong, Vu Khanh Phat 29
Oreta, Andres Winston 79, 110
Othman, F S 435
Othman, N. 428, 507
Penaredondo, Suzette 110
Ponraj, M 418
Pratama, A 257
Pratiwi, E P A 59
Purba, L D A 441, 490
Quan, L M 561
Quinay, P E 49
Quinay, Pher Errol B 191
Rafi, S. B. Mohd. 507
Rahajeng, S S 328
Rahardjo, A P 216, 499
Rahayu, Rita Laksmitasari 182
Rakhmania 585
Ramadhani, E. E. 370
Rambat, Shuib Bin 398
Ramos, J A 355
Ramos, Jamie Alea B 191
Razali, M. 451
Razip, N A Mohd 1
Rifa'I, A 274, 363
Riyadi, F A 348, 596
Sabri, Nsa 348
Sabtu, N. 164
Sahabudin, E 428
Said, Nf. 304
Saleh, A 164
Salleh, M S N 603, 613
Satyarno, I 257
Sauri, S 363
Seman, N 603, 613
Sembuil, Azila 534
Sethi, S 316
Setiawan, A 499
Sevieri, Giacomo 79
Sha'Arani, S. 348, 507
Sharani, Noor Atiqah 551
Shimizu, K 441
Sok, Ty 284, 294
Solidum, Renato U. 69
Soriano, C M 49
Sugeng, Triyadi S 182
Sujono, J. 370
Sukri, S N A Mohamad 603
Suswanto, B. 174
Suzuki, I 428
Syafitri, R. A. W. D. 134
Taib, S H M 603
Takemura, J 119
Tambunan, Lily 182
Tanavarotai, Karn 464
Tarmizi, Z I 603, 613
Taryono 370
Teo, Y C 435
Thao, N D 227
Tharima, A. F. 39
Tobe, H. 451
Tra, Nga N Q 154
Tran, Kim T 154
Trance, Rene 110

Trang, D D T 227
Try, Sophal 284
Ujang, Zaini 544
Warmadewanthi, I 561
Wignyosukarto, B S 499, 515
Wignyosukarto, Budi 338
Wijaya, I P K 100
Windianita, K 59
Yakub, Fitri 380
Yakubson, Yakubson 515
Yokota, Y 451
Yusof, N 596
Yusoff, I. N. 451
Yuzir, A 471, 480
Zahir, H. 39
Zahra, S A 490
Zaki, Sheikh Ahmad 380
Zhe, J C 603
Zuhairi, Ainaa Hanis 380

Institute of Physics
Dirac House, Temple Back
Bristol BS1 6BE UK

ISSN: 1755-1307
ISBN 978-1-7138-6437-0
9 781713 864370